Hyperspectral Remote Sensing

Fundamentals and Practices

Remote Sensing Applications

Series Editor

Qihao Weng

Indiana State University
Terre Haute, Indiana, U.S.A.

Hyperspectral Remote Sensing: Fundamentals and Practices, *Ruiliang Pu*

Integrating Scale in Remote Sensing and GIS, *Dale A. Quattrochi, Elizabeth A. Wentz, Nina Siu-Ngan Lam, and Charles W. Emerson*

Remote Sensing for Sustainability, *Qihao Weng*

Remote Sensing Applications for the Urban Environment, *George Z. Xian*

Remote Sensing of Impervious Surfaces in Tropical and Subtropical Areas, *Hui Lin, Yuanzhi Zhang, and Qihao Weng*

Global Urban Monitoring and Assessment through Earth Observation, *edited by Qihao Weng*

Remote Sensing of Natural Resources, *edited by Guangxing Wang and Qihao Weng*

Remote Sensing of Land Use and Land Cover: Principles and Applications, *Chandra P. Giri*

Remote Sensing of Protected Lands, *edited by Yeqiao Wang*

Advances in Environmental Remote Sensing: Sensors, Algorithms, and Applications, *edited by Qihao Weng*

Remote Sensing of Coastal Environments, *edited by Yeqiao Wang*

Remote Sensing of Global Croplands for Food Security, *edited by Prasad S. Thenkabail, John G. Lyon, Hugh Turral, and Chandashekhar M. Biradar*

Global Mapping of Human Settlement: Experiences, Data Sets, and Prospects, *edited by Paolo Gamba and Martin Herold*

Hyperspectral Remote Sensing: Principles and Applications, *Marcus Borengasser, William S. Hungate, and Russell Watkins*

Remote Sensing of Impervious Surfaces, *edited by Qihao Weng*

Multispectral Image Analysis Using the Object-Oriented Paradigm, *Kumar Navulur*

Urban Remote Sensing, *edited by Qihao Weng and Dale A. Quattrochi*

Hyperspectral Remote Sensing

Fundamentals and Practices

Ruiliang Pu

CRC Press is an imprint of the
Taylor & Francis Group, an **informa** business

CRC Press
Taylor & Francis Group
6000 Broken Sound Parkway NW, Suite 300
Boca Raton, FL 33487-2742

Printed on acid-free paper

International Standard Book Number-13: 978-1-4987-3159-1 (Hardback)
International Standard Book Number-13: 978-1-1387-4717-3 (Paperback)

Library of Congress Cataloging-in-Publication Data

Names: Pu, Ruiliang, 1956- author.
Title: Hyperspectral remote sensing : fundamentals and practices / Ruiliang Pu.
Description: Boca Raton : Taylor & Francis, [2017]
Identifiers: LCCN 2016059563 | ISBN 9781498731591 (hardback : alk. paper)
Subjects: LCSH: Remote sensing. | Hyperspectral imaging. | Geology--Remote sensing. | Land use--Remote sensing. | Natural resources--Remote sensing.
Classification: LCC G70.4 .P838 2017 | DDC 621.36/78--dc23
LC record available at https://lccn.loc.gov/2016059563

Visit the Taylor & Francis Web site at
http://www.taylorandfrancis.com

and the CRC Press Web site at
http://www.crcpress.com

This book is dedicated to my wife, Guoling, and two sons, William and Wilson, for their constant support during the course of my preparing this book.

Contents

Foreword

Hyperspectral remote sensing has been a frontier of geospatial technology since the late 1980s, first with the launch of airborne sensors, and later, in the 1990s, with the addition of spaceborne sensors. Today, we can witness the maturity of hyperspectral remote sensing from its wide range of applications, readily available software for image processing, and large volume of pertinent publications. With the expectation of future satellite missions of hyperspectral imaging with global coverage and increasing improvement in image-processing efficiency and information extraction effectiveness, hyperspectral remote sensing will usher in an era of reinvention with the integrated use of high spatial and temporal resolution, LiDAR, and radar sensing. In this context, I am pleased that Dr. Ruiliang Pu has the vision and energy to have written a book on this important topic.

The book provides an exhaustive review of the characteristics of airborne and spaceborne hyperspectral sensors, systems, and missions; it discusses in detail the algorithms, techniques, and methods for processing and analyzing hyperspectral data. Further, the author assesses the features and modules of operational tools and software for processing hyperspectral data, followed by a comprehensive investigation of hyperspectral remote sensing applications to geology, soils, vegetation, and environments. Dr. Pu instills his research of more than two decades and rich teaching experience across the Pacific Ocean into the writing of this book. It is a one-of-a-kind textbook, as well as research monograph. I plan to use this book for my multiple remote sensing classes.

This book is the sixteenth volume in the Taylor & Francis Series in Remote Sensing Applications, and 2017 marks the tenth anniversary of this series. As envisioned in 2007, the books in the series contribute to advancements in theories, methods, techniques, and applications of remote sensing in various fields. In fact, as seen today, many of the books in the series have served well as references for professionals, researchers, and scientists, and as textbooks for teachers and students throughout countries all over the world. I hope that the publication of this book will promote a wider and deeper appreciation and application of hyperspectral remote sensing technology. Finally, my hearty congratulations go to Dr. Pu for creating a new milestone in the history of remote sensing.

Qihao Weng, PhD
Series Editor
Hawthorn Woods, Indiana

Preface

Remote sensors can be seen widely in the biological world, and they have existed for at least hundreds of millions of years. Eyes, ears, and noses of animals are such examples. Even the skin of a living animal contains sensing cells for heat. Remote sensing devices made by humans have a much shorter history—less than 200 years—but at their first appearance, they have shown a unique capability not possessed by biological sensors: recording. From the first photographing device invented by Joseph Nicéphore Niépce in the 1820s to the first phonograph device invented by Thomas Edison in 1877, all have recording capabilities. These expanded the memorizing capability of images and sounds of biological systems. The technology has since evolved through electrical, magnetic, and into the present digital age. In the 1950s, multispectral scanners were first put on board airplanes to take multispectral images of the land. Such images contain 4 to 12 spectral bands, with some bands expanding human vision from the visible spectral range to the shorter wavelength range of ultraviolet and the longer wavelength range of near infrared, both not visible to human eyes. Although it was not until 1962 that remote sensing as a term came into being, the history of remote sensing can be traced back to the advent of photography. In the past 200 years or less, remote sensing has brought waves of excitement to human society, among which is the invention of an airborne imaging spectrometer by Alex Goetz and his colleagues in the 1980s. The principle of an airborne imaging spectrometer is similar to a multispectral scanner but can produce images in several dozen to several hundred spectral bands. Essentially, imaging spectrometers made continuous spectroscopic airborne imaging possible. To make this type of data and associated analysis distinguishable from traditional multispectral remote sensing, it was natural to choose hyperspectral remote sensing.

The strength of hyperspectral remote sensing is the detailed recording of spectroscopic properties of the imaged area. The large number of spectral bands, however, also limited the level of detail obtainable in the spatial aspects. In satellite remote sensing, a tradeoff must be done between the spectral and spatial level of details. Since hyperspectral data emphasizes the spectral aspects, either the spatial resolution or the spatial coverage cannot be great. This is why we have not yet been able to enjoy a full coverage of the world with hyperspectral data in relatively high resolution. Since technologies of data storage capacity and bandwidth of data transmission have sufficiently advanced, I think the remote sensing community will not need to wait long for the time in which we are able to make use of hyperspectral images with global coverage and sufficiently high spatial resolution.

With a background in forestry, Ruiliang Pu began hyperspectral remote sensing research when he visited Professor John Miller at the Institute of Space and Terrestrial Science located at York University in 1990. As a physicist, Professor Miller had been collaborating with foresters such as Professor Barry Rock at the University of New Hampshire and Professor Dick Waring at Oregon State University. This gave Dr. Pu a great opportunity to learn from those pioneers at a time that the application of hyperspectral remote sensing in the field of forestry had just started. He participated in a NASA project led by Dick Waring and David Peterson, the Oregon Transect Ecosystem Research. As a result, he and his colleagues published the first paper on the use of hyperspectral images in the estimation of forest leaf area index in 1992. Over the past 25 years, Dr. Pu has worked continuously on the application of hyperspectral remote sensing to various environmental problems. He has become an authority in this field.

I am pleased to be among the first few readers of this book. It first introduces the various field spectrometer devices, airborne and spaceborne hyperspectral sensors. It then introduces the radiometric processing of hyperspectral images. A large collection of various data processing algorithms and corresponding software packages is then presented. The remainder of this book introduces the application of hyperspectral remote sensing in studies of geology, vegetation, soils, water, and atmosphere. This book is well written, with detailed coverage of methods of data analysis. It will be particularly useful to students and researchers who wish to use hyperspectral remote sensing data. I hope readers will enjoy this book as much as I do.

Peng Gong
Department of Earth System Sciences
Tsinghua University
Beijing, China

Acknowledgments

Like most books, the successful completion of this book is due to many people's contributions, support, and assistance. First of all, I must express my gratitude to Professor Peng Gong at University of California at Berkeley (UCB), USA, and Tsinghua University, China, for his invaluable supervisorship, guidance, collaboration, brotherliness, and financial support to all my research projects on HRS theoretical and application studies at UCB. I greatly appreciate Dr. Qihao Weng, Professor at Indiana State University, USA, and Series Editor for Remote Sensing Applications (Taylor & Francis Publishing Group), for his consistent encouragement and recommendation. I would also like to thank my former supervisor Professor John R. Miller at York University, Canada, for his guidance of my research on HRS in the early 1990s. I am grateful to University of South Florida for granting me a sabbatical leave in fall of 2015 to complete part of the book writing process and for the financial support to publish the book. I gratefully acknowledge all assistance from Irma Shagla Britton (senior editor) and Claudia Kisielewicz (editorial assistant), as well as all other staff at Taylor & Francis Publishing Group. Finally, I would like to express my most sincere appreciation to my lovely wife, Guoling, for her endless support and encouragement.

Author

Dr. Ruiliang Pu is currently an associate professor in the School of Geosciences at the University of South Florida (USF). He earned his MSc in forest management from Nanjing Forestry University (NFU), China, in 1985. He earned a PhD in cartography and geographic information systems in 2000, conducted at University of California (UC) at Berkeley, from the Chinese Academy of Sciences. His research experience and interests are in remote sensing, GIS and spatial statistics with direct applications to natural hazard monitoring, land use/cover change detection, biophysical and biochemical parameters extraction, and coastal and terrestrial ecosystems modeling. Dr. Pu's research projects have been funded mostly by the Natural Science Foundation of China and NASA. He has published more than 100 journal papers, book chapters, and segments in English; and more than 25 journal papers and books in Chinese. He has been a reviewer of NASA (United States), NSERC (Canada), and Belgian Science Policy research proposals and of papers for more than 20 journals. Dr. Pu is currently an editorial board member/academic editor of *Remote Sensing* and *Geosciences* of MDPI.

Introduction

Hyperspectral remote sensing (HRS), frequently known as imaging spectroscopy, can provide subtle imaging spectral information. Given the characteristics of HRS, combining imaging with spectroscopy and possessing individual absorption features of materials due to specific chemical bonds in a solid, liquid, or gas, researchers and scientists have investigated and applied imaging spectroscopy techniques for the detection, identification, and mapping of minerals on land and in waters and atmosphere. Hence, HRS technology, as an advanced remote sensing tool, has been studied for many applications, such as in geology, geomorphology, limnology, pedology, hydrology, vegetation and ecosystems, and atmospheric sciences. To well investigate and apply HRS technology, it is necessary to systematically introduce and extensively review and summarize existing HRS technologies. This book intends to achieve that purpose.

The book systematically introduces concepts, theories, principles, and applications of HRS and imaging spectral technologies. It is built on my research experience with HRS and on extensive review of literature items cited in the references at the end of each chapter. After discussing the basic concepts, fundamentals, and history of imaging spectral technologies, the book gives full and authoritative introductions to characteristics and principles of field spectrometers and biological instruments as well as airborne and spaceborne hyperspectral sensors, systems, and missions. While ground-based instruments (field spectrometers and bio-instruments) are important for studying underlying mechanisms and vegetative applications of HRS, it is necessary for us to understand the underlying imaging principles of both airborne and spaceborne hyperspectral sensors and systems to successfully apply sensor data in different application areas. Given the sensitivity of hyperspectral data to atmospheric effects, conducting radiometric and atmospheric correction is a key step for successful utilization of hyperspectral imaging data. Therefore, the book extensively introduces and deeply discusses empirical and radiative transfer–based atmospheric correction algorithms, techniques, and methods. Different from a set of techniques and methods for processing and analyzing traditional remote sensing data, this book fully and authoritatively introduces and discusses the techniques and methods developed specially for processing and analyzing hyperspectral data sets. To efficiently and successfully conduct various HRS projects, including both theoretical and applicative studies, operational tools and software for processing various hyperspectral data are necessary. Thus, the book introduces relevant features and modules of contemporary hyperspectral image processing tools and packages in detail. After describing the spectral characteristics of various materials and discussing analysis techniques and methods applicable to different application purposes of HRS, a comprehensive literature review is conducted for different application areas of HRS: geology and soil, vegetation and ecosystem, environments and other areas; the book then comments on the applicability of HRS in these areas. In short, the book can be used as a reference or professional book for graduate students, professionals, and researchers for their studies and applications of HRS technologies.

There are a total of nine chapters in this book. The book is organized with an overview of HRS in Chapter 1; an introduction to field spectrometers, bio-instruments and imaging spectrometers, and sensors and systems onboard airborne and space-borne platforms in Chapters 2 and 3; an introduction to, and review of, techniques/models and algorithms of radiometric and atmospheric correction to hyperspectral data in Chapter 4; an introduction to, and discussion of, techniques and methods used for processing hyperspectral data in Chapter 5; a brief description of features and models of relevant software and tools for processing hyperspectral images in Chapter 6; and an overview on HRS research and applications in geology/soils, vegetation, environments, and other disciplines in Chapters 7 through 9.

More specifically, in Chapter 1, I introduce basic concepts of imaging spectroscopy and HRS and discuss differences between multispectral remote sensing and HRS. Then the development stages of

HRS with some representative missions and sensor systems in particular periods are reviewed, and an overview of HRS applications in geology, vegetation, atmosphere, hydrology, urban, and other environments is presented. Finally, a perspective of HRS is provided and discussed.

In Chapter 2, work principles, structures, operations in the field, and application areas of frequently used modern spectrometers and plant biology instruments are briefly introduced. In Chapter 3, working principles, technical characteristics, and status of current/operational and future airborne and spaceborne hyperspectral sensors, systems, and missions are briefly introduced. The knowledge about field spectrometers is capable of not only helping one to understand the mechanism of imaging spectral technology but is also necessary to calibrate various airborne and spaceborne imaging sensors and systems. Introduction to several key bio-instruments in this book is critical for HRS applied in ecosystems and plants. To reasonably utilize and analyze the various data of hyperspectral sensors, it is necessary for users to know and understand the working principles and technical characteristics of existing and future airborne and satellite hyperspectral sensors and systems.

Chapters 4 and 5 present and discuss techniques/methods and algorithms for radiometric and atmospheric corrections and hyperspectral image processing. Radiometric correction for HRS data is necessary for most HRS application cases. Systematical introduction and discussion of both empirical and radiative transform algorithms and models for atmospheric correction can benefit most readers to efficiently utilize contemporary atmospheric correction algorithms and techniques to process their HRS data. Such improved spectral quality via accurate radiometric correction can increase the accuracy of HRS application. Hyperspectral data contain a wealth of spectral information about the material content. However, such a wealth of interesting information usually hides in a huge data volume that hyperspectral imagery can represent, and hyperspectral data also has a high dimensionality and high correlation of adjacent bands. Therefore, it is necessary to introduce and discuss a collection of techniques and algorithms, commonly used and newly developed, which are suitable for processing HRS data in Chapter 5. For readers to conveniently be able to find, and to be familiar with, hyperspectral processing software and tools/systems currently available in the market, Chapter 6 introduces the main functionalities and features of a collection of major and minor software tools, programs, and systems for processing and analyzing various hyperspectral data sets.

In the last three chapters (Chapters 7 through 9), a comprehensive review of HRS application studies in geology and soil sciences, vegetation and ecosystems, and other environmental areas is conducted. In Chapter 7, spectral characteristics and properties of various mineral/rock species or classes are introduced and discussed, the relevant techniques and methods are summarized and their applications with various hyperspectral data for estimating and mapping minerals and rocks are reviewed and discussed, and the spectral characteristics of soils and hyperspectral technology applied in estimating and mapping properties of soils are reviewed and summarized. In Chapter 8, spectral characteristics of typical green plants including green leaf structure and plant spectral reflectance curve are introduced, nine types of analytical techniques and methods suitable for extracting and estimating plant biophysical and biochemical parameters and vegetation mapping with hyperspectral data are introduced and reviewed, and an overview on application cases of various hyperspectral data in estimating and mapping a set of plant physical and chemical parameters is presented. In the last chapter (Chapter 9), an overview of application studies of various hyperspectral data sets to other environments and disciplines is presented. The other environments and disciplines include an estimation of atmospheric parameters (water vapor, clouds, aerosols, and carbon dioxide), snow and ice hydrology, coastal environments and inland waters, environmental hazards and disasters, and urban environments.

Ruiliang Pu
School of Geosciences
University of South Florida
Tampa, Florida

1 Overview of Hyperspectral Remote Sensing

Remote sensing is an advanced technology used for obtaining information about a target through the analysis of data acquired by a sensor from the target at a distance. Compared to traditional remote sensing (multispectral remote sensing, MRS), hyperspectral remote sensing (HRS) has a relatively short history of development, spanning about three decades. In this chapter, basic concepts of imaging spectroscopy and HRS are introduced, and differences between MRS and HRS are discussed. Then, the development stages of HRS with some representative missions and sensor systems in particular periods are described, and an overview of HRS applications is presented. And, finally, a perspective of HRS is provided and discussed.

1.1 CONCEPTS OF IMAGING SPECTROSCOPY

1.1.1 Spectroscopy

Spectroscopy (or spectrography) refers to the measurement of radiation intensity as a function of wavelength from an object and is often used to describe experimental spectroscopic methods (Wikipedia 2014a). Spectroscopic studies focus on the interaction between matter and radiated energy. Daily observations of color are related to spectroscopy. Historically, spectroscopy originated with the study of visible light dispersed according to its wavelength, as dispersed through a prism. Analyzing white light by dispersing it through a prism is an example of spectroscopy (Figure 1.1). The constituent colors consist mainly of red (620–780 nm), orange (585–620 nm), yellow (570–585 nm), green (490–570 nm), blue (440–490 nm), indigo (420–440 nm), and violet (400–420 nm). Spectroscopic data are often represented by spectrum as a function of wavelength or frequency. More recently, the definition of spectroscopy has been expanded to include the study of interactions among particles, including electrons, protons, and ions, as well as their interactions with other particles as a function of their collision energy associated with wavelength or frequency (Encyclopedia Britannica 2014). Spectroscopy also includes the study of the absorption and emission of light and other radiation by matter, as related to the dependence of these processes on the wavelength of the radiation.

Spectroscopy has been employed by physicists and chemists in laboratories for over a century (Skoog et al. 1998). Nowadays, spectroscopic techniques have been applied in virtually all technical fields of science and technology. For example, microwave spectroscopy is used to discover the so-called three-degree blackbody radiation. Optical spectroscopy is routinely used to identify the chemical composition of matter, and to determine its physical structure (Encyclopedia Britannica 2014). Spectral measurement devices, including spectrometers, spectrophotometers, spectrographs, and spectral analyzers, are frequently used to measure spectroscopic data in practice (Wikipedia 2014a). As a result, these spectral measurements can be used to detect, identify, and quantify information about atoms and molecules. Spectroscopic techniques are also used in astronomy and remote sensing on Earth. The measured spectra are used to determine the chemical composition and physical properties (e.g., temperature and velocity) of objects and matter in space and on Earth.

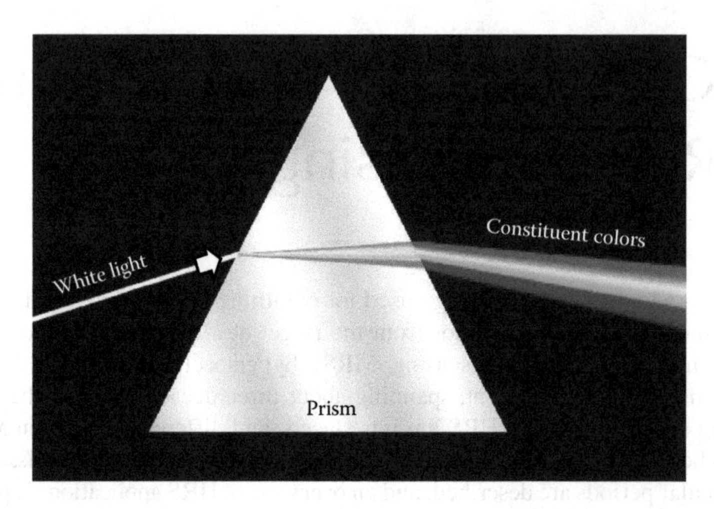

FIGURE 1.1 An example of spectroscopy illustrated by analyzing a white light by dispersing it with a prism.

1.1.2 Imaging Spectroscopy

Imaging spectrometry (IS; also called hyperspectral imaging) refers to the art and science of designing, fabricating, evaluating, and applying instrumentation capable of simultaneously capturing spatial and spectral attributes of a scene with enough fidelity to preserve the fundamental spectral features that provide for object detection, classification, identification, and characterization (Eismann 2012). Compared with Eismann's comprehensive definition of IS (2012), the U.S. Geological Survey (USGS) Speclab gives a relatively straightforward definition of IS: The main objective of IS is to measure the spectral signatures and chemical compositions of all features within the sensor's field of view; IS data contain both spatial and spectral information from materials within a given scene; and each pixel across a sequence of continuous, narrow spectral bands contains both spatial and spectral properties (2014). Both definitions of IS are derived from the classic definition given by Goetz et al. (1985), which defines IS as an acquisition of images in many narrow contiguous spectral bands throughout the visible and solar-reflected infrared spectral bands, simultaneously. Nowadays, the IS definition covers all spectral regions (i.e., visible, near infrared, shortwave infrared, midwave infrared, and longwave infrared), all spatial domains (microscopic to macroscopic), and all targets (solid, liquid, and gas) (Ben-Dor et al. 2013). Further, an imaging spectrometer typically can collect several dozens to a few hundreds of bands of data, which enable the construction of an effectively continuous and complete reflectance spectrum for every pixel in an image (Goetz et al. 1985, Vane and Goetz 1988, Lillesand and Kiefer 1999). Figure 1.2 shows the basic concept of imaging spectroscopy. In this figure, each picture element (pixel) in an image scene has associated with it a large number of spectral data points, which allows the reconstruction of a complete reflectance or radiance spectrum. Figure 1.3 presents an "image cube" in which pixels are sampled across many narrowband images at particular spatial locations, resulting in a one-dimensional spectrum that is a plot of wavelength versus radiance or reflectance. The spectrum can be used to identify and characterize a particular feature within the scene, based on unique spectral signatures. Image spectral data can be acquired using both airborne and spaceborne platforms, and typically involves scanning many narrowband images simultaneously, while using some type of dispersion grating to produce the spectrum (USGS Speclab 2014).

The value of the IS technique lies in its ability to acquire a complete spectrum for each pixel in the image (Goetz et al. 1985, Vane and Goetz 1988), and the fact that its data allow the diagnostic narrowband spectral features that are present in most natural materials to be uniquely identified (Vane and Goetz 1993). Most natural materials on the Earth's surface have diagnostic absorption features in the 0.4 to 2.5 μm range of the reflectance spectrum. Since these diagnostic features are

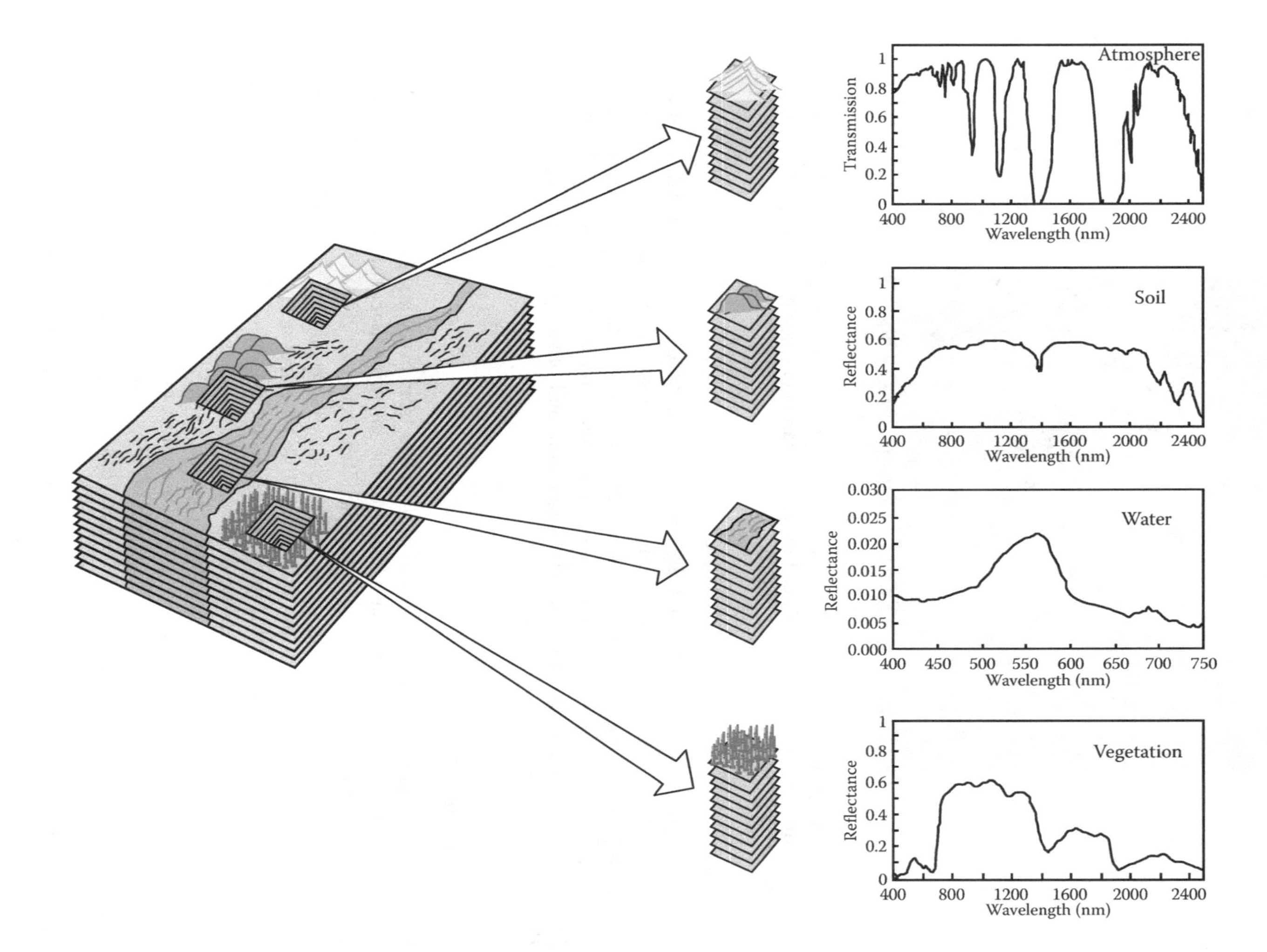

FIGURE 1.2 The concept of imaging spectroscopy illustrated using NASA's AVIRIS sensor with a spectrum measured for each spatial element in an image. The complete spectra can be analyzed in science research and applications in a variety of disciplines. (Courtesy of NASA/JPL-Caltech, http://aviris.jpl.nasa.gov/html/aviris.concept.html.)

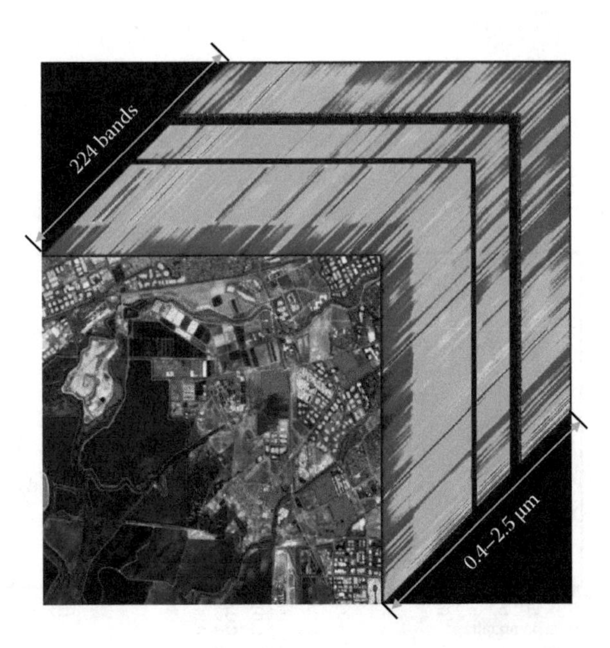

FIGURE 1.3 Hyperspectral image cube. The AVIRIS imagery was acquired from Moffett Field, San Francisco Bay, California, on June 20, 1997, and the surface false-color composite image of the cube was composed with a typical NIR/Red/Green vs. R/G/B.

of very narrow spectral appearance, typically 20–40 nm in width at the half-band depth (Hunt 1980), surface materials can be directly identified only if the spectrum is sampled at sufficiently high resolution (Figures 1.2 and 1.3). The IS technique meets the resolution requirements of spectral sampling, and, therefore, was developed for mineral mapping in the early 1980s (e.g., Goetz et al. 1985). It has since been employed for marine, vegetation, and other applications (e.g., Wessman et al. 1988, Gower and Borstad 1990).

1.1.3 Hyperspectral Remote Sensing

HRS technology can provide detailed spectral information from every pixel in an image. Whereas HRS refers mostly to remote sensing (from a distance), the emerging IS technique covers all spatial–spectral domains, from microscopic to macroscopic (Ben-Dor et al. 2013). The goal of HRS is to obtain the spectrum for each pixel in the image of a scene, with the purpose of finding objects, identifying materials, or detecting processes (Wikipedia 2014b). Besides acquiring the electromagnetic spectrum using the IS technique, HRS, like general traditional remote sensing technology, also considers hyperspectral imaging data processes (including image processing, and useful information extraction and presentation) and applications (Jensen 2005, Eismann 2012). In HRS, the actual capability in identifying materials is dependent on a number of factors. These factors include the abundance of the material of interest; the strength of absorption features for that material in the wavelength region measured; and the spectral coverage, spectral resolution, and the signal-to-noise ratio of the hyperspectral sensor or spectrometer. In spite of these factors, and since HRS was developed for improved identification of materials and quantitative determination of physical and chemical properties in areas of interest such as minerals, water, vegetation, soils, and man-made materials, HRS technology is well accepted in remote sensing as a tool for many applications. Some such applications include geology, ecology, geomorphology, limnology, pedology, and atmospheric sciences—especially in cases where other remote sensing means have failed or are incapable of obtaining additional information (Ben-Dor et al. 2013).

Although HRS technology has many advantages over traditional remote sensing, it presents several challenges. For example, acquiring high-quality spectral data in airborne and spaceborne HRS cases is not like acquiring data in a laboratory, where conditions are constant, optimal, and well controlled. Significant interference is encountered, such as the short dwell time of data acquisition over a given pixel, and hence leading to a lower signal-to-noise ratio (SNR); atmospheric attenuation of gases and aerosols (scatterings and absorptions); and the uncontrolled illumination conditions of the source and objects. Such interferences could result in undesirable HRS data for application purposes and present challenges to many disciplines, including atmospheric science, electro-optical engineering, aviation, computer science, statistics, applied mathematics, and more (Ben-Dor et al. 2013). Unlike conventional panchromatic imagery or even multispectral imagery, the information content in hyperspectral imagery does not readily lend itself to be easily visualized and extracted by computer processing if using MRS image processing techniques and algorithms. Thus, significant attention in the field of HRS research and application should be paid to developing algorithmic techniques to detect, classify, identify, quantify, and characterize objects and features of interest in captured HRS data (Eismann 2012). The general goal of developing HRS is to extract physical information from raw HRS data across a spectrum (radiance), similar to that which is collected by a spectrometer in a laboratory. Given the spectral measurements taken under laboratory conditions, the spectral information across all spectral regions can be quantitatively analyzed for all natural and artificial Earth materials, such as vegetation, water, gases, artificial material, soils, minerals and rocks, with many already available in spectral libraries (Ben-Dor et al. 2013). It is expected that if an ideal HRS sensor that can result in high SNR spectral data is used, a spectral analytical technique can be incorporated to yield new spectral products never before sensed by traditional remote sensing means (Clark et al. 1990, Krüger et al. 1998).

1.1.4 Differences between Hyperspectral and Multispectral Imaging

The major differences between hyperspectral and multispectral (or traditional) imaging include two aspects: (1) A hyperspectral sensor can acquire image data in several dozens to a few hundred narrow and contiguous spectral bands covering certain spectral ranges, whereas multispectral sensor measures image data in a few wide and discrete spectral bands within the certain spectral ranges; and (2) more importantly, the hyperspectral sensor's data can be used to extract diagnostic spectral features that are unique absorption bands in 20–40 nm for most natural materials (Hunt 1980), and are not possible for the multispectral imaging technology. For instance, in Figure 1.4, four Thematic

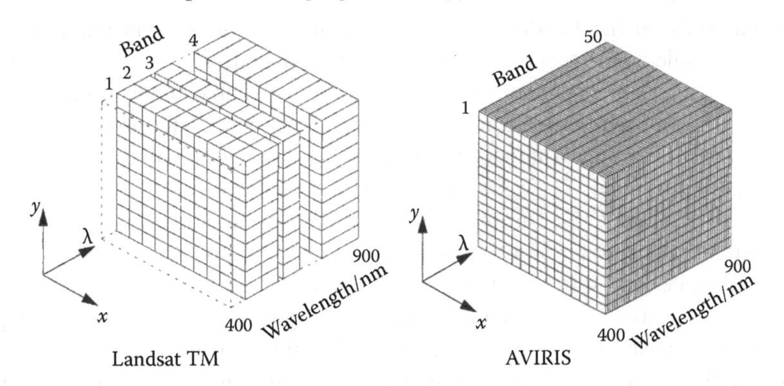

FIGURE 1.4 Visualization of the spatial and spectral resolutions of the Landsat TM and AVIRIS in the VNIR spectral range. The relative proportions between the two sensors are correct along each axis, and each small rectangular box represents one image pixel. The TM samples the spectral dimension incompletely and with relatively broad spectral bands. In comparison, AVIRIS represents almost a continuous spectral sampling. (Modified from Schowengerdt, R. A., *Remote Sensing Models and Methods for Image Processing*, 2nd ed., pp. 1–522, New York, Academic Press, 1997.)

Mapper (TM) bands cover the wavelength range of 0.4 to 0.9 μm but are not continuous; while with Airborne Visible/Infrared Imaging Spectrometer (AVIRIS), the spectral information in the same range is recorded in more than 50 narrow but contiguous bands. Multispectral sensors usually produce three to ten different band measurements in each pixel of the images they produce. Their image data are in a few discrete bands in a bandwidth (in full width at half maximum) of >50 nm. The bandwidth of multispectral remote sensing (e.g., TM, SPOT) is generally between 50 and 200 nm, which is much wider than the absorption bandwidths of most surface materials (Goetz et al. 1985, Vane and Goetz 1988). Hyperspectral sensors measure energy in narrower and more numerous bands than multispectral sensors. Hyperspectral images can contain as many as 200 (or more) contiguous spectral bands. The numerous narrow bands of hyperspectral sensors provide a continuous spectral measurement across the entire electromagnetic spectrum and therefore are more sensitive to subtle variations in reflected and emitted energy. Images produced from hyperspectral sensors contain much more spectral information than images from multispectral sensors and have a greater potential for detecting differences among materials on land and in water and the atmosphere. For example, multispectral imagery can be used to map forested areas, while hyperspectral data can be used to map tree species and other biophysical parameters within the forest. It is important to remember that HRS does not simply increase the number of bands from that multispectral remote sensing and that there is not necessarily a difference in spatial resolution between hyperspectral and multispectral data, but rather in their spectral resolutions (Govender et al. 2007).

When using remote sensing image processing to extract useful information for various application purposes, there are different natures of image information to focus on for hyperspectral and multispectral image data. In this case, Chang (2013) summarized such different natures of image information. Multispectral image processing must rely on image spatial information and correlation to make up insufficient spectral resolution as it results from a few discrete spectral bands. However, with the very high spectral resolution hyperspectral imaging sensors, many material substances that cannot be detected by multispectral imaging sensors can now be uncovered by hyperspectral images for data analysis and exploration. Therefore, an early development of traditional multispectral image processing has focused on spatial domain-based techniques (e.g., traditional supervised classification methods), and more recently developed hyperspectral data processing techniques have centered on efficiently using image spectral information (e.g., various spectral unmixing approaches).

1.1.5 Absorption Features and Diagnostic Spectral Features

Most natural materials on the Earth's surface have diagnostic absorption features in the 0.4 to 2.5 μm range on the reflectance spectrum. This is because, in general, there are two types of physical and chemical processes in most natural materials: (1) In the visible and near-infrared wavelength ranges of the spectra of minerals, electronic transition and charge transfer processes (via changes in energy states of electrons bound to atoms or molecules), in association with the presence or absence of transition metal ions (e.g., Fe, Ti, Cr, Co, Ni), mostly determine positions of diagnostic absorption features (Adams 1974, Hunt 1977, Burns 1989); (2) in the mid- to short-wave infrared part of the spectrum, vibrational processes in H_2O and OH^- associated with the presence or absence of water and hydroxyl, carbonate, and sulfate determine the absorption features (Hunt 1977, van der Meer 2004). According to the study by van der Meer (2004), diagnostic absorption features parameters/variables that include position, shape, depth, and width are controlled by the particular crystal structure in which the absorbing species is contained and by the chemical structure of the material. Consequently, the parameters characterizing the diagnostic absorption features can be directly correlated with the chemistry and structure of the sample (e.g., a mineral or vegetation species sample).

As one of the unique characteristics of hyperspectral imaging, HRS data allow the diagnostic spectral features (or narrow absorption bands) that are present in most natural materials to be uniquely identified (Vane and Goetz 1993). Since most natural materials on the Earth's surface possess diagnostic absorption features in the 0.4 to 2.5 μm spectral range and these diagnostic features typically have a very narrow spectral appearance (around 20–40 nm bandwidth; Hunt 1980), surface materials can be directly identified by using high-spectral resolution data (<10 nm bandwidth) as produced by hyperspectral imaging techniques. For example, Figure 1.5 presents the laboratory-measured reflectance spectra for a number of common minerals over a wavelength range between 2.0 and 2.5 μm. The Landsat TM obtains only one data point in this wavelength range (TM band 7), while an imaging spectrometer is able to acquire data points in continuous narrow bands at a bandwidth of approximately 10 nm. In Figure 1.5, the diagnostic spectral features of a number of common minerals are clearly present, such as Alunite at 2.14 μm and Kaolinite at 2.20 μm. Consequently, the parameters of different minerals' absorption features can be directly extracted from the hyperspectral data. Figure 1.6 illustrates hyperspectral imaging spectra of healthy green vegetation where some absorption features (bands) of different plant species are also shown, especially those present in the right part (b) of the figure. In practice, quantitative estimates of mineralogical composition and chemical analysis on the basis of hyperspectral image data have been demonstrated by many researchers (e.g., Resmini et al. 1997, Galvão et al. 2008). In vegetation and ecosystem studies, hyperspectral reflectance spectra have been used to estimate foliar biochemistry for more than two decades (e.g., Pu, Ge et al. 2003, Huber et al. 2008).

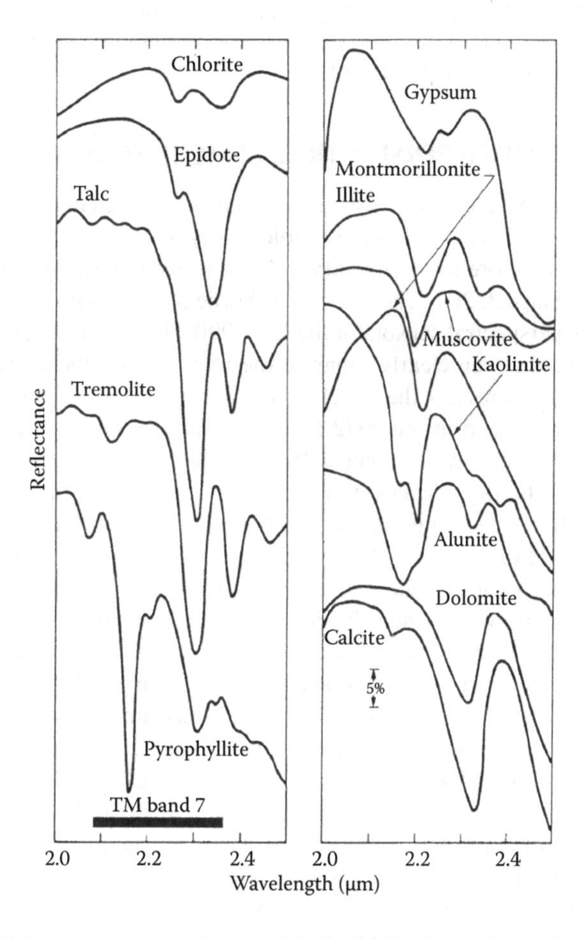

FIGURE 1.5 Selected laboratory spectra of minerals' diagnostic absorption and reflectance features. The spectra are displaced vertically to avoid overlap. The bandwidth of TM band 7 is also shown. (From Goetz, A. F. H., G. Vane, J. E. Solomon, and B. N. Rock, *Science*, 228(4704), 1985, 1147–1153. Reprinted with permission from AAAS.)

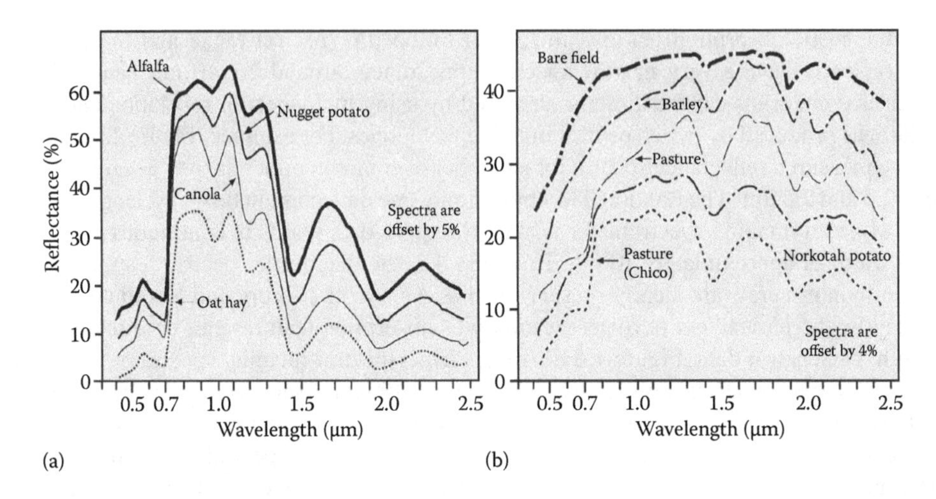

FIGURE 1.6 Hyperspectral data of healthy green vegetation in the San Luis Valley of Colorado obtained on September 3, 1993, using Airborne Visible Infrared Imaging Spectrometer (AVIRIS). (a) 224 channels, each 10 nm wide with 20 m × 20 m pixels. (b) Absorption features of some vegetation types. (From Clark, R. N., T. V. V. King, C. Ager, and G. A. Swayze. *Summaries of the Fifth Annual JPL Airborne Earth Science Workshop*, JPL Publication 95–1, 1995, pp. 35–38.)

1.2 DEVELOPMENT OF HYPERSPECTRAL REMOTE SENSING

Based on the history and representative instruments and sensors' characteristics of imaging spectroscopy, the development of HRS could be chronologically introduced in four development stages: Stage 1, HRS preparation before the 1980s; Stage 2, 1st generation of airborne hyperspectral sensors in the early 1980s; Stage 3, 2nd generation of airborne hyperspectral sensors in the late 1980s; and Stage 4, satellite hyperspectral sensors in the late 1990s. However, in fact, there is not an exact chronological time table that can clearly separate the four stages—these groupings are meant to provide a convenient way to review the development stages of HRS. The imaging spectrometry of the Earth could be initialized in the 1972 launch of ERST-1, later renamed Landsat-1 (Goetz 2009). The Landsat-1 multispectral scanner (MSS) data could be used in agricultural work and geologic interpretation of taking advantage of the synoptic view, as well as the spectral coverage that extended beyond the visible range of the spectrum (Goetz 1992). For geological application, by comparing enhanced MSS images with ground observations in field studies, geologists quickly found that subtle color variations on the images were difficult to identify in the field and spectral reflectance measurements of undisturbed, *in situ* surface samples would be necessary to interpret the image colors properly (Goetz 2009). Thus an early field instrument called the *portable field reflectance spectrometer* (PFRS) was developed to cover the spectral range (0.4–2.5 μm) of solar reflected radiance (Goetz 1975). As the demand for field instruments in the research community grew, other more technically advanced and easier-to-carry field spectrometers were developed, such as those created by Geophysical Environmental Research (GER) of Millbrook, New York, and Analytical Spectral Devices (ASD) of Boulder, Colorado (Goetz 2009). All *in situ* and laboratory spectra of minerals supported the notion that reflectance spectroscopy in the visible and shortwave infrared (SWIR) range would provide rich information about the composition of the Earth's surface (Hunt 1977). In the mid-1970s, although the imaging spectroscopic technique we know today was not available, field spectrometry and airborne multispectral imaging had proved the rationale for acquiring spectral information of minerals in SWIR (Goetz 1992). In 1976, the Shuttle Multispectral Infrared Radiometer (SMIRR), a ten-channel spinning filter wheel radiometer was tested, and, in 1981, the SMIRR flew on the second flight of the shuttle. It showed that a direct identification of surface mineralogy could be made from orbit with contiguous, narrowband spectral measurements.

The SMIRR radiometer acquired data with three channels only 10 nm wide and centered 10 nm apart around 2.2 μm. The data from the instrument provided the first direct identification of clay, kaolinite, and limestone from orbit (Goetz et al. 1982).

Due to the success of the SMIRR radiometer, a proposal was submitted to NASA that put forth the idea to build an airborne HRS sensor (airborne imaging spectrometer, or AIS), which would be sensitive to capturing mineral information across the SWIR region (Ben-Dor et al. 2013). The AIS with 2D detector arrays consisting of HgCdTe detectors (32 × 32 elements) enabled, for the first time, the generation of images at a wavelength longer than 1.1 μm. The array detector did not need a scan and provided sufficient improvement in the SNR to suit airborne applications. AIS had two versions, with two modes used in each: "tree mode," from 0.9–2.1 μm; and "rock mode," from 1.2–2.4 μm. The first image from AIS, acquired in late 1982, presented the scientific community with a fundamentally new class of data that required new approaches to information handling and extraction (Goetz et al. 1985). Although research on building this imaging spectrometer began at the end of 1970s (Chiu and Collins 1978), acquisition of the first AIS image kicked off the new generation of hyperspectral sensors. The first generation of imaging spectrometers included the AIS-I and AIS-II (Vane and Goetz 1988, Vincent 1997). The AIS-I was flown from 1982 to 1985; the AIS-II system was flown in 1986, collecting 128 bands of data, approximately 9.3 nm bandwidth. The AIS-I yielded an image with a narrow swath of 32 pixels while the AIS-II had a swath of 64 pixels, all with a ground pixel size of approximately 8 × 8 m^2 (for a detailed description, see the sensor systems in Chapter 3). Since then, AIS images had been successfully applied in several geological study areas. Because of the limitation of the swath-width of the two-dimension solid array detectors at that time, the commercial development of this type of imaging spectrometer was limited. Nevertheless, these developments did initiate a new era of hyperspectral remote sensing. The proceedings of conferences that summarized the activity and first set of results of the AIS missions were published by NASA Jet Propulsion Laboratory (JPL) AIS workshops in 1985–1987 (Vane and Goetz 1985, 1986; Vane 1987). Some important research work related to the analysis of AIS data was also published in a special issue of *Remote Sensing of Environment* (vol. 24, Febuary 1988).

Funded by NASA and proposed by the JPL, development of the second generation of airborne imaging spectrometers, represented by Airborne Visible/Infrared Imaging Spectrometer (AVIRIS), began in 1984, and the image first flew aboard the NASA ER-2 aircraft at 20 km altitude in 1987. Despite being a relatively low-quality SNR instrument (compared with today's hyperspectral sensors, and especially to the current upgraded AVIRIS sensor), the first AVIRIS demonstrated excellent performance relative to the AIS. It was a whiskbroom sensor with an SNR of around 100 and was the first imaging spectrometer to cover the solar-reflected spectrum from 0.4 μm to 2.5 μm with a swath of 614 pixels. AVIRIS collects upwelling radiance through 224 contiguous spectral bands at approximately 10 nm intervals across the spectrum (Green et al. 1998). The AVIRIS system consists of a sensor flown on the NASA ER-2 high altitude research aircraft, a dedicated ground data-processing facility, a dedicated calibration facility, and a small full-time operation team (Vane and Goetz 1993). Since 1987, the AVIRIS sensor has undergone upgrades, and today, the instrument is significantly different from the original. The major differences are its SNR (100 in 1987 compared with >1000 today), spectral coverage (0.4–2.5 μm compared with 0.35–2.5 μm) and spatial resolution (20 m compared with 2 m). The instrument can fly on different platforms at lower altitudes and has opened up new capabilities for potential users in many applications. The AVIRIS program has established an active HRS community in the United States that has rapidly matured (Ben-Dor et al. 2013). This continuous effort is to meet the requirements of investigators using AVIRIS spectral images for scientific research and applications. The AVIRIS has acquired and provided a large number of hyperspectral images for scientific research and applications every year since 1987 (Vane and Goetz 1993, Green et al. 1998). Most important research work related to analyses and applications of AVIRIS data was published in two special issues of *Remote Sensing of Environment* (vol. 44, May–June 1993; vol. 65, September 1998). In the late 1980s, several airborne commercial hyperspectral imaging sensors were also developed to acquire corresponding airborne hyperspectral images. For example, the fluorescence line imager (FLI; Hollinger et al. 1988), advanced solid state array spectrometer (ASAS) (Huegel 1988), and compact airborne spectrographic imager (CASI) also provided researchers and practitioners with a large number of hyperspectral images. In particular, CASI is the first commercially available, programmable airborne hyperspectral scanner. It uses along-track scanning with a 578-pixel CCD linear array to collect data in up to 288 bands between 400 and 900 nm at a 1.8-nm spectral interval. The precise number of bands, their locations, and bandwidths could all be programmed in flight. In addition, the hyperspectral digital image collection experiment (HYDICE) came into being in 1994 (Basedow and Zalewski 1995). It has characteristics similar to the AVIRIS, but acquires images with a push broom CCD-linear array. Another like the AVIRIS sensor is the airborne hyperspectral mapper (HyMap) series of airborne hyperspectral scanners, which have been deployed in a large number of countries in support of a wide variety of remote sensing applications, ranging from mineral exploration to defense research to satellite simulation on a global basis (Goetz 2009). The HyMap series continues to provide data with hyperspectral coverage across the solar wavelengths (0.4–2.5 μm) and in the thermal infrared (8–12 μm). The spectral bandwidths of HyMap are 10–20 nm in the spectral range of 0.4–2.5 μm and 100–200 nm in the thermal infrared range (see http://www.hyvista.com).

In the late 1990s, and in addition to the airborne hyperspectral systems mentioned previously, NASA and the European Space Agency (ESA) started development on the first generation of spaceborne hyperspectral sensor systems. Earth Observing-1 (EO-1; see http://eo1.usgs.gov) was the first satellite in NASA's New Millennium Program Earth Observing series. EO-1 was launched on November 21, 2000. The three primary EO-1 instruments are the advanced land imager (ALI), the Hyperion, and the atmospheric corrector (LAC). Among the three sensors, Hyperion and LAC are both hyperspectral sensors. The Hyperion instrument provides a new class of Earth observation data for improving Earth surface characterization. The Hyperion mission, designed for three years, is still operational today with a healthy sensor and data, although the SNR is poor in SWIR range. It provides a high-spectral resolution imager capable of resolving 220 spectral bands (from 0.4 to 2.5 μm) with a 30-meter spatial resolution. By 2014, Hyperion had produced a large number of satellite hyperspectral images for scientific validation and application studies. Initial scientific validation results of Hyperion images were published in a special issue of *IEEE Transactions on Geoscience and Remote Sensing* (vol. 41, issue 6, 2003). The compact high resolution imaging spectrometer (CHRIS) is a new imaging spectrometer, carried on board the ESA's PROBA satellite, launched on October 22, 2001. The CHRIS acquires 13 km^2 scenes at 17 m spatial resolution in 18 user-selected visible and near-infrared wavelengths. This agile satellite can also deliver up to five different viewing angles (nadir, +/– 55^0, and +/– 36^0). It is sensitive to the VIS–NIR region (410–1059 nm), and the number of bands is programmable, with up to 63 spectral bands. Although limited in its spectral region, the instrument provides a first view of the bidirectional reflectance distribution function (BRDF) effects for vegetation and water applications. Nearly 20,000 environmental science images have been acquired (https://earth.esa.int/web/guest/missions/esa-operational-eo-missions/proba). The hyperspectral and multi-angle capability of CHRIS makes it an important resource for studying BRDF phenomena of vegetation. Other applications also include coastal and inland waters, wild fires, education, and public relations. Moderate-resolution imaging spectroradiometer (MODIS) onboard the NASA EOS Terra satellite was launched on December 18, 1999. The MODIS sensor can also be considered part of the HRS activities in space due to its several spectral bands with bandwidth of 10 nm. Other orbital satellite hyperspectral sensors also include the Indian Space Research Organization (ISRO) VNIR hyperspectral imager (HySI), on board the Indian Microsatellite 1 and the Chinese VNIR HJ-1A satellite sensor (hyperspectral imager, HSI) both launched in 2008 (Staenz 2009, Miura and Yoshioka 2012). The former sensor has 64 bands with a spectral resolution of approximately 10 nm while the latter has 110–128 bands with a spectral resolution of 5 nm. All airborne and spaceborne hyperspectral sensor systems mentioned previously have provided a large amount of valuable hyperspectral image data for various research and applications.

1.3 OVERVIEW OF HYPERSPECTRAL REMOTE SENSING APPLICATIONS

HRS technology provides an innovative way to study many spatial phenomena on the Earth's surfaces and in the atmosphere by the merging of spectral and spatial information acquired by hyperspectral imaging systems and sensors. If the hyperspectral image data are of high quality, they can potentially allow near-laboratory level spectral sensing of targets from a distance. Thus, the information and knowledge gathered in the laboratory environment can be used to process the HRS data on a pixel-by-pixel basis (Ben-Dor et al. 2013). The initial motivation for the development of imaging spectrometry was for mineral identification, although early experiments were also conducted in botanical remote sensing (Goetz et al. 1985). However, since the late 1988s, imaging spectroscopy technology (i.e., HRS technology) has been successfully applied to a wide range of disciplines and areas, including geology, ecology, forestry, snow and ice, soil, environment, hydrology, disaster management, urban mapping, atmospheric study, agriculture, fisheries, oceans, and even national security—and these only constitute a few of the applications for HRS technology today. Since the advent of the first generation of airborne hyperspectral imaging sensors (AIS) in the early 1980s, thousands of journal articles, conference/workshop proceedings, books, and special issues of remote sensing journals have published results and findings of HRS research and applications with various levels of hyperspectral data acquired at different platforms, including laboratory, *in situ*, airborne, and spaceborne. Major publications in HRS research and applications literature include the following:

1. Workshop proceedings of NASA JPL Airborne Earth Science related to research and applications of airborne hyperspectral sensors AIS and AVIRIS (e.g., Vane and Goetz 1985, 1986 for AIS workshops; Green 1991, 1992 for AVIRIS workshops).
2. HRS books, including *Imaging Spectrometry: Basic Principles and Prospective Applications* (edited by van der Meer and de Jong 2001)—application areas include land degradation, soil, vegetation science, agriculture, minerals identification and mapping, petroleum geology, urban, and water; *Hyperspectral Remote Sensing of Vegetation* (edited by Thenkabail, Lyon, and Huete 2012); and *Hyperspectral Remote Sensing of Tropical and Sub-Tropical Forests* (edited by Kalacska and Sanchez-Azofeifa 2008).
3. Four special issues published in *Remote Sensing of Environment* (vol. 24, Feb. 1988; vol. 44. May–June 1993; vol. 65, Sept. 1998; vol. 113, Sept. 2009) and one special issue in *IEEE Transactions on Geosciences and Remote Sensing* (vol. 41, no. 6, June 2003).

An overview of literature in application studies of HRS imaging data in different disciplines and areas, including geology and soils, vegetation, ecosystems (agriculture and forestry), the atmosphere, coastal and inland waters, snow and ice hydrology, environmental hazards, and urban environment, will be given throughout this chapter.

1.3.1 Geology and Soils

Hyperspectral data have been available to researchers since the early 1980s, and their use for geologic applications is well documented. HRS was first tested for mineral identification and geological mapping (Goetz et al. 1985, Vane and Goetz 1993, Resmini et al. 1997). The application of hyperspectral imaging spectra to geology and soil investigations is based on the large number of minerals with unique spectral absorption bands in the solar-reflected spectrum. A variety of hyperspectral imaging sensor data have been used to determine the point, local, and regional distribution of minerals for a range of geology and soil science investigations (Green et al. 1998). The reflectance spectra acquired with fine spectral resolution hyperspectral sensors in the solar-reflected spectrum can be used to identify a large range of surface cover materials that cannot be identified with traditional broadband, low-spectral resolution data, such as Landsat TM and SPOT HRV sensors.

Diagnostic spectral features of various minerals and rocks in this spectral region allow us to determine their chemical composition and relative abundance (Crosta et al. 1997). Figure 1.5 presents selected laboratory spectra of several minerals' diagnostic absorption and reflectance features. The diagnostic absorption features were first extracted and applied for geological and soil mapping and mineral identification from various hyperspectral images. These features could include absorption-band wavelength position at the band minimum, absorption depth, width at half the band depth, and area (Schowengerd 1997). In addition, to sufficiently utilize the spectral and spatial information of various hyperspectral imaging data in geology and soils, investigators have also applied different analysis methods and techniques developed during the last three decades, including spectral matching, spectral unmixing, etc.

Since diagnostic absorption features generally have a 20–40 nm bandwidth (Hunt 1980) and are available from various hyperspectral data, researchers can directly extract diagnostic absorption features, such as absorption depth, width, and wavelength position, from hyperspectral images to identify and map some minerals and rock components. For example, Goetz et al. (1985) indicated that remote mineral identification was possible with AIS data from Cuprite, Nevada. They obtained similar characteristic spectra as those measured in the laboratory or in the field for kaolinite, alunite, and secondary quartz, containing an overtone Si-OH absorption feature from AIS image. *In situ* ground spectral measurements and laboratory spectra of field-collected samples verified the absorption features of alunite (2.17 μm) and kaolinite (2.20 μm) extracted from the AIS imaging image. In the study of hydrothermally altered rocks, Kruse (1988) mapped areas of a quartz–sericite–pyrite alteration zone by identifying sericite absorption features at 2.21, 2.25, and 2.35 μm, areas of argillic alteration containing montmorillonite based on a single absorption feature at 2.21μm, and calcite and dolomite by identifying their sharp diagnostic features at 2.34 and 2.32 μm with three combined flight lines of AIS images from the northern Grapevine Mountains, Nevada/California. The mapped areas using the AIS data agreed well with the areas identified by field mapping techniques. Crowley (1993) used high-spectral resolution data to study evaporate minerals in Death Valley playas. Evaporite minerals possess spectral absorption features in the AVIRIS spectral range that have been used to map these sedimentary units (Crowley 1993). In the study, eight different saline minerals were remotely identified, including three borates, hydroboracite, pinnoite, and revadavite, which previously had not been reported from the Death Valley evaporite crusts. By using relative absorption band-depth (RBD) images, Crowley et al. (1989) demonstrated that a number of rock and soil units were distinguished in the Ruby Mountains, including weathered quartz–feldspar pegmatites, marbles of several compositions, and soils developed over poorly exposed mica achists. They reported that the RBD images are both highly specific and sensitive to the presence of particular mineral absorption features. Baugh et al. (1998) also reported that a linear relation was found between ammonium concentration and the depth of a 2.12 μm ammonium absorption feature in buddingtonite in the southern Cedar Mountains, Nevada, with AVIRIS data.

To sufficiently utilize spectral information of hyperspectral images, spectral matching is an effective technique to identify and map different minerals and rock components. In the application of HRS images to geology and soils, two spectral matching approaches are frequently used: spectral angle mapper (SAM) approach and cross-correlogram spectral matching (CCSM). For instances, Baugh et al. (1998) employed AVIRIS data and the SAM approach to map ammonium minerals (buddingtonite) in hydrothermally altered volcanic rocks in the southern Cedar Mountains, Nevada. Their study indicated that AVIRIS high-spectral resolution data could be used for geochemical mapping and the SAM approach could be extended to quantitatively map other minerals that have absorption features in the short-wave infrared range. Kruse et al. (2003) also used the SAM mapping technique (as well as mixture-tuned matched filtering [MTMF]) with AVIRIS and satellite hyperspectral sensor Hyperion to successfully estimate and map mineral abundances for carbonates, chlorite, epidote, kaolinite, alunite, buddingtonite, muscovite, hydrothermal silica, and zeolite. The CCSM mapping technique, developed by van der Meer and Bakker (1997), was first applied to surface mineralogical mapping by using 1994 AVIRIS data from Cuprite, Nevada. Accurate mapping of kaolinite, alunite, and buddingtonite was achieved using the CCSM technique and AVIRIS data, assessed by using three parameters from the cross correlograms that were constructed on a pixel-by-pixel basis: the correlation coefficient at match position zero, the moment of skewness, and the significance (based on a Student's *t*-test of the validity of the correlation coefficients) (van der Meer and Bakker 1997).

Given the fact that there are a large amount of imaging pixels mixed with more than one mineral or component, spectral unmixing technique with hyperspectral data can help improve the accuracy of identifying and mapping various minerals. Although the technique also can be used to extract fractions or abundances of end-members from a mixed pixel in multispectral remotely sensed data, it is possible to simultaneously retrieve fraction information from more end-members with hyperspectral data. The spectral unmixing of each pixel is a process of inverting the linear mixing of those end-members to derive their proportion of surface coverage in each pixel. For example, Mustard (1993) used a linear spectral mixing model and AVIRIS data to investigate an area of soil, grass, and bedrock associated with the Kaweah serpentinite mélange in the foothills of the Sierra Nevada Mountains in California. With five spectral end-members plus one for shade in the spectral mixing model, he was able to account for almost all the spectral variability within the data set and derive information on the areal distribution of several surface cover types. Three of the end-members were shown to accurately model green vegetation, dry grass, and illumination. For the other three end-members, the spatial distributions in mixed pixels were still interpretable and coherent although their spectral distinction was very small. Kruse et al. (1993) used AVIRIS images and other classical geological information for the Death Valley region of California to produce detailed geological maps for subsequent use in a regional synthesis. A linear spectral unmixing of the AVIRIS data was used to determine relative mineral abundance and identification of mineral assemblages and mixtures. The mineral maps derived from the imaging spectrometer data were validated through subsequent ground spectral measurements and field mapping. Similar application studies using hyperspectral imaging data and spectral unmixing technique in mapping geology were also done by Adams et al. (1986), Kruse (1995), Ferrier and Wadge (1996), and Resmini et al. (1997).

The value of hyperspectral imaging data in discriminating subtle differences in soil composition and mapping other natural surface material is well demonstrated (Vane and Goetz 1993). As the materials on the Earth's surface, the absorption and scattering characteristics of soils are expressed in the solar spectral range. The absorption features in soils tend to be subtle due to particle size, scattering, and coating effects. Much research work with hyperspectral remote sensing images has been done for identification and distribution mapping of soils (e.g., Krishnan et al. 1980, Stoner and Baumgardner 1981, de Jong 1992, Palacios-Orueta and Ustin 1996, and Coops et al. 1998). AVIRIS spectra have been successfully used to map the distribution and relationship between soil sequences and other surface materials. For example, Palacios-Orueta and Ustin (1996) demonstrated that two soils belonging to the same series and a third soil belonging to a different but related series can be discriminated at a high level of accuracy using reflectance data from AVIRIS. In reviewing HRS technology used for soil science applications, Ben-Dor et al. (2009) discussed some cases of studies regarding soil degradation (salinity, erosion, and deposition), soil mapping and classification, soil genesis and formation, soil contamination, soil water content, and soil swelling using different hyperspectral imaging data. Based on the accumulated knowledge presented in the review paper and on the promising results of the reviewed studies, they feel strongly that the HRS technology has a strong capability for soil applications, some of which are yet to be identified.

Compared with those techniques and methods reviewed previously, some modeling methods (e.g., the modified Gaussian model [MGM] by Sunshine et al. 1990 and Sunshine and Piepers 1993) and artificial neural network algorithms (e.g., Benediktsson et al. 1995, Yang et al. 1997) are also developed but are relatively less-applied in geological mapping and mineral identification with hyperspectral data. For those readers who are interested in the techniques and methods and their applications, please see Chapter 5 and Chapter 7.

1.3.2 Vegetation and Ecosystems

Ecology and the study of terrestrial vegetation are important application fields for hyperspectral remote sensing (Green et al. 1998). A number of forest ecosystem variables, including leaf area index (LAI), absorbed fraction of photosynthetically active radiation (fPAR), canopy temperature, and community type are correlated with remotely sensed data or their derivatives (Johnson et al. 1994). Plant leaf water, chlorophyll, ancillary pigments, cellulose, lignin, and other constituents combine with the leaf and canopy structure to produce the reflectance of vegetation that could be measured by AVIRIS (Green et al. 1998). Figure 1.7 presents a vegetation reflectance curve across which several absorption and reflectance features caused by biophysical and biochemical parameters (bioparameters) and leaf internal structure are illustrated. However, sensors in common use, such as the Landsat TM, integrating radiance data over wide bands of the electromagnetic spectrum, have limited value in studying the dominant canopy reflectance features, such as the red spectral absorption band, the near-infrared reflectance band, and the mid-infrared water absorption band (Wessman et al. 1989). Moreover, extraction of red-edge and other "slops" optical parameters (e.g., Miller et al. 1990, 1991, Pu et al. 2004) that are related to plant stress or senescence is impossible with broadband sensors.

Hyperspectral sensors on board different types of platforms have made it possible to acquire higher spectral resolution data that contain more information on subtle spectral features of plant canopies. The use of narrow (1 to 10 nm) instead of broad (50 to 200 nm) spectral bands could offer new potentials for remote sensing applied to vegetation (Guyot et al. 1992). Plant biochemical

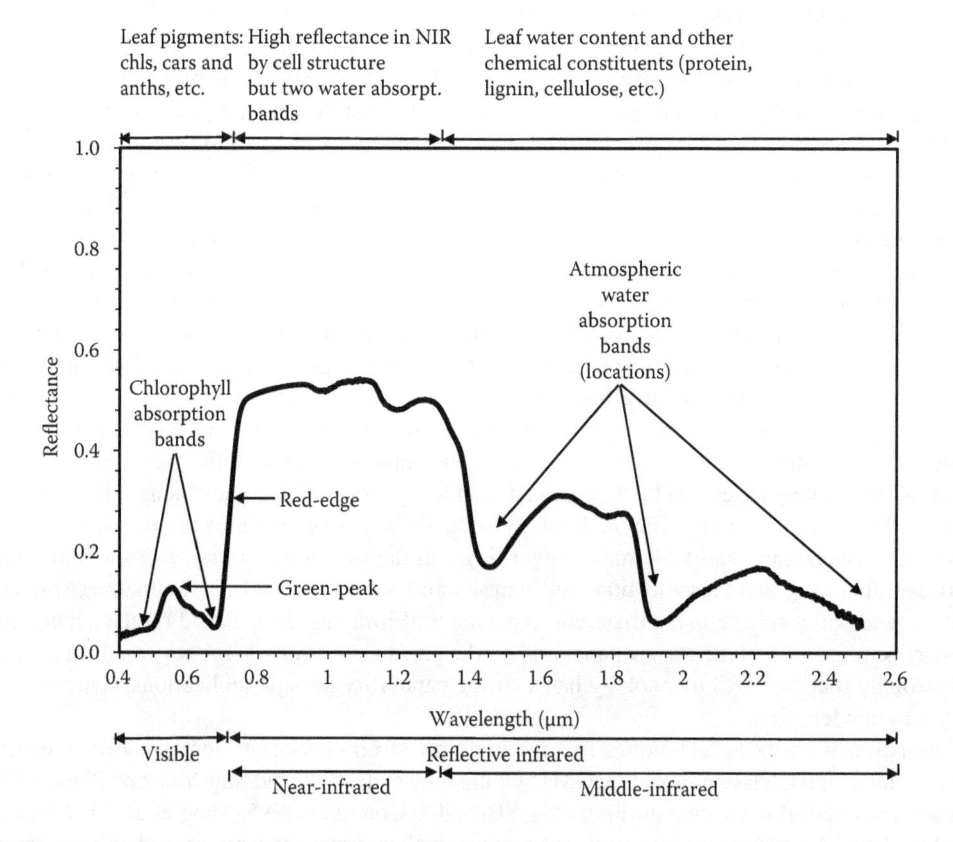

FIGURE 1.7 An oak tree leaf spectrum in details to illustrate major absorption and reflectance features and locations caused by pigments, water, other chemical constituents, and plant cell structure. Leaves from different species may have different strength of absorption and reflectance.

constituents include chlorophyll, water, protein, cellulose, sugar and lignin content, and nutrients, while biophysical parameters consist of LAI, APAR, canopy temperature and structure, and community type. Hyperspectral data have been proven to be more useful in estimating biochemical parameters' content and concentration at both leaf and canopy levels (e.g., Peterson et al. 1988, Johnson et al. 1994, Smith et al. 2003, Townsend et al. 2003, Darvishzadeh et al. 2008, Asner and Martin 2009, Ustin et al. 2009) and some other ecosystem structural (biophysical) parameters such as LAI, plant species composition, and biomass (e.g., Gong et al. 1997, Martin et al. 1998, le Maire et al. 2008, Pu et al. 2008) than traditional remotely sensed data. Therefore, besides classification and identification of vegetation types in terrestrial ecosystem study, hyperspectral remote sensing can also be applied to estimation of bioparameters and to evaluation of ecosystem functions.

Many minerals found on the Earth's surface have unique and diagnostic spectral reflectance signatures. Plants, on the other hand, are composed of the same few compounds and therefore should have similar spectral signatures (Vane and Goetz 1993). Indeed, major features of "peaks and valleys" along the spectral reflectance curve of a plant are due to the presence of pigments (e.g., chlorophyll), water, and other chemical constituents (Figure 1.7). Therefore, characterizing diagnostic absorption features in plant spectra with hyperspectral data as done in geological mapping and mineral identification can also be used for extraction of bioparameters of plants (e.g., Wessman et al. 1989, Johnson et al. 1994, Curran et al. 1995, Jacquemoud et al. 1996, Gong et al. 2003, Pu, Ge et al. 2003, Pu and Gong 2004, Cheng et al. 2006, Asner and Martin 2009). For example, Galvão et al. (2005) successfully used some absorption features extracted with a continuum-removal technique and other spectral indices from EO-1 Hyperion data to discriminate the five sugarcane varieties in southeastern Brazil. Huber et al. (2008) also estimated foliar biochemistry (the concentrations of N and carbon and the content of water) from hyperspectral data HyMap in mixed forest canopy using the continuum-removal technique.

Most researchers have employed the statistical analysis methods to correlate bioparameters (as dependent variables) with spectral reflectance, vegetation indices (VIs), or derivatives (as explanatory/independent variables) of spectra in the visible, NIR and SWIR wavelengths of hyperspectral data at leaf-, canopy-, or plant community-level (e.g., Peterson et al. 1988, Wessman et al. 1988, Smith et al. 1991, Johnson et al. 1994, Matson et al. 1994, Pinel et al. 1994, Yoder and Waring 1994, Gastellu-Etchegorry et al. 1995, Gamon et al. 1995, Gong et al. 1997, Yoder and Pettigrew-Crosby 1995, Grossman et al. 1996, Zagolski et al. 1996, Gitelson and Merzlyak 1997, Martin and Aber 1997, Blackburn 1998, Chen et al. 1998, Datt 1998, Martin et al. 1998, Serrano et al. 2002, Galvão et al. 2005, Colombo et al. 2008, Hestir et al. 2008, Huber et al. 2008, Pu 2012, Thenkabail et al. 2013). For example, Johnson et al. (1994) determined predictive relationships for biochemical concentrations using regressions between the chemical composition of forest canopy and the AVIRIS reflectance. After correlating vegetation indices of R_{NIR}/R_{700} and R_{NIR}/R_{550} with chlorophyll content, Gitelson and Merzlyak (1996, 1997) demonstrated that the indices for chlorophyll assessment were important for two deciduous species, maple and chestnut. With EO-1 Hyperion hyperspectral image data, Galvão et al. (2005) successfully discriminated five sugarcane varieties in southeastern Brazil using the multiple discriminant analysis method, which produced a classification accuracy of 87.5%. With multiple linear regression models, continuum-removal technique, and normalized HyMap spectra, Huber et al. (2008) estimated foliar concentrations of N and carbon, and the content of water in a mixed forest canopy. Recently, researchers have increasing interest in applying the partial least square regression (PLSR) approach to calibrate relationships between spectral variables, often derived from hyperspectral data and a set of bioparamters (e.g., Hansen and Schjoerring 2003, Asner and Martin 2008, Darvishzadeh et al. 2008, Martin et al. 2008, Weng et al. 2008, Prieto-Blanco et al. 2009). For instance, using spectral measurements taken from leaves and bioparameter data (chlorophyll-a, -b, carotenoids, anthocyanins, water, N, P, and specific leaf area) collected from 162 Australian tropical forest species, along with PLSR approach and canopy radiative transfer modeling, Asner and Martin (2008) concluded that a suite of leaf properties among tropical forest species can be estimated using full-range leaf spectra of fresh foliage collected in the field.

Some spectral position variables, rather than the amplitude of the spectral feature, were investigated in estimating and mapping bioparameters. For example, the red edge of vegetation between 670 nm and 780 nm has been widely modeled by a number of researchers. Based on the spectral properties of the pigments, some researchers have used red edge optical parameters to estimate plant, leaf, and canopy chlorophylls (Chls) content and concentration. The red edge optical parameters can be used for estimating Chls concentration (e.g., Belanger et al. 1995, Curran et al. 1995), nutrient constituent concentrations (e.g., Gong et al. 2002, Cho et al. 2008), leaf relative water content (e.g., Pu et al. 2004), and forest LAI (e.g., Pu, Gong et al. 2003).

Physically based models rest on a theoretical basis consisting of developing a leaf or canopy scattering and absorption model that involves biochemistry and biophysics. In the context of the remote sensing of bioparameters, such models have been used in the forward mode to calculate leaf or canopy reflectance and transmittance and in inversion to estimate leaf or canopy chemical and physical properties from remote sensing data, especially from hyperspectral imaging data. During the last three decades, based on literature searched and analyzed by Jacquemoud et al. (2009), the most popular and important radiative transfer (RT) models on leaf, canopy, and leaf–canopy coupled optical properties are PROSPECT, SAIL, and PROSAIL, as well as their modified versions. Many investigators employed the physically based models at leaf or canopy level to retrieve biochemical parameters including leaf pigments from either simulated spectra or hyperspectral image data (e.g., Asner and Martin 2008, Feret et al. 2008, Zhang et al. 2008a,b). Such modeling approaches have also been used in defining predictive relationships that have been applied to hyperspectral imagery to generate maps of Chl (Haboudane et al. 2002; Zarco-Tejada et al. 2004, 2005). In addition, geometric–optical (GO) models belong to one type of RT models developed to capture the variation of remote sensing signals of the Earth's surface with illumination and observation angles. Since GO models emphasize the effect of canopy architecture, they are very effective in capturing the angular distribution pattern of reflected radiance; thus, they are used widely in hyperspectral remote sensing applications (e.g., Chen and Leblanc 2001) as aforementioned RT models.

1.3.3 The Atmosphere

Since there are absorption and scattering of molecular and particle constituents in the atmosphere, the atmosphere has a profound impact on the signal recorded by airborne or spaceborne imaging systems operating in the solar-reflected portion of the spectrum (Vane and Goetz 1993). These atmospheric constituents generally include water vapor, carbon dioxide, oxygen, clouds, aerosols, ozone, and other atmospheric gases (Green et al. 1998). Water vapor is of interest because it is the most variable component in the atmosphere and the primary absorber across the solar-reflected spectrum. It is also a key driver in global circulation and hence is also of interest in studies involving weather and climate modeling (Vane and Goetz 1993) and in the hydrologic cycle. A number of algorithms and techniques have been developed for derivation and analysis of water vapor from hyperspectral data (e.g., Gao and Goetz 1990, Green et al. 1991, Gao et al. 1993, Gao and Davis 1998, Schläpfer et al. 1998). Generally, these algorithms evaluate the strengths of the water vapor absorption bands at 0.94 μm and 1.14 μm and relate the strengths to the total column water abundance in the atmosphere. For example, Gao et al. (1993) used a three-channel ratioing technique (with a few channels near 0.94 μm and 1.14 μm) in their derivation of the water vapor on a pixel-by-pixel basis from AVIRIS radiance data. Carrère and Conel (1993) performed a sensitivity analysis on the two simple techniques to retrieve path precipitable water from the AVIRIS high-spectral resolution radiance data (with continuum interpolated band ratio and narrow/wide ratio), using the 0.94 μm water absorption band. Amounts retrieved using the N/W approach match more closely to *in situ* measurements. Schläpfer et al. (1998) applied a technique called atmospheric precorrected differential absorption (APDA), which is derived directly from simplified radiative transfer equations to two AVIRIS images acquired in 1991 and 1995. The technique combines a partial atmospheric correction with a differential absorption technique. The accuracy of the measured water vapor columns from the AVIRIS data is within a range of 65% compared to ground truth radiosonde data.

Cirrus clouds play an important role in the global energy balance and are often an undetected confounding factor in remotely sensed data. AVIRIS images in strong water vapor absorption bands at 1.38 μm and 1.85 μm provide a new approach to detect the presence and distribution of cirrus clouds (e.g., Gao and Goetz 1992, Hutchison and Chloe 1996). Non–cirrus-cloud fraction and cloud shadow analyses have also been studied with AVIRIS data (e.g., Kuo et al. 1990, Feind and Welch 1995). Detection of various aerosols in the atmosphere is very important because they are a type of scattering sources to solar-reflected spectrum and people are interested in understanding their impact on the local and global atmosphere and climate. Knowledge on aerosols is required for atmospheric correction of imaging spectroscopy and other remotely sensed data. Isakov et al. (1996) employed AVIRIS data to derive aerosol information. High-contrast natural surface and artificial surface were taken as two different study sites in order to obtain aerosol optical depth. Their preliminary results suggest that background aerosol optical depth (i.e., $\tau_{aerosol} < 0.1$) could not be retrieved from space with adequate accuracy; however, the aerosol optical depth of the most polluted atmosphere (i.e., $\tau_{aerosol} < 0.2$) could be retrieved with adequate accuracy.

1.3.4 Coastal and Inland Waters

The number and abundance of absorbing and scattering components found in the coastal ocean, inland lakes, and rivers support the use of spectroscopy to measure and map the constituents in these environments. These constituents include chlorophyll-a, a variety of planktonic species, dissolved organics, suspended sediments with local and distant sources, bottom composition and substrates, submerged aquatic vegetation, and water depth (Green et al. 1998). Hyperspectral data and *in situ* measurements have been used to investigate the distribution and concentration (or content) of those constituents in coastal environments (e.g., Carder et al. 1993, Richardson et al. 1994, Brando and Dekker 2003, Pu et al. 2012, Pu and Bell 2013) and inland lakes (e.g., O'Neill et al. 1987, Hamilton et al. 1993, Jaquet et al. 1994). They have also been used in bathymetric mapping in coastal environments and inland lakes (e.g., Clark et al. 1997, George 1997, Kappus et al. 1998, Sandidge and Holyer 1998). The increased signal-to-noise ratio and absolute calibration of AVIRIS in the wavelength range of 400–1000 nm of the spectrum is supporting new investigations in the coastal and inland water environments. For instance, Carder et al. (1993) used AVIRIS data to study dissolved and particulate constituents in Tampa Bay, Florida. Images of the absorption coefficient at 415 nm and backscatter coefficient at 671 nm were used to show the dissolved and particulate constituents of the Tampa Bay plume. The results were verified at three sites within the bay using *in situ* ship-based data measured at the time of the AVIRIS overflight. Hamilton et al. (1993) used AVIRIS data to study chlorophyll concentration and bathymetry in Lake Tahoe and demonstrated that AVIRIS data is useful for a variety of oceanographic and limnological applications. Using a CASI image, Clark et al. (1997) mapped tropical coastal environments and most components related to reef, seagrass habitats, coastal wetlands, and mangroves. The CASI data can provide detailed quantitative information on habitat extent and composition, water depth, seagrass biomass, and mangrove canopy cover, etc. Holden and LeDrew (1999) mapped coral reef in tropical areas. Brando and Dekker (2003) tested satellite hyperspectral Hyperion data for their capabilities over a range of water targets in Eastern Australia, including Moreton Bay in southern Queensland. This region was selected due to its spatial gradients in optical depth, water quality, bathymetry, and substrate composition. Their experimental results demonstrated that Hyperion imagery, after preprocessing and noise reduction strategies, had sufficient sensitivity to detect, and could be used to map, optical water quality concentrations of colored dissolved organic matter, chlorophyll, and total suspended matter simultaneously in the complex waters of estuarine and coastal systems in Moreton Bay. Pu and Bell (2013) evaluated a protocol that utilizes image optimization algorithms followed by atmospheric and sunglint corrections to the three satellite sensors (Landsat TM, EO-1 Advanced

Land Imager, and Hyperion) and a fuzzy synthetic evaluation technique to map and assess seagrass abundance in Pinellas County, Florida. The experimental results indicate that the Hyperion sensor produced the best results of the five-class seagrass cover classification and better multiple regression models for estimating the three biometrics (seagrass cover percentage, LAI, and biomass) for creating seagrass abundance maps along with two environmental factors.

1.3.5 Snow and Ice Hydrology

Snow albedo plays an important role in the regional and global energy balance. The study of seasonal snow pack is important because of its relevance to the aquatic ecology of alpine watersheds and to the Earth's radiation budget (Pu and Gong 2000). Properties of snow and ice recorded in the solar spectral range include fractional cover, grain size, surface liquid water content, impurities, and shallow depth (Green et al. 1998). Derivation of snow grain size has been pursued with AVIRIS spectra (Nolin and Dozier 1993). Models based on radiative transfer theory have been pursued to derive the snow grain size and liquid water content simultaneously from spectra measured by AVIRIS (Green and Dozier 1996). Painter et al. (1998) used AVIRIS data and advanced methods of spectral mixture analysis to derive accurate grain size parameters. The sensitivity of snow spectral reflectance to grain size translates these grain size gradients into spectral gradients. They performed numerical simulations to demonstrate the sensitivity of nonlinear mixture analysis to grain size for a range of sizes and snow fractions. Analysis of fraction under/overflow and residuals confirmed mixture analysis sensitivity to grain-size gradients. From imaging spectrometer data (AVIRIS and Hyperion), Dozier et al. (2009) estimated the fraction of each pixel that is covered by snow, the liquid water content in the surface layer, and the amount of radiative forcing caused by absorbing impurities. Their experience with imaging spectrometer data has allowed extension of the fractional snow cover and albedo estimates to multispectral sensors, particularly MODIS.

1.3.6 Environmental Hazards

During the last three decades, hyperspectral data measured by various imaging spectrometers have been used to determine and assess surface compositions that are directly or indirectly related to hazard environments. For example, Swayze et al. (1996) used AVIRIS data over the EPA Superfund site at Leadville, Colorado, to map the distribution of acid-generating minerals. Acidic water and mobilized heavy metals are an environmental hazard at Leadville. Farrand and Harsanyi (1995) also used AVIRIS imaging spectra to assess the transport of hazardous mine waste downstream through alluvial processes. CASI data have been acquired over opencast mining residual lakes in central Germany. The CASI data combined with water sampling of hydrochemical and hydrobiological properties were used to map different water qualities in lakes (Boine et al. 1999). Ferrier et al. (2007) used the data from two airborne hyperspectral sensors, AVIRIS and CASI, to map the release of hazardous mine waste from abandoned gold mining areas. Their results for the study of an abandoned gold mine at Rodalquilar in southern Spain have demonstrated the potential of HRS technology to resolve the distributions of mine waste and secondary iron species at mine sites and in adjacent rivers. The pH value is one of the major chemical parameters affecting results of remediation programs carried on abandoned mines and dumps and one of the major parameters controlling heavy metal mobilization and speciation. Kopačková et al. (2012) demonstrated the feasibility of mapping surface pH on the basis of airborne hyperspectral sensor HyMap datasets. A natural hazard related to volcanoes in the Cascade Mountains of the western United States was investigated with AVIRIS data in 1996. The spectra derived from the AVIRIS data were used to map the spatial distribution of specific alteration minerals. These mapped alteration minerals from AVIRIS images are associated with weak zones in the volcano-derived slopes and are indicators of regions of slope instability and potential collapse (Crowley and Zimbelman 1997).

Biomass burning is one of the major sources of trace gases and aerosol particles, with significant ramifications for atmospheric chemistry, cloud properties, and radiation budget (Kaufman, Justice et al. 1998). In the smoke, clouds, and radiation–Brazil (SCAR-B) experiment, Kaufman, Hobbs et al. (1998) used remote sensing data (AVIRIS, NOAA-AVHRR, and MODIS simulator data) to study biomass burning, emphasizing the biomass burning associated with measurements of surface biomass, fires, smoke aerosol and trace gases, clouds, radiation, and their climate effects. Their experimental results demonstrated that remote sensing data including high spectral resolution data proved to be useful for monitoring smoke properties, surface properties, and the impacts of smoke on radiation and climate. Characteristics of the fuel, fire process, combustion products, and post-fire regrowth of biomass burning are expressed in the spectra measured by imaging spectrometers. For example, Roberts et al. (2003) evaluated the performance of satellite hyperspectral sensor Hyperion relative to airborne hyperspectral sensor AVIRIS for fire danger assessment in the southern California Chaparral. They focused on the Santa Barbara area, a region that has experienced a number of recent catastrophic fires. They compared reflectance, measures of fuel moisture/live biomass, fuel condition, and fuel type derived from spatially overlapping Hyperion and AVIRIS datasets acquired in early June 2001. The comparative results indicate that while neither sensor exceeded 85% accuracy, AVIRIS came close to this requirement and produced a map significantly more accurate than Hyperion. However, Hyperion was capable of mapping three critical land-cover classes at high accuracy that are of importance due to fire danger: bare soil, senesced grasslands, and chamise.

1.3.7 Urban Environments

In terrestrial urban environment, two major aspects can be remotely sensed: natural targets (e.g., soil, water, vegetation, and gases) and man-made targets (e.g., buildings, pools, roads, and vehicles) (Ben-Dor 2001). Urban areas provide a complex material environment, both in the number of materials present and in the spatial scale of material variation. Whereas HRS technology produces mixed spectral information regarding complex urban areas, use of sophisticated classification techniques and well-known knowledge about the spectral behavior of pure urban components are the basic keys for the successful application of HRS technology in the urban environment (e.g., Ridd et al. 1997, Huqqani and Khurshid 2014). For example, Ridd et al. (1997) used AVIRIS data acquired over Pasadena, California, and showed that a combination of the hyperspectral data with neural network analysis improved the classification results. Kalman and Bassett III (1997) made use of HYDICE hyperspectral data in classification and material identification in an urban environment. Based on a comprehensive set of material spectral library, they developed a procedure for land cover classification that can be automated and performed with little or no prior knowledge of objects in the scene. Comparison of the material identifications produced using the technique with ground truth observations indicated a relatively high agreement. The type of product could potentially provide input to urban analysis models which require accurate assessment of impervious surfaces. Imaging spectroscopy provides a uniform synoptic approach to mapping surface materials in the urban and adjacent environments to support urban characterizing, monitoring, and planning (Green et al. 1998). A demonstration of this has been given by Fiumie and Marino (1997) over the city of Rome, using MIVIS (consisting of 102 bands across the 0.43–8.18 μm spectral region, with 4-m spatial resolution) data. They concluded that it is possible to differentiate between paving materials made with basalt and those made with marble, which are products found in close proximity of Rome. An alternate urban application example is measurement of the composition of roofs in fire hazard zones. For instance, Zhao (2001) used PHI data to distinguish the difference among four roof-cover materials that are painted blue and difficult to differentiate with regular band compositing. Another example is that one mission to map asbestos roofs over an urban area has shown remarkable results using the MIVIS sensor data integrated with GIS data (Marino et al. 2000). During the mission they were able to distinguish between aluminum, tiles, bitumen, and plain concrete using the spectral information derived from the MIVIS imagery.

To improve the application effectiveness of HRS data in urban environments, many studies have demonstrated that hyperspectral data measured by various imaging spectrometers combined with other remotely sensed data may be a good idea. For example, Lehmann et al. (1998) merged the data from high-spectral resolution sensor HyMap and high-resolution digital camera sensors obtained over Berlin to improve urban recognition. Hepner et al. (1998) combined AVIRIS data with IF-SAR data over the Westwood neighborhood in the city of Los Angeles, demonstrating this capability toward optimal sensing of the complex urban environment. In addition, the creation of a 3D urban GIS database for more efficient urban planning will obviously benefit from the combinations of hyperspectral data with LiDAR or IF-SAR data (Wicks and Campos-Marquetti 2010). Hence, it is widely accepted that HRS technology has a great capacity for providing more useful information and will attain its maximum potential regarding the urban environment if HRS data are combined with data acquired simultaneously by other advanced sensors (Ben-Dor 2001).

1.4 PERSPECTIVE OF HYPERSPECTRAL REMOTE SENSING

NASA's Mission to Planet Earth (MTPE) and Earth Observing System (EOS) program, ESA's PRoject for On-Board Autonomy (PROBA) platform, and India and China's spaceborne hyperspectral missions will continue through 2017 and beyond. The final objectives of these missions and programs are to permit assessment of various Earth system processes, including hydrological processes, biogeochemical processes, atmospheric processes, ecological processes, and geophysical processes. Imaging spectrometry will be a key technology in realizing these missions and programs. For this case, a large number of existing airborne hyperspectral sensors/systems, such as AVIRIS, CASI, HyMap, HYDICE, etc., will continue to provide airborne hyperspectral imaging data to meet requirements of investigations and applications of HRS technology and for the needs of calibration and the development of new satellite hyperspectral sensors. Current EO-1 is the first satellite in NASA's New Millennium Program (NMP) Earth Observing series. The EO-1 and other NMP missions will develop and validate instruments and technologies for space-based Earth observations with unique spatial, spectral, and temporal characteristics not previously available. CHRIS/PROBA is an important multiangular hyperspectral sensor whose potential has not been fully unfolded. With ISRO's VNIR Hyper-Spectral Imager (HySI) on board the Indian Microsatellite 1 (IMS-1) and the Chinese VNIR HJ-1A satellite sensors, new opportunities will arise for the use of hyperspectral imaging data in various application and research areas as the larger ground sampling distance (GSD) (≥100 m) combined with a larger swath width (≥50 km) of these sensors (Staenz 2009). In the future, all these existing airborne and spaceborne hyperspectral sensors/systems will continue to offer hyperspectral images in the solar-reflected spectral range from all over the Earth with multitemporal coverage.

Future spaceborne satellite sensors will not only continue to measure solar-reflected spectra as images in the manner executed by AVIRIS and Hyperion, but they will also sense some emitted thermal infrared radiance. The satellite hyperspectral sensors/missions developed and launched in the near future may include the environmental mapping program (EnMAP), the hyperspectral precursor of the application mission (PRISMA), the hyperspectral portion of the multi-sensor microsatellite imager (MSMI), hyperspectral infrared imager (HyspIRI), hyperspectral imager suite (HISUI), and fluorescence explorer visible (FLEX-VIS; Buckingham and Staenz 2008, Staenz 2009, Miura and Yoshioka 2012, Matsunaga et al. 2013). Table 1.1 briefly summarizes the future spaceborne hyperspectral sensors/missions. EnMAP and PRISMA, which are very similar in the spatial and spectral characteristics, are scheduled to operate in space during the 2018/2018 time frame. Both sensors, plus the hyperspectral portion of MSMI, will have a much higher data acquisition capacity than their technology precursors, Hyperion and CHRIS, which will allow for better data provision capabilities for the civilian community to feed initial operational applications on a regional basis. The HyspIRI mission is one of the Tier 2 missions recommended for launch in the 2020 time frame (NRC's Decadal Survey Report 2007). This global survey mission provides an

TABLE 1.1
Future Spaceborne Hyperspectral Sensors/Missions

Sensor/Mission Country	EnMAP Germany	PRISMA Italy	MSMI South Africa & Belgium	HyspIRI USA	HISUI Japan	FLEX-VIS ESA
GSD (m)	30	30	15	60	30	300
Swath at nafir (km)	30	30	15	150	30	390
Wavelength coverage (nm)	420–2450	400–2500	400–2350	380–2500	400–2500	400–1000
Number of bands	218	>200	200	212	185	>60
Spectral resolution (nm @ FWHM)	5/10 for VNIR	<10	10	10	10 for VNIR 12.5 for SWIR	5–10
Expected launch year	2018	2018	2010 (TBD)	>2023	>2018	>2016

Source: Buckingham, R., and K. Staenz, *Canadian Journal of Remote Sensing* 34(S1), S187–S197, 2008; Staenz, K., Terrestrial imaging spectroscopy, some future perspectives. In *Proceedings of the 6th EARSEL SIG IS*, Tel Aviv, Israel. 2009; Miura, T., and H. Yoshioka, Hyperspectral data in long-term, cross-sensor continuity studies, in *Hyperspectral Remote Sensing of Vegetation*, eds. P. S. Thenkabail, J. G. Lyon, and A. Huete, pp. 611–633. Boca Raton, FL: CRC Press/Taylor & Francis Publishing Group. 2012; Matsunaga, T. et al., Current status of Hyperspectral Imager Suite (HISUI). In *IGARSS 2013*, Melbourne, Australia, pp. 3510–3513, 2013.

unprecedented capability to assess how ecosystems respond to natural and human-induced changes, and it will help us assess the status of elements of biodiversity around the world and the role of different biological communities on land, coastal zones, and the surface of the deep ocean (NASA 2009 HyspIRI Science Workshop Report 2010). The HyspIRI mission has its roots in the imaging spectrometer Hyperion on EO-1 and in ASTER with the multispectral thermal IR instrument flown on EOS/Terra. Although similar to HyspIRI in spectral characteristics, the hyperspectral mission of the HISUI is a targeting mission with spatial and spectral characteristics similar to MSMI (Staenz 2009). The ESA's FLEX-VIS hyperspectral instrument is sensitive to the VNIR only, which was proposed to complement the Earth Explorer mission FLEX. The FLEX mission represents a fully innovative way of addressing vegetation monitoring at a global scale by providing information on terrestrial vegetation fluorescence never available before from space observations (Bézy et al. 2008).

Since the advent of HRS, many technical difficulties have been overcome in areas such as sensor development and calibration, data processing, and mining. However, there are still several main issues or challenges today that require solutions in order to move HRS technology toward more frequent operational use. These issues or challenges are addressed as follows:

1. The altitude and flight velocity of satellites cause significant challenges to the measurement of high-quality spectra from space. Accurate spectral, radiometric, and spatial calibration of spaceborne imaging spectrometers is difficult and requires new techniques and capabilities (Green et al. 1998). Therefore, development of a robust atmospheric correction algorithm is very important for spaceborne imaging spectrometer data (e.g., Hyperion data). Also, it is necessary to improve or upgrade the properties of existing airborne and spaceborne imaging systems and sensors (e.g., improving SNR of Hyperion data in SWIR range).
2. An operational AVIRIS-like spaceborne mission is needed to provide 20/30 m GSD data. Such a mission must cover an adequate area on the ground with a daily data acquisition capacity or revisit time of less than five days to ensure that data are readily available and can fulfill the requirements of users, such as large-scale mapping (Staenz 2009). With PRISMA and HyspIRI currently under development, some of these requirements will be met in the 2018/2023 time frame. However, the data quality level of these sensors does not quite match that of AVIRIS, since its SNR is lower and has a relatively large GSD (>30 m).

Therefore, it is necessary to continue improvements on imaging spectrometers in order to further advance imaging spectroscopy, especially toward its operational use.

3. Hyperspectral systems and sensors need to be improved on capabilities of onboard processing and data compression. For this purpose, automated procedures are required to successfully create onboard products. For instance, raw data can be calibrated onboard to generate an at-sensor radiance data product, which can be further converted into surface reflectance using an automatic atmospheric correction procedure. It is also necessary to directly move away from traditional data sets to higher-level products, such as land cover, vegetation indices, and other thematic maps, so that these high-quality spectral data products can be used directly for many application purposes, such as emergency events, since onboard processing will cut the response time significantly.
4. Hyperspectral data processing algorithms need to be improved in order to increase data processing efficiency and capability of automatic target detection. The demand for efficient algorithms on data reduction, feature extraction, fusion, target detection, and quantitative interpretation of hyperspectral data acquired from different sources at different times will continue to grow. For example, spectral unmixing remains an important information extraction task for hyperspectral data analysis, especially to improve the end-member selection from scenes with only mixed pixels. The use of principal component analysis, mathematical programming, and factor analysis needs to be further assessed in solving the linear mixing problem. Radiative transfer modeling in vegetation canopy should emphasize improving the model inversion, bridging different scaling levels (e.g., leaf to canopy), and real-time target detection. Inversion of radiative transfer models with hyperspectral data assisted by analysis of multiangular data will be useful tools for solving nonlinear spectral mixing problems for angular data, and they can be used to retrieve structural information of vegetation on the Earth's surface.
5. Although hyperspectral data analytical tools are now readily available, there is a general lack of robust automated procedures to process data quickly with a minimum of user intervention. Therefore, a powerful information extraction tool that is easily portable and easily used will allow rapid expansion in the use of imaging spectrometry. Moreover, not only individual tools, but entire processing chains need to be automated and implemented into dedicated systems in order to be effective. An example of such a system would be one that monitors coastal waters. These systems need to be intelligent and have a learning capability built in to improve their capability over time (e.g., expert systems) (Staenz 2009). This will help ease user interaction.
6. Attention needs to incorporate the spectral and spatial information inherent in the hyperspectral data and to emphasize procedures for fusion of these data with data from various other sources. For instance, a data fusion procedure is hyperspectral and LiDAR data, a powerful combination, which is currently used to better characterize the surface properties and canopy structure. Moreover, such procedures need more development in the future to fully utilize this capability. In the future, the richness of information available in the continuous spectral coverage afforded by both airborne and spaceborne imaging spectrometers makes it possible to more accurately and correctly address questions in water quality, snow/ice hydrology, vegetation ecosystem, atmospheric science, oceanic and surface mineralogy.

1.5 SUMMARY

HRS technology provides detailed spectral information from every pixel in an image and offers an innovative way to detect and map many spatial phenomena on the Earth's surface and in the atmosphere via subtle and contiguous spectral information acquired by hyperspectral imaging systems and sensors. In the beginning of this chapter, fundamentals of spectroscopy and imaging spectroscopy were introduced, and then the definition and concept of HRS as well as differences between multispectral remote sensing and HRS were explained. In the next section, the history and development of HRS were chronologically reviewed, following the four development stages: HRS preparation before the 1980s; the first generation of airborne hyperspectral sensors in the early 1980s (representative HRS sensor, AIS); the second generation of airborne hyperspectral sensors in the late 1980s (AVIRIS, CASI, HyMap, and HYDICE); and satellite hyperspectral sensors in the late 1990s (EO-1/Hyperion and Probe/CHRIS). After reviewing HRS development, an overview of applications and investigations of HRS technology to different disciplines and areas, which included geology and soils, vegetation and ecosystems, the atmosphere, coastal and inland waters, snow and ice hydrology, environmental hazards, and urban environment was reviewed. This overview placed an emphasis on the current status of applications and investigations of HRS imaging data to these disciplines and fields. Finally, a perspective of HRS based on NASA, the ESA, and other countries' spaceborne hyperspectral imaging missions and programs that are being implemented and developed in the near future was delivered and discussed. Some main issues and challenges associated with HRS applications that we must solve were also discussed.

REFERENCES

Adams, J. B. 1974. Visible and near-infrared diffuse reflectance: Spectra of pyroxenes as applied to remote sensing of solid objects in the solar system. *Journal of Geophysical Research* 79(32):4829–4836.

Adams, J. B., M. O. Smith, and P. E. Johnson. 1986. Spectral mixture modeling: A new analysis of rock and soil types at the Viking Lander 1 site. *Journal of Geophysical Research* 91:8098–8112.

Asner, G. P., and R. E. Martin. 2009. Airborne spectranomics: Mapping canopy chemical and taxonomic diversity in tropical ecosystems. *Frontiers in Ecology and the Environment* 7(5):269–276.

Basedow, R. W., and E. Zalewski. 1995. Characteristics of the HYDICE sensor. In *Summaries of the Fifth Annual JPL Airborne Earth Science Workshop*, JPL Publication 95–1, vol. 1:9.

Baugh, W. M., F. A. Kruse, and W. W. Atkinson. 1998. Quantitative geochemical mapping of ammonium minerals in the southern Cedar Mountains, Nevada, using the Airborne Visible/Infrared Imaging Spectrometer (AVIRIS). *Remote Sensing of Environment* 65:292–308.

Belanger, M. J., J. R. Miller, and M. G. Boyer. 1995. Comparative relationships between some red edge parameters and seasonal leaf chlorophyll concentrations. *Canadian Journal of Remote Sensing* 21(1):16–21.

Ben-Dor, E. 2001. Imaging spectroscopy for urban applications, in *Imaging Spectrometry: Basic Principles and Prospective Applications*, eds. F. van der Meer and S. M. de Jong, pp. 243–281. Dordrecht, the Netherlands: Kluwer Academic Press.

Ben-Dor, E., T. Malthus, A. Plaza, and D. Schläpfer. 2013. Hyperspectral remote sensing, in *Airborne Measurements for Environmental Research: Methods and Instruments*, eds. M. Wendisch and J.-L. Brenguier, pp. 413–456. Weinheim, Germany: Wiley-VCH.

Ben-Dor, E., S. Chabrillat, J. A. M. Demattê, G. R. Taylor, J. Hill, M. L. Whiting, and S. Sommer. 2009. Using Imaging Spectroscopy to study soil properties. *Remote Sensing of Environment* 113:S38–S55.

Benediktsson, J. A., J. R. Sveinsson, and K. Arnason. 1995. Classification and feature extraction of AVIRIS data. *IEEE Transactions on Geoscience and Remote Sensing* 33(5):1194–1205.

Bézy, J.-L., S. Delwart, and M. Rast. 2000. MERIS: A new generation of ocean-colour sensor onboard Envisat. *ESA Bulletin* 103:48–56.

Blackburn, G. A. 1998. Quantifying chlorophylls and caroteniods at leaf and canopy scales: An evaluation of some hyperspectral approaches. *Remote Sensing of Environment* 66:273–285.

Boine, J., K. Kuka, C. GlaBer, C. Olbert, and J. Fischer. 1999. Multispectral investigations of acid mine lakes of lignite open cast mines in Central Germany. *IGARSS1999 Proceedings*, Hamburg, Germany, June 28–July 2, 1999, pp. 855–857.

Brando, V. E., and A. G. Dekker. 2003. Satellite hyperspectral remote sensing for estimating estuarine and coastal water quality. *IEEE Transactions on Geoscience and Remote Sensing* 41(6):1378–1387.

Buckingham, R., and K. Staenz. 2008. Review of current and planned civilian space hyperspectral sensors for EO. *Canadian Journal of Remote Sensing* 34(S1):S187–S197.

Burns, R. G. 1989. Spectral mineralogy of terrestrial planets: Scanning their surfaces remotely. *Mineralogical Magazine* 53:135–151.

Carder, K. L., P. Reinersman, R. F. Chen, F. Muller-Karger, C. O. Davis, and M. Hamilton. 1993. AVIRIS calibration and application in coastal oceanic environments. *Remote Sensing of Environment* 44:205–216.

Carrère, V., and J. E. Conel. 1993. Recovery of atmospheric water vapor total column abundance from imaging spectrometer data around 940 nm-sensitivity analysis and application to Airborne Visible/Infrared Imaging Spectrometer (AVIRIS) data. *Remote Sensing of Environment* 44:179–204.

Chang, C. I. 2013. Overview and Introduction, in *Hyperspectral Data Processing: Algorithm Design and Analysis*, pp. 1–30, 164 p. Hoboken, NJ: John Wiley & Sons, Inc.

Chen, J. M., and S. G. Leblanc. 2001. Multiple-scattering scheme useful for geometric optical modeling. *IEEE Transactions on Geoscience and Remote Sensing* 39:1061–1071.

Chen, Z., C. D. Elvidge, and D. P. Groeneveld. 1998. Monitoring seasonal dynamics of arid land vegetation using AVIRIS data. *Remote Sensing of Environment* 65:255–266.

Cheng, Y.-B., P. J. Zarco-Tejada, D. Riaño, C. A. Rueda, and S. L. Ustin. 2006. Estimating vegetation water content with hyperspectral data for different canopy scenarios: Relationships between AVIRIS and MODIS indexes. *Remote Sensing of Environment* 105:354–366.

Chiu, H. Y., and W. Collins. 1978. A spectroradiometer for airborne remote sensing. *PE & RS* 44:507–517.

Cho, M. A., A. K. Skidmore, and C. Atzberger. 2008. Towards red-edge positions less sensitive to canopy biophysical parameters for leaf chlorophyll estimation using properties optique spectrales des feuilles (PROSPECT) and scattering by arbitrarily inclined leaves (SAILH) simulated data. *International Journal of Remote Sensing* 29(8):2241–2255.

Clark, C. D., H. T. Ripley, E. P. Green, A. J. Edwards, and P. J. Mumby. 1997. Mapping and measurement of tropical coastal environments with hyperspectral and high spatial resolution data. *International Journal of Remote Sensing* 18(2):237–242.

Clark, R., A. Gallagher, and G. Swayze. 1990. Material absorption band depth mapping of imaging spectrometer data using a complete band shape least-squares fit with library. In *Proceedings of the Second Airborne Visible/Infrared Imaging Spectrometer (AVIRIS)*, Pasadena, California, pp. 176–186, JPL Publication 90–54.

Clark, R. N., T. V. V. King, C. Ager, and G. A. Swayze. 1995 (January 23–26). Initial vegetation species and senescence/stress mapping in the San Luis Valley, Colorado using imaging spectrometer data. *Summaries of the Fifth Annual JPL Airborne Earth Science Workshop*, ed. R. O. Green, JPL Publication 95–1, pp. 35–38.

Colombo, R., M. Meroni, A. Marchesi, L. Busetto, M. Rossini, C. Giardino, and C. Panigada. 2008. Estimation of leaf and canopy water content in poplar plantations by means of hyperspectral indices and inverse modeling. *Remote Sensing of Environment* 112:1820–1834.

Coops, N., P. Ryan, and A. Bishop. 1998. Investigating CASI responses to soil properties and disturbance across an Australian Eucalypt forest. *Canadian Journal of Remote Sensing* 24(2):153–168.

Crosta, A. P., and C. R. de F. Souza. 1997 (November 17–19). Evaluating AVIRIS hyperspectral remote sensing data for geological mapping in Laterized Terranes, Central Brazil. In *Proceedings of the Twelfth International Conference and Workshops on Applied Geologic Remote Sensing*, Denver, Colorado, vol. II:II-430–II-437.

Crowley, J. K. 1993. Mapping playa evaporite minerals with AVIRIS data. *Remote Sensing of Environment* 44(2–3):337–356.

Crowley, J. K., and D. R. Zimbelman. 1997. Mapping hydro thermally altered rocks on Mount Rainier, Washington, with Airborne Visible/Infrared Imaging Spectrometer (AVIRIS) data. *Geology* 25(6): 559–562.

Crowley, J. K., D. W. Brickey, and L. C. Rowan. 1989. Airborne imaging spectrometer data of the Ruby Mountains, Montana: Mineral discrimination using relative absorption band-depth images. *Remote Sensing of Environment* 29:121–134.

Curran, P. J., W. R. Windham, and H. L. Gholz. 1995. Exploring the relationship between reflectance red edge and chlorophyll content in slash pine leaves. *Tree Physiology* 15:203–206.

Darvishzadeh, R., A. Skidmore, M. Schlerf, C. Atzberger, F. Corsi, and M. Cho. 2008. LAI and chlorophyll estimation for a heterogeneous grassland using hyperspectral measurements. *ISPRS Journal of Photogrammetry and Remote Sensing* 63:409–426.

Datt, B. 1998. Remote sensing of chlorophyll a, chlorophyll b, chlorophyll a+b, and total carotenoid content in Eucalyptus leaves. *Remote Sensing of Environment* 66:111–121.

De Jong, S. M. 1992. The analysis of spectroscopical data to map soil types and soil crusts of Mediterranean eroded soils. *Soil Technology* 5:199–211.

Dozier, J., R. O. Green, A. W. Nolin, and T. H. Painter. 2009. Interpretation of snow properties from imaging spectrometry. *Remote Sensing of Environment* 113:S25–S37.

Eismann, M. 2012. *Hyperspectral Remote Sensing*. Bellingham, WA: SPIE Press.

Encyclopedia Britannica. 2014. Spectroscopy. http://www.britannica.com/EBchecked/topic/558901/spectroscopy, accessed on December 1, 2014.

Farrand, W. H., and J. C. Harsanyi. 1995. Minerologic variations in fluvial sediments contaminated by mine tailings as determined by AVIRIS data, Coeur D'Alene River Valley, Idaho. In *Summaries of the Fifth Annual JPL Airborne Earth Science Workshop*, JPL Publication 95–1, vol. 1:47–50.

Feind, R. E., and R. M. Welch. 1995. Cloud fraction and cloud shadow property retrievals from coregistered TIMS and AVIRIS imagery: The use of cloud morphology for registration. *IEEE Transactions on Geoscience and Remote Sensing* GE-33(1):172–184.

Feret, J.-B., C. François, G. P. Asner, A. A. Gitelson, R. E. Martin, L. P. R. Bidel, S. L. Ustin, G. le Maire, and S. Jacquemoud. 2008. PROSPECT-4 and 5: Advances in the leaf optical properties model separating photosynthetic pigments. *Remote Sensing of Environment* 112:3030–3043.

Ferrier, G., and G. Wadge. 1996. The application of imaging spectrometry data to mapping alteration zones associated with gold mineralization in southern Spain. *International Journal of Remote Sensing* 17(2):331–350.

Ferrier, G., B. Rumsby, and R. Pope. 2007. Application of hyperspectral remote sensing data in the monitoring of the environmental impact of hazardous waste derived from abandoned mine sites, in *Mapping Hazardous Terrain Using Remote Sensing*, ed. R. M. Teeuw, Geological Society of London, Special Publications, 283:107–116.

Fiumie, L., and C. M. Marino. 1997. Airborne hyperspectral MIVIS data for the characterization of urban historical environments. In *Proceedings of the Third International Airborne Remote Sensing Conference and Exhibition II*, Copenhagen, Denmark, pp. 770–771.

Galvão, L. S., A. R. Formaggio, and D. A. Tisot. 2005. Discrimination of sugarcane varieties in Southeastern Brazil with EO-1 Hyperion data. *Remote Sensing of Environment* 94:523–534.

Galvão, L. S., A. R. Formaggio, E. G. Couto, and D. A. Roberts. 2008. Relationships between the mineralogical and chemical composition of tropical soils and topography from hyperspectral remote sensing data. *ISPRS Journal of Photogrammetry & Remote Sensing* 63:259–271.

Gamon, J. A., C. B. Field, M. L. Goulden, K. L. Griffin, A. E. Hartley, G. Joel, J. Penuelas, and R. Valentini. 1995. Relationships between NDVI, canopy structure, and photosynthesis in three California vegetation types. *Ecological Applications* 5(1):28–41.

Gao, B., and A. F. Goetz. 1990. Determination of total column water vapor in the atmosphere at high spatial resolution from AVIRIS data using spectral curve fitting and band ratioing techniques. In *Proceedings of SPIE: Imaging Spectroscopy of the Terrestrial Environment*, 1298:138–149.

Gao, B. C., and C. O. Davis. 1998. Examples of using imaging spectrometry for remote sensing of the atmosphere, land, and ocean. In *Proceedings of SPIE: Hyperspectral Remote Sensing and Applications*, 3502:234–242.

Gao, B. C., and A. F. H. Goetz. 1992. Separation of cirrus clouds from clear surface from AVIRIS data using the 1.38mm water vapor band. In *Summaries of the Third Annual JPL Airborne Geoscience Workshop*, JPL Publication 92–14, vol. 1:98–100.

Gao, B.-C., K. B. Heidebrecht, and A. F. H. Goetz. 1993. Derivation of scaled surface reflectances from AVIRIS data. *Remote Sensing of Environment* 44:165–178.

Gastellu-Etchegorry, J. P., F. Zagolski, E. Mougin, G. Marty, and G. Giordano. 1995. An assessment of canopy chemistry with AVIRIS case study in the Landes Forest, Southwest France. *International Journal of Remote Sensing* 16(3):487–501.

George, D. G. 1997. Bathymetric mapping using a compact airborne spectrographic imager (CASI). *International Journal of Remote Sensing* 18(10):2067–2071.

Gitelson, A. A., and M. N. Merzlyak. 1996. Signature analysis of leaf reflectance spectra: Algorithm development for remote sensing of chlorophyll. *Journal of Plant Physiology* 148:494–500.

Gitelson, A. A., and M. N. Merzlyak. 1997. Remote estimation of chlorophyll content in higher plant leaves. *International Journal of Remote Sensing* 18:2691–2697.

Goetz, A. 1992. Imaging spectroscopy for Earth remote sensing, in *Imaging Spectroscopy: Fundamentals and Prospective Applications*, eds. F. Toselli and J. Bodechtel, pp. 1–19. London: Kluwer Academic Publishers.

Goetz, A. 2009. Three decades of hyperspectral remote sensing of the Earth: A personal view. *Remote Sensing of Environment* 113:5–16.

Goetz, A. F. H. 1975. Portable field reflectance spectrometer. Appendix E in Application of ERTS images and image processing to regional geologic problems and geologic mapping in Northern Arizona. *JPL Technical Report*, pp. 32–1597.

Goetz, A. F. H., L. C. Rowan, and M. J. Kingston. 1982. Mineral identification from orbit: Initial results from the shuttle multispectral infrared radiometer. *Science* 218:1020–1024.

Goetz, A. F. H., G. Vane, J. E. Solomon, and B. N. Rock. 1985. Imaging spectrometry for earth remote sensing. *Science* 228(4704):1147–1153.

Gong, P., R. Pu, and B. Yu. 1997. Conifer species recognition: An exploratory analysis on in situ hyperspectral data. *Remote Sensing of Environment* 62:189–200.
Gong, P., R. Pu, and R. C. Heald. 2002. Analysis of in situ hyperspectral data for nutrient estimation of giant sequoia. *International Journal of Remote Sensing* 23(9):1827–1850.
Gong, P., R. Pu, G. S. Biging, and M. Larrieu. 2003. Estimation of forest leaf area index using vegetation indices derived from Hyperion hyperspectral data. *IEEE Transactions on Geoscience and Remote Sensing* 41(6):1355–1362.
Govender, M., K. Chetty, and H. Bulcock. 2007. A review of hyperspectral remote sensing and its application in vegetation and water resource studies. *Water SA* 33(2):145–152.
Gower, J. F. R., and G. A. Borstad. 1990. Mapping of phytoplankton by solar-stimulated fluorescence using an imaging spectrometer. *International Journal of Remote Sensing* 11(2):313–320.
Green, R. O., ed. 1991. *Proceedings of the Third Airborne Visible/Infrared Imaging Spectrometer (AVIRIS) Workshop*, JPL Publication 91–28, Jet Propulsion Laboratory, Pasadena, California, 326 pp.
Green, R. O., ed. 1992. *Summaries of the Third Annual JPL Airborne Geoscience Workshop*, vol 1. AVIRIS, JPL Publication 92–14, Jet Propulsion Laboratory, Pasadena, California, 159 pp.
Green, R. O., and J. Dozier. 1996. Retrieval of surface snow grain size and melt water from AVIRIS spectra. In *Summaries of the Sixth Annual JPL Airborne Earth Science Workshop*, JPL Publication 96–4, vol. 1:127–134.
Green, R. O., J. E. Conel, J. S. Margolis, C. J. Bruegge, and G. L. Hoover. 1991. An inversion algorithm for retrieval of atmospheric and leaf water absorption from AVIRIS radiance with compensation for atmospheric scattering. In *Proceedings of the Third AVIRIS Workshop*, JPL Publication, 91–28, pp. 51–61.
Green, R. O., M. L. Eastwood, C. M. Sarture, T. G. Chrien, M. Aronsson, B. J. Chippendale, J. A. Faust, B. E. Pavri, C. J. Chovit, M. Solis, M. R. Olah, and O. Williams. 1998. Imaging spectroscopy and the airborne visible/infrared imaging spectrometer (AVIRIS). *Remote Sensing of Environment* 65:227–248.
Grossman, Y. L., S. L. Ustin, S. Jacquemoud, E. W. Sanderson, G. Schmuck, and J. Verdebout. 1996. Critique of stepwise multiple linear regression for the extraction of leaf biochemistry information from leaf reflectance data. *Remote Sensing of Environment* 56:182–193.
Guyot, G., F. Baret, and S. Jacquemoud. 1992. Imaging spectroscopy to vegetation studies, in *Imaging Spectroscopy: Fundamentals and Prospective Applications*, eds. F. Toselli, and J. Bodechtel, pp. 145–165. The Netherlads: Springer.
Haboudane, D., J. R. Miller, N. Tremblay, P. J. Zarco-Tejada, and L. Dextraze. 2002. Integrated narrow-band vegetation indices for prediction of crop chlorophyll content for application to precision agriculture. *Remote Sensing of Environment* 81(2–3):416–426.
Hamilton, M. K., C. O. Davis, W. J. Rhea, S. H. Pilorz, and K. L. Carder. 1993. Estimating chlorophyll content and bathymetry of Lake Tahoe using AVIRIS data. *Remote Sensing of Environment* 44:217–230.
Hansen, P. M., and J. K. Schjoerring. 2003. Reflectance measurement of canopy biomass and nitrogen status in wheat crops using normalized difference vegetation indices and partial least squares regression. *Remote Sensing of Environment* 86:542–553.
Hepner, G. F., B. Houshmand, I. Kulikov, and N. Brayant. 1998. Investigation of the Integration of AVIRIS and IFSAR for urban analysis. *Photogrammetric Engineering and Remote Sensing* 64:813–820.
Hestir, E. L., S. Khanna, M. E. Andrew, M. J. Santos, J. H. Viers, J. A. Greenberg, S. S. Rajapakse, and S. L. Ustin. 2008. Identification of invasive vegetation using hyperspectral remote sensing in the California Delta ecosystem. *Remote Sensing of Environment* 112:4034–4047.
Holden, H., and E. Ledrew. 1999. Hyperspectral identification of coral reef features. *International Journal of Remote Sensing* 20(13):2545–2563.
Hollinger, A. B., L. H. Gray, J. F. R. Gower, and H. R. Edel. 1988. The fluorescence line imager: An imaging spectrometer for ocean and remote sensing. In *Proceedings of SPIE: Imaging Spectroscopy II*, 834:2–11.
Huber, S., M. Kneubühler, A. Psomas, K. Itten, and N. E. Zimmermann. 2008. Estimating foliar biochemistry from hyperspectral data in mixed forest canopy. *Forest Ecology and Management* 256:491–501.
Huegel, F. G. 1988. Advanced solid state array spectroradiometer: Sensor and calibration improvements. In *Proceedings of SPIE: Imaging Spectroscopy II*, 834:12–21.
Hunt, G. R. 1977. Spectral signatures of particulate minerals in the visible and near-infrared. *Geophysics* 42:501–513.
Hunt, G. R. 1980. Electromagnetic radiation: The communication link in remote sensing, in *Remote Sensing in Geology*, eds. B. Siegal and A. Gillespia, pp. 5–45. New York: Wiley.
Huqqani, I., and K. Khurshid. 2014. Comparative study of supervised classification of urban area hyperspectral satellite imagery. *Journal of Space Technology* 4(1):7–15.
Hutchison, K. D., and N. J. Choe. 1996. Application of 1 center dot 38 mm imagery for thin cirrus detection in daytime imagery collected over land surface. *International Journal of Remote Sensing* 17(17):3325–3342.

Isakov, V. Y., R. E. Feind, O. B. Vasilyev, and R. M. Welch. 1996. Retrieval of aerosol spectral optical thichmess from AVIRIS data. *International Journal of Remote Sensing* 17(11):2165–2184.

Jacquemoud, S., S. L. Ustin, J. Verdebout, G. Schmuck, G. Andreoli, and B. Hosgood. 1996. Estimating leaf biochemistry using the PROSPECT leaf optical properties model. *Remote Sensing of Environment* 56:194–202.

Jacquemoud, S., W. Verhoef, F. Baret, C. Bacour, P. J. Zarco-Tejada, G. P. Asner, C. François, and S. L. Ustin. 2009. PROSPECT+SAIL models: A review of use for vegetation characterization. *Remote Sensing of Environment* 113:S56–S66.

Jaquet, J. M., F. Schanz, P. Bossard, K. Hanselmann, and F. Gendre. 1994. Measurements and significance of biooptical parameters for remote sensing in 2 sub alpine lakes of different trophic state. *Aquatic Sciences* 56(3):263–305.

Jensen, J. R. 2005. *Introductory Digital Image Processing: A Remote Sensing Perspective*, 3rd edition. Upper Saddle River, NJ: Prentice Hall.

Johnson, L. F., C. A. Hlavka, and D. L. Peterson. 1994. Multivariate analysis of AVIRIS data for canopy biochemical estimation along the Oregon transect. *Remote Sensing of Environment* 47:216–230.

Kalacska, K., and G. A. Sanchez-Azofeifa, eds. 2008. *Hyperspectral Remote Sensing of Tropical and Sub-Tropical Forests*. Boca Raton, FL: CRC Press/Taylor & Francis Publishing Group.

Kalman, L. S., and E. M. Bassett III. 1997. Classification and material identification in an urban environment using HYDICE hyperspectral data. In *Proceedings of SPIE: Imaging Spectrometry III*, 3118:57–68.

Kappus, M. E., C. O. Davis, and W. J. Rhea. 1998. Bathymetry from fusion of airborne hyperspectral and laser data. In *Proceedings of SPIE: Imaging Spectrometry IV*, 3438:40–51.

Kaufman, Y. J., C. O. Justice, L. P. Flynn, J. D. Kendall, E. M. Prins, L. Giglio, D. E. Ward, W. P. Menzel, and A. W. Setzer. 1998. Potential global fire monitoring from EOS-MODIS. *Journal of Geophysical Research* 103(D24):32215–32238.

Kaufman, Y. J., P. V. Hobbs, V. W. J. H. Kirchhoff, P. Artaxo, L. A. Remer, B. N. Holben, M. D. King, D. E. Ward, E. M. Prins, K. M. Longo, L. F. Mattos, C. A. Nobre, J. D. Spinhirne, Q. Ji, A. M. Thompson, J. F. Gleason, S. A. Christopher, and S.-C. Tsay. 1998. Smoke, clouds, and radiation-Brazil (SCAR-B) experiment. *Journal of Geophysical Research* 103(D24):31783–31808.

Kopačková, V., S. Chevrel, A. Bourguignon, and P. Rojík. 2012. Mapping hazardous low-pH material in mining environment: Multispectral and hyperspectral approaches. *IGARSS 2012*, 2695–2698.

Krishnan, P., J. D. Alexander, B. J. Butler, and J. W. Hummel. 1980. Reflectance technique for predicting soil organic matter. *Soil Science Society of America Journal* 44:1282–1285.

Krüger, G., H. Erzinger, and H. Kaufmann. 1998. Laboratory and airborne reflectance spectrometric analyses of lignite overburden dumps. *Journal of Geochemical Exploration* 64:47–65.

Kruse, F. A. 1988. Use of Airborne Imaging Spectrometer data to map minerals associated with hydrothermally altered rocks in the Northern Grapevine Mountains, Nevada and California. *Remote Sensing of Environment* 24:31–51.

Kruse, F. A. 1995. Mapping spectral variability of geologic targets using Airborne Visible/Inrared Imaging Spectrometer (AVIRIS) data and a combined spectral feature/unmixing approach. In *Proceedings of SPIE: Imaging Spectrometry*, 2480:213–224.

Kruse, F. A., A. B. Lefkoff, and J. B. Dietz. 1993. Expert system-based mineral mapping in northern Death Valley, California/Nevada using the Airborne Visible/Infrared Imaging Spectrometer (AVIRIS). *Remote Sensing of Environment* 44:309–336.

Kruse, F. A., J. W. Boardman, and J. F. Huntington. 2003. Comparison of airborne hyper-spectral data and EO-1 Hyperion for mineral mapping. *IEEE Transactions on Geoscience and Remote Sensing* 41:1388–1400.

Kuo, K. S., R. M. Welch, B. C. Gao, and A. F. H. Goetz. 1990. Cloud identification and optical thickness retrieval using AVIRIS data. In *Proceedings of the Second AVIRIS Workshop*, JPL Publication 90–54, 149–156.

Le Maire, G., C. François, K. Soudani, D. Berveiller, J.-Y. Pontailler, N. Bréda, H. Genet, H. Davi, and E. Dufrêne. 2008. Calibration and validation of hyperspectral indices for the estimation of broadleaved forest leaf chlorophyll content, leaf mass per area, leaf area index and leaf canopy biomass. *Remote Sensing of Environment* 112:3846–3864.

Lehmann, F., T. Buchert, S. Hese, A. Hofmann, A. Mayer, F. Oschuts, and Y. Zhang. 1998. Data fusion of HyMAP hyperspectral with HRSC-A multispectral stereo and DTM data: Remote sensing data validation and application in different disciplines. In *First EARSeL Workshop on Imaging Spectroscopy*, Zurich, Switzerland, 105–117.

Lillesand, T. M., and R. W. Kiefer. 1999. *Remote Sensing and Image Interpretation*, 4th ed. New York: John Wiley & Sons.

Marino, C. M., C. Panigada, A. Galli, L. Boschetti, and L. Buseto. 2000. Environmental applications of airborne hyperspectral remote sensing: Asbestos concrete sheeting identification and mapping. In *Proceedings of the International Conference on Applied Geologic Remote Sensing*, Las Vegas, Nevada, pp. 607–610.

Martin, M. E., and J. D. Aber. 1997. High spectral resolution remote sensing of forest canopy lignin, nitrogen, and ecosystem processes. *Ecological Applications* 7(2):431–443.

Martin, M. E., S. D. Newman, J. D. Aber, and R. G. Congalton. 1998. Determining forest species composition using high spectral resolution remote sensing data. *Remote Sensing of Environment* 65:249–254.

Martin, M. E., L. C. Plourde, S. V. Ollinger, M.-L. Smith, and B. E. McNeil. 2008. A generalizable method for remote sensing of canopy nitrogen across a wide range of forest ecosystems. *Remote Sensing of Environment* 112:3511–3519.

Matson, P. A., L. F. Johnson, J. R. Miller, C. R. Billow, and R. Pu. 1994. Seasonal changes in canopy chemistry across the Oregon transect: Patterns and spectral measurement with remote sensing. *Ecological Applications* 4(2):280–298.

Matsunaga, T., A. Iwasaki, S. Tsuchida, J. Tanii, O. Kashimura, R. Nakamura, H. Yamamoto, T. Tachikawa, and S. Rokugawa. 2013 (July 21–26). Current status of Hyperspectral Imager Suite (HISUI). In *IGARSS 2013*, Melbourne, Australia, pp. 3510–3513.

Miller, J. R., E. W. Hare, and J. Wu. 1990. Quantitative characterization of the vegetation red edge reflectance. 1. An Inverted-Gaussian reflectance model. *International Journal of Remote Sensing* 11:1775–1795.

Miller, J. R., J. Wu, M. G. Boyer, M. Belanger, and E. W. Hare. 1991. Season patterns in leaf reflectance red edge characteristics. *International Journal of Remote Sensing* 12(7):1509–1523.

Miura, T., and H. Yoshioka. 2012. Hyperspectral data in long-term, cross-sensor continuity studies, in *Hyperspectral Remote Sensing of Vegetation*, eds. P. S. Thenkabail, J. G. Lyon, and A. Huete, pp. 611–633. Boca Raton, FL: CRC Press/Taylor & Francis Publishing Group.

Mustard, J. F. 1993. Relationships of soil, grass, and bedrock over the Kaweah serpentinite melange through spectral mixture analysis of AVIRIS data. *Remote Sensing of Environment* 44:293–308.

NASA 2009 HyspIRI Science Workshop Report. 2010 (June). JPL Publication 10-3, HyspIRI Group, Jet Propulsion Laboratory. Pasadena, CA: California Institute of Technology, pp. 1–73.

Nolin, A. W., and J. Dozier. 1993. Estimating snow grain size using AVIRIS data. *Remote Sensing of Environment* 44(2–3):231–238.

NRC's Decadal Survey Report. 2007. *Earth Science and Applications from Space: National Imperatives for the Next Decade and Beyond*. Accessed from http://www.nap.edu/catalog/11820.html.

O'Neill, N. T., A. R. Kalinauskas, G. A. Borstad, H. Edel, J. F. Gower, and H. Van der Piepen. 1987. Imaging spectrometry for water applications. In *Proceedings of SPIE: Imaging Spectroscopy II*, 834:129–135.

Painter, T. H., D. A. Roberts, R. O. Green, and J. Dozier. 1998. Improving mixture analysis estimates of snow-cover area from AVIRIS data. *Remote Sensing of Environment* 65:320–332.

Palacios-Orueta, A., and S. L. Ustin. 1996. Multivariate statistical classification. *Remote Sensing of Environment* 57:108–118.

Peterson, D. L., J. D. Aber, D. A. Matson, D. H. Card, N. Swanberg, C. Wessman, and M. Spanner. 1988. Remote sensing of forest canopy and leaf biochemical contents. *Remote Sensing of Environment* 24:85–108.

Pinel, V., F. Zagolski, J. P. Gastellu-Ethchgorry, G. Giordano, J. Romier, G. Marty, E. Mougin, and R. Joffre. 1994. An assessment of forest chemistry with ISM. *SPIE* 2318:40–51.

Prieto-Blanco, A., P. R. J. North, M. J. Barnsley, and N. Fox. 2009. Satellite-driven modelling of Net Primary Productivity (NPP): Theoretical analysis. *Remote Sensing of Environment* 113:137–147.

Pu, R. 2012. Mapping leaf area index over a mixed natural forest area using ground-based measurements and Landsat TM imagery. *International Journal of Remote Sensing* 33(20):6600–6622.

Pu, R., and S. Bell. 2013. A protocol for improving mapping and assessing of seagrass abundance along the West Central Coast of Florida using Landsat TM and EO-1 ALI/Hyperion images. *ISPRS Journal of Photogrammetry and Remote Sensing* 83:116–129.

Pu, R., and P. Gong. 2000. *Hyperspectral Remote Sensing and Its Applications* [in Chinese]. Beijing, China: Higher Education Press.

Pu, R., and P. Gong. 2004. Wavelet transform applied to EO-1 hyperspectral data for forest LAI and crown closure mapping. *Remote Sensing of Environment* 91:212–224.

Pu, R., L. Foschi, and P. Gong. 2004. Spectral feature analysis for assessment of water status and health level of coast live oak (*Quercus Agrifolia*) leaves. *International Journal of Remote Sensing* 25(20):4267–4286.

Pu, R., P. Gong, and Q. Yu. 2008. Comparative analysis of EO-1 ALI and Hyperion, and Landsat ETM+ data for mapping forest crown closure and leaf area index. *Sensors* 8:3744–3766.

Pu, R., S. Ge, N. M. Kelly, and P. Gong. 2003. Spectral absorption features as indicators of water status in Quercus Agrifolia leaves. *International Journal of Remote Sensing* 24(9):1799–1810.

Pu, R., P. Gong, G. S. Biging, and M. R. Larrieu. 2003. Extraction of red edge optical parameters from Hyperion data for estimation of forest leaf area index. *IEEE Transactions on Geoscience and Remote Sensing* 41(4):916–921.

Pu, R., S. Bell, C. Meyer, L. Baggett, and Y. Zhao. 2012. Mapping and assessing seagrass habitats using satellite imagery. *Estuarine, Coastal, and Shelf Science* 115:234–245.

Resmini, R. G., M. E. Kappus, W. S. Aldrich, J. C. Harsanyi, and M. Anderson. 1997. Mineral mapping with HYperspectral Digital Imagery Collection Experiment (HYDICE) sensor data at Cuprite, Nevada, U.S.A. *International Journal of Remote Sensing* 18(7):1553–1570.

Richardson, L. L., D. Buisson, C. J. Liu, and V. Ambrosia. 1994. The detection of algal photosynthetic accessory pigments using airborne visible-infrared imaging spectrometer (AVIRIS) spectral data. *Marine Technology Society Journal* 28(3):10–21.

Ridd, M. K., N. D. Ritter, and R. O. Green. 1997. Neural network analysis of urban environments with airborne AVIRIS data. In *Proceedings of the Third International Airborne Remote Sensing Conference and Exhibition*, Copenhagen, Denmark, pp. 197–203.

Roberts, D. A., P. E. Dennison, M. E. Gardner, Y. Hetzel, S. L. Ustin, and C. T. Lee. 2003. Evaluation of the potential of Hyperion for fire danger assessment by comparison to the airborne visible/infrared imaging spectrometer. *IEEE Transactions on Geoscience and Remote Sensing* 41(6):1297–1310.

Sandidge, J. C., and R. J. Holyer. 1998. Coastal bathymetry from hyperspectral observation of water radiance. *Remote Sensing of Environment* 65:341–352.

Schläpfer, D., C. C. Borel, J. Keller, and K. I. Itten. 1998. Atmospheric precorrected differential absorption technique to retrieve columnar water vapor. *Remote Sensing of Environment* 65:353–366.

Schowengerdt, R. A. 1997. *Remote Sensing Models and Methods for Image Processing*, 2nd ed. New York: Academic Press.

Serrano, L., J. Peñuelas, and S. L. Ustin. 2002. Remote sensing of nitrogen and lignin in Mediterranean vegetation from AVIRIS data: Decomposing biochemical from structural signals. *Remote Sensing of Environment* 81:355–364.

Skoog, D. A., E. J. Holler, and T. A. Nieman. 1998. *Principles of Instrumental Analysis*, 5th ed. Philadelphia, PA: Saunders College Publishers.

Smith, M.-L., M. E. Martin, L. Plourde, and S. V. Ollinger. 2003. Analysis of hyperspectral data for estimation of temperate forest canopy nitrogen concentration: Comparison between an airborne (AVIRIS) and a spaceborne (Hyperion) sensor. *IEEE Transactions on Geoscience and Remote Sensing* 41(6):1332–1337.

Smith, N. J., G. A. Boratad, D. A. Hill, and R. C. Kerr. 1991. Using high-resolution airborne spectral data to estimate forest leaf area and stand structure. *Canadian Jounral of Forest Research* 21:1127–1132.

Staenz, K. 2009 (March 16–18). Terrestrial imaging spectroscopy, some future perspectives. In *Proceedings of the 6th EARSEL SIG IS*, Tel Aviv, Israel.

Stoner, E. R., and M. F. Baumgardner. 1981. Characteristic variations in reflectance of surface soils. *Soil Science Society of America Journal* 45:1161–1165.

Sunshine, J. M., and C. M. Pieters. 1993. Estimating modal abundances from the spectra of natural and laboratory pyroxene mixture using the modified gaussian model. *Journal of Geophysical Research* 98(E5):9075–9087.

Sunshine, J. M., C. M. Pieters, and S. F. Pratt. 1990. Deconvolution of mineral absorption bands: An improved approach. *Journal of Geophysical Research* 95(B5):6955–6966.

Swayze, G. A., R. N. Clark, R. M. Pearson, and K. E. Livo. 1996. Mapping acid generating minerals at the California gulch superfund site in Leadville, Colorado, using imaging spectroscopy. In *Summaries of the Sixth Annual JPL Airborne Earth Science Workshop*, Jet Propulsion Laboratory, Pasadena, California, pp. 231–234.

Thenkabail, P. S., J. G. Lyon, and A. Huete, eds. 2012. *Hyperspectral Remote Sensing of Vegetation*. Boca Raton, FL: CRC Press/Taylor & Francis Publishing Group.

Thenkabail, P. S., I. Mariotto, M. K. Gumma, E. M. Middleton, D. R. Landis, and K. F. Huemmrich. 2013. Selection of hyperspectral narrowbands (HNBs) and composition of hyperspectral two band vegetation indices (HVIs) for biophysical characterization and discrimination of crop types using field reflectance and Hyperion/EO-1 Data. *IEEE Journal of Selected Topics in Applied Earth Observations and Remote Sensing* 6(2):427–439.

Townsend, P. A., J. R. Foster, R. A. Chastain, Jr., and W. S. Currie. 2003. Application of imaging spectroscopy to mapping canopy nitrogen in the forests of the central Appalachian Mountains using Hyperion and AVIRIS. *IEEE Transactions on Geoscience and Remote Sensing* 41(6):1347–1354.

USGS Speclab. 2014. What is imaging spectroscopy? Accessed on December 6, 2014, from http://speclab.cr.usgs.gov/aboutimsp.html.

Ustin, S. L., A. A. Gitelson, S. Jacquemoud, M. Schaepman, G. P. Asner, J. A. Gamon, and P. Zarco-Tejada. 2009. Retrieval of foliar information about plant pigment systems from high resolution spectroscopy. *Remote Sensing of Environment* 113:S67–S77.

Van der Meer, F. 2004. Analysis of spectral absorption features in hyperspectral imagery. *International Journal of Applied Earth Observation and Geoinformation* 5:55–68

Van der Meer, F., and W. Bakker. 1997. Cross correlogram spectral matching: Application to surface mineralogical mapping by using AVIRIS data from Cuprite, Nevada. *Remote Sensing of Environment* 61:371–382.

Van der Meer, F. D., and S. M. de Jong, eds. 2001. *Imaging Spectrometry: Basic Principle of Prospective Applications.* The Netherlands: Springer.

Vane, G., ed. 1987. *Proceedings of the Third Airborne Imaging Spectrometer Data Analysis Workshop*, JPL Publication 87–30, Jet Propulsion Laboratory, Pasadena, California.

Vane, G., and A. F. H. Goetz, eds. 1985. *Proceedings of the Airborne Imaging Spectrometer Data Analysis Workshop*, JPL Publication 85–41, Jet Propulsion Laboratory, Pasadena, California.

Vane, G., and A. F. H. Goetz, eds. 1986. *Proceedings of the Second Airborne Imaging Spectrometer Data Analysis Workshop*, JPL Publication 86–35, Jet Propulsion Laboratory, Pasadena, California.

Vane, G., and A. F. H. Goetz. 1988. Terrestrial imaging spectroscopy. *Remote Sensing of Environment* 24:1–29.

Vane, G., and A. F. H. Goetz. 1993. Terrestrial imaging spectrometry: Current status, future trends. *Remote Sensing of Environment* 44:117–126.

Vincent, R. K. 1997. *Fundamentals of Geological and Environmental Remote Sensing.* Prentice Hall Series in Geographic Information Science. Upper Saddle River, NJ: Prentice Hall.

Weng, Y. L., P. Gong, and Z. L. Zhu. 2008. Soil salt content estimation on the Yellow River Delta with satellite hyperspectral data. *Canadian Journal of Remote Sensing* 34(3):259–270.

Wessman, C. A., J. D. Aber, and D. L. Peterson. 1989. An evaluation of imaging spectrometry for estimating forest canopy chemistry. *International Journal of Remote Sensing* 10:1293–1316.

Wessman, C. A., J. D. Aber, D. L. Peterson, and J. M. Melillo. 1988. Remote sensing of canopy chemistry and nitrogen cycling in temperate forest ecosystems. *Nature* 335:154–156.

Wicks, D., and A. R. Campos-Marquetti. 2010. *Creation of a 3D Urban GIS Database: Data Fusion Approach Technical Session on Photogrammetry and 3D Visualization.* Accessed on December 23, 2014, from http://www.geospatialworld.net/Paper/Technology/ArticleView.aspx?aid=2474.

Wikipedia. 2014a. Accessed on December 6, 2014, from http://en.wikipedia.org/wiki/Spectroscopy.

Wikipedia. 2014b. Accessed on December 7, 2014, from http://en.wikipedia.org/wiki/Hyperspectral_imaging.

Yang, H., F. van der Meer, and W. Bakker. 1997 (July 7–10). A back-propagation neural network for mineralogical mapping from AVIRIS data. In *Proceedings of the Third International Airborne Remote Sensing Conference and Exhibition*, Copenhagen, Denmark, vol. I:265–272.

Yoder, B. J., and R. E. Pettigrew-Crosby. 1995. Predicting nitrogen and chlorophyll content and concentration from reflectance spectra (400–2500 nm) at leaf and canopy scales. *Remote Sensing of Environment* 53:199–211.

Yoder, B. J., and R. H. Waring. 1994. The normalized difference vegetation index of small Douglas fir canopies with varying chlorophyll concentrations. *Remote Sensing of Environment* 49:81–91.

Zagolski, F., V. Pinel, J. Romier, D. Alcayde, J. Fontanari, J. Gasstellu-Etchegorry, G. Giodano, G. Marty, E. Mougin, and R. Joffre. 1996. Forest canopy chemistry with high spectral resolution remote sensing. *International Journal of Remote Sensing* 17(6):1107–1128.

Zarco-Tejada, P. J., J. R. Miller, A. Morales, A. Berjón, and J. Agüera. 2004. Hyperspectral indices and model simulation for chlorophyll estimation in open-canopy tree crops. *Remote Sensing of Environment* 90:463–476.

Zarco-Tejada, P. J., A. Berjón, R. López-Lozano, J. R. Miller, P. Martín, V. Cachorro, M. R. González, and A. Frutos. 2005. Assessing vineyard condition with hyperspectral indices: Leaf and canopy reflectance simulation in a row-structured discontinuous canopy. *Remote Sensing of Environment* 99:271–287.

Zhang, Y., J. M. Chen, J. R. Miller, and T. L. Noland. 2008a. Leaf chlorophyll content retrieved from airborne hyperspectral remote sensing imagery. *Remote Sensing of Environment* 112:3234–3247.

Zhang, Y., J. M. Chen, J. R. Miller, and T. L. Noland. 2008b. Retrieving chlorophyll content in conifer needles from hyperspectral measurements. *Canadian Journal of Remote Sensing* 34(3):296–310.

Zhao, Y. C. 2001. An analysis of spectral characteristics of typical ground targets and information extraction strategies in hyperspectral remote sensing—A research report on some critical problems. Institute of Remote Sensing Applications, Beijing, China (Postdoc Report).

2 Field Spectrometers and Plant Biology Instruments for HRS

The development of hyperspectral remote sensing needs supports from non-imaging field spectroscopy techniques and plant biology instruments for purposes of calibrating and validating HRS sensors, data, and applications. The *in situ* spectral measurements taken by using frequently used modern field spectrometers are important to better understanding spectral characteristics of Earth's surface and atmospheric materials/constitutes. Biophysical and biochemical parameter (bioparameter) measurements taken by a set of plant biology instruments are necessary in calibrating and validating empirical and physically processed models for retrieving and mapping bioparameters from hyperspectral imaging data. Therefore, in this chapter, work principles, structures, operations in the field, and application areas of frequently used modern spectrometers and plant biology instruments are introduced.

2.1 NON-IMAGING FIELD SPECTROMETERS

In this section, the definition, history, and roles of development of field spectroscopy are first introduced. The basic theories and principles of field spectroscopy and general field operational guidelines of spectroradiometers are then addressed. Finally, a list of frequently used modern field spectroradiometers is briefly introduced, associated with a summary of major technical characteristics and parameters in a table as well as necessary illustrations of each instrument.

2.1.1 Introduction

Given that field spectroscopy is a fundamental technique in multi-/hyperspectral remote sensing, both at the level of primary research and in operational applications (Milton 1987), field spectrometers have long been used to characterize the multi-angle reflectance properties of various relatively homogeneous surface targets (e.g., soils, water, snow and ice fields, vegetation canopies), generally under solar illumination. Many disciplines associated with research and applications of remote sensing technology are interested in the measurement of light reflected off, or radiation energy emitted from, various targets in the natural environment, although the paralleled laboratory spectroscopic technique is also necessary for extensive research and applications of multi-/hyperspectral remote sensing. To better introduce various field spectrometers, several terms must be clarified prior to describing the role of field spectroscopy in Earth observation and principles of field spectroscopy. *Field spectroscopy* is the term used to describe spectroscopy undertaken in the natural environment and to emphasize spectroscopic principles and mechanisms of measuring reflectance properties from targets, particularly where the reflectance properties of vegetation, soils and minerals, and water bodies are measured under solar illumination. Field spectrometry is the term used to describe the application technique of field spectroscopy to make spectral measurements, whereas the actual instruments used to measure point spectra (reflectance or radiance) in the laboratory and in the field are referred to as spectrometers. The term field spectrometry is frequently called by other terms, such as handheld radiometry, ground radiometry, field radiometry, reflectance spectrometry, field spectroradiometry, etc.

Field spectrometers were first used to investigate human color vision, such as the color of the Earth's surface from the air (Penndorf 1956), whereas the development of the first instruments that could make accurate measurements of spectral reflectance in the field environment was spurred by the development of airborne multispectral scanners in the 1960s (Milton et al. 2009). However, the first truly portable field reflectance spectrometer (PFRS), which could cover the spectral range of solar reflected radiance, 0.4–2.5 μm, didn't appear until 1975 (Goetz 1975). The motivation behind developing PFRS during the 1970s was to make accurate measurements in the short-wave infrared region (SWIR, 1.1–2.4 μm), which had been proved by laboratory measurements to be a very important spectral range for geological applications such as spectrally identifying minerals on the Earth's surface. The ability that PFRS could accurately measure the SWIR spectrum finally resulted in the addition of band 7 to the Landsat Thematic Mapper (TM) and encouraged the development of more capable field spectrometers such as the Jet Propulsion Laboratory PIDAS field spectroradiometer (Goetz 1987, 2009). Later on, companies such as Geophysical Environmental Research (GER) of Millbrook, New York, and Analytical Spectral Devices (ASD) of Boulder, Colorado, developed more technically advanced and easier-to-carry field spectrometers in response to the demand of field spectral measurement from targets. For instance, the ASD field instrument can acquire a full spectrum in 100 ms. Laboratory and field spectra of minerals, measured using such advanced spectrometers, supported the notion that reflectance spectroscopy in the visible and SWIR regions would be a source of rich information about the composition of the Earth's surface (Hunt 1977). Spectra Vista Corporation (SVC), another leading manufacturer of field spectroradiometers, can also trace its origins back to this period through its predecessor, GER Corporation. However, compared with ASD and GRE field spectrometers, SVC spectrometers (e.g., SVC HR-1024) have higher spectral resolution.

As a means of scaling up the understanding of electromagnetic radiation energy–matter interactions from a finer scale of individual spectral measurement elements, such as tree leaf to much coarser tree canopy-scale studies, and calibrating airborne and satellite systems and sensors, the key role of field spectroscopy (FS) in Earth observation has been demonstrated clearly during the last three decades (Gamon et al. 2006). The role of field spectroscopy is briefly explained in the following three aspects:

1. Spectral measurements of individual scene elements measured by using the FS technique can help us understand the spectral characteristics of individual scene elements such as tree leaves, minerals and soils, snow and ice covers, water bodies, etc., and scaling-up issues in remote sensing. For example, the spectral reflectance spectra measured from individual plant leaves/branches and minerals using an FS instrument in the natural environment can help us understand and locate the diagnostic absorption features (bands) in visible and NIR spectral ranges for plant species and those in the SWIR spectral region for minerals. This information of diagnostic absorption features in different spectral ranges for different scene elements is beneficial for designing airborne and spaceborne remote sensing sensors, such as band setting and determination of optimal spectral resolution for each band. It is easy to understand that the *in situ* spectral measurements taken using the FS instrument from the scene elements can be scaled up to a pixel or landscape level to directly help analyze remote sensing imaging data through some scaling-up modeling techniques and approaches.
2. The FS technique can be used for various calibration of remote sensing sensor systems and atmospheric correction of airborne and satellite remote sensing data. The reflectance-based and radiance-based "vicarious calibration" methods have been extensively adopted as the means to assure the quality of remotely sensed data acquired from airborne and spaceborne sensors (Milton et al. 2009). Usually the reflectance-based method uses near Lambertian reflectance measured from a flat, featureless, and extensive area of bare ground in a radiative transfer model together with data on the state of the atmosphere at the time when satellite data are acquired to predict the radiance at the top of the atmosphere within the satellite sensors' bands. For this case, similar Lambertian reflectance can also be measured from high-altitude dry lake beds with the FS technique. The radiance-based vicarious calibration method collects radiance with field spectrometers onboard a helicopter or aircraft platform from the ground with a similar Lambertian surface. For example, Slater et al. (1987) used

the reflectance- and radiance-based methods for the in-flight absolute calibration of the Coastal Zone Color Scanner (CZCS) and the Thematic Mapper (TM) on Landsat 4. Their experimental results indicate that the five reflectance-based calibrations of the Landsat 5 TM at White Sands, New Mexico, in the period of July 1984 to November 1985 show a ±2.8% standard deviation (1 SD) for the six solar-reflective bands. The measurement of reflected spectral radiance using a spectrometer onboard a helicopter at 3000 m mean sea level (MSL) has been shown experimentally and theoretically to offer a precise alternative procedure to reflectance-based calibration for bands in the visible and NIR. Thome et al. (1998) used the reflectance measured from Lunar Lake Playa, Nevada, in June 1996 to evaluate the accuracy of reflectance-based vicarious calibrations of Earth Observing Systems. In their experiment, four groups participated in the campaign and made independent measurements of surface reflectance and atmospheric transmittance on five different days. Then each group predicted top-of-the-atmosphere radiance for several bands in the 400 nm to 2500 nm spectral range. Analysis of the data showed differences in the order of 5% to 10% throughout the spectral region under study. With such vicarious calibration methods, similar calibration research includes calibrations of NASA AVIRIS (Green et al. 1991), SPOT HRV instruments (Santer et al. 1992), and ESA's ENVISAT MERIS sensor (Kneubühler et al. 2003). The empirical line calibration (ELC) is a typical atmospheric correction method in which ground-based reflectance is measured by well-characterized field spectrometers and can be directly used to calibrate airborne and satellite imaging data (in either radiance or digital number) (Jensen 2005). For instance, Pu et al. (2015) evaluated three atmospheric correction methods (an ELC method, a radiative transfer modeling approach, and a combination of both methods) in identifying urban tree species/groups with high resolution WorldView-2 (WV2) imagery in the city of Tampa, Florida. Their experimental results demonstrate that the reflectance-based ELC methods were more effective than a radiative transfer–based atmospheric correction model to atmospherically correct the image data, due to the latter's lacking accurate and reliable atmospheric parameters on which to run the radiative transfer model.

3. The FS provides a tool for the development, refinement, testing, and verifying of models (either empirical/statistical or physically processed), correlating remotely sensed attributes or features with a set of biophysical and biochemical parameters. For examples, *in situ* hyperspectral data were used for nutrient estimation (total nitrogen, total phosphorus, and total potassium) of a conifer tree species, the giant sequoia (*Sequoiadendron giganteum*) in California, using the empirical modeling technique (Gong et al. 2002). Pu (2009) used *in situ* reflectances measured using the ASD full-range spectrometer from the branches/canopies of eleven urban tree species, including the American elm (*Ulmus americana*), bluejack oak (*Quercus incana*), crape myrtle (*Lagerstroemia indica*), laurel oak (*Q. laurifolia*), live oak (*Q. virginiana*), southern magnolia (*Magnolia grandiflora*), persimmon (*Diospyros virginiana*), red maple (*Acer rubrum*), sand live oak (*Q. geminata*), American sycamore (*Platanus occidentalis*), and turkey oak (*Q. laevis*) to identify them using statistical models. Zarco-Tejada et al. (2001) utilized the physically processed modeling method and *in situ* spectral reflectance measurements and airborne compact airborne spectrographic imager (CASI) hyperspectral image data collected from twelve sites of sugar maple (*Acer saccharum M.*) in the Algoma Region, Canada, to estimate the chlorophyll content of closed forest canopies. In all such cases, the FS technique has a prominent role to play in the modeling processes and in the testing of models subsequently derived.

2.1.2 Principles of Field Spectroscopy and General Guidelines on Field Techniques

2.1.2.1 Principles of Field Spectroscopy

Natural targets (e.g., soils and minerals, vegetation types, snow and ice layers, water bodies, etc., in the natural environment) are usually illuminated by the whole hemisphere of the sky, and thus receive direct solar flux and scattered sky light. Interactions of materials with incident radiation (i.e., direct solar flux plus scattered sky light) on the Earth's surface result in a proportion of the total incident radiation energy being reflected, either directly from the surface or after multiple interactions within the surface (Milton 1987). In general, natural targets are not perfectly diffuse (Lambertian) reflectors, and thus the intensity of the reflected radiation varies with the angle from which it leaves the surface. Consequently, the radiation environment consists of two hemispherical distributions of electromagnetic radiation: incoming and outgoing distributions. And it presents an interaction between these two distributions, which constitutes a focus of interest in field spectroscopy under discussion.

The simplified radiation geometry of the field environment is illustrated in Figure 2.1 with two sets of angles describing the zenith and azimuth positions of the primary source of irradiance (solar light) and the sensor. In the geometry of Figure 2.1, the angle from the vertical (the zenith angle, θ) and the angle measured in the horizontal plane from a reference direction (the azimuth angle, ϕ) are considered, but the scattering skylight is not considered. Referring to Figure 2.1, the bidirectional reflectance distribution function (BRDF) describes the scattering of a parallel beam of incident light from one direction in the hemisphere into another direction in the hemisphere (Schaepman-Strub et al. 2006). Note that for reasons of clarity, all wavelength dependencies have been omitted in the equations below. Indirect illumination (i.e., ignoring skylight), incident, and reflected can be regarded as confined to two slender elongated cones. If the solid angles of the two cones measured in steradians (sr) are infinitesimally small, the reflectance of the target can be defined as a function:

$$f(\theta_i,\phi_i,\theta_r,\phi_r)=\frac{dL(\theta_r,\phi_r)}{dE(\theta_i,\phi_i)} \tag{2.1}$$

where dL is the reflected radiance per unit solid angle, dE is the irradiance per unit solid angle, and the subscripts i and r denote incident and reflected rays, respectively. Both the radiance and the

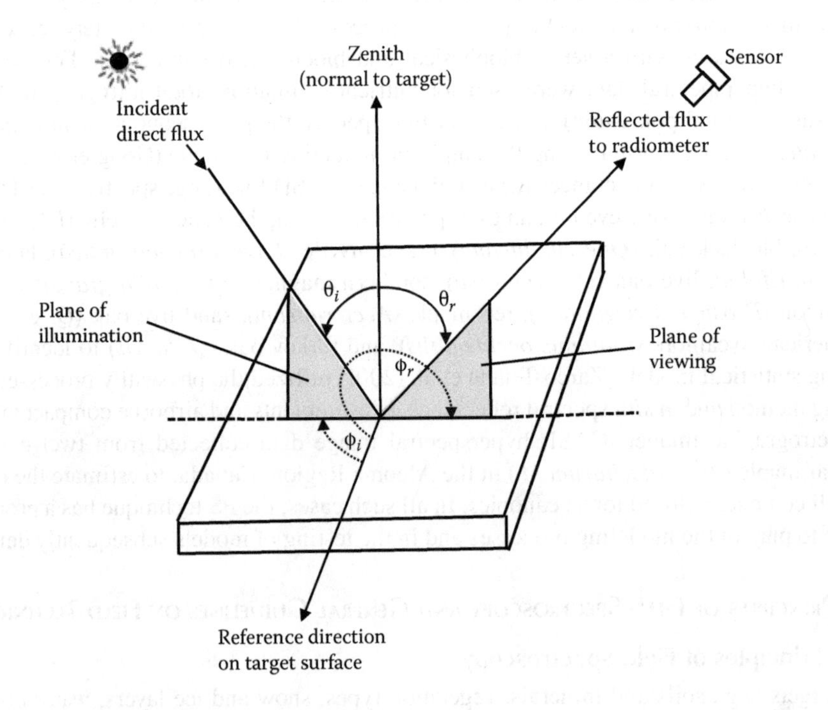

FIGURE 2.1 Definition of the angles involved in the field spectroscopy.

irradiance vary in zenith and azimuth, hence to specify completely the reflectance field at the target, the reflectance must be measured at all possible source/sensor positions, resulting in the bidirectional reflectance distribution function (BRDF). Although the BRDF is a useful theoretical concept, it has drawbacks for practical applications and cannot be directly measured (Nicodemus et al. 1977), notably the requirement that dE be measured at the target surface and that both dE and dL be measured over infinitesimally small solid angles. A solution is simplifying the BRDF theoretical model (Equation 2.1 in considering all possible source/sensor positions) to bidirectional reflectance factor (BRF), which describes reflectance for parallel beams of radiance and irradiance.

For practical purposes, therefore, an alternative measure is needed to represent the directional reflectance of natural surfaces. There are two simplification methods (i.e., two practical measurement geometries in the field) made to the concept of BRDF: Cos-conical method and Bi-conical method. For the Cos-conical method, the solid angle is increased to be large enough to contain measurable quantities of energy. For the Bi-conical method, dE at the target is estimated, either from the global irradiance measured at a short distance above the target, or by estimating it from the amount of energy reflected by a calibrated reflectance panel.

With the Cos-conical method, an upward-looking spectrometer with a cosine-corrected receptor is used, which is that a sensor shows no dependence upon the zenith or azimuth angle of the incident flux. This measurement configuration may be termed *Cos-conical geometry* to indicate that the target is measured using an aperture receptor. In this case the BRF (R) from a natural target can be expressed as

$$R(\theta_i,\phi_i,\theta_r,\phi_r) = \frac{dL_t(\theta_r,\phi_r)}{dE} k(\theta_i,\phi_i,\theta_r,\phi_r) \tag{2.2}$$

where dL_t is the radiance of the target and dE is the irradiation as measured by the upward-looking sensor, and k is a correction factor relating the signal from the cosine-corrected receptor to that expected from a perfectly diffuse white panel.

With the Bi-conical method, the BRF may also be determined by comparing the reflected radiance from the target to that from a reflectance panel specified to be perfectly diffuse, completely reflecting and viewed under the same irradiation conditions and in the same geometry as the target. Because in practice a perfect reflecting panel does not exist, a correction is made to account for the spectral reflectance of the panel. Thus,

$$R(\theta_i,\phi_i,\theta_r,\phi_r) = \frac{dL_t(\theta_r,\phi_r)}{dL_p(\theta_r,\phi_r)} k(\theta_i,\phi_i,\theta_r,\phi_r) \tag{2.3}$$

where dL_t is the radiance of the target and dL_p is the radiance of the panel under the same specified conditions of illumination and viewing, and k is the panel correction factor. Figure 2.2 presents a spectral reflectance spectrum of a live oak leaf measured in the laboratory (top) and an *in situ* unprocessed reflectance spectrum of a dense live oak canopy measure under natural solar light (bottom) using the Bi-conical method. Both noise bands locating around 1.4 μm and 1.9 μm in Figure 2.2(b) are caused by atmospheric water vapor in the natural environment. Note that k in Equation 2.3 is also dependent upon the angular configuration, as perfectly Lambertian standard panels are impossible to achieve in practice. For a perfectly diffuse surface, the BRF (R) may be related to the BRDF as follows:

$$R(\theta_i,\phi_i,\theta_r,\phi_r) = \pi f(\theta_i,\phi_i,\theta_r,\phi_r) \tag{2.4}$$

An ideal Lambertian surface reflects the same radiance in all view directions, and its BRDF is $1/\pi$. Thus, the BRF (unitless) of any surface can be expressed as its BRDF [sr^{-1}] timing π (Equation 2.4).

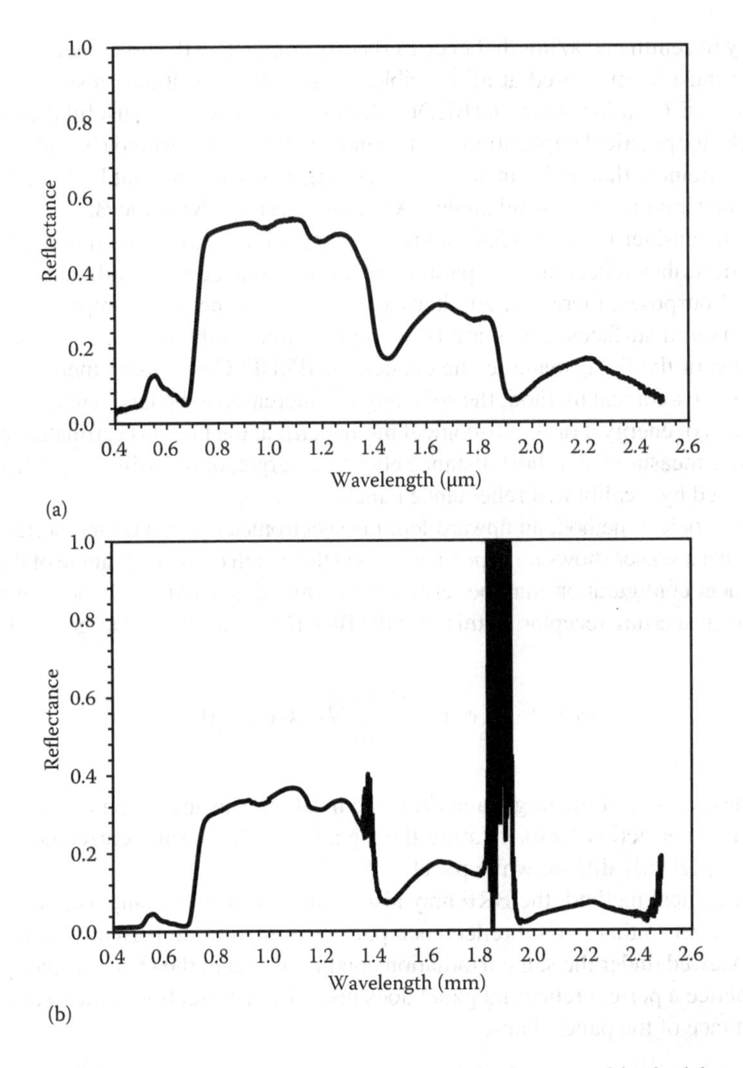

FIGURE 2.2 A spectral reflectance spectrum of (a) a live oak leaf measured in the laboratory and (b) an *in situ* unprocessed reflectance spectrum of a dense live oak canopy measured under natural solar light. Both reflectance spectra were measured using an Analytical Spectral Devices Inc. FieldSpec Pro spectrometer with a field of view of 25°.

When the BRF (Equation 2.2 and Equation 2.3) is used instead of the BRDF to represent the spectral reflectance of natural targets, several assumptions in reflectance field spectroscopy need to be considered (Milton 1987) as follows:

- The angular field-of-view of the sensor is as small as possible (<20 degrees).
- The reflectivity panel must fill the field-of-view of the sensor.
- There should be no change in the irradiation amount or its spectral distribution between measurement of dL_t and dL_p (or dE).
- Direct solar flux dominates the irradiation field. That is, the sun is assumed to shine out of a black sky, and skylight is ignored.
- The sensor responds in a linear fashion to changes in radiant flux.
- The reflectance properties of the standard panel are known and invariant over the course of the measurements.
- The sensor is sufficiently distant from targets.

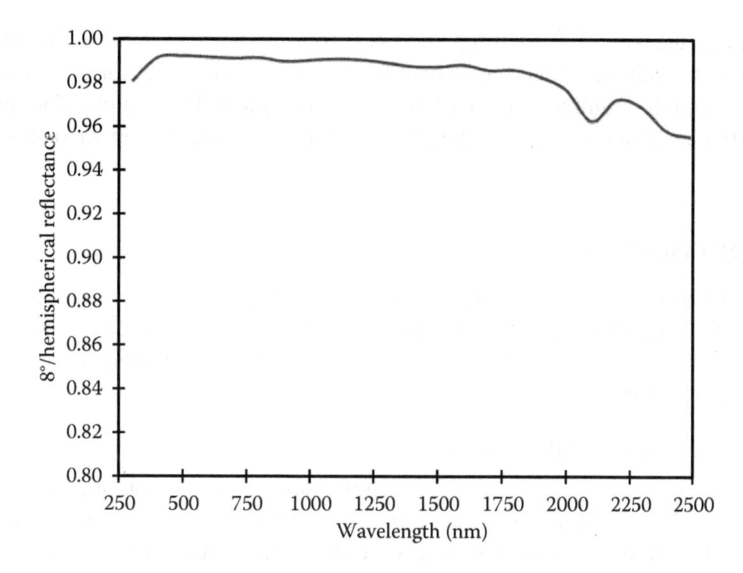

FIGURE 2.3 Reflectance curve of a Spectralon®. (Courtesy of Labsphere, Inc., USA.)

In practice, an ideal reflecting panel (or spectralon) is required to have high and flat reflectance over the optical range when clean (>99% reflectance over 400–1500 nm and >95% reflectance over 1500–2500 nm). It is supposed to be Lambertian or closed Lambertian and it is highly water-resistant and thermally stable. Figure 2.3 shows a sample of reflectance of spectralon that is necessary to measure reflectance spectra in the field.

2.1.2.2 General Guidelines on Field Technique

To improve the consistency and accuracy of spectral reflectance data collected using field spectrometers in the natural environment, several researchers have published practical suggestions. This is important if the methodology of field spectroscopy is to be refined and standardized between different research groups. According to Milton (1987) and Maracci (1992), the most important points may be summarized as follows:

- Use a mast or tripod to ensure a fixed geometry among the sensor, the standard panel, and the target, if at all possible. Handheld measurements are less precise due to the variable geometry involved and because the operator should be close to the target and the target should be close to the spectrometer.
- Ensure that the sensor is an appropriate distance (e.g., one meter) above the upper surface of the target. The "appropriate distance" is determined when a typical and stable reflectance spectrum can be seen from the upper surface of the target.
- Unless variations in reflectance factor with azimuth are being studied, be consistent in always orientating the sensor horizontal support and positioning the other field equipment (including people) in the same positions relative to the sun. This is most easily achieved by pointing the sensor support directly toward the sun.
- Check that the standard panel (spectralon) fills the field-of-view of all bands of the sensor and that it is not shaded by the sensor or the operator.
- It is useful to position a continuously recording solarimeter centrally within the field area during the measurement time. It provides three benefits: (1) It allows any anomalies in the data from the primary sensor to be screened and possibly corrected; (2) it allows the atmospheric variability to be quantified on a range of time scales; (3) it can provide data to correct for the effects of variable cloud cover (also see Richardson 1981).

- Operators should wear dark clothing and kneel some distance away from the target during measurements. Vehicles should be kept at least 3 m from the target for the same reason.
- Take an adequate number of measurements in a single field (or plot). The average has to be statistically significant for the pixel seen by an airborne sensor or a spaceborne sensor.

2.1.3 Field Spectrometers

Table 2.1 lists important technical specifications of frequently used modern spectrometers. In this section, substantial working principles, operating procedures, spectral range covered, spectral characteristics, and appropriate application areas of several important field spectroradiometer series products are briefly introduced.

2.1.3.1 ASD Field Spectroradiometers

ASD FieldSpec® series spectroradiometers are manufactured by Analytical Spectral Devices (ASD), Inc., United States. ASD FieldSpec series spectroradiometers can generate high-quality field spectra at the required illumination and viewing geometry for many applications and studies including hyperspectral remote sensing. Two types of ASD FieldSpec are introduced below. They are carry-on (ASD FieldSpec 3 and 4) and handheld 2 field spectroradiometers. Figure 2.4 shows the exterior view of both types of ASD field spectroradiometers, whereas Figure 2.5 presents the operation of the two types of instruments in the field. Referring to Table 2.1, readers can learn some important technical parameters for ASD field spectroradiometers.

ASD FieldSpec 3 is a user friendly, field portable, precision instrument with a full spectral range (350–2500 nm), rapid data collection (0.1 s per spectrum), and high spectral resolution full width at half maximum [FWHM] 3 nm at 700 nm and 10 nm at 1400/2100 nm. The improved instrument design combined with the new wireless connectivity takes portability and flexibility to a higher level, allowing users to perform applications never before possible. The wireless connection allows for remote control of your data collection up to 50 meters away. The FieldSpec 3 is ready for the field, mountains, the desert, the arctic, etc. With exceptional portability and flexibility, the FieldSpec 3 allows people to collect more data from more sites in less time.

ASD FieldSpec 4 series spectroradiometers include FieldSpec 4 Standard-Res, FieldSpec 4 Hi-Res, and FieldSpec 4 Wide-Res. FieldSpec 4 Standard-Res offers dramatically improved speed, performance, and portability over previous models. This full-range Vis/NIR (350–2500 nm) instrument features enhance capabilities in the SWIR 1 and 2 regions, as well as double the signal-to-noise ratio performance. While collecting spectral data in the field, it possesses high sensitivity and low noise without increasing scan time. Improvements in the SWIR 1 and 2 regions allow researchers to cover twice the area in half the time as the FieldSpec 3. The new spectrometer configuration provides double the signal-to-noise ratio performance as previous models. A new ruggedized cable protects the fiber optics, nearly eliminating fiber breakage. An expanded wireless range adds flexibility by helping users capture spectra farther from the instrument controller.

The FieldSpec 4 Hi-Res spectroradiometer has been designed for faster, more precise spectral data collection in remote sensing applications. The enhanced spectral resolution with 3 nm at 700 nm and 8 nm at 1400/2100 nm offered by the portable, ruggedized Hi-Res model is well suited to geologic applications and other fields of research that requires definition of narrow spectral features, especially in the longer wavelengths. Faster spectral collection speed allows users to measure more targets at more sites in a smaller timeframe. Within the full-range (350–2500 nm) the FieldSpec 4 Hi-Res provides the highest spectral resolution available in a portable, ruggedized spectroradiometer. The FieldSpec 4 Hi-Res has increased signal throughput, improved spectral resolution and vastly improved signal-to-noise ratio over previous models. For remote sensing, this means faster spectrum capture in the field with improved spectrum quality. Researchers looking at

TABLE 2.1
Summary of Modern Field Spectrometers

Spectrometer	Date	Spectral Range (nm)	Resolution (FWHM, nm)	Number of Bands	FOV (degree)	Display	Weight (kg)
ASD PS II™	1989	350–1050	3	512	$1/15/2\pi$	Real time	3.5
ASD FieldSpec	1994	350–1050	3	512 or 1024	$1/24/2\pi$	Real time	8
ASD FieldSpec NIR™	1994	1000–2500	10	750	$25/1/2\pi$	Real time	8
ASD FieldSpec-FR™	1994	350–2500	3 @ 700 10 @ 1400/2100	512 or 1024, 750	$1/25/2\pi$	Real time	8
ASD FieldSpec® 3	2006	350–2500	3 @ 700 10 @ 1400/2100	2151	25	Real time	5.6
ASD FieldSpec 4 Standard-Res	2014	350–2500	3 @ 700 10 @ 1400/2100	2151	25	Real time	5.44
ASD FieldSpec 4 Hi-Res	2014	350–2500	3 @ 700 8 @ 1400/2100	2151	25	Real time	5.44
ASD FieldSpec 4 Wide-Res	2014	350–2500	3 @ 700 30 @ 1400/2100	2151	25	Real time	5.44
ASD FieldSpec HandHeld™	1994	325–1075	3.5	512	25	Real time	1.2
ASD FieldSpec HandHeld 2	2014	325–1075	<3 @ 700	512	25	Real time	1.2
GER IRIS Mk IV™	1986	300–3000	2, 4	<1000	6 × 4	Real time	11
GERSIRIS™	1988	300–3000	2, 4	<1000	15 × 5	Real time	11
GER 1500™	1994	300–1100	3	512	3 × 1	Real time	4
GER 2100™	1994	400–2500	10, 24 or 8	140	3 × 15	Real time	11
GER 2600™	1994	400–2500	3, 24 or 8	512 + 128 or 64	3 × 15	Real time	11
GER 3700™	1994	400–2500	3, 4.8, 6.25, 8	512 + 192	1.4 × 0.3	Real time	12
Ocean Optic USB2000+	2014	200–1100	0.1–10	2048	–	Real time	0.19
Ocean Optic USB2000+UV-VIS	2014	200–850	1.5	2048	–	Real time	0.19
Ocean Optic HR2000+CG	2014	200–1100	1	2048	–	Real time	0.57
Spectral Evolution PSR+3500	2012	350–2500	3 @ 700 8 @ 1500 6 @ 2100	2151	25	Real time	3.31
Spectral Evolution PSR-2500	2012	350–2500	3.5 @ 700 22 @ 1500/2100	2151	25	Real time	3.31
Spectral Evolution PSR-1900	2012	350–1900	3.5 @ 700 10 @ 1500	1551	25	Real time	3.31
Spectral Evolution PSR-1100	2012	320–1100	3.2	512	4	Real time	1.8
Spectral Evolution PSR-1100F	2012	320–1100	3.2	512	1–8	Real time	1.8
SpectraScan® PR-655	2011	380–780	<3.5	128	7	Near real time	1.7
SpectraScan® PR-670	2011	380–780	<2	128	7	Near real time	1.7

(*Continued*)

TABLE 2.1 (CONTINUED)
Summary of Modern Field Spectrometers

Spectrometer	Date	Spectral Range (nm)	Resolution (FWHM, nm)	Number of Bands	FOV (degree)	Display	Weight (kg)
SpectraScan® PR-680	2011	380–780	<2	256	7	Near real time	2.04
SpectraScan® PR-680L	2011	380–780	<2	256	7	Near real time	2.04
Spectron SE590™	1984	370–1110	11	256	1/15/2π	Near real time	4.5
SVC HR-512i	2005	350–1050	3.2 @ 700	512	25	Real time	3.1
SVC HR-768si	2005	350–1880	3.5 @ 700 9.5 @ 1500	768	25	Real time	3.8
SVC HR-768i	2005	350–2500	3.5 @ 700 16 @ 1500 14 @ 2100	768	25	Real time	3.9
SVC HR-1024i	2005	350–2500	3.5 @ 700 9.5 @ 1500 6.5 @ 2100	1024	25	Real time	3.9

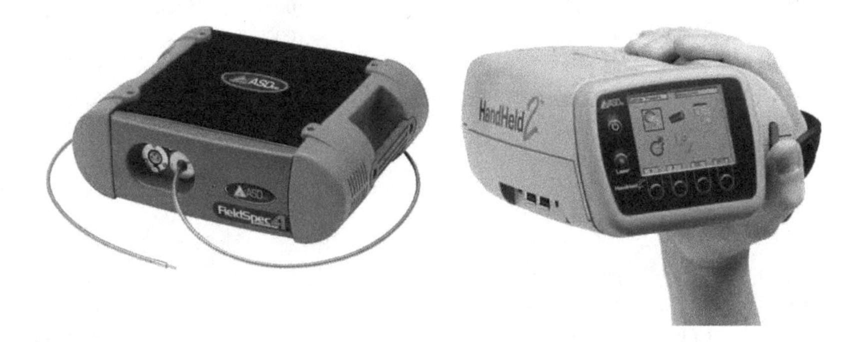

FIGURE 2.4 Exterior view of ASD FieldSpec® spectroradiometers: FieldSpec® 4 (left) and FieldSpec® Handheld 2 (right) spectroradiometers. (Courtesy of Analytical Spectral Devices, Inc., USA.)

materials at longer wavelengths—such as carbonates, clays, and chlorites—will benefit from a 2× radiometric performance increase in the SWIR 1 and 2 regions.

The FieldSpec 4 Wide-Res spectroradiometer is suited for support of multispectral remote sensing studies and analysis of materials with broad spectral features. The instrument offers dramatically improved speed, performance, and portability over previous models. The Wide-Res spectroradiometer can cover twice the ground in half the time without any loss of data quality.

The FieldSpec HandHeld 2 minimizes field data collection time while maximizing the quality of spectral results. The durable and flexible HandHeld 2 produces laboratory-quality results in remote measurement and analysis applications, in a range of orientations and environments, and without the need for an external controlling computer in the field. Leveraging proprietary technology, the HandHeld 2 produces high signal-to-noise ratio spectra in under a second using a highly sensitive detector array, low stray light grating, built-in shutter, DriftLock™ dark current compensation, and second-order filtering. The HandHeld 2 provides accurate, quickly derived reflectance, radiance, and irradiance spectra in a variety of settings. The HandHeld 2 Pro offers even greater sensitivity for applications with low reflectance. The Pro incorporates a photodiode array with an

FIGURE 2.5 Operation of ASD FieldSpec® spectroradiometers: FieldSpec® 4 (left) and FieldSpec® Handheld 2 (right) spectroradiometers. (Courtesy of Analytical Spectral Devices, Inc., USA.)

area five times larger than the standard HandHeld 2, allowing for spectrum collections up to three times faster than the standard model, with a 5× improvement in the signal-to-noise ratio. With a wavelength range of 325–1075 nm, an accuracy of ±1 nm and a resolution of <3 nm at 700 nm, the HandHeld 2 delivers an excellent performance. The ergonomic design, light weight, accessory handle, laser targeting, and tilting color LCD display all contribute to the convenience and ease-of-use in a range of orientations and environments.

Overall, the FieldSpec series field spectroradiometers are designed to collect solar reflectance, radiance, and irradiance measurements. They are ideal for applications in remote sensing, oceanography, ecology, forestry, plant physiology, geology, etc.

2.1.3.2 SVC (GER) Field Spectroradiometers

Geophysical Environmental Research (GER) series field spectroradiometers, including GER 1500, GER 2100, GER 2600, and GER 3700, are predecessors of current Spectra Vista Corporation (SVC) series field spectroradiometers, including SVC HR-512i, SVC HR-768si, SVC HR-768i, and SVC HR-1024i (SVC 2013). The GER series field spectrometers were discontinued in the early 2000s and were replaced with SVC series instruments, such as GER3700, a predecessor of SVC HR-1024i. Figure 2.6 shows an exterior view of the front and top of SVC HR-1024i field spectroradiometer, whereas Figure 2.7 presents the working principles of GER 3700, which can help readers to understand the working principles of SVC new generation spectroradiometers, including SVC HR-1024i. Readers can refer to Table 2.1 to learn some important technical parameters for SVC (GER) field spectroradiometers. All SVC (GER) field spectroradiometers allow measurements in a broad range of areas, from ground truthing of airborne campaigns to a number of industrial, military, agricultural, and environmental applications.

As illustrated by Figure 2.7, after a light beam enters the instrument through a lens, it first passes an aperture controlled by a shutter and regulated by a slit. The entered light then passes through the first beam splitter to be separated into two parts: Vis-NIR and SWIR spectral energies in two ways. In the first way, the Vis-NIR energy passes one more aperture/lens to reach the Vis-NIR diffraction grating, then finally projects onto the Vis-NIR detector array. In the second way, the SWIR energy continues to pass the second beam splitter to separate into SWIR 1 and SWIR 2 energies, then

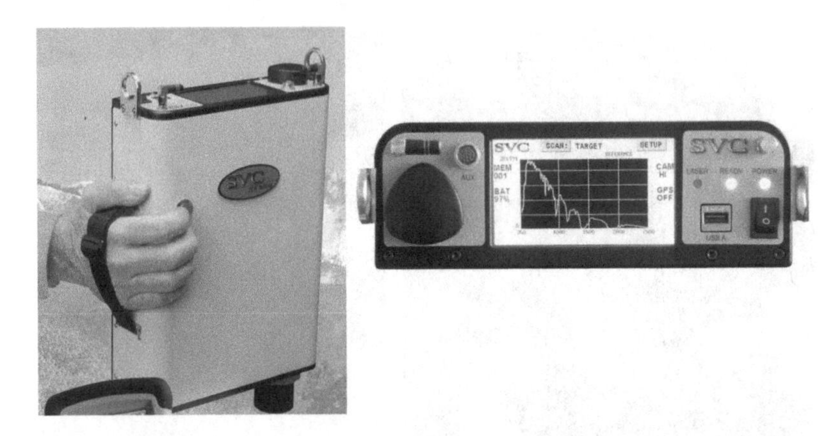

FIGURE 2.6 Exterior view of SVC HR-1024i: Front view (left) and top view (right). (Courtesy of Spectra Vista Corporation, Inc., USA.)

both SWIR lights reach their corresponding diffraction gratings and finally are projected onto their corresponding SWIR detector arrays. SVC (GER) instruments applied similar working principles as those shown in Figure 2.7 to measure spectral measurement covering a full spectral range from 3500 nm to 2500 nm.

The SVC HR-1024i combines the latest technology required to produce exceptional spectral data while capturing digital photographic, GPS, and external sensor data (SVC 2013). The instrument covers a full spectral range from 350 to 2500 nm. All data streams are gathered coincidentally and written to a single measurement file in order to provide the important spectral, positional, and visual data for analysis. The 32-bit instrument processor and internal memory allow operation without the use of an external computer while displaying the data graphically on the QVGA sunlight readable touch screen for immediate confirmation. Measurements are easily acquired by one person by first setting up instrument parameters via the touch screen display and then initiating a measurement. The high spectral resolution and low noise ensure that the collected data are of the highest quality. This high-quality data can be stored internally, along with scene photos and GPS coordinates while operating in stand-alone mode. The SVC HR-1024i includes a second Bluetooth device, allowing the instrument to receive data from an external sensor suite containing up to eight separate sensors. The sensor suite can include downwelling sensors supplying instantaneous broad- or narrowband solar response, so that the instrument can help researchers address a problem that is frequently encountered in taking field spectral measurements, which is a spectral measurement error caused by the possible illumination condition change between taking reference radiance and the radiance from a target. The sensor data is stored with the spectral data file, allowing researchers to understand changes in solar irradiance and to assist in corrections. The use of 100% linear array detectors ensures excellent wavelength stability, while the cooled InGaAs and extended InGaAs detectors provide superior radiometric stability. Fixed fore optics and hard-mounted internal spectrometer elements provide a robust optical path (similar to Figure 2.7a).

SVC HR-512i is a lightweight portable spectroradiometer with most features the same as SVC HR-1024i, except it operates in the VNIR spectral range (350 nm–1050 nm). SVC HR-512i takes spectral measurements from both terrestrial and marine targets in seconds as the internal CPU sets the appropriate integration based on current lighting conditions, while dark current is automatically measured and subtracted. The internal computer applies the selected radiometric calibration and the graphic data is promptly displayed on the sunlight-readable LCD touch screen. The SVC HR-512i also includes a second Bluetooth® radio for communication with up to eight optional external sensors as the SVC HR-1024i.

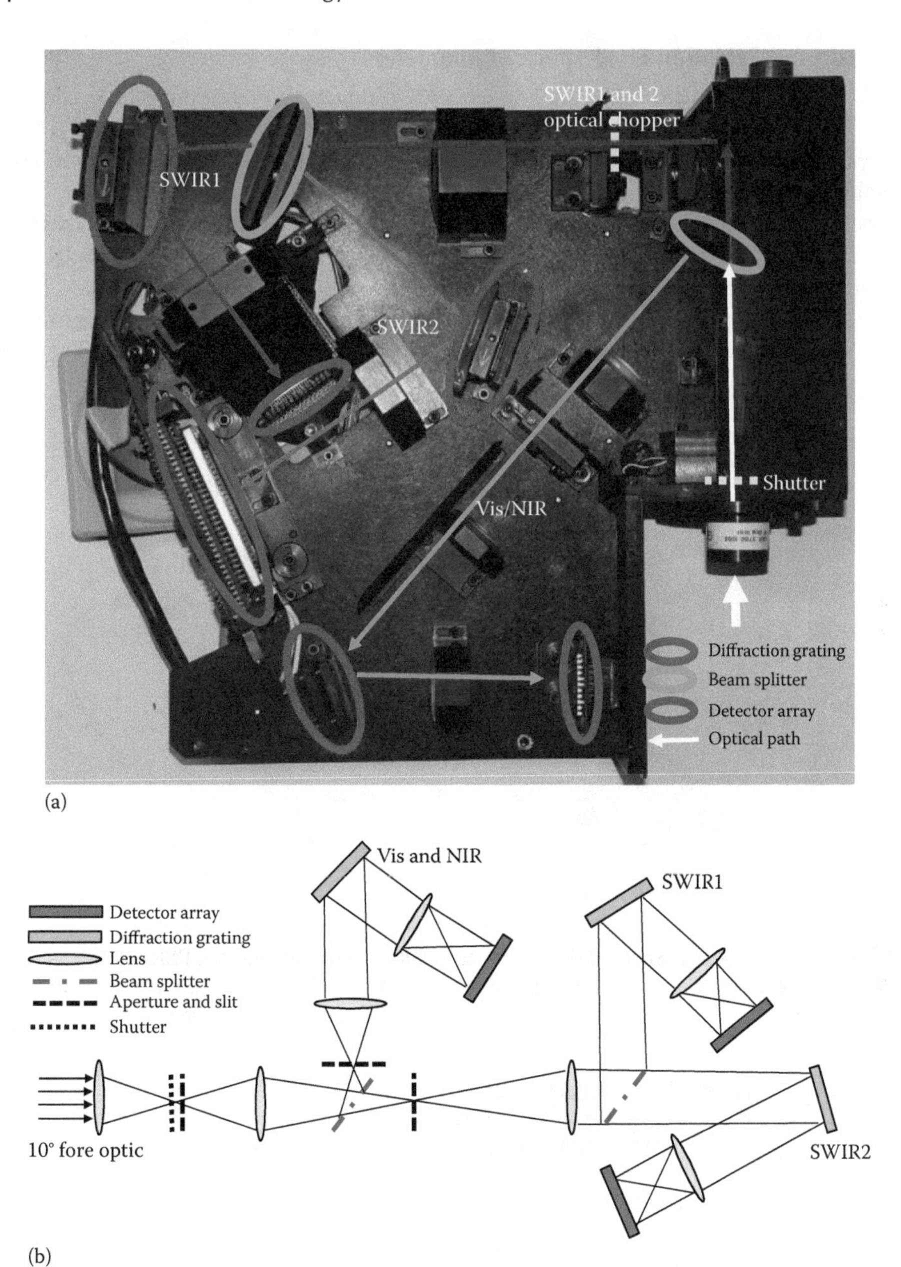

FIGURE 2.7 Illustration of working principle of GER 3700: (a) internal optical paths and (b) diagrammatic form.

SVC HR-768si field spectroradiometer produces the same high-quality data and high spectral resolution as the SVC HR-1024i, while covering the spectral range from 350 nm–1900 nm. The onboard LCD provides the researcher with the instantaneous graphic display of the measurement without need for a separate computer or PAD. The internal digital camera captures the image of the target area for reference during the data analysis, while the built-in GPS acquires the geolocation of the instrument and writes the coordinates to memory. Therefore, the SVC HR-768si is engineered to be the central device integrating target images, GPS location, time, and external sensor inputs with high-resolution spectral data.

2.1.3.3 Spectral Evolution Field Spectroradiometers

Spectral Evolution series field spectroradiometers are products of Spectral Evolution, Inc., United States (PSR 2014). Its new models are PSR+ and PSR-1100, which are introduced below. (See Table 2.1 for their important technical parameters.) Figure 2.8 presents a side-view of PSR series instruments while Figure 2.9 shows their general operation in the field.

The spectrometer component in each PSR Series spectroradiometer is a crossed Czerny–Turner configuration using ruled gratings as the dispersive elements. Energy enters the spectrometer and is collimated before being reflected off the gratings and refocused onto the detectors. Depending on the model PSR Series instruments, there are two or three detectors in use. All systems incorporate a 512-element silicon array covering the spectral range from 350–1000 nm. The PSR-1900

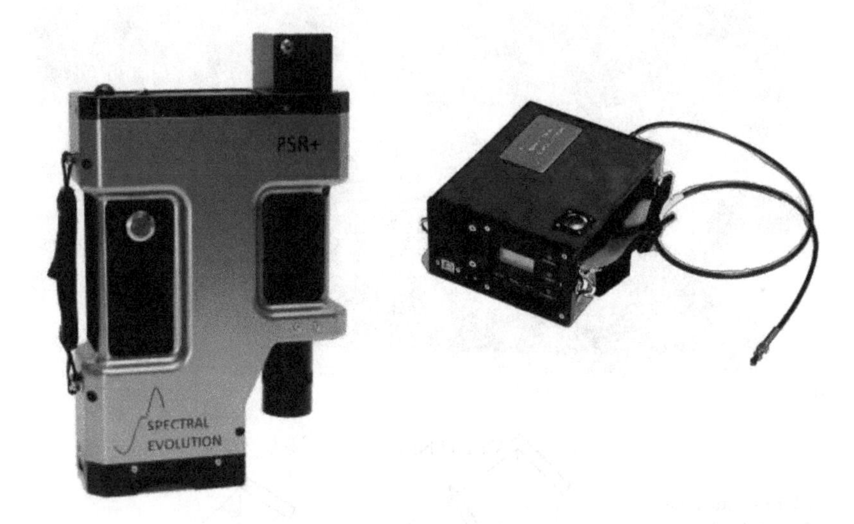

FIGURE 2.8 Exterior view of Spectral Evolution series field spectroradiometers: PSR+ (left) and PSR-1100F (right) spectroradiometers. (Courtesy of Spectral Evolution, Inc., USA.)

FIGURE 2.9 Operation of Spectral Evolution series field spectroradiometers: PSR+ (left) and PSR-1100F (right) spectroradiometers. (Courtesy of Spectral Evolution, Inc., USA.)

and PSR-2500 contain a single 256-element thermoelectrically cooled InGaAs array that extends spectral range into the NIR (to 1900 nm and to 2500 nm, respectively). The PSR+3500 model uses two InGaAs arrays: a 256-element array covering 1000–1900 nm and another array 256-element that extends from 1900–2500 nm with finer spectral resolution. The PSR+ series have the following features: (1) high spectral resolution (3nm at 700 nm, ≤8 nm at 1500 nm, and ≤6 nm at 2100 nm); (2) proprietary Sotex™ filter technology for improved order sorting, smoother transitions, and enhanced stray light performance; (3) no moving optical parts and improved optical path for reliable, superior operation no matter what the conditions; (4) best-in-class signal-to-noise ratio with industry leading resolution; (5) auto-shutter, auto-exposure, and auto-dark correction for one-touch operation; (6) direct attach 4°, 8°, and 14° lenses, 25° fiber optic, diffuser or integration sphere; (7) fiber mount: 1°, 2°, 3°, 4°, 5°, 8°, and 10° lenses; and (8) optional GETAC microcomputer with digital camera, GPS–tags photos, GPS coordinates, and voice notes to scans. The PSR+ series spectroradiometers are lightweight, battery operated, easy to move from project to project, and are suitable for many applications.

Spectral Evolution offers two budget-minded, lightweight, handheld PSR-1100 spectroradiometers to meet the rigorous demands of the remote sensing community. The PSR-1100 features a durable, fixed 4° FOV optic and laser targeting for field operation while the PSR-1100F provides greater optical flexibility, which includes a detachable fiber optic cable with a wide array of optional fiber mount FOV lenses (1°, 2°, 3°, 4°, 5°, and 8°) and irradiance diffusers. The PSR-1100F can be equipped with a GETAC PS336 microcomputer in the field, with a sunlight-readable display or viewing spectra, the ability to store an almost limitless number of scans, and the capability of tagging scans with voice notes, images from its digital camera, GPS coordinates, and altimeter reading.

The PSR+ and PSR-1100 spectroradiometers allow users to perform a wide variety of tasks, from ground truthing for hyperspectral and multispectral imagery from satellites like Worldview 3 to scanning a single leaf for measuring plant health. Some of the applications where the PSR+ is being used include ground truthing, including confirming/interpreting hyperspectral and multispectral data; estimation of crop and grass chlorophyll; environmental research and atmospheric/climate research; crop health–measuring photosynthesis efficiency; forestry research and canopy studies; plant species identification, water body studies, and soil analysis, including topsoil fertility and erosion tests; radiometric calibration transfer, geological remote sensing, and mapping, including surveying, mineral identification, and geomorphology; and forage analysis and precision agriculture.

2.1.3.4 SpectraScan Spectroradiometers

SpectraScan® spectroradiometers (PR-6xx series) are products made by Photo Research, Inc., United States, and are designed to measure radiance utilizing the patented Pritchard optics to collect the incoming optical radiation. They can be used for measuring spectral reflectance and transmittance for remote sensing. Their ease-of-use, accuracy, and reliability have made this product family the most widely used. Figure 2.10 shows the exterior view of PR®-6xx series spectroradiometers.

When measuring small character stroke widths or miniscule samples in a large field, an accurate and precise optical system is essential to insure proper alignment. Pritchard Optics is such a system, widely accepted as the most accurate and versatile in use today. The PR-6xx series are mounted with such an optical system. In the Pritchard system, the user sees a bright and magnified image, in the center of which is a black dot (aperture). Since only the light passing through the aperture is measured, the aperture accurately and unambiguously defines the measuring field within the field of view. The PR-6xx is as easy to use as point-and-shoot. The Pritchard optics system makes target alignment as easy as aim and focus. The PR-670, -680, and -680L are all supplied with four automated measuring apertures (1°, 1/2°, 1/4°, and 1/8°). The multi-apertures are ideal for measuring both large and small targets without the need for any additional lenses or accessories, and with minimal repositioning of the instrument. If the spot coverage with the standard lens is too large, the PR-6xx instruments can be optionally equipped with a series of magnification lenses to achieve spot coverage as small as 0.0036 mm.

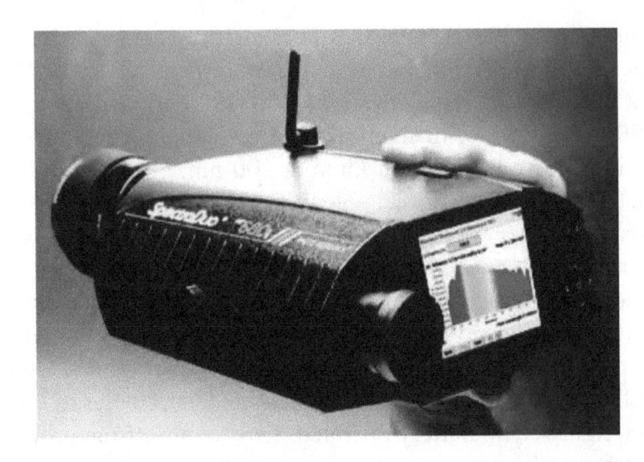

FIGURE 2.10 Exterior view of Photo Research, PR® series spectroradiometers. (Courtesy of Photo Research, Inc., USA.)

The PR-6xx series design provides standalone operation (no PC required). Menus are accessed via the on-board, 3.5-inch high-resolution, full-color touch screen LCD display, and 5-way navigation keypad. Following a measurement, the PR-6xx displays data and color spectral graphs on the system display. In addition, the PR-6xx series of instruments takes portability to the next level. All instruments come standard with a Li-ion battery which, on a full charge, can last for more than 12 hours of continuous use. Secure Digital (SD) storage allows for over 80,000 measurements on a 512 MB card.

The PR-680 and -680L SpectraDuo® are the first and only instruments to contain a fast-scanning 256 detector element spectroradiometer and an ultra-sensitive, low-noise photo multiplier tube based filter photometer in one. Available operating modes are (1) as a fast scanning spectroradiometer; (2) as a highly sensitive photometer; or (3) in auto select mode, which automatically chooses a detector based on the available signal. The unique design of the SpectraDuo® series makes tasks such as spectrally based colorimetry and high speed, low-level luminance—required for display metrology—possible with a single instrument. Since the Pritchard viewing and measuring optics are shared by both the spectroradiometer and photometer, instrument realignment is not necessary when switching between spectral and photometer modes. The SpectraDuo® can optionally be equipped with analog output capability, using the photo multiplier tube as a high-speed sensor for characterizing display response (3 μsec response) or waveform analysis of flash sources.

2.1.3.5 Ocean Optical Spectrometers

USB2000+ (HR2000+) series custom modular spectrometers are made by Ocean Optics, Inc., United States, mostly working in UV-Vis and Vis-NIR spectral ranges. Figure 2.11 shows an exterior view of three Ocean Optic custom modular spectrometers. The USB2000+ Miniature Fiber Optic Spectrometer is a unique combination of technologies: a powerful 2-MHz analog-to-digital (A/D) converter, programmable electronics, a 2048-element CCD-array detector, and a high-speed USB 2.0 port. This innovative combination produces the fastest spectrometer yet and provides resolution to 0.35 nm (FWHM). The USB2000+ allows users to capture and store a full spectrum into memory every millisecond (or 1,000 full spectra every second) when the spectrometer is interfaced to a computer via a USB 2.0 port. The USB2000+ can be sensitive to weak light sources. Components of the USB2000+ series and its working principle are shown in Figure 2.12, with an explanation of each component's function summarized in Table 2.2. Figure 2.12 shows how light moves through the optical bench of a USB2000+ spectrometer. The optical bench has no moving parts that can wear out or break; all the components are fixed in place at the time of manufacture.

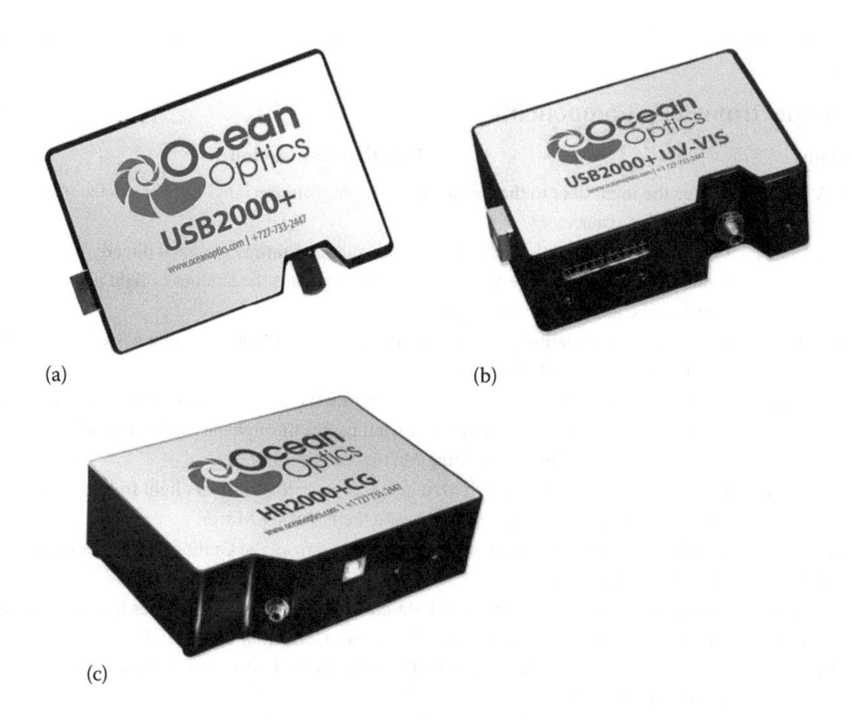

FIGURE 2.11 Exterior view of Ocean Optic custom modular spectrometers. (a) Ocean Optics USB2000+ Fiber Optic Spectrometer; (b) Ocean Optics USB2000+UV-VIS Fiber Optic Spectrometer; and (c) Ocean Optics HR2000+CG High-Speed Fiber Optic Spectrometer. (Courtesy of Ocean Optics, Inc., USA.)

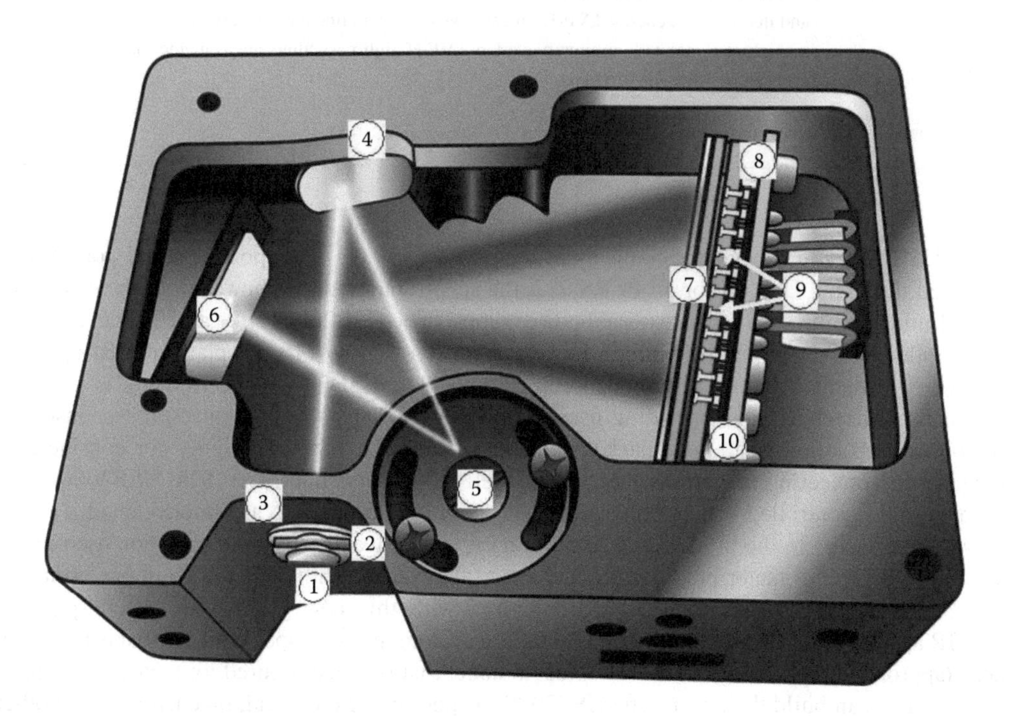

FIGURE 2.12 USB2000+ with components to show how the spectrometer works. (Courtesy of Ocean Optics, Inc., USA; see Table 2.2 for an explanation of the function of each numbered component in the USB2000+ spectrometer shown in this figure.)

TABLE 2.2
USB2000+ Spectrometer's Components

Item	Name	Function Description
1	SMA 905 Connector	Secures the input fiber to the spectrometer. Light from the input fiber enters the optical bench through this connector.
2	Slit	A dark piece of material containing a rectangular aperture is mounted directly behind the SMA Connector. The size of the aperture (200 μm) regulates the amount of light that enters the optical bench and controls spectral resolution.
3	Filter	Restricts optical radiation to pre-determined wavelength regions. Light passes through the filter before entering the optical bench.
4	Collimating Mirror	A SAG+, Ag-coated mirror focuses light entering the optical bench toward the Grating of the spectrometer. Light enters the spectrometer, passes through the SMA Connector, Slit, and Filter, and then reflects off the Collimating Mirror onto the Grating.
5	Grating	A #3 (600 lines per millimeter, blazed at 500 nm) grating diffracts light from the Collimating Mirror and directs the diffracted light onto the Focusing Mirror.
6	Focusing Mirror	A SAG+, Ag-coated mirror receives light reflected from the Grating and focuses first-order spectra onto the detector plane.
7	L2 Detector Collection Lens	Attaches to the Detector to increase light-collection efficiency. It focuses light from a tall slit onto the shorter Detector elements. The L2 Detector Collection Lens should be used with large diameter slits or in applications with low light levels. It also improves efficiency by reducing the effects of stray light.
8, 10	Detector	Collects the light received from the Focusing Mirror or L2 Detector Collection Lens and converts the optical signal to a digital signal. Each pixel on the Detector responds to the wavelength of light that strikes it, creating a digital response. The spectrometer then transmits the digital signal to the software application.
9	LVF Filters	Optional Linear Variable Filters (LVF) construct systems with excellent separation of excitation and fluorescence energy. LVF-L Linear low-pass filters fine tune the excitation source for maximum signal with minimum overlap. LVF-H high-pass filters are available for the detection side. These filters are optional.

Source: Courtesy of Ocean Optics, Inc., USA.

The USB2000+ Spectrometer connects to a computer via the USB port or serial port. When connected through a USB 2.0 or 1.1 or RS-232, the spectrometer draws power from the host computer, eliminating the need for an external power supply. The USB2000+, like all USB devices, can be controlled by OceanView software, a Java-based spectroscopy software platform that operates on Windows, Macintosh, and Linux operating systems. The USB2000+ spectrometer is a unique combination of technologies providing users with both an unusually high spectral response and high optical resolution in a single package. The electronics have been designed for considerable flexibility in connecting to various modules as well as external interfaces. The USB2000+ is a microcontroller-controlled spectrometer, thus all operating parameters are implemented through software interfacing to the unit. A special 500 lines/mm groove density grating option used in the USB2000+XR spectrometer provides broader spectral coverage with no sacrifice in performance. This extended-range spectrometer is preconfigured with this new grating for general-purpose UV-NIR applications. Therefore, the USB2000+ is a custom modular spectrometer that is relatively cheap (approximately $2000). Users can select a model that is preconfigured for a particular application or they can build their own. The USB2000+ is perfect for chemical, biochemical, and other applications (e.g., measuring *in situ* hyperspectral data from interesting targets) where fast reactions need to be monitored.

2.2 PLANT BIOLOGY INSTRUMENTS FOR HRS

2.2.1 Introduction

One application area of developing HRS is extracting and mapping biophysical (e.g., plant canopy leaf area index [LAI] plant species, and biomass) and biochemical (e.g., plant chlorophyll content, nitrogen concentration) parameters using the HRS data. The frequently used modern field spectrometers introduced in Section 2.1 are for calibrating and validating airborne and spaceborne multi-/hyperspectral imaging data and modeling relationships between *in situ* hyperspectral data and a set of bioparameters. To better understand and apply HRS imaging data to vegetation, including bioparameter estimation and mapping, we need some plant biology instruments to measure important bioparameters to calibrate and validate a set of empirical/statistical models and physically based models that can finally be used to retrieve and map such bioparameters based on HRS data (acquired onboard different platforms: *in situ*, airborne, and spaceborne). Such bio-instruments frequently use spectral radiance (or reflectance) and light photon density measurements taken from the surface of plant canopy or individual leaves or flowers to convert to a set of bioparameter measurements, such as plant canopy leaf area, LAI, photosynthesis rate and fraction of photosynthetically active radiation (fPAR), chlorophyll and nitrogen contents, etc. Such bioparameter observation values measured by bio-instruments can then be used to calibrate and validate both statistical models and physically processed models, and finally such calibrated/validated models can be utilized for retrieving and mapping the bioparameters using hyperspectral imaging data. Therefore, in the following section, some frequently used plant biology instruments will be briefly introduced. Such bio-instruments may include those for measuring plant individual leaf area and plant canopy LAI, plant photosynthesis rate and fPAR, and chlorophyll and nitrogen content.

2.2.2 Plant Biology Instruments

Table 2.3 summarizes plant biology instruments that can be used for calibrating a set of bioparameter models using hyperspectral data. In this section, working principles, structure, and operating procedures in the field of three categories of bio-instruments are briefly introduced.

2.2.2.1 Instruments for Measuring Leaf Area and Leaf Area Index

2.2.2.1.1 LAI-2200C Plant Canopy Analyzer (PCA)

The LAI-2200C PCA is an instrument made by LI-COR, Inc., and is used for measuring leaf area index (LAI) of plant canopies. Figure 2.13 presents two parts of the LAI-2200C and its operation in the field. The LAI-2200C consists of the LAI-2250 optical sensor and LAI-2270 control unit. Of them, the LAI-2250 optical sensor is considered the "heart" of the LAI-2200C PCA. The instrument calculates LAI and other canopy structure attributes from radiation measurements made with a fisheye optical sensor (148° field of view). Measurements made above and below the canopy are used to determine canopy light interception at five angles, from which LAI is computed using a model of radiative transfer in vegetative canopies (LAI-2200C PCA, 2014). The LAI-2200C PCA can be used to measure *in situ*, non-destructive plant LAI, and to provide quick and accurate LAI measurements. It is ideal for studies of canopy growth, canopy productivity, forest vigor, canopy fuel load, air pollution deposition modeling, insect defoliation, remote sensing, and the global carbon cycle.

The LAI-2200C PCA consistently outperforms other methods in terms of flexibility, advanced features, accuracy, and ease of use. The LAI-2200C can be used to measure LAI of a variety of plant canopies, from very short row crops to isolated trees. The low-profile sensor head in the LAI-2200C allows LAI measurements in short canopies, such as grasslands. In forests, automated logging can collect above-canopy readings in a clearing while a second LAI-2250 optical sensor is used for below-canopy readings. When ceptometry methods (Finzel et al. 2012) are used in tall canopies, it is not possible to use the sun-fleck or gap-fraction mode because shadows are blurred due to the penumbra effect. LAI-2200C measures the LAI of small plots, gapped, and non-uniform canopies simply and easily by restricting the

TABLE 2.3
Summary of Plant Bio-Instruments that May Be Used for Hyperspectral Remote Sensing

Category	Instrument	Year	Measuring Characteristics and Accuracy	Application	Manufacture	Weight (kg)
Leaf area, leaf area index (LAI)	LAI-2200C Plant Canopy Analyzer (PCA)	2014	Quickly and accurately measuring LAI across a wide variety of canopy types from row crops to isolated trees, under any sky conditions, and with GPS measurement.	Measuring effect LAI of plant canopy (crops, forests, etc.)	LI-Cor, Inc., USA	1.30
	LI-3000C Portable Area Meter (PAM)	2014	Measuring leaf area of living plants or detached leaves in the field or laboratory; 1 mm^2 resolution; displays and stores individual and accumulated leaf area.	Measuring plant leaves area (crops, trees, etc.)	LI-Cor, Inc., USA	2.00
	LI-3100C Area Meter (AM)	2004	Rapid, precise area measurement of large or small leaves; adjustable resolution: 0.1 or 1 mm^2; high accuracy and repeatability; individual or cumulative area.	Measuring plant leaves area (crops, trees, etc.)	LI-Cor, Inc., USA	43.00
Photosynthesis rate and fPAR	LI-6400XT Portable Photosynthesis System	2013	Open system design, automatic and independent controls of leaf chamber CO_2, H_2O concentrations, temperature, and light, and operation control by user's programming.	Measuring plant (crops and trees, etc.) photosynthesis, fluorescence and respiration	LI-Cor, Inc., USA	N/A
	TRAC	2002	Measuring canopy 'gap size' distribution in addition to canopy 'gap fraction' along a transect in the forests; data collected at 32 readings per sensor per second.	Measuring LAI, fPAR absorbed by plant canopies and other forest struture parameters	3rd Wave Engineering, Canada	N/A
Chlorophyll content	SPAD-502 Plus Chlorophyll Meter	2010	Based on absorption amounts at red and NIR regions by chlorophyll the SPAD-502 measurement can be used to determine chlorophyll content.	Measuring chlorophyll content of plants (crops and forests, etc.)	Spectrum Technologies, Inc., USA	0.20
	SPAD-502DL Plus Chlorophyll Meter	2014	The same as SPAD-502 Plus but with a built-in data logger (Item 2900PDL).	Measuring chlorophyll content of plants (crops and forests, etc.)	Spectrum Technologies, Inc., USA	0.20

(Continued)

TABLE 2.3 (CONTINUED)
Summary of Plant Bio-Instruments that May Be Used for Hyperspectral Remote Sensing

Category	Instrument	Year	Measuring Characteristics and Accuracy	Application	Manufacture	Weight (kg)
	Field Scout CM-1000 Chlorophyll Meter	2013	Based on absorption at red by chlorophyll and reflectance at NIR by leaf physical structure to determine chlorophyll content. Repeatability: ± 5% of measurement.	Measuring chlorophyll content of plants (crops and forests, etc.)	Spectrum Technologies, Inc., USA	0.70
	MC-100 Chlorophyll Concentration Meter	2016	Measuring a CCI (= %Transmittance at 931 nm / %Transmittance at 653 nm) selative chlorophyll value. ± 0.1 Chlorophyll Concentration Index (CCI) unit; sample acquisition time: 2–3 s.	Measuring chlorophyll content/ concentration of plants (crops and forests, etc.)	Apogee Instruments, Inc., USA	0.21

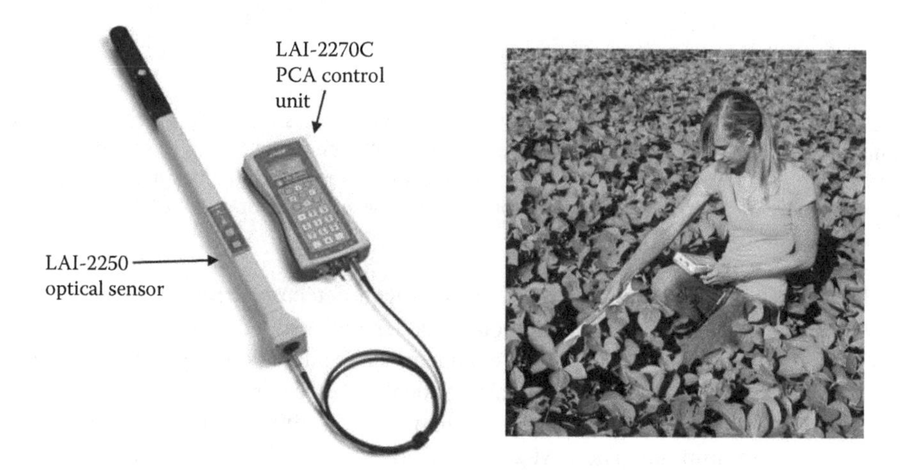

FIGURE 2.13 LAI-2200C plant canopy analyzer (PCA) (left) and its operation (right). (Copyright LI-COR Inc. Used with permission.)

view in terms of both azimuth (with view-restricting caps) and zenith angles (by excluding one or more of the view rings). With ceptometry methods, it is not possible to restrict either the azimuth or the zenith view angles. Foliage density (m^2 foliage area per m^3 canopy volume) is the result when measuring an isolated tree with the LAI-2200C. The File Viewer 2200 software package calculates canopy volume from simple measurements describing the average shape of the tree's crown, and then calculates foliage density. The LAI-2200C can work in most daylight conditions, unlike other methods that require specific sun angles or cloud cover. The instrument with an internal Global Positioning System (GPS) module integrates location data into the LAI file for easy mapping LAI.

The LAI-2250 optical sensor is designed specially with a fisheye lens with hemispheric field-of-view that ensures that LAI calculations are based on a large sample of the foliage canopy. The fisheye lens takes in a hemispherical image, which the optical system focuses onto the five-ring

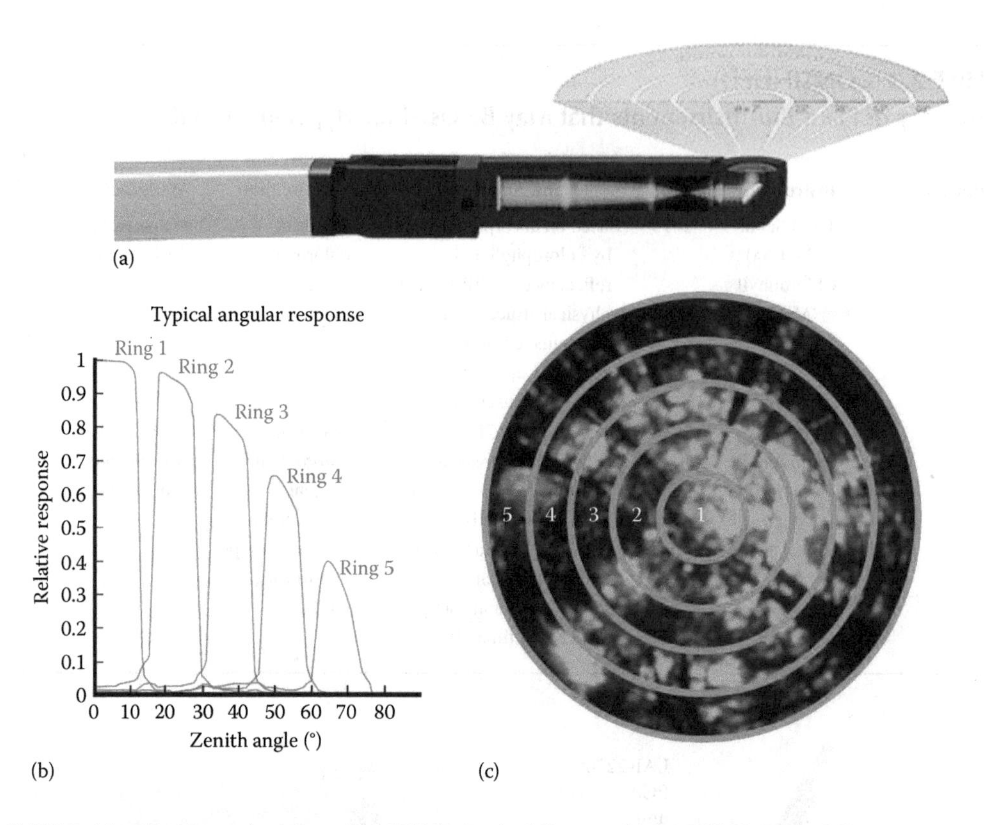

FIGURE 2.14 Working principles of LAI-2250 Optical Sensor: (a) LAI-2250 Optical Sensor measures light from five different zenith angles with one reading; (b) each detector ring responds over a different range of zenith angles; and (c) the approximate field-of-view of each ring is simulated by the fisheye photo (above), where concentric circles represent the five view angles. (Copyright LI-COR Inc. Used with permission.)

photodiode optical sensor (see Figure 2.14a). Five silicon detectors arranged in concentric rings ensure measuring light from five different zenith angles with one reading (Figure 2.14a,b). Each detector ring views a different portion of the canopy or sky centered on one of the five view angles (Figure 2.14c). An internal filter in the LAI-2250 optical sensor can reduce errors from transmitted and reflected light. The optical sensor receives radiation below 490 nm only, where leaf reflectance and transmittance are minimal. The LAI-2250 optical sensor with restricting caps can allow for LAI measurements of small plots and hedges by blocking undesired objects from the sensor's view, such as the operator or a neighboring plot. Its light scattering correction feature can correct light scattering measurements throughout the daylight hours, even under clear skies. The LAI-2200C PCA computes LAI from measurements made above and below the canopy, which are used to determine canopy light interception at five angles. These data are fit to a well-established model of radiative transfer inside vegetative canopies to compute LAI, mean tilt angle, and canopy gap fraction.

Given a plant canopy condition, factors including the height, size, and structure of the canopy, size and orientation of leaves, sky condition, and field of view of the optical sensor can affect the result of LAI measurements. If not paying attention to them, such factors all may affect the accuracy of LAI measurement. The LAI measurement taken by the LAI-2200C PCA is called "effective" LAI (Chen and Cihlar 1995).

2.2.2.1.2 LI-3000C Portable Area Meter (PAM)

LI-3000C PAM is designed for measuring the leaf area of both living plants and detached leaves in the field or laboratory (LI-3000C PAM, 2014). Similar to the LAI-2200C PCA, the LI-3000C

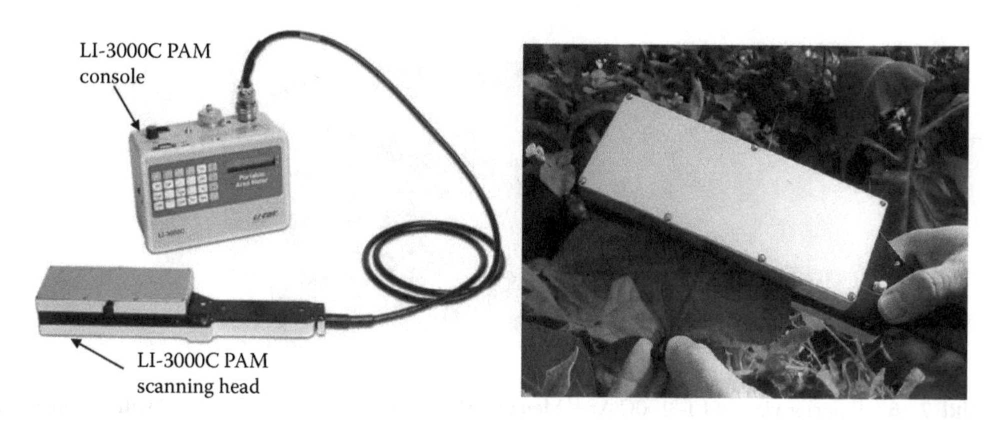

FIGURE 2.15 LI-3000C Portable Area Meter (PAM) (left) and its operation (right). (Copyright LI-COR Inc. Used with permission.)

PAM also consists of two components: LI-3000C scanning head and LI-3000C readout console (Figure 2.15). The LI-3000C provides a non-destructive, precise method to measure leaf area. Leaf area development or reduction associated with conditions of insect infestation, drought, disease, or air pollution can be measured in the field or in growth chambers. Leaf area can be monitored without damaging the plants. This allows precise evaluation of plant canopy development throughout the growing season. This technique allows area data collection in the field before the detached leaves appreciably shrink or curl. For many experiments, the need to transport samples away from the plot site is eliminated. The remaining area of insect-damaged leaves can also be measured. Data can be collected from the same plant throughout its life cycle. For severely damaged leaves, a transparent sheath can be used to support the remaining tissue. Forest canopy or urban plant foliage development can be recorded for purposes such as evaluating air pollution effects or pest damage.

For each leaf area measurement, the LI-3000C records values of leaf area, leaf length, average width, and maximum width. Area data are recorded by the readout console as the scanning head is passed over the leaf. Leaves with irregular margins or with holes, as in cases of insect damage, are correctly measured by the LI-3000C. As the portion of leaf with the hole passes through the scanning head, the lens-photodiode system senses the LED light. That particular LED location does not contribute to the accumulated area on the display until the LEDs are once again masked by a portion of leaf without a hole. The area is integrated and displayed as the scanning head is drawn over the leaf. Each of these values can be shown on the display. Individual leaf area values can also be added to a secondary summing register to collect accumulated leaf area data for leaf area index or whole plant leaf area. After a measurement, the area data are stored using a convenient file system along with a time stamp from the real time clock and an alphanumeric remark to identify the data. The LI-3000C PAM utilizes an electronic method of rectangular approximation providing 1 mm^2 resolution.

2.2.2.1.3 LI-3100C Area Meter (AM)

The LI-3100C AM is designed for efficient and exacting measurement of both large and small leaves (individual or cumulative area) (LI-3100C AM, 2004). Figure 2.16 presents the exterior view of LI-3100C AM. Per the LI-3100C, its adjustable resolution settings (0.1 or 1 mm^2) provide versatility for diverse project requirements. A wide variety of leaves can be measured, ranging from larger samples, such as corn, tobacco, and cotton, to smaller samples, such as wheat, rice, or alfalfa. Small leaves or leaf discs are measured with the same precision as larger leaves. The LI-3100C can also handle conifer needles, perforated leaves, and leaves with irregular margins. This is especially important in determining leaf damage and insect feeding trails.

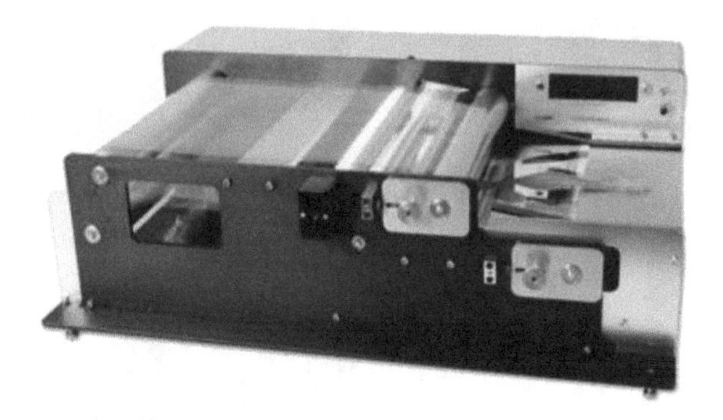

FIGURE 2.16 Exterior view of LI-3100C Area Meter (AM). (Copyright LI-COR Inc. Used with permission.)

The working principle of the LI-3100C AM can be described as follows. Samples are placed between the guides on the lower transparent belt and allowed to pass through the LI-3100C. As the sample travels under the fluorescent light source, the projected image is reflected by a system of three mirrors to a scanning camera. This unique optical design results in high accuracy and dependability. An adjustable press roller flattens curled leaves and feeds them properly between the transparent belts. This provides for accurate measurement of small grasses, legumes, aquatic plants, and similar types of leaves. As samples pass under the light source, the accumulating area in mm^2 is shown on the LED display or on a computer screen.

One of major application areas with LAI-2200C PCA, LI-3000C PAM, and LI-3100C AM is for multi-/hyperspectral remote sensing. The measured LAI is a major biophysical parameter and can be directly used for calibrating remote sensing modeling for estimating and mapping LAI (crops and forests, for example). Quickly and accurately measuring sampling leaf area is a vital step toward estimating and mapping biochemical content/concentration at canopy even landscape scale with *in situ* or hyperspectral imaging data (Johnson et al. 1994).

2.2.2.2 Instruments for Measuring Photosynthesis and fPAR

2.2.2.2.1 LI-6400/LI-6400XT Portable Photosynthesis System

The LI-6400XT is LI-COR's newest photosynthesis system, which is an ideal instrument for measuring plant photosynthesis rates (LI-6400XT System, 2013). The LI-6400 is its last version and has become the world leader in portable photosynthesis measurement systems. The LI-6400XT is its latest version. The advanced features that lead most researchers with expertise in remote sensing, biological, and ecological studies to choose the LI-6400XT include

1. Proven technology that places gas analyzers in the sensor head to provide rapid response and eliminate time delays
2. An open system design that allows complete control over environmental variables of interest
3. A flexible, open source software language in the LI-6400XT console that can be modified to write user own equations or AutoPrograms, providing an unprecedented level of automation
4. Powerful networking capability via Ethernet connectivity, providing a world of data output, file-sharing, and training possibilities
5. A variety of leaf chambers and light sources, a leaf chamber fluorometer, and soil CO_2 flux chamber that are interchangeable with the same LI-6400XT Sensor Head

Figure 2.17 presents the LI-6400XT system associated with two major components: LI-6400XT system sensor head and LI-6400XT system console, and its basic operation in the field.

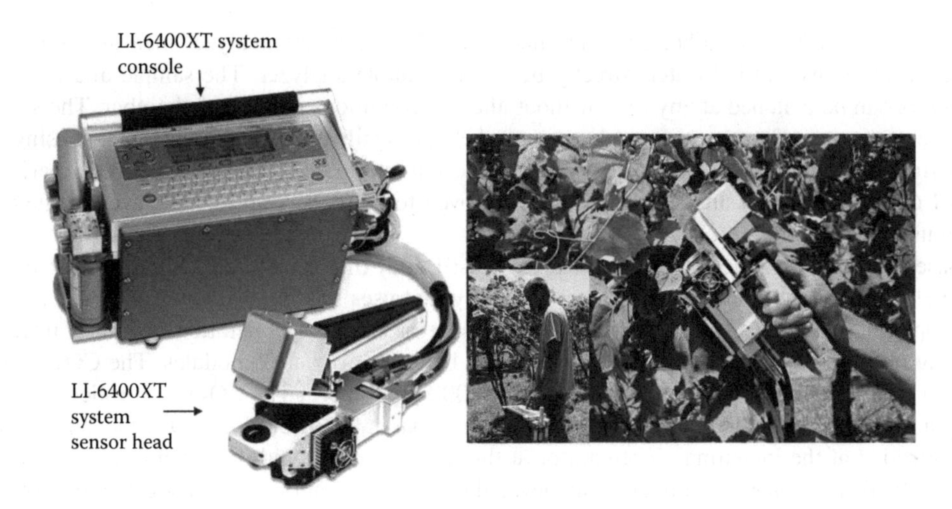

FIGURE 2.17 LI-6400XT System (left) and its operation (right). (Copyright LI-COR Inc. Used with permission.)

The LI-6400XT System can also be used for measuring plant fluorescence and soil respiration simultaneously, besides measuring plant photosynthesis rate (LI-6400XT System 2013). However, only the working flowchart (the system console) and operation principle (the sensor head) of the instrument and its operation related to measuring photosynthesis rate are briefly introduced.

The LI-6400XT has two absolute CO_2 and two absolute H_2O non-dispersive infrared (IR) analyzers in the sensor head (see Figure 2.18). Based on LI-6400XT System (2013), the operation principle of the sensor head can be briefly described as follows. When the IR radiation from the sample analyzer source passes into the leaf chamber mixing volume, it is twice reflected 90° by gold mirrors. The mirrors are gold-plated to enhance IR reflection and provide long-term stability. After being reflected through the leaf chamber mixing volume where IR absorption occurs, IR radiation passes through a chopping filter wheel and into the sample analyzer detector. The chopping filter wheel has four filters that pass light in absorption and optical reference wavelengths for CO_2 and H_2O. These filters provide excellent rejection of IR radiation outside the wavelengths of interest,

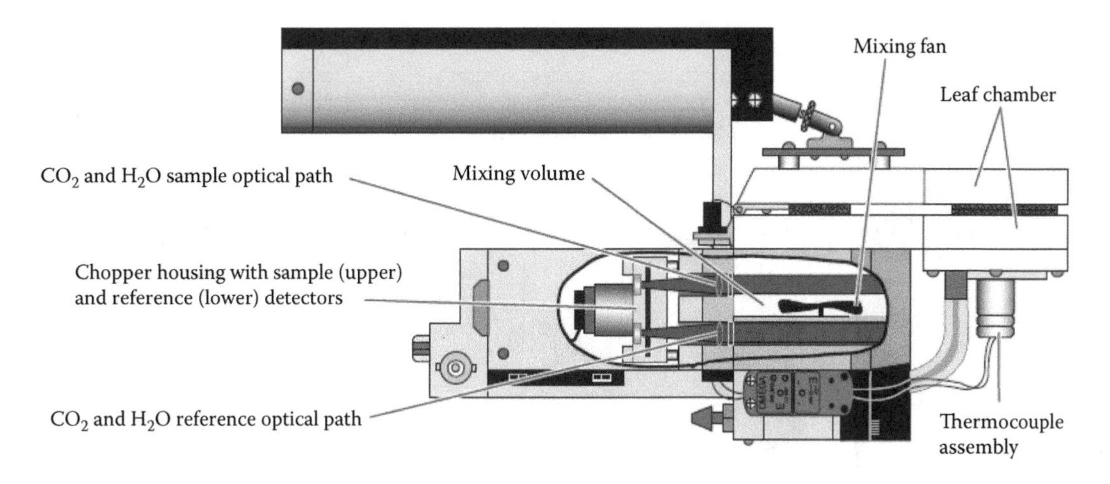

FIGURE 2.18 Profile view of LI-6400XT System Sensor Head. (Copyright LI-COR Inc. Used with permission.)

eliminating the effects of other IR absorbing gases. The reference analyzer measures incoming gas concentrations and is located directly below the sample analyzer. The sample and reference analyzers can be matched at any time without altering conditions in the leaf chamber. The sample analyzer detector, reference analyzer detector, and chopping filter wheel are sealed in a housing that is continuously purged of CO_2 and water vapor to prevent interference. Through years of experience, the LI-6400XT analyzer and sensor head have proven to be robust and reliable, even in the most rigorous field conditions.

Based on LI-6400XT System (2013), the 6400-01 CO_2 Injector System (Figure 2.19) consists of an electronic controller, a CO_2 Source Assembly that uses readily available cartridges for portable operation, and a CO_2 tank fitting for greenhouse or laboratory operation. All parts integrate directly into the standard console with no external batteries or control modules. The CO_2 Injector System provides a constant CO_2 input from 50 to 2000 μmol mol^{-1}. The CO_2 is controlled by delivering a precisely controlled pure CO_2 stream into air that is CO_2-free. The CO_2 concentration can be controlled at the incoming air stream or at the leaf surface to within 1 ppm of a target value. The 6400-01 facilitates measurements at elevated CO_2 concentrations and easy generation of CO_2 response curves. The CO_2 injector is under complete software control, allowing you to manually set CO_2 levels from the console, or use AutoPrograms to make measurements at a series of concentrations. The LI-6400XT controls chamber humidity by automatically varying the flow rate to null-balance at the chamber humidity level you specify in software; the input flow rate can also be held constant. Flow rate is controlled by pump speed in the standard system. With the LI-6400-01 CO_2 Injector System, pump speed is constant and flow rate to the chamber is controlled by redirecting excess flow. This "shunt regulation" allows flow to be controlled smoothly and quickly across a broad range. Whether the controller in the 6400-01 is used, air supplied to the chamber may be dry or moist. Supplying the chamber with moist air allows higher flow rates to be used to balance low transpiration rates, which provides more stable control and more accurate measurements. Inaccuracies and time delays due to water sorption on the air lines between the console and the sensor head are eliminated by measuring the reference and sample water vapor concentrations in the sensor head.

In general, under a natural light condition, the LI-6400XT System can be used directly for measuring the photosynthesis rate. The major characteristics of the LI-6400XT System include open system design; automatic and independent controls of leaf chamber CO_2 H_2O concentrations, temperature, and light; and operation control by user programming, etc. In short, the LI-6400XT System intends first to accurately measure changes of CO_2 and H_2O concentrations that flow through the

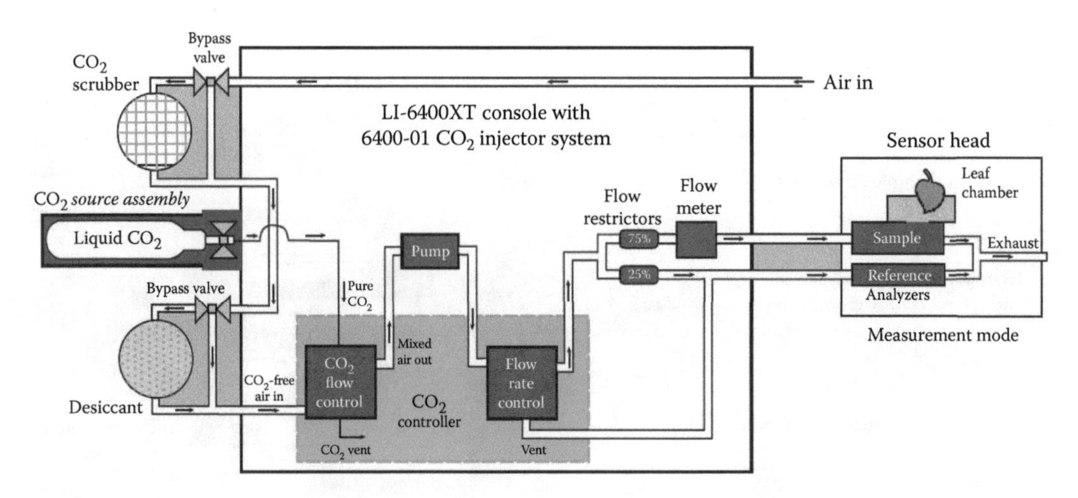

FIGURE 2.19 Working flowchart of LI-6400XT Console with 6400-01 CO_2 Injector System. (Copyright LI-COR Inc. Used with permission.)

chamber that contains plant leaves, and then calculates photosynthesis rate and evapotranspiration rate based on the relationships between changes of CO_2 and H_2O concentrations, flow rate, and measured plant leaf area.

2.2.2.2.2 *Tracing Radiation and Architecture of Canopies (TRAC)*

The TRAC instrument was designed by Dr. Jing M. Chen (a former research scientist at the Canada Centre for Remote Sensing) and manufactured by 3rd Wave Engineering, Canada. It is a new optical instrument for measuring the LAI and the fraction of photosynthetically active radiation absorbed by plant canopies (fPAR). The TRAC measures canopy gap size distribution in addition to canopy gap fraction. *Gap fraction* is the percentage of gaps in the canopy at a given solar zenith angle. It is usually obtained from radiation transmittance. *Gap size* is the physical dimension of a gap in the canopy. For the same gap fraction, gap size distributions can be quite different (Leblanc et al. 2002). The observation values measured by the TRAC instrument along a transect line allow one not only to calculate mean values (canopy gap fraction and fPAR) but also to extract foliage clumping index and calculate canopy gap size distribution. Figure 2.20 shows the structure of the TRAC instrument and its operation in the field. TRAC consists of three PAR sensors (400 to 700 nm) and amplifiers, an analog-to-digital converter, a microprocessor, a battery-backed memory, and a clock and serial I/O circuitry. A power switch controls the power to all components of the system except the memory. A control button controls the operating mode when the power is on. This button is also used to insert distance/time markers.

TRAC is hand-carried by a person (see Figure 2.20b) walking at a steady pace (about 0.3 m/s). Using the solar beam as a probe, TRAC records the transmitted direct light at a high frequency. Figure 2.21 shows an example of such measurements where each spike, large or small, in the time trace represents a gap in the canopy in the sun's direction. The measured photosynthetic photon flux density (PPFD) along a 20 m transect (a small portion of the original 200 m record) shows large flat-topped spikes corresponding to large canopy gaps between tree crowns and small spikes resulting from small gaps within tree crowns. The baseline is the diffuse irradiance under the canopy measured using the shaded sensor. These individual spikes are then converted into gap size values to obtain a gap size distribution. A gap size distribution curve like this reveals the composition of the gap fraction and contains much more information than the conventional gap fraction measurements

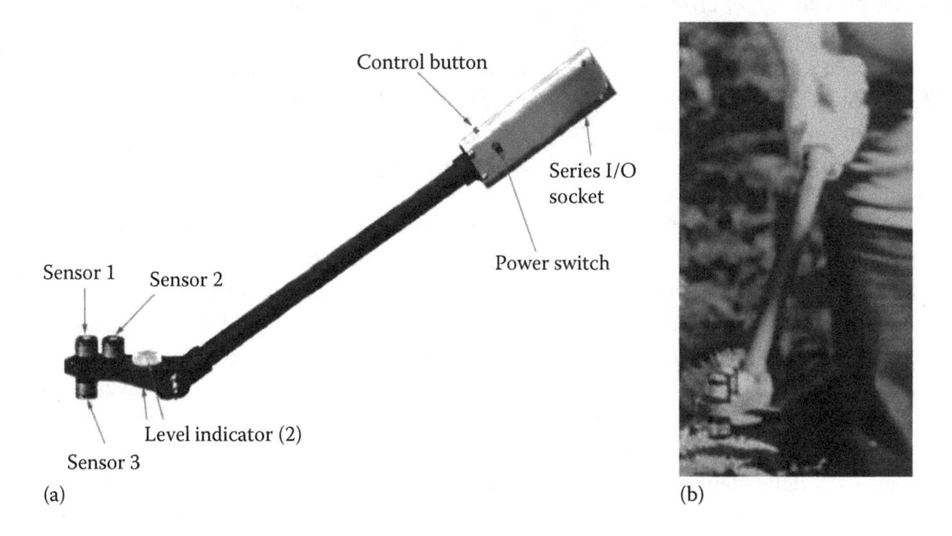

FIGURE 2.20 (a) A view of TRAC structure. (b) TRAC measures light under the canopy while the operator holds it levelled and walks at a constant slow pace. (From Leblanc, S. G. et al., Tracing Radiation and Architecture of Canopies, TRAC Manual, version 2.1.3, Natural Resources Canada, 2002.)

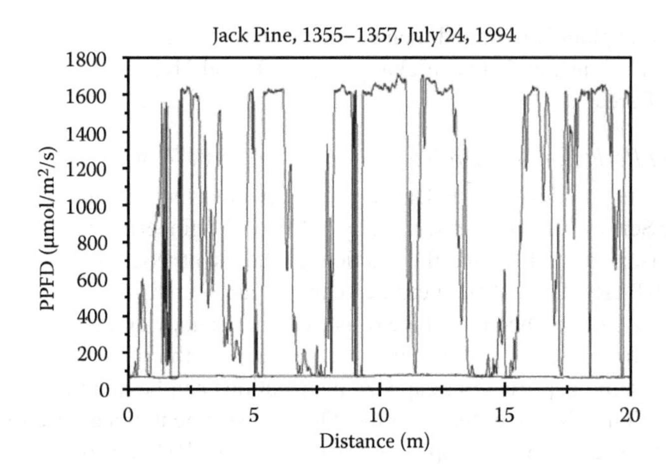

FIGURE 2.21 An example of TRAC measurements in a mature jack pine stand near Candle Lake, Saskatchewan, Canada. (From Leblanc, S. G. et al., Tracing Radiation and Architecture of Canopies, TRAC Manual, version 2.1.3, Natural Resources Canada, 2002.)

(Leblanc et al. 2002). In addition to measuring canopy LAI and fPAR, consequently, the gap size distribution measured by the TRAC instrument can be also used to (1) estimate several canopy architectural parameters, including foliage clump size and area, as well as foliage element size (Chen and Cihlar 1995); and (2) model the hotspot and the bidirectional reflectance distribution function (BRDF) of the optical remote sensing signal from plant canopies (Chen and Leblanc 1997).

2.2.2.3 Instruments for Measuring Chlorophyll Content

2.2.2.3.1 SPAD-502 Plus Chlorophyll Meter

Chlorophyll Meter SPAD-502 Plus is a compact, lightweight meter that can be used to quickly determine the amount of chlorophyll present in plant leaves without damaging the leaves (SPAD-502 PCM 2010). It is manufactured by Spectrum Technologies, Inc. Chlorophyll content is one indicator of plant health, and can be used to optimize the timing and quantity of applying additional fertilizer to provide larger crop yields of higher quality with lower environmental load. Figure 2.22 illustrates the exterior view of SPAD-502 PCM and its operation in the field. The SPAD 502DL Plus meter is with a built-in data logger (SPAD-502DL PCM, 2014). The data logging version (Item 2900PDL) allows the user to compile readings for statistical analysis and include an RS-232 port for communicating with a PC or portable GPS receiver. Use geo-referenced data to correlate *N* measurements to yield maps or download the data into mapping software.

The values measured by the Chlorophyll Meter SPAD-502 Plus correspond to the amount of chlorophyll present in the plant leaf. The values are calculated based on the amount of light transmitted by the leaf in two wavelength regions in which the absorbance of chlorophyll is different. Figure 2.23a shows the spectral absorbance characteristics of chlorophyll extracted from two leaves using 80% acetone. The chlorophyll content of leaf B is less than that of leaf A. The graph also shows that the peak absorbance areas of chlorophyll are in the blue and red regions, with low absorbance in the green region and almost no absorbance in the NIR region. Based on this, the wavelength ranges chosen to be used for measurement are the red area (where absorbance is high and unaffected by carotene) and the NIR area (where absorbance is extremely low). Since the chlorophyll content in plant leaves is dependent on the nutrient state of the leaf, the SPAD-502 Plus can be used for determining when and how much nitrogen fertilizer should be provided to plants (Figure 2.23b). Optimizing fertilization in this way not only leads to greater yields, but also results in less over-fertilization, reducing environmental contamination due to the leaching of excess fertilizer into the soil and underground water.

FIGURE 2.22 Exterior view of SPAD-502P Chlorophyll Meter and its operation. (Courtesy of Spectrum Technologies, Inc., USA.)

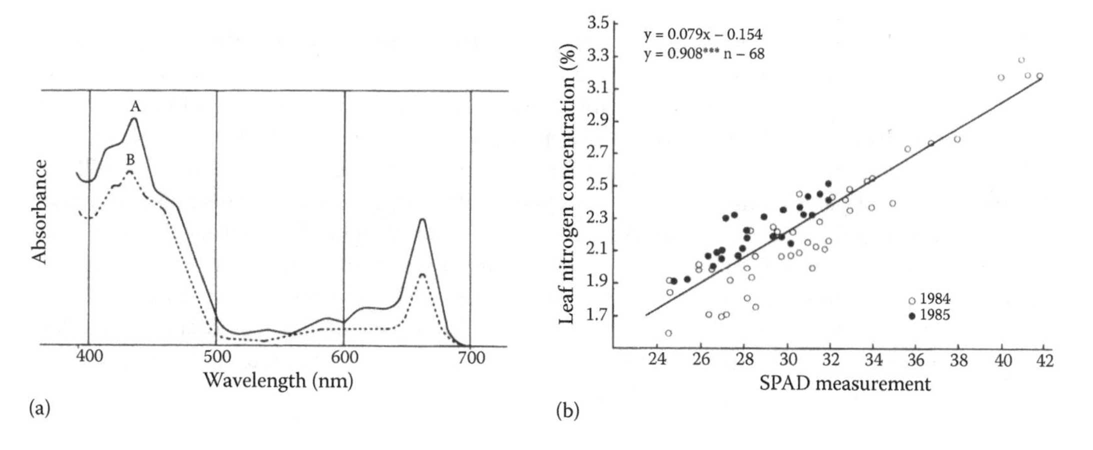

FIGURE 2.23 (a) Absorbance spectra of two leaf samples caused by chlorophyll; (b) a relationship between SPAD-502 measurements and leaf N concentration. (Courtesy of Spectrum Technologies, Inc., USA.)

2.2.2.3.2 CM-1000 Chlorophyll Meter

The Field Scout CM-1000 Chlorophyll Meter senses light at wavelengths of 700 nm and 840 nm to estimate the quantity of chlorophyll in leaves (CM-1000 2013). The ambient and reflected light at each wavelength is measured. Chlorophyll-a absorbs red band energy at 700 nm significantly. NIR light at 840 nm is almost unaffected by leaf chlorophyll content and serves as an indication of how much light is reflected due to leaf structural characteristics such as the presence of a waxy or hairy leaf surface. The CM-1000 chlorophyll index reading is calculated on a scale of 0 to 999 from the measured ambient and reflected light data. The index value is a measure of the relative greenness of the leaf with higher chlorophyll content leading to a higher index value. The CM-1000 is also a product of Spectrum Technologies, Inc. Figure 2.24 shows the instrument's exterior view and its operation in the field.

FIGURE 2.24 Exterior view of Field Scout CM-1000 Chlorophyll Meter and its operation. (Courtesy of Spectrum Technologies, Inc., USA.)

The instrument uses lasers to define the target as the trigger is pressed. At a distance of 28.4 cm, the field of view is 1.10 cm in diameter, while at a distance of 183 cm, the field of view increases to 18.8 cm in diameter. The number of samples taken and a running average of chlorophyll index values can be displayed. The response of the ambient light sensor is displayed as a brightness index value (BRT) from 0–9. A BRT value of 1 or greater indicates that there are at least 250 to 300 $\mu mol \cdot m^{-2} \cdot s^{-1}$ of photosynthetically active radiation (PAR) light available. This is the minimum light level at which the meter is useful. At low levels of ambient light, the chlorophyll index reading may be suspect. Full sun should return a BRT value of seven to eight. Because of the way the CM1000 estimates chlorophyll content, higher light levels enable greater resolution in the chlorophyll content index. Natural sunlight is the best source of light for measuring chlorophyll by reflectance because both wavelengths are present in approximately equal quantities and the quantity of light remains relatively constant (CM-1000 2013).

2.2.2.3.3 MC-100 Chlorophyll Concentration Meter

The MC-100 chlorophyll concentration meter (CCM) is handheld and designed for the rapid, nondestructive, determination of chlorophyll concentration in intact leaf samples. It is manufactured by Apogee Instruments, Inc. Figure 2.25 presents the exterior view of MC-100 CCM and its operation in the field. Chlorophyll concentration or content is a direct indication of plant health and condition. The data collected by the instrument can then be applied to a multitude of crop production and research initiatives such as nutrient and irrigation management, pest control, environmental stress evaluation, and crop breeding (MC-100 2016), as well as calibrating chlorophyll estimation models using hyperspectral data.

Chlorophyll has several distinct optical absorbance characteristics that the MC-100 CCM exploits in order to determine relative chlorophyll concentration. Strong absorbance bands are present in the blue and red but not in the green or infrared bands. The MC-100 CCM uses absorbance to estimate the chlorophyll content in leaf tissue. Two wavelengths are used for absorbance determinations. One wavelength falls within the chlorophyll absorbance range while the other serves to compensate for

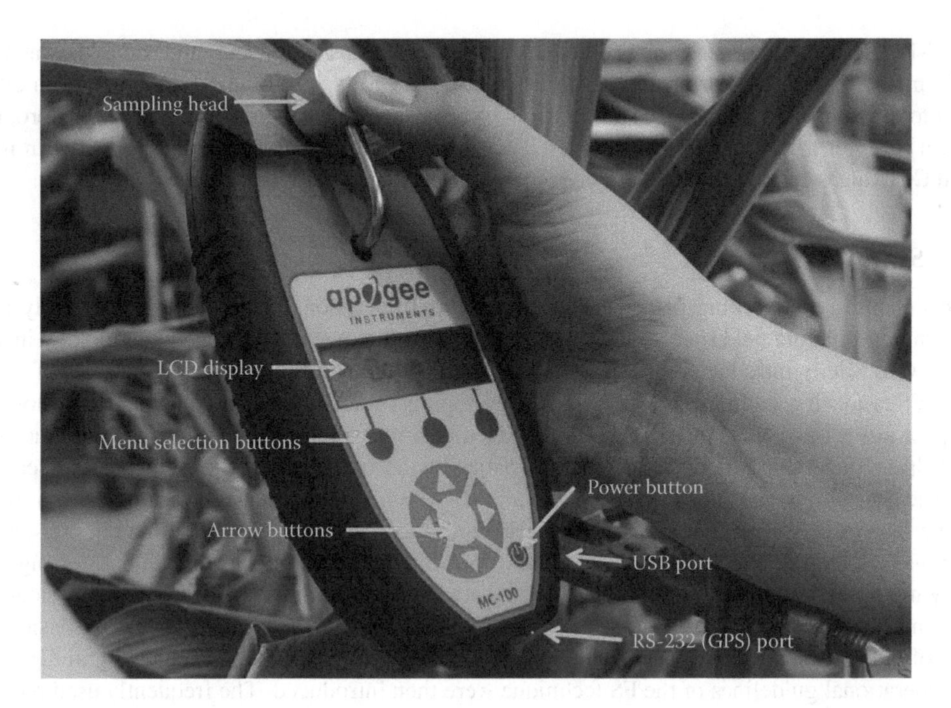

FIGURE 2.25 Exterior view of MC-100 Chlorophyll Concentration Meter and its operation. (Courtesy of Apogee Instruments, Inc., USA.)

mechanical differences such as tissue thickness. The meter measures the absorbance of both wavelengths and calculates a chlorophyll content index (CCI = %Transmittance at 931 nm/%Transmittance at 653 nm) value that is proportional to the amount of chlorophyll in a sample (MC-100 2016). Note that the CCI value is a relative chlorophyll value. There is a nonlinear relationship between CCI and chlorophyll concentration, and the relationship is different among different plant species. The MC-100 CCM has included equations with species-specific coefficients for 22 different species (see the 22 species

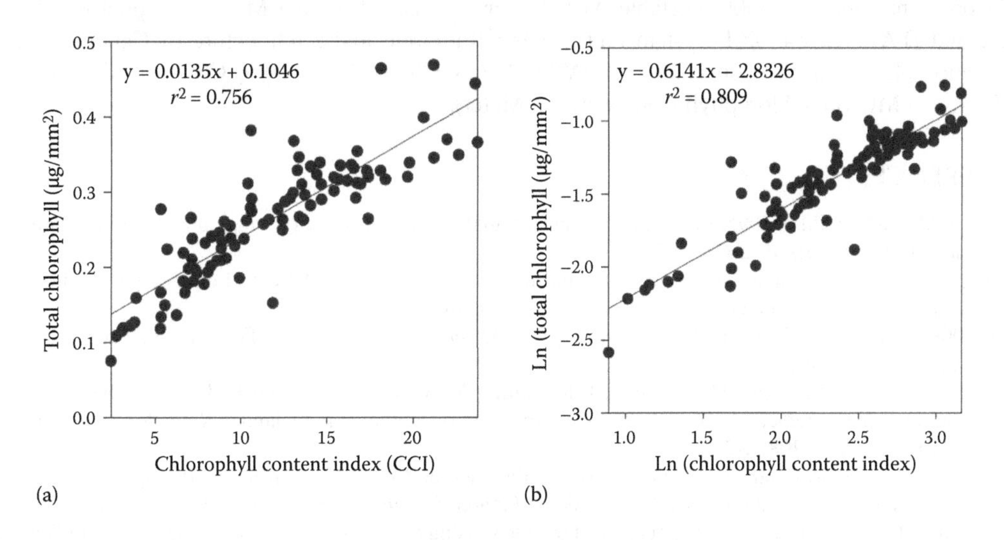

FIGURE 2.26 (a) Linear regression of chlorophyll content index (CCI) against total extractable chlorophyll (mg/mm²) with 98 sugar maple leaves; (b) linear regression of the natural log of CCI values against the natural log of total extractable chlorophyll (mg/mm²) with 98 sugar maple leaves. (From van den Berg, A. K., and T. D. Perkins, *Forest Ecology Management*, 200, 113–117, 2004.)

in MC-100 [2016]). Users can create additional species' equations/coefficients and add to the database in the meter by Apogee. For example, van den Berg and Perkins (2004) used the old version of the meter to measure and calculate CCI values from 98 sugar maple leaves and also extracted chlorophyll content (μg/mm^2) from the 98 leaf samples. Their experimental result of relationship between measured CCI values and extracted chlorophyll content is presented in Figure 2.26.

2.3 SUMMARY

In this chapter, the working principles, structures, and operations in the field of frequently used modern non-imaging field spectroscopy techniques (or spectrometers, FS) and plant biology instruments were briefly introduced. For the non-imaging FSs, the definition, history, and roles of developing FS were first addressed. The roles of developing FS technique may include the following: (1) spectral measurements of individual scene elements measured by using the FS techniques that can help us understand the spectral characteristics of the individual scene elements and scaling-up issues in remote sensing; (2) the FS technique, which can be used for various calibration of remote sensing sensor systems and atmospheric correction of airborne and satellite remote sensing data; and (3) the FS, which can also provide a tool for the development, refinement, testing, and verifying of models' correlating remotely sensed attributes or features with a set of biophysical and biochemical parameters. The basic theories and principles of FS technique, including basic concepts of bidirectional reflectance distribution function, bidirectional reflectance factor, and general field operational guidelines of the FS technique were then introduced. The frequently used modern field spectroradiometers that were briefly introduced in this chapter include ASD FieldSpec® series spectroradiometers, Spectra Vista Corporation (GER) series field spectroradiometers, Spectral Evolution series field spectroradiometers, SpectraScan® spectroradiometers (PR-6×× series), and Ocean Optics USB2000+ (HR2000+) series custom modular spectrometers. Per the plant biology instruments, we need to use them to measure some important bioparameters in order to calibrate and validate a set of empirical/statistical models and physical-based models that can finally be used to retrieve and map such bioparameters based on HRS data. The basic working principles, field operation, and applications of the three categories of plant biology instruments were introduced, and they include those used for measuring (1) plant leaf area and canopy LAI (LAI-2200C Plant Canopy Analyzer, LI-3000C Portable Area Meter, and LI-3100 Area Meter); (2) photosynthesis rate and fPAR (LI-6400XT System and Tracing Radiation and Architecture of Canopies); and (3) chlorophyll and nitrogen content (SPAT-502 Plus Chlorophyll Meter, CM-1000 Chlorophyll Meter, and MC-100 Chlorophyll Concentration Meter).

REFERENCES

Chen, J. M., and J. Cihlar. 1995. Plant canopy gap size analysis theory for improving optical measurements of leaf area index. *Applied Optics* 34:6211–6222.

Chen, J. M., and S. Leblanc. 1997. A 4-scale bidirectional reflection model based on canopy architecture. *IEEE Transactions on Geoscience and Remote Sensing* 35:1316–1337.

CM-1000. 2013. *Field Scout CM-1000 Chlorophyll Meter Product Manual.* Fort Worth, TX: Spectrum Technologies, Inc.

Finzel, J. A., M. S. Seyfried, M. A. Weltz, J. R. Kiniry, M.-V. V. Johnson, and K. L. Launchbaugh. 2012. Indirect measurement of leaf area index in sagebrush-steppe rangelands. *Rangeland Ecology & Management* 65:208–212.

Gamon, J. A., A. F. Rahman, J. L. Dungan, M. Schildhauer, and K. F. Huemmrich. 2006. Spectral Network (SpecNet)—What is it and why do we need it? *Remote Sensing of Environment* 103:227–235.

Goetz, A. 2009. Three decades of hyperspectral remote sensing of the Earth: A personal view. *Remote Sensing of Environment* 113:5–16.

Goetz, A. F. H. 1987. The portable instant display and analysis spectrometer (PIDAS). *Proceedings of the Third Airborne Imaging Spectrometer Data Analysis Workshop*, JPL Publication, vol. 87-30 (pp. 8–17).

Goetz, A. F. H. 1975. Portable field reflectance spectrometer, Appendix E in Application of ERTS images and image processing to regional geologic problems and geologic mapping in Northern Arizona. *JPL Technical Report* (pp. 32–1597).

Gong, P., R. Pu, and R. C. Heald. 2002. Analysis of in situ hyperspectral data for nutrient estimation of giant sequoia. *International Journal of Remote Sensing* 23(9):1827–1850.

Green, R. O., J. E. Conel, C. J. Bruegge, J. S. Margolis, V. Carrere, G. Vane, and G. Hoover. 1991. *In-Flight Calibration of the Spectral and Radiometric Characteristics of ARIVIS in 1991.* Accessed January 8, 2015, from http://ntrs.nasa.gov/archive/nasa/casi.ntrs.nasa.gov/19940012194.pdf.

Hunt, G. R. 1977. Spectral signatures of particulate minerals in the visible and near-infrared. *Geophysics* 42:501–513.

Jensen, J. R. 2005. *Introductory Digital Image Processing: A Remote Sensing Perspective*, 3rd ed. Upper Saddle River, NJ: Prentice Hall.

Johnson, L. F., C. A. Hlavka, and D. L. Peterson. 1994. Multivariate analysis of AVIRIS data for canopy biochemical estimation along the Oregon transect. *Remote Sensing of Environment* 47:216–230.

Kneubühler, M., M. E. Schaepman, K. J. Thome, and D. R. Schläpfer. 2003. MERIS/ENVISAT vicarious calibration over land. *Proceedings of SPIE—The International Society for Optical Engineering* 5234: 614–623.

LAI-2200C PCA. 2014. *The Brochure of LAI-2200C Plant Canopy Analyzer.* Lincoln, NE: LI-COR Inc.

Leblanc, S. G., J. M. Chen, and M. Kwong. 2002. Tracing Radiation and Architecture of Canopies. *TRAC Manual*, Version 2.1.3. Natural Resources Canada.

LI-3000C PAM. 2014. *The Brochure of LI-3000C Portable Area Meter.* Lincoln, NE: LI-COR Inc.

LI-3100C AM. 2004. *The Brochure of LI-3100C Area Meter.* Lincoln, NE: LI-COR Inc.

LI-6400XT System. 2013. *The Brochure of LI-6400XT System.* Lincoln, NE: LI-COR Inc.

Maracci, G. 1992. Field and laboratory narrow band spectrometers and the techniques employed, in *Imaging Spectroscopy: Fundamentals and Prospective Applications*, eds. F. Toselli and J. Bodechtel, pp. 33–46. London: Kluwer Academic Publishers.

MC-100. 2016. *MC-100 Chlorophyll Concentration Meter Owner's Manual.* Logan, UT: Apogee Instruments, Inc.

Milton, E. J. 1987. Principles of field spectroscopy. *International Journal of Remote Sensing* 8(12):1807–1827.

Milton, E. J., M. E. Schaepman, K. Anderson, M. Kneubuehler, and N. Fox. 2009. Progress in field spectroscopy. *Remote Sensing of Environment* 113:S92–S109.

Nicodemus, F. F., J. C. Richmond, J. J. Hsia, I. W. Ginsberg, and T. L. Limperis. 1977. Geometrical considerations and nomenclature for reflectance. National Bureau of Standards Monograph, vol. 160 (pp. 20402). Washington, DC: United States Government Printing Office.

Penndorf, R. 1956. Luminous and spectral reflectance as well as colors of natural objects. Bedford, MA: U.S. Air Force Cambridge Research Center.

PSR. 2014. *PSR Series Spectroradiometer Operator's Manual*, Revision: 1.07. Lawrence, MA: Spectral Evolution, Inc.

Pu, R. 2009. Broadleaf species recognition with in situ hyperspectral data. *International Journal of Remote Sensing* 30(11):2759–2779.

Pu, R., S. Landry, and J. Zhang. 2015. Evaluation of atmospheric correction methods in identifying urban tree species with WorldView-2 imagery. *IEEE Journal of Selected Topics in Applied Earth Observations and Remote Sensing* 8(5):1886–1897.

Richardson, A. J. 1981. Measurement of reflectance factors under daily and intermittent irradiance variations. *Applied Optics* 20(19):3336–3340.

Santer, R., X. F. Gu, G. Guyot, C. Deuzé, E. Vermote, and M. Verbrugghe. 1992. SPOT calibration at the La Crau Test Site (France). *Remote Sensing of Environment* 41:227–237.

Schaepman-Strub, G., M. E. Schaepman, T. H. Painter, S. Dangel, and J. V. Martonchik. 2006. Reflectance quantities in optical remote sensing—Definitions and case studies. *Remote Sensing of Environment* 103:27–42.

Slater, P. N., S. F. Biggar, R. G. Holm, R. D. Jackson, Y. Mao, M. S. Moran, J. M. Palmer, and B. Yuan. 1987. Reflectance- and radiance-based methods for the in-flight absolute calibration of multispectral scanners. *Remote Sensing of Environment* 22:11–37.

SPAD-502 PCM. 2010. *SPAD-502 Plus Chlorophyll Meter Product Manual.* Fort Worth, TX: Spectrum Technologies, Inc.

SPAD-502DL PCM. 2014. *SPAD-502DL Plus Chlorophyll Meter Product Manual.* Fort Worth, TX: Spectrum Technologies, Inc.

SVC. 2013. Spectra Vista Corporation, SVC HR-1024i/SVC HR-768i/SVC HR-640i/SVC HR-768si/SVC HR-512i. *User's manual*, revision 1.12.

Thome, K. J., S. Schiller, J. E. Conel, K. Arai, and S. Tsuchida. 1998. Results of the 1996 Earth Observing System vicarious calibration joint campaign at Lunar Lake Playa, Nevada (USA). *Metrologia* 35:631–638.

van den Berg, A. K., and T. D. Perkins. 2004. Evaluation of a portable chlorophyll meter to estimate chlorophyll and nitrogen contents in sugar maple (*Acer saccharum Marsh.*) leaves. *Forest Ecology and Management* 200:113–117.

Zarco-Tejada, P. J., J. R. Miller, T. L. Noland, G. H. Mohammed, and P. H. Sampson. 2001. Scaling-up and model inversion methods with narrowband optical indices for chlorophyll content estimation in closed forest canopies with hyperspectral data. *IEEE Transactions on Geoscience and Remote Sensing* 39(7):1491–1507.

3 Imaging Spectrometers, Sensors, Systems, and Missions

It has been demonstrated that hyperspectral remote sensing technology has been successfully applied in different areas such as monitoring and management of natural resources and environments on the land and in the ocean and atmosphere. To efficiently and reasonably utilize and analyze the various hyperspectral sensors' data, it is necessary for us to know and understand the working principles and technical characteristics of current/operational and future airborne and spaceborne hyperspectral sensors, systems, and missions. Therefore, in this chapter, the working principles, technical characteristics, and status of current/operational and future airborne and spaceborne hyperspectral sensors, systems, and missions are briefly introduced.

3.1 WORKING PRINCIPLES OF IMAGING SPECTROMETRY

Imaging spectrometers can typically use a 2D matrix array (e.g., a charge couple device [CCD]) to produce a 3D data cube (spatial dimensions and a third spectral dimension). These progressive data cubes are built by either sequentially recording one full spatial image after another, each at a different wavelength, or sequentially recording one narrow image (one pixel wide, multiple pixels long) swath after another with the corresponding spectral signature for each pixel in the swath (Ortenberg 2012, Ben-Dor et al. 2013). Depending upon the mechanism of scanning and data acquisition, there are two commonly used types of imaging spectrometry: (1) whiskbroom imaging spectrometry and (2) pushbroom imaging spectrometry (see Figure 3.1a and b) (Goetz 1992, Gupta 2003, Ortenberg 2012).

3.1.1 WHISKBROOM IMAGING SPECTROMETRY

Figure 3.1a shows the working principle of a whiskbroom line-array imaging spectrometer. It basically belongs to an electro-mechanical device, but produces image data in contiguous spectral bands. A rotating scan mirror sweeps from one edge of the swath to the other. It directs radiations from different parts of the ground onto the radiation-sorting optics. The radiation-sorting optics consists of a dispersion device such as a grating plane or a prism. The radiation dispersed and separated wavelength-wise is focused onto a line array. The heart of the device is the photo-detectors, which are arranged along the CCD-linear array (see spec. dim. λ in Figure 3.1a). The radiations corresponding to different wavelength ranges, collected by the mirror, are passed onto different elements of the CCD-linear array. Thus, each pixel corresponding to each ground instantaneous field of view (IFOV), is simultaneously sensed in as many spectral bands as there are detector elements in the linear array. In this way, imaging is carried out pixel by pixel and line by line as the sensor-craft keeps moving forward. This leads to the construction of images in numerous contiguous bands. Since the dwell time for each pixel must be very short in this imaging principle, it leads to a relatively low signal-to-noise ratio (SNR). This is because the increase in SNR is proportional to the square root of integration time (Goetz 1992). Therefore, with this type of imaging spectrometer it is more difficult to further improve spectral resolution and increase radiative sensitivity. Whiskbroom imaging spectrometers tend to be large and complex. However, a significant advantage of the whiskbroom imaging spectrometers is that they have a single detector (or fewer detectors) subject to calibration, which is vastly less complicated than calibrating an area array imaging spectrometer which may have tens of thousands of detectors, such as the pushbroom imaging spectrometers introduced later in this chapter. Well-known examples working in the whiskbroom linear array

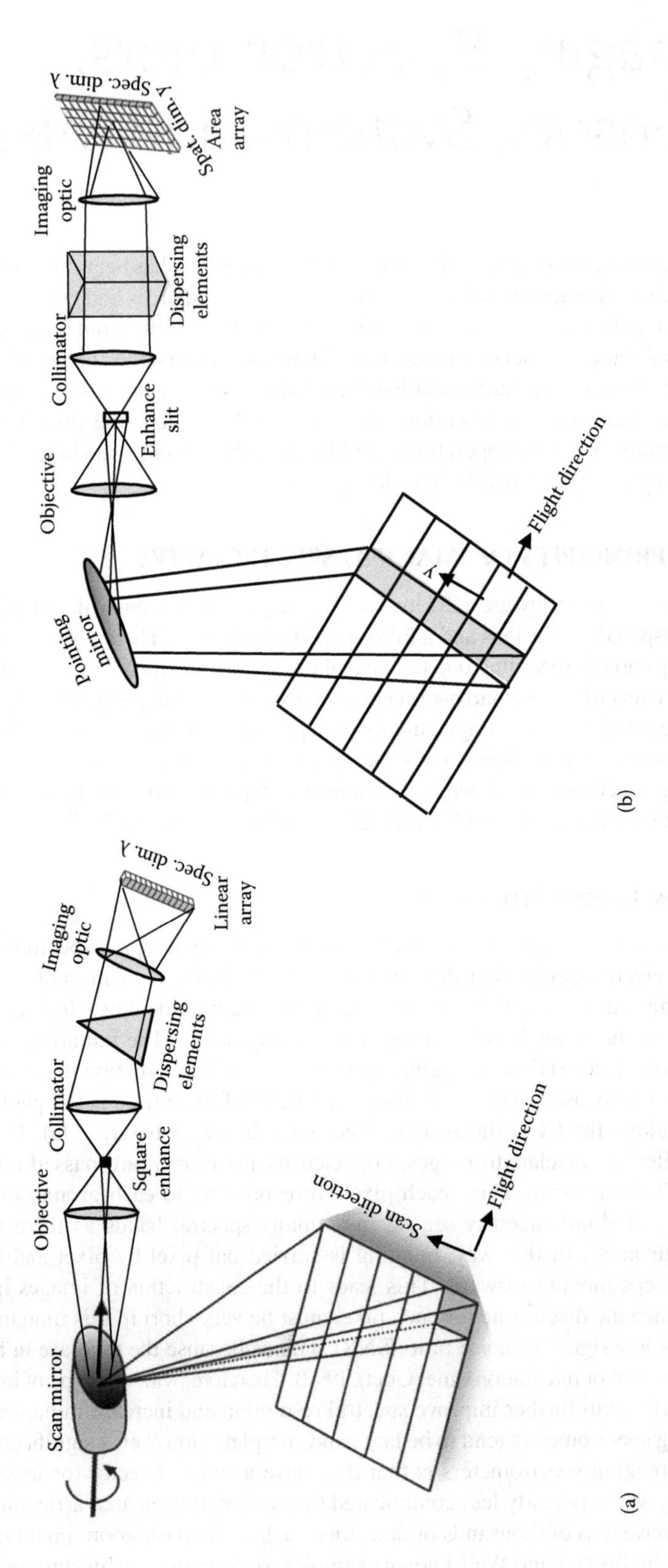

FIGURE 3.1 The basic types of imaging principles: (a) Whiskbroom linear array perpendicular flight direction; (b) pushbroom two-dimension area array, one used as spectrometer (spectral dimension) and the other as pushbroom linear array (spatial dimension).

include AVIRIS, Hyperspectral Mapper (HyMap), and Geographical and Environmental Research Corporation (GER) Daedalus systems.

3.1.2 Pushbroom Imaging Spectrometry

The pushbroom imaging sensor uses a 2D CCD-area array of detectors, one dimension for spectral and the other for spatial (see spec. dim. λ and spat. dim. y in Figure 3.1b) at the focal plane of the spectrometer. The number of pixels equals that of ground cells for a given swath. The motion of carrier aircraft or satellite realizes the scan in the along-track direction, thus the inverse of the line frequency equals the pixel dwell time (Ortenberg 2012). The optical design and major components are shown in Figure 3.1b. Imaging is carried out, in pushbroom mode, one line at a time without a rotating mirror. The radiations are collected by a pointing mirror, pass through a slit, and are collimated onto a dispersing element. They are separated according to wavelengths and focused onto a CCD-area array of detectors. In this manner, the imaging device measures radiative intensity of each band (spectral dimension, λ) for each pixel (spatial dimension, y) for a $\lambda \times y$ area array of detector elements. The rest of the pushbroom scanning mechanism remains the same as in the linear array mode, i.e., the radiations are integrated for a certain period (dwell time) by a frontal photogate after which the charge is quickly transferred to the shift register, where readout is carried out with more leisure (Gupta 2003). Therefore, the characteristics of this type of imaging spectrometer are (1) spatial scanning completed by solid scanning of device and longer time of pixel-photography so that the SNR is much higher than the whiskbroom scanner and the sensitivity and spatial resolution of the system can be improved; (2) in the visible region, the spectral resolution can be increased up to 1–2 nm due to mature technology of CCD elements and high integration time; (3) pushbroom spectrometers are lighter, smaller, and less complex than their whiskbroom counterparts because they have fewer moving parts. However, their major disadvantage is that a large number of detectors need to be calibrated, which is very complicated and time consuming. The AIS is the first imaging spectrometer to acquire ground data in the mode of a pushbroom area array. In addition, CASI, FLI, AISA and HYDICE were designed in this working principle. Since the pushbroom imaging spectrometers use electronic scanning only, they are ideally suited for use as spaceborne imaging spectrometers.

3.2 AIRBORNE HYPERSPECTRAL SENSORS/SYSTEMS

Table 3.1 presents a summary of existing airborne hyperspectral sensors and systems in terms of basic technical parameters, characteristics, application, date of development, and manufacturers. A survey of a total of 23 selected important airborne hyperspectral sensors and systems is summarized in this section by referring to Green et al. (1998), Pu and Gong (2000), Kramer (2002), Lucas et al. (2004), Qi et al. (2012), and Ben-Dor et al. (2013).

3.2.1 Advanced Airborne Hyperspectral Imaging Sensor (AAHIS)

The AAHIS sensor was built by Science Application International Corporation (SAIC) of San Diego, California, and is owned and operated by SETS Technology, Inc., of Hawaii (Kramer 2002). AAHIS was initially developed in 1993 and the first test flights were conducted in 1994. It is currently in its third generation (AAHIS 3). The AAHIS sensor gathers hyperspectral data over the entire visible region and beyond into the NIR. The AAHIS sensor features a proprietary, state-of-the-art, three-axis navigation and stabilization system with reliable, georegistered data, independent of aircraft movement. A cooled CCD detector array of 576 × 384 pixels is employed, but half of the active array is used for image storage. The system provides a fully stabilized 40° view from the aircraft, assuring a large area coverage rate.

TABLE 3.1
Summary of Existing Airborne Hyperspectral Sensors and Systems

Sensor (Country)	Number of Bands	Spectral Range (nm)	Band Width (FWHM, nm)	IFOV (mrad)	FOV (°)	Application	Operation	Manufacturer
AAHIS (U.S.A.)	288	440–880	3	1	40	Littoral environment/vegetation monitoring, effluent and military target detection	1994	SAIC of San Diego, CA
AIS-I	128	990–2100 1200–2400	9.3	1.91	3.7	Geochemistry, identification of minerals, altered rock, plant infected	1983–1985	NASA/JPL
AIS-II (U.S.A.)	128	800–1600 1200–2400	10.6	2.05	7.3		1986–1987	
AISA EAGLE	Up to 488	400–970	1.56–9.36	1	21	Hydrology, geology, agriculture and forestry	Since 1993	SPECIM, Spectral Imaging, Ltd
AISA HAWK	254	970–2500	8.5		17.8, 24, 35.5			
AISA DUAL	Up to 500	400–2500	3.3, 12		24, 35.5, 37.7			
AISA OWL (Finland)	Up to 96	7600–12300	100		24 or 32.3			
ASAS	29	455–873	15	0.8	25	Measuring bidirectional radiance of terrestrial targets	1987–1991	GSFC of NASA
upgrade ASAS (U.S.A.)	62	400–1060	11.5	0.8	25		Since 1992	
AVIRIS (U.S.A.)	224	380–2500	9.7–12.0	1	30	Ecology, oceanology, geology, atmosphere, ice, snow, clouds	Since 1987	NASA/JPL
CASI-2	288	400–1000	1.8	1.3, 1.6	37.8	Ecosystem, terrestrial features observation	Since 1990	ITRES Research
CASI-1500	288	380–1050						
SASI-600	100	950–2450						
MASI-600	64	3000–5000						
TASI-600 (Canada)	32	8000–1150						
CHRISS (U.S.A.)	40	430–860	11	0.05	10	Petroleum seepage monitoring, vegetation, forestry survey, ocean color and environmental monitoring	1992	SAIC of San Diego, CA

(Continued)

TABLE 3.1 (CONTINUED)
Summary of Existing Airborne Hyperspectral Sensors and Systems

Sensor (Country)	Number of Bands	Spectral Range (nm)	Band Width (FWHM, nm)	IFOV (mrad)	FOV (⁰)	Application	Operation	Manufacturer
CIS (China)	64	400–1040	10	1.2 × 3.6	80	Terrestrial surface observation	Since 1993	Shanghai Institute of Technical Physics
	24	2000–2480	20	1.2 × 1.8				
	1	3530–3490	410	1.2 × 1.2				
	2	10500–12500	1000	1.2 × 1.2				
C_2VIFIS (U.S.A.)	96	420–870	10	Program	Program	Agriculture and forestry, terrestrial, ocean environmental monitoring	Since 1994	Flight Landata, Inc.
DAIS-7915 (U.S.A.)	32	400–1010	10~16	3.3, 2.5 or 5.0	64–78	Terrestrial, ocean ecological environmental monitoring, agriculture and forestry, geological mapping	Since 1994	GER Corporation
	8	1500–1788	36					
	32	1970–2540	36					
	1	3000–5000	2000					
	6	8700–12700	600					
DAIS-16115 (U.S.A.)	76	400–1000	8	3	78	Terrestrial, ocean ecological environmental monitoring, agriculture and forestry, geological mapping	Since 1994	GER Corporation
	32	1000–1800	25					
	32	2000–2500	16					
	6	3000–5000	333					
	12	8000–12000	333					
	2	400–1000	Stereo					
DAIS-21115 (U.S.A.)	76	400–1000				Terrestrial, ocean ecological environmental monitoring, agriculture and forestry, geological mapping	Since 1994	GER Corporation
	64	1000–1800						
	64	2000–2500						
	1	3000–5000						
	6	8000–12000						
EPS-H (Germany)	76	430–1050						GER Corporation
	32	1500–1800						
	32	2000–2500						
	12	8000–12500						

(*Continued*)

TABLE 3.1 (CONTINUED)
Summary of Existing Airborne Hyperspectral Sensors and Systems

Sensor (Country)	Number of Bands	Spectral Range (nm)	Band Width (FWHM, nm)	IFOV (mrad)	FOV (⁰)	Application	Operation	Manufacturer
FLI (U.S.A.)	288	430–805	2.5	1.3	70	Ecosystem, terrestrial surface observation	1984–1990	Moniteq Ltd and Itres Ltd
GER-63 (U.S.A.)	24	400–1000	25	2.5, 3.3 or 4.5	90	Environmental monitoring, geological study	Since 1986	GER Corporation
	4	1500–2000	125					
	29	2000–2500	17.2					
	6	8000–12500	750					
HYDICE (U.S.A.)	210	400–2500	3, 10–20	0.5	8.94	Military and civilian use, agriculture and forestry, environmental monitoring, and resources/disaster	Since 1995	Hughes-Danbury Optical System, Inc.
HyMAP (Australia)	32	450–890	15–16	2.5, 2.0	61.3	Mineral exploration and environmental monitoring	1996	Australian Integrated Spectronics Company
	32	890–1350	15–16	2.5, 2.0	61.3			
	32	1400–1800	15–16	2.5, 2.0	61.3			
	32	1950–2480	18–20	2.5, 2.0	61.3			
	32	8000–12000	100–200	2.5, 2.0	61.3			
HySpex (Norway)	108	400–1000	5.4	0.28, 0.56	16	Hyperspectral imaging systems for a multitude of applications, designed in a close collaboration with customers	Since 1995	Norsk Elektro Optikk
	160	400–1000	3.7	0.18, 0.36	17			
	182	400–1000	3.26	0.16, 0.32	17			
	288	1000–2500	5.45	0.73, 0.73	16			
	427	400–2500	3, 6.1	0.25, 0.25	15			
ISM (France)	64	800–1600	12.5	3.3 × 11.7	40 (selectable)	Geology, clouds, ice snow, vegetation canopy chemistry	Since 1991	DESPA of Paris-Meudon Observatory
	64	1600–3200	25.0					
MAIS (China)	32	450–1100	20.0	3	90	Geological and environmental survey	1987–1990	Shanghai Institute of Technical Physics of CAS
	32	1400–2500	30.0	3	90			
	7	8200–12200	400–800	3	90			

(*Continued*)

TABLE 3.1 (CONTINUED)
Summary of Existing Airborne Hyperspectral Sensors and Systems

Sensor (Country)	Number of Bands	Spectral Range (nm)	Band Width (FWHM, nm)	IFOV (mrad)	FOV (⁰)	Application	Operation	Manufacturer
MISI (U.S.A.)	70	400–1000	10.0	2	±45	Environmental protection, water quality assessment	Since 1996	The Rochester Institute of Technology (RIT) of Rochester, NY
MIVIS (U.S.A.)	20 8 64 10	433–833 1150–1550 2000–2500 8200–12700	20 50 8 400–500	2	70	Geological and environmental study	1993	Sensys Technologies Inc.
MUSIC (U.S.A.)	90 90	2500–7000 6000–14500	25–70 60–1400	0.5	1.3, 2.6	Chemical vapor sensing, plume disgnostics, spectral	Since 1989	Lockheed Palo Alto Research Laboratory, CA
PHI (China)	244	400–850	<5		1.1	Terrestrial ecosystem, and resources inventory	Since 1997	Shanghai Institute of Technical Physics of CAS
PROBE-1 (U.S.A.)	128	400–2500	11–18		60	Geology survey		Earth Search Science Inc.
ROSIS (Germany)	84	430–830	(4–12)	0.56	± 16	Fluorescence of chlorophyll in coastal zones	Since 1992	Jointly by DASA/MBB, CKSS and DLR (Institute of Optoelectronics), Germany
SFSI (Canada)	122	1200–2400	10	0.4	11.7	Geology survey and mineral mapping	Since 1994	Canada Centre for Remote Sensing
SMIFTS (U.S.A.)	100	1000–5200		0.66	9.7	Geology survey and general terrestrial surface observation	1993	University of Hawaii, Honolulu

(Continued)

TABLE 3.1 (CONTINUED)
Summary of Existing Airborne Hyperspectral Sensors and Systems

Sensor (Country)	Number of Bands	Spectral Range (nm)	Band Width (FWHM, nm)	IFOV (mrad)	FOV (º)	Application	Operation	Manufacturer
TRWIS-B	90	450–880	4.8	0.4–2	(5–25)	Vegetation or mineral	Since 1991	TRW of Redondo
TRWIS-II	128	900–1800 or	12	0.45	6	identification, or		Beach, CA
		1500–2500				quantification		
TRWIS-III	128	370–1040	5.25	0.9	13.1			
(U.S.A.)	256	890–2450	6.25					
VIFIS	64	420–870	10–18	1	31.5	A variety of environmental	Since 1991	Univerity of
(U.K.)						monitoring tasks		Dundee, UK
WIS-FDU	64	400–1030	7.2–18.5	1.36	10 and 15	Earth's surface observation	1992	Hughes Santa
WIS-VNIR	129+265	400–1000	5.4–14.4	0.66	19.1		1994	Barbara Research
WIS-SWIR	81+90	1000–2500	18.0–37.8	0.66	12		1995	Center
(U.S.A.)								

Source: Pu, R., and P. Gong, *Hyperspectral Remote Sensing and Its Applications*. Beijing, China: Higher Education Press of China, 2000; Kramer, H. J., *Observation of the Earth and Its Environment: Survey of Missions and Sensors*, 4th ed. Berlin: Springer-Verlag, 2002; Lucas, R. et al., Hyperspectral sensor and applications, in *Advanced Image Processing Techniques for Remotely Sensed Hyperspectral Data*, eds. P. K. Varshney and M. K. Arora, pp. 11–45. Berlin and Heidelberg: Springer-Verlag, 2004; Qi, J. et al., Hyperspectral remote sensing in global change studies, in *Hyperspectral Remote Sensing of Vegetation*, eds. P. S. Thenkabail, J. G. Lyon, and A. Huete, pp. 69–89. Boca Raton, FL: CRC Press/Taylor & Francis Publishing Group, 2012; Ben-Dor, E. et al., Hyperspectral remote sensing, in *Airborne Measurements for Environmental Research: Methods and Instruments*, eds. M. Wendisch, J.-L. Brenguier, pp. 413–456. Weinheim, Germany: Wiley-VCH, 2013.

A software-controlled variable aperture, spectral shaping filters, and a high-quantum efficiency, back-illuminated CCD contribute to the excellent sensitivity of the sensors. The system is compact and carried in most fixed- and rotary-wing aircraft with installed optical windows, achieving ground-sampling distances ranging from 6.5 cm to 2 m. Optimized for operation in littoral areas, the AAHIS-3 operated successfully during a number of diverse remote sensing applications, including environmental monitoring, vegetation discrimination, and effluent detection. Other applications of this system have been used in military target detection, classification, and identification on land and under water.

3.2.2 Airborne Imaging Spectrometer (AIS)

AIS was designed and built by NASA Jet Propulsion Laboratory (JPL) in the early 1980s as one of first types of airborne imaging spectrometers ever built. It is a pushbroom imaging spectrometer with a 2D area array of detectors that were tested for the first hybrid infrared area array (Goetz 1995). The AIS-I made use of a 32 × 32 element HgCdTe array bonded to a silicon multiplexer and was tested and applied during the period from 1983 to 1985. This design of a 32 × 32 element array permitted a pushbroom-style scan in which the image of each crosstrack pixel was detected by a specific line of detectors in the area array, eliminating the need for cross rack mechanical scanning. The AIS-I collects 128 bands at approximately 9.3 nm wide between 0.9 μm and 2.1 μm in its "tree mode" and between 1.2 μm and 2.4 μm in its "rock mode." A later version of the instrument, AIS-II, incorporated a second generation 64 × 64 element array, covered the 0.8–2.4 μm spectral range, and was tested and used between 1986 and 1987. The bandwidth of AIS-II is 10.6 nm. Spatial resolution and swath width are 12 m and 365 m for AIS-I, and 12 m and 787 m for AIS-II, respectively, if both flight altitudes reach 6 km AGL. A comparison of performance parameters between AIS-I and AIS-II is presented in Table 3.2. AIS provided the first direct identification of minerals on the Earth's surface from a remote imaging system. AIS can also be used in botanical studies, such as by extracting biophysical and biochemical parameters from vegetation canopies (e.g., Peterson et al. 1988, Wessman et al. 1988). AIS has several shortcomings. Among them are the lack of full spectral coverage in the visible and shortwave infrared and a very narrow FOV (Goetz 1995).

TABLE 3.2
Comparison of AIS-I and AIS-II Imaging Spectrometers' Performance Parameters

Parameter	AIS-I	AIS-II
Preparation (year)	1983–85	1986–87
Imaging principle	pushbroom	pushbroom
Flight altitude (km)	6	6
FOV(^{0})	3.7	7.3
Swath width (m)	365	787
Spatial resolution (m)	11.5	12.3
Spectral sampling interval (nm)	9.3	10.6
Spectral range (nm)	990–2400	800–2400
Number of bands	128	128
Detector array	area array	area array

Source: Vane, G., and A. F. H. Goetz, *Remote Sensing of Environment*, 24, 1–29, 1998; Vincent, R. K., *Fundamentals of Geological and Environmental Remote Sensing*, Prentice Hall, Upper Saddle River, NJ, 1997.

3.2.3 Airborne Imaging Spectrometer for Different Applications (AISA)

AISA is a commercially available hyperspectral imaging spectrometer from Spectral Imaging Ltd. (Specim) of Finland. The objective of developing the AISA is to provide users with a wide variety of application configurations. It operated at the beginning of 1993. AISA is a programmable pushbroom imaging spectrometer based on CCD technology (Kramer 2002). Application areas of AISA include hydrology, geology, forestry, and agriculture. The AISA can be flown on a light aircraft. A nominal flight altitude is 1000 m, and this configuration can provide a spatial resolution of 1 m across track. The AISA family consists of several models, including AISA EAGLE, AISA HAWK, AISA DUAL, and AISA OWL, etc. AISA EAGLE is a high-performance airborne VNIR pushbroom hyperspectral system that works in a spectral range of 400 nm to 1000 nm. AISA EAGLE has become the reference in the hyperspectral imaging mapping thanks to its extraordinary spectral resolution of 3.3 nm and up to 488 spectral channels that allow detecting the finest spectral signatures. It provides the highest performance in terms of signal dynamic range, signal-to-noise ratio (SNR), image rate, and resolution. To fulfill each application's need of FOV, AISA EAGLE is available with various lenses that can be easily exchanged by the sensor user. AISA HAWK is the first small and low maintenance SWIR (970–2500 nm) hyperspectral sensor, which provides high-speed data acquisition at high sensitivity. The major advantages of the AISA HAWK are its distinctive size, high quality of data, and low investment cost. It can be quickly and easily installed in any aircraft. AISA DUAL is a high-performance hyperspectral sensor system for simultaneous acquisition of VNIR and SWIR data. AISA DUAL combines the AISA EAGLE and AISA HAWK sensors in a dual-sensor bracket mount and provides an economic imaging tool for remote sensing companies and researchers. Specim's thermal airborne hyperspectral sensor AISA OWL covers the contiguous spectral range from 7.6 to 12.3 μm in 96 channels and even has the sensitivity to detect and classify gases. The AISA OWL is a fast pushbroom hyperspectral system designed to provide the remote sensing market with a high performance LWIR hyperspectral imager in the smallest achievable size. Thanks to its design and sensitivity, the AISA OWL captures all its spectral bands for the whole swath-width of 384 pixels simultaneously, and sustains the 100 frames per second storage speed of the whole data cube continuously. As a result, high sensor or target dynamics are possible without degradation of image quality.

3.2.4 Advanced Solid-State Array Spectroradiometer (ASAS)

ASAS is a hyperspectral, multi-angle, airborne remote sensing instrument maintained and operated by the Laboratory for Terrestrial Physics at the NASA Goddard Space Flight Center (GSFC), Maryland. ASAS is an off-nadir-pointing imaging sensor with an objective of acquiring bidirectional radiance data from targets. The original instrument has been modified for off-nadir pointing by GSFC of NASA in order to study the directional anisotropy of solar radiance reflected from terrestrial surface. A nine-bit shaft encoder is used to measure the sensor pointing angle. The maximum pointing angle is 45°. The pointing capability was first utilized in 1987 (Huegel 1988). ASAS acquires data in the range 455–871 nm with 29 bands at approximately 15 nm wide. ASAS employs a cooled 512 × 32 element silicon charge-injection-device (CID) detector array to generate multispectral digital image data in the pushbroom mode. The first row and last two rows of the array are blacked-out, with the remaining 29 rows intended for digital imaging data acquisition. The long dimension of the array is reserved for the spatial resolution of 512 crosstrack ground elements. For a typical flight at a 5000 m altitude for data acquisition, the crosstrack field of view is 2200 m and the corresponding spatial resolution is 4.25 m (Irons et al. 1990, Kramer 2002). The scan rate is selectable (3, 6, 12, 24, 48, and 64 frames/s). The quantization is 12-bit. The ASAS was upgraded in 1992. A new tilting system for the optical head was installed to allow tilting angles up to 75° forward and up to 60° aft. A new CCD array replaced the old one, along with a new data acquisition system to accommodate the new array. The new ASAS system can acquire imaging data in 62 spectral bands ranging from 400 to 1060 nm with a spectral resolution of 11.5 nm. Some specific applications of ASAS imagery have included validation of forest canopy reflectance models, estimation of biophysical parameters used in forest ecosystem models, derivation of vegetation indices for mapping vegetative vigor, land cover classification, testing geometric registration algorithms, and retrieval of atmospheric properties.

3.2.5 Airborne Visible/Infrared Imaging Spectrometer (AVIRIS)

The AVIRIS sensor was developed by the JPL as a NASA facility instrument to fly on an ER-2 aircraft. The instrument can measure transmitted, reflected, and scattered solar energy from the Earth's surface and atmosphere. AVIRIS is regarded as the first operational hyperspectral instrument that acquires data using whiskbroom scanning mode. The spectral data measured by AVIRIS are used to determine constituent composition through the physics and chemistry of spectroscopy for science research and applications over the regional scale of the image. Research areas include ecology, oceanography, geology, snow hydrology, and cloud and atmospheric study (Green et al. 1998). AVIRIS data are also used for satellite calibration, modeling, and algorithm development and validation. Research with AVIRIS is predominantly directed toward understanding processes related to the global environment and climate change (Elvidge and Portigal 1990, Melack and Pilorz 1990, Gao and Davis 1998). AVIRIS was flown for the first time in 1986, and it was fully operational in 1989. AVIRIS uses scanning optics and a group of four spectrometers to image 614-pixel swath width simultaneously in 224 contiguous spectral bands in a spectral range of 380–2500 nm at a spectral resolution of approximately 10 nm. A standard scene of AVIRIS image consists of 614 pixels × 512 lines × 224 bands. Since 1989, over 4000 scenes have been acquired in different countries. AVIRIS is of modular construction, consisting of six optical subsystems and five electrical subsystems, and its principal sensor and data characteristics are listed in Table 3.3. The recorded data set forms an image cube of which two axes represent spatial dimensions, and the third represents a spectral dimension. AVIRIS has been upgraded since 1992.

TABLE 3.3
AVIRIS Sensor and Data Characteristics

Imaging principle	Whiskbroom scanner
Scan rate	12 Hz
Dispersion	Four grating spectrometers (A, B, C, D)
Detectors	224 detectors (32, 64, 64, 64) Si and InSb
Digitization	12 bits
Data rate	20.4 mbits/s
Spectrum rate	7300 spectra/s
Data capacity	>10 GB (>8000 km^2)
Detector array	Line array
Wavelength range	400–2500 nm
Spectral resolution (FWHM)	10 nm
Calibration accuracy	1 nm
FOV	30° (11 km at 20 km altitute)
IFOV	1.0 mrad (20 m at 20 km altitute)
Flight line length	800 km total

Source: Green, R. O. et al., *Remote Sensing of Environment*, 65, 227–248, 1998.

3.2.6 Compact Airborne Spectrographic Imager (CASI)

CASI has been a commercially available instrument since 1990. It is manufactured by Itres Research Limited of Calgary, Alberta, Canada. The CASI instrument covers visible and near-infrared spectrum from 385 to 900 nm, with 1.8 nm sampling interval (spectral resolution approximately 3–4 nm), with a total of 288 bands and 512 spatial pixels (see more CASI-2 technical parameters in Table 3.4). CASI is a lightweight spectrometer/multispectral pushbroom imaging system for airborne remote sensing applications. CASI employs an imaging spectrograph with an area solid-state image sensor (CCD) to capture both spatial (in spatial mode) and spectral (in spectral mode) information from each line image. CASI can operate either as a multispectral imager or as a high-speed multipoint spectrometer with co-registered monochromatic imagery. In spatial mode, the spectral band configurations are defined interactively with a graphical user interface; there are 512 cross-track pixels, FOV = 34.2°, pixel width = 0.0012 × AGL; and pixel length is a function of altitude, ground speed, number of spectral bands, and integration time. In spectral mode, high-resolution spectra are recorded for the full spectral range of the instrument for up to 39 regularly spaced points in every data frame; there are up to 288 bands, 2 nm bandpass, and noncontiguous spatial coverage; and pixel size is a function of the number of look directions, altitude, and ground speed. CASI can be provided with optional roll correction and calibration systems. It is suitable for a variety of remote sensing applications due to the programmability of its spectral band sets and the easy tradeoffs achievable between spectral and spatial information during data acquisition. Water applications include pollution and algae bloom monitoring, benthic weed survey, and bathymetry and subsurface feature investigations. In the vegetation area, applications include vitality analysis, species identification, cover estimation, right of way surveillance, illicit crop detection and soil evaluation (Anger et al. 1990). Currently, over 20 CASI systems are in use by government and educational institutes, private service companies, international space agencies, and the military, including the CASI-1500, SASI-600, MASI-600, and TASI-600, etc. The CASI-1500 is a visible near infrared (VNIR) sensor which offers an impressive 1500 pixels across its field of view, allowing users to image a vast area with a single pass, or achieve spatial resolutions as high as 25 cm using a standard fixed-wing aircraft. The SASI-600 is one of the only commercially available pushbroom SWIR imagers that offers this feature. It is possible with the SASI-600 to consistently produce the best hyperspectral-SWIR imagery. The MASI-600 is the first commercially available midwave hyperspectral sensor designed specifically for airborne use. The MASI-600 has custom-designed optics which provides

TABLE 3.4
CASI-2 Sensor Technical Parameters (ITRES, Canada)

FOV (cross-track)	37.8°
IFOV (mrad)	1.3 for standard optics, 1.6 for custom lens
Spatial sampling	512 pixels
Spectral sampling	288 bands
Spectral range	385–900 nm
Spectral resolution	2.2 nm at wavelength 650 nm
Spectral sampling interval	1.8 nm
SNR (peak value)	420:1
Aperture	f/2.8 ~ f/16.0
Radiation precision	470–800 nm, absolute ±2%; 385–900 nm, absolute ±5%
Power support	28VDC/13A
Temperature range	5–40°C
Weight	55 kg
Integration times	30 ms (in spatial mode), 100 ms (in spectral mode)

extremely sharp imagery and precise spectral resolution. This sensor also has a high signal-to-noise ratio due to the Stirling cycle cooled mercury cadmium telluride (MCT) detector. With a spectral range of 3 to 5 μm and 64 bands, this 600-pixel sensor provides the advanced midwave infrared imagery available today. The TASI-600 is the commercially available pushbroom hyperspectral thermal sensor system designed specifically for airborne use. Available in a 32-spectral channel configuration, with a spectral range of 8 to 11.5 μm, and featuring 600 spatial pixels, it is well suited to numerous critical applications, including mining exploration, geophysical mapping, and landmine and ordnance detection.

3.2.7 Compact High-Resolution Imaging Spectrograph Sensor (CHRISS)

CHRISS is a commercial high-resolution hyperspectral imaging spectrometer developed by Science Applications International Corporation (SAIC) of San Diego, California, for SETS Technology Inc., Hawaii, and was first developed during 1991–1992. CHRISS is a nadir-pointing CCD pushbroom two-dimension imaging spectrometer. It is capable of imaging contiguous spectral range from ultraviolet to infrared region with a traditional blazed diffraction grating. The instrument is configurable with regard to spatial and spectral resolutions. In the spatial dimension, between 192 and 385 pixels may be chosen; in the spectral dimension, the number of bands may range from 79 to 144 with a bandwidth of 11 nm. Spatial/spectral sampling varies according to the format of the CCD camera selected for a particular application. The frame rate of the CCD and noise characteristics can also be matched to a particular mission and platform in order to provide the optimum resolution and SNR (Kramer 2002). Possible applications of the instrument include surveying for petroleum seepage, vegetation identification, forestry inventory, ocean color monitoring, and environmental monitoring.

3.2.8 Digital Airborne Imaging Spectrometers (DAIS 7915, 16115)

DAIS-7915 is a 79-channel/15-bit quantization electronical–mechanical scanner developed and built by Geographical Environmental Research Corporation (GER) of Millbrook, New York. The DAIS-7915 sensor covers the spectral range from the visible to the thermal infrared wavelengths at variable spatial resolution from 3–20 m, depending on the carrier aircraft flight altitude (1.5–4.5 km AGL). DAIS is accessible through DLR (German Aerospace Center) to serve European Airborne Remote Sensing Capabilities (EARSEC), as well as for other contractual application worldwide (Kramer 2002). Image data are measured with a radiometric resolution of 15 bits/pixels and coregistered bands. The DAIS scan mechanism is of the Kennedy type, where a cubic polygon mirror scans the terrain below through the opened window hatch in the bottom of the aircraft. The scan mirror rotates anticlockwise with respect to the aircraft heading to provide a ground element crosstrack scanning motion while the forward motion of the aircraft provides a line-by-line scan. DLR's DAIS-7915 is operated from a Dornier DO-228 aircraft and has a swath angle of ±26°, which is covered with 512 pixels per scanline. The IFOV currently used is 3.3 mrad, giving a typical pixel size of 10 m from a flight altitude of 3 km AGL (Carrère et al. 1995, Schaepman et al. 1998). The DAIS-7915 has flown since 1994. The DAIS-16115 has 161 channels but is still at 15-bit quantization. All spectrometer channels are spatially registered (image cube) and its potential applications are the same as DAIS-7915. The DAIS imaging spectrometers can be used in the environmental monitoring of land and marine ecosystems, vegetation stress research, agriculture and forestry resources mapping, geological mapping, mineral exploration, and provision of data for geographic information systems.

3.2.9 Fluorescence Line Imager (FLI)

The FLI (sometimes also called the Programmable Multispectral Imager [PMI]) is an airborne imaging sensor developed and built by Moniteq Ltd. and Itres Ltd. for the Canadian Department of

Fisheries and Oceans. The instrument was in operation from 1984–1990. The imaging spectrometer makes use of a 2D multi-element array (with CCD pushbroom technology) in the focal plane of a dispersive optical system. The FLI can also operate in two models as its successor, CASI: spatial mode and spectral mode. In spatial mode, high spatial resolution imaging data are collected in 8 selectable spectral bands; in spectral mode, high spectral resolution imaging data are acquired in 288 spectral bands (spatial resolution is reduced to 40 look directions) (Kramer 2002). The system consists of five separate optical camera modules, aligned to provide an FOV of about 70°, with 1925 detectors elements and 1.3 mrad resolution (IFOV). The objectives of developing the system are not only imaging ocean chlorophyll fluorescence and spectral reflectance changes in water caused by phytoplankton in the ocean but also operating in missions for applications related to water quality and aquatic and terrestrial vegetation monitoring.

3.2.10 Hyperspectral Digital Imagery Collection Experiment (HYDICE)

The HYDICE is an airborne, hyperspectral imaging system that provides high spatial and spectral resolution images of the Earth. It was built by Hughes-Danbury Optical System, Inc., Connecticut, and was operated by the Naval Research Laboratory (NRL). The HYDICE became operational in January of 1995. The HYDICE is a nadir-viewing pushbroom imaging spectrometer. Reflected solar energy is measured along a ground swath approximately 1 km wide based on a design flight altitude of 6 km. The ground sampling distance varies from 1 to 4 meters depending on aircraft altitude AGL. The full spectral range of the instrument covers from the visible into the short wave infrared (0.4 to 2.5 μm). HYDICE's spectral resolution ranges from 3 nm for the short wavelengths to 10–20 nm for the longer wavelengths (Shen 1996; see more technical parameters in Table 3.5). The optics consist of an off-axis, f/3, 27-mm aperture diameter Paul Baker telescope and a Schmidt double pass prism spectrometer. Light enters the HYDICE spectrometer through a narrow slit parallel to the 320-CCD focal-plane array (FPA) and perpendicular to the line of flight. For each band, a scene image is built up line by line in a pushbroom scan by the forward motion of the aircraft. Dispersion of the incoming light is accomplished with a prism, and the dispersed light is detected by FPA detectors simultaneously in 210 spectral channels by 320 spatial pixels and subject to 12-bit

TABLE 3.5
HYDICE Instrument Technical Characteristics

Imager type	Pushbroom scanner
Detector array (SBRC)	320 × 210 element InSb
Integration time	1.0–42.5 ms
Spectral bands	210
Digitization	12 bits
Aperture diameter	27 mm
Objective lens	f/3.0
Detector array	area array
Wavelength range	400–2500 nm
Spectral resolution (FWHM)	3 for VNIR, 10–20 for SWIR
Calibration accuracy	1 nm
FOV	8.94°
IFOV	0.5
SNR	250–280 in VNR, 100–107 in SWIR, 50–80 in MWIR
Altitude	8–10 km AGL

Source: Kramer, H. J., *Observation of the Earth and Its Environment: Survey of Missions and Sensors*, 4th ed., Springer-Verlag, Berlin, 2002.

quantization. The heart of the HYDICE is single indium-antimonide (InSb) 320 × 210 elements, cryo-cooled, CCD FPA detectors (Kramer 2002). The hyperspectral data can be used in the following fields (Kramer 2002):

1. Agriculture: For crop analysis, pest control, and stress analysis.
2. Forestry: For the monitoring of inventory, habitat mapping, pest control, and reforestation.
3. Environment: For the monitoring of toxic waste, acid rain, air pollution, eutrophication, soil conservation, and water pollution.
4. Resource management: For land use and mineral identification.
5. Mapping: For area classification, bathymetry, wetland, and critical habitats.
6. Disaster management: For damage assessment and search and rescue support.
7. Law enforcement: For a host of applications.

3.2.11 Hyperspectral Mapper (HyMap)

The HyMap Imaging Spectrometer was built by Australian Integrated Spectronics Ltd. and was operated by HyVista Corp. It is a "whiskbroom-style" instrument. The HyMap systems have set the benchmark for commercially available hyperspectral sensors in terms of SNR, image quality, stability, adaptability, and ease of use. The instrument offers an extremely high SNR of more than 500:1 which is more than most existing imaging spectrometers including AVIRIS can realize. The evolution of the HyMap series continues with the development of a system providing hyperspectral coverage across the solar wavelengths (0.4–2.5 μm) with 126 bands and 32 bands in the thermal infrared (8–12 μm). The spectral bandwidths of HyMap hyperspectral data are 15–20 nm in the spectral range of 0.4–2.5 μm and 100–200 nm in the thermal infrared range (see more technical parameters in Table 3.6). The design of the HyMap series of airborne hyperspectral sensors features opto-mechanically scanned foreoptics combined with modular, high-efficiency spectrographs and optimized detector arrays. The HyMap sensor is mounted on a stabilized platform. Position and orientation during the flight are measured continuously by means of Differential-GPS and INS (Bucher and Lehmann 2000). The HyMap records an image by using a rotating scan mirror, which allows the image to build line by line as the aircraft flies forward. The reflected sunlight collected by the scan mirror is then dispersed into different wavelengths by four spectrometers in the system. The HyMap offers rapid and efficient wide-area imaging, especially for mineral exploration and environmental monitoring applications. It has been operated in Europe, Australia, the United States, and Africa in 24 specific campaigns (Ben-Dor 2013).

TABLE 3.6
HyMap Sensor Technical Characteristics

Imager type	Whiskbroom scanner
Detector array	Line array
Wavelength range	450–2500 nm, 8000–12000 nm
Spectral resolution (FWHM)	15–16 nm (0.4–2.5 μm), 100–200 nm (8.0–12.0 μm)
SNR	>500:1
FOV	60° (512 pixels across track)
IFOV	2.5 mrad (alon track), 2.0 mrad (across track)
Spatial resolution	3–10 m
Altitudes	1500–5000 m AGL

Source: Kramer, H. J., *Observation of the Earth and Its Environment: Survey of Missions and Sensors*, 4th ed., Springer-Verlag, Berlin, 2002.

3.2.12 HyperSpectral Cameras (HySpex)

HySpex sensors are compact, high-performance, and versatile instruments built and developed by Norsk Elektro Optikk, Norway, for a multitude of applications ranging from airborne to laboratory and industrial use of imaging spectroscopy (see http://www.hyspex.no). The line of HySpex hyperspectral cameras comprises VNIR models, operating in the spectral range from 400–1000 nm, and SWIR models operating in the spectral range from 900–2500 nm. Basically, the working principle of the HySpex camera can be described as follows: The camera foreoptic images the scene onto a slit which only passes light from a narrow line in the scene. After collimation, a dispersive element (a transmission grating) separates the different wavelengths, and the light is then focused onto a CCD detector array. The net effect of the optics is that for each pixel interval along the line defined by the slit, a corresponding spectrum is projected on a column of detectors on the array. The data readout from the array thus contains a slice of a hyperspectral image, with spectral information in one direction and spatial information in the other direction. By scanning over the scene, the HySpex camera collects slices from adjacent lines, forming a hyperspectral image or "data cube" (i.e., two spatial dimensions and one spectral dimension). The objectives of developing HySpex hyperspectral cameras are for applications ranging from online industrial monitoring/sorting/classification to laboratory measurements, clinical instruments for medical diagnostic, and airborne- and satellite-based remote sensing tools.

3.2.13 Infrared Imaging Spectrometer (ISM)

The ISM French airborne spectrometer was developed by the Space Research Department (DESPA) of the Paris-Meudon Observatory as part of a planetary mission in collaboration with Institut d'Astrophysique Spatiale (IAS), France. The first test flight occurred in 1991. It operates in the NIR and MWIR spectral regions, and provides 128 spectral channels between 0.8 and 3.2 μm. The ISM adopts the electromechanical scanner with a whiskbroom imaging principle to acquire hyperspectral images. The spectrometer provides a fixed grating system, which measures the radiation wavelengths in three ranges. The ISM has two 64-PbS detectors located in parallel rows, which provide acquisitions with 128 contiguous spectral bands. The two CCD arrays are cooled by passive cryogenics, and the ISM provides internal calibration from an incandescent lamp (Kramer 2002). The ISM hyperspectral imagery can be used in geology, clouds, ice and snow, and vegetation (including forest and agriculture), with a special emphasis on the mid-wave infrared (MWIR) region for estimating and analyzing plant canopy chemistry, including lignin, nitrogen, cellulose, etc.

3.2.14 Modular Airborne Imaging Spectrometer (MAIS)

The MAIS Airborne Imaging Spectrometer was designed and operated by the Shanghai Institute of Technical Physics (SITP) of the Chinese Academy of Sciences (CAS). The instrument is the first real imaging spectrometer in China (Tong et al. 2014). It provides 71 spectral bands; the first 32 bands having a bandwidth of 20 nm in the spectral range 0.44–1.08 μm, another 32 having a bandwidth of 30 nm in the spectral range 1.50–2.50 μm, and the last 7 bands covering the thermal infrared range 8.2–12.2 μm. The MAIS features a modular design. Its design is based on the linear array detectors added to an optical mechanical scanner frame. The whole system consists of a tilted 45° rotation mirror optical mechanical scanner unit, main optical system, VNIR spectrometer, SWIR spectrometer, TIR spectrometer, data acquisition system, motor drive, and power supply system (Wang and Xue 1998). Therefore, separate calibration of each module is possible. The MAIS was first used in 1990 and thereafter in a number of remote sensing applications by SITP for geological and environmental survey. During the period of September–October 1991, the MAIS was installed on a Citation S/II aircraft of the CAS and flown successfully in a joint Sino-Australian remote sensing campaign near Darwin and at several other test sites in Western Australia.

3.2.15 Modular Imaging Spectrometer Instrument (MISI)

The MISI is a hyperspectral imager manufactured by the Rochester Institute of Technology (RIT of Rochester, New York). It provides 70 bands covering visible and NIR regions from 400–1000 nm and 5 LWIR bands. The MISI features a line scanner design collecting two lines per scan for the 70 high-resolution channels and one line per scan for remaining channels (5 LWIR bands). The coveraging image is split onto four slightly off-axis (< 2°) focal planes by a four-sided pyramid mirror (Kramer 2002). The total FOV of the MISI scanner is ±45° with calibration standards viewed every rotation. The imager is flown at altitudes ranging from 0.3 to 3 km AGL. The purpose of developing the MISI sensor was for (1) airborne laboratory for EO research; (2) under flight system for airborne/spaceborne sensor performance evaluation; (3) versatile data collection platform for acquiring imagery to be used for algorithm development, reconnaissance evaluation, and environmental applications; and (4) survey instrument for studies of image analysis methods in areas such as energy conservation, water quality assessment, hazardous waste site management, etc.

3.2.16 Multispectral Infrared Camera (MUSIC)

The MUSIC was developed and built at the Lockheed Palo Alto Research Laboratory, California. The current version of the MUSIC airborne hyperspectral imager has been operated since 1989 and has flown on NASA research aircraft at up to 20 km AGL. The MUSIC measures infrared image data simultaneously in the MWIR (2.5–7.0 μm) and TIR (6.0–14.5 μm) spectral ranges (Kramer 2002). The spatial resolution is 0.5 mrad and each infrared range is covered with a total of 90 spectral bands. The sensor consists of two parallel optical telescopes, each with its own set of optical filters and detector arrays. Each detector array has a 45 × 90 pixel matrix for a total 4050 detectors in each array and all detectors collect data in parallel at up to 80 frames/s. The detector arrays are cooled with liquid helium. Due to cooling, the internal background temperature is very low and detectors operate close to the BLIP limit as defined by the external signal through the cooled spectral filters (Kramer 2002). To provide a continuously variable narrow bandpass ($\lambda/\Delta\lambda \sim 100$), in addition to the several spectral bandpass filters in each telescope, a circular variable filter is installed in the MUSIC sensor. The system also offers a quicklook capability. The objective of developing the MUSIC sensor is for chemical vapor sensing, plume diagnostics, and extraction of spectral signatures.

3.2.17 Probe-1

The Probe-1 airborne hyperspectral remote sensing system was manufactured by Earth Search Science, Inc., and can record high-resolution spectral reflectance from the Earth's surface. The Probe-1 imager covers a spectral range from 0.4–2.5 μm. It records 512 crosstrack pixels with 128 bands each. Pixel size is typically between 5 and 10 m on a side as determined by the aircraft's altitude AGL. Probe-1 images are built up by collecting successive scan lines as the aircraft moves forward. The imaging FOV of Probe-1 (60°) allows Earth Search to collect data over large areas quickly and has an important advantage with respect to cost effectiveness when compared with other hyperspectral sensors. The system has relatively high SNR characteristics, which offer Earth Search's customers with the highest quality hyperspectral data. The hyperspectral system flew over desert regions in northern Chile and over the Canadian Arctic, proving to be a useful tool in defining exposed geology, alterations, and gossans.

3.2.18 Reflective Optics System Imaging Spectrometer (ROSIS)

ROSIS is a compact airborne imaging spectrometer and was developed jointly by German industry and research organizations (i.e., jointly by DASA/MBB, CKSS and DLR [Institute of

Optoelectronics]). After the initial test phase of the modified sensor, it was planned to make the sensor available to a large customer community worldwide within a suitable environmental unitization program (Kramer 2002). The sensor makes use of a 2D CCD array for simultaneously imaging 115 spectral bands of 512 pixels along a crosstrack direction. The spatial resolution is 0.56 mrad. The selection of bands can be pre-programmed per orbit or aircraft flight out of the available 115 bands. The radiometric resolution is defined as 0.05% of the apparent albedo. The total FOV covers ± 16° per optics module. The basic working procedure can be described as follows: The primary optics focus the image into an entrance slit for the spectrometer part, which reduces the Earth's image to a single scanline. The subsequent larger collimator optics forms a parallel beam to the reflective echelette grating, which disperses the scan line into a continuum of spectral lines within the preselected waveband range. The dispersed image passes the collimator optics in a reverse mode again and is focused via a small fixed-fold mirror onto the matrix CCD detector array. Per the CCD array, one direction represents the spatial scanning of the scan line, while the other direction lines correspond to discrete narrowband spectral channels (Kunkel et al. 1991). The objective for the design of ROSIS was for application in detecting spectral fine structures in coastal and inland waters, which determined the selection of spectral range, bandwidth, number of channels, radiometric resolution, and tilt capability for sun glint avoidance. However, ROSIS also can be used for monitoring spectral features above land and within the atmosphere. In view of the ESA (and NASA) planning for Polar Platform Missions, ROSIS could represent a promising candidate of multipurpose remote sensing instruments.

3.2.19 SWIR Full Spectrographic Imager (SFSI)

The SFSI was developed by the Canada Centre for Remote Sensing (CCRS) for research purposes, and it has undergone recent modifications. The first test flights with the SFSI were carried out in October 1994. The SFSI imager was designed for focus on the shortwave infrared region of the electromagnetic spectrum. The pushbroom design sensor has the ability to simultaneously acquire the full spectrum at high spatial (up to 20 cm) and spectral (10.4 nm) resolution. It collects image data in 115 contiguous bands over the range from 1219 nm to 2405 nm, with a nominal band spacing of 10 nm. A 2D detector array technology is used, providing a full image cube consisting of 496 pixels by 580 lines by 115 spectral bands (Kramer 2002). The instrument utilizes a platinum silicide (PtSi) detector array, refractive optics, and a transmission grating. The special feature of this sensor is its ability to achieve a spatial ground resolution down to 0.2 m using the pitch-scan technique. The sensor can be flown in a small twin engine aircraft at an altitude appropriate for any desired sample spacing. The control electronics, data processing, and data recording system have been developed to store the image cube. During a pilot project flown in Nevada in June 1995 over desert terrain, the SNR was measured under the operational conditions with a solar angle of 27° with the SNR peaked at 120:1 in the 1.20–1.32 μm region, and at 80:1 in the 1.50–1.79 μm and 2.0–2.4 μm regions (Kramer 2002). The imager can be employed in measuring the depth of absorption bands and in identifying most of the Earth's surface components in the regions of 1.2–1.3 μm and 1.5–1.7 μm. During the first flights of the sensor, hyperspectral imagery was acquired over a calcite quarry and a dolomite quarry. Both these minerals show distinctive carbonate absorption features in the 1.7–2.5 um region (Rowlands and Neville 1996).

3.2.20 Spatially Modulated Imaging Fourier Transform Spectrometer (SMIFTS)

The SMIFTS is a cryogenically cooled, spatially modulated imaging Fourier transform interferometer-spectrometer and was designed and manufactured in the Department of Geology and Geophysics at the University of Hawaii, Honolulu. The instrument utilizes a 256 × 256 element InSb detector array covering an infrared spectral range from 1.0–5 μm. The crosstrack IFOV of the sensor is 0.5 mrad and has a swath width of 256 pixels. The along-track IFOV is variable in order to maximize the

SNR for specific application. It was intended to demonstrate the feasibility of a highly sensitive IR hyperspectral imager based upon new technology. The first successful test flights with the prototype of SMIFTS occurred in the summer of 1993 over the lava fields of the Kilauea volcano on the island of Hawaii. The SMIFTS system consists of three major subsystems: a Sagnac interferometer, which produces the spatially modulated interferogram; a Fourier transform lens, which frees the spectral properties from dependence on aperture geometry and allows the wide FOV; and a cylindrical lens, which reimages on axis of the input aperture onto the detector array providing one dimension of imaging (Kramer 2002). Using a Michelson interferometer, the proven Fourier transform spectroscopy serves a basis to obtain the sampled interference pattern from the input source. The SMIFTS has had the following characteristics: broad wavelength range, wide FOV, simultaneous measurement of all spectral channels, moderate spectral resolution, and one-dimension imaging. On the 2D detector array used by the SMIFTS, one axis contains the spectral information, and the other axis contains spatial information along one axis of the source. The spatial axis is retained for a wide field monitoring purpose.

3.2.21 TRW Imaging Spectrometers (TRWIS)

A set of commercially available airborne hyperspectral imaging spectrometers has been developed and built by TRW of Redondo Beach, California, operational since 1990. Since then, TRW has designed and built four imaging spectrometers: TRWIS-A, TRWIS-B, TRWIS-II, TRWIS-III, and SSTI HSI (NASA's Small Satellite Technology Initiative, Hyperspectral Imager, was not successful due to loss of control of the satellite) (Kramer 2002). Their characteristics and technical parameters are compared in Table 3.7. The three TRWIS-B instruments and one TRWIS-II instrument were in operation in 1994. The instruments provide real-time image data for vegetation and mineral identification and quantification (e.g., crop health, biomass, and algae contents.).

TRWIS-A was initially a laboratory model. The first sensor array used was an intensified CCD type TV camera. The spatial and spectral resolution is limited by the fiber optics coupler between the intensifier and the CCD. TRWIS-B used an unintensified CCD camera. It features improved performance and lower cost and has been used extensively for several years. TRWIS-II was the

TABLE 3.7
Performance Comparison of TRWIS Family with HIS

Parameter	TRWIS-A	TRWIS-B	TRWIS-II	TRWIS-III	SSTI HIS
Spectral range (μm)	0.43–0.85	0.46–0.88	1.5–25	0.4–2.5	0.4–2.5
Spectral channels	128	90	108	384	384
Spectral sampling interval (nm)	3.3	4.8	12	5.0 (VNIR) 6.25 (SWIR)	5 (VNIR) 6.38 (SWIR)
Spatial pixels	240	240	240	256	256
IFOV (mrad)	1.0	1.0	0.5/1.0	0.9	0.06
TFOV (mrad)	240	240	120/240	230	15.4
Aperture (mm)	1.5	5	17.5/8.5	20	125
Focal length (mm)	25	25	70/34	70	1048
Focal ratio	f/16	f/5	f/5.3; f/4.8	f/3.3	f/8.3
Detectors	intensified CCD	Si CCD	INSb CCD	CCD/HCT	CCD/HCT
Quantization (bits)	8	8	8	12	12
Recoding median	vodeotape	videotape	videotape	digital	digital
Operation (year)	1990	1991	1992	1995	1996

Source: Kramer, H. J., *Observation of the Earth and Its Environment: Survey of Missions and Sensors*, 4th ed., Springer-Verlag, Berlin, 2002.

first attempt to make measurements in the SWIR region. It uses a custom IR lens foreoptics system, a SPEX 270M spectrometer (repackaged), and a modified commercial InSb camera. The IR focal plane array hybrid is an FLIR camera-type array that uses a direct injection readout and has a very large integration capacitor. The CCD detector array is cooled with liquid nitrogen. The data are calibrated using spectrally flat and spatially uniform calibration standards (Kramer 2002). The TRWIS-III consists of two imaging spectrometers, one visible-near infrared (VNIR) covering from 370 nm to 1040 nm at a 5.25 nm resolution and the other shortwave infrared (SWIR) covering from 890 nm to 2450 nm at a 6.25 nm resolution. Both spectrometers are co-aligned and matched in crosstrack field of view to allow coregistration across the 384 spectral bands. An image is generated by moving the instrument across a scene in a pushbroom fashion, perpendicularly to the instrument's slits, and recording frames of spectral and spatial information detected by the VNIR and SWIR focal plane arrays. The image cubes generated in this manner contain 256 pixels in the crosstrack direction and 384 spectral bands per pixel. TRWIS-III offers very good SNR images due to relatively long integration times and low thermal background because the CCD detector array of the SWIR spectrometer is cooled to 115 K using a TRW-developed pulse tube cryo-cooler. The system is designed to interface with a variety of aircraft platforms. TRWIS-III's variable frame rate allows for imaging at altitudes from 0.6 km to 12 km. The IFOV of 0.9 mrad of the instrument results in spatial resolution ranging from 0.5 m–11 m. The system consists of a piece of onboard navigation equipment, including a global positioning system (GPS) with a differential correction receiver and an inertial navigation sensor (INS). The navigation data from these systems are used in TRWIS-III data postprocessing for image rectification and geolocation (Sandor-Leahy et al. 1998). Unlike many other aircraft-based scanners, the TRW Imaging Spectrometer has in-flight calibration resources to permit radiometric corrections without the need of spectral reference targets located on the ground.

3.2.22 Variable Interference Filter Imaging Spectrometer (VIFIS)

The VIFIS imaging spectrometer was developed by the Department of Applied Physics and Electronic & Manufacturing at the University of Dundee, United Kingdom. The imager covers the visible-near infrared range from 450 nm to 870 nm. While earlier test flights with a single-module prototype instrument took place in August 1991, later test flights with the three-module hybrid VIFIS instrument started in May 1994. The three synchronized CCD-imager modules, aligned to a common field of view, consist of two modules with variable interference filters (VIF): one VIS-range filter and one NIR-range filter and a third module of a normal videograph unit without the visible filter. The analysis of the design and test results demonstrates that such a hybrid three-CCD array instrument has the potential to acquire both wavelength spectral image data and directional spectral image data in a single pass. All VIFIS CCD-imaging modules are full-frame imagers that create a sequence of instantaneous 2D images as synchronized snapshots. The video images can, of course, be displayed on a screen in real-time mode during data acquisition (Kramer 2002). The VIFTS also has a third panchromatic video camera so that a combination of the panchromatic data channel with two spectrally filtered data channels offers an even-better reference and increased spectral information content. The VIFIS is flexible for acquiring a hyperspectral image that can be used for a variety of environmental monitoring tasks, such as mapping shallow water bottom materials, growth and distribution of phytoplankton, crop discrimination, oil spill detection and mapping, etc.

3.2.23 Wedge Imaging Spectrometer (WIS)

The WIS is a hyperspectral imaging spectrometer developed by Hughes Santa Barbara Research Center (SBRC) on Corporate Independent R&D. Flight demonstrations of the instrument took place in 1992. The WIS sensor provides a novel spectral separation technique in which the spectral separation filters are mated to the detector array to achieve a 2D sampling of the combined spatial/spectral

TABLE 3.8
Performance Comparison of Three WIS Imaging Spectrometers

Parameter	Early VNIR (WIS-FDU)	New VNIR Instrument (WIS-VNIR)		New SWIR Instrument (WIS-SWIR)	
Filter	1	1	2	1	2
Spectral range (μm)	0.40–1.03	0.40–0.60	0.60–1.0	1.0–1.80	1.80–2.50
Spectral bands	64	129	265	81	90
Spectral resolution (nm)	7.2–18.5	9.6–14.4	5.4–8.6	20 0–37.8	18.0–2.50
Detector material and type	Si CCD	Si CCD		InSb	
Spatial pixels	128	512		320	
IFOV (mrad)	1.36	0.66		0.66	
FOV (swath width) (°)	10, 15	19.1		12.0	
Telescope focal length (mm)	55, 108	27.4		61.0	
Spatial resolution (m) at 1.5 km AGL	0.5–5	1.0		1.0	
Quantization (bits)	12	12		12	

Source: Kramer, H. J., *Observation of the Earth and Its Environment: Survey of Missions and Sensors*, 4th ed., Springer-Verlag, Berlin, 2002.

information passed by the filter. This innovation is based on using a linear spectral wedge filter mated directly to an area detector array, avoiding the use of bulky and complex aft optics required for imaging spectrometers based on grating or prism concepts. The wedge filter is a thin-film optics device that transmits light at a center wavelength depending on the spatial position of the illumination. When a detector array is placed behind the wedge filter device, each detector in the "spectral" dimension will receive radiation from the scene at a different center wavelength and the array output is the sampled spectrum of the scene. The detected scene information will vary spatially in one direction and spectrally in the other direction by using an array of detectors. Scanning the filter/array assembly along the spectral dimension (provided by forward motion of the platform) will then build a spatial image (2D) in each of the detected spectral bands. Thus, it became an imaging spectrometer (Demro et al. 1995). Therefore, the WIS sensor concept does not provide simultaneous spectral sampling of all spatial locations within the sensor's FOV. Instead, the sensor samples each ground point in all spectral bands over a short period of time (in the order of 1 s) by using the forward motion of the aircraft to "pushbroom" the ground image across the WIS detector array. The data are then registered in post-processing to superimpose the spectral bands (Kramer 2002). A comparison of performance characteristics of the three WIS imaging spectrometers is presented in Table 3.8.

In terms of several solar-light covered spectral regions, the WIS image data can be applied as follows: (1) detect spectral differences in most vegetation applications in visible spectral portion (0.4–0.6 μm) with a spectral resolution of 15 nm; (2) detect the "red edge" shifts due to stress in vegetation in the red-NIR spectral range (0.6–1.0 μm) with a spectral resolution of 6 nm; and (3) detect most terrestrial materials and several water and water vapor absorption features by using broad reflectance peaks from the spectral range 1.2–2.4 μm with a spectral resolution of 30 nm. Thirty nanometer spectral resolution is adequate for most applications in this region, and the far end of this region covers the range of diagnostic absorption features of hydroxyl-bearing minerals and carbonates (Demro et al. 1995).

3.3 SPACEBORNE HYPERSPECTRAL SENSORS/MISSIONS

In this section, a survey of fifteen (nine for current/operational and six for future/planned) selected important spaceborne hyperspectral sensors and missions is summarized by referring to

Lucas et al. (2004), Staenz (2009), Miura and Yoshioka (2012), Qi et al. (2012), and Ben-Dor et al. (2013). Table 3.9 presents a summary of current and future spaceborne hyperspectral sensors and missions in terms of their launch year, actual or potential platform/satellite and orbit altitude, operational spectral range and spatial/spectral resolution, swath width and instantaneous field of view (IFOV), features of sensors and missions, etc.

3.3.1 Advanced Responsive Tactically Effective Military Imaging Spectrometer (ARTEMIS), TacSat-3 Satellite

ARTEMIS onboard U.S. Air Force TacSat-3 satellite launched on May 19, 2009, and was developed by Raytheon Space and Airborne Systems of El Segundo, California. The spectrometer uses a single HgCdTe Focal Plane Array covering the entire V/NIR/SWIR spectral range from 400 nm to 2500 nm at a uniform resolution of 5 nm. ARTEMIS first measures the spectral information at each point on the ground in 400 spectral channels at a spatial resolution of 4 m, which enables it to detect and identify tactical targets. It is the first hyperspectral satellite with the ability to provide reconnaissance within 10 minutes after passing overhead (ARTEMIS 2015). The instrument consists of a telescope (35 cm diameter), an Offner imaging spectrometer known as a hyperspectral imager (HSI), a high-resolution imager (HRI) and a real-time processor referred to as a hyperspectral imaging processor (HSIP). ARTEMIS also provides HSI observations with panchromatic data. The HSI design also relies upon a single substrate-removed HgCdTe focal plane array (FPA) that extends its sensitivity into the blue wavelengths to cover the full spectral range (VNIR and SWIR). The HRI is adapted from off-the-shelf hardware for simplicity and cost savings (ARTEMIS 2015). The objective of developing ARTEMIS is to demonstrate tactically significant hyperspectral imagery collection and processing sufficient to meet militarily relevant detection thresholds (30 min.).

3.3.2 Compact High-Resolution Imaging Spectrometer (CHRIS), PROBA Satellite

The CHRIS is a satellite imaging spectrometer on board the European Space Agency (ESA) Project for On-Board Autonomy (PROBA) satellite, launched on October 22, 2001. CHRIS is the prime instrument of the PROBA-1 mission (CHRIS-PROBE 2015). The sensor was developed by the United Kingdom Company with support from the British National Space Center. CHRIS operates over the visible/near infrared range from 400 nm to 1050 nm and can operate in 63 spectral bands at a spatial resolution of 36 m, or with 18 bands at full spatial resolution of 18 m. From a 580 km orbit, CHRIS can image the Earth in a 17.5 km swath width with a spatial resolution of 18 m (this is somewhat variable as the altitude varies around the orbit). CHRIS design is capable of providing up to 150 channels over the spectral range of 400–1050 nm. Using PROBA's agile steering capabilities in along- and crosstrack directions enables observation of selectable targets well outside the nominal field of view of 1.3°. This agile satellite can also deliver up to five different viewing angles (nadir, ± 55° and ± 36°) along the track direction. Spectral sampling varies from 2–3 nm at the blue end of the spectrum, to about 12 nm at 1050 nm. Sampling is about 7 nm near the red edge (~690–740 nm). The instrument is very flexible and different sets of bands can be used for different applications. PROBA, by virtue of its agile pointing capability, enables CHRIS to acquire images from five different angles over a selected site. The hyperspectral and multiangle capability of CHRIS makes it an important resource for studying bidirectional reflectance distribution function (BRDF) of vegetation. Other applications include coastal and inland waters, wild fires, education, and public relations. An effective data acquisition planning procedure has been implemented and, since mid-2002, users have received data for analysis. A cloud prediction routine has been adopted that maximizes the image acquisition capacity of CHRIS-PROBA. Although the mission was designed for a 1-year life, the sensor has been in operation for more than 10 years now (CHRIS-PROBE 2015).

TABLE 3.9
Summary of Current and Future Spaceborne Hyperspectral Sensors and Missions

Sensor/Mission (Country)	Launch	Platform/Altitude (km)	Pixel Size (m)	No. of Bands	Spectral Range (nm)	Spectral Resolution (FWHM, nm)	IFOV (μrad)	Swath (km)	Features
Current/operational spaceborne hyperspectral sensors/missions:									
ARTEMIS (USA)	May 2009	TacSat-3	4	400	400–2500	5			Identify tactical targets within 10 minutes
CHRIS (ESA)	Oct. 2001	ESA PROBE (580)	18 –36	up to 63	410–1050	1.25–11.0	43.1	up to17.5	Variable spatial/spectral resolution, multi-angle imaging, BRDF
FTHSI (USA)	Jul. 2000	Mighty SatII (575)	30	150	475–1050	1.7–9.7	50	13	Identify military targets
GLI (Japan)	Dec. 2002	NASDA ADEOS-II (803)	250 – 1000	36	380–11950	10–1000	313–1250	1600	Monitoring the carbon cycle in the ocean and biological processes
HJ-1A/HIS (China)	Sep. 2008	HJ-1A (649)	100	115, 128	450–950	5		50, 60	Fourier Transform HIS, and environmental and disaster monitoring
Hyperion (USA)	Nov. 2000	EOS/EO-1 (705)	30	220	400–2500	10	42.5	7.5	1st satellite hyperspectral sensor, advancing Earth observation
HySI (India)	Apr. 2008	Indian Microsatellite 1 (635)	505.6	64	450–950	8		128	Measure vegetation type and resource characterization
MERIS (ESA)	Mar. 2002	ENVISAT (800)	300	15	390–1040	1.8		1150 (FOV of 68.5°)	Spectral bands programmable in width and position
MODIS (USA)	Dec. 1999	Terra/Acqu (705)	250–1000	36	405–14385	VNIR-SWIR: 10–50; TIR: 30–360		2330	Global coverage 1 to 2 days, variable spectral/spatial resolutions

(Continued)

TABLE 3.9 (CONTINUED)
Summary of Current and Future Spaceborne Hyperspectral Sensors and Missions

Sensor/Mission (Country)	Launch	Platform/Altitude (km)	Pixel Size (m)	No. of Bands	Spectral Range (nm)	Spectral Resolution (FWHM, nm)	IFOV (μrad)	Swath (km)	Features
Future/planned spaceborne hyperspectral sensors/missions:									
EnMAP/HSI (Germany)	Schedule 2017	Polar, sun-synchronous platform (652)	30 30	~89 ~155	420–1000 900–2450	5–10 10–20	30 30	30 30	Monitoring and characterizing the Earth's environment
FLEX (ESA)	Schedule 2016	LEO sun-synchronous platform (800)	300	> 60	400–1000	5–10		390	Monitoring health of Earth's vegetation
HISUI (Japan)	2018 or later	ALOS-3 (620)	5 30	4 185	450–900 400–2500	60–110 10–12.5		90 30	Both hyper/multispectral imagers operated independently or simultaneously
HyspIRI (USA)	Schedule 2023	LEO sun-synchronous platform (626)	60	>200	380–2500 3000–12000	10		90 600	Disasters, ecological forecasting, health and air quality and water resources
MSMI (South Africa)		Sunsat (660)	15	200	400–2350	10	22	15	Imager suitable for microsatellites
PRISMA (Italy)	Schedule 2018	Sun-synchronous platform (620)	30	250	400–2500	10	48.34	30	Europe and the Mediterranean region observation

Source: Lucas, R. et al., Hyperspectral sensor and applications, in *Advanced Image Processing Techniques for Remotely Sensed Hyperspectral Data*, eds. P. K. Varshney and M. K. Arora, pp. 11–45. Berlin and Heidelberg: Springer-Verlag, 2004; Staenz, K., Terrestrial imaging spectroscopy: Some future perspectives. *Proceedings of the 6th EARSEL SIG IS*. Tel Aviv, Israel, 2009; Miura, T., and H. Yoshioka, Hyperspectral data in long-term, cross-sensor continuity studies, in *Hyperspectral Remote Sensing of Vegetation*, eds. P. S. Thenkabail, J. G. Lyon, and A. Huete, pp. 611–633. Boca Raton, FL: CRC Press/Taylor & Francis Publishing Group, 2012; Qi, J., Hyperspectral remote sensing in global change studies, in *Hyperspectral Remote Sensing of Vegetation*, eds. P. S. Thenkabail, J. G. Lyon, and A. Huete, pp. 69–89. Boca Raton, FL: CRC Press/Taylor & Francis Publishing Group, 2012; Ben-Dor, E. et al., Hyperspectral remote sensing, in *Airborne Measurements for Environmental Research: Methods and Instruments*, eds. M. Wendisch, J.-L. Brenguier, pp. 413–456. Weinheim, Germany: Wiley-VCH, 2013.

3.3.3 Fourier Transform Hyperspectral Imager (FTHSI), MightySat II Satellite

FTHSI on MightySat II, the first mission of the U.S. Air Force program initiated in 1995, was successfully launched in July 2000. The sensor operates in a spectral waveband of 470–1050 nm and has 30 m spatial resolution, 150 spectral bands, and an FOV of 3°. FTHIS is operated in pushbroom mode, which provides individual 1D frames consisting of 1 by 1024 samples. The FTHSI payload consists of two components: The hyperspectral instrument (HSI), which is an imager; and the hyperspectral instrument interface card (HII), which allows the user to control the amount and parameters of the data collect (Yarbrough et al., 2002). FTHSI is the only Department of Defense space-based HSI to use state-of-the-art Fourier Transform technique and may provide improvement over traditional dispersive- or grating-type sensors, particularly for long-wave infrared applications. This instrument may also provide the means to detect and identify military targets, despite camouflage or other concealment, categorize terrain, and assess traffic ability for ground troop movement. Commercial applications include classification of environmental/crop damage and many others (Freeman et al. 2000).

3.3.4 Global Imager (GLI), NASDA ADEOS-II Satellite

GLI is an optical NASDA (National Space Development Agency of Japan) core sensor on Advanced Earth Observing Satellite 2 (ADEOS II) launched by NASDA, NASA, and CNES in December 2002. GLI operates in spectral range from 0.38–11.95 μm with a total of 36 spectral bands. The GLI imaging spectrometer is an opto-mechanical instrument. It features a crosstrack mirror and an off-axis parabolic mirror as the collecting optics and focal planes. The detectors in the instrument are arrayed in the along-track direction with spectral interference (dichroic) filters. GLI can tilt the scan mirror ±20° from nadir in order to avoid sun glitter. GLI has five focal planes, two for VNIR, two for SWIR, and one for MWIR/TIR. Two VNIR focal planes have detector arrays for 13 and 10 bands, respectively. Two SWIR focal planes have detector arrays for 4 and 2 bands, while the MWIR/TIR regions have one focal plane with a detector array for 7 bands. One SWIR and the MWIR/TIR focal planes are cooled to 220 K and 80 K by a multistage Peltier element and Stirling cycle mechanical cooler, respectively. The VNIR detector material is Si, and the SWIR is InGaAs, while the MWIR/TIR material is CMT (GLI 2015).

GLI is for studying and monitoring the carbon cycle in the ocean, principally for biological processes. Multispectral observations from the near-UV to the near-IR reflected solar radiation from the Earth's surface include land, ocean, and clouds. Out of the 36 observation bands of GLI (up to about 10) extending 380 nm to 865 nm bands could be used for ocean color remote sensing with 1 km spatial resolution at nadir. Substantially, the GLI sensor's data can be used for determination of chlorophyll pigment, phycobilin, and dissolved organic matter in the ocean, as well as for the classification of phytoplankton according to their pigment. The GLI data can also be utilized for measuring sea surface temperature, cloud distribution, land coverage, vegetation index, etc.

3.3.5 HJ-A/HSI (Hyperspectral Imager, HJ-1A Satellite)

The Chinese environment satellite HJ-1A was launched successfully on September 6, 2008. The HJ-1A satellite carries the first Chinese Earth observation spaceborne hyperspectral sensor: the hyperspectral imager (HSI) (Gao et al. 2010). The HSI instrument operates in visible-near infrared spectral range from 450 nm to 950 nm with a ground swath width of 50 km and a spatial resolution of 100 m. It has 110 –128 spectral bands and ± 30° side observing ability and calibration function on satellite. HSI is a Fourier Transform HyperSpectral Imager built by the Xian Institute of Optics and Precision Mechanics (XIOPM) of the Chinese Academy of Sciences (CAS). Since the Fourier Transform HSI has its theoretical advantages and its stable capability, this new spectral imaging technology was developed and boarded on a satellite fewer than 10 years after its emergence (Zhao et al. 2010). The HSI is composed of an interferometer, a Fourier mirror, a calibration system, a swing mirror, a detector array, and some other components. Usually, the data from a dispersive imaging spectrometer can be normally processed through relative radiometric correction and absolute radiometric correction, but the data of HSI are Fourier transformed and must be processed through a series of special processes which include data preprocessing, FFT, absolute corrections, and image combination (Zhao et al. 2010). The HSI Fourier transformed data can be reconstructed successfully by GDPS in a timely way every day. The primary mission of HJ-1A is to validate new instrument technologies in flight and to provide remotely sensed data to the user community for environment and disaster monitoring, including fields of water quality, aerosol pollution, flood, earthquake, etc.

3.3.6 Hyperion (Hyperspectral Imager, EO-1 Satellite)

The first satellite hyperspectral sensor, Hyperion, is one of three primary Earth Observing-1 (EO-1) (see http://eo1.gsfc.nasa.gov) instruments: Advanced Land Imager (ALI), the Hyperion hyperspectral imager, and the Linear etalon imaging spectrometer array Atmospheric Corrector (LAC). Among the three sensors, Hyperion and LAC are hyperspectral sensors. EO-1 is the first satellite in NASA's New Millennium Program (NMP) Earth Observing series. The EO missions will develop and validate instruments and technologies for space-based Earth observations with unique spatial, spectral, and temporal characteristics not previously available. EO-1 was launched on November 21, 2000. EO-1 is flying in a 705 km circular, sun-synchronous orbit at a 98.7° inclination. This orbit allows EO-1 to match within one minute the Landsat 7 orbit and collect identical images for later comparison on the ground. Once or twice a day, sometimes more, both Landsat 7 and EO-1 will image the same ground areas (scenes). The Hyperion instrument provides a new class of Earth observation data for improving Earth surface characterization. It provides a science-grade instrument with quality calibration based on heritage from the LEWIS Hyperspectral Imaging Instrument (HSI) and a high-resolution hyperspectral imager capable of resolving 220 spectral bands (from 0.4 to 2.5 μm) with a 30-meter spatial resolution. The instrument can image a 7.5 km by 100 km land area per image and can provide detailed spectral mapping across all 220 bands with high radiometric accuracy.

The Hyperion acquires each frame of imagery with a pushbroom fashion by capturing the spectrum of a line 30 m align in the along-track direction by 7.5 km in the crosstrack direction. The Hyperion has a single telescope and two spectrometers, one visible/near infrared (VNIR) spectrometer and one short-wave infrared (SWIR) spectrometer. The Hyperion instrument consists of three physical units (Folkman et al. 2001): (1) the Hyperion Sensor Assembly (HSA), (2) the Hyperion Electronics Assembly (HEA), and (3) Cryocooler Electronics Assembly (CEA). The HSA includes the optical systems, cryocooler, in-flight calibration system, and the high-speed focal plane electronics. The HEA contains the interface and control electronics for the instrument while the CEA controls cryocooler operation. These units are placed on the deck of the spacecraft with the viewing direction along the major axes of the spacecraft. A dichroic filter in the system reflects solar light from 400–1000 nm to one spectrometer (VNIR) and that from 900–2500 nm to the other spectrometer (SWIR). The VNIR and SWIR spectrum overlap from 900–1000 nm will allow cross calibration between the two spectrometers. The VNIR spectrometer has an array of 60 mm pixels created by aggregating 3 × 3 sub-arrays of a 20 mm CCD detector array. It uses a 70 (spectral) by 256 (spatial) pixel array, which provides a 10 nm spectral bandwidth over a range of 400–1000 nm. The SWIR spectrometer has 60 mm, HgCdTe detectors in an array of 172 spectral channels by 256 spatial pixels similar to the VNIR. The SWIR also has a spectral bandwidth of 10 nm. The HgCdTe detectors in the SWIR spectrometer are cooled by an advanced TRW cryocooler and maintained at 110 K during data collection (Hyperion Summary 2015). A common on-board calibration system is provided for the two spectrometers. Solar and in-flight calibration data will be used as the primary sources for monitoring radiometric stability, with ground site (vicarious) and lunar imaging treated as secondary calibration data. The hyperspectral imaging data acquired with Hyperion have wide-ranging applications in mining, geology, vegetation, forestry, agriculture, and environmental management.

3.3.7 HySI (HyperSpectral Imager, IMS-1 Satellite)

The HySI onboard IMS-1(Indian Microsatellite 1) is a low-cost microsatellite imaging mission of ISRO (Indian Space Research Organization). The HySI operates in the visible-near infrared (VNIR) spectra range from 450 nm to 950 nm with a total of 64 spectral bands at a spectral resolution of 8 nm. Spectral separation is realized using the wedge filter technique. The Sun's reflected light from Earth's surface is being collected through a telecentric refractive optics system and is being focused onto an Active Pixel Sensor (APS) area detector (HySI 2015). The area detector images 260 km in along-track and 128 km of cross-track area on the ground in an integration time of 78.45 ms. The APS area detector has 256 × 512 pixels of 50 μm pixel size; 256 elements are in the cross-track direction and 512 elements are in the along-track direction. The overall objective for developing HySI is to provide free medium-resolution imagery for developing countries. The HySI data may be used for resource characterization and detailed studies.

3.3.8 Medium-Resolution Imaging Spectrometer (MERIS), ESA ENVISAT Satellite

MERIS was launched by the European Space Agency (ESA) onboard its polar orbiting ENVISAT Earth Observation Satellite in March 2002. MERIS is an FOV of 68.5° pushbroom imaging spectrometer that measures the solar radiation reflected by the Earth in the VNIR part of the spectrum during daytime, at a ground spatial resolution of 300 m, in 15 spectral bands, programmable in width and position, and in the spectral range of 390 nm to 1040 nm. MERIS allows global coverage of the Earth in three days. The swath of 1150 km is divided into five segments covered by five identical cameras that have corresponding fields of view with a slight overlap between adjacent cameras. Each camera images an along-track stripe of the Earth's surface onto the entrance slit of an imaging optical grating spectrometer. This entrance slit is imaged through the spectrometer onto a 2D CCD array, thus providing spatial and spectral information simultaneously. The spatial information along-track is determined by the pushbroom principle via successive readouts of the CCD-array. The scene is imaged simultaneously across the entire spectral range, through a dispersing system, and onto the CCD array. Signals read out from the CCD pass through several processing steps in order to achieve the required image quality. These CCD processing tasks include dumping of spectral information from unwanted bands, and spectral integration to obtain the required bandwidth. Onboard analog electronics perform pre-amplification of the signal and correlated double sampling and gain adjustment before digitization (MERIS 2015). Full spatial resolution data (i.e., 300 m at nadir) are transmitted over coastal zones and land surfaces. Reduced spatial resolution data, achieved by an onboard combination of 4 × 4 adjacent pixels across-track and along-track resulting in a resolution of approximately 1200 m at nadir, are generated continuously (MERIS 2015).

The primary mission of MERIS is the measurement of sea color in the oceans and in coastal areas. Measurements of sea color can be converted into a measurement of chlorophyll pigment concentration, suspended sediment concentration, and of aerosol loads over the marine domain. The secondary mission of MERIS is also the capable acquiring of atmospheric parameters associated with clouds, water vapor, and aerosols, in addition to land surface parameters and, in particular, vegetation processes.

3.3.9 Moderate-Resolution Imaging Spectroradiometer (MODIS), Terra/Aqua Satellites

Of the Mission to Planet Earth (MTPE) programs (Wharton and Myers 1997), the Earth Observation System (EOS) program is regarded as the principal element, providing systematic and continuous observation from low Earth orbit for a minimum time period of 15 years. MODIS is NASA's flagship sensor system of the sensors on board Terra (EOS AM) and Aqua (EOS PM) satellites. Terra's orbit around the Earth is timed so that it passes from north to south across the equator in the morning, while Aqua passes south to north over the equator in the afternoon. Terra MODIS and Aqua MODIS are viewing the entire Earth's surface, and MODIS can be regarded as a hyperspectral sensor with 36 bands covering visible, near infrared, shortwave infrared, and thermal infrared spectral ranges (see MODIS groups of wavelength at http://modis.gsfc.nasa.gov). EOS/Terra was successfully launched on December 18, 1999, while the EOS/Aqua was launched on May 4, 2002.

The MODIS sensor system provides high radiometric sensitivity (12 bit) in 36 spectral bands ranging in wavelength from 0.4 μm–14.4 μm. Among the 36 MODIS spectral bands, two bands are imaged at a nominal resolution of 250 m at nadir, with five bands at 500 m and the remaining 29 bands at 1000 m. A ±55° scanning pattern at the EOS orbit of 705 km achieves a 2330 km swath and provides global coverage every one to two days. The Scan Mirror Assembly uses a continuously rotating double-sided scan mirror to scan ±55 ° driven by a motor encoder built to operate at 100 percent duty cycle throughout the six-year instrument design life. The optical system consists of a two-mirror off-axis and a focal telescope which directs energy to four refractive objective assemblies; one for each of the VIS, NIR, SWIR/MWIR, and LWIR spectral regions covering a total spectral range of 0.4–14.4 μm. A high-performance passive radiative cooler provides cooling to 83 K for the 20 IR spectral bands on two HgCdTe Focal Plane Assemblies (FPAs). Novel photodiode-silicon readout technology for the VNIR provides unsurpassed quantum efficiency and low-noise readout with exceptional dynamic range. MODIS system consists of three dedicated electronics modules: The Space-Viewing Analog Module (SAM), the Forward-Viewing Analog Module (FAM), and the electronics module (MEM). The third MEM module provides power, control systems, command and telemetry, and calibration electronics (see http://modis.gsfc.nasa.gov).

The MODIS systems offer unprecedented image data at terrestrial, atmospheric, and ocean phenomenology for a wide and diverse community of users throughout the world. These data will improve our understanding of global dynamics and processes occurring on the land, in the oceans, and in the lower atmosphere. The MODIS sensor plays a vital role in the development of validated, global, interactive Earth system models able to predict global change accurately enough to assist policymakers in making sound decisions concerning the protection of our environment (see http://modis.gsfc.nasa.gov).

3.3.10 Environmental Mapping and Analysis Program (EnMAP)

EnMAP is a German hyperspectral satellite mission, scheduled for launch in 2017, which aims at monitoring and characterizing the Earth's environment at a global scale. EnMAP will operate in an imaging pushbroom principle and will monitor the Earth's surface with a ground sampling distance (GSD) of 30 m × 30 m (30 km × 5000 km per day) measured in the 420–2450 nm spectral range by means of two separate spectrometers covering the visible to near-infrared (VNIR) and short-wave infrared (SWIR) spectral regions with 244 contiguous bands. The mean spectral sampling distance and resolution is of 6.5 nm at the VNIR, and of 10 nm at the SWIR. Accurate radiometric and spectral responses are guaranteed by a defined SNR of ≥400:1 in the VNIR and ≥170:1 in the SWIR, a radiometric calibration accuracy better than 5 percent and a spectral calibration uncertainty of 0.5 nm in the VNIR and 1 nm in the SWIR. An off-nadir pointing capability of up to 30° enables a target revisit time of 4 days (EnMAP 2015). The primary goal of EnMAP is to offer accurate, diagnostic information on the state and evolution of terrestrial ecosystems on a timely and frequent basis, and to allow for a detailed analysis of surface parameters with regard to the characterization of vegetation canopies, rock/soil targets and coastal waters at a global scale. EnMAP is designed to record biophysical, biochemical, and geochemical variables to increase our understanding of biospheric/geospheric processes and to ensure the sustainability of our resources (Kaufmann et al. 2008).

3.3.11 Fluorescence Explorer (FLEX)

The FLEX mission proposes to launch a satellite for the global monitoring of steady-state chlorophyll fluorescence in terrestrial vegetation (Rascher et al. 2008). Fluorescence is a highly specific signal of vegetation function, stress, and vitality. Solar-induced fluorescence, a very weak signal, is detectable using the Fraunhofer lines of the solar spectrum, allowing the observation from a satellite. The FLEX mission, selected by ESA, will operate in a three-instrument array for measurement of the interrelated features of fluorescence, hyperspectral reflectance, and canopy temperature. The FLEX scientific payload, from a sun-synchronous low Earth orbit (LEO), will measure the fluorescence of the H_α line (656.3 nm) and at least one other Fraunhofer line in the blue-UV region with a CCD matrix type imaging spectrometer (Stoll et al. 1999). A thermal infrared imaging radiometer, optional although highly desirable, will measure the vegetation temperature. An additional CCD camera will provide cloud detection and scene identification. The spatial resolution will be better than 0.5×0.5 km^2, with a nominal FOV of 8.4°; a steering mirror will allow plus or minus 4° across track depointing, while allowing, if necessary, freezing of the image to increase the SNR (Stoll et al. 1999). Processing of the data will require atmospheric characteristics (aerosols) from other missions. Interpretation of the fluorescence signal will require the following:

Reflectance data over the 400–800 nm region.
Biome characteristics (LAI, architecture, biomass density factors, APAR, etc.) from space mission providing high spectral resolution.
Directional reflectance measurements in the visible domain.
Ground data on environmental factors and plant physiology.
In situ fluorescence measurements for satellite signal validation.

The FLEX mission is to provide global coverage of chlorophyll fluorescence of vegetation canopies and new data in the field of Earth observation, addressing big-scale screening of terrestrial vegetation in relation to agricultural, forestry, and global change issues, such as solar irradiance, ozone depletion, water availability, air temperature, and pollution in air/water/soils.

3.3.12 Hyperspectral Imager Suite (HISUI)

HISUI is under development in Japan and is planned to be launched onboard ALOS-3 in 2018 or later (Kashimura et al. 2013). The HISUI mission consists of hyperspectral and multispectral imagers that will be operated independently or simultaneously. The hyperspectral imager has 30 m spatial resolution with a swath width of 30 km covering the spectral range of visible to shortwave-infrared region (0.4–2.5 μm). The multispectral imager has a 5 m spatial resolution and 90 km swath width. In order to fill the gap of the swaths of two imagers, the hyperspectral imager has a crosstrack pointing mechanism up to ±2.75° (Kashimura et al. 2013). To satisfy high SNR ratio, the diameter of the telescope is designed to 30 cm for the ground sampling distance of 30 m. The ground footprint is projected to the slits with a gap of 30 mm. The designed SNR ratio is sufficiently high in order to be at 450 and 300 for visible and near infrared (VNIR) and shortwave infrared (SWIR), respectively. The light entering the slits is introduced to two spectrometers, one for VNIR radiometer and the other for SWIR radiometer. The two spectrometers adopt a reflective grating system. Since the smile and keystone phenomena distort the spectrogram, which is difficult to correct by data processing on the ground, a fine spectrogram is obtained at the interval of 2.5 nm for VNIR and 6.75 nm for SWIR that are decimated on board (Iwasaki et al. 2011). The optical components are arranged to minimize stray light caused by reflection at the elements. By coupling with the multi- and hyperspectral radiometers, the HISUI system provides users with data at both the high spatial and spectral resolutions, which are useful for precise landcover management, such as a classification of ground target and change detection based on spectral response. The HISUI dataset will be applied in monitoring global energy and resources, environment, agriculture, and forestry.

3.3.13 Hyperspectral Infrared Imager (HyspIRI)

The HyspIRI mission, being developed by JPL/NASA, will study the world's ecosystems and provide critical information on natural disasters such as volcanoes, wildfires, and drought (HyspIRI 2015). The mission was recommended in the National Research Council Decadal Survey requested by NASA, NOAA, and USGS (NRC's Decadal Survey Report 2007). The HyspIRI system includes two instruments mounted on a satellite in low Earth orbit. An imaging spectrometer measures from visible to short-wave infrared (VSWIR: 380 nm to 2500 nm) in 10 nm contiguous bands while a multispectral imager measures from 3 to 12 μm in the mid and thermal infrared (TIR). Both instruments (VSWIR and TIR spectral ranges) have a spatial resolution of 60 m at nadir. Both instruments have a different revisit day: 19 and 5 days for the VSWIR and the TIR instruments, respectively. The HyspIRI mission has its roots in the imaging spectrometer Hyperion on EO-1 and ASTER with the multispectral thermal IR instrument flown on EOS/Terra. The mission will provide a benchmark on the state of the world's ecosystems against which future changes can be assessed. The mission will also assess the pre-eruptive behavior of volcanoes and the likelihood of future eruptions as well as carbon and other gases released from wildfires. The data from HyspIRI will be used for a wide variety of studies, primarily in the areas of carbon cycle, ecosystem, Earth's surface, and interior focus (HyspIRI 2015).

3.3.14 Multisensor Microsatellite Imager (MSMI)

The MSMI, developed by SunSpace in Stellenbosch, South Africa, is an imager suitable for microsatellites, with multi/hyperspectral and video detectors on the same focal plane of a single telescope that could be used to capture variations in surface features at different scales. The multispectral data acquired by the MSMI are with a resolution of around 5 m and swath width about 25 km and three bands (red, NIR, and a water band of 950–970 nm). The hyperspectral array with the spectral range from 400 nm to 2350 nm (see Table 3.9) will provide data for detecting crop health, nutrition, water stress, etc. The MSMI has standardized and modularized components and communication systems, which allow the reuse of sensors, processors, and mass memory on imagers for other small satellites produced by SunSpace in association with Stellenbosch University (Schoonwinkel et al. 2005). The MSMI imager has substantial advantages over the existing sensors, including the sensor's pointing and high temporal revisit abilities (Mutanga et al. 2009). The pointing ability of the platform allows sensors to acquire imagery of the same target area at different viewing angles which, in turn, allow for the assessment of bidirectional reflectance distribution function (BRDF) effects. The multiangular viewing capabilities of the MSMI enable capturing of such off-nadir variations. The high temporal resolution of microsatellites' sensors when compared with the current spaceborne sensors provides high multi-temporal data that can facilitate constant monitoring of dynamic variables, such as crop growth and climate change effects on vegetation. The MSMI mission makes the possibility of affordable satellite constellations that can deliver remote sensing data to commercial and science users more frequently in the future, including food security applications, invasive species mapping, mineral mapping, etc. Although the MSMI sensor is built and ready for launch, the launch situation is not clear because the mission has been descoped (Staenz and Held 2013).

3.3.15 Hyperspectral Precursor and Application Mission (PRISMA)

PRISMA (PRecursore IperSpettrale della Missione Applicativa) is a medium-resolution hyperspectral imaging mission of the Italian Space Agency (ASI) with a projected launch in 2018 (Lopinto and Ananasso 2013). The PRISMA project is conceived as a pre-operational and technology demonstrator mission, focused on the development and delivery of hyperspectral products and the qualification of the hyperspectral payload in space (PRISMA 2015). PRISMA (the name of the mission and the name of the sensor are identical) is an advanced hyperspectral instrument including a panchromatic (Pan) camera at a medium resolution. The design is based on a pushbroom-type observation concept providing hyperspectral imagery (~250 bands) at a spatial resolution of 30 m on a swath of 30 km. The spectral resolution is better than 12 nm in a spectral range of 400–2500 nm (VNIR and SWIR regions). In parallel, Pan image is provided at a spatial resolution of 5 m; the Pan data are co-registered with the hyperspectral data to permit testing of image fusion techniques (PRISMA 2015). The overall instrument spectral radiation (400–2505 nm) is split into two regions (VNIR and SWIR), adopting two different focal plane arrays. The VNIR covers the spectral range (400–1010 nm) with 66 spectral bands, while the SWIR spectral range is from 920 nm to 2505 nm with 171 bands. Due to the adopted optical design, the spectral dispersion is not constant with wavelength. The Pan detector is a linear array with 6000 pixels and a pitch of 6.5 μm with a spatial resolution of 5 m. The overall objective of the PRISMA mission is to provide a global observation capability and the specific areas of interest to be covered include Europe and the Mediterranean region. The images acquired by the PRISMA will be used for detecting land degradation and vegetation status, product development for agricultural areas, and management and monitoring of natural and induced hazards (Lopinto and Ananasso 2013).

3.4 SUMMARY

In this chapter, the working principles, performance, and technical characteristics of current and future airborne and spaceborne hyperspectral sensors, systems, and missions were briefly introduced. Based on the mechanism of scanning and data acquisition, two types of commonly used scanning principles of imaging spectrometry—whiskbroom imaging spectrometry and pushbroom imaging spectrometry—were first described. Then, in Section 3.2, a survey of selected important airborne hyperspectral sensors and systems was summarized in terms of basic technical parameters, characteristics, date of development, and application areas. In such a survey, a total of 23 selected airborne hyperspectral sensors and systems were summarized. These included

- Advanced Airborne Hyperspectral Imaging Sensor (AAHIS)
- Airborne Imaging Spectrometer (AIS)
- Airborne Imaging Spectrometer for Different Applications (AISA)
- Advanced Solid-State Array Spectroradiometer (ASAS)
- Airborne Visible/Infrared Imaging Spectrometer (AVIRIS)
- Compact Airborne Spectrographic Imager (CASI)
- Compact High-Resolution Imaging Spectrograph Sensor (CHRISS)
- Digital Airborne Imaging Spectrometers (DAIS 7915, 16115)
- Fluorescence Line Imager (FLI)
- Hyperspectral Digital Imagery Collection Experiment (HYDICE)
- Hyperspectral Mapper (HyMap)
- Hyperspectral cameras (HySpex)
- Infrared Imaging Spectrometer (ISM)
- Modular Airborne Imaging Spectrometer (MAIS)
- Modular Imaging Spectrometer Instrument (MISI)
- MultiSpectral Infrared Camera (MUSIC)
- PROBE-1
- Reflective Optics System Imaging Spectrometer (ROSIS)
- SWIR Full Spectrographic Imager (SFSI)
- Spatially Modulated Imaging Fourier Transform Spectrometer (SMIFTS)
- TRW Imaging Spectrometers (TRWIS)

- Variable Interference Filter Imaging Spectrometer (VIFIS)
- Wedge Imaging Spectrometer (WIS)

In Section 3.3, a survey of selected important spaceborne hyperspectral sensors and missions was summarized in terms of their launch year, actual or possible platform/satellite and orbit altitude, operational spectral range and spatial/spectral resolution, swath width and IFOV, features of sensors and missions. The nine current/operational spaceborne hyperspectral sensors and missions include the following:

- Advanced Responsive Tactically Effective Military Imaging Spectrometer (ARTEMIS), TacSat-3 satellite
- Compact High Resolution Imaging Spectrometer (CHRIS), PROBA satellite
- Fourier Transform Hyperspectral Imager (FTHSI), MightySat II satellite
- Global Imager (GLI), NASDA ADEOS-II satellite
- HJ-A/HSI (Hyperspectral Imager, HJ-1A satellite)
- Hyperion (Hyperspectral Imager, EO-1 satellite)
- HySI (HyperSpectral Imager, IMS-1 satellite)
- Medium Resolution Imaging Spectrometer (MERIS), ESA ENVISAT satellite
- Moderate Resolution Imaging Spectroradiometer (MODIS), Terra/Aqua satellites

Next, six future/planned spaceborne hyperspectral sensors and missions were summarized. These include the following:

- Environmental Mapping and Analysis Program (EnMAP)
- Fluorescence Explorer (FLEX)
- Hyperspectral Imager Suite (HISUI)
- Hyperspectral Infrared Imager (HyspIRI)
- Multisensor Microsatellite Imager (MSMI)
- Hyperspectral Precursor and Application Mission (PRISMA)

REFERENCES

Anger, C. D., S. K. Babey, and R. J. Adamson. 1990. A new approach to imaging spectroscopy. In *Proceedings of SPIE: Imaging Spectroscopy of the Terrestrial Environment* 1298:72–86.

ARTEMIS. 2015. ARTEMIS, Earth Observation Portal. Accessed March 19, 2015, from https://directory.eoportal.org/web/eoportal/satellite-missions/t/tacsat-3.

Ben-Dor, E., T. Malthus, A. Plaza, and D. Schläpfer. 2013. Hyperspectral remote sensing, in *Airborne Measurements for Environmental Research: Methods and Instruments*, eds. M. Wendisch, J.-L. Brenguier, pp. 413–456. Weinheim, Germany: Wiley-VCH.

Bucher, T., and F. Lehmann. 2000. Fusion of HyMap hyperspectral with HRSC-A multispectral and DEM data. *Proceedings of Geoscience and Remote Sensing Symposium 2000 (IGARSS 2000)* 7:3234–3236.

Carrère, V., D. Oertel, J. Verdebout, G. Maracci, G. Schmuck, and A. J. Sieber. 1995. Optical component of the European airborne remote sensing capabilities (EARSEC). *Proceedings of SPIE: Imaging Spectrometry* 2480:186–194.

CHRIS-PROBE. 2015. CHRIS, ESA Earth Online. Accessed March 15, 2015, from https://earth.esa.int/web/guest/missions/esa-operational-eo-missions/proba/instruments/chris.

Demro, J. C., R. Hartshorne, L. M. Woody, P. A. Levine, and J. R. Tower. 1995. Design of a multispectral, wedge filter, remote-sensing instrument incorporating a multiport, thinned, CCD area array. *Proceedings of SPIE: Imaging Spectrometry* 2480:280–286.

Elvidge, C. D., and F. P. Portigal. 1990. Change detection in vegetation using 1989 AVIRIS data. *Proceedings of SPIE: Imaging Spectroscopy of the Terrestrial Environment* 1298:178–189.

EnMAP. 2015. EnMAP hyperspectrak sensor, Earth Observation Center. Accessed March 19, 2015, from http://www.enmap.org.

Folkman, M. A., J. Pearlman, L. B. Liao, and P. J. Jarecke. 2001. EO-1/Hyperion hyperspectral imager design, development, characterization, and calibration. *Proceedings of SPIE: Hyperspectral Remote Sensing of the Land and Atmosphere* 4151:40–51.

Freeman, L. J., C. C. Rudder, and P. Thomas. 2000. MightySat II: On-orbit lab bench for Air Force Research Laboratory. *Proceedings of the 14th Annual AIAA/USU Conference on Small Satellites*. Accessed March 17, 2015, from http://digitalcommons.usu.edu/smallsat/2000.

Gao, H. L., X. F. Gu, T. Yu, H. Gong, J. Li, and X. Li. 2010. HJ-1A HSI on-orbit radiometric calibration and validation research. *Science China Technological Sciences* 53:3119–3128.

Gao, B. C., and C. O. Davis. 1998. Examples of using imaging spectrometry for remote sensing of the atmosphere, land, and ocean. *Proceedings of SPIE: Hyperspectral Remote Sensing and Applications* 3502:234–242.

GLI. 2015. GLI, eoPortal News. Accessed March 18, 2015, from https://directory.eoportal.org/web/eoportal/satellite-missions/a/adeos-ii.

Goetz, A. 1992. Principles of narrow band spectrometry in the visible and IR: Instruments and data analysis, in *Imaging Spectroscopy: Fundamentals and Prospective Applications*, eds, F. Toselli and J. Bodechtel, pp. 21–32. London: Kluwer Academic Publishers.

Goetz, A. F. H. 1995. Imaging spectrometry for remote sensing: Vision to reality in 15 years. *Proceedings of SPIE: Imaging Spectrometry* 2480:2–13.

Green, R. O., M. L. Eastwood, C. M. Sarture, T. G. Chrien, M. Aronsson, B. J. Chippendale, J. A. Faust, B. E. Pavri, C. J. Chovit, M. Solis, M. R. Olah, and O. Williams. 1998. Imaging spectroscopy and the airborne visible/infrared imaging spectrometer (AVIRIS). *Remote Sensing of Environment* 65:227–248.

Gupta, R. P. 2003. *Remote Sensing Geology*. Berlin and Heidelberg: Springer-Verlag.

Huegel, F. G. 1988. Advanced solid state array spectroradiometer: Sensor and calibration improvements. *Proceedings of SPIE: Imaging Spectroscopy II*, 834:12–21.

Hyperion Summary. 2015. Accessed March 15, 2015, from http://eo1.gsfc.nasa.gov/new/baseline/techVal/readMoreInstTech.html.

HySI. 2015. HySI, eoPortal News. Accessed March 19, 2015, from https://directory.eoportal.org/web/eoportal/satellite-missions/i/ims-1.

HyspIRI. 2015. HyspIRI Mission Study. Accessed March 20, 2015, from http://hyspiri.jpl.nasa.gov.

Irons, J. R., P. W. Dabney, J. Paddon, R. R. Irish, and C. A. Russell. 1990. Advanced solid-state array spectroradiometer support of 1989 field experiments. *Proceedings of SPIE: Imaging Spectroscopy of the Terrestrial Environment* 1298:2–10.

Iwasaki, A., N. Ohgi, J. Tanii, T. Kawashima, and H. Inada. 2011. Hyperspectral imager suite (HISUI): Japanese hyper-multi spectral radiometer. *IGARSS 2011* 1025–1028.

Kashimura, O., K. Hirose, T. Tachikawa, and J. Tanii. 2013. Hyperspectral space-borne sensor hisui and its data application. Accessed March 21, 2015, from http://www.a-a-r-s.org/acrs/administrator/components/com_jresearch/files/publications/SC01-0579_Full_Paper_ACRS2013_Osamu_Kashimura.pdf.

Kaufmann, H., K. Segl, L. Guanter, S. Hofer, K.-P. Foerster, T. Stuffler, A. Mueller, R. Richter, H. Bach, P. Hostert, and C. Chlebek. 2008 (July 7–11). Environmental Mapping and Analysis Program (EnMAP)—Recent advances and status. *Geoscience and Remote Sensing Symposium 2008 (IGARSS 2008)*, Boston, Massachusetts, IV:109–112.

Kramer, H. J. 2002. *Observation of the Earth and Its Environment: Survey of Missions and Sensors*, 4th ed. Berlin: Springer-Verlag.

Kunkel, B., F. Blechinger, D. Viehmann, H. Van Der Piepen, and R. Doerffer. 1991. ROSIS imaging spectrometer and its potential for ocean parameter measurements (airborne and space-borne). *International Journal of Remote Sensing* 12(4):753–761.

Lopinto, E., and C. Ananasso. 2013. *EARSeL*, eds. R. Lasaponara, N. Masini, and M. Biscione, pp. 135–146.

Lucas, R., A. Rowlands, O. Niemann, and R. Merton. 2004. Hyperspectral sensor and applications, in *Advanced Image Processing Techniques for Remotely Sensed Hyperspectral Data*, eds. P. K. Varshney and M. K. Arora, pp. 11–45. Berlin and Heidelberg: Springer-Verlag.

Melack, J. M., and S. H. Pilorz. 1990. Radiative-transfer-based retrieval of reflectance from calibration radiance imagery measured by an imaging spectrometer for lithological mapping of the Clark Mountains, California. *Proceedings of SPIE: Imaging Spectroscopy of the Terrestrial Environment* 1298:202–212.

MERIS. 2015. MERIS, ESA Earth Online. Accessed March 16, 2015, from https://earth.esa.int/web/guest/missions/esa-operational-eo-missions/envisat/instruments.

Miura, T., and H. Yoshioka. 2012. Hyperspectral data in long-term, cross-sensor continuity studies, in *Hyperspectral Remote Sensing of Vegetation*, eds. P. S. Thenkabail, J. G. Lyon, and A. Huete, pp. 611–633. Boca Raton, FL: CRC Press/Taylor & Francis Publishing Group.

Mutanga, O., J. van Aardt, and L. Kumar. 2009. Imaging spectroscopy (hyperspectral remote sensing) in southern Africa: An overview. *South African Journal of Science* 105:193–198.

NRC's Decadal Survey Report. 2007. *Earth Science and Applications from Space: National Imperatives for the Next Decade and Beyond*. Accessed March 26, 2015, from http://www.nap.edu/catalog/11820.html.

Ortenberg, F. 2012. Hyperspectral sensor characteristics: Airborne, spaceborne, hand-held, and truck-mounted; intergration of hyperspectral data with LIDAR, in *Hyperspectral Remote Sensing of Vegetation*, eds. P. S. Thenkabail, J. G. Lyon, and A. Huete, pp. 39–68. Boca Raton, FL: CRC Press/Taylor & Francis Publishing Group.

Peterson, D. L., J. D. Aber, D. A. Matson, D. H. Card, N. Swanberg, C. Wessman, and M. Spanner. 1988. Remote sensing of forest canopy and leaf biochemical contents. *Remote Sensing of Environment* 24:85–108.

PRISMA. 2015. PRISMA, Earth Observation Portal. Accessed March 20, 2015, from https://directory.eoportal.org/web/eoportal/satellite-missions/p/prisma-hyperspectral.

Pu, R., and P. Gong. 2000. *Hyperspectral Remote Sensing and Its Applications*. Beijing, China: Higher Education Press of China.

Qi, J., Y. Inoue, and N. Wiangwang. 2012. Hyperspectral remote sensing in global change studies, in *Hyperspectral Remote Sensing of Vegetation*, eds. P. S. Thenkabail, J. G. Lyon, and A. Huete, pp. 69–89. Boca Raton, FL: CRC Press/Taylor & Francis Publishing Group.

Rascher, U., B. Gioli, and F. Miglietta. 2008. FLEX–Fluorescence Explorer: A remote sensing approach to quantify spatio-temporal variations of photosynthetic efficiency from space, in *Photosynthesis. Energy from the Sun: 14th International Congress on Photosynthesis*, eds. J.F. Allen, E. Gantt, J.H. Golbeck, and B. Osmond (eds.), 1387–1390. Copyright Springer, 2008.

Rowlands, N., and R. A. Neville. 1996 (November 13). Calcite and dolomite discrimination using airborne SWIR imaging spectrometer data. *Proceedings of SPIE 2819, Imaging Spectrometry II*, 36, doi:10.1117/12.258085.

Sandor-Leahy, S. R., D. Beiso, M. A. Figueroa, M. A. Folkman, D. A. Gleichauf, T. R. Hedman, P. J. Jarecke, and S. Thordarson. 1998. TRWIS III hyperspectral imager: Instrument performance and remote sensing applications. *Proceedings of SPIE: Imaging Spectrometry IV* 3438:13–22.

Schaepman, M. E., M. Kneubuehler, E. H. Meier, A. Muller, P. Strobl, R. Reulke, and R. Horn. 1998. Fusion of hyperspectral (DAIS 7915), wide-angle (WAAC), and SAR (E-SAR) data acquistion methods: The multi Swiss'97 campaign. *Proceedings of SPIE: Imaging Spectrometry IV* 3438:84–95.

Schoonwinkel, A., H. Burger, and S. Mostert. 2005. Integrated hyperspectral, multispectral and video imager for microsatellites. *The 19th Annual AIAA/USU Conference on Small Satellites*. Accessed March 21, 2015, from http://digitalcommons.usu.edu/smallsat/2005.

Shen, S. S. 1996. Relative utility of HYDICE and multispectral data for object detection, identification, and abundance estimation. *Proceedings of SPIE: Hyperspectral Remote Sensing and Applications* 2821:256–267.

Staenz, K. 2009 (March 16 –18). Terrestrial imaging spectroscopy: Some future perspectives. *Proceedings of the 6th EARSEL SIG IS*. Tel Aviv, Israel.

Staenz, K., and A. Held. 2013. Summary of current and future terrestrial civilian hyperspectral spaceborne systems. Accessed March 19, 2015, from http://www.grss-ieee.org/wp-content/uploads/2013/07/Staenz_Held_2607.pdf.

Stoll, M.-P., A. Court, K. Smorenburg, H. Visser, L. Crocco, J. Heilimo, and A. Honig. 1999. FLEX—Fluorescence Explorer. In *Part of the EUROPTO Conference on Remote Sensing for Earth Science Applications*, Florence, Italy, *SPIE*, vol. 3868.

Tong, Q., Y. Xue, and L. Zhang. 2014. Progress in hyperspectral remote sensing science and technology in China over the past three decades. *Journal of Selected Topics in Applied Earth Observations and Remote Sensing* 7(1):70–91.

Vane, G., and A. F. H. Goetz. 1988. Terrestrial imaging spectroscopy. *Remote Sensing of Environment* 24:1–29.

Vincent, R. K. 1997. *Fundamentals of Geological and Environmental Remote Sensing*. Prentice Hall, Upper Saddle River, NJ, 400p.

Wang, J., and Y. Xue. 1998. Airborne imaging spectrometers developed in China. *Proceedings of SPIE: Hyperspectral Remote Sensing and Applications* 3502:12–22.

Wessman, C. A., J. D. Aber, D. L. Peterson, and J. M. Melillo. 1988. Remote sensing of canopy chemistry and nitrogen cycling in temperate forest ecosystems. *Nature* 335:154–156.

Wharton, S. W., and M. F. Myers. 1997. *MTPE EOS Data Products Handbook* 1:1–266.

Yarbrough, S., T. R. Caudill, E. T. Kouba, V. Osweiler, J. Arnold, R. Quarles, J. Russell, L. J. Otten III, B. A. Jones, A. Edwards, J. Lane, A. D. Meigs, R. B. Lockwood, and P. S. Armstrong. 2002. MightySat II.1 hyperspectral imager: Summary of on-orbit performance. *Proceedings of SPIE 4480, Imaging Spectrometry VII* 186–197. doi: 10.1117/12.453339.

Zhao, X., Z. Xiao, Q. Kang, Q. Li, and L. Fang. 2010 (July 25–30). Overview of the Fourier transform hyperspectral imager (HSI) boarded on the HJ-1 Satellite. *Proceedings of IGARSS (IEEE International Geoscience and Remote Sensing Symposium) 2010*, Honolulu, Hawaii.

4 Hyperspectral Image Radiometric Correction

Usually, radiance recorded at a sensor is not fully representative of Earth's surface features; it suffers from radiometric errors caused by remote sensors/systems themselves and atmospheric effects and terrain surface effects, as well. In order to accurately and reliably make use of remotely sensed data, especially hyperspectral imaging data, it is necessary to conduct imaging radiometric correction, especially correcting various atmospheric effects to retrieve surface reflectance before applying the imaging data for various research and application purposes. Therefore, in this chapter, the reasons why it is necessary to perform hyperspectral imaging data correction (Section 4.1); various atmospheric effects (Section 4.2); some sensors/systems induced radiometric errors and correction (Section 4.3); and methods, principles, and algorithms of atmospheric correction (Section 4.4) are concisely introduced and discussed. In addition, several techniques and methods for estimating atmospheric total column water vapor amounts and retrieving aerosol optical thickness or loading are also reviewed and discussed.

4.1 INTRODUCTION

In general, the radiance recorded at a sensor is not fully representative of Earth's surface features within the sensor's field of view. This is because some radiometric errors are introduced in remotely sensed data by the sensor system itself and the radiance recorded in remote sensing imagery is usually altered by the atmosphere through which electromagnetic energy has to pass (Teillet 1986, Moses and Philpot 2012). Several of the radiometric errors introduced by multi-/hyperspectral sensors include line or column striping, spatial and spectral misregistration across spatial axis and spectral axis in pushbroom imaging systems, etc. The solar radiation on the Sun-surface–sensor ray path is subject to molecular and aerosol scattering and absorption by gases (Liang 2004, Gao et al. 2006, 2009, see Figure 4.1). The atmospheric scattering (e.g., Rayleigh and aerosol) with the shorter wavelength region below 1.00 μm modifies the radiances of adjacent fields of different reflectance (adjacency effect). Therefore, dark areas surrounded by bright areas appear to have a higher reflectance than the intrinsic reflectance (Figure 4.1a). In Figure 4.1a, the thick curve represents atmospheric path radiance caused by atmospheric molecular and aerosol scattering. The absorptions can be caused by water vapor absorption bands centered at approximately 0.94, 1.14, 1.38, and 1.88 μm, an oxygen absorption band at 0.76 μm, and a carbon dioxide absorption band near 2.08 μm (Figure 4.1b). Sometimes image preprocessing can recover the miscalibrated spectral information and make it relatively compatible with the correctly acquired data in the scene. To remove and/or suppress the atmospheric effects on target spectra, there are many ways to atmospherically correct remotely sensed data. Relatively straightforward methods include the earlier empirical line method, flat field method, and image/scene-based method; more complex methods include recently developed rigorous radiative transfer modeling approaches (Lu et al. 2002, Gao et al. 2009). For example, Staben et al. (2012) used an empirical line method to calibrate WorldView-2 (WV2) imagery to surface reflectance. They compared the calibrated image reflectance values against the surface reflectance values of nineteen independent field targets and obtained RMSE of eight WV2 bands between 0.94% and 2.14% (reflectance) with the greatest variation in NIR bands. Their results show that the empirical line method can be used to successfully calibrate WV2 imagery to surface reflectance. Perry et al. (2000) also compared the performance of an empirical line method with a radiative transfer method, the ATmospheric

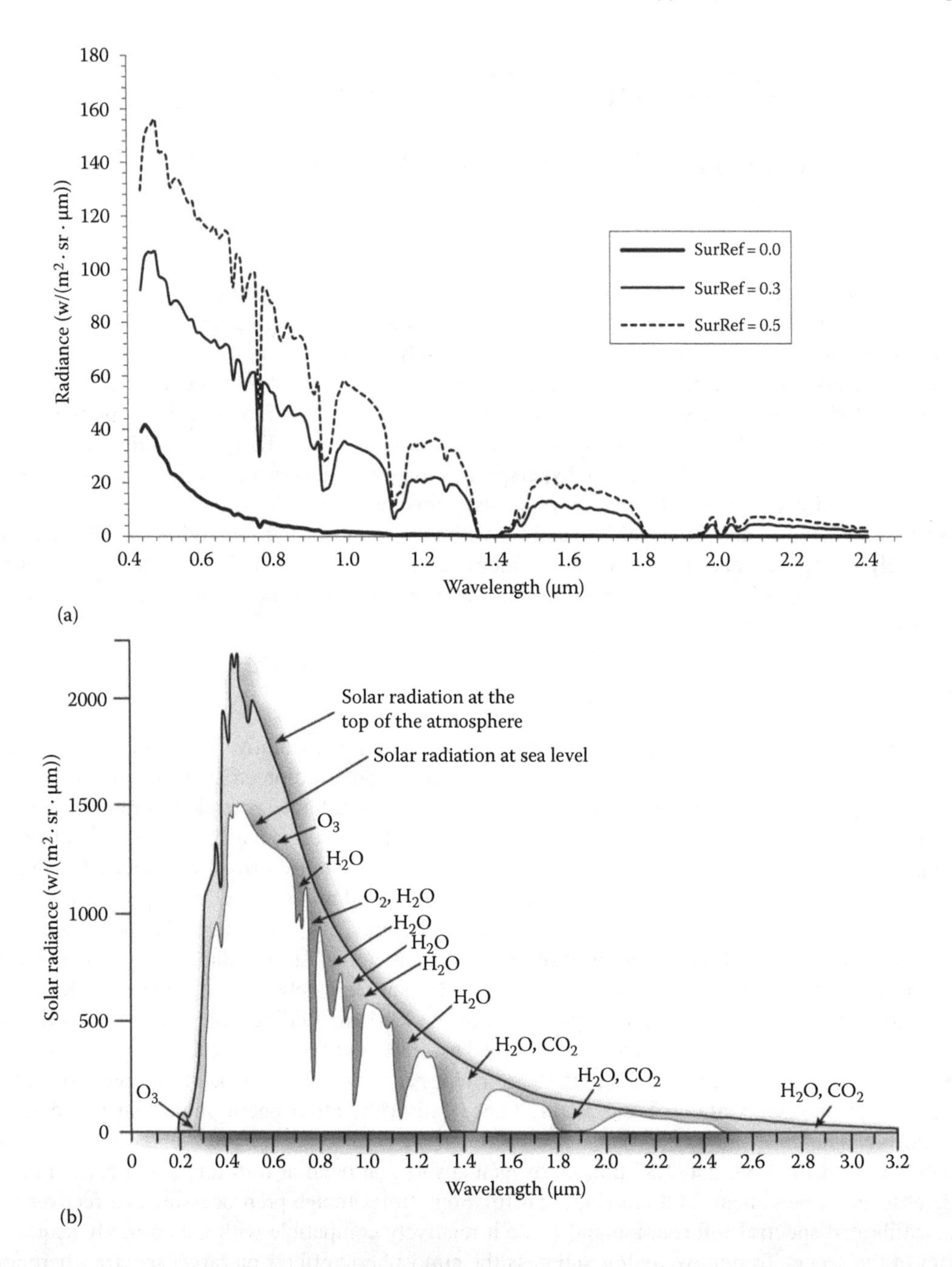

FIGURE 4.1 (a) Total radiances simulated using MODTRAN4 with inputs of surface reflectances 0.0, 0.3, and 0.5 and water vapor 0.7; (b) the integration effects of atmospheric absorption, scattering, and reflectance on reducing the solar irradiance reaching the Earth's surface at sea level compared with solar radiance at the top of atmosphere. (From Jensen, J. R., *Remote Sensing of the Environment: An Earth Resource Perspective*, 2nd ed., Pearson Education, Inc., New York, 2007. Reprinted by permission of Pearson Education, Inc.)

REMoval algorithm (ATREM) (Gao et al. 1993) and found both atmospheric correction (AC) methods had reduced the differences in apparent reflectance for the same targets in the two overlapping HYperspectral Digital Imaging Collection Experiment (HYDICE) frames. San and Suzen (2010) compared four radiative transfer modeling approaches, the Atmospheric CORrection Now (ACORN), the Fast Line-of-sight Atmospheric Analysis of Spectral Hypercubes (FLAASH), and

ATmospheric CORrection (ATCOR 2-3), to atmospherically correct EO-1 Hyperion imagery for lithological and mineralogical mapping. As a result of their study, all four AC methods were found to be successful in overall evaluation, with ACORN performing slightly better than others. In AC, molecular scattering and absorptions by ozone, oxygen, and other gases are relatively easy to correct as their concentrations are fairly stable over both space and time. However, the most difficult and challenging task is to estimate the aerosol and water vapor contents directly from imagery over both space and time (Liang 2004).

Although accurate removal of atmospheric absorption and scattering effects is required for some application purposes, AC may not be necessary in some studies that examine surface properties using multispectral data. For example, AC is not always necessary for certain types of classification and change detection. Theoretical analysis and empirical results indicate that atmospheric correction may only be necessary for image classification and many types of change detection when training data from one time or place must be extended through space or time (Song et al. 2001). For instance, it is not generally necessary to perform AC on a single date of remote sensing data that will be classified using a maximum likelihood classification algorithm (Jensen 2005). However, when applications involve multitemporal data, some AC and radiometric normalization of the multitemporal images are required in order to account for the different atmospheric effects on the images (Nielsen et al. 1998, Canty et al. 2004, Moses and Philpot 2012). It is also necessary to atmospherically correct the remote sensing data even over a single date image if quantitative analyses by combining field survey data with spectral image data for applications such as extracting biophysical parameters from vegetation (e.g., biomass, leaf area index, chlorophyll, and crown closure) are required (Haboudance et al. 2002, Pu et al. 2003). Since the effects of atmospheric scattering and absorption on hyperspectral remote sensing data are critical, the atmospheric correction is particularly important to imaging spectrometer data (Vane and Goetz 1993, Green et al. 1998) for retrieval of surface reflectance and atmospheric constituents. If the data are not atmospherically corrected, the subtle spectral difference in reflectance (or emittance) among the important components may be lost (Jensen 2005).

Given the fact that hyperspectral imaging data have been collected with different types of imaging spectrometers from aircraft and satellite platforms since just the mid-1980s, there is a relatively short history for hyperspectral remote sensing. However, atmospherically correcting remotely sensed imagery quantitatively has a relatively long history. Therefore, representative methods of atmospherically correcting hyperspectral imaging data are briefly introduced and discussed in this chapter after introducing atmospheric effects in detail and methods for correcting general radiometric errors induced by sensors/systems.

4.2 ATMOSPHERIC EFFECTS

Electromagnetic radiation (EMR) can travel at a speed of light of 3×10^8 m s^{-1} in a vacuum in which the EMR is not affected. However, solar radiation, as an EMR form, on the sun–surface–sensor path is affected by atmosphere because the atmosphere is not a vacuum medium. Atmospheric effects on EMR mainly include absorption and scattering from atmospheric gases and aerosols. Due to such effects on EMR, the atmosphere may affect not only the speed of radiation but other properties of EMR, such as wavelength, intensity, and spectral distribution. In the atmosphere, the solar radiation may also be diverted from its original direction due to refraction. Since a remote sensor or system onboard an airborne or spaceborne platform records the total radiance that is originally passing through the atmosphere twice (once only for recording emitted radiance), the recorded radiance is always contaminated by the atmospheric effects, especially scattering and absorption. Therefore, in this section, to better understand and utilize hyperspectral remote sensing data, the major atmospheric effects including radiation refraction, scattering, and absorption and atmospheric transmittance are briefly introduced and discussed from basic concepts and processing mechanisms associated with the atmosphere.

4.2.1 Atmospheric Refraction

Refraction can be defined as the bending of light when it passes from one medium to another with different density. The refraction phenomenon usually occurs when the media are of differing densities and the speed of the EMR traveling varies in each medium layer with different density. Since atmospheric density is variable from the top of the atmosphere to the Earth's surface with a trend of density increasing when approaching the Earth's surface, the refraction phenomenon may take place in the atmosphere. The radiation refraction can be described using an *index of refraction*, n, of the optical density of a substance. The index is defined as a ratio of the speed of light in a vacuum, c, to the speed of light in a substance such as the atmosphere or water, c_n (Mulligan 1980):

$$n = \frac{c}{c_n} \tag{4.1}$$

According to Equation 4.1, since the speed of light in a substance can never reach the speed of light in a vacuum, the index of refraction in the substance must always be greater than 1. For instance, the n for atmosphere and water is 1.0002926 and 1.33, respectively (Jensen 2005). Due to difference of substance densities, light travels more slowly in a substance with higher n (higher density) than that with lower n (lower density).

In practice, *Snell's law* can be used to describe a relationship between two indices of refraction in two media n_1 and n_2 and their corresponding angles (θ_1 and θ_2) of incidence of the radiation to media 1 and 2. The relationship can be described as

$$n_1 \sin\theta_1 = n_2 \sin\theta_2 \tag{4.2}$$

Therefore, if we know the indices of refraction of two adjacent media n_1 and n_2 and the angle of incident radiation to medium 1, we can use Snell's law (Equation 4.2) to predict the amount of refraction that will take place in medium 2 using the trigonometric relationship. For example, in Figure 4.2, if we know n_3, n_4, and θ_3, we can easily predict θ_4 using Snell's law (Equation 4.2).

4.2.2 Atmospheric Scattering

Atmospheric scattering is the result of diffuse multiple reflections of EMR by gas molecules and suspended particles (aerosols) in the atmosphere (Gupta 2003). Scattering is unpredictable and differs from refraction that is predictable (Equation 4.2). Based on the length of the wavelength of incident EMR relative to the diameter of gases, water vapor, and suspended particles with which the EMR interacts, the atmospheric scattering can be divided into two major types: Rayleigh scattering and Mie scattering (sometimes called nonselective scattering).

Rayleigh scattering occurs when the diameter of gas molecules (e.g., O_2 and N_2) and tiny particles in the atmosphere is much smaller than the wavelength of incident radiation and can be extended to scattering from particles up to about a tenth of the wavelength of the incident radiation. Rayleigh scattering, named after British physicist Lord Rayleigh (Sagan 1994), does not change the state of material; hence, it is a parametric process. Rayleigh scattering applies to the case when the scattering particle is very small (i.e., with a particle size $< 1/10$ incident wavelength) and the whole surface re-radiates with the same phase (Barnett 1942). In detail, the intensity I of light Rayleigh scattered by any one of the small spheres of diameter d and refractive index n from a beam of unpolarized light of wavelength λ and intensity I_0 is given by Seinfeld and Pandis (2006):

$$I = I_0 \frac{1+\cos^2\theta}{2R^2}\left(\frac{2\pi}{\lambda}\right)^4\left(\frac{n^2-1}{n^2+2}\right)^2\left(\frac{d}{2}\right)^6 \tag{4.3}$$

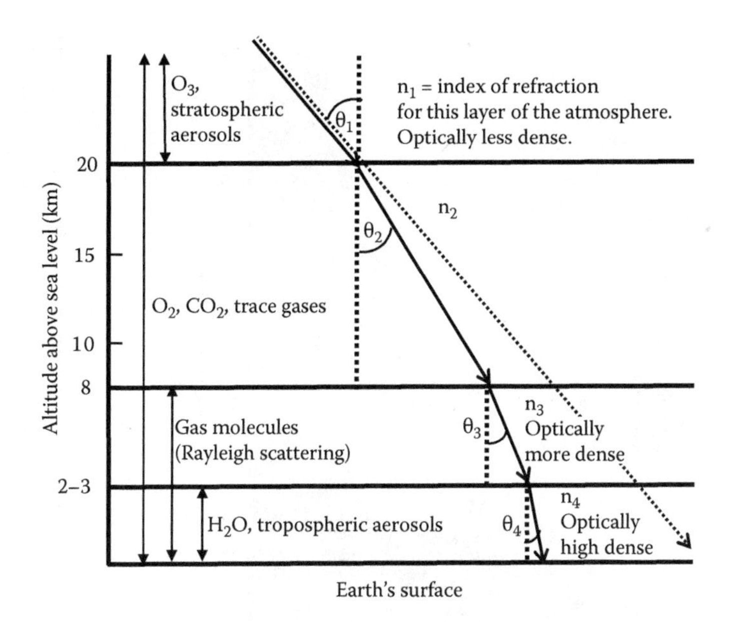

FIGURE 4.2 Major subdivisions of the atmosphere and the types of molecules and aerosols found in each layer and atmospheric refraction as well. Note that the index of refraction for each layer of the atmosphere and the density of corresponding atmospheric layer are assumed and for illustration only. (Modified from Miller, S. W., and Vermote, E., NPOESS Visible/Infrared Imager/Radiometer: Algorithm Theoretical Basis Document, Version 5, Raytheon, Lanham, Maryland, 2002.)

where R is the distance to the particle and θ is the scattering angle. Therefore, averaging this over all angles gives the Rayleigh scattering cross-section, σ_s, (Cox et al. 2002) as follows:

$$\sigma_s = \frac{2\pi^5 d^6}{3\lambda^4}\left(\frac{n^2-1}{n^2+2}\right)^2 \tag{4.4}$$

The Equation 4.3 can also be rewritten for individual molecules by expressing the dependence on refractive index in terms of the molecular polarizability, α, proportional to the dipole moment induced by the electric field of the light. In this case, the Rayleigh scattering intensity, I, for a single molecule is given in CGS-units by Blue Sky (2015):

$$I = I_0 \frac{8\pi^4\alpha^2}{\lambda^4 R^2}(1+\cos^2\theta) \tag{4.5}$$

The Rayleigh scattering intensity at right angles is half the forward scattering intensity (Figure 4.3a) (Blue Sky 2015). According to Equation 4.4 and Equation 4.5, the Rayleigh scattering effect decreases rapidly with increasing wavelength (λ^{-4}) (Figure 4.3b). For example, the blue light at 0.4 μm is scattered about 5 times more than red light at 0.6 μm (i.e., $(0.6/0.4)^{-4} = 5.06$). The short wavelength region between 0.4 and 0.7 μm is strongly affected by molecular scattering (Rayleigh scattering), whereas the aerosol scattering effect also decreases with increasing wavelength, but at a slower rate (typically λ^{-2} to λ^{-1}) (Gao et al. 2009). Most Rayleigh scattering by gas molecules occurs in the atmosphere 2 to 8 km above ground level (Figure 4.2). The shorter violet and blue wavelengths are more efficiently scattered than longer orange and red wavelengths. The blue sky is caused by the Rayleigh scattering of sunlight off the molecules of the atmosphere because the Rayleigh scattering is more effective at short wavelengths (the blue end of the visible spectrum) (Figure 4.3a).

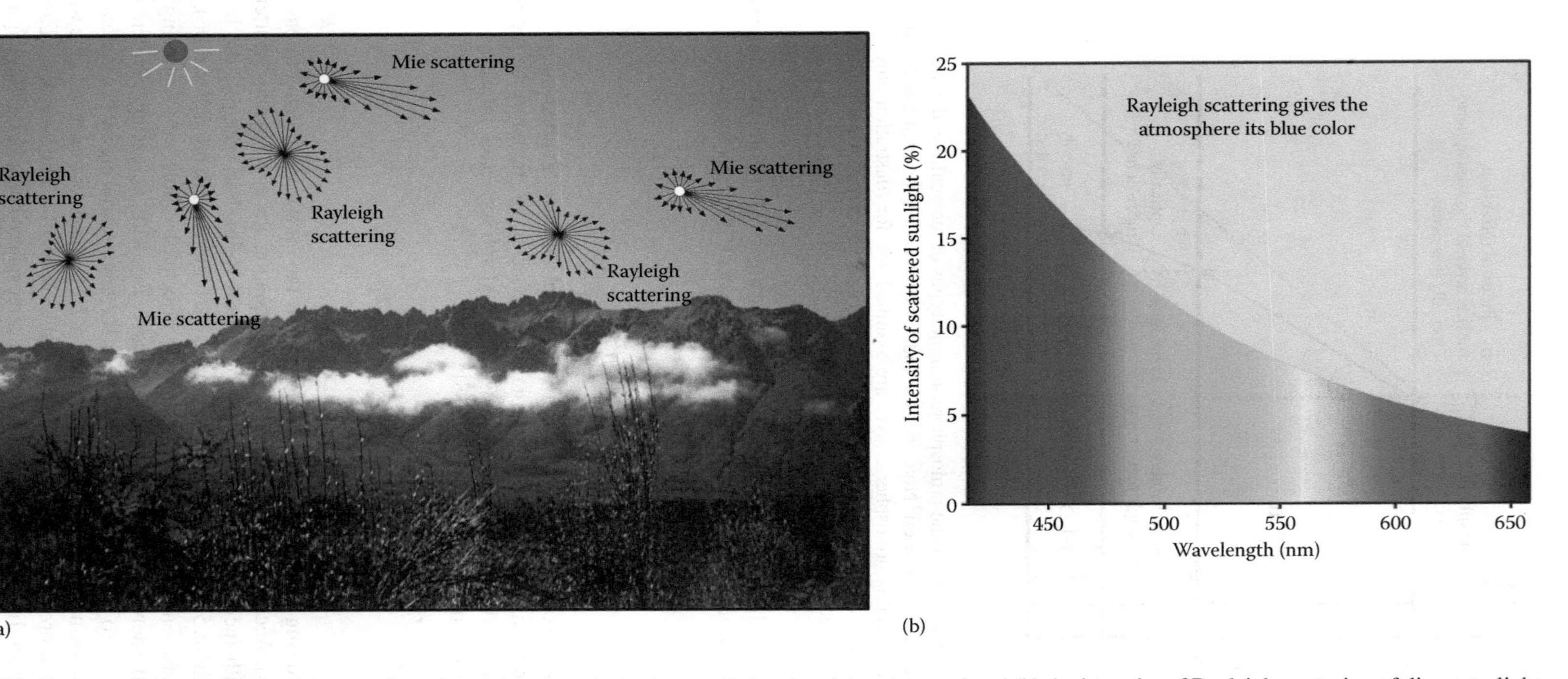

FIGURE 4.3 (a) Atmospheric scattering (Rayleigh and Mie scattering) caused by molecular and aerosol and (b) the intensity of Rayleigh scattering of direct sunlight (%) varying inversely with the fourth power of the incident wavelength (λ^4). The Rayleigh and Mie scattering symbols in part (a) and the Rayleigh scattering curve in part (b) are referred to in Blue Sky (2015).

Mie scattering takes place when the diameter of large particles in sphere in the atmosphere may range from 0.1 to 10 times the wavelength of an incident radiation; thus sometimes Mie scattering is also called *aerosol particle scattering* or *nonselective scattering* (Gupta 2003) if we also consider particle size > 10 times the wavelength of incident EMR, which takes place in the lowest portions of the atmosphere. Suspended dust particles and water vapor molecules are main Mie scatters, which are important in low altitudes of the atmosphere (Figure 4.2). The Mie scattering produces a pattern like an antenna lobe, with a sharper and more intense forward lobe for larger particles (Figure 4.3a) (Blue Sky 2015). Compared to Rayleigh scattering, Mie scattering is not strongly wavelength dependent and produces the almost white glare around the sun when a lot of particulate material is present in the air. It also gives us the white light from mist, fog, and cloud when all wavelengths of white light are scattered equally. Mie scattering influences the entire spectral region from near-UV up to and including the near-IR and has a greater effect on longer wavelengths than Rayleigh scattering (Gupta 2003). Due to Rayleigh scattering and Mie scattering, air pollution also contributes to beautiful sunsets and sunrises. This is because the more the smoke and dust particles in the atmospheric column, the more the violet and blue light will be scattered away and only the longer orange and red wavelength light in the visible region will reach our eyes (Jensen 2005).

Both Rayleigh scattering and Mie scattering are considered the main contributors to atmospheric path radiance. Atmospheric path radiance can severely reduce the information content of hyperspectral remotely sensed data to the point where the imagery loses contrast among surface features and thus it becomes difficult to differentiate one feature from another.

4.2.3 Atmospheric Absorption

Atmospheric absorption is a process by which incident radiant energy is retained by the atmosphere. In the process, when the atmosphere absorbs energy, the result is an irreversible transformation of radiation into another form of energy. There are many different gases and particles in the atmosphere, which absorb and transmit many different wavelengths of EMR energy that passes the atmosphere. Among the approximately thirty atmospheric gases, only eight gases, namely water vapor (H_2O), carbon dioxide (CO_2), ozone (O_3), nitrous oxide (N_2O), carbon monoxide (CO), methane (CH_4), oxygen (O_2), and nitrogen dioxide (NO_2) cause observable absorption features in imaging spectrometer data over the spectral range 0.4 to 3.0 μm with a spectral resolution between 1 and 20 nm (Gao et al. 2009). An *absorption band* is a range of wavelengths (or frequency) in the electromagnetic spectrum within which the radiant energy is absorbed by a substance. Selective absorption by particular gases in the atmosphere as a whole is shown in Figure 4.4, which summarizes the effects of H_2O, CO_2, O_2, O_3, and N_2O on transmission of EMR through the atmosphere. Approximately half of the spectral region between 0.1 and 30 μm is affected by atmospheric water vapor absorption. The cumulative effect of the absorption by the various substances in the atmosphere can cause the atmosphere to "close down" completely in certain wavelength regions (see the bottom portion of Figure 4.4), whereas in other regions of the spectrum, the EMR can be transmitted through the Earth's atmosphere. Those wavelength regions of the spectrum that allow the radiation effectively to pass through the atmosphere are called *atmospheric windows*. In atmospheric windows, there is very little attenuating of the radiation by the atmosphere it passes through.

As a result, only the wavelength regions outside the main absorption bands of the atmospheric gases can be used for remote sensing because they minimize the atmospheric absorption effects. These windows are found in the visible, NIR, certain bands in thermal infrared and the microwave regions. From the bottom portion of Figure 4.4, it is clear that the visible part of the spectrum is an excellent atmospheric window, and then there are several narrow windows in the shortwave infrared region. In the thermal-IR region, there are two important atmospheric windows available at 8.0–9.2 μm and 10.2–12.4 μm, which are separated by an absorption band due to ozone present in the upper atmosphere (Figure 4.2). If considering aerial platforms, the thermal region can be used as 8–14 μm.

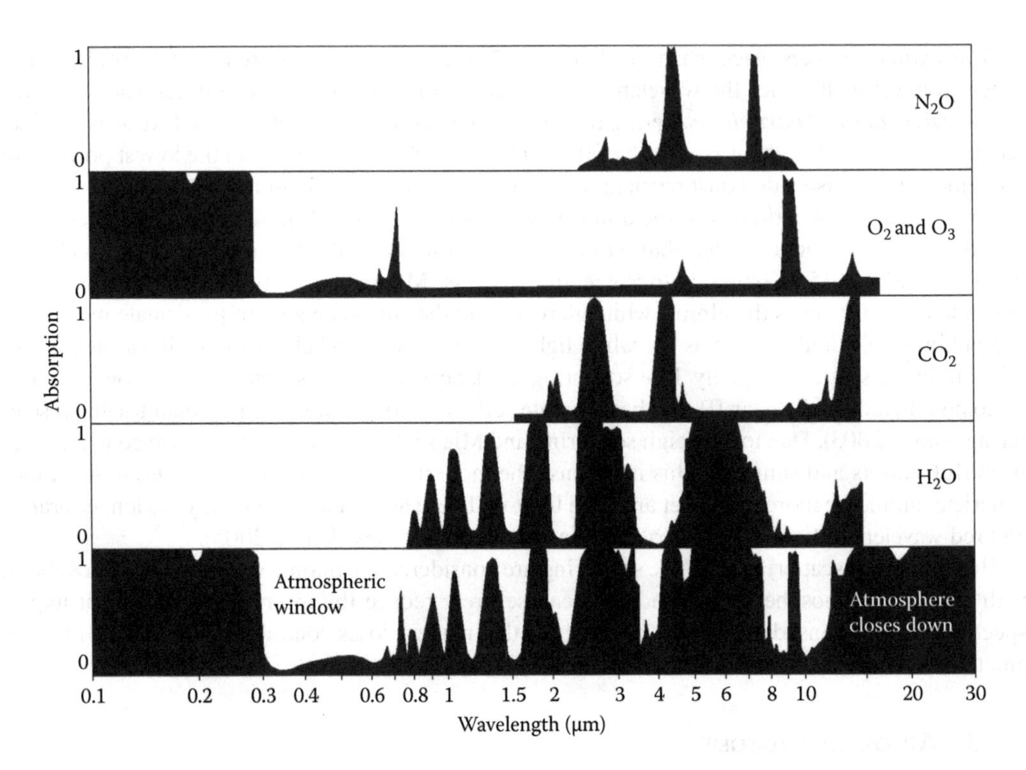

FIGURE 4.4 Atmospheric absorption caused by a set of atmospheric gases: N_2O, O_2 and O_3, CO_2, and H_2O across solar incident electromagnetic energy in the region from 0.1 to 30 μm, while the final graphic illustrates the cumulative result of all these constituents being in the atmosphere at one time. (From Jensen, J. R., *Remote Sensing of the Environment: An Earth Resource Perspective*, 2nd ed., Pearson Education, Inc., New York, 2007. Reprinted by permission of Pearson Education, Inc.)

In addition, microwaves of wavelength longer than 20 mm are transparent to the atmosphere with least attenuation (Gupta 2003), thus used by microwave remote sensing.

4.2.4 Atmospheric Transmittance

If there were no atmosphere over the Earth, the transmittance of solar radiant energy to the Earth's surface would be 100 percent. However, because of scattering and absorptions by the atmosphere, not all of the solar radiant energy reaches the Earth's surface at 100 percent. The amount of energy that does reach the ground, relative to that at the top of the atmosphere, is called *transmittance. Atmospheric transmittance*, T_λ, is defined as a ratio of radiant flux density arriving at a distance l from the origin (Sturm 1992):

$$T_\lambda = \frac{\phi_\lambda(l)}{\phi_\lambda(0)} = e^{-\int_0^l K_\lambda(z)dl} \tag{4.6}$$

where $K_\lambda(z)$ is the extinction coefficient at height z. Then an optical thickness τ_λ at a height z_0 can be defined as

$$\tau_\lambda(z_0) = \int_0^{z_0} K_\lambda(z)dz \tag{4.7}$$

and because of $dl = dz/\cos\,\theta$, the transmittance T_λ for a light beam between point 0 and at-sensor flying at height z_0 becomes

$$T_\lambda(z_0,\ \cos\theta) = e^{-\int_0^{z_0} K_\lambda(z)dz/\cos\theta} = e^{-\tau_\lambda(z_0)/\cos\theta} \tag{4.8}$$

where θ represents either view or sun zenith angle. According to Sturm (1992) and Jensen (2005), the optical thickness of the atmosphere at certain wavelength λ, τ_λ, equals the sum of all extinction coefficients mainly due to Rayleigh ($\tau_{R\lambda}$) and aerosol ($\tau_{A\lambda}$) scattering and absorption by gaseous components of which the most important are water vapor ($\tau_{W\lambda}$) and ozone ($\tau_{OZ\lambda}$):

$$\tau_\lambda = \tau_{R\lambda} + \tau_{A\lambda} + \tau_{W\lambda} + \tau_{OZ\lambda} \tag{4.9}$$

Each of these optical thicknesses in the atmosphere has its proper dependence on the height z and wavelength λ. For example, most attenuation of light in visible-NIR regions (wavelength < 0.8 μm) is caused by Rayleigh and aerosol (Mie) scattering, and thus its optical thickness is derived from ($\tau_{R\lambda}$) and ($\tau_{A\lambda}$) primarily while in NIR-SWIR (> 0.8 μm up to 2.5 μm), its corresponding optical thickness can be considered to be caused mostly by absorptions ($\tau_{W\lambda} + \tau_{OZ\lambda}$).

As discussed above, the atmosphere has selective scattering and absorption characteristics, which mean that the atmospheric effects depend on the wavelength of a given a remote sensor or system. In the visible and NIR regions, gas molecules and aerosol particles scattering effects, which produce upwelling atmospheric radiance or path radiance (Slater 1980), are additive to remotely recorded data. Since the gas molecules are stable with wavelengths < 0.7 μm and obey the Rayleigh scattering rule, this part of the scattering effect is relatively easy to remove.

However, since the character of aerosol particles is often variable, their scattering effect on remotely sensed data is difficult to estimate and remove (Lu et al. 2002, Liang 2004). In the visible region, the absorption caused by water vapor and other gases is very weak and can be ignored. Therefore, in visible and partial NIR regions for radiometerically correcting remote sensing data, we should mainly concentrate on correcting atmospheric scattering effects caused by gas molecules and aerosol particles. However, in the shortwave infrared region, the influence of air molecules and aerosol particle scattering can be negligible, and the atmospheric absorption caused by H_2O, CO_2, O_2, O_3, N_2O, etc., is significant. The absorption effect with multiplicative characteristics is difficult to estimate and remove from remotely sensed data. Among the several absorption substances in the atmosphere, the contents of CO_2, O_2, O_3, N_2O, etc., are relatively stable, but the H_2O is much more variable over space and time. Thus, how to accurately estimate atmospheric column water vapor amounts usually is a key to developing an atmospheric correction algorithm to correct atmospheric absorption effect for multi-/hyperspectral remote sensing data.

4.3 CORRECTING RADIOMETRIC ERRORS INDUCED BY SENSORS/SYSTEMS

4.3.1 Introduction to Radiometric Errors Caused by Sensors/Systems

Before a remote sensing instrument is launched into space or installed in an aircraft, its spectral and radiometric calibration is usually performed in a laboratory (this is called pre-launch calibration). However, since some processes such as outgassing, aging, degradation of optical or electronic components, and misalignment due to mechanical vibrations are inevitable after the remote sensing instrument is in flight (onboard satellite or aircraft), the instrument performance may be changed compared to its laboratory-calibrated version (Guanter et al. 2006), and thus the spectral and radiometric errors may be created in the flight. Per hyperspectral remote sensing, such system-induced errors may consist of two types: (1) hyperspectral image striping artifacts caused by bad pixels and inaccuracies in the precision of the radiometric calibration across the detector array (Goetz et al. 2003), and (2) hyperspectral imaging sensor artifacts that are caused by spectral and spatial misregistrations across the detector array along the spectral and spatial dimensions, respectively (Yokoya et al. 2010). Per the first error, the result would be an image with systematic, noticeable columns that are brighter than adjacent columns (see striping effects in Figure 4.5a,c). Such individual detectors induced maladjusted columns contain valuable information, but should be corrected to have approximately the same radiometric scale as the data collected by the appropriate calibrated detectors in association with the same band (Jensen 2005). Per the second error, both spectral and spatial misregistrations are generated by optical aging, degradations, aberrations, and misalignments in pushbroom systems, especially in hyperspectral pushbroom systems where crosstrack and spectral pixels are continuously recorded at the same time using an area detector array (Figure 4.6). Spectral misregistration, also called a "smile" or "frown" curve, is a spectral shift in the spectral domain. The spectral shift is a function of the crosstrack pixel number. The "smiling" error originates a nonlinear

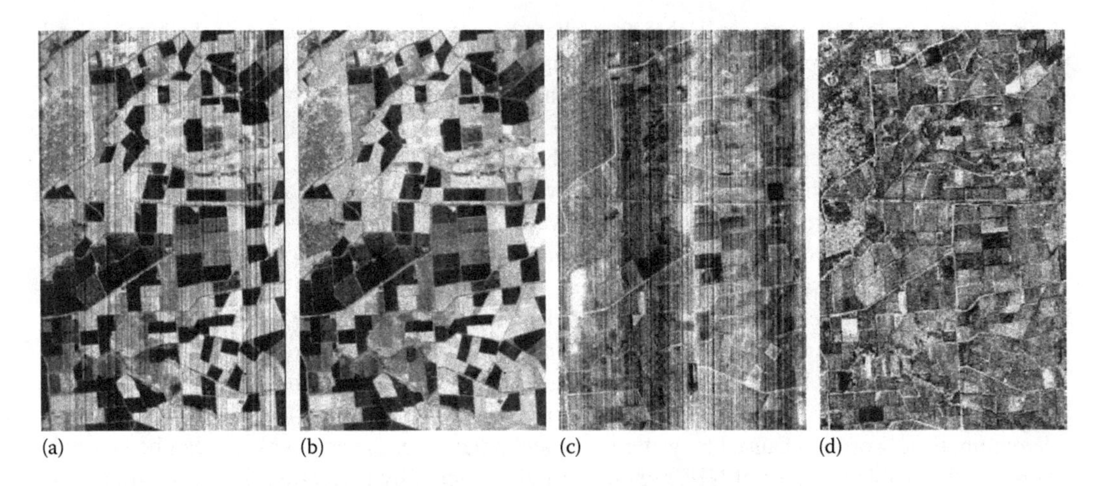

FIGURE 4.5 (a) Hyperion band 128 of Coleambally, NSW, Australia, on January 11, 2002, showing effects of striping (before destriping); (b) band 128 after destriping; (c) MNF-15 before destriping; and (d) MNF 15 after destriping. (©2003 IEEE. Reprinted from Goetz, A. F. H., Kindel, B. C., Ferri, M., and Qu, Z., *IEEE Transactions on Geoscience and Remote Sensing*, 41, 1215–1221, 2003. With permission.)

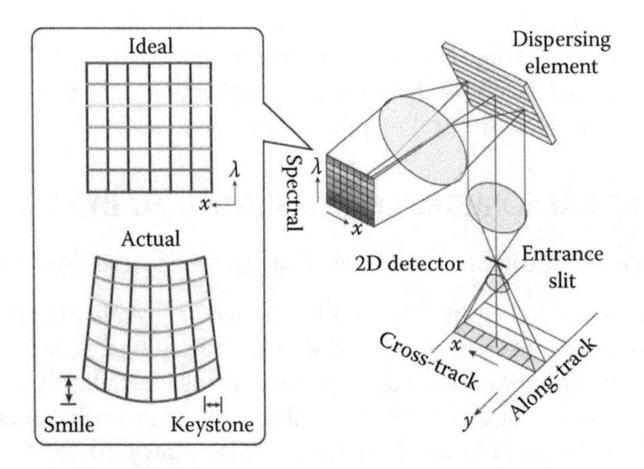

FIGURE 4.6 Hyperspectral system induced spectral and spatial errors, commonly called "smile" and "keystone" effects, respectively. (Image courtesy of N. Yokoya, http://naotoyokoya.com/Research.html. See Yokoya, N. et al., *Applied Optics*, 49(24), 4568–4575, 2010.)

variation of the spectral shift with the crosstrack position of the different detectors compounding the sensor. Spatial misregistration, also known as a "keystone," corresponds to band-to-band misregistration. Both errors of spectral and spatial misregistrations distort the spectral signatures and thus might lower image classification analysis (Yokoya et al. 2010).

4.3.2 De-Striping

Correcting hyperspectral image striping artifacts that are caused by bad pixels and inaccurate radiometric calibration across the detector array is called de-striping. Here, two hyperspectral imaging de-striping techniques are introduced, one more suitable for land and the other for water areas.

The de-striping technique for hyperspectral imaging data acquired on land is briefly introduced based on Goetz et al. (2003). The technique assumes that the mean and standard deviation of each column in a band image match the grand mean and standard deviation of the spectral band image. Mathematically, the procedure can be summarized as follows. Let X_{ijk} be the value (e.g., DN or reflectance) of a pixel in row i, column j, and band k. And let μ_k and σ_k be the grand mean and standard deviation of band k, and μ_{jk} and σ_{jk} be the mean and standard deviation of column j and band k before de-striping. Each vector column X_{jk} can be replaced by a corresponding destriped vector column X'_{jk} with the linear function:

$$X'_{jk} = a_{jk} X_{jk} + b_{jk} \tag{4.10}$$

so that the destriped X'_{jk} has the same mean and standard deviation of band k, i.e., $\mu'_{jk} = \mu_k$ and $\sigma'_{jk} = \sigma_k$. In Equation 4.10, a_{jk} is the gain and b_{jk} is offset value. From Equation 4.10, we have

$$\mu'_{jk} = a_{jk}\mu_{jk} + b_{jk} \tag{4.11}$$

$$\sigma'_{jk} = a_{jk}\sigma_{jk} \tag{4.12}$$

Therefore, in considering the assumption, it is easy to have

$$\mu_k = a_{jk}\mu_{jk} + b_{jk} \tag{4.13}$$

$$\sigma_k = a_{jk}\sigma_{jk} \tag{4.14}$$

then

$$a_{jk} = \frac{\sigma_k}{\sigma_{jk}} \tag{4.15}$$

$$b_{jk} = \mu_k - \frac{\sigma_k}{\sigma_{jk}}\mu_{jk} \tag{4.16}$$

Therefore, any de-spriped X'_{ijk} pixel value in band k can be obtained by modifying the DN or reflectance of the pixel in column j according to

$$X'_{ijk} = \frac{\sigma_k}{\sigma_{jk}} X_{ijk} + \mu_k - \frac{\sigma_k}{\sigma_{jk}} \mu_{jk}, \quad i = 1, 2, \ldots, N \tag{4.17}$$

where N is the number of rows.

This technique was applied to Hyperion hyperspectral image in radiance or reflectance to destrip processing and the destriped image quality was determined by inspecting the image striping effects and higher-order maximum noise fraction (MNF) images for striping artifacts. Figure 4.5 demonstrates the de-striping effects by utilizing the de-striping technique (Equation 4.17) to a Hyperion band 128, acquired on January 11, 2002, in Coleambally, NSW, Australia. The clear striping effects show on the band image (Figure 4.5a) and on an MNF image (Figure 4.5c) and their corresponding destriped results present on Figure 4.5b,d.

The other de-striping technique is based on the destripe method called Vertical Radiance Correction (VRadCor), developed by Zhao et al. (2013), which fits water areas. According to the study by Zhao et al. (2013) on Hyperion image de-striping over a water area, they found that most existing approaches were developed and validated to improve image quality for large terrestrial areas, rather than aquatic environments. For example, there are two algorithms applied in ENVI software (RSI 2005) that can be used to assist for de-striping: Cross Track Illumination Correction (CTIC) and the de-striping function derived from solving the 8-element-sensor stripes of TM, which work well on processing images over land. In addition, the stripes in the Hyperion Level 1R images (Zhao et al. 2013) were the residues after the regular relative calibration and destripe optimization (e.g., a similar techniques addressed above; Goetz et al. 2003). These residual stripes are relatively enhanced compared to the low signal of water area, and cannot be adequately removed by the conventional methods (Zhao et al. 2013). Therefore, an ideal destripe method should remove/compress both high and low frequency stripes in a limited, uneven area while retaining spectral features over the water area. Accordingly, the VRadCor destriping technique is concisely summarized as follows (for a detailed description of VRadCor, see the paper by Zhao et al. 2013):

> VRadCor is a technique to compress the cross-track radiance abnormity to assess both additive and multiplicative correction factors and is accommodated for the water area with an uneven background. It is based on the spectral statistics of multi-/hyperspectral images along the track, and is a relative calibration algorithm of sensors/images. In considering some causes that are additive while the others are multiplicative, and so to the stripe effect generally, a universal destripe algorithm can be expressed as:
>
> $$R_i(\lambda_j) = GAIN_i(\lambda_j) Ro_i(\lambda_j) + OFFSET_i(\lambda_j) \tag{4.18}$$
>
> where R_i is the corrected image value of pixel sample i; Ro_i is its original value, and λ_j is the wavelength of band j; $GAIN_i$ and $OFFSET_i$ are the multiplicative and additive linear correction factors for sample i, respectively.

Based on a general inference of the correlation simulating analysis model (CSAM) for spectral analysis (Zhao et al. 2005), VRadCor provides a method to use both spatial and spectral information assuming that adjacent pixels with statistical expectation values along the track have similar spectra. Thus, it may not require the crosstrack with an even precondition. The general algorithm of VRadCor consists of the following steps:

1. Select a multi/hyperspectral image R_{mn}, with a number of bands $n \geq 3$ and a number of samples $m \geq 2$.
2. Select an area for the along-track spectral statistics. A criterion of selection considered is that the statistical expectation spectra from the selected area represent similar mean of features' spectra in the image.

3. Calculate the mean from the selected area of image R_{mn} along the track for each band, and record the crosstrack spectral mean as S_{mn} (i.e., each column mean).
4. Smooth the crosstrack spectral mean S_{mn} in the direction of the crosstrack to remove/compress information of real differentiation among adjacent pixels.
5. Calculate the gain and offset factors based on the assumption that the optimal statistic result, S_{mn}, has similar spectral curves in adjacent pixels and the CASM inference requires these curves to have a linear relationship, using the following expression:

$$S_i = GAIN_i S_0 + OFFSET_i \tag{4.19}$$

   where S_i is the mean spectrum of sample i in S_{mn}; S_0 is the optimized base spectrum to which the image will be corrected. Using Equation 4.19, VRadCor can obtain both gain and offset correction coefficients (matrices).
6. Calculate the base spectrum S_0 along with S_{mn} to correct the different parts of crosstrack radiance abnormity. To calculate the base spectrum S_0, it can be fixed to the grand mean of image band j if an even expectation is true for the whole image, while it represents the interpolation of adjacent pixels in an uneven area (e.g., the study area in Zhao et al. [2013]). Usually, to obtain the most optimized S_0 estimation, an iterative optimizing algorithm needs to be run (Zhao et al. 2005).
7. Correct the image using the coefficient matrices calculated from Equation 4.19.

In general, VRadCor can be used not only to correct the low frequency component with a higher applicability than CTIC, but also to correct the high frequency stripes by behaving like a relative calibration method (Zhao et al. 2005). Figure 4.7 presents a comparison of before (left) and after

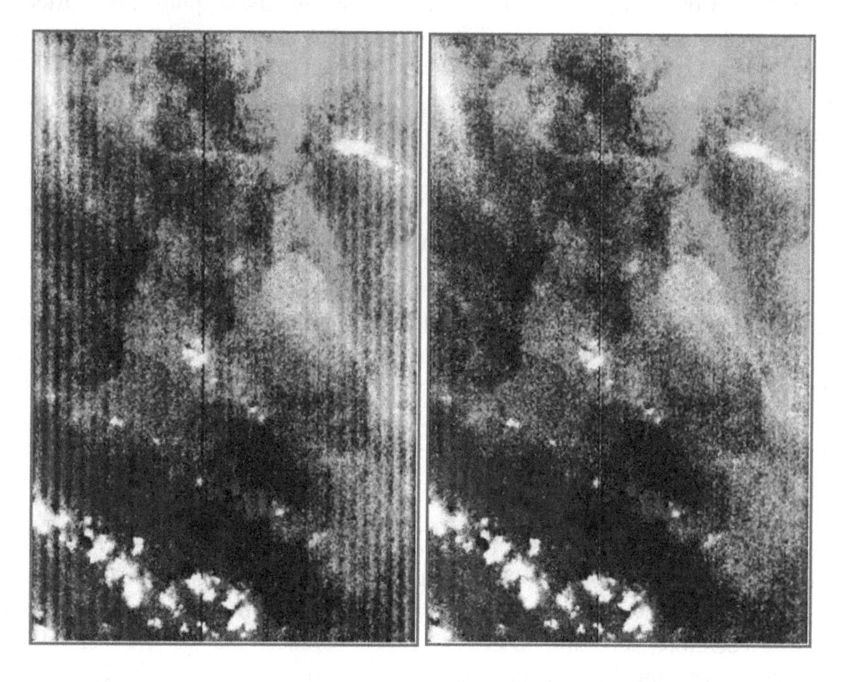

FIGURE 4.7 A comparison of before (left) and after (right) striping removal (destripe). These results were produced after equalization enhancement. Note that the stripes after processing were significantly reduced while the uneven (heterogeneous) phenomena in patches were still retained. The figure was produced with Hyperion in Clearwater, Florida, and was acquired on October 8, 2009. (©2013 IEEE. Reprinted from Zhao, Y., R. Pu, S. Bell, C. A. Meyer, L. Baggett, X. Geng, *IEEE Transactions on Geoscience and Remote Sensing*, 51(2), 1025–1036, 2013. With permission.)

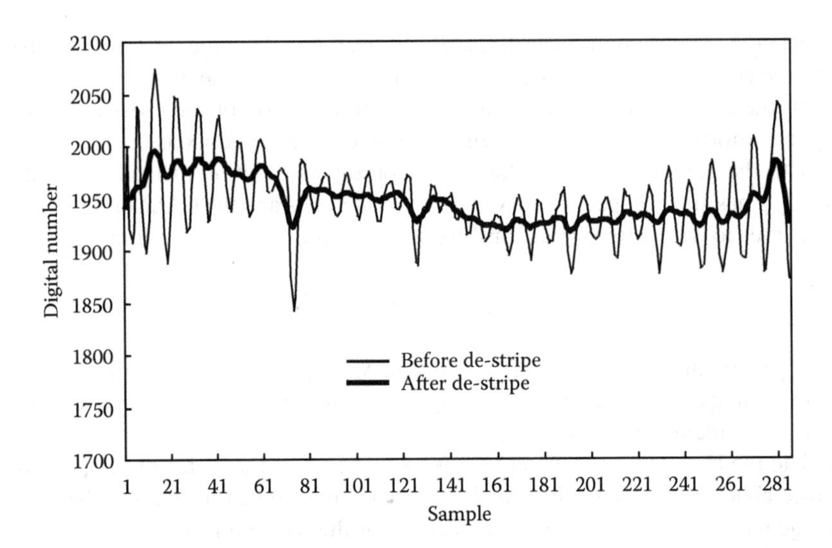

FIGURE 4.8 The crosstrack distribution of the mean statistics calculated across the track from Hyperion image band 8. The thin and thick curves represent values before and after destriping, respectively. The plot shows that stripes with an interval of approximately eight pixels were mostly removed from the uneven background. (©2013 IEEE. Reprinted from Zhao, Y., R. Pu, S. Bell, C. A. Meyer, L. Baggett, X. Geng, *IEEE Transactions on Geoscience and Remote Sensing*, 51(2), 1025–1036, 2013. With permission.)

(right) destriping processing of the Hyperion image, Clearwater, Florida, acquired on October 8, 2009, using the VRadCor algorithm (Zhao et al. 2013). Note that the striping effects after processing were significantly reduced while the uneven (heterogeneous) phenomena in patches were still retained. To visually see the effectiveness of before and after destriping processing, these results were shown after equalization enhancement. Figure 4.8 provides a crosstrack distribution of the mean statistics calculated across the track from Hyperion image band 8 and shows that stripes with an interval of approximately eight pixels were mostly removed from the uneven background (Zhao et al. 2013).

4.3.3 Correcting Smile- and Keystone-Induced Errors

Based on previous studies, very high sensor's accuracy in hyperspectral remote sensing is required to obtain spectral signatures useful for scientific research and applications (e.g., Green 1998, Mouroulis et al. 2000). Synthetic data with a systematic spectral shift of 1 nm for channels with Full Width at Half Maximum (FWHM) of 10 nm showed associated errors in the measured radiance of up to ±25% in strong water vapor absorption bands (Guanter, Estellés et al. 2007). Moreover, such spectral shifting errors will be higher as the spectral resolution becomes finer. For instance, the Hyperion hyperspectral sensor has an up to 2 nm shift in spectral wavelength calibration across the array (Ungar et al. 2003). Figure 4.9 shows that the magnitude of the errors associated with smile in the Hyperion instrument as an image of the detector array (Goetz et al. 2003). In the figure, the initial surface reflectance was set at 50% and the radiances were generated and then retrieved using MODTRAN4. As a result, the maximum spectral errors around the water vapor absorption bands are ±10%. Therefore, according to Green's (1998) detailed analysis of the errors in at-sensor radiances caused by spectral shifts, a spectral uncertainty (i.e., smile error) of less than 1% of the FWHM throughout the spectral response function is necessary. Furthermore, the maximum spatial misregistration error (i.e., keystone error) in high hyperspectral sensor calibration is required to be less than 5% of the pixel size (Mouroulis et al. 2000).

To detect and correct the spectral shift (i.e., smile) errors, most methods use the atmospheric absorption features/bands of O_2, H_2O, CH_4, and CO_2 at difference spectral regions (Qu et al. 2003,

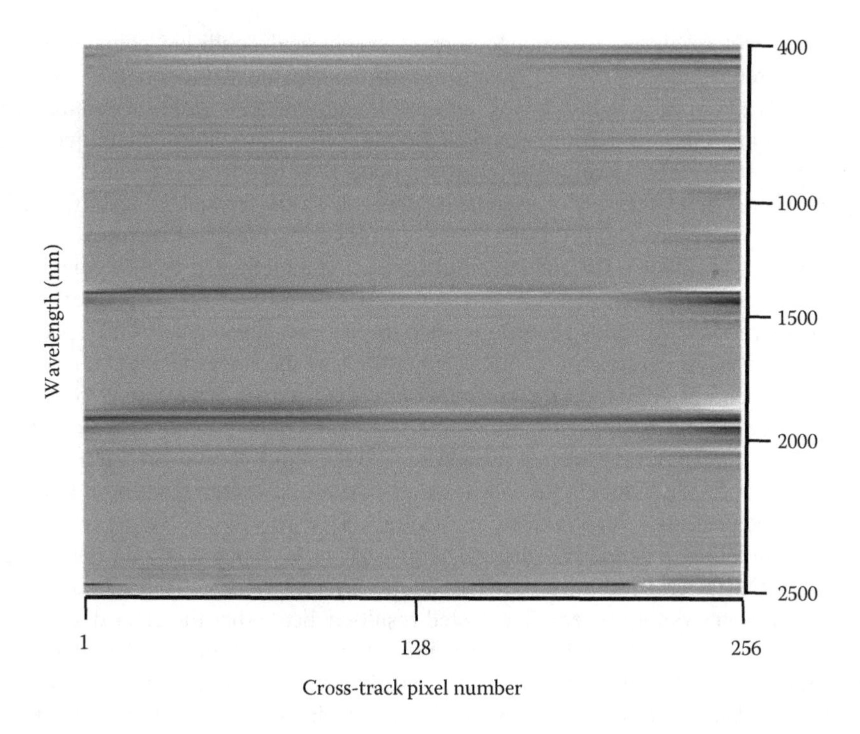

FIGURE 4.9 Image of the errors expected from smile-induced spectral calibration error. The initial reflectance was set at 50% and radiances were generated and then retrieved using MODTRAN4. The grayscale white-to-black around the water vapor bands is ±10% error. (©2003 IEEE. Reprinted from Goetz, A. F. H., Kindel, B. C., Ferri, M., and Qu, Z., *IEEE Transactions on Geoscience and Remote Sensing*, 41, 1215–1221, 2003. With permission.)

Guanter, Estellés et al. 2007). For instance, O_2 band around 0.76 μm is used for spectral calibration for wavelengths between 0.6–0.86 μm, while the 1.14 μm water vapor absorption region is used for spectral calibration for wavelengths between 1.06–1.25 μm, etc. (Qu et al. 2003). Usually when the O_2 absorption around 0.76 μm is present in the spectral range registered by the sensor, it is preferred to use it for the spectral calibration. If not, H_2O absorptions in the NIR wavelengths are considered. The reason for preferring the O_2 absorption is that it is less variable in temporal and spatial terms than the water vapor one (Guanter, Estellés et al. 2007).

In the remainder of this section, three methods used for detection and correction of imaging sensors induced errors will be summarized:

1. The *smoothness technique* developed by Goetz et al. (2003) and Qu et al. (2003) looks for the spectral shift that generates the smoothest surface reflectance spectrum after atmospheric correction. The smoothness technique is embedded in a version of High-Accuracy Atmospheric Correction for Hyperspectral data (HATCH) called HATCH-2d that was developed to make a column-by-column atmospheric correction to minimize the errors associated with different model–data calibration effects (Goetz et al. 2003, Qu et al. 2003). Per the technique, the spectral shifts are assumed to be linear across the array. If they are not matched to 1 nm or less, major overshoots occur at the edges of strong atmospheric H_2O and O_2 absorptions (Goetz et al. 2003). By shifting the central wavelengths up to a maximum of ±3 nm and FWHM response of the spectrometer up to ±2 nm in the strong H_2O absorption regions and the O_2 absorption band, HATCH searches for the smoothest spectrum. HATCH-2d is capable of speeding up the procedure by utilizing the fact that the response function shift varies gradually across the entire detector array, because

adjacent detectors always experience very close physical condition changes (Qu et al. 2003). Therefore, using different spectral absorption features (including O_2, H_2O, CH_4, and CO_2), HATCH-2d is able to handle possible spectral miscalibration separately for each spectral region. Qu et al. (2003) demonstrated that this smoothness technique produces reflectance spectra with lower amplitude overshoots than that from ATREM.

2. A method for the evaluation of systematic spectral shifts from hyperspectral data was developed by Guanter et al. (2006) using atmospheric absorption features. According to Guanter et al. (2006), the fundamental basis of the method is the calculation of the value of the spectral shift that minimizes the error in estimating surface reflectance in the neighborhood of gaseous absorption spectral features. The value of the spectral shift can be found by an iterative procedure, which leads to the smoothest surface reflectance spectrum when it is applied to the convolution of the hyperspectral output from radiative transfer codes in the atmospheric correction procedure. This procedure is performed for both the assessment of the spectral calibration and the atmospheric correction of the data. Therefore, the final product is the surface reflectance that is free from both atmospheric influence and possible errors induced by spectral shifts. To validate the method and determine its limitation, a sensitivity analysis was performed using synthetic data generated with the MODTRAN4 radiative transfer code for several values of the spectral shift and the column water vapor content. The tested results indicate that the error detected in the retrieval is less than ± 0.2 nm for spectral shifts smaller than 2 nm, and less than ± 1.0 nm for extreme spectral shifts of 5 nm. A low sensitivity to uncertainties in the estimation of water vapor content demonstrates the usefulness of the algorithm. Using other both airborne instruments (HyMap, AVIRIS, and ROSIS) and spaceborne (PROBA-CHRIS) real data to test the method also proved that in all the cases errors around atmospheric absorption bands were successfully reduced or removed, further demonstrating the robustness of the method to different sensors for spectral calibrations and to different hyperspectral images for atmospheric correction (Guanter et al. 2006).
3. An integrative method was developed by Yokoya et al. (2010) for detecting and correcting spectral and spatial misregistrations for hyperspectral data by using subpixel image registration and cubic spline interpolation techniques. According to Yokoya et al. (2010), two subpixel image registration techniques based on normalized cross correlation (NCC) and phase correlation (PC) are used to detect the artifacts of spectral and spatial misregistrations, and their effectiveness is evaluated from the viewpoints of accuracy and robustness with Hyperion visible and near-IR (VNIR) imaging data. Cubic spline interpolation is adopted to correct the artifacts because the co-registration of measured radiance spectra is distinctly improved (Feng and Xiang 2008). The two subpixel image registration techniques can detect smile and keystone properties. The smile is detected by estimating the distortion of the atmospheric absorption line in the spectrum image whereas the keystone is detected by estimating the band-to-band misregistration at all pixel positions in a whole image. Since the cubic spline interpolation has been used for estimation of spectra and can effectively correct the errors in the radiance measurement and mitigate the measured spectral radiance misregistration (Feng and Xiang 2008), it may be used to modify the spectral curvatures distorted by smile and keystone effects. Figure 4.10 shows the result of keystone property as estimated and corrected using the method proposed by Yokoya et al. (2010). The detailed NCC, PC, and cubic spline interpolation algorithms and procedures were described in detail in Yokoya et al. (2010). The validation result of the method demonstrates that the Hyperion VNIR characteristics were properly evaluated and corrected with 0.01-pixel accuracy. In addition, when the method was applied to other scenes of the hyperspectral images, consistent results of detecting and correcting the errors induced by smile and keystone properties were obtained, which demonstrate the robustness of the integrative method.

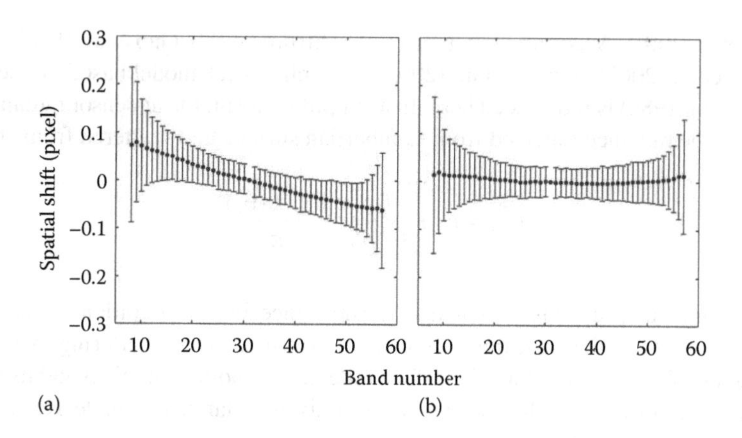

FIGURE 4.10 Average keystone property estimated (a) before and (b) after correction. Error bars represent 1 standard deviation. The detected spatial misregistration (keystone) examined was using the desert scene from Hyperion imagery in Chile acquired on December 25, 2002. (From Yokoya, N. et al., *Applied Optics*, 49(24), 4568–4575, 2010. Used with permission from OSA.)

4.4 ATMOSPHERIC CORRECTION METHODS

4.4.1 Introduction to Atmospheric Correction

As addressed in previous sections, the radiance recorded at-sensor with hyperspectral remote sensing (HRS) instruments is not accurate to represent Earth's surface features within the sensor's field of view. Apart from the HRS data with the possible HRS instrument-induced radiometric errors (addressed in Section 4.3), most HRS remote sensing data are contaminated by the interaction of electromagnetic radiation with the atmosphere, caused by the atmospheric gas molecules and aerosol scatterings and gaseous substance absorptions. For HRS to provide accurate measurements of surface inherent properties from at-sensor radiance acquired with HRS instruments, atmospheric correction (AC) is usually necessary. Most AC methods are to convert the at-sensor raw radiance data recorded by HRS imaging spectrometers to scaled surface reflectance by correcting for atmospheric effects so that the corrected HRS data is a data set in which each pixel can be represented by a reflectance spectrum which can be directly compared to reflectance spectra of surface features measured either in the field or in the laboratory (van der Meer et al. 2001). There are many AC ways that can be used to correct atmospheric effects. Some are relatively straightforward, whereas others are very complicated. In general, two categories of AC methods can be identified, that is, the absolute atmospheric correction and relative atmospheric correction (normalization). The absolute correction method converts remotely sensed digital numbers into surface reflectance or radiance by removing the effects caused by atmospheric attenuation, topographic condition, and other parameters. Common models/methods for absolute AC include MODTRAN (Moderate Resolution Atmospheric Radiance and Transmittance Model), 6S (Second Simulation of the Satellite Signal in the Solar Spectrum), the Fast Line-of-sight Atmospheric Analysis of Spectral Hypercubes (FLAASH), Empirical Line Calibration (ELC) models, etc. The relative AC methods are used to remove or normalize the variation within a scene and normalize the intensities between images of the same study area collected at different dates. The methods for relative correction can be histogram adjustment and multi-date normalization using a regression model approach, etc. (Lu et al. 2002).

Per a physically-processed AC method (belonging to the absolute AC category), usually a Radiative Transfer (RT) model (with different versions available depending on its degree of complication) is needed to model the atmosphere's optical behavior given user-defined boundary conditions. In a RT model, the total radiance reaching the sensor can be split into four components: path radiance, reflected diffuse radiance, reflected direct radiance, and reflected radiance from neighborhood (van der Meer et al. 2001). According to Green (1992), Strum (1992), Gao et al. (1993), van der Meer et al. (2001), Pu et al. (2003), and Qu et al. (2003), a simplified RT model based on the same principles (see Tanré et al. 1986) is introduced here. In a simplified form, the at-sensor radiance L_{obs} can be a combination of the radiance reflected from Lambertian surface and scattered from atmosphere by

$$L_{obs} = L_a + \frac{T_2\rho}{1-\rho S} \cdot \frac{E_s Cos(\theta_s)}{\pi} \tag{4.20}$$

where T_2 is the sun–surface–sensor two-way transmittance in consideration of the total gaseous transmittance effect; L_a is the path radiance caused by atmospheric scattering; S is the spherical albedo of the atmosphere; ρ is the Earth's surface reflectance (both from pixel and its neighborhood areas); E_s is the exoatmospheric solar irradiance; and θ_s is the solar zenith angle. Rewriting Equation 4.20 for retrieval of surface reflectance, ρ, yields

$$\rho = \frac{L_{obs} - L_a}{(L_{obs} - L_a)S + T_2 \cdot \dfrac{E_s \cdot Cos(\theta_s)}{\pi}} \tag{4.21}$$

By simulating the atmospheric quantities T_2, L_a, and S with radiative transfer models such as MODTRAD4, the surface reflectance (assumed for the horizontal Lambertian surface), ρ, can be retrieved from the measured radiance, L_{obs}, using Equation 4.21. Pixel-based L_{obs} value can be directly converted from DN from HRS raw data using its corresponding metadata provided by an image provider. For example, Pu et al. (2003) used an atmospheric RT code MODTRAN4; three at-sensor total radiances were first simulated with three surface reflectance values (e.g., 0.0, 0.3, 0.5) as inputs. Other parameter values necessary for calculating the total radiance, such as water vapor (0.7 in this experiment), aerosols, and atmospherically geographical–seasonal model were also estimated. In simulating the total at-sensor radiance with MOTRAN4, the two key input parameters are water vapor and aerosol values because both are variable in space and time and all other input parameters are relatively stable and easy to determine. The simulated output radiance was used as L_{obs} in Equation 4.20. With the three simulated at-sensor total radiances, we can solve the RT model (Equation 4.20) for parameters T_2, L_a, and S. Then Pu et al. (2003) used AVIRIS image pixel value (radiance) as at-sensor radiance and solved T_2, L_a, and S to calculate pixel-based surface reflectance from AVIRIS image radiances by Equation 4.21. A high accurate atmospheric correction code, HATCH, also adopts Equation 4.21 to realize the retrieval of surface reflectance from an HRS image such as AVIRIS and Hyperion (Goetz et al., 2003).

According to the AC model characteristics and complexity, three general types of AC methods or models can be classified. The first type of AC methods is empirical/statistical methods, including ELC and image-/scene-based techniques, Internal Average Reflectance (IAR), etc. Such a type of AC method does not require any *in situ* atmospheric information, and inputs required to implement these methods are mainly based on image and ground spectrum measurements. The second type of AC methods are physically based (RT-based) models, which require many simulation submodels that use many atmospheric parameters. Theoretically, such models can produce a high accuracy of retrieved surface reflectance from HRS data. These methods/models, such as ATREM, HATCH, and FLAASH, are often very complex and require many input parameters from the *in situ* field atmospheric information acquired at the time of remote sensing data acquisition (Lu et al. 2002). The third type of method focuses on relative radiometric correction and normalization and develops methods/models based on histogram matching and regression equations. In the following subsections, all three types of AC methods/models are introduced and discussed in detail, including their principles, algorithms, operating procedures, and key input parameters. Among them, some corresponding operational software and tools to process HRS images including AC can be seen in Chapter 6.

4.4.2 Empirical/Statistical Methods

The empirical/statistical and image-/scene-based AC methods are developed to remove atmospheric effects from HRS imaging data for deriving surface reflectance spectra. Such types of AC methods include Empirical Line Calibration (ELC), Internal Average Reflectance (IAR), and Flat Field Correction (FFC). IAR and FFC actually are used for removing atmospheric effects from HRS data for retrieving relative surface reflectance spectra.

4.4.2.1 The Empirical Line Calibration (ELC)

The Empirical Line Calibration (ELC) is a direct application of the linear regression model in Equation 4.22 to estimate surface reflectance in the VNIR/SWIR spectral regions from the image digital number (DN) in a hyperspectral image (Conel et al. 1987). ELC requires field or laboratory reflectance spectra for at least one bright target and one dark target. The image DN values over the surface targets are linearly regressed against the field or laboratory spectra to derive gain (slope) and offset (intercept) curves (i.e., $a(\lambda)$ and $b(\lambda)$ in Equation 4.22). The gain and offset curves are then applied to the whole image to derive surface reflectance for the entire scene. This method produces spectra that are most comparable to reflectance spectra measured in the field, or in the laboratory. The basic ELC band-specific linear regression model is as

$$L_{obs}(\lambda) = a(\lambda)\rho(\lambda) + b(\lambda) \tag{4.22}$$

where L_{obs} (λ) is at-sensor pixel radiance (or pixel DN) of band λ in an image; $a(\lambda)$ and $b(\lambda)$ are gain and offset for band λ; $\rho(\lambda)$ is surface reflectance of band λ. If considering a case with only two samples (pixels) to simulate Equation 4.22, for which reflectance spectra are known to be $\rho_1(\lambda)$ (e.g., from a bright target) and $\rho_2(\lambda)$ (e.g., from a dark target) and corresponding at-sensor radiance $L_1(\lambda)$ and $L_2(\lambda)$ (or DNs) for image band λ, the unknown parameters ($a(\lambda)$ and $b(\lambda)$) can be estimated from simple slope–intercept relationship for a line as

$$\hat{a}(\lambda) = \frac{L_2(\lambda) - L_1(\lambda)}{\rho_2(\lambda) - \rho_1(\lambda)} \tag{4.23}$$

and

$$\hat{b}(\lambda) = \frac{L_1(\lambda)\rho_2(\lambda) - L_2(\lambda)\rho_1(\lambda)}{\rho_2(\lambda) - \rho_1(\lambda)} \tag{4.24}$$

According to these estimates from Equations 4.23 and 4.24, the surface reflectance $\hat{\rho}(\lambda)$ corresponding to the image pixel at-sensor radiance $L_{obs}(\lambda)$ or DN can be estimated by

$$\hat{\rho}(\lambda) = \frac{L_{obs}(\lambda) - \hat{b}(\lambda)}{\hat{a}(\lambda)} \tag{4.25}$$

If considering a case with multiple instances (pixels) to simulate the Equation 4.22, a linear least-square regression analysis is adopted. For this case, multiple reflectance spectra are known to be $\rho_1(\lambda)$, $\rho_2(\lambda)$,..., $\rho_N(\lambda)$ (measured from ground or laboratory) and corresponding at-sensor radiance $L_1(\lambda)$, $L_2(\lambda)$,..., $L_N(\lambda)$ (or DNs) for image band λ. The unknown parameters ($a(\lambda)$ and $b(\lambda)$) can be estimated from the set of linear equations of the form (Eismann 2012):

$$L_i(\lambda) = a(\lambda)\rho_i(\lambda) + b(\lambda) \tag{4.26}$$

where $i = 1, 2, \ldots, N$. Equation 4.26 can be replaced with a standard matrix-vector form:

$$\mathbf{AX} = \mathbf{B} \tag{4.27}$$

where

$$\boldsymbol{A} = \begin{bmatrix} \rho_1(\lambda) & 1 \\ \rho_2(\lambda) & 1 \\ \vdots & \\ \rho_N(\lambda) & 1 \end{bmatrix}, \ \boldsymbol{X} = \begin{bmatrix} a(\lambda) \\ b(\lambda) \end{bmatrix}, \text{ and } \boldsymbol{B} = \begin{bmatrix} L_1(\lambda) \\ L_2(\lambda) \\ \vdots \\ L_N(\lambda) \end{bmatrix}$$

The linear least-squares estimates can be independently computed on a band-by-band basis for each spectral band using the pseudo-inverse (Eismann 2012):

$$\mathbf{X} = [\mathbf{A}^{\mathrm{T}}\mathbf{A}]^{-1}\mathbf{A}^{\mathrm{T}}\mathbf{B}. \tag{4.28}$$

The major limitation of the ELC is the need to have a scene with identifiable materials that have known reflectance spectra. The materials must be sufficiently large to ensure that identified image pixels can cover only the reference target and not any surrounding background. Eismann (2012) suggested that an application of the ELC is to identify natural materials in the scene, such as uniform areas of a very homogeneous land cover type, for which the reflectance spectra can be measured by a calibrated spectrometer. To measure reflectance spectra from the natural materials that appear in the scene of image, an alternative method is bringing samples of the materials to a laboratory where their reflectance spectra can be measured by a spectrometer. The method assumes the entire scene is topographically flat, with no scan-angle or view-angle effects (Clark et al. 1995, Perry et al. 2000). In addition, if changes occur in the atmospheric properties outside the area used for the ELC method, the spectral reflectance data will contain atmospheric features (Gao et al. 2009).

The ELC is demonstrated by applying it to CASI hyperspectral data for which the ELC results are depicted in Figure 4.11 (Pu et al. 2008). In the study, the CASI imagery with 48 spectral bands with band width of approximately 11 nm and a spatial resolution of 2 m was acquired on July 2, 2002. Ground spectral data from light and dark targets located within the CASI scene were collected using a full-range Analytical Spectral Device (ASD) (FieldSpec®ProFR) on July 23, 2002. The targets included a white target, a parking lot, dry grasses, dense shrubs, a water ditch, an asphalt road surface, and a dense coast live oak canopy. Using the *in situ* spectral measurements taken from parking lot and water ditch as $\rho_1(\lambda)$ and $\rho_2(\lambda)$ and corresponding pixel DNs from the CASI scene as $L_2(\lambda)$ and $L_1(\lambda)$, both slope $a(\lambda)$ and intercept $b(\lambda)$ were calculated with Equations 4.23 and 4.24, and then the entire CASI image in DN were converted into surface reflectance using Equation 4.25. The six targets were selected because they represent nearly the whole range of spectral variation in the study area (from bright to dark targets). Figure 4.11a,b shows the six CASI DN curves that were extracted from six targets and corresponding six ASD spectral reflectance curves. Figure 4.11c shows the R^2 between CASI DN and ASD reflectance data for each band, which were calculated from six samples (i.e., the six curves in Figure 4.11a,b) for each band. The converted reflectance curves presented in Figure 4.11d are very similar to the original ASD reflectance curves shown in Figure 4.11b, with the exception of the band near 700 nm. The cause of the low R^2 value at that point (Figure 4.11c) was undetermined, possibly caused by the spectral shift by the CASI sensor.

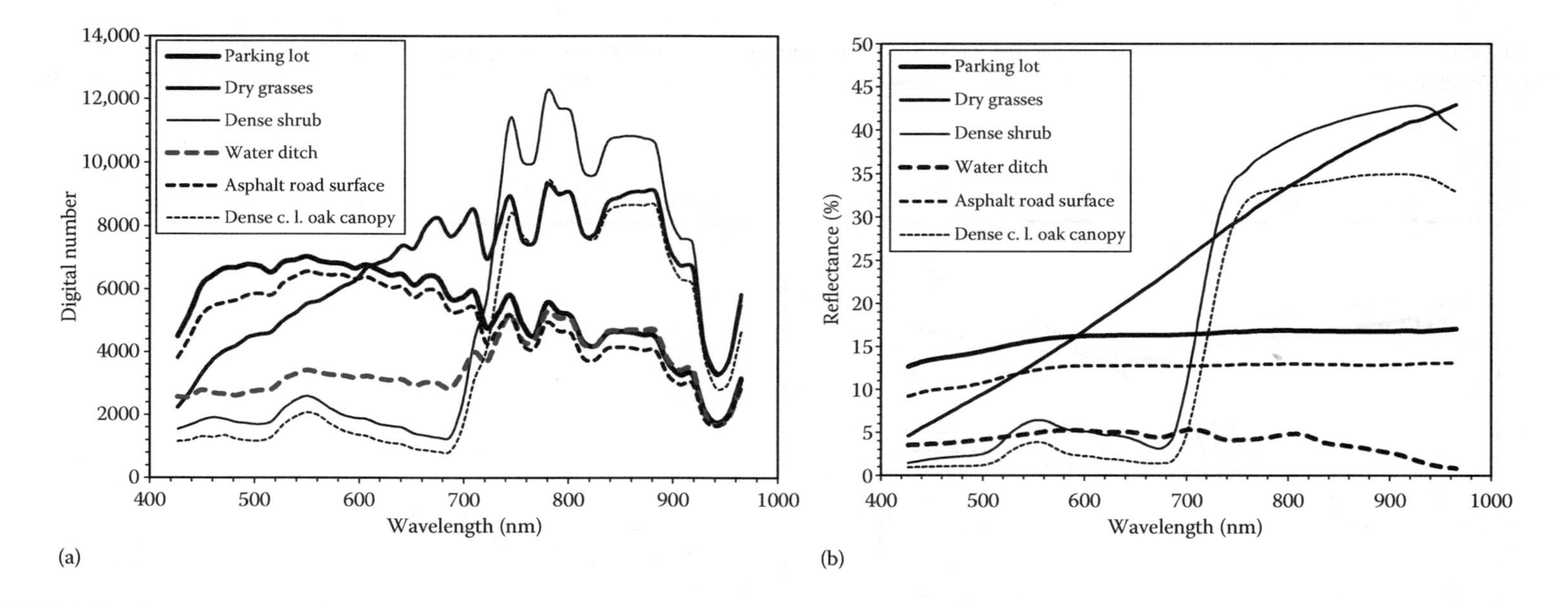

FIGURE 4.11 CASI hyperspectral data calibrated from digital number to reflectance using the empirical line calibration method. (a) CASI raw data (DN); (b) ASD spectral measurements taken from the same targets on the ground as (a). *(Continued)*

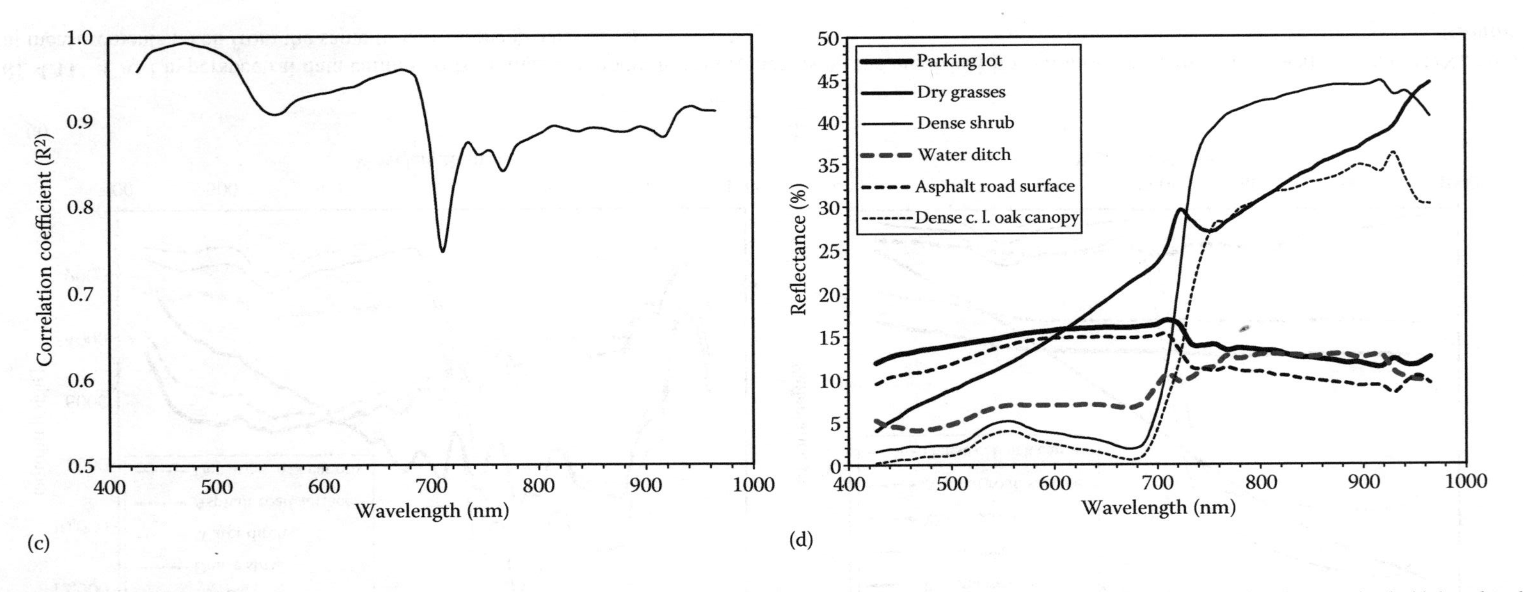

FIGURE 4.11 (CONTINUED) CASI hyperspectral data calibrated from digital number to reflectance using the empirical line calibration method. (c) band to band correlation between ASD spectral reflectances and CASI raw data; and (d) CASI converted to surface reflectances. (Modified from Pu, R., Kelly, M., Anderson, G. L., and Gong, P., *PE&RS*, 74(1), 65–75, 2008.)

4.4.2.2 Internal Average Reflectance (IAR) and Flat Field Correction (FFC)

Several scene-based empirical approaches were developed to remove atmospheric effects from hyperspectral imaging data for the derivation of relative surface reflectance spectra during the mid-1980s. The Internal Average Reflectance (IAR) approach developed by Kruse (1988) and the Flat Field Correction (FFC) approach created by Roberts et al. (1986) are two examples of such scene-based empirical approaches. According to Kruse (1988), the IAR approach first normalizes the data by scaling the sum of the DNs in each pixel spectrum in a hyperspectral image to a constant value. The effect of the normalization is to shift all the spectra (of pixels in DN) to nearly the same overall relative brightness. After the albedo differences in the spectra were removed, the next step is to convert the data to reflectance so that individual spectra could be compared directly with laboratory data for mineral identification. To do so, the approach uses the properties of the data themselves to calculate an approximation of the reflectance which is called IRA reflectance. The IRA reflectance is calculated by determining an average spectrum for a single scene or for all scenes acquired on an individual mission. Each spectrum (normalized pixel with a total number of bands in a hyperspectral image) in the scene is then divided by the average spectrum. The resulting spectra (i.e., the IRA reflectances) represent reflectance relative to the average spectrum and resemble laboratory spectra acquired of the same materials. Per the IRA reflectance approach, as noted by Kruse (1988), if the average spectrum used for calculating the IAR reflectance spectrum may itself have spectral characteristics associated with mineral absorption features, this can adversely influence the quality of the IAR reflectance spectra and thus significantly limit their usefulness in matching laboratory spectra. Without requirements of onboard calibration data and a priori knowledge of each scene, the IAR reflectance technique has an additional advantage of effectively removing the majority of atmospheric effects. This is because the average spectrum is supposed to contain the same or similar contributions from the atmosphere (so that, for instance, the same multiplicative atmospheric effect can be removed after a ratio process as in creating the IRA reflectance spectra). However, if the imaging scene covers an area with a wide variation in ground elevation or if the atmosphere is not uniform across the scene, the IAR technique will not effectively remove the atmosphere effects (Kruse 1988). In addition, this approach works best for hyperspectral imaging data acquired over arid areas without vegetation cover (Gao et al. 2009).

The FFC method (Roberts et al. 1986) relies on the existence of a neutral, homogeneous area, called *flat field*, in an image and normalizes all pixel spectra in the image to the flat field spectrum by dividing every pixel spectrum in the image by the mean spectrum of the flat field. With an appropriate flat field spectrum, the method can effectively convert hyperspectral data to relative surface reflectance and largely remove atmospheric multiplicative effects. The neutral, homogeneous flat field within the image is required to be (1) topographically flat, (2) spectrally flat (spectrally neutral reflectance, i.e., uniform spectral reflectance at all wavelengths without significant absorption features or vegetation), and (3) bright in order to reduce the effects of random image noise on the correction. The average flat field spectrum is supposed to be almost pure solar irradiance signature associated with atmospheric scattering and absorption effects.

Both the IAR approach and the FFC approach do not need any field measurements of reflectance spectra of surface targets. However, it is worth noting that the derived relative reflectance spectra with both approaches often have absorption features that are not present in reflectance spectra of corresponding materials measured in the field or laboratory. This is because the mean spectrum of a scene or the flat field often contains absorption effects of surface materials and is not 100 percent spectrally neutral (Gao et al. 2009). For the FFC approach, if noise is restricted to the flat field, the noise will then be introduced into the derived relative reflectance spectra in the scene. Therefore, caution must be used when applying both approaches, as spurious features can be introduced into the converted spectra if the mean spectrum contains absorption features associated with surface materials or vegetation (Clark and King 1987). In addition, caution may be needed on artifacts introduced by approaches including shifting wavelength for some features (Carrère and Abrams 1988) and changing intensity or distorting other spectral features of interest (Clark and King 1987).

4.4.3 Radiative Transfer Methods

After developing the empirical- and scene-based AC methods to correct atmospheric effects and convert HRS raw data to scaled surface (or relative) reflectance in the mid-1980s, many physically based (i.e., based on a radiative transfer [RT] model) AC models/methods were developed specifically for correcting atmospheric effects caused by scattering and absorptions for HRS imaging data. The physically based AC models/methods can overcome limitations of empirical approaches for surface reflectance retrievals. They can be used in different atmospheric conditions associated with seasonal and geographic variations, atmospheric scattering, and absorption, and can provide highly accurate surface reflectance with a spectral range covering most of the solar radiation spectrum. Usually, to run a physically based AC model, many atmospheric characteristic parameters are used to convert the HRS raw data into scaled surface reflectance. However, fortunately, many free variables in an atmospheric model have a similar influence on downwelling and path characteristics, and other parameters are relatively stable over space and time and can be adequately modeled using nominal values. Therefore, actually, the atmospheric correction problem can be reduced to an estimation problem over a reasonable set of key underlying parameters, such as aerosol optical thickness and column water vapor content (Eismann 2012). Many physically based AC models/methods need to use robust atmospheric RT codes such as Moderate Resolution Atmospheric Radiance and Transmittance Model 4 (MODTRAN4) and Second Simulation of the Satellite Signal in the Solar Spectrum (6S) to obtain necessary atmospheric scattering and absorption information. To perform an RT model-based AC algorithm to an HRS image, the following information is usually provided (Jensen 2005):

- Latitude and longitude of the image scene
- Date and time of the image acquisition
- Altitude of acquiring the image (e.g., 705 km AGL)
- Mean elevation of the image scene (e.g., 100 m ASL)
- A standard atmospheric model (e.g., mid-latitude summer, mid-latitude winter, tropical)
- Radiometrically calibrated image radiance data (e.g., a radiance data form in $\mu W\ cm^{-2}\ nm^{-1}\ sr^{-1}$ in ENVI/FLAASH)
- Information about specific band (i.e., band mean and FWHM)
- Local atmospheric visibility at the time of the image acquisition (e.g., 20 km)

In this subsection, six important AC models/methods, especially for correcting atmospheric effects on HRS imaging data, are reviewed and discussed. They are all RT model-based methods/algorithms, including ACORN, ATCOR, ATREM, FLAASH, HATCH, and ISDAS.

4.4.3.1 Atmospheric Correction Now (ACORN)

ACORN is a commercially available package developed by Analytical Imaging and Geophysics, LLC, which offers a range of options for atmospherically correcting hyperspectral and multispectral remote sensing imageries in the 0.4–2.5 μm spectral range (ImSpec 2002). ACORN is based on MODTRAN4 (Berk et al. 1999) radiative transfer (RT) code to produce high-quality surface reflectance without ground measurements. The RT model-based ACORN uses both the calibrated data and provided parameters to derive and model the absorption and scattering characteristics of the atmosphere. These derived and modeled atmospheric characteristics are then used to invert the image radiance to apparent surface reflectance (surface reflectance if terrain is flat).

The ACORN program for atmospheric correction is based on radiative transfer equations developed by Chandrasekhar (1960). In simplified terms, Equation 4.29 gives the relationship from contributions of the exo-atmospheric solar source, the atmosphere, and the Lambertian surface to the radiance measured by an Earth-looking sensor for a homogeneous plane parallel atmosphere (ImSpec 2002):

$$L_{obs}(\lambda) = \frac{E_s(\lambda)}{\pi}\left(E_{du}(\lambda) + \frac{T_{\theta_s}(\lambda)\rho(\lambda)T_{\theta_v}(\lambda)}{1 - E_{dd}(\lambda)\rho(\lambda)}\right) \tag{4.29}$$

where $L_{obs}(\lambda)$ is the total radiance arriving at the sensor; $E_s(\lambda)$ is the top of the atmospheric solar irradiance; $E_{du}(\lambda)$ and $E_{dd}(\lambda)$ are the upward and downward reflectances of the atmosphere, respectively; $T_{\theta_s}(\lambda)$ and $T_{\theta_v}(\lambda)$ are the downward and upward transmittances of the atmosphere, respectively; $\rho(\lambda)$ is the scaled spectral reflectance of the surface; and λ is spectral wavelength. Equations 4.29 and 4.20 are equivalent when considering $L_a = \frac{E_s(\lambda)E_{du}(\lambda)}{\pi}$, $E_s Cos(\theta_s) = E_s(\lambda)$, $S = E_{dd}(\lambda)$, $T_2 = T_{\theta_s}(\lambda)T_{\theta_v}(\lambda)$, and wavelength λ omitted.

When all the appropriate parameters in Equation 4.29 are given, the apparent (scaled) surface reflectance $\rho(\lambda)$ can be calculated using the equation as

$$\rho(\lambda) = \frac{1}{\left[\frac{\left(T_{\theta_s}(\lambda)E_s(\lambda)T_{\theta_v}(\lambda)\right)/\pi}{L_{obs}(\lambda) - \left(E_s(\lambda)E_{du}(\lambda)\right)/\pi}\right] + E_{dd}(\lambda)} \tag{4.30}$$

ACORN uses look-up-tables calculated with the MODTRAN4 radiative transfer code to model atmospheric gas absorption as well as molecular and aerosol scattering effects, converting the calibrated at-sensor radiance measurements from an HRS image to apparent surface reflectance (ImSpec 2002). The hyperspectral data must be spectrally and radiometrically calibrated to be suitable for ACORN. The ACORN user controls the strategy for water vapor estimation, artifact suppression, and visibility constraint and estimation. Differing water vapor derivations and constraints are possible. If the hyperspectral data set measures the region between 0.78 and 1.25 μm at 5 to 20 nm spectral resolution, then ACORN should be able to derive the water vapor from the calibrated hyperspectral data set. Water vapor can be estimated from the hyperspectral data set on a pixel-by-pixel basis using the water vapor absorption bands at 0.94 and/or 1.15 μm. A lookup table for a range of column water vapor densities is generated using MODTRAN4 and then fitted in a least-squares sense against the imaging spectrometer data (Kruse 2004). ACORN estimates the visibility by analyzing the spatial and spectral content of the hyperspectral data in the region from 0.40 to 1.00 μm. The two-way transmitted radiance and atmospheric reflectance are calculated for each pixel using MODTRAN4 and the derived water vapor, pressure elevation, and aerosol optical depth estimations. As an example over a geologically interesting region, calibrated hyperspectral radiance data are shown in Figure 4.12a for minerals from Cuprite, Nevada. The atmosphere is suppressed and mineral absorption features near 0.9, 2.2, and 2.3 μm are clearly identifiable (Figure 4.12b). For an ecologically interesting and urban example, using an AVIRIS image acquired over the Jasper Ridge Ecological Preserve, Stanford, California, and adjacent urban areas, Figure 4.12c,d presents calibrated radiance spectra from the AVIRIS hyperspectral instrument from five target areas as well as derived reflectance spectra after use of ACORN. After atmospheric correction the spectral absorption feature inherent to the vegetation and man-made surface materials are revealed (ImSpec 2002).

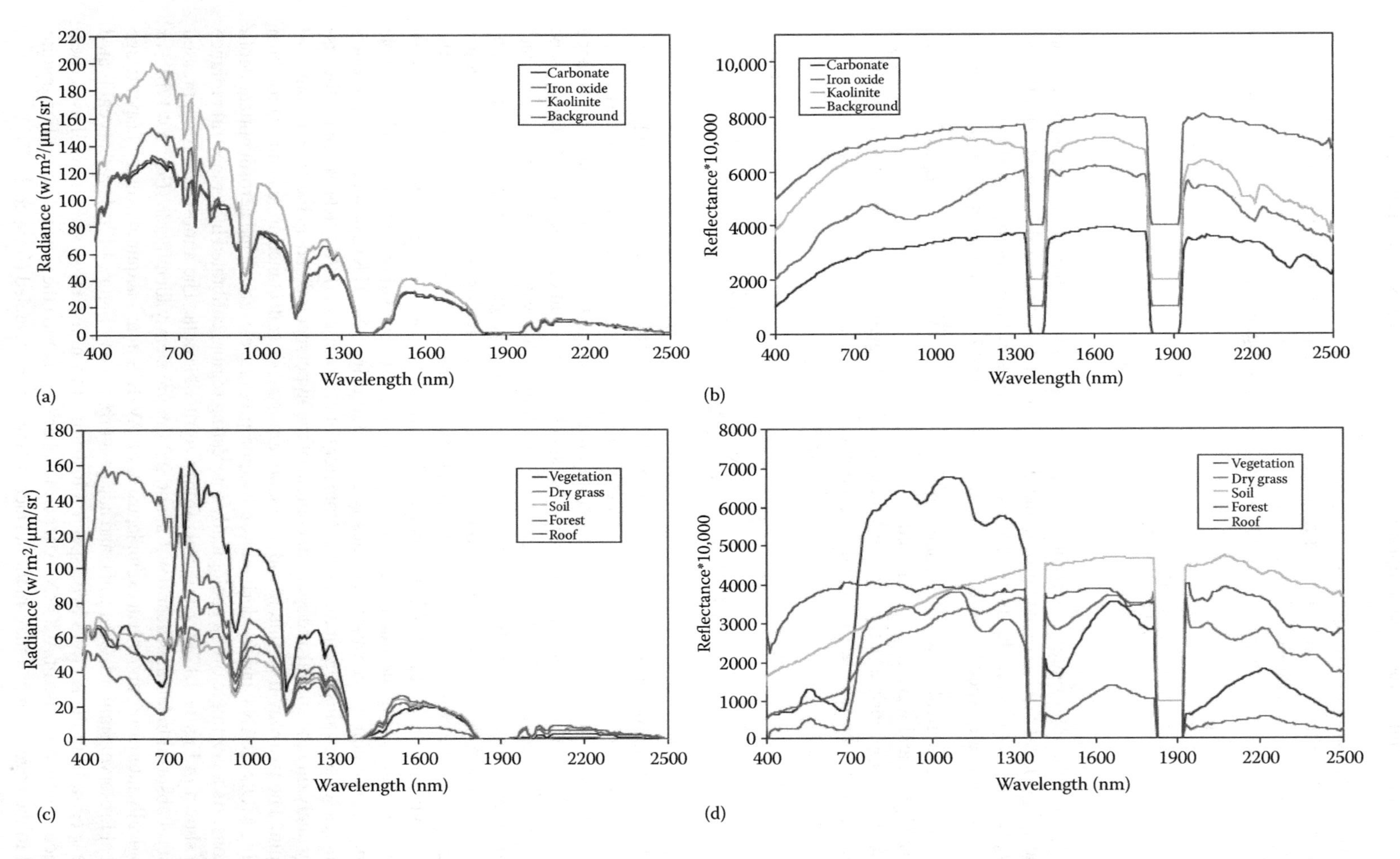

FIGURE 4.12 (a) Calibrated measured AVIRIS radiance spectra from Cuprite, Nevada; (b) the same spectra after ACORN atmospheric correction from (a); (c) calibrated measured AVIRIS radiance spectra from Jasper Ridge Ecological Preserve, Stanford, California; and (d) the same spectra after ACORN atmospheric correction from (c). (Courtesy of ImSpec LLC.)

A key feature of ACORN uses a full spectral fitting technique to model the overlap of absorptions between water vapor and liquid water in surface vegetation (Kruse 2004). ACORN also has a feature that is single spectrum enhancement of a hyperspectral atmospheric correction. Per the feature, ACORN uses a spectrum derived from an atmospherically corrected hyperspectral data and an accurately known spectrum measured at the same target. Using the two spectra, the full atmospherically corrected hyperspectral image is then corrected (improved) to the accuracy of the known spectrum. In addition, ACORN contains several other functions, including artifact suppression options (types), automated wavelength adjustment, elimination of noisy channels, and correction (or "polishing") of residual reflectance spectral errors (Gao et al. 2009). ACORN is designed to process data acquired under typical remote sensing conditions. However, if the atmospheric condition is too poor, such as if it's very hazy or smoky, some of the modes/features of ACORN may not work properly in fully correcting for atmosphere. Although ACORN can also be applied to water targets such as lakes, rivers, and the ocean, the quality of the atmospheric correction will vary depending on the quality of the calibration and the quality of the MODTRAN4 calculations and their implementation in ACORN (ImSpec 2002).

4.4.3.2 Atmospheric Correction (ATCOR)

A series of ATCOR atmospheric and topographic correction codes have been continually updated and developed throughout the 1990s and 2000s (Richter 1990, 1996, 1998; Richter and Schläpfer 2002), with the latest version, ATCOR4, using a large database containing results of radiative transfer calculations based on the MODTRAN5 radiative transfer model (Berk et al. 2005, Black et al. 2014). ATCOR is a method for computing a ground reflectance image for the reflective spectral bands in a spectral range of 0.4–2.5 μm, and emissivity images for the thermal bands in a spectral range 8–14 μm. ATCOR algorithm was originally developed at the DLR—German Aerospace Center and currently is in three different types (ATCOR2, ATCOR3, and ATCOR4). For a nearly horizontal surface or flat terrain, ATCOR2 ("2" = two geometric degrees of freedom [DOF] of the flat plane) is a spatially adaptive fast atmospheric correction algorithm; ATCOR3 ("3" = three DOFs, x, y, z) is designed for rugged topographical surface, hence a Digital Elevation Model (DEM) is used in the ATCOR3 algorithm for atmospheric correction; and ATCOR4 ("4," indicating up to four geometric DOFs: x, y, z, and scan angle) is used with remotely sensed data acquired by suborbital (airborne) remote sensing system. All versions of ATCOR codes allow the corrections for adjacency effect, and 3D codes additionally include a correction of topographical effects and bidirection reflectance distribution function (BDRF) in addition to haze and low cirrus cloud removal. All ATCOR versions share the following key features (Richter 2004):

- A fast numerical correction algorithm accounting for the adjacency effect.
- Capabilities for constant or spatially varying atmospheric conditions (visibility or aerosol optical depth).
- Inflight/vicarious calibration to obtain/verify the radiometric instrument calibration.
- Solar reflective spectral bands: Surface reflectance is calculated from calibrated imagery.
- Thermal spectral band(s): Surface emissivity and temperature are calculated.
- Combined atmospheric/topographic correction in rugged terrain. A DEM is required and surface topography (slope/aspect orientation) is taken into account.
- Atmospheric water vapor retrieval on a per-pixel basis, water vapor map (0.94/1.13 μm bands required).
- Haze removal.
- Cloud shadow removal.
- Maps of atmospheric optical depth, haze, cloud, and cloud shadow masks.

ATCOR uses a database of look-up tables (LUTs) for atmospheric correction for calculations of surface reflectance for optical imaging data and surface temperature for thermal data. The LUTs

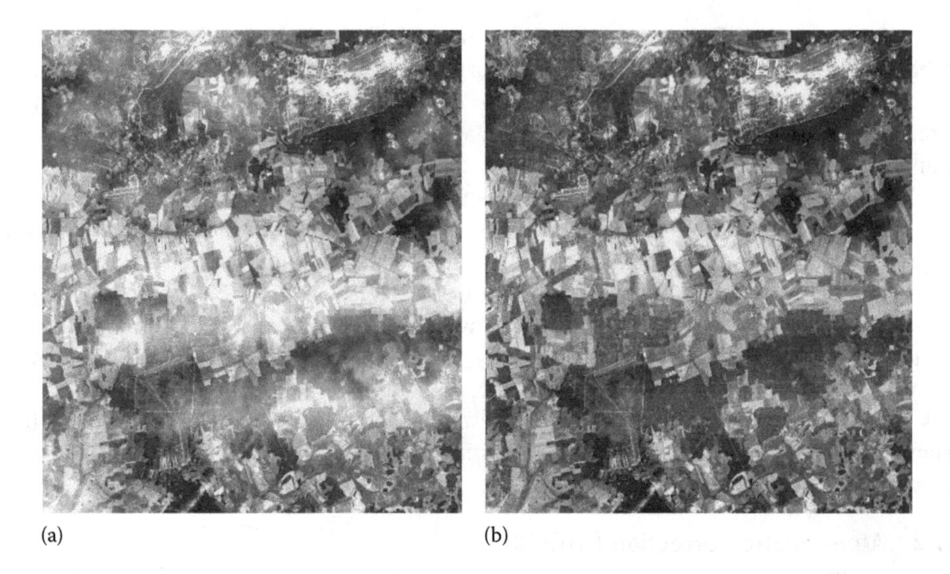

FIGURE 4.13 (a) Image containing substantial haze prior to atmospheric correction; (b) image after atmospheric correction using ATCOR. (Courtesy of Leica Geosystems and DLR—the German Aerospace Center.)

of atmospheric correction functions (path radiance, atmospheric transmittance, direct and diffuse solar flux, etc.) are calculated with a radiative transfer code (MODTRAN4+), which depends on scan angle, relative azimuth angle between scan line and solar azimuth, and terrain elevation. Atmospheric water vapor retrieval uses the band ratio technique on central absorption bands at 0.94 and 1.13 μm (Gao and Goetz 1990). Additionally, the terrain shape obtained from a DEM is taken into account. An image-based atmospheric/topographic correction in the latest version (ATCOR4) needs to be performed iteratively employing the LUTs and to consider three major steps for the solar spectral region: (1) neglecting the influence of the adjacency effect, (2) approximating correction of the adjacency effect, and (3) enhancing the result achieved at step 2 (see Richter 1998 and Richter and Schläpfer 2002 for the detailed algorithm). A large database is adopted in the program to avoid time-consuming iterations with a radiative transfer code for cases that are already included in the database. For a specific user, only a resampling with the channel-specific spectral response function of the selected remote sensing sensors is required in these cases before processing the image data (Richter and Schläpfer 2002). More specifically, the airborne version of ATCOR (ATCOR4) allows the user to include new airborne hyperspectral instruments including along-track (fore/aft) viewing sensors. A standard list of supported hyperspectral imaging sensors includes AVIRIS, HyMap, DAIS-7915, CASI, SASI, and Daedalus (Richter 2004). Panchromatic, multispectral, and hyperspectral imagery from the solar spectral region and thermal region can be processed to obtain surface reflectance and temperature images by the ATCOR4. An example of before and after AC with the ATCOR method is presented in Figure 4.13. Per the figure, the AC effectiveness is visually significant.

4.4.3.3 Atmosphere Removal (ATREM)

The ATREM program was developed by the Center for the Study of Earth from Space (CSES) at the University of Colorado in the early 1990s for retrieving scaled surface reflectance from hyperspectral data using a radiative transfer model (Gao et al. 1993, 2009). It allows the retrieval of scaled surface reflectance spectra from VNIR/SWIR spectral regions from hyperspectral imaging data, such as AVIRIS data, without ground-based spectral measurements (Green et al. 1993, Gao et al. 2009). The ATREM code is based on the 5S (later on the 6S) radiative transfer code (Tanré et al. 1986) and retrieves atmospheric characteristics by first estimating three key characteristics of the atmosphere: atmospheric water vapor, aerosol optical depth, and surface pressure elevation. These characteristics may determine the modeling parameters for estimating atmospheric transmittance and path radiance that are then coupled with the known solar zenith angle to estimate scaled surface reflectance from measured at-sensor radiance data (Eismann 2012). The underlying models (Equations 4.31–4.33) are for retrieval of scaled surface reflectance from image-based at-sensor radiance. The at-sensor radiance measured by imaging spectrometer, including atmospheric path radiance and surface reflected solar radiance, can be often converted to *apparent reflectance*. According to Gao et al. (1993) and Tanré et al. (1986), the apparent reflectance, $\rho^*_{obs}(\lambda, \theta_v, \varphi_v, \theta_s, \varphi_s)$, is defined by

$$\rho^*_{obs}(\lambda, \theta_v, \varphi_v, \theta_s, \varphi_s) = \frac{\pi L_{obs}(\lambda, \theta_v, \varphi_v, \theta_s, \varphi_s)}{E_s(\lambda) Cos(\theta_s)} \tag{4.31}$$

where θ_s and φ_s are the solar zenith and azimuth angles, respectively; θ_v and φ_v are the sensor zenith and azimuth angles, respectively; L_{obs} is the at-sensor radiance; E_s is the solar flux at the top of the atmosphere when the solar zenith angle is equal to zero; and λ is the spectral wavelength. As the apparent reflectance is measured at the top of atmosphere, according to Tanré et al. (1986), when the surface is assumed to be Lambertian and the adjacency effect is neglected, the apparent reflectance, $\rho^*_{obs}(\lambda, \theta_v, \varphi_v, \theta_s, \varphi_s)$, can be approximated as:

$$\rho^*_{obs}(\lambda, \theta_v, \varphi_v, \theta_s, \varphi_s) = \left[\rho^*_{atm}(\lambda, \theta_v, \varphi_v, \theta_s, \varphi_s) + \frac{T_d(\lambda, \theta_s) T_u(\lambda, \theta_v) \rho(\lambda)}{1 - S(\lambda)\rho(\lambda)}\right] T_g(\lambda, \theta_v, \theta_s) \tag{4.32}$$

where, ρ^*_{atm} is the path reflectance; T_d and T_u are the downward scattering transmittance and upward scattering transmittance, respectively; S is the spherical albedo of the atmosphere; ρ is the surface reflectance; and T_g is the total gaseous transmittance in the sun–surface–sensor path. Per Equation 4.32, the first term in the bracket in Equation 4.32, ρ^*_{atm}, is the contribution from atmospheric scattering caused by gas molecules and aerosols, whereas the second term in the bracket, $T_d T_u \rho/(1 - s\rho)$, is the contribution directly from surface reflection. The term T_g addresses all absorptions caused by atmospheric gases, as shown in Figure 4.4. Clearly, Equation 4.32 treats the atmospheric scattering and gaseous absorption processes as two independent processes. According to Gao et al. (2009), when coupling effects between the two terms are neglected and the notations for relevant quantities are simplified, Equation 4.32 can be solved for surface reflectance ρ as follows:

$$\rho = \frac{\frac{\rho^*_{obs}}{T_g} - \rho^*_{atm}}{\left[T_d T_u + S\left(\frac{\rho^*_{obs}}{T_g} - \rho^*_{atm}\right)\right]} \tag{4.33}$$

In the ATREM program, after the atmospheric quantities in Equation 4.33: T_g, ρ^*_{atm}, T_d, T_u, and S are simulated with radiative transfer models (e.g., 5S, 6S, or MOATRAN), along with the at-sensor measured radiance (L_{obs} converted from hyperspectral imaging data) and considering Equations 4.31 and 4.33, the scaled surface reflectance, ρ, can be retrieved.

The ATREM program derives the integrated water vapor amount on a pixel-by-pixel basis from the 0.94 and 1.13 μm water vapor absorption bands using a band ratio technique (a ratio of the band center to the two band shoulders) (Gao et al. 1993). Atmospheric transmittance spectra are simulated for each of seven atmospheric gases (water vapor [H_2O], carbon dioxide [CO_2], ozone [O_3], nitrous oxide [N_2O], carbon monoxide [CO], methane [CH_4], and oxygen [O_2]) using the Malkmus narrowband model (Malkmus 1967) based on the solar and observational geometry and the derived water vapor values (Kruse 2004). The atmospheric scattering effect due to atmospheric molecules and aerosols is modeled using the 6S radiative transfer code (Tanré et al. 1986). An aerosol model and a surface visibility need to be determined to model the aerosol effect (optical depth). To estimate aerosol optical depth, at-sensor radiance data are fit over a 0.40–0.60 μm spectral range to model radiance data using a nonlinear least-squares-fitting algorithm. Surface pressure elevation parameter is estimated by ATREM from the strength of a 0.76 μm oxygen absorption line in measured radiance data. This is also performed by a nonlinear least-squares fit of the measured-to-modeled data in the 0.76 μm region. Both aerosol optical depth and surface pressure elevation are simulated in a radiative transfer model-generated lookup table using a simplex search method (Eismann 2012). Apparent reflectance spectra are obtained by dividing the measured radiances (from hyperspectral imaging data) by solar irradiances above the atmosphere. The scaled surface reflectance spectra are then derived from the apparent reflectance spectra using the simulated atmospheric gaseous transmittances and the simulated molecular and aerosol scattering data in the ATREM program.

The final output results from the ATREM program are a water vapor image and a surface reflectance data cube (two dimensions for spatial imaging and one dimension for spectral information; assuming that the imaging area is over a horizontal Lambertian surface). If the imaging area is not true for the assumption, the retrieved surface reflectance is called *scaled surface*

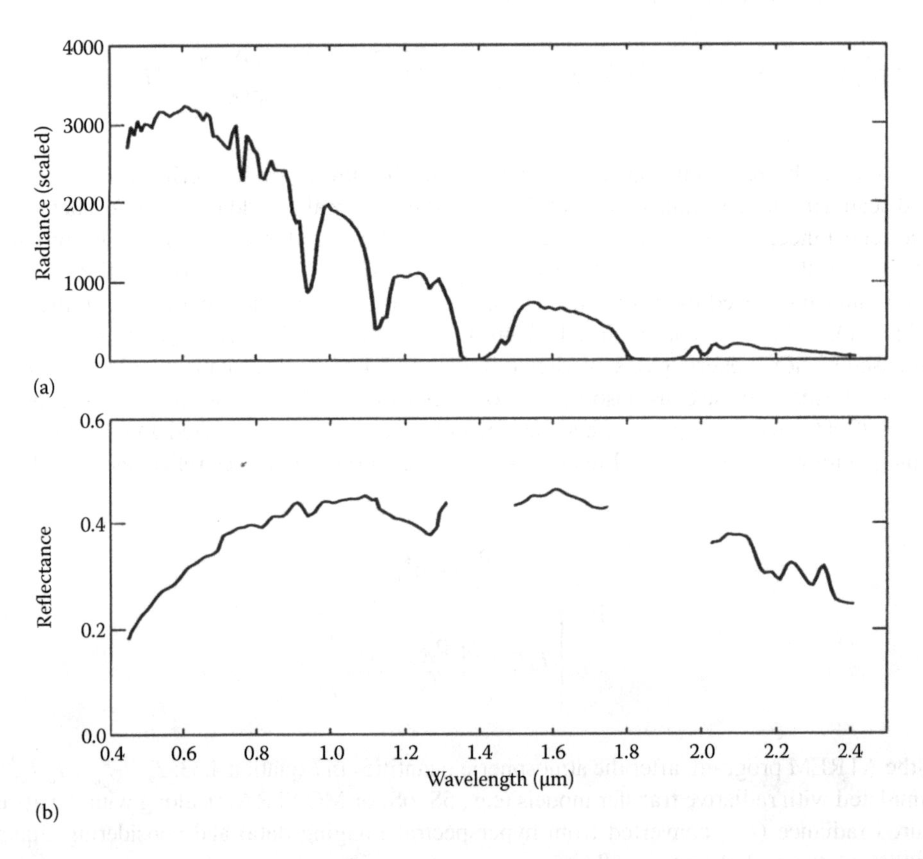

FIGURE 4.14 (a) An AVIRIS spectrum acquired on July 23, 1990, over an area covered by the mineral kaolinite in Cuprite, Nevada; (b) the corresponding retrieved surface reflectance spectrum using the ATREM code, which clearly shows the characteristic kaolinite doublet feature near 2.17 μm. (Reprinted from Gao, B.C., Montes, M. J., Davos, C. O., and Goetz, A. F. H., *Remote Sensing Environment*, 113, S17–S24, 2009. Copyright 2009 with permission from Elsevier.)

reflectance that can be converted to *surface reflectance* after further correcting the effects of terrain surface (slope and aspect) and BRDF property of the surface (Gao et al. 2009). Figure 4.14 presents an example of surface reflectance retrieved from AVIRIS data using the ATREM program. Several mineral absorption features in the 2.1–2.4 μm spectral region, especially the kaolinite doublet absorption features near 2.2 μm, are recovered after atmospherically correcting the AVIRIS data (Gao et al. 2009).

The early version of the ATREM source code, called band model version (Version 3.1), was developed in the early 1990s and marked a major advance in imaging spectrometry at that time. This is because the early version made it possible to retrieve surface reflectance spectra from imaging spectrometer data for a variety of research and application activities in a large land remote sensing community, in particular, the HRS community. During the late 1990s and early 2000s, the ATREM code was upgraded. The major upgrades include (1) replacing the band model with a line-by-line atmospheric transmittance model (Gao and Davis 1997) and the HITRAN2000 line database (Rothman et al. 2003); (2) upgrading the 5S radiative transfer code to the newer 6S code for modeling atmospheric characteristics including scattering effects; and (3) adding a new module for modeling absorption effects caused by atmospheric NO_2 in the 0.4–0.8 μm spectral region. The updated ATREM algorithm has been released to a small number of users for further improving derivation of surface reflectance spectra, especially for spectral regions that have atmospheric gaseous absorption features (Gao et al. 2009).

4.4.3.4 Fast Line-of-Sight Atmospheric Analysis of Spectral Hypercubes (FLAASH)

The FLAASH atmospheric correction program was developed jointly by the Air Force Research Laboratory, Space Vehicles Directorate (AFRL/VS), Hanscom AFB, and Spectral Sciences, Inc. (SSI) over the last 20 years to provide accurate and fast first-principles atmospheric correction of VNIR–SWIR (0.4–2.5 μm) hyperspectral and multispectral imagery (Adler-Golden et al. 1999, Cooley et al. 2002, Perkins et al. 2012). FLAASH draws on existing spectral analysis methods and codes that have been developed for research and applications (e.g., Richter 1996, Moses and Philpot 2012). It is designed as a general code and has been generated in parallel with upgrades to MODTRAN, the radiative transfer (RT) model that offers the physical understanding behind the mathematical assumptions in FLAASH, in order to take advantage of latest improvements in accuracy and speed (Cooley et al. 2002). FLAASH provides accurate, physics-based derivation of scaled surface reflectance through derivation of atmospheric properties such as surface albedo, surface elevation, column water vapor, and aerosol and cloud optical depths from hyperspectral imaging data (Kruse 2004). Apart from finally outputting scale surface reflectance image from both hyperspectral and multispectral imaging data, FLAASH also produces a column water vapor image, a cloud map, and a visibility range value for the scene.

FLAASH solves a RT model for the pixel surface reflectance ρ using a standard at-sensor radiance, L_{obs}, equation (similar to Equations 4.20 and 4.29) when the surface is assumed to be flat and Lambertian, which may be written as (Cooley et al. 2002, Richter and Schläpfer 2002, Perkins et al. 2012)

$$L_{obs} = L_a + \frac{a\rho}{1-\rho_e S} + \frac{b\rho_e}{1-\rho_e S} \tag{4.34}$$

where ρ_e is the spatially averaged surface reflectance; S is the spherical albedo of the atmosphere from the ground; L_a is the atmosphere path radiance; and a and b are coefficients that solely vary with atmospheric and geometric conditions. Wavelength dependence of these quantities is omitted for notational convenience. The second term in Equation 4.34 responds to the radiance reaching the surface (from both sky light and direct solar illumination) that is backscattered directly in the sensor, whereas the third term corresponds to the radiance from the surface that is diffusely transmitted into the sensor, giving rise to the *adjacency effect*. The spatially averaged reflectance, ρ_e, is used to account for the adjacency effect (spatial mixing of radiance among nearby pixels) caused by atmospheric scattering. To ignore the adjacency effect correction, set $\rho_e = \rho$. However, this correction can result in significant reflectance errors at short wavelengths, especially under hazy conditions and when strong contrasts occur among the materials in the scene (ACM User's Guide 2009).

The values of a, b, S, and L_a in Equation 4.34 are determined from MODTRAN calculations with inputs of the viewing and solar angles and the mean surface elevation of the measurement. These atmospheric characteristic quantities assume a certain model atmosphere, aerosol type, and visible range. The variable atmospheric quantities that have the greatest impact on VNIR–SWIR wavelengths are the column water vapor and the aerosol amount (Perkins et al. 2012). To determine unknown and variable column water vapor and aerosol optical thickness, the FLAASH built-in MODTRAN calculations are iterated over a series of varying column water vapor amounts, and the water vapor is estimated using the method described later on in the text. A visibility estimate associated with aerosol optical thickness for the scene is retrieved using "dark" pixels, as described later on.

The spatial averaging implied in ρ_e is a convolution with a spatial point spread function (PSF) (Perkins et al. 2012). The PSF describes the relative contributions to the pixel radiance from points on the ground at different distances from the direct line of sight. To be accurate averaging, cloud-contaminated pixels must be removed prior to averaging. Strictly speaking, the ρ_e in the numerator and the denominators in Equation 4.34 are not identical. However, since $\rho_e S$ is generally very small, it is possible to approximate the denominator PSF with the numerator PSF, which describes the upward diffuse transmittance. Therefore, ρ_e can be estimated from an approximate form of Equation 4.34:

$$L_e = L_a + \frac{(a+b)\rho_e}{1-\rho_e S} \tag{4.35}$$

where L_e is the radiance image convolved with the PSF (i.e., the spatial averaged radiance image). With the values of a, b, S, and L_a estimated by MODTRAN calculations along with image-based L_e and L_{obs}, Equation 4.34 is then solved for ρ.

To retrieve the column water vapor, the FLAASH uses radiance averages for two sets of spectral bands: an "absorption" band or set of bands centered at a water absorption band (typically the 1.13 μm band) and a reference set of bands taken from both edges of the band. The default band selections may be overridden by the user in FLAASH. For performing rapid water retrieval, a 2D look-up table is constructed from the MODTRAN outputs. One dimension of the table is the ratio of "reference" to "absorption" and the other is the "reference" radiance (Perkins et al. 2012). To estimate aerosol amount, the FLAASH code includes a method for retrieving an estimated aerosol/haze amount based on selected dark land pixels in the scene. The dark land pixels are dominated by green vegetation and soil with a characteristic reflectance ratio of 0.5 at ~0.66 μm versus ~2.10 μm (Kaufman et al. 1997). According to observations by Kaufman et al. (1997), there is a nearly fixed ratio between reflectances for dark pixels at 0.66 μm and 2.10 μm. FLAASH uses the method to retrieve the aerosol amount by iterating Equations 4.34 and 4.35 over a series of visible ranges, such as 17 km to 200 km. For each specific visible range, FLAASH calculates the ratio of the average reflectance of the dark pixels at 0.66 μm to that at 2.10 μm in the scene, and it interpolates to search for the best estimate of the visible range by matching the ratio to the average reflectance of ~0.45 that was observed by Kaufman et al. (1997). FLAASH derives surface pressure elevations by applying the same method to the oxygen 0.762 μm absorption band (similar for ATREM). FLAASH also offers several additional options, including (1) correcting for light scattered from adjacent pixels (a key feature); (2) recalibrating hyperspectral sensor wavelengths using sharp molecular absorption features in the atmosphere, such as water vapor, oxygen, and carbon dioxide bands; (3) correcting spectral "smile" error; and (4) removing artifacts from hyperspectral reflectance retrievals using the data itself only (Perkins et al. 2012). FLAASH supports hyperspectral sensors (such as HyMAP, AVIRIS, HYDICE, Hyperion, Probe-1, CASI, and AISA) and multispectral sensors (such as ASTER, IRS, Landsat, RapidEye, and SPOT). Water vapor and aerosol retrieval are only possible when the image contains bands in appropriate wavelength positions (e.g., narrow bands at 0.66, 0.76, 1.13, and 2.10 μm). In addition, FLAASH can correct images collected in either vertical (nadir) or slant-viewing geometries (AMC User's Guide 2009). Figure 4.15 presents an example of surface reflectance retrieved from Hyperion imagery using the FLAASH program. The retrieved reflectance of Hyperion imagery shows the rich spectral information of objects. For

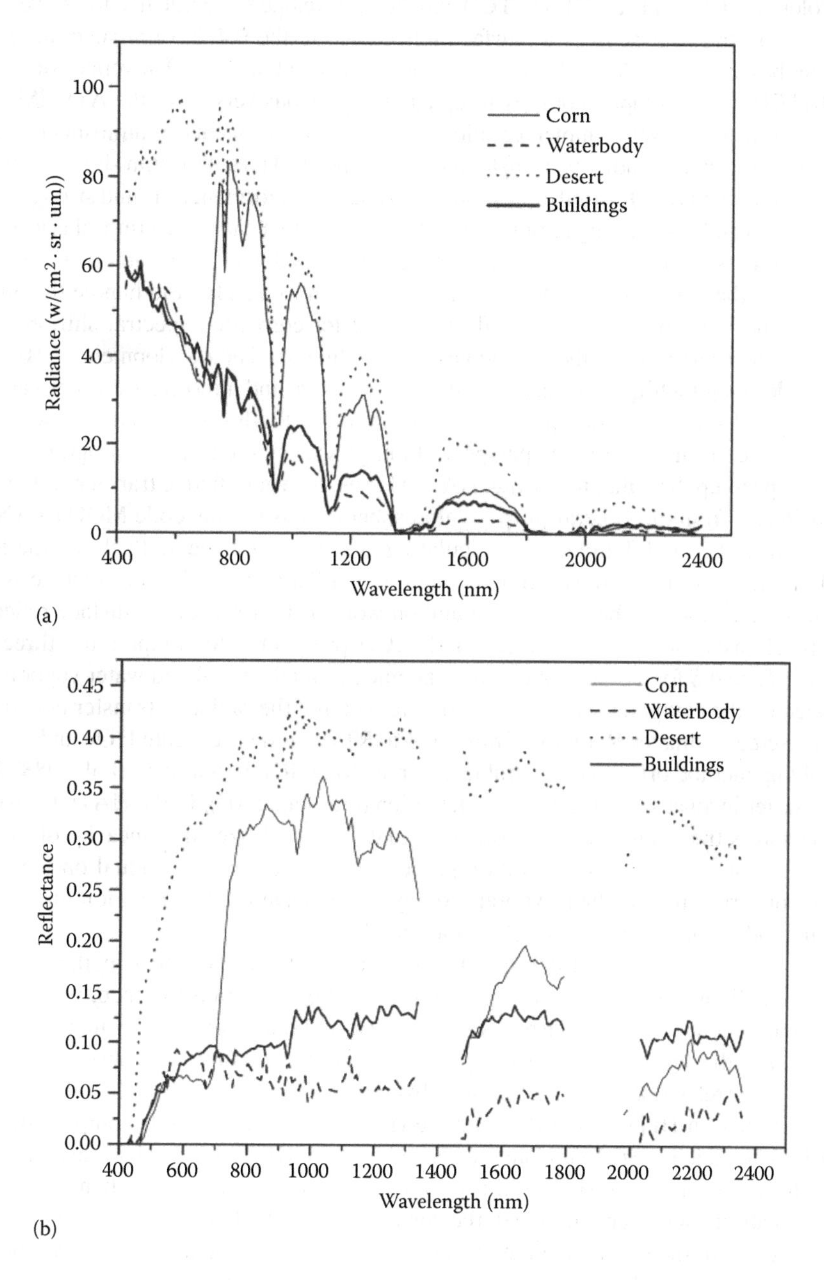

FIGURE 4.15 (a) At-sensor radiance of four typical objects extracted from Hyperion imagery (427–2396 nm) acquired in the City of Zhangye, Gansu Province, China, on September 10, 2007; (b) reflectance of the four typical objects from (a) after FLAASH atmospheric correction. (©2008 IEEE. Reprinted from Yuan, J., and Niu, Z., *Proceedings of 2008 IEEE International Workshop on Earth Observation and Remote Sensing Applications*, 6 pp., 2008. With permission.)

instance, after FLAASH atmospheric correction, corn typical reflectance spectrum curve presents with high reflectance at the green band and low reflectance at the red band, but the depth of the red band was not obvious due to its poor growth because of infection (Yuan and Niu 2008).

4.4.3.5 High-Accuracy Atmospheric Correction for Hyperspectral Data (HATCH)

The atmospheric correction program, HATCH, was developed at the University of Colorado in Boulder, Colorado (Qu et al. 2003). HATCH has been developed specifically to convert radiance from imaging spectrometer sensors to surface reflectance in the 0.4–2.5 μm spectral region on a pixel-by-pixel basis. It is available through a license agreement with the University of Colorado at Boulder. HATCH was developed not only to update the previous version of the ATREM model but also to add new features. With implementation of recent advancements in atmospheric correction techniques and atmospheric radiative transfer modeling, the HATCH algorithm shows improvements in many aspects over ATREM. Briefly, they include (1) better performance around strong water vapor absorption regions and overlapping regions for different gases that cause the atmospheric absorptions using the smoothness test technique and the correlated-*k* method, and (2) the automatic spectral calibration capability that has proved to be a promising function for HATCH to handle the problematic residual atmospheric feature in derived reflectance and for correction spectral shift errors due to pushbroom imaging mode using the smoothness test technique. The development of HATCH program was for deriving high-quality surface reflectance spectra and atmospheric water vapor images from hyperspectral imaging data (Qu et al. 2003). Figure 4.16 shows a sample of scaled surface reflectance retrieved from Hyperion hyperspectral imagery using the HATCH program.

In order to speed up the data processing, HATCH uses its own radiative transfer (RT) model that is slightly different from the general-purpose atmospheric transmission code MODTRAN (Berk et al. 1999). In HATCH, the RT model for calculating at-sensor radiance and solving the RT model for ground surface reflectance retrieval associated with a flat and Lambertian surface is the same as Equations 4.20 and 4.21. The major computation workload for retrieving surface reflectance on a pixel-by-pixel basis using the RT model in HATCH program is to compute the three radiative quantities, L_a, T_2, and S for a given solar-sensor geometry and the tabulated water vapor amount. In order to speed up the computation, a new method for solving the radiative transfer equation, called Multi-Grid discrete ordinates Radiative Transfer (MGRT) method is executed (Qu and Goetz 1999). The new solving method offers a comparable accuracy to DISORT (Stamnes et al. 1988), but is five to ten times faster in an at-sensor radiance calculation (Qu et al. 2003). In the HATCH program, the new method solves the integral form of the RT equation and utilizes the linearity of the equation to speed up the convergence of the iteration process by computing the residual on fewer streams. It generally converges in less than five iteration cycles to solve the RT equation due to using the multigrid method in the angular space (Qu et al. 2003).

According to Qu et al. (2003), the HATCH program uses the correlated-*k* method (Goody et al. 1989, Lacis and Oinas 1991) to calculate gaseous absorption and transmittance, which transforms the line-by-line integration over a narrow spectral band of a radiative quantity (e.g., transmittance) to integration over the cumulative probability distribution function of the gas absorption coefficient. HATCH uses a correlated-*k* lookup table built from the line-by-line code LBLRTM (Clough and Iacono 1995) based on HITRAN 2000 database (Rothman et al. 2003) to compute the transmittance. The lookup table is easy to update once a new version HITRAN database becomes available (it is independent of the HATCH algorithm). In addition to better accuracy than band models for computing gas absorption (transmittance), the correlated-*k* method also provides an explicit way to accurately account for the interaction between multiple scattering and absorption. A more accurate account for this interaction by using the correlated-*k* method is thus expected to improve HATCH performance in the short wavelength regions (Qu et al. 2003).

Water vapor amount is one of the major uncertain factors in the atmospheric constituents that significantly influences radiation in the 0.4–2.5 μm spectral region. In the HATCH program, a new technique, called the *smoothness test*, was developed to avoid the linearity assumption for the

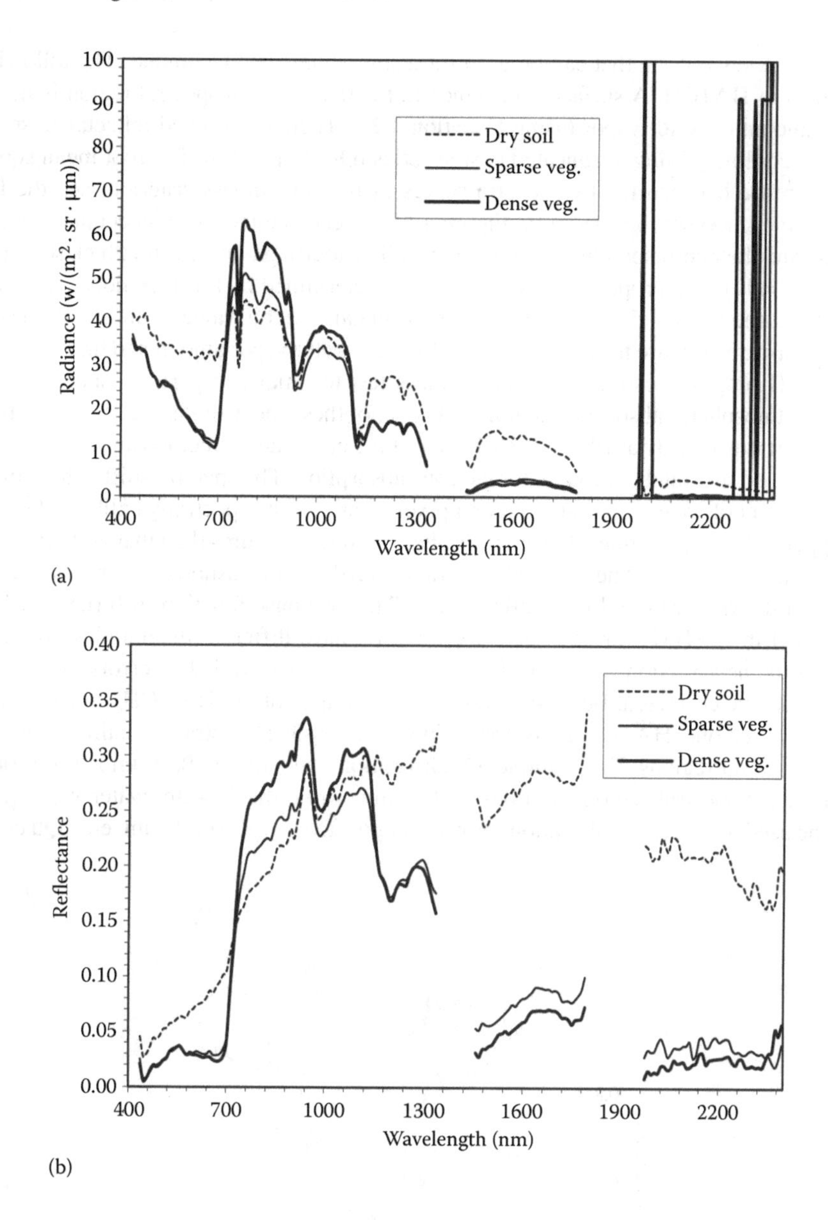

FIGURE 4.16 Representation of effectiveness of atmospheric correction with Hyperion hyperspectral imagery acquired at Blodgett Forest Research Station, University of California, Berkeley, California, on October 9, 2001: (a) Original at-sensor radiance and (b) surface reflectance retrieved from (a) with an atmospheric correction code: High Accuracy Atmospheric Correction for Hyperspectral Data (HATCH). (Modified from Pu, R., and Gong, P., *Remote Sensing of Environment,* 91, 212–224, 2004.)

surface reflectance (see the three-band ratioing technique from Gao et al. [1993] for the linearity assumption for retrieving water vapor). According to Qu et al. (2003), the smoothness test technique is briefly summarized as follows. The smoothness technique is based on the principle often used by hyperspectral data analysts that either under- or overestimation of water vapor amount results in irregularities in the retrieved surface reflectance. Features and reflectance curves generated from poor atmospheric corrections (e.g., using an improper transmittance) are generally rougher than the inherent surface spectral features. Therefore, the best water vapor estimation may be expected to create the smoothest retrieved surface reflectance curve in the water vapor absorption regions.

There are quite a few criteria that can be used for a smoothness test technique. The following criterion is adopted in HATCH. A surface reflectance in the 0.8–1.25 μm spectral region is first derived for a given amount of water vapor (using Equation 4.21). Then a smoothed reflectance spectrum is constructed accordingly using a truncated cosine series (Qu et al. 2003). The root mean square error (RMSE) difference between the two spectra serves as the smoothness criterion, i.e., the lower the RMSE value, the smoother the spectrum. Figure 4.17 presents the retrieved reflectance spectra using AVIRIS data and their corresponding smoothed ones. The fourth pair of spectra (thick solid line) from the top corresponds to the proper water vapor amount determined by this technique. Once the water vapor amount at the first pixel is derived, the next pixel can use this value as an initial guess and the smoothness testing procedure needs using a few different water vapor amounts to find the proper one.

The HATCH program also uses the smoothness test technique to perform spectral calibration using known atmospheric absorption features. The smoothest spectrum associated with the proper wavelength shift is primarily attributable to the use of the correlated-*k* method and the new HITRAN database that more accurately accounts for gaseous absorption. The spectral shift determined in this way is independent of water vapor amount and pixel location. The spectrally calibrated spectrum in HATCH apparently contains much fewer atmospheric residual features than that without the spectral calibration (Qu et al. 2003). The spectral calibration process for pushbroom sensors needs to correct spectral shift errors caused by so-called "smile," the response function shift (i.e., shift in center wavelength and in FWHM) for each detector that can have different magnitudes and directions. Therefore, in the new version of HATCH (HATCH-2d), the spectral shift errors on a column-by-column basis can be corrected. Because of absorption features of O_2, H_2O, CH_4, and CO_2 located at various spectral regions, HATCH-2d is able to process possible spectral miscalibration separately for different spectral regions. For example, O_2 absorption band around 0.76 μm is used for spectral calibration for the spectral region from 0.6 to 0.86 μm, while the 1.14 μm water vapor absorption region can be used for spectral calibration for wavelengths from 1.06 to 1.25 μm, etc. (Qu et al. 2003).

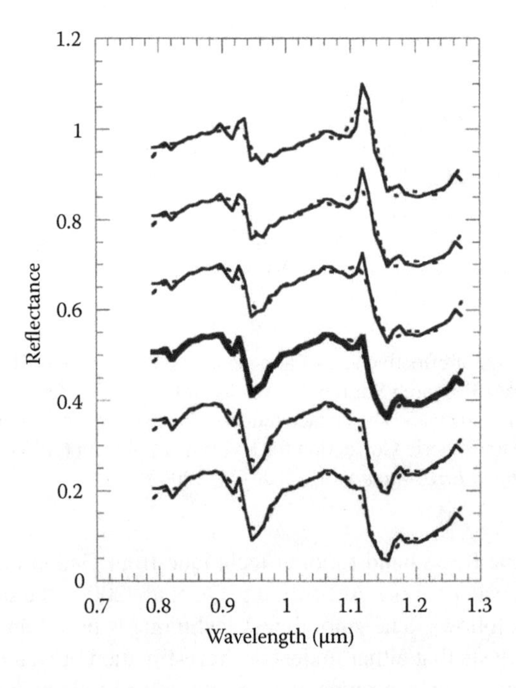

FIGURE 4.17 "Smoothness test" for water vapor retrieval. From the figure, the fourth pair of spectra (thick solid line) from the top corresponds to the proper water vapor amount determined by this smoothness technique. Adjacent reflectance spectra offset: 0.1. (©2003 IEEE. Reprinted from Qu, Z., Kindel, B. C., and Goetz, A. F. H., *IEEE Transactions on Geoscience and Remote Sensing*, 41, 1223–123, 2003. With permission.)

4.4.3.6 Imaging Spectrometer Data Analysis System (ISDAS)

ISDAS is an image and spectral analysis system developed by the Canada Centre for Remote Sensing in conjunction with industry for efficient processing and analysis of hyperspectral imaging data starting in the late 1990s (Staenz et al. 1998). It is built on top of the Application Visualization System (AVS), a commercially available graphics data processing software package. ISDAS processes hyperspectral imaging data (both airborne and spaceborne hyperspectral sensors) covering spectral regions from VNIR and SWIR up to 2.7 μm. Tools included in the ISDAS system allow users to process hyperspectral and associated ancillary data; remove sensor, calibration, and atmospheric modeling artifacts; convert at-sensor radiances to scaled surface reflectances; and extract information from imaging data associated with interactive viewing and analyzing of data.

The ISDAS tools can consist of four main categories: Data input and output, data preprocessing, data visualization, and data information extraction. An overview of the tools implemented in ISDAS conducted by Staenz et al. (1998) is summarized below, with an emphasis placed on data preprocessing (including atmospheric correction). The data input/output tools are geared to ingest/output data cubes and associated ancillary data from a variety of sources (in different media and formats). ISDAS visualization tools help display image cube data and related data for exploratory analysis. These tools provide functionalities to view efficiently hyperspectral data in 1D, 2D, and 3D display and manipulate environments. One of the essential functionalities for a hyperspectral data processing system is extracting and manipulating spectra from a data cube. Further manipulations of spectra through the visualization tools are available in the post-processing tool as addressed below. Information extraction includes tools for classification of data cubes and for the mapping of specific quantitative parameters. These tools include spectral matching, spectral unmixing, automatic end-member selection, interactive end-member selection, quantitative estimation, calculator, etc.

The data preprocessing category includes eight tools that are geared to remove major sensor, calibration, and atmospheric modeling artifacts in order to support the surface reflectance tool that is the major data preprocessing tool. The eight tools include the following:

1. The *roll correction* tool removes the most significant aircraft motion effects from the image cube.
2. The *transform tool* transforms the image cube in the spectral domain in order to enhance information by reducing the spectral dimensionality of the cube. Currently implemented transformations involve the minimum/maximum autocorrelation factor (MAF) (Switzer and Green 1984), the minimum noise fraction (MNF), the principal component analysis (PCA), and the band-moment analysis (BMA) (Staenz 1992). The transformations are also used during classification processes, as in the selection of end-members.
3. The *noise removal* tool removes noise from the image cube in the spectral/spatial domain. MAP, MNF, and PCA can be used in combination with the noise removal tool to correct the striping, etc., effects of imaging data.
4. The *SPECAL* tool is used for assessing and correcting spectral calibration errors caused by spectral shift (spectral smile–frown detection and correction) based on atmospheric gas absorption features. The user can select wavelength range in the radiance domain for a given image cube to correct spectral shift errors. Since the SPECAL tool is linked to the surface reflectance retrieval tool, the assessment/correction of a wavelength shift can be conducted iteratively until the user is satisfied with the surface reflectance retrieved result.
5. The *keystone detection and correction* tool in ISDAS can be used to detect and correct the interband spatial misregistration errors in imaging spectrometers (e.g., the Hyperion sensor). The technique used is based on interband correlation local image matching of spatial features, such as roads and field boundaries, and is described in detail by Neville et al. (2004).
6. The *surface reflectance retrieval* tool retrieves surface reflectances based on a look-up table approach, which is based on the Canadian Advanced Modified 5S (CAM5S RT) model that incorporates most of the 6S (O'Neill et al. 1996). This tool plays a central role

in the analysis of hyperspectral data and provides the physical underpinning for quantitative information extraction. Based on the LUT approach, the surface reflectance retrieval procedure (Staenz and Williams 1997) can significantly reduce the number of RT code runs. The five-dimensional LUTs with tunable break points are generated with the RT code for providing additive and multiplicative coefficients for the removal of atmospheric scattering and absorption effects. The dimensions include wavelength, pixel position, atmospheric water vapor, aerosol optical depth, and terrain elevation. The tool also includes these features:

a. Altitude-dependent layering of atmospheric gases and aerosols;
b. scene-based estimation of atmospheric water vapor;
c. first-order removal of adjacency effects;
d. Lambertian correction for slope and aspect effects; and
e. choice of exo-atmospheric solar irradiance functions from different sources (Staenz et al. 1998).

Both individual spectra and entire image cubes can be processed in both forward (at-sensor radiance) and backward (surface reflectance) computation modes.

7. The *post-processing* tool allows removing artifacts caused during calibration and atmospheric modeling from the reflectance spectra of a cube. Flat targets may be used to create a best-fit function between the modeled and observed reflectance, which produces gains and offsets for band-to-band correction (Staenz and Williams 1997). Several alternative techniques based on least-square smoothing and Gaussian, triangular, exponential, and box car convolution kernels are available to apply to the retrieved reflectances (Staenz et al. 1998).
8. The *spectral simulation* tool spectrally simulates data for a variety of predefined/user-defined sensors, including existing and future sensor data. This tool can simulate band spectral characteristics as long as the band's central wavelength, bandwidth (at FWHM), and relative spectral response profile are available to use.

Figure 4.18 presents an example of processed results from Hyperion hyperspectral data using several tools from the data preprocessing category of the ISDAS system (Khurshid et al. 2006). Figure 4.18a illustrates the comparison of original (contaminated with smile–frown errors in dashed line) and gain–offset corrected (the errors corrected in solid line) at-sensor radiance for a spectrum of vegetation and soil pixels, whereas Figure 4.18b shows the results of the retrieved surface reflectance spectra from Figure 4.18a before (dashed line) and after (solid line) post-processing.

In short, the ISDAS system incorporates tools to provide the functionalities that include ingesting hyperspectral and associated ancillary data, removing sensor and calibration artifacts, transferring at-sensor radiance to surface reflectance, interactively viewing and analyzing data, extracting qualitative and quantitative information products, and outputting results. They have been applied to hyperspectral data from a variety of sources (such as AVIRIS, CASI, Probe-I, and SFSI) for applications in precision agriculture, forest monitoring, geology exploration, environmental monitoring, etc. (Staenz et al. 1999, Nadeau et al. 2002, Khurshid et al. 2006).

4.4.3.7 Comparison

After introducing the six commonly used RT model-based atmospheric correction methods/algorithms, their RT model type, major characteristics and key features, estimate techniques and approaches of atmospheric total-column water vapor content and aerosol optical thickness, and developers as well as key references have been summarized in Table 4.1. Based on limited studies on evaluations and comparisons of performances of the six AC methods applied to hyperspectral imaging data for retrieving scaled surface reflectance and estimating atmospheric water vapor and aerosol optical thickness/visibility, it seems that the retrieval results produced using most of the

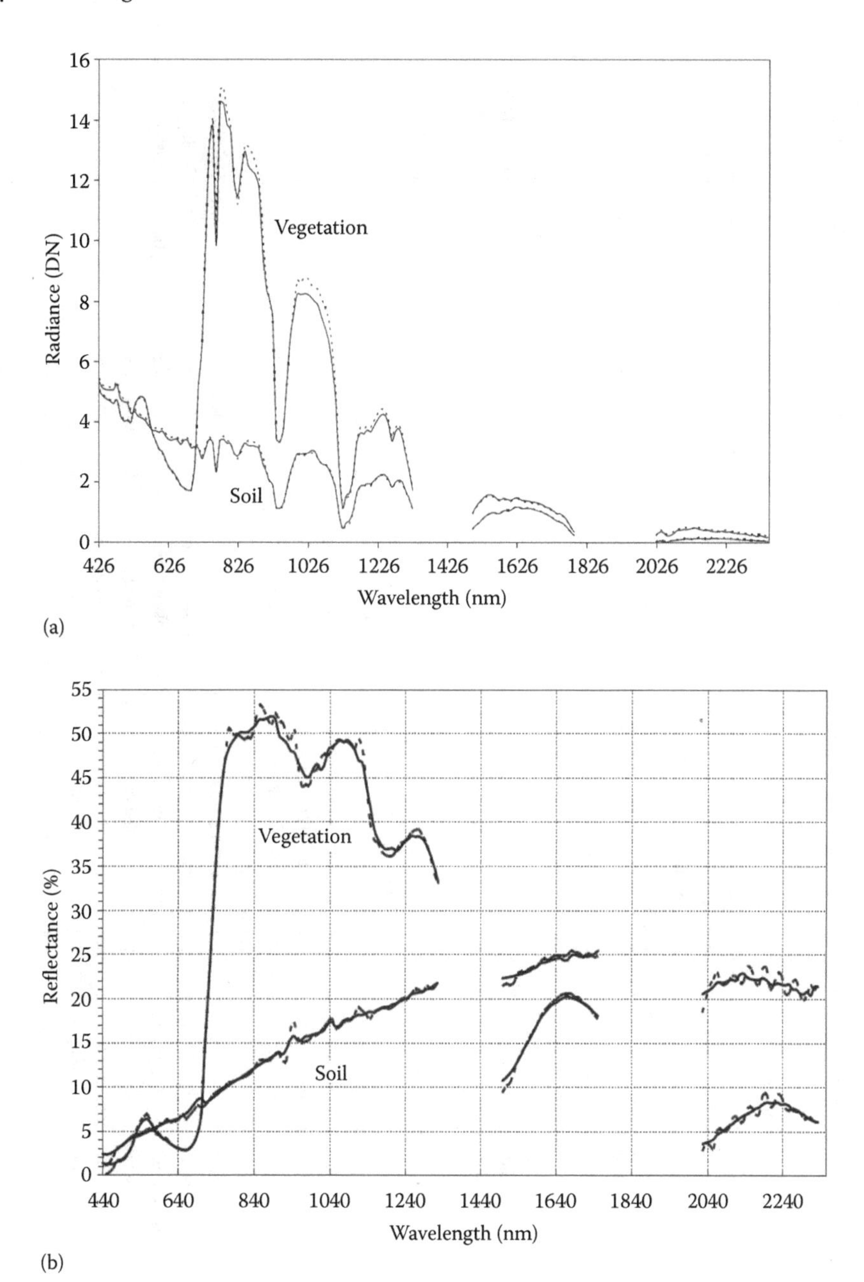

FIGURE 4.18 Comparison of single-pixel radiance spectra extracted from vegetation and soil before (a) and after (b) atmospheric correction using the ISDAS program. Hyperion imagery was acquired over Indian Head, Saskatchewan, on June, 30, 2002. Using the ISDAS, the results with (solid line) and without (dashed line) gain and offset correction for smile–frown errors are presented in (a), whereas the results before (dashed line) and after (solid line) post-processing are presented in (b). (From Khurshid, K. S. et al., *Canadian Journal of Remote Sensing*, 32(2), 84–97, 2006. Used with the permission of the Canadian Aeronautics and Space Institute.)

six AC methods are similar and comparable when the same or similar input parameters are used, although some differences also occur. Table 4.2 summarizes a list of limited studies on evaluations and performance comparisons between two AC methods or among three through six AC methods for their utilization in retrieving surface reflectance and estimating water vapor and aerosol optical thickness, etc., from the data of imaging spectrometers.

TABLE 4.1
Summary of Six Radiative Transfer Model–Based Atmospheric Correction Methods Suitable for Correcting Hyperspectral Imaging Data

Atmospheric Corrrection Method	Characteristics and Key Features	Water Vapor/Aerosol Estimation Approach/Technique	Developer/Sponsor	References
ACORN (Atmospheric CORrection Now)	Based on MODTRAN4+ to retrieve surface reflectance without ground measurements. A full spectral fitting technique to model the overlap of absorptions between water vapor and liquid water in surface vegetation. Single spectrum enhancement and artifact suppression available.	Water vapor can be estimated from HRS data on a pixel-by-pixel basis using the water vapor absorption bands at 940 and/or 1150 nm. Estimate visibility by analyzing the spatial and spectral content of the hyperspectral data in the region from 400 to 1000 nm.	Analytical Imaging and Geophysics LLC., Boulder, CO, USA	Chandrasekhar 1960; ImSpec 2002
ATCOR (ATmospheric CORrection)	Based on MODTRAN4+, both atmospheric and topographic correction codes. A large database is adopted in the program to avoid time-consuming iterations. Correct adjacency effect and topographic and BRDF effects.	Water vapor retrieval on a per-pixel basis using a band ratio technique at 940/1130 nm as central bands. Use aerosol model in MODTRA4+ to estimate aerosol optical depth.	DLR - German Aerospace Center.	Richter 1990, 1996,1998; Richter and Schläpfer 2002
ATREM (ATmosphere REMoval)	Based on 5S and 6S to retrieve scaled surface reflectance and a water vapor image from HRS data without field measurements. New version uses a line-by-line atmospheric transmittance model and the HITRAN2000 line database.	Water vapor amount on a per-pixel basis from the 0.94 and 1.13 μm water vapor absorption bands using a three-band ratioing technique. At-sensor radiance data are fit over a 0.40–0.60 μm spectral range to model radiance data using a nonlinear least-squares-fitting algorithm for estimating aerosol optical depth.	The Center for the Study of Earth from Space (CSES), the University of Colorado, USA	Gao et al. 1993, 2009
FLAASH (Fast Line-of-sight Atmospheric Analysis of Spectral Hypercubes)	Based on MODTRAN4+ to retrieve surface reflectance without ground measurements. Correct light scattering from adjacent pixels in the FOV. Recalibrating hyperspectral sensor wavelengths using sharp molecular absorption features. Correcting spectral "smile" error; and removing artifacts from retrieved reflectance using the data themselves only.	Radiance averages for two sets of spectral bands: an "absorption" band or set of bands centered at 1.13 μm and a reference set of bands taken from both edges of the band. The ratio of average reflectance of the dark pixels at 0.66 μm to that at 2.10 μm to match the average reflectance of ~0.45 for estimating aerosol optical thickness.	The Air Force Research Laboratory, Space Vehicles Directorate (AFRL/VS), Hanscom AFB, and Spectral Sciences, Inc. (SSI), USA	Adler-Golden et al. 1999; Cooley et al. 2002; Perkins et al. 2012

(Continued)

TABLE 4.1 (CONTINUED)

Summary of Six Radiative Transfer Model–Based Atmospheric Correction Methods Suitable for Correcting Hyperspectral Imaging Data

Atmospheric Corrrection Method	Characteristics and Key Features	Water Vapor/Aerosol Estimation Approach/Technique	Developer/Sponsor	References
HATCH (High-accuracy ATmospheric Correction for Hyperspectral data)	Using HATCH own RT model to retrieve scaled surface reflectance and water vapor image from HRS imagery, the correlated-*k* method is used to calculate gaseous absorption and transmittance. Spectral calibration using known atmospheric absorption features. Correcting spectral shift error caused by pushbroom mode's smile phenomenon.	Water vapor amount is estimated using a new spectral-matching technique, called "smoothness test." HATCH uses AFGL standard aerosol data for tropospheric aerosols. One innovative function in HATCH is to allow different aerosol types to be mixed externally, e.g., a mixture of oceanic and urban aerosols can be used for coastal regions.	University of Colorado in Boulder, CO, USA	Qu and Goetz 1999; Qu et al. 2003
ISDAS (Imaging Spectrometer Data Analysis System)	Based on MODTRAN4+ and CAM5S to retrieve surface reflectance without ground measurements. Built on top of the Application Visualization System (AVS). Correct light scattering from adjacent pixels in the FOV. Recalibrating hyperspectral sensor wavelengths using gas absorption features. Correcting spectral "smile" error and spatial "keystone" interband misregistration error; and removing artifacts from calibration and retrieved reflectances.	The convolved sensor-specific look-up tables are used in combination with a curve-fitting technique in the 1130 nm water vapor absorption region to estimate the atmospheric water vapor content on a pixel-by-pixel basis from the data themselves. Use aerosol model in MODTRA4+ to estimate aerosol optical depth.	Canada Centre for Remote Sensing, Canada	Staenz and Williams 1997; Staenz et al. 1998

TABLE 4.2
Summary of Evaluations and Performance's Comparisons between Two or among Three through Six AC Methods for Retrieving Surface Reflectances and Water Vapor Content from Hyperspectral Imaging Data

AC Methods to Be Compared	Surface Features Covered by Imaging Data	Hyperspectral Sensor/ Data and Reference Spectra	Results of Atmospheric Correction and Derived Products	Evaluation Approach and Result of AC Methods' Performances	Reference
FLAASH vs. ATCOR2	The hydrothermal alteration area north of Gümüshane province (NE Turkey). Phyllic, argillic, propylitic, iron oxidation, and silicification are some of the hydrothermal alteration types observed in the region.	EO-1 Hyperion hyperspectral image data. ASD field spectral measurements taken from laboratory.	The retrieval of surface reflectance by the two RT model AC methods from the Hyperion data.	The measured reflectance data of six minerals obtained from these hydrothermal alteration zones were used to investigate the retrieved surface reflectance from Hyperion image's single pixels. Five spectral parameters extracted from both ASD and imaging spectra are (i) general shape and symmetry of normal spectral curves, (ii) continuum-removed band depths, (iii) variations at wavelength positions of the diagnostic absorption features at continuum-removed spectra, (iv) normalized absorption depths, and (v) reflectance intensity of continuum-removed spectra. According to the results of the five parameters used in the analyses of the spectral features of the image reflectance data, the surface reflectance data that are similar, suitable for, and best fitting to the ground truth reflectance data, were produced by FLAASH and ATCOR2, respectively. However FLAASH is slightly better than ATCOR2 due to FLAASH's higher scores.	Kayadibi and Aydal 2013
ACORN vs. FLAASH vs. ATCOR2-3	Sparse vegetation cover with a group of metamorphic rocks, central Anatolia, Turkey	EO-1 Hyperion hyperspectral image data. ASD field spectral measurements.	Surface reflectance retrieved from Hyperion data and absorption features/ bands extracted from the atmospheric correction spectra.	The two comparison schemes, cross correlation and comparison of absorption features in specific wavelengths between reference spectra and AC corrected spectra, were used to evaluate the success of four different atmospheric correction methods. As a result of the study, all four AC methods were found to be successful in overall evaluation with ACORN performing slightly better than others for natural Earth materials through lithological and mineralogical mapping needs.	San and Suzen 2010

(*Continued*)

TABLE 4.2 (CONTINUED)

Summary of Evaluations and Performance's Comparisons between Two or among Three through Six AC Methods for Retrieving Surface Reflectances and Water Vapor Content from Hyperspectral Imaging Data

AC Methods to Be Compared	Surface Features Covered by Imaging Data	Hyperspectral Sensor/ Data and Reference Spectra	Results of Atmospheric Correction and Derived Products	Evaluation Approach and Result of AC Methods' Performances	Reference
ACORN vs. ATCOR4 vs. ATREM vs. CAM5S (ISDAS) vs. FLAASH vs. HATCH	The surface spectral information for input to create the synthetic data sets was obtained from two field spectra of soil and vegetation acquired in Morgan County, Colorado that were convolved to an actual AVIRIS 2001 wavelength file (band centers and FWHM)	Synthetic AVIRIS data. The input data (surface reflectance and water vapor content) for running MODTRAN4 to create the synthetic ARIVIS data as reference reflectance spectra and water vapor content.	Retrieving surface reflectance and water vapor content from the synthetic AVIRIS data with the six RT model-based AC methods.	The synthetic AVIRIS image was run by each selected AC method and the results (retrieved surface reflectance and water vapor content) were compared to the initial spectral information (for creating the synthetic AVIRIS data). To judge the performance of six AC methods to retrieve the reflectance values, a ratio calculation technique in which a selected target is divided by its corresponding true (input) reflectance data and the deviation from a unity is then calculated and squared-sum across selected wavelengths using Average Sum of Deviations Squared (ASDS). There is no a single method that optimally performs (retrieving surface reflectance) across the entire spectral regions and thus a combination between available methods is required. To check the performances of AC methods in water vapor retrieval, a 2D scatter plot between the retrieved and true water vapor images in each scenario was constructed. The results in water vapor retrieval, a sequence to select the relatively best AC method for water vapor content retrieval as follows: ACORN=ATCOR (best) > HATCH>CAM5S>ATREM> >> FLAASH (worst).	Ben-Dor et al. 2005

(Continued)

TABLE 4.2 (CONTINUED)
Summary of Evaluations and Performance's Comparisons between Two or among Three through Six AC Methods for Retrieving Surface Reflectances and Water Vapor Content from Hyperspectral Imaging Data

AC Methods to Be Compared	Surface Features Covered by Imaging Data	Hyperspectral Sensor/ Data and Reference Spectra	Results of Atmospheric Correction and Derived Products	Evaluation Approach and Result of AC Methods' Performances	Reference
ACORN vs. ATREM vs. FLAASH	Several AVIRIS flightlines cover the City of Boulder, Colorado, USA, distributed by a typical Boulder, Colorado gravel.	AVIRIS airborne hyperspectral data. ASD field spectral measurements.	Several other products were produced in addition to the corrected surface reflectance data: A water vapor image produced by each of all three AC methods; ACORN also optionally produces a liquid water image, while FLAASH also produces a cloud mask.	AVIRIS data were corrected using similar parameters and options for each software. With the three AC methods all corrected reflectance spectra were similar for specific materials and generally matched known spectra, as judged by comparison with field spectral measurements. All three models produce similar surface reflectance spectra when the same parameters are used, though some differences occur. Absolute reflectances for the 3.8 m resolution AVIRIS data were within about 5% of field spectral measurements acquired during the AVIRIS flight. In summary, ATREM, ACORN, and FLAASH produce comparable atmospheric correction results.	Kruse 2004
ATREM vs. FLAASH	Two scenes of AVIRIS imagery cover Moffett Field and Jasper Ridge, California, and both scenes consist primarily of a hilly ridge and an urban area. One scene of HYDICE imagery covers Keystone (Able), Pennsylvania, with a dense vegetation cover and in a typical warm, moist and hazy day.	Airborne AVIRIS and HYDICE sensors' data. No reference spectra are needed in this study.	In addition to the retrieved surface reflectance from the AVIRIS and HYDICE data, secondary products produced by the two AC methods include a map of the integrated column water vapor and scene aerosol type and visibility.	The AVIRIS and HYDICE airborne sensors' spectra were used to compare retrieved values of column water vapor and surface reflectance obtained from the AC models. Both models retrieved similar column water vapor and surface reflectances under dry, clear conditions but differ under moist, hazy conditions. By assessing the sensitivity to incomplete knowledge or inaccurate estimation of specific input parameters in terms of the degree of error to the retrieval of the surface reflectance, certain variations in the inputs to the AC models produce retrieved reflectance differences near 0.1. The computational budget for processing hyperspectral data cubes showed that optimized versions of the more complex FLAASH model are similar in computational costs to those of ATREM model.	Griffin and Burke 2003

Theoretically, it is ideal that the RT model-based AC methods are designed to remove the atmospheric effects caused by gaseous molecular and particulate scattering and absorption from the radiance at the sensor and to obtain accurate reflectance at the surface. However, these physically based models require *in situ* atmospheric measurements and radiative transfer codes to correct for atmospheric effects. In practice, such models suffer from a main limitation in which it is often impossible to collect the *in situ* atmospheric parameters for many applications, especially using historical remotely sensed data. Accordingly, many researchers just use existing general atmospheric models and parameters, usually provided by commercial software providers (e.g., FLAASH in ENVI). Results of the atmospherically corrected image data may not be desirable and the AC processing may result in a reduction of signal-to-noise ratio compared with its raw data. For example, in the study on evaluation of AC methods in identifying urban tree species with WorldView-2 imagery by Pu et al. (2015), although they deliberately selected those necessary input parameters (but no *in situ* measurements were available) for running FLAASH, their comparison result indicated that FLAASH's performance was worse than the ELC-based AC methods. This is because using the empirical line methods does not require a priori knowledge of the surface characteristics and atmospheric conditions.

4.4.4 Relative Correction Methods

As introduced and discussed previously, an absolute atmospheric correction (AC) using either empirical/statistical or physically processing–based AC methods and models makes it possible to convert HRS imaging data in DN to image cube in scaled surface reflectance. Such retrieved results, including surface reflectance and other additional products (e.g., water vapor content and aerosol optical thickness), after performing AC processing are useful and necessary to most HRS research and applications. However, not each case of research or application needs such absolute AC correction to HRS imaging data. Some relative AC correction and normalization methods can also meet the requirements of some cases of research and applications, such as using a single scene of an HRS image for surface feature classification and multiple-date images for some feature change detection, etc. (Song et al. 2001). Other reasons for choosing not to use the absolute AC methods may be that specific knowledge and field data required by the absolute AC methods for a specific scene and date and for both the sensor spectral profile and atmospheric properties at the time of imaging data collection will not be available (Du et al 2002). Therefore, in this section, two relative correction methods are reviewed. These are (1) single-image correction using histogram adjustment, and (2) multiple-date image normalization using regression model.

The histogram adjustment method for correcting a single image is based on the assumption that NIR and MIR image data (> 0.7 μm) are free of atmospheric scattering effects, while the visible image data (0.4–0.7 μm) is strongly influenced by atmospheric effects. The method includes two steps. The first is to evaluate the histograms of multi-/hyper-spectral remotely sensed data to identify minimum values for every visible band. The second is to use a simple algorithm: *output DN = input DN – bias*, where *bias* usually equals the scattering value of a dark object, to correct the atmosphere-induced additive scattering effects (path radiance). Normally, the imaging data collected in the visible range (e.g., TM bands 1–3) have a higher minimum DN because of the atmosphere-induced additive effects on these visible images. Therefore, if we can determine the minimum values that are caused by the scattering effects for the visible images using some ways, we can deduct the atmospheric effects from the visible images mostly through performing the second step. There are different ways that can be used to identify the minimum values (i.e., equivalent scattering effects) for visible bands, such as by evaluating the histograms of visible bands to determine values of very dark objects (e.g., deep and clear waters) (Lu et al. 2002). In practice, the dark pixel subtraction is often used. It assumes that the pixel of lowest DN in each visible band should be zero, and hence, its radiometric value (in DN) is the result of atmospheric scattering or haze effects

(Jensen 2005). Thus, subtraction of the dark pixel value from the image may achieve a purpose of relative correction of atmospheric effects.

The multiple-date image normalization method using regression modeling normalizes a date of one image into another so that the multiple dates of images have approximately the same radiometric characteristics. Multitemporal remote sensing data are frequently required in land use/cover change and other surface feature change detection. However, multiple-date image data are usually not captured on anniversary dates, thus sun angle, atmospheric conditions, and soil moisture likely vary from date to date (Lu et al. 2002). The normalization method involves selecting a reference image from the multiple-date images and the radiometric scale of remaining images needs to be normalized to that of the reference image, usually by using a linear regression model. Such a relative atmospheric normalization method is inherently empirical and based on the assumption of a simple linear relationship among images across time and the dominance of stable features in the scene. In the normalized process, one date of image acts as the reference image and another date of image as the predicted/normalized image. To establish a regression model, pseudo-invariant features (targets) (PIFs) need to be identified from both reference and normalized images, based on the assumption that the values of the normalization PIFs are constant and any changes of these values in other images are caused by the satellite sensor, atmospheric conditions, and soil conditions. To be of value, the identified PIFs should have at least the following two characteristics (Song et al. 2001; Jensen 2005): (1) the spectral characteristics of a PIF should change very little over time, such as deep nonturbid water bodies, bare soil, large rooftops, etc.; (2) PIFs should normally contain only minimal amounts of vegetation because vegetation can change over time due to environmental stress and plant phenology. However, an extremely stable, undisturbed dense mature forest canopy imaged on near-anniversary dates might be considered. Therefore, existing relative atmospheric normalization approaches rely on the ability to identify PIFs from the images. There are several approaches to help identify desirable PIFs. Among them, the two-dimension scatterplot approach (Song et al. 2001, Du et al. 2002) can help locate the PIFs pixels in the "ridge" area. The scatterplot is formed with all the pixels in a scene with one axis being the DN value of date 1 and the other being the DN value of date 2 image. In such a scatterplot, DN values of all stable features form a ridge, with the straight line that passes along the ridge defining the relationship between dates of imagery. Figure 4.19 shows the ridge

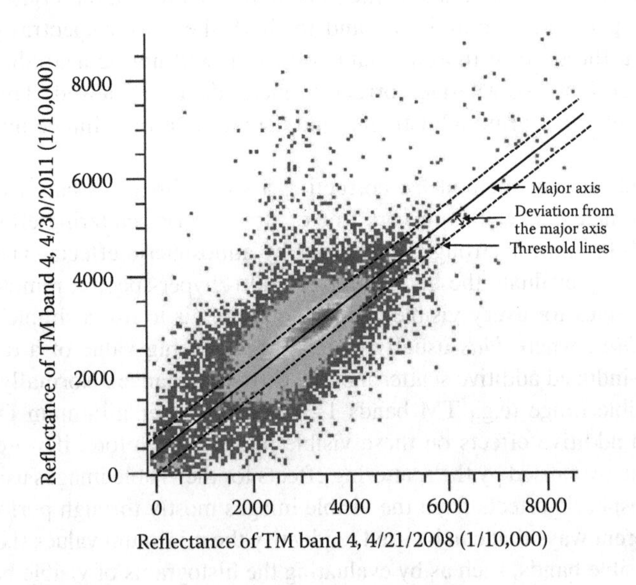

FIGURE 4.19 Scatterplot of Landsat TM band 4 from two images acquired on April 21, 2008, and April 30, 2011, to illustrate the major axis and "ridge area" between the two threshold lines. The perpendicular deviation distance from the major axis can be used to determine the thresholds based on substantial application case.

area from which some PIFs pixels may be located. After the desirable PIFs are identified, a linear regression model is developed by correlating the values of the PIFs in both the image (DN_x) being normalized and the reference image (DN_y). This linear regression equation can be described as: $DN_y = a \cdot DN_x + b$, where the regression slope a is a multiplicative component (gain), which can normalize differences in changes of sun angle, atmospheric conditions, etc., between images acquired at different times; the intercept b is an additive component (offset), which can normalize differences in atmospheric path radiance between dates of image data. After obtaining the a and b parameters, all pixels in DN in the scene of the image being normalized can be normalized to the reference image using the regression equation. Since the a and b parameters are different for different image bands, each spectral band must be corrected (normalized) separately. Note that the pixel values in PIFs and both images can be DN, radiance and reflectance.

4.5 TECHNIQUES FOR ESTIMATING ATMOSPHERIC WATER VAPOR AND AEROSOLS

As discussed previously, atmospheric water vapor content and aerosol loading are variable in the space and time domain and thus are considered two key factors in atmospheric correction for hyperspectral imaging data. In this section, several retrieval techniques, which are used for estimating water vapor content and aerosol optical thickness from hyperspectral imaging data and have theoretical and practical significance, are briefly introduced and discussed. These retrieval techniques can be used not only for accurately performing atmospheric correction to retrieve scaled surface reflectance, but also for creating the products of column water vapor content and aerosol optical thickness or loading for other applications.

4.5.1 Atmospheric Water Vapor

To retrieve atmospheric water vapor (WV) from hyperspectral imaging data, there are two general types of retrieval techniques: differential absorption techniques and spectral fitting techniques. The differential absorption techniques are a relatively simple and practical way to determine WV content from a spectrum of absorption bands at a low-computing time cost. In general, a ratioing is performed in this type of technique between the radiance at bands within the absorption feature (measurement bands) and a reference radiance of bands (reference bands) in the vicinity of the absorption feature to detect the relative strength of the absorption (Schläpfer et al. 1998). The ratioing value is then used to directly estimate the WV content using linear or nonlinear models. This type of technique includes narrow/wide (N/W; Frouin et al. 1990), continuum interpolated band ratio (CIBR; Green et al. 1989, Bruegge et al. 1990), three-band ratioing (3BR; Gao et al. 1993), linear regression ratio (LIRR; Schläpfer et al. 1996), and atmospheric pre-corrected differential absorption (APDA; Schläpfer et al. 1998). The second type of technique relies on radiative transfer (RT) and spectral models to simulate or fit the spectral curve to compare with sensor-measured spectrum to determine the WV content from hyperspectral imaging data. Typically, such spectral-fitting techniques include the curve-fitting algorithm (Gao and Goetz 1990), the smoothness test (Qu et al. 2003), and the band-fitting technique (Guanter, Estellés et al. 2007). Since these types of spectral-fitting techniques usually require users to have knowledge and experience on atmospheric radiative transfer, spectral molding, and linear and nonlinear least square fitting, and also need high computing time cost, they are not introduced here. For readers who are interested in these techniques, refer to their corresponding references.

4.5.1.1 Narrow/Wide (N/W) Technique

Frouin et al. (1990) developed a new technique for estimating total WV amount from space (from multi-/hyper-spectral imaging data). The technique is called narrow/wide (N/W) channel technique and consists of two spectral bands, one narrow and the other wide, both centered on the same wavelength at the water absorption maximum near 0.94 μm. Figure 4.20a shows the two bands' locations and their relative band width. The ratio, $R_{N/W}$, of the narrow band to the wide band (Equation 4.36) is independent of the surface reflectance and yields a direct estimate of WV integrated along the optical path. The ratio equation is expressed as

$$R_{N/W} = \frac{L_{narrow}}{L_{wide}} \tag{4.36}$$

where L_{narrow} is the (averaged) radiance in a narrow measurement band; L_{wide} is the (averaged) radiance in a wide measurement band at the same central wavelength as L_{narrow}. The technique is mostly applicable to multispectral data, but in a hyperspectral case, the channel sets need to be averaged into a narrow and a wide band. The ratio has an appropriate exponential relationship with radiosonde total WV amount measurement with a 10–15% error (which is a commonly acceptable error range) (Frouin et al. 1990).

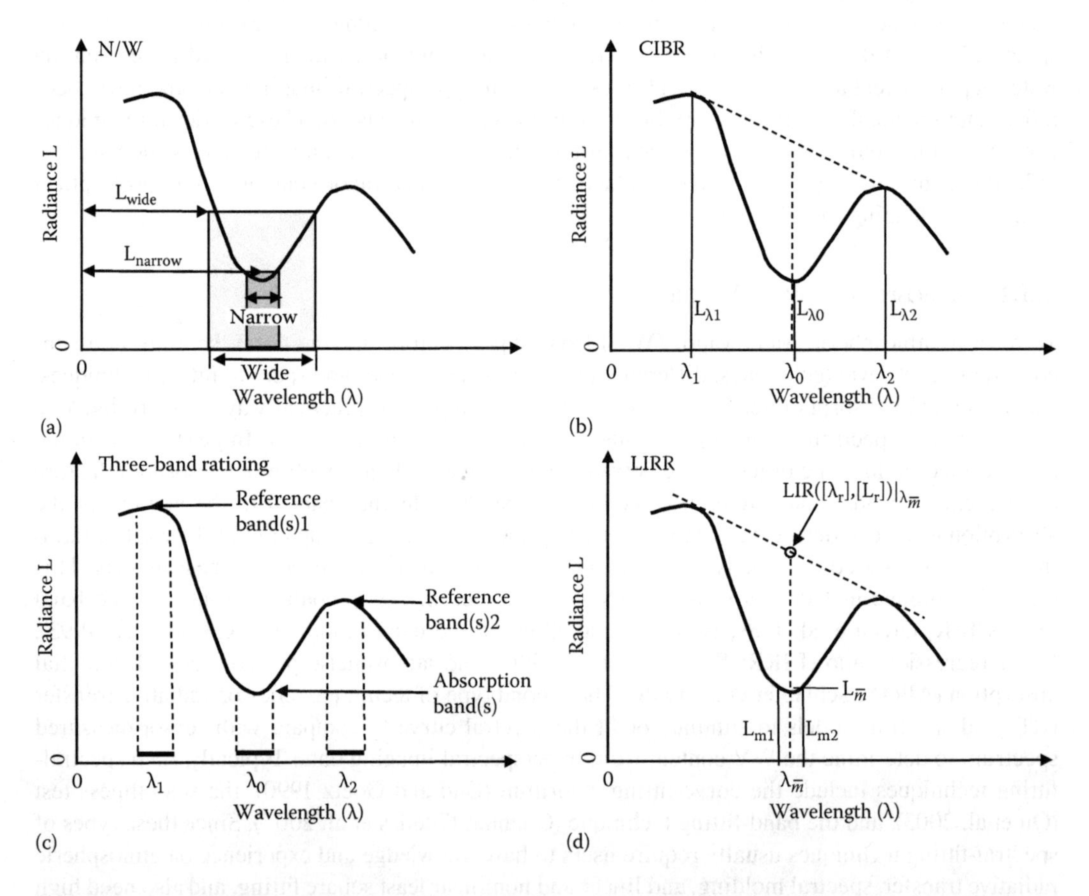

FIGURE 4.20 A partial simulated spectrum in radiance illustrating the four differential absorption techniques for retrieving water vapor content. (a) N/W technique is shown with a narrow (a relatively narrow band contains the measured absorption feature) and a wide (a relatively wide band also contains the measured absorption feature) band; (b) CIBR technique is illustrated with the three bands (middle one as measured absorption band and left and right ones as reference bands); (c) three-band ratioing technique is shown with the three band locations (the middle one located at the measured absorption band and both adjacent bands being reference bands); and (d) LIRR is demonstrated with its linear regression line and the average measured absorption band location.

Airborne experiments carried out by Frouin et al. (1990) demonstrated the technique's feasibility under clear-sky conditions over both sea and land. In the presence of thick aerosol layers, however, the WV amounts derived over the ocean may be underestimated by as much as 20%. In general, compared to satellite microwave remote sensing techniques that are frequently used under most weather conditions, the N/W technique has the advantage of simplicity and is a promising alternative over land, especially where microwave radiometry is inappropriate (Frouin et al. 1990).

4.5.1.2 Continuum Interpolated Band Ratio (CIBR)

The CIBR technique (Green et al. 1989, Bruegge et al. 1990) uses radiance measurements at the center of an absorption band at 0.94 μm, together with values of the continuum radiances on either side of the absorption band (Figure 4.20b). Green et al. (1989) first used the CIBR technique to retrieve the total-column atmospheric WV from AVIRIS data. Then the accuracy of this technique was established with respect to radiosonde and other spectroscopic water recoveries (Bruegge et al. 1990). A value of the continuum radiance at the wavelength of maximum water band absorption is estimated by linear interpolation between two adjacent continuum values (Green et al. 1989). The CIBR ratio, R_{CIBR}, is formed between the interpolated continuum radiances at wavelengths λ_1 and λ_2 and water absorption band radiance at wavelength λ_0:

$$R_{CIBR} = \frac{L_{\lambda_0}}{w_{\lambda_1} L_{\lambda_1} + w_{\lambda_2} L_{\lambda_2}} \tag{4.37}$$

where L_{λ_0}, L_{λ_1}, and L_{λ_2} represent at-sensor radiances at wavelengths λ_0, λ_1, and λ_2, and

$$w_{\lambda_1} = \frac{\lambda_2 - \lambda_0}{\lambda_2 - \lambda_1} \tag{4.38}$$

and

$$w_{\lambda_2} = \frac{\lambda_0 - \lambda_1}{\lambda_2 - \lambda_1} \tag{4.39}$$

Per Figure 4.20b, it is easy to see that L_{λ_0} decreases when WV content increases while the radiances at both shoulders at λ_1, and λ_2 do not change significantly. Both simulations and experiments reveal the ratio R_{CIBR} is well-correlated with the total column water vapor amount of the atmosphere (W) in the following approximate form (Carrère and Conel 1993, Liang 2004):

$$R_{CIBR} = e^{-\alpha W^{\beta}} \tag{4.40}$$

where, α and β are two coefficients.

Carrère and Conel (1993) applied and compared the two simple techniques (CIBR and N/W) to retrieve total-column WV content from ARIVIS imaging data acquired over Salton Sea, California, using the 0.94 μm water absorption band. Analysis of systematic and random errors based on the radiative transfer code LOWTRAN7 showed that the CIBR proved to be the technique less sensitive to perturbing effects, except for errors in visibility estimate. Validation was carried out through comparison between an independent estimate of water vapor from concurrent Reagan sunphotometer measurements and AVIRIS estimates. The validation result demonstrated that WV amounts retrieved using the N/W approach match more closely *in situ* measurements, even after adjusting model parameters for background reflectance, viewing geometry, and type of aerosol at the site.

4.5.1.3 Three-Band Ratioing (3BR)

The three-band ratioing (3BR) method actually is a special type of the CIBR technique when $w_{\lambda 1} = w_{\lambda 2} = 0.5$. Gao et al. (1993) first used the 3BR technique to estimate the integrated WV amount on a pixel-by-pixel basis that is derived from the 0.94 μm and 1.14 μm WV absorption features from AVIRIS hyperspectral imaging data. The retrieval of WV values from AVIRIS data with the 3BR technique is mainly based on two reasons. One is that the surface reflectance curves for common soils and rocks vary nearly linearly with wavelengths in the 0.94 μm and 1.14 μm WV band absorption regions (Gao et al. 1993). The other is that under typical atmospheric conditions, the transmittances of the 0.94 μm and 1.14 μm WV absorption bands are sensitive to the changes in the amount of water vapor (Gao and Goetz 1990). The first reason means that the 3BR described in Figure 4.20c may remove the surface reflectance effects and may give WV transmittances of the two absorption channels. The total WV amounts are then derived from the transmittances. The accuracy of WV derivations from the 0.94 μm and 1.14 μm WV bands in AVIRIS data depends not only on the spectral model used, but also on the assumed atmospheric temperature, pressure, and WV volume mixing ratio profiles (Gao and Goetz 1990).

For example, in retrieval of WV amounts from AVIRIS data, Gao et al. (1993) first averaged apparent reflectances of five AVIRIS bands near 0.945 μm to give a mean apparent reflectance of the 0.94 μm WV band. Then apparent reflectances of three bands near 0.865 μm were averaged to give a mean apparent reflectance for the left region of WV absorption band at 0.94 μm. Apparent reflectances of three bands near 1.025 μm were averaged to obtain a mean apparent reflectance for the right region of WV absorption band at 0.94 μm. Next, the mean apparent reflectance at the water vapor center was divided by the averaged mean apparent reflectances at both the left and right regions to obtain the three-band ratio value. Finally, by comparing the mean observed transmittance (equivalent to the ratio) with theoretically calculated mean transmittances using atmospheric and spectral models, the amount of WV in the sun–surface–sensor path is obtained. Gao et al. (1993) used similar three-band ratioing procedures to derive another WV amount in the sun–surface–sensor path from the 1.14 μm WV band. The average of water vapor values from the 0.94 μm and 1.14 μm bands was considered as the best estimate of the water vapor value corresponding to the pixel. Note that in practice, in implementation of the 3BR technique, the center positions and widths of both side regions and water vapor absorption bands are all allowed to vary to reduce any possible errors.

4.5.1.4 Linear Regression Ratio (LIRR)

To extend CIBR technique to include more suitable measurement bands in order to reduce the total noise of the sensor through considering a maximum number of selected bands, Schläpfer et al. (1996) developed a linear regression ratio (LIRR) technique to retrieve the WV amount. The linear regression through a set of reference channels can be taken as an interpolation line at the center wavelength of the measurement channels around a water vapor absorption band (e.g., 0.94 μm). With a similar continuum interpolated regression line (LIR([λ_r], [L_r]) as for CIBR approach and referring to Figure 4.20d, the LIRR ratio R_{LIRR} can be defined as follows:

$$R_{LIRR} = \frac{L_{\bar{m}}}{\mathrm{LIR}([\lambda_r],[L_r]|_{\lambda_{\bar{m}}}} \tag{4.41}$$

where $L_{\bar{m}}$ represents the mean radiance of all measurement bands containing a water vapor absorption band with the corresponding mean wavelength $\lambda_{\bar{m}}$; $\mathrm{LIR}([\lambda_r],[L_r]|_{\lambda_{\bar{m}}}$ represents estimated value (see Figure 4.20d) by the linear regression line LIR([λ_r], [L_r]) when wavelength = $\lambda_{\bar{m}}$.

According to the experimental results of comparing performances of several differential absorption techniques including N/W, CIBR, and LIRR techniques for retrieving WV amount from AVIRIS imaging data, conducted by Schläpfer et al. (1996), over water, only the LIRR method can be evaluated. It calculates a linear regression ratio in the weak WV absorption band at 0.73 μm. The noise propagation error is about 7%, whereas all the techniques in the 0.94 μm band show errors of 30% or more due to the lack of ground reflected radiance (lake water) at that wavelength. However, much better results are achieved over vegetation. All methods can be evaluated, and CIBR and LIRR can be quantified with a good accuracy. Their relative noise propagation error over vegetation was 6.7% (CIBR) and 2.6%(LIRR), respectively.

4.5.1.5 Atmospheric Pre-Corrected Differential Absorption (APAD)

Experimentally and theoretically simulated results have demonstrated that the CIBR is proportional to WV amount only for high background reflectances and creates greater error if retrieving the WV amount over low background albedo (Schläpfer et al. 1998). Over the area with low background reflectances, the WV retrieval gets less accurate, resulting in an underestimation over dark surfaces (Gao and Goetz 1990). The N/W technique has a similar problem as the CIBR to retrieve WV content from hyperspectral data with low background reflectance. Such error could be corrected using a differential absorption procedure, which depends on the apparent reflectance at-sensor level (Schläpfer et al. 1998). For this case, Schläpfer et al. (1998) developed a new differential absorption technique called the Atmospheric Pre-corrected Differential Absorption (APAD) technique to correct this effect (i.e., WV retrieval less accurate with low background reflectance) based on the physical RT model. In fact, the "pre-corrected" here means that the at-sensor radiance values L of each band (including both reference bands and measurement WV absorption bands) are reduced by the corresponding atmospheric path radiance term L_{atm}. Then, the pre-corrected at sensor radiance values are inserted in the LIRR equation (Equation 4.41) to form the APDA ratio index R_{APAD} that is analogous to the form of R_{LIRR}:

$$R_{APAD} = \frac{\left[L_{\overline{m}} - L_{atm,\overline{m}}\right]_i}{\mathrm{LIR}\left(\left[\lambda_r\right]_j, \left[\mathrm{L}_r - \mathrm{L}_{atm,r}\right]_j\right)\Big|_{\lambda_{\overline{m}}}} \tag{4.42}$$

where $\mathrm{LIR}([x],[y])|_a$ denotes a linear regression line (similar to the one in Figure 4.20d) through the points (x, y) evaluated for y at the point $x = a$. The parameters in brackets are the central wavelengths and atmospheric pre-corrected radiances of measurement bands (i) and reference bands (j), respectively.

As we know, the atmospheric path radiance L_{atm} does not depend on ground reflectance but is sensitive to atmospheric composition, especially the aerosol loading and the WV content. The atmospheric pre-correction term $L_{atm,i}$ is a function of terrain height, aerosol profile and contents, band position, and water vapor contents (Schläpfer et al. 1998). With a radiative transfer model (e.g., MODTRAN4), the $L_{atm,i}$ can be estimated by calculating the total radiance of the sensor at ground zero albedo under varying ground altitude and water vapor contents (see Schläpfer et al. 1998 for details). If the pre-corrected at sensor radiance values use only one measurement and two reference bands, the form of R_{APAD} is analogous to the form of R_{CIBR} (Equation 4.37). According to Schläpfer et al. (1998), when the APDA technique was applied to two AVIRIS images acquired in 1991 and 1995, the accuracy of the retrieved total-column WV amounts was within a range of ±5% compared to ground truth radiosonde measurements.

4.5.2 Atmospheric Aerosols

Atmospheric aerosols (dust, smoke, and air pollution particles) have a significant effect on remote sensing through scattering and absorbing radiation energy passing through the atmosphere. Correction for the aerosol effect on remote sensing needs to know aerosol loading or aerosol optical thickness when a scene of remote sensing image is acquired. In this section, two aerosol retrieval techniques that can be used to estimate atmospheric aerosol loading or optical thickness from hyperspectral data are introduced and discussed. One is called dark dense vegetation technique

(DDV; Kaufman and Tanré 1996) and the other is called aerosol optical thickness at 550 nm technique (AOT at 550 nm; Guanter, González-Sampedro et al. 2007; Guanter, Estellés et al. 2007).

4.5.2.1 Dark Dense Vegetation (DDV) Technique

The aerosol effect is strongest for low background surface reflectance. Therefore it is appropriate to use the darkest pixels in the image to estimate the aerosol loading. Equation 4.32 should be an approximation good enough for dark surfaces used to retrieve the aerosol path radiance and optical thickness. The effect of the path radiance on apparent reflectance is larger for shorter wavelengths (e.g., visible range), and for low values of the surface reflectance (e.g., $\rho < 0.05$). For example, the smoke has a large effect in the visible range of the spectrum, decreasing with wavelength from the blue to red regions. Its effect is smaller than variation in the high background surface reflectance in the NIR region, and is small in the MIR (2.2 μm) region. Many surface covers (vegetation, water, and some soils) are dark in the red (0.60–0.68 μm) and blue (0.4–0.48 μm) wavelengths. In addition, some longer wavelengths (2.2 μm or 3.7 μm) are less sensitive to aerosol scattering (since the wavelengths are much larger than most aerosol particles), but still sensitive to surface characteristics. Therefore, detection of aerosol in the blue and red channels may be based on the assumption that the aerosol effect is much smaller or negligible at 2.2 μm versus that at the blue and red channels. Based on these assumptions and atmospheric radiative transfer principle, Kaufman and Tanré (1996) developed a dark pixel approach (i.e., DDV technique) to directly estimate aerosol loading or aerosol optical thickness from multi-/hyperspectral image with some visible bands (e.g., 0.41 μm, 0.47 μm, and 0.66 μm) and MIR band (e.g., 2.2 μm) available.

According to Kaufman and Tanré (1996), over dark or dense vegetated areas, the DDV technique is based on the relationships between the surface reflectance at 2.2 μm and those at 0.47 μm and 0.66 μm, derived from Landsat TM and aircraft AVIRIS images over the mid-Atlantic United States (Figure 4.21). The images covered forested area and crop lands, as well as exposed soil, residential areas, and water. The images were corrected for the atmospheric effect, using sunphotometer measurements from the ground of the aerosol optical thickness and optical properties. The uncertainty in the estimate of the surface reflectance in the visible bands (0.47 μm and 0.66 μm) from the 2.2 μm

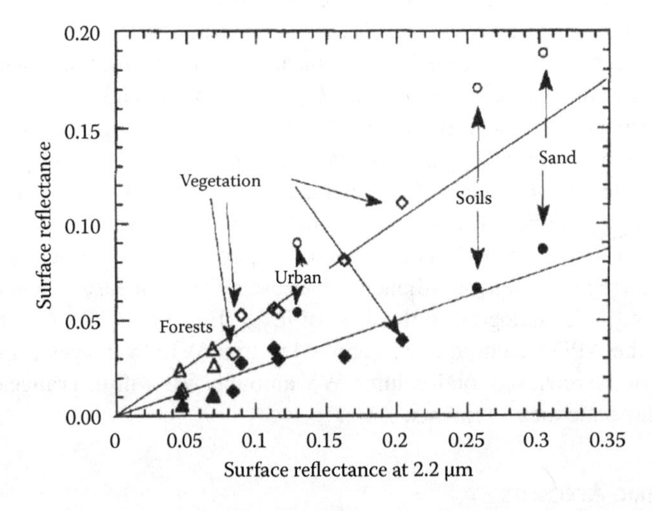

FIGURE 4.21 Scatterplot between average values of the surface reflectance at 0.47 μm (filled symbols) and 0.66 μm (empty symbols) and that at 2.2 μm. Several of the surface types are identified. The relationships $\rho_{0.47}/\rho_{2.23} = 0.25$ and $\rho_{0.66}/\rho_{2.23} = 0.50$ are also plotted (solid lines). The data were taken from Landsat TM and aircraft AVIRIS data over the mid-Atlantic region of the United States. The images were corrected for the atmospheric effect using sunphotometer measurements from the ground of the aerosol optical thickness and optical properties. (Reprinted from Kaufman, Y., and Tanré, D., *Remote Sensing Environment*, 55, 65–79, 1996. ©1996, with permission from Elsevier.)

band reflectance is + 0.005 to + 0.01 in the red and blue bands, respectively, for dark targets (reflectance at 2.2 μm < 0.10) (Kaufman et al. 1997). Using the DDV technique, the aerosol effect $\Delta\rho^*$ that is directly related to aerosol optical thickness can be calculated as:

$$\Delta\rho^* = \rho^*(aerosol) - \rho^*(no\ aerosol) \tag{4.43}$$

where ρ^* (*aerosol*) is the apparent reflectance at the top of the atmosphere (Equation 4.32), converted from imaging data. ρ^* (*no aerosol*) at 0.47 μm and 0.66 μm can be estimated by using relationships between $\rho^*_{2.2}$ and $\rho^*_{0.47}$ (*no aerosol*) or $\rho^*_{0.66}$ (*no aerosol*) (e.g., $\rho_{0.47} = (0.25 + 0.08)\ \rho_{2.23}$ and $\rho_{0.66} = (0.50 + 0.11)\ \rho_{2.23}$ in Figure 4.21) because $\rho^*_{2.23}$ is almost not influenced by aerosol effect. According to Kaufman et al. (1997), for the continental model, even for the heavy aerosol loading (optical thickness of 0.5 at 0.55 μm) the aerosol effect, $\Delta\rho^*$, on the reflectance observed from space at 2.2 μm, $\rho^*_{2.2}$, is very small. It is close to zero for forests and most dense vegetation types $\left(\Delta\rho^*_{2.2} \le -0.002\right)$ and increases to $\Delta\rho^*_{2.2} = -0.01$ for soils and sand. At the same time the aerosol effect on the blue and red bands is $\Delta\rho^*_{0.47} \sim 0.03$ and $\Delta\rho^*_{0.66} \sim 0.02$. The aerosol effect in the blue and red channels is smaller for brighter surfaces (soils) than for darker surfaces (dense vegetation) and therefore aerosol detection is more accurate for darker surfaces, so the aerosol retrieval method is called the *dark dense vegetation technique* (Kaufman et al. 1997). More detailed discussion of the dependence of the aerosol effect $\Delta\rho^*$ on the surface reflectance and the aerosol properties is given by Fraser and Kaufman (1985). In short, the DDV direct technique identifies dark pixels (dark dense vegetation pixels) using the MIR bands, estimates their reflectance in the red and blue bands using reflectance from the MIR bands, and determines the aerosol optical thickness.

4.5.2.2 Aerosol Optical Thickness at 550 nm (AOT at 550 nm)

The retrieval technique of AOT at 550 nm is described by Guanter, González-Sampedro et al. (2007) and Guanter, Estellés et al. (2007) in detail. The final total aerosol loading may be parameterized by the AOT at 550 nm retrieved by the technique. The technique assumes that the AOT is constant over a 30 km × 30 km area. The AOT at 550 nm in the area is retrieved from a set of five land reference pixels with a high spectral contrast within this area, from the purest vegetation to the purest bare soil, by means of a multiparameter inversion of the at-sensor spectral radiances. A perfect choice for the set of five reference pixels would be a pure vegetation pixel, a pure bare soil pixel, and three intermediate ones, mixed with different proportions of vegetation and soil (Guanter, González-Sampedro et al. 2007). To perform the procedure, the following four steps are required, as according to Guanter, Estellés et al. (2007).

The first step is to generate a set of synthetic at-sensor radiances by following Equation 4.20 for the inversion of the measured (imaging) radiances. To do so, one single look-up table (LUT) consisting of five breakpoints with varying horizontal visibility (as equivalent to AOT at 550 nm, see their relationship from Figure 4.22 for a sample) is built. In Figure 4.22, the particular transfer function between AOT at 550 nm and horizontal visibility for a target located at sea level with rural aerosol model and atmospheric summer profile is displayed (Guanter, Estellés et al. 2007). The view zenith angle (VZA) is set to 0°, the surface elevation (SE) to the mean SE as provided by the digital elevation model (DEM), and the column water vapor (CWV) is set to 2.0 g cm^{-2} (as the CWV amount has no relevant influence on aerosol scattering). According to these settings, the selection of five reference pixels is constrained to those pixels with VZA between 0° and 5° and with SE deviating only ±10% from the mean SE in the area. These simplifications allow reducing considerably the total computation time (Guanter, Estellés et al. 2007).

The next step is to provide an estimation of the surface reflectance each of the five reference pixels so that the five synthetic at-sensor radiances can be simulated using Equation 4.20. The surface reflectance ρ of each of the five reference pixels may be modelled as a linear combination of two artificial end-members:

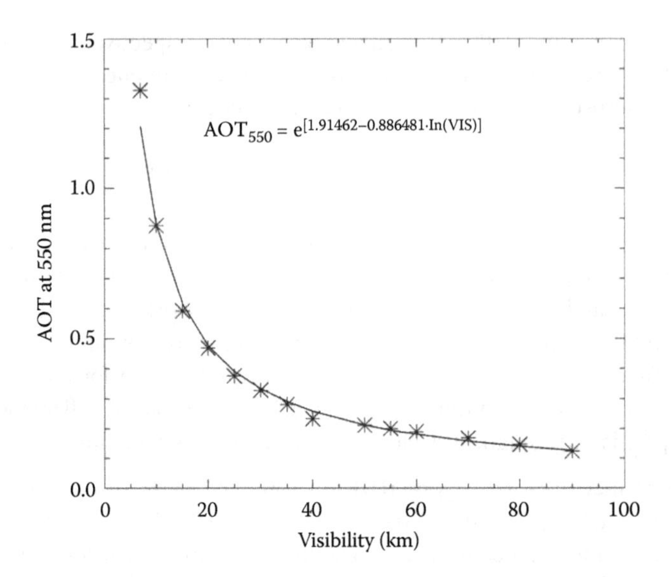

FIGURE 4.22 A transfer function between AOT at 550 nm and horizontal visibility at sea level for the MODTRAN4 rural aerosol model and summer profile. (Reprinted from Guanter, L., Estellés, V., and Moreno, J., *Remote Sensing of Environment*, 109, 54–65, 2007. ©2007, with permission from Elsevier.)

$$\rho = C_v \rho_{veg} + C_s \rho_{soil} \quad C_{v,s} > 0, \rho \in [0,1] \tag{4.44}$$

where ρ_{veg} and ρ_{soil} are representative vegetation and bare soil reflectance spectra, and $C_{v,s}$ corresponds to weighting coefficients. Note that the end-members are called "artificial" because their role is to provide a reflectance basis for simulating at-sensor radiances, reproducing real targets present in the scene. As a result, the $C_{v,s}$ coefficients are not real abundances (called effective abundances of the end-members) so the sum of $C_{v,s}$ may be > 1.0. The hypothesis that any common land pixel can be represented by this kind of linear combination works well in the VIS/NIR spectral ranges, but further effort is needed when the SWIR is also included (Guanter, Estellés et al. 2007).

The third step is to invert five reference pixels for providing the visibility (which will finally be used for estimating AOT at 550 nm) and the five couples of $C_{v,s}$ coefficients as a by-product. To do so, Powell's Minimization Method (Press et al. 2007) is carried out, based on separate 1D minimizations in each of the eleven directions comprising the parameter space (one for AOT at 550 nm and ten for $C_{v,s}$ coefficients in the five reference pixels). Therefore, the inversion of the five reference pixels is performed by using the minimization of a merit function δ^2 specifically designed for this problem (Press et al. 2007),

$$\delta^2 = \sum_{pix=1}^{5} \sum_{\lambda_i} \frac{1}{\lambda_i^2} \left[L^{SIM}\Big|_{pix,\lambda_i} - L^{SEN}\Big|_{pix,\lambda_i} \right]^2 \tag{4.45}$$

where L^{SIM} is the set of simulated at-sensor radiances using Equation 4.20; λ_i corresponds to the central wavelength (in μm) of band i; and L^{SEN} represents the at-sensor radiances measured by the sensor. To run Equation 4.45, it is initialized using linear correlations (see Figure 4.23) between the NDVI and the coefficients $C_{v,s}$ that were found from several simulations and the visibility of 23 km as suggested by Guanter, Estellés et al. (2007).

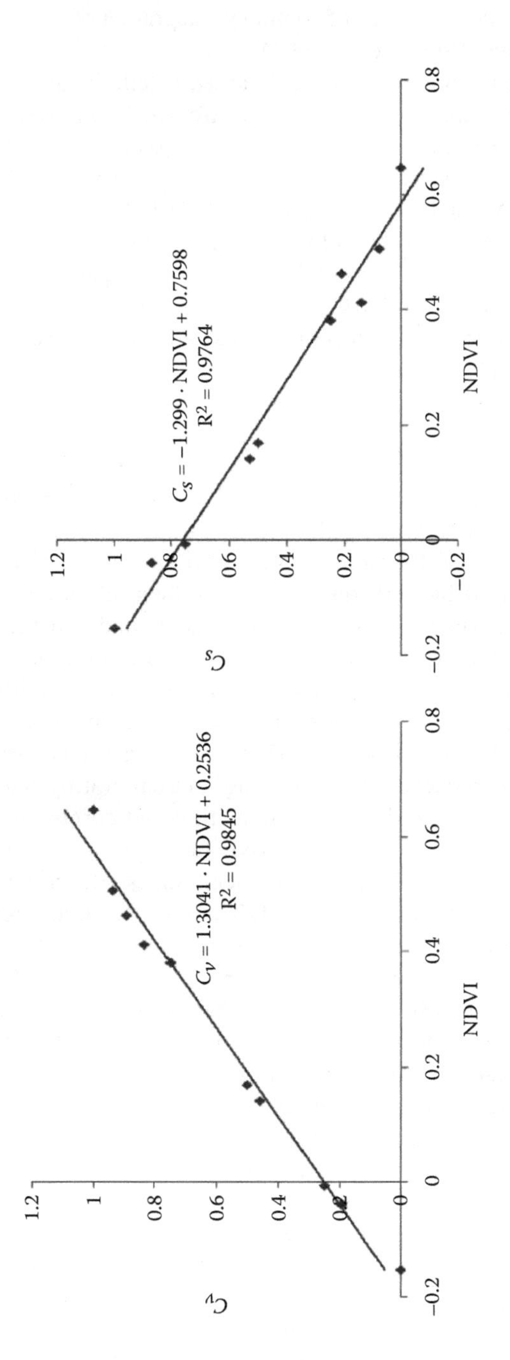

FIGURE 4.23 Correlation functions between the NDVI and the $C_{v,s}$ coefficients used to initialize Powell's minimization method. (From Guanter, L., González-Sampedro, M. C., and Moreno, J., *International Journal of Remote Sensing*, 28(3–4), 709–728, 2007.)

The final step is to convert the visibility derived from the five reference pixels to AOT at 550 nm through a particular transfer function (e.g., a conversion in Figure 4.22 as a sample). Visibility is the MODTRAN4 variable to express aerosol loading, while AOT is the magnitude that is measured by a sunphotometer at the ground. Note that the modifications performed over the MODTRAN4 original code also included the calculation of AOT at 550 nm by integration of the aerosol extinction coefficient vertical profiles (Guanter, Estellés et al. 2007).

Guanter, González-Sampedro et al. (2007) and Guanter, Estellés et al. (2007) applied the technique introduced above to the data of two imaging spectrometers to retrieve AOT at 550 nm and found that the technique is feasible after an extensive validation exercise with ground-based measurements. With the data of the medium resolution imaging spectrometer (MERIS) onboard the European Space Agency (ESA) Environmental Satellite (ENVISAT) platform, Guanter, González-Sampedro et al. (2007) found that the values of the RMSE of AOT retrieved from the MERIS data were 0.085 at 440 nm, 0.065 at 550 nm, and 0.048 at 870 nm, compared with the Aerosol Robotic Network measurements from two sites. With CASI-1500 data, Guanter, Estellés et al. (2007) demonstrated a good correlation between ground measurements and CASI-derived AOT with a Pearson's correlation coefficient R^2 up to 0.71.

4.6 SUMMARY

At the beginning of this chapter, the importance and necessity of HRS image radiometric correction and atmospheric correction for most HRS applications were discussed. This is because the remotely sensed data are not fully representative of the Earth's surface features and suffer from radiometric errors caused by both remote sensors themselves and atmospheric effects, and sometimes also terrain surface effects. To better understand and accurately correct various atmospheric effects, the theoretical concepts and principles of atmospheric characteristics (including transmittances) and effects were first introduced before a substantial discussion of various atmospheric correction (AC) methods and algorithms. The atmospheric effects introduced include Rayleigh and Mie (aerosol) scattering, various gaseous absorptions, and refraction. Some techniques and approaches for correcting and calibrating remote sensors/systems induced radiometric errors including destriping, detecting and correcting spectral and spatial mis-registration errors (caused by smile–frown curve and keystone) were introduced and discussed. In Section 4.4, various methods, principles, techniques, algorithms, and models of AC were introduced, reviewed, and discussed in detail. The introduced and reviewed AC methods and models consist of three categories, including (1) empirical/statistical and image/scene-based AC methods (ELC, IAR, and FFC); and (2) physically-based AC methods (ACORN, ATCOR, ATREM, FLAASH, HATCH, and ISDAS) and relative/normalization AC methods (single-image correction using histogram adjustment, and multiple-date image normalization using regression model). Particularly, characteristics, key features, and references, etc., of the six physically based AC methods were summarized in Table 4.1 and the comparison of evaluation and performance between two and among three through six AC methods were highlighted in Table 4.2. Finally, the five techniques for estimating atmospheric total-column water vapor (WV) content and two techniques for retrieving aerosol optical thickness (AOT) or loading from HRS imaging data were introduced and reviewed, given the fact that the two atmospheric parameters are the most difficult and challenging to estimate for AC to retrieve surface reflectance. The five techniques for estimating WV content include narrow/wide (N/W), continuum interpolated band ratio (CIBR), three-band ratioing (3BR), linear regression ratio (LIRR), and atmospheric pre-corrected differential absorption (APDA). The two techniques for retrieving AOT consist of dark dense vegetation technique (DDV) and aerosol optical thickness at 550 nm technique (AOT at 550 nm).

REFERENCES

ACM User's Guide. 2009. *Atmospheric Correction Module: QUAC and FLAASH User's Guide*. ENVI, Version 4.7, August 2009 Edition, 44p.

Adler-Golden, S. M., M. W. Matthew, L. S. Bernstein, R. Y. Levine, A. Berk, S. C. Richtsmeier, P. K. Acharya, G. P. Anderson, G. Felde, J. Gardner, M. Hoke, L. S. Jeong, B. Pukal, J. Mello, A. Ratkowski, and H.-H. Burke 1999. Atmospheric correction for short-wave spectral imagery based on MODTRAN4, in *Summaries of the Eighth JPL Airborne Earth Science Workshop JPL Publication*, vol. 99–17, in ed. R. O. Green, pp. 21–29. Pasadena, California: Jet Propulsion Lab 1.

Barnett, C. E. 1942. Some application of wavelength turbidimetry in the infrared. *The Journal of Physical Chemistry* 46(1):69–75.

Ben-Dor, E., B. C. Kindel, and K. Patkin. 2005. A comparison between six model-based methods to retrieve surface reflectance and water vapor content from hyperspectral data: A case study using synthetic AVIRIS data. *International Conference on Optics & Optoelectronics—ICOL 2005*, Dehradun, India.

Berk, A., G. P. Anderson, P. K. Acharya, J. H. Chetwynd, L. S. Bernstein, E. P. Shettle, M. W. Matthew, and S. M. Adler-Golden. 1999. *MODTRAN4 User's Manual*. Air Force Research Laboratory, Hanscom AFB, Maryland.

Berk, A., G. P. Anderson, P. K. Acharya, L. S. Bernstein, L. Muratov, J. Lee, M. J. Fox, S. M. Adler-Golden, J. H. Chetwynd, M. L. Hoke, R. B. Lockwood, T. W. Cooley, and J. A. Gardner. 2005 (January 20). MODTRAN5: A reformulated atmospheric band model with auxiliary species and practical multiple scattering options. *Proceedings of SPIE 5655, Multispectral and Hyperspectral Remote Sensing Instruments and Applications II*, 88. doi:10.1117/12.578758.

Black, M., A. Fleming, T. Riley, G. Ferrier, P. Fretwell, J. McFee, S. Achal, and A. U. Diaz. 2014. On the atmospheric correction of Antarctic airborne hyperspectral data. *Remote Sensing* 6:4498–4514.

Blue Sky. 2015. Atmospheric scattering. Accessed May 14, 2015, from http://hyperphysics.phy-astr.gsu.edu/hbase/atmos/blusky.html#c5.

Bruegge, C. J., J. E. Conel, J. S. Margolis, R. O. Green, G. C. Toon, V. Carrere, R. G. Holm, and G. Hoover. 1990. In situ atmospheric water-vapor retrieval in support of AVIRIS validation. *Imaging Spectroscopy of the Terrestrial Environment, SPIE* 1298:150–163.

Canty, M. J., A. A. Nielsen, and M. Schmidt. 2004. Automatic radiometric normalization of multitemporal satellite imagery. *Remote Sensing of Environment* 91:441–451.

Carrère, V., and M. J. Abrams. 1988. An assessment of AVIRIS data for hydrothermal alteration mapping in the Goldfield Mining District, Nevada. *Proceedings of the Airborne Visible/Infrared (AVIRIS) Performance Evaluation Workshop, JPL Publication* 88–38, pp. 134–154.

Carrère, V., and J. E. Conel. 1993. Recovery of atmospheric water vapor total column abundance from imaging spectrometer analysis and application to Airborne Visible/Infrared Imaging Spectrometer (AVIRIS) data. *Remote Sensing of Environment* 44:179–204.

Chandrasekhar, S. 1960. *Radiative Transfer*. Colorado: Dover Publications.

Clark, R. N., and T. V. V. King. 1987. Causes of spurious features in spectral reflectance data, *Proceedings of the Third Airborne Imaging Spectrometer Data Analysis Workshop*, JPL Publication 87–30, pp. 132–137. Retrieved from http://speclab.cr.usgs.gov.

Clark, R. N., G. A. Swayze, K. B. Heidebrecht, R. O. Green, and A. F. H. Goetz. 1995. Calibration to surface reflectance of terrestrial imaging spectrometry data: Comparison of methods, ed. R. O. Green, *Summaries of the Fifth Annual JPL Airborne Earth Science Workshop*, JPL Publication 95–1, pp. 41–42.

Clough, S. A., and M. J. Iacono. 1995. Line-by-line calculation of atmospheric fluxes and cooling rates. 2. Application to carbon-dioxide, ozone, methane, nitrous-oxide and the halocarbons. *Journal of Geophysical Research* 100(D8):16519–16535.

Conel, J. E., R. O. Green, G. Vane, C. J. Bruegge, and R. E. Alley. 1987. AIS-2 radiometry and a comparison of methods for the recovery of ground reflectance, ed. G. Vane, *Proceedings of the 3rd Airborne Imaging Spectrometer Data Analysis Workshop*, JPL Publication 87–30, pp. 18–47.

Cooley, T., G. P. Anderson, G. W. Felde, M. L. Hoke, A. J. Ratkowski, J. H. Chetwynd, J. A. Gardner, S. M. Aldler-Golden, M. W. Matthew, A. Berk, L. S. Bernstein, P. K. Acharya, D. Miller, and P. Lewis. 2002. FLAASH: A Modtran4-based atmospheric correction algorithm: Its application and validation. *Proceedings of IGARSS*, vol. III, pp. 1414–1418. IEEE, Toronto, Canada.

Cox, A. J., A. J. DeWeerd, and J. Linden. 2002. An experiment to measure Mie and Rayleigh total scattering cross sections. *American Journal of Physics* 70:620–625.

Du, Y., P. M. Tiellet, and J. Cihlar. 2002. Radiometric normalization of multitemporal high-resolution satellite images with quality control for land cover change detection. *Remote Sensing of Environment* 82:123–134.

Eismann, M., 2012. *Hyperspectral Remote Sensing*. Bellingham, WA: SPIE Press.

Feng, Y., and Y. Xiang. 2008. Mitigation of spectral mis-registration effects in imaging spectrometers via cubic spline interpolation. *Optics Express* 16:15366–15374.

Fraser, R. S., and Y. J. Kaufman. 1985. The relative importance of scattering and absorption in remote sensing. IEEE Trans. *Geoscience and Remote Sensing* 23:625–633.

Frouin, R., P.-Y. Deschamps, and P. Lecomte. 1990. Determination from space of atmospheric total water vapor, amounts by differential absorption near 940 nm: Theory and airborne verification. *Journal of Applied Meteorology* 29:448–459.

Gao, B., and A. F. H. Goetz. 1990. Column atmospheric water vapor and vegetation liquid water retrievals from airborne imaging spectrometer data. *Journal of Geophysical Research* 95(D4):3549–3564.

Gao, B. C., and C. O. Davis. 1997. Development of a line-by-line-based atmosphere removal algorithm for airborne and spaceborne imaging spectrometers. *SPIE Proceedings* 3118:132–141.

Gao, B. C., C. O. Davis, and A. F. H. Goetz. 2006. A review of atmospheric correction techniques for hyperspectral remote sensing of land surfaces and ocean color. *IEEE International Geoscience and Remote Sensing Symposium (IGARSS 2006)*, pp. 1979–1981. Denver, Colorado.

Gao, B. C., K. B. Heidebrecht, and A. F. H. Goetz. 1993. Derivation of scaled surface reflectances from AVIRIS data. *Remote Sensing of Environment* 44:165–178.

Gao, B. C., M. J. Montes, C. O. Davos, and A. F. H. Goetz. 2009. Atmospheric correction algorithms for hyperspectral remote sensing data of land and ocean. *Remote Sensing Environment* 113(supplement 1): S17–S24.

Goetz, A. F. H., B. C. Kindel, M. Ferri, and Z. Qu. 2003. HATCH: Results from simulated radiances, AVIRIS and HYPERION. *IEEE Transactions on Geoscience and Remote Sensing* 41:1215–1221.

Goody, R., R. West, L. Chen, and D. Crisp. 1989. The correlated-*k* method for radiation calculations in nonhomogeneous atmospheres. *Journal of Quantitative Spectroscopy and Radiation Transfer* 42:539–550.

Green, R. O. 1992. Retrieval of reflectance from calibrated radiance imagery measured by the airborne visible infrared imaging spectrometer (AVIRIS) for lithological mapping, in *Imaging Spectroscopy: Fundamentals and Prospective Application, ECSC EEC EAEC*, eds. F. Toselli and J. Bodechtel, pp. 61–71. Springer, Netherlands: Brussels and Luxembourg.

Green, R. O. 1998. Spectral calibration requirement for Earth looking imaging spectrometers in the solar-reflected spectrum. *Applied Optics* 37:683–690.

Green, R. O., V. Carrère, and J. E. Conel. 1989. Measurement of atmospheric water vapor using the Airborne Visible/Infrared Imaging Spectrometer, in *Workshop Imaging Processing*. Sparks, NV: American Society for Photogrammetry and Remote Sensing.

Green, R. O., J. E. Conel, and D. A. Roberts. 1993. Estimation of aerosol optical depth, pressure elevation, water vapor, and calculation of apparent surface reflectance from radiance measured using the Airborne Visible/Infrared Imaging Spectrometer (AVIRIS) using a radiative transfer code. *Proceedings of SPIE* 1937, 2–11. doi:10.1117/12.157054.

Green, R. O., M. L. Eastwood, C. M. Sarture, T. G. Chrien, M. Aronsson, B. J. Chippendale, J. A. Faust, B. E. Pavri, C. J. Chovit, M. Solis, M. R. Olah, and O. Williams. 1998. Imaging spectroscopy and the airborne visible/infrared imaging spectrometer (AVIRIS). *Remote Sensing of Environment* 65:227–248.

Griffin, M. K., and H. K. Burke. 2003. Compensation of hyperspectral data for atmospheric effects. *Lincoln Laboratory Journal* 14(1):29–54.

Guanter, L., V. Estellés, and J. Moreno. 2007. Spectral calibration and atmospheric correction of ultra-fine spectral and spatial resolution remote sensing data application to CASI-1500 data. *Remote Sensing of Environment* 109:54–65.

Guanter, L., M. C. González-Sampedro, and J. Moreno. 2007. A method for the atmospheric correction of ENVISAT/MERIS data over land targets. *International Journal of Remote Sensing* 28(3–4):709–728.

Guanter, L., R. Richter, and J. Moreno. 2006. Spectral calibration of hyperspectral imagery using atmospheric absorption features. *Applied Optics* 45:2360–2370.

Gupta, R. P. 2003. *Remote Sensing Geology*, 2nd ed. Berlin, NY: Springer-Verlag.

Haboudance, D., J. R. Miller, N. Tremlaly, P. J. Zarco-Tajada, and L. Dextraze. 2002. Integrated narrow-band vegetation indices for prediction of crop chlorophyll content for application to precision agriculture. *Remote Sensing of Environment* 81:416–426.

ImSpec. 2002. *ACORN 4.0 User's Guide*. Boulder, CO: Analytical Imaging and Geophysics LLC.

Jensen, J. R. 2005. *Introductory Digital Image Processing: A Remote Sensing Perspective*, 3rd ed. Upper Saddle River, NJ: Prentice Hall.

Kaufman, Y., and D. Tanré. 1996. Strategy for direct and indirect methods for correction the aerosol effect on remote sensing: From avhrr to EOS-MODIS. *Remote Sensing of Environment* 55:65–79.

Kaufman, Y. J., A. E. Wald, L. A. Remer, B.-C. Gao, R.-R. Li, and L. Flynn. 1997. The MODIS 2.1-μm channel-correlation with visible reflectance for use in remote sensing of aerosol. *IEEE Transactions on Geoscience and Remote Sensing* 35:1286–1298.

Kayadibi, Ö., and D. Aydal. 2013. Quantitative and comparative examination of the spectral features characteristics of the surface reflectance information retrieved from the atmospherically corrected images of Hyperion. *Journal of Applied Remote Sensing* 7(1), 073528.

Khurshid, K. S., K. Staenz, L. Sun, R. Neville, H. P. White, A. Bannari, C. M. Champagne, and R. Hitchcock. 2006. Preprocessing of EO-1 Hyperion data. *Canadian Journal of Remote Sensing* 32(2):84–97.

Kruse, F. A. 1988. Use of airborne imaging spectrometer data to map minerals associated with hydrothermally altered rocks in the northern Grapevine Mountains, Nevada and California. *Remote Sensing of Environment* 24:31–51.

Kruse, F. A. 2004. Comparison of ATREM, ACORN, and FLAASH atmospheric corrections using low-altitude AVIRIS data of Boulder, CO, in *Summaries of 13th JPL Airborne Geoscience Workshop.* Pasadena, CA: Jet Propulsion Lab.

Lacis, A. A., and V. Oinas 1991. A description of correlated k distribution method for modeling nongray gaseous absorption, thermal emission, and multiple scattering in vertically inhomogeneous atmospheres. *Journal of Geophysical Research* 96:9027–9063.

Liang, S. 2004. Atmospheric correction of optical imagery, in *Quantitative Remote Sensing of Land Surfaces*, ed. S. Liang, pp. 196–230. Hoboken, NJ: John Wiley & Sons.

Lu, D., P. Mausel, E. Brondizio, and E. Moran. 2002. Assessment of atmospheric correction methods for Landsat TM data applicable to Amazon basin LBA research. *International Journal of Remote Sensing* 23(13):2651–2671.

Malkmus, W. 1967. Random Lorentz band model with exponential-tailed S line intensity distribution function. *Journal of the Optical Society America* 57:323–329.

Miller, S. W., and E. Vermote. 2002. NPOESS Visible/Infrared Imager/Radiometer: Algorithm Theoretical Basis Document, Version 5. Lanham, MD: Raytheon.

Moses, W. J., and W. D. Philpot. 2012. Evaluation of atmospheric correction using bi-temporal hyperspectral images. *Israel Journal of Plant Sciences* 60:253–263.

Mouroulis, P., R. O. Green, and T. G. Chrien. 2000. Design of pushbroom imaging spectrometers for optimum recovery of spectroscopic and spatial information. *Applied Optics* 39:2210–2220.

Mulligan, J. F. 1980. *Practical Physics: The Production and Conservation of Energy.* New York: McGraw-Hill.

Nadeau, C., R. A. Neville, K. Staenz, N. T. O'Neill, and A. Royer. 2002. Atmospheric effects on the classification of surface minerals in an arid region using Short-Wave Infrared (SWIR) hyperspectral imagery and a spectral unmixing technique. *Canadian Journal of Remote Sensing* 28(6):738–749.

Neville, R. A., L. Sun, and K. Staenz. 2004. Detection of keystone in imaging spectrometer data. In *Proceedings of the International Symposium on Algorithms and Technologies for Multispectral, Hyperspectral, and Ultraspectral Imagery X*, eds. S. S. Shen and P. E. Lewis. Orlando, Florida. Bellingham, WA: International Society for Optical Engineering. *Proceedings of SPIE* 5425:208–217.

Nielsen, A. A., K. Conradsen, and J. J. Simpson. 1998. Multivariate alteration detection (MAD) and MAF post-processing in multispectral, bitemporal image data: New approaches to change detection studies. *Remote Sensing of Environment* 64:1–19.

O'Neill, N. T., A. Royer, and M. N. Nguyen. 1996. *Canadian Advanced Modified 5S (CAM5S) Internal Report*, CARTEL-1996-0202, Centre d'application et de recherches en teledetection (CARTEL). Quebec: Universite Sherbrooke.

Perkins, T., S. Adler-Golden, M. W. Matthew, A. Berk, L. S. Bernstein, J. Lee, and M. Fox. 2012. Speed and accuracy improvements in FLAASH atmospheric correction of hyperspectral imagery. *Optical Engineering* 51(11):111707.

Perry, E. M., T. Warner, and P. Foote. 2000. Comparison of atmospheric modelling versus empirical line fitting for mosaicking HYDICE imagery. *International Journal of Remote Sensing* 21(4):799–803.

Press, W. H., S. A. Teukolsky, W. T. Vetterling, and B. P. Flannery. 2007. *Numerical Recipes: The Art of Scientific Computing,* 3rd ed. Cambridge, NY: Cambridge University Press.

Pu, R., and P. Gong. 2004. Wavelet transform applied to EO-1 hyperspectral data for forest LAI and crown closure mapping. *Remote Sensing of Environment* 91:212–224.

Pu, R., P. Gong, and G. S. Biging. 2003. Simple calibration of AVIRIS data and LAI mapping of forest plantation in southern Argentina. *International Journal of Remote Sensing* 24(23):4699–4714.

Pu, R., S. Landry, and J. Zhang. 2015. Evaluation of atmospheric correction methods in identifying urban tree species with WorldView-2 imagery. *Journal of Selected Topics in Applied Earth Observations and Remote Sensing*, 8(5):1886–1897.

Pu, R., M. Kelly, G. L. Anderson, and P. Gong. 2008. Using CASI hyperspectral imagery to detect mortality and vegetation stress associated with a new hardwood forest disease. *PE&RS* 74(1):65–75.

Qu, Z., and A. H. Goetz. 1999. A fast algorithm for radiative intensity calculation in plane parallel scattering-absorbing atmospheres. *Proceedings of the 22nd Annual Review of Atmospheric Transmission Models*, Hanscom AFB, Massachusetts.

Qu, Z., B. C. Kindel, and A. F. H. Goetz. 2003. The high-accuracy atmospheric correction for hyperspectral data (HATCH) model. *IEEE Transactions on Geoscience and Remote Sensing* 41:1223–1231.

Richter, R. 1990. A fast atmospheric correction algorithm applied to Landsat TM images. *International Journal of Remote Sensing* 11:159–166.

Richter, R. 1996. Atmospheric correction of DAIS hyperspectral image data. *Computers & Geosciences* 22:785–793.

Richter, R. 1998. Correction of satellite imagery over mountainous terrain. *Applied Optics* 37:4004–4015.

Richter, R. 2004. ATCOR: Atmospheric and Topographic Correction. DLR—German Aerospace Center. Accessed June 3, 2015, from http://www.dlr.de/caf/en/Portaldata/36/Resources/dokumente/technologie/atcor_flyer_march2004.pdf.

Richter, R., and D. Schläpfer. 2002. Geo-atmospheric processing of airborne imaging spectrometry data. Part 2: Atmospheric/topographic correction. *International Journal of Remote Sensing* 23:13, 2631–2649.

Roberts, D. A., Y. Yamaguchi, and R. Lyon. 1986. Comparison of various techniques for calibration of AIS data, in eds. G. Vane and A. F. H. Goetz. *Proceedings of the 2nd Airborne Imaging Spectrometer Data Analysis Workshop*, JPL Publication, vol. 86–35, pp. 21–30. Pasadena, California.

Rothman, L. S., A. Barbe, D. C. Benner, L. R. Brown, C. Camy-Peyret, M. P. Carleer et al. 2003. The HITRAN molecular spectroscopic database: Edition of 2000 including updates through 2001. *Journal of Quantitative Spectroscopy and Radiative Transfer* 82:5–44.

Research Systems Inc. (RSI). 2005. ENVI User's Guide. Online manual of software ENVI.

Sagan, C. 1994. *Pale Blue Dot A Vision of the Human Future in Space*, 1st ed. New York: Random House.

San, B. T., and M. L. Suzen. 2010. Evaluation of different atmospheric correction algorithms for eo-1 Hyperion imagery. *International Archives of the Photogrammetry, Remote Sensing and Spatial Information Science*, volume XXXVIII, part 8, pp. 392–397, Kyoto, Japan.

Schläpfer, D., J. Keller, and K. I. Itten. 1996. Imaging spectrometry of tropospheric ozone and water vapor, ed. E. Parlow. In *Proceedings of the 15th EARSeL Symposium Basel*, pp. 439–446. Rotterdam, A. A. Balkema. Accessed from http://www.geo.unizh.ch/dschlapf/paper.html.

Schläpfer, D., C. C. Borel, J. Keller, and K. I. Itten. 1998. Atmospheric precorrected differential absorption technique to retrieve columnar water vapor. *Remote Sensing of Environment* 65:353–366.

Seinfeld, J. H., and S. N. Pandis. 2006. Interaction of aerosols with radiation, in *Atmospheric Chemistry and Physics: From Air Pollution to Climate Change*, 2nd ed., pp. 691–719. Upper Saddle River, NJ: John Wiley & Sons.

Slater, P. N. 1980. *Remote Sensing: Optics and Optical System*. Reading, MA: Addison-Wesley.

Song, C., C. E., Woodcock, K. C. Seto, M. P. Lenney, and S. A. Macomber. 2001. Classification and change detection using Landsat TM data: When and how to correct atmospheric effects. *Remote Sensing of Environment* 75:230–244.

Staben, G. W., K. Pfitzner, R. Bartolo, and A. Lucieer. 2012. Empirical line calibration of WorldView-2 satellite imagery to reflectance data: Using quadratic prediction equations. *Remote Sensing Letters* 3(6): 521–530.

Staenz, K. 1992. Imaging spectrometer data analyzer (ISDA): A software package for analyzing high spectral resolution data. *Canadian Journal of Remote Sensing* 18(2):90–101.

Staenz, K., and D. J. Williams. 1997. Retrieval of surface reflectance from hyperspectral data using a look-up-table approach. *Canadian Journal of Remote Sensing* 23:354–368.

Staenz, K., T. Szeredi, and J. Schwarz. 1998. ISDAS: A system for processing/analyzing hyperspectral data. *Canadian Journal of Remote of Sensing* 24:99–113.

Staenz, K., R. A. Neville, J. Levesque, T. Szeredi, V. Singhroy, G. A. Borstad, and P. Hauff. 1999. Evaluation of casi and SFSI hyperspectral data for environmental and geological applications: Two case studies. *Canadian Journal of Remote Sensing* 25:311–322.

Stamnes, K., S.-C. Tsay, W. Wiscombe, and K. Jayaweera. 1988. A numerically stable algorithm for discrete-ordinate-method radiative transfer in multiple scattering and emitting layered media. *Applied Optics* 27:2502–2509.

Sturm, B. 1992. Atmospheric and radiometric corrections for imaging spectroscopy, in *Imaging Spectroscopy: Fundamentals and Prospective Application*, eds. F. Toselli and J. Bodechtel, pp. 47–60. Brussels and Luxembourg: ECSC EEC EAEC.

Switzer, P., and A. A Green. 1984. *Min/Max Autocorrelation Factors for Multivariate Spatial Imagery*. Technical Report No. 6. Stanford, CA: Department of Statistics at Stanford University.

Tanré, D., C. Deroo, and P. Duhaut, M. Herman, J. J. Morcrette, J. Perbos, and P. Y. Deschamps. 1986. *Simulation of the Satellite Signal in the Solar Spectrum (5S), User's Guide*. Laboratory d'Optique Atmospherique, U. S. T. de Lille, 59655 Villeneuve D'ascq, France.

Teillet, P. M. 1986. Image correction for radiometric effects in remote sensing. *International Journal of Remote Sensing* 7:12, 1637–1651.

Ungar, S. G., J. S. Pearlman, J. Mendenhall, and D. Reuter. 2003. Overview of the Earth Observing 1 (EO-1) mission. *IEEE Transactions of Geoscience and Remote Sensing* 41:1149–1159.

Van der Meer, F. D., S.M. de Jong, and W. Bakker. 2001. Imaging spectrometry: Basic analytical techniques, in *Imaging Spectrometry: Basic Principles and Prospective Applications*, eds. F. van der Meer and S. M. de Jong, Springer, The Netherlands. pp. 17–61.

Vane, G., and A. F. H. Goetz. 1993. Terrestrial imaging spectrometry: Current status, future trends. *Remote Sensing of Environment* 44:117–126.

Yokoya, N., N. Miyamura, and A. Iwasaki. 2010. Detection and correction of spectral and spatial misregistrations for hyperspectral data using phase correlation method. *Applied Optics* 49(24):4568–4575.

Yuan, J., and Z. Niu. 2008. Evaluation of atmospheric correction using FLAASH. In *Proceedings of 2008 IEEE International Workshop on Earth Observation and Remote Sensing Applications.*

Zhao, Y., Z. Meng, L. Wang, S. Miyazaki, X. Geng, G. Zhou, R. Liu, N. Kosaka, and M. Takahashi. 2005. A new cross-track radiometric correction method (VRadCor) for airborne hyperspectral image of Operational Modular Imaging Spectrometer (OMIS). *IEEE IGARSS2005*, pp. 3553–3556.

Zhao, Y., R. Pu, S. Bell, C. A. Meyer, L. Baggett, and X. Geng. 2013. Hyperion image optimization in coastal waters. *IEEE Transactions on Geoscience and Remote Sensing* 51(2):1025–1036.

5 Hyperspectral Data Analysis Techniques

Hyperspectral data contain a wealth of spectral information about the material content that presents in remotely sensed scenes. However, such a wealth of, interesting information usually hides in a huge data volume that hyperspectral imagery can represent, and hyperspectral data has a high dimensionality and high correlation of adjacent bands. Consequently, while many existing analytical methods are effective for processing multispectral data, they generally face an overall challenge that they are less effective to process, or do not work on, hyperspectral data. Moreover, due to the nature of hyperspectral data, some data analytical techniques are unique to processing hyperspectral data, such as derivative spectral analysis and diagnostic spectrum extraction. Therefore, in this chapter, a collection of general hyperspectral image processing techniques and algorithms, commonly used and newly developed, are briefly introduced and discussed. Particularly, spectral derivative analysis, spectral similarity measuring, spectral absorption feature extraction and wavelength position variable determination, and hyperspectral vegetation indices are introduced and summarized in Sections 5.2 through 5.5. Relevant hyperspectral data transform methods and feature extraction algorithms in Section 5.6, various spectral mixture analysis algorithms especially linear spectral unmixing models in Section 5.7, and different segment-based traditional classifiers and advanced neural networks and support vector machines algorithms in Section 5.8 are introduced and reviewed. In addition, many specific analysis techniques/methods used in different application areas for analyzing hyperspectral data are reviewed and discussed in the book's last three chapters: applications to geology and soil, vegetation and plants, and environment and others.

5.1 INTRODUCTION

After hyperspectral data are acquired by imaging spectrometers, the imaging data first need to be calibrated and corrected for spectral precision and radiometric accuracy. Methods and techniques dealing with spectral and radiometric calibration and correction are found in Chapter 4. Although hyperspectral data contain wealthy spectral information about the material content of remotely sensed scenes based on the unique spectral characteristics of the materials within them, such interesting information usually hides in a huge data volume that hyperspectral imagery can represent. Consequently, while analytical methods are unique to a particular remote sensing application, they generally face an overall challenge of sorting through the huge data volume of hyperspectral imagery to detect specific materials of interest, classify materials into groups with similar and more meaningful spectral properties, or estimate and retrieve specific physical properties based on their subtle spectral characteristics (Eismann 2012). The special characteristics of hyperspectral datasets also pose different processing problems, which must be tackled under specific mathematical formalisms (Plaza et al. 2009). Therefore, over the last three decades, many image-processing techniques specially for processing hyperspectral image data have been developed and applied to process and extract relevant information from hyperspectral data (Schaepman et al. 2009). Such processing techniques and algorithms include spectral derivative analysis (e.g., Tsai and Philpot 1998), spectral matching (e.g., van der Meer and Bakker 1997a), spectral absorption features and wavelength position variable extraction (e.g., Clark and Roush 1984), spectral unmixing (Adams et al. 1986), spectral index analysis (e.g., Pu and Gong 2011), spectral transform and feature extraction (e.g., Eismann 2012), and image segmentation and classification (e.g., Jia et al. 1999, Xu and Gong et al. 2007, Plaza et al. 2009). In the following, these imaging data processing and information extraction techniques and algorithms developed primarily for hyperspectral data are introduced and discussed. Performance of these data analysis and information extraction techniques requires specific dedicated processing software and hardware platforms (Plaza and Chang 2007). Relevant processing software and tools are introduced in Chapter 6.

5.2 SPECTRAL DERIVATIVE ANALYSIS

The derivative spectrum is the normalized spectral difference of two continuous/neighbor narrowbands with their wavelength interval. Since *in situ* or imaging spectrometer data obtained in the field are rarely from a single object, they are usually contaminated by illumination variations caused by terrain background, atmosphere, and viewing geometry (Pu and Gong 2011). Therefore, spectral derivative analysis are considered a desirable tool to remove or compress the effect of such illumination variations with low frequency on target spectra (Demetriades-Shah et al. 1990, Tsai and Philpot 1998). For the first- and second-order derivative spectra, a finite approximation (Tsai and Philpot 1998) can be applied to calculate them from hyperspectral data:

$$\rho'(\lambda_i) \approx [\rho(\lambda_{i+1}) - \rho(\lambda_{i-1})]/\Delta\lambda \tag{5.1}$$

and

$$\begin{aligned} \rho''(\lambda_i) &\approx [\rho'(\lambda_{i+1}) - \rho'(\lambda_{i-1})]/\Delta\lambda \\ &\approx [\rho(\lambda_{i+1}) - 2\rho(\lambda_i) + \rho(\lambda_{i-1})]/\Delta\lambda^2 \end{aligned} \tag{5.2}$$

where $\rho'(\lambda_i)$ and $\rho''(\lambda_i)$ are the first and second derivatives, respectively; $\rho(\lambda_i)$ is the reflectance at a wavelength (band) i; $\Delta\lambda$ is the wavelength interval between λ_{i+1} and λ_{i-1} and equals twice the bandwidth for this case. When implementing the spectral derivative analysis, a spectral resolution better than 10 nm is required and spectral bands are continuous. Figure 5.1 shows a raw reflectance, first and second order derivative spectrum of an *in situ* hyperspectral spectrum measured from a red maple (*Acer rubrum*) tree canopy.

It is believed that the accuracy of derivative analysis is sensitive to the signal-to-noise ratio (SNR) of hyperspectral data and thus higher order spectral derivative processing is susceptible to the noise (Cloutis 1996). Lower order derivatives (e.g., the first order derivative) are less sensitive to noise and hence more effective in operational remote sensing. For example, Gong et al. (1997, 2001) and Pu (2009) reported that first derivative of tree spectra could considerably improve the accuracy in recognizing six conifer species commonly found in northern California and eleven urban tree species in the City of Tampa, Florida, respectively.

5.3 SPECTRAL SIMILARITY MEASURES

The simplest spectral similarity measure is binary coding of hyperspectral data for classification. The resultant binary (coding) chain is regarded to contain curve shape information that can support material recognition. A set of reference curves are also encoded into such binary chains and used for comparison with the binary chain (i.e., test curves) to be classified. The smallest difference resulting from the comparison between two binary chains (one reference binary chain and the other test [target] binary chain) indicates the greatest similarity between them. Although this encoding method was designed to minimize computation requirement, clearly it does not use detailed differences in magnitude and thus cannot differentiate spectral curves that have similar shapes but different absolute magnitudes. Thus, this method has not been widely used. Currently, there are two groups of spectral matching or spectral similarity measures that have been developed, which all efficiently utilize the detailed spectral differences in magnitude between two spectra (one used as a reference and the other as a test spectrum) to measure the spectral similarity of the two spectra.

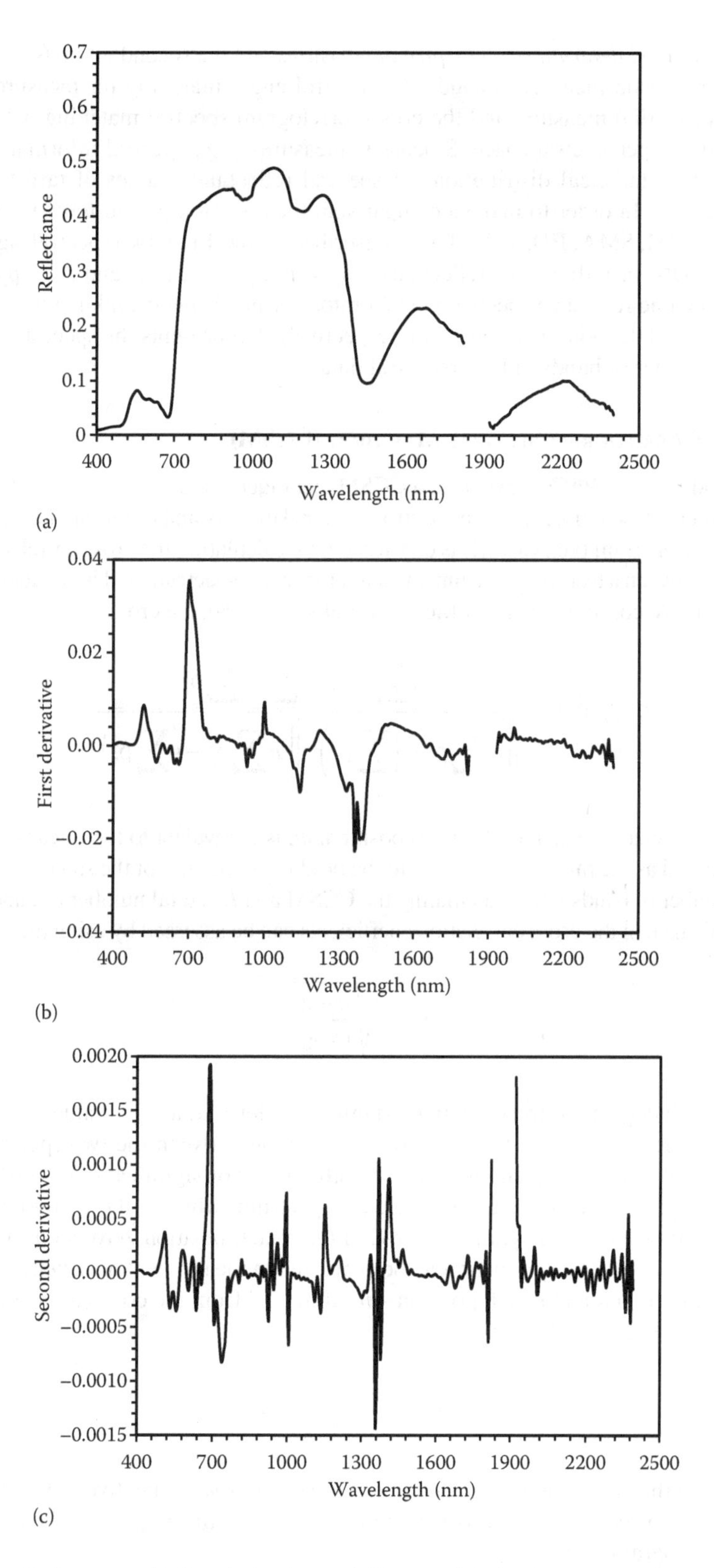

FIGURE 5.1 (a) Reflectance, (b) first-derivative, and (c) second-derivative spectra of red maple tree (*Acer rubrum*). The spectrum was measured in Tampa, Florida, with an ASD spectrometer.

The first group is called *deterministic–empirical measures* and the second group is called *stochastic measures*. Deterministic measures include the spectral angle matching (or measure) (SAM), the Euclidean distance (ED) measure, and the cross-correlogram spectral matching (CCSM) of spectral vectors in the hyperspectral space. Stochastic measures (e.g., spectral information divergence [SID]) evaluate the statistical distributions of spectral reflectance values of target end-members (van der Meer 2006). In order to make a comparison, the commonly used four spectral similarity measures (i.e., CCSM, SMA, ED, and SID) are described below. First, two spectral signature curves (pixel vectors in DN, or radiance or reflectance): $\rho_r = [\rho_{r1}, \rho_{r2}, \ldots, \rho_{rL}]^T$ and $\rho_t = [\rho_{t1}, \rho_{t2}, \ldots, \rho_{tL}]^T$ are assumed to be one reference spectrum (a laboratory or pixel spectrum known to characterize a target of interest) and the other test spectrum, respectively. L represents the spectral dimensionality and equals the number of bands of hyperspectral data.

5.3.1 Cross-Correlogram Spectral Matching (CCSM)

van der Meer and Bakker (1997a) developed a CCSM technique under consideration of the correlation coefficient between a test and a reference spectrum, their skewness and criterion of correlation significance. A cross-correlogram (i.e., CCSM) is constructed by calculating the cross correlation at different match positions between a test (ρ_t) spectrum and a reference (ρ_r) spectrum and is suitable for processing hyperspectral data. According to van der Meer and Bakker (1997a), the cross correlation is defined as

$$r_m(\rho_r, \rho_t) = \frac{n\sum \rho_r\rho_t - \sum \rho_r \sum \rho_t}{\sqrt{\left[n\sum \rho_r^2 - \left(\sum \rho_r\right)^2\right]\left[n\sum \rho_t^2 - \left(\sum \rho_t\right)^2\right]}} \tag{5.3}$$

where the cross correlation, r_m, at each match position, m, is equivalent to the linear correlation coefficient and is defined as the ratio of covariance to the product of the sum of the standard deviations; n is the effective number of bands when calculating the CCSM and L is total number of bands ($n < L$). The statistical significance of the cross correlation coefficients can be assessed by using the following t-test:

$$t = r_m\sqrt{\frac{n-2}{1-r_m^2}} \tag{5.4}$$

Based on (n – 2) degrees of freedom and a significance level α, a $t_{\alpha(n-2)}$ value can be found from a t-distribution table. If $t > t_{\alpha(n-2)}$, then the cross correlation between the two spectra at a specific match position m is statistically significant at α; otherwise not significant. To further help judge whether a target spectrum belongs to the reference spectrum, whether the correlogram curve calculated from the two spectra is symmetric around the match position zero is considered (van der Meer and Bakker 1997b). The symmetry can be expressed in the moment of skewness, which can be calculated as the correlation at match position $-m$ subtracted from the correlation at match position $+m$, called adjusted skewness (AS_{ke}):

$$AS_{ke} = 1 - \frac{\left|r_{+m} - r_{-m}\right|}{2} \tag{5.5}$$

where r_{+m}, r_{-m} are the correlations at match position $+m$ and $-m$, respectively. When $AS_{ke} = 1$, the peak value of the correlogram curve is symmetric; the closer the $AS_{ke} \to 0$, the skewer the peak value of the correlogram curve.

Figure 5.2 demonstrates the potential of using the CCSM for mineral mapping through spectral shape matching, using the results of calculating CCSM for spectra resampled to the AVIRIS band

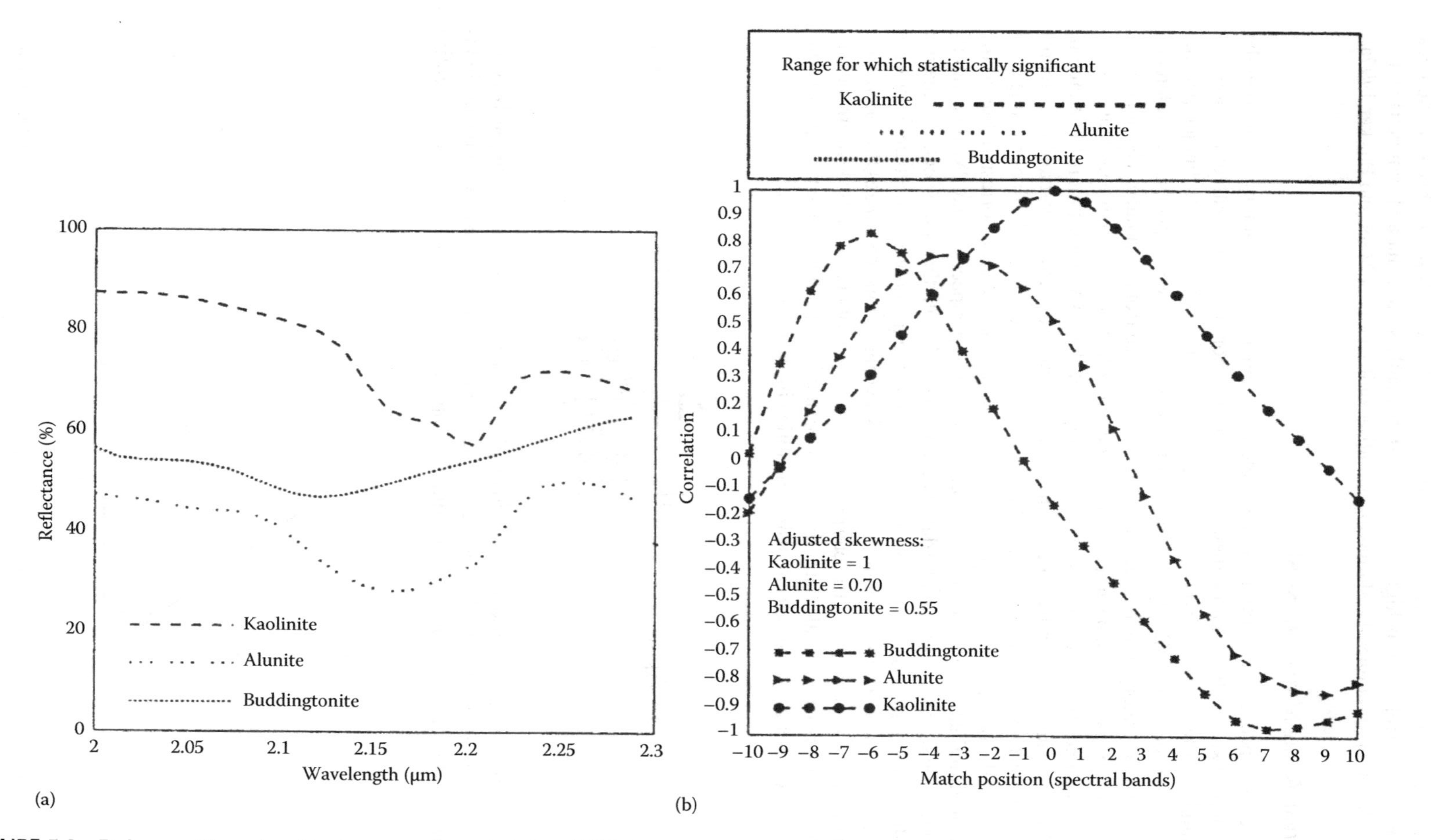

FIGURE 5.2 Reference (library) and test spectra (a) resampled to AVIRIS band passes and corresponding cross-correlograms (b) for kaolinite (as a reference spectrum) versus kaolinite, alunite, and buddingtonite (as test spectra). (a) The dash and dotted lines above the diagram in (b) show the match positions for which the cross correlation calculated was found to be statistically significant at alpha = 0.05. Per (b), it is clear that a perfect match is found for kaolinite, whereas the skew peak values for both alunite and buddingtonite are very different from the kaolinite. (Reprinted from van der Meer, F., and W. Bakker, *Remote Sensing Environment*, 61, 371–382, 1997. ©1997, with permission from Elsevier.)

passes. Per Figure 5.2b, the test spectrum (pixel) should be labeled as *kaolinite* because the peak value of the kaolinite correlogram is higher and more statistically significant and symmetric when compared with the other two minerals: alunite and buddingtonite (van der Meer and Bakker 1997a).

5.3.2 Spectral Angle Matching (SAM)

SAM determines the spectral similarity between a reference and a test spectrum by calculating the "angle" between the two spectra, treating them as spectrum vectors in a space with dimensionality equal to the number of bands (L) of hyperspectral data (Kruse et al. 1993). The technique permits rapid mapping of the spectral similarity of image (test) spectra to reference spectra. The reference spectra can be either laboratory or field spectra or directly extracted from the image. This algorithm assumes that the imaging data have been reduced to apparent reflectance, which means that all dark current and path radiance biases are removed. The principle of the SAM algorithm is illustrated in Figure 5.3, which considers a reference spectrum and a test spectrum from two-band data represented on a 2D plot as two points (empty dots). According to Kruse et al. (1993), lines passing each spectrum point and the origin contain all possible positions for that material, corresponding to the range of possible illuminations recorded in different bands of hyperspectral data. Weakly illuminated pixels may fall closer to the origin (the dark point) than pixels with the same spectral signature but stronger illumination. In addition, it is worth noting that the angle between spectrum vectors is the same regardless of their length, which means that the SAM algorithm generalizes this geometric interpretation to L dimensional space. To determine spectral similarity between two spectrum vectors, the SAM (in radian) actually calculates the arccosine of the dot product of the two spectra (one as reference and the other as test spectrum) using the following equation:

$$SAM(\rho_t, \rho_r) = \alpha = \cos^{-1}\left[\frac{\sum_{i=1}^{L}\rho_{ti}\rho_{ri}}{\left(\sum_{i=1}^{L}\rho_{ti}^2\right)^{\frac{1}{2}}\left(\sum_{i=1}^{L}\rho_{ri}^2\right)^{\frac{1}{2}}}\right] \tag{5.6}$$

This spectral similarity measure is insensitive to gain factors because the angle between two spectrum vectors is invariant with respect to the lengths of the vectors. As a result, laboratory spectra can be directly compared to imaging apparent reflectance spectra to identify

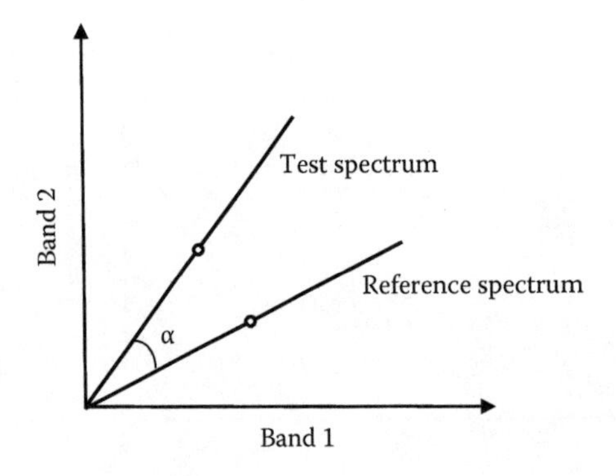

FIGURE 5.3 Plot of a reference spectrum and test spectrum for a two-band image to illustrate the principle of the SAM. The same materials with varying illumination are represented by the spectrum vectors connecting the origin (dark) and projected through the points representing the actual spectra.

materials, which inherently have an unknown gain factor related to topographic illumination effects (Kruse et al. 1993).

5.3.3 Euclidian Distance (ED)

ED measure between two pixels or between a reference spectrum and a test spectrum in the L-dimensional feature space is given as

$$ED(\rho_t, \rho_r) = \left[\sum_{i=1}^{L} (\rho_{ti} - \rho_{ri})^2 \right]^{\frac{1}{2}} \tag{5.7}$$

The ED measure is a common geometric-based vector-distance measure extended to L- dimensions (Kong et al. 2010), and the smaller measure value indicates that the two pixel spectra are more similar. The ED measure between two pixels in the L-dimensional spectral feature space can be derived from the spectral angle measure (SAM) if both ρ_t and ρ_r are normalized to unity as follows (Chang 2003):

$$\begin{aligned} ED(\rho_t, \rho_r) &= \sqrt{2 - 2\sum_{i=1}^{L} \rho_{ti}\rho_{ri}} = \sqrt{2(1 - \cos(SAM(\rho_t, \rho_r)))} \\ &= 2\sqrt{(1 - \cos(SAM(\rho_t, \rho_r)))/2} = 2\sin(SAM(\rho_t, \rho_r)/2) \end{aligned} \tag{5.8}$$

The major difference of the ED measure compared to the SAM is that the ED takes into account the brightness difference between the two vectors (spectra), whereas the SAM (and also the CCSM) is invariant with brightness (van der Meer 2006).

5.3.4 Spectral Information Divergence (SID)

The SID measure calculates the distance between the probability distributions produced by the spectral signatures of two pixels defined as (Chang 2000)

$$SID(\rho_t, \rho_r) = D(\rho_r \parallel \rho_t) + D(\rho_t \parallel \rho_r) \tag{5.9}$$

where

$$D(\rho_t \parallel \rho_r) = \sum_{l=1}^{L} q_l D_l(\rho_t \parallel \rho_r) = \sum_{l=1}^{L} q_l [I_l(\rho_r) - I_l(\rho_t)] \tag{5.10}$$

and

$$D(\rho_r \parallel \rho_t) = \sum_{l=1}^{L} p_l D_l(\rho_r \parallel \rho_t) = \sum_{l=1}^{L} p_l [I_l(\rho_t) - I_l(\rho_r)] \tag{5.11}$$

derived from the probability vectors $p = (p_1, p_2, \ldots, p_L)^T$ and $q = (q_1, q_2, \ldots, q_L)^T$ for the spectral signatures of vectors; ρ_r and ρ_t, where $p_k = \rho_{rk} / \sum_{l=1}^{L} \rho_{rl}$, $q_k = \rho_{tk} / \sum_{l=1}^{L} \rho_{tl}$, and $I_l(\rho_t) = -\log q_l$; and, similarly, $I_l(\rho_r) = -\log p_l$. Measures $I_l(\rho_t)$ and $I_l(\rho_r)$ are referred to as the self-information of ρ_t for band l. Note that Equations 5.10 and 5.11 represent the relative entropy of ρ_t with respect to ρ_r (indicated with the || symbol). It is also worth noting that the spectral similarity between the spectral signatures of two pixel vectors measured by SID is based on the discrepancy between their corresponding spectral signature–derived probability distributions. Therefore, compared to SAM and ED that extract geometric features—angle and spectral distance between two pixel vectors—the SID measures the discrepancy of probability distributions between two pixel vectors (Chang 2003). Accordingly, SID is expected to be more effective than ED and SAM in capturing spectral variability.

van der Meer (2006) compared the performance of four spectral similarity measures (CCSM, SAM, ED, and SID) using synthetic hyperspectral and real (i.e., AVIRIS) hyperspectral data of a (artificial and real) hydrothermal alteration system characterized by the minerals alunite, kaolinite, montmorillonite, and quartz to measure spectral similarity between a known reference and unknown target spectrum. Results from the study of AVIRIS data show that SAM yields more spectral confusion (i.e., class overlap) than SID and CCSM. In turn, SID is more effective in mapping the four target minerals than CCSM as it clearly outperforms CCSM when the target mineral coincides with the mineral phase on the ground.

5.4 SPECTRAL ABSORPTION FEATURES AND WAVELENGTH POSITION VARIABLES

Identification and analysis of spectral absorption (i.e., diagnostic) features is one step toward recognizing some essential properties of a target of interest. Quantitative characterization of absorption features allows for abundance estimation and classification of materials from hyperspectral data. Spectral absorption features are caused by a combination of factors inside and outside matter surface, including electronic processes, molecular vibrations, abundance of chemical constituents, granular size and physical structure, and surface roughness relative to electromagnetic wavelength. Figures 1.5 through 1.7 in Chapter 1 show the major absorption and reflectance features of minerals and plants on selected laboratory and imaging spectra.

Quantitative measures of these absorptions may be determined from each absorption peak after the normalization of the raw spectral reflectance curve. This can be done using a *continuum removal technique* as proposed by Clark and Roush (1984). As illustrated by Figure 5.4, a continuum is defined for each spectral curve by finding the high points (local maxima) along the curve and fitting straight-line segments between these points. This can be done either manually or automatically. The normalized curve is obtained by dividing the original spectral value at each band location with the value on the straight line segments at the corresponding wavelength location (Pu, Ge et al. 2003). An asymmetric term can also be defined by subtracting area A from area B (Figure 5.4) (Kruse et al. 1993). The quantitative measures shown in Figure 5.4 can be used to determine abundances of certain compounds in a pixel. For instance, Pu, Ge et al. (2003) explored the effectiveness of these absorption parameters in correlation with leaf water content of oak trees at various stages of disease infection.

Some absorption and reflectance features can also be modeled. For this case, some feature variables coupled with the modeling of absorption and reflectance features are related to change of wavelength position. For example, the extraction of red edge optical parameters from a plant spectrum between 670 nm and 780 nm has been widely modeled by a number of researchers. The red edge position (REP) is a wavelength position variable. There are several methods/algorithms used for extracting the red edge optical parameters from hyperspectral data, such as the four-point interpolation method (Guyot et al. 1992), high-order polynomial fitting (Pu, Gong et al. 2003), Lagrangian interpolation (Dawson and Curran 1998), invert Gaussian model fitting techniques (Miller et al. 1990), and linear extrapolation technique (Cho and Skidmore 2006). In the following, principles and algorithms of the five methods of extracting red edge optical parameters are introduced and discussed. All the extracted optical parameters may be used to estimate biophysical and biochemical parameters of vegetation (e.g., Pu, Gong et al. 2003).

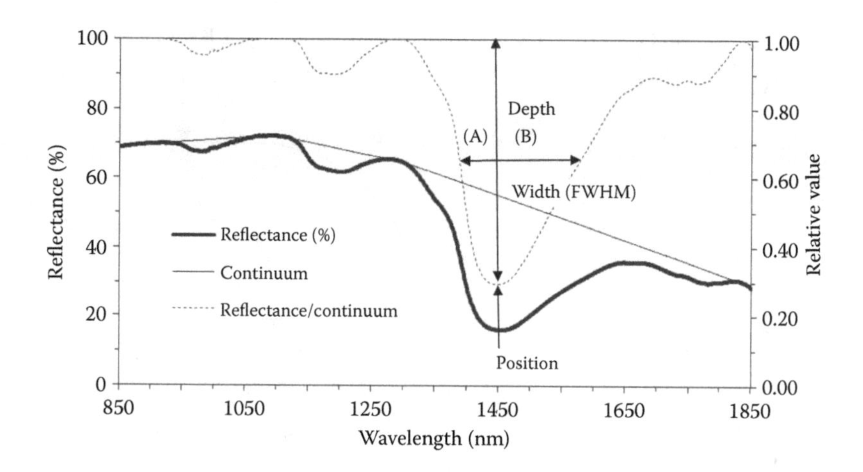

FIGURE 5.4 A portion of coast live oak leaf spectrum adjusted by a continuum removal technique (Clark and Roush 1984) and the definitions of three absorption features: depth, width, and position. Depth measures the deepest absorption. Width measures full width half maximum (FWHM). Position marks the wavelength at the deepest absorption and areas (A or B) on each side of the deepest absorption are used for calculating asymmetry. (Modified from Pu, R., Ge, S., Kelly, N. M., and Gong, P., *International Journal of Remote Sensing*, 24(9), 1799–1810, 2003.)

5.4.1 Four-Point Interpolation

Guyot et al. (1992) proposed a four-point interpolation method to find the wavelength position of the inflection point on the red edge and the red well position. The REP is defined as the wavelength position of the greatest first-order derivative spectrum. The linear interpolation as described by Guyot et al. (1992) assumes that the reflectance curve at the red edge can be simplified to a straight line centered near a midpoint between 670 nm and 780 nm (Figure 5.5). The REP interpolated locates in the range between 700 and 740 nm. To interpolate REP, there are two steps: (1) Calculating the reflectance at the inflection point along the red edge of reflectance (ρ_i, as in Equation 5.12); and (2) calculating the wavelength of the reflectance at the inflection point (λ_i, as in Equation 5.13):

$$\rho_i = (\rho_1 + \rho_4)/2 \tag{5.12}$$

$$\lambda_i = \lambda_2 + (\lambda_3 - \lambda_2)\frac{\rho_i - \rho_2}{\rho_3 - \rho_2}, \tag{5.13}$$

where λ_1, λ_2, λ_3, and λ_4 are the wavelengths of 670, 700, 740, and 780 nm respectively; and ρ_1, ρ_2, ρ_3, and ρ_4 are the reflectances at corresponding wavelengths, respectively. Accordingly, another red edge parameter, RWP (λ_0), can also be interpolated from Equation 5.14:

$$\lambda_0 = \lambda_1 + (\lambda_2 - \lambda_1)\frac{\rho_i - \rho_2}{\rho_3 - \rho_2} \tag{5.14}$$

In practice, we may not have the four bands with the exact same wavelengths as in Equation 5.13. For this case, we can choose the four bands with similar wavelengths in Equation 5.13. For example, four Hyperion bands—671.62, 702.12, 742.80, and 783.48 nm—were selected for extracting the red edge parameters in Pu, Gong et al. (2003). The REP interpolated with this method was originally used for estimating leaf chlorophyll content and LAI of the plant canopy.

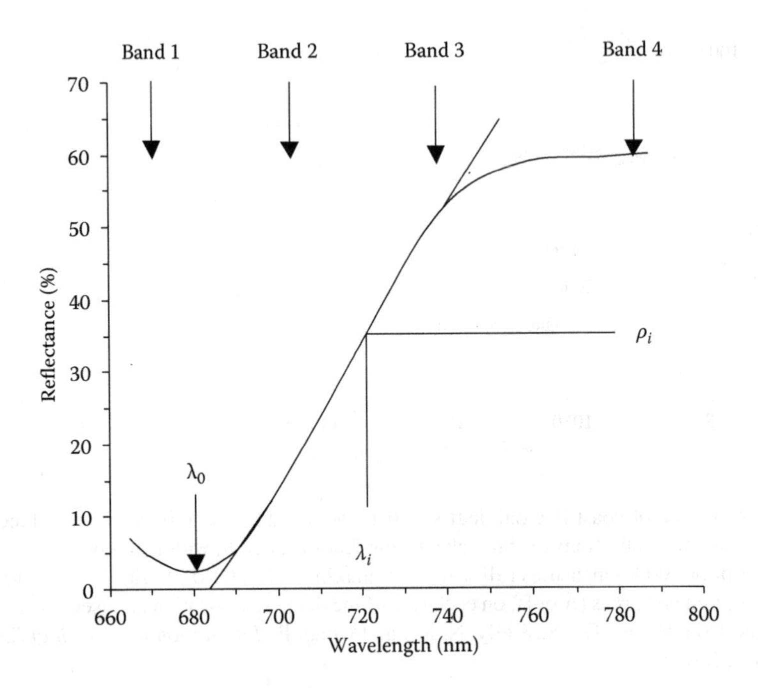

FIGURE 5.5 Illustration of the 4-point interpolation method. The ideal wavelengths for the four bands (points) are 670, 700, 740, and 780 nm, respectively. (Modified from Guyot, G., Baret, F., and Jacquemoud, S., in *Imaging Spectroscopy: Fundamentals and Prospective Application*, eds. F. Toselli and J. Bodechtel, Kluwer, Dordrecht, pp. 145–165, 1992.)

5.4.2 Polynomial Fitting

A red edge reflectance curve between the wavelengths corresponding to the minimum reflectance in red and the maximum NIR "shoulder" reflectance can be fitted with a fifth-order polynomial equation (Pu, Gong et al. 2003) as

$$\rho = a_0 + \sum_{i=1}^{5} a_i \lambda^i \qquad (5.15)$$

where λ represents band wavelengths locating within a spectral region (660–780 nm). After a red edge reflectance curve is fitted with the high-order polynomial equation, the red edge optical parameters are then calculated based on the definitions of the parameters. For example, Pu, Gong et al. (2003) selected 13 band wavelengths of Hyperion total from 661.45 nm to 783.48 nm to run a fifth-order polynomial fitting, and the red edge reflectance curve is presented in Figure 5.6. The R^2 of the fifth-order polynomial model reaches greater than 0.998 for the 13 Hyperion bands. An advantage of this method is that a continuous derivative curve can be created by taking the derivative of the 5th order polynomial curve, instead of taking differences and ratios band by band.

5.4.3 Lagrangian Technique

Dawson and Curran (1998) developed a fast technique based upon a three-point Lagrangian interpolation for determining the REP in spectra that are sampled coarsely. With the three

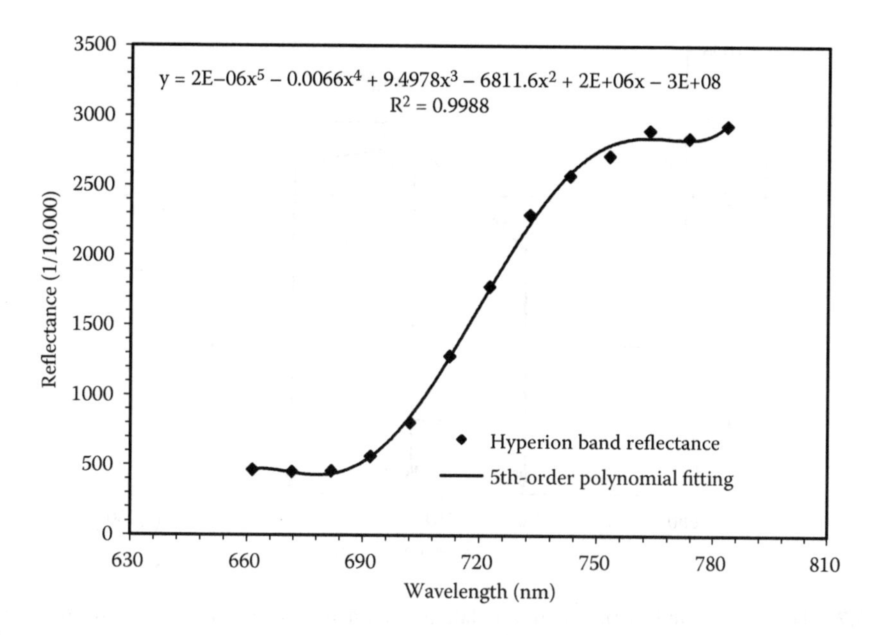

FIGURE 5.6 Illustration of the polynomial fitting (in fifth order) for determining red-edge optical parameters. The fifth-order polynomial fitting was made by using 13 band wavelengths of Hyperion from 661.45 nm to 783.48 nm. (Modified from Pu, R., Gong, G. S., Biging, S., and Larrieu, M. R., *IEEE Transactions on Geoscience and Remote Sensing*, 41(4), 916–921, 2003.)

wavebands centered around the maximum first derivative of a vegetation reflectance red edge, the technique fits a parabola through the first-derivative values. Then a second derivative is performed upon the Lagrangian equation to determine the maximum slope position (i.e., REP) (see Figure 5.7):

$$REP = \frac{A(\lambda_i + \lambda_{i+1}) + B(\lambda_{i-1} + \lambda_{i+1}) + C(\lambda_{i-1} + \lambda_i)}{2(A + B + C)} \tag{5.16}$$

where

$$A = \frac{D_{\lambda(i-1)}}{(\lambda_{i-1} - \lambda_i)(\lambda_{i-1} - \lambda_{i+1})}, \quad B = \frac{D_{\lambda(i)}}{(\lambda_i - \lambda_{i-1})(\lambda_i - \lambda_{i+1})}, \text{ and } C = \frac{D_{\lambda(i+1)}}{(\lambda_{i+1} - \lambda_{i-1})(\lambda_{i+1} - \lambda_i)} \tag{5.17}$$

where $D_{\lambda(i-1)}$, $D_{\lambda(i)}$, and $D_{\lambda(i+1)}$ are the first derivative values around the red edge position of the maximum slope at wavelengths λ_{i-1}, λ_i, and λ_{i+1}, respectively. The technique assumes no *a priori* knowledge of the spectrum and its simplicity, coupled with the flexibility that results from the fact that the wavebands need not to be equally spaced, makes it very useful for observing red edge shifts using the latest generation of airborne and spaceborne imaging spectrometers (Dawson and Curran 1998). To test this technique, Pu, Gong et al. (2003) used two methods to obtain the three bands of the first derivatives. One method uses the three first-derivative bands (713, 720, and 727 nm) calculated from the polynomial fitting curve (the 2nd approach) and the other uses the three first-derivative bands (712.29, 722.46, and 732.63 nm) calculated from five continuous Hyperion bands from 702.12 to 742.80 nm.

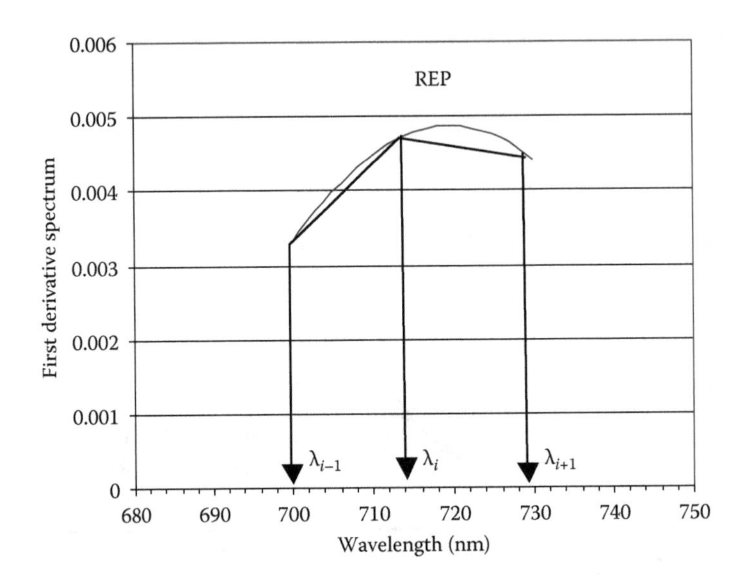

FIGURE 5.7 The three-point Lagrangian interpolation technique to determine the red edge position. (From Dawson, T. P., and Curran, P. J., *International Journal of Remote Sensing*, 19, 2133–2139, 1998.)

5.4.4 IG Modeling

Based on Bonhan-Carter (1988) and Miller et al. (1990), the spectral shape of the red edge reflectance can be approximated by one-half of an Inverted-Gaussian (IG) function (Figure 5.8). Accordingly, the IG model represents the red edge as follows:

$$R(\lambda) = R_s - (R_s - R_0)\exp\left(\frac{-(\lambda_0 - \lambda)^2}{2\sigma^2}\right) \tag{5.18}$$

where R_s is the maximum or "shoulder" spectral reflectance; R_0 and λ_0 are the minimum spectral reflectance and corresponding wavelength (i.e., RWP), respectively; λ is the wavelength; and σ^2 is the Gaussian function variance parameter. A fifth related parameter, λ_p (i.e., REP), is calculated as follows:

$$\lambda_p = \lambda_0 + \sigma \tag{5.19}$$

The IG model has been fitted to laboratory spectral reflectance data (Miller et al. 1990) and airborne imaging spectrometer data (Patel et al. 2001). The IG model can be fitted using a standard numerical procedure, an iterative optimization fitting procedure and a linearized fitting approach (Miller et al. 1990) with spectral reflectance data acquired in laboratory or from airborne platforms. The best estimates of R_0 and R_s can be used with the transformation equation (Miller et al. 1990):

$$B(\lambda) = \left\{-\ln\left[\frac{R_s - R(\lambda)}{R_s - R_0}\right]\right\}^{\frac{1}{2}} \tag{5.20}$$

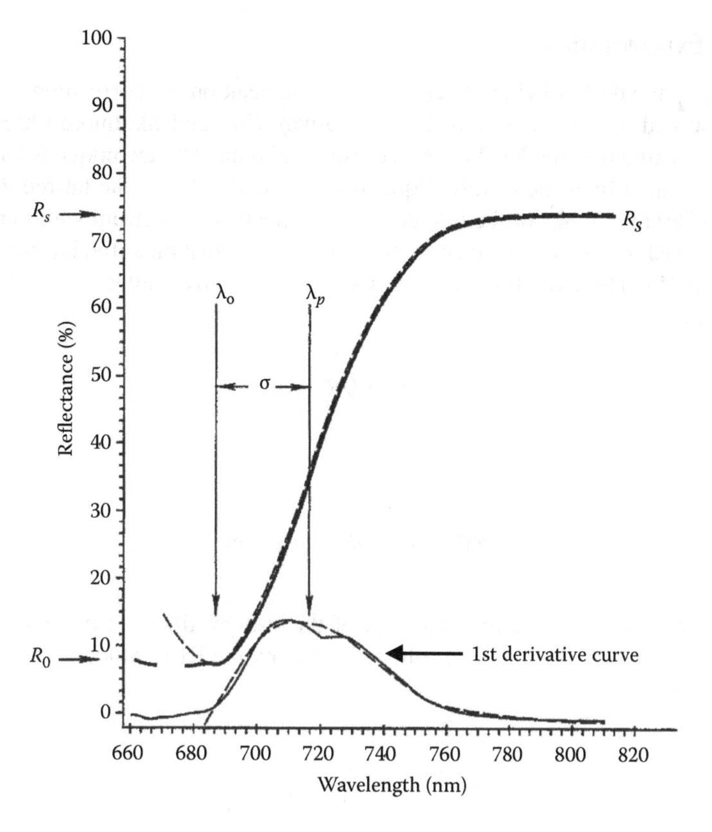

FIGURE 5.8 A measured red edge reflectance spectrum (solid curve) for a seven-leaf stack of bur oak (*Quercus macrocarpa*) and its best-fit IG model representation (dashed curve) illustrating the model parameters. The lower curves represent the measured and IG model first derivative curves. (Modified from Miller, J. R., Hare, E. W., and Wu, J., *International Journal of Remote Sensing*, 11, 1775–1795, 1991.)

to yield $B(\lambda)$ as a linear function of λ. Thus, a linear regression ($B = a_0 + a_1\lambda$) between B and λ for reflectance data in the spectral range of 685 nm to 780 nm yields best-fit coefficients a_0 and a_1 that are related to the IG model spectral parameters:

$$\lambda_0 = \frac{-a_0}{a_1} \tag{5.21}$$

and

$$\sigma = \frac{1}{\sqrt{2}\, a_1} \tag{5.22}$$

For example, Pu, Gong et al. (2003) used a linearized fitting approach with Hyperion data to determine and test the two red edge parameters. Following this approach, mean reflectances were determined for spectral regions 670 to 685 nm and 780 to 795 nm to provide estimates of R_0 and R_s. In the study, they also used reflectance from 13 Hyperion bands to fit the IG model in order to extract the two red edge parameters for each spectral sample.

5.4.5 Linear Extrapolation

In order to mitigate the destabilizing effect of the double peak on the correlation between nitrogen and REP (determined as the maximum first derivative), Cho and Skidmore (2006) developed a new technique for extracting the REP from hyperspectral data. The technique is based on a linear extrapolation of straight lines, defined by Equations 5.23 and 5.24, on the far-red (680 to 700 nm) and NIR (725 to 760 nm) flanks of the first derivative reflectance spectrum with in red edge region (Figure 5.9). The REP is then determined by the wavelength position at the intersection of the two lines (Equation 5.25). The two straight lines with first derivative reflectance (FDR) spectra are defined as follows:

$$\text{Far-red line: } FDR = m_1\lambda + c_1 \tag{5.23}$$

and

$$\text{NIR line: } FDR = m_2\lambda + c_2 \tag{5.24}$$

where m and c represent the slope and intercept of the straight lines. At the intersection, the two lines have equal λ (wavelength) and FDR values. Therefore, the REP, which is the λ at the intersection, can be given by:

$$REP = \frac{-(c_1 - c_2)}{(m_1 - m_2)}. \tag{5.25}$$

In short, the technique needs only four coordinate points (or wavebands) for calculating the REP. For instance, two bands near 680 and 700 nm to calculate m_1 and c_1 for the far-red line and the

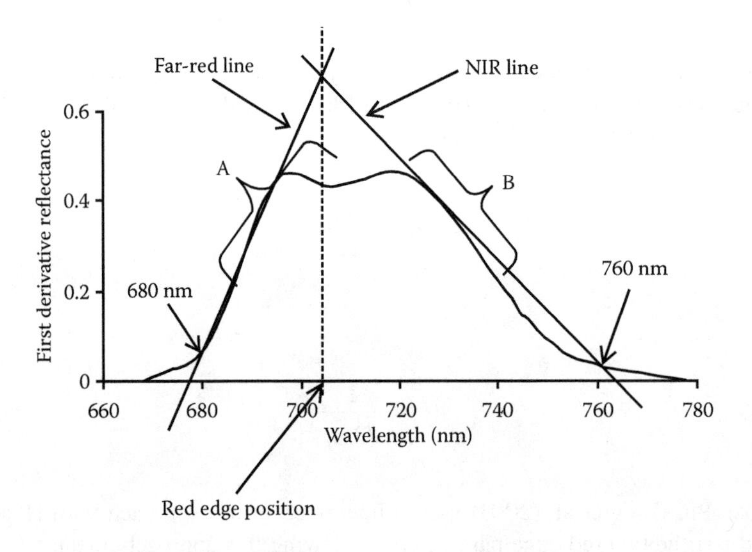

FIGURE 5.9 Schematic representation of the linear extrapolation technique for extracting the red edge position (REP)—the wavelength position of meeting point between two straight lines extrapolated on the far-red and NIR flanks of the first derivative spectrum. (Reprinted from Cho, M. A., and Skidmore, A. K., *Remote Sensing Environment*, 101, 181–93, 2006. ©2006, with permission from Elsevier.)

other two bands near 725 and 760 nm to calculate m_2 and c_2 for the NIR line are needed. To test the sensitivity of the technique for estimating nitrogen concentration of plant leaves, Cho and Skidmore (2006) identified far-red wavebands at 679.65 and 694.30 nm and NIR wavebands at 732.46 and 760.41 nm or at 723.64 and 760.41 nm as the optimal combinations for calculating nitrogen-sensitive REPs for three spectral data sets (rye canopy, maize leaf, and mixed grass/herb leaf stack spectra). Their experimental results demonstrated that REPs extracted using this new technique showed high correlations with a wide range of foliar nitrogen concentrations for both narrow and wider bandwidth spectra. The results derived with the new technique were comparable with the results obtained using the traditional four-point interpolation, polynomial, and inverted Gaussian fitting techniques.

To evaluate the performance of the first four techniques used for determining red edge optical parameters, Pu, Gong et al. (2003) conducted a correlation analysis between forest leaf area index (LAI) and two red edge parameters: red edge position (REP) and red well position (RWP). The two red edge parameters were extracted from reflectance images retrieved from Hyperion data with the four methods: four-point interpolation, polynomial fitting, Lagrangian technique, and IG modeling. Experimental results indicated that the four-point approach was the most practical and suitable method for extracting the two red edge parameters from Hyperion data because only four bands and a simple interpolation computation are needed. The polynomial fitting method is a direct method and has its practical value if hyperspectral data are available. However, the method requires more computation time. The Lagrangian technique is applicable only if the first derivative spectra are available whereas the IG approach needs further testing and refinement for Hyperion data.

In addition to the five techniques used for extracting the two red edge optical parameters, Pu et al. (2004) also proposed to extract 20 spectral variables (10 maximum 1st derivatives [1D] plus 10 corresponding wavelength-position [WP] variables) from "10 slopes" defined across a reflectance curve from 0.4 to 2.5 μm for estimating oak leaf relative water content (see the "10 slopes" in Figure 5.10). The definitions of the 20 spectral variables are similar to those defining red edge 1D and REP

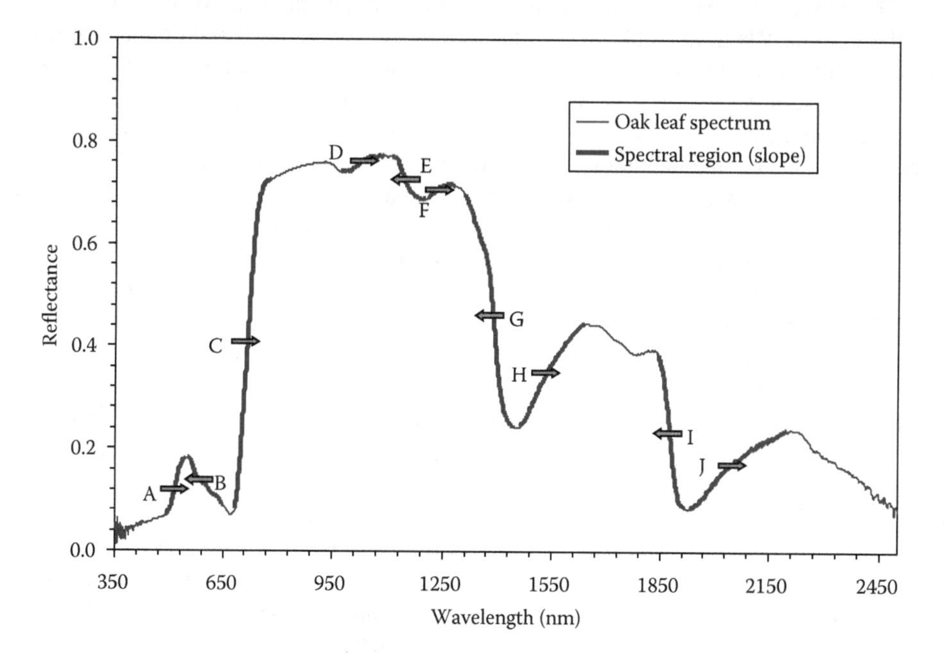

FIGURE 5.10 A spectral reflectance curve showing 10 spectral slopes (A–J). (Modified from Pu, R., Foschi, L., and Gong, P., *International Journal of Remote Sensing*, 25(20), 4267–4286, 2004.)

(Pu et al. 2004). The analysis results indicated that high correlations exist between some 1Ds and WPs and the RWC of oak leaves:

With all 306 leaf spectra covering a wide range of RWC of oak leaves (including healthy, infected, and newly dead), 1D values at both sides of the absorption valleys near 1200 nm, 1400 nm, and 1940 nm and the WPs of the right side of the 1200 nm valley and of both sides of the 1400 nm and 1940 nm valleys were found to have high correlations with RWC.

With a selection of 260 samples of only green and green-yellowish leaves (including only healthy and infected levels), the WPs at the right side of the 1400 nm valley, at the left side of the 1940 nm valley, and at the red well have a stable relationship with leaf RWC.

5.5 SPECTRAL VEGETATION INDICES

While hyperspectral data support for spectral derivative analysis, like multispectral data, they can also be used to build various forms of difference and/or ratio spectral indices. With multispectral data to construct various spectral vegetation indices (VIs), the advantages of these VIs include easy usability and that they normalize background effects on target (vegetation) spectra, e.g., reducing atmosphere and soil background effect on target spectra, and hence the VIs usually perform much better than their individual bands in remote sensing applications in vegetation and other areas. When using hyperspectral data to conduct spectral VI analysis, the hyperspectral remote sensing has the additional advantage of increased chance and flexibility to choose more sensitive spectral bands to construct various VIs. For example, with multispectral data, one may have the choice of using only red and NIR bands. However, with hyperspectral data, one can choose many of such red and NIR narrow band combinations (Gong et al. 2003). Spectral VIs constructed with hyperspectral data are called narrowband VIs (Zarco-Tejada et al. 2001, Eitel et al. 2006, He et al. 2006). Table 5.1 lists a set of 82 VIs that are mostly updated from a collection by Pu and Gong (2011) to reflect newly developed narrow band VIs from hyperspectral data. The purpose of developing these VIs is frequently for extracting and evaluating plant biophysical and biochemical parameters from hyperspectral data. As those in Pu and Gong (2011), the total 82 VIs in the table are organized into five categories based on the characteristics and functions of the VIs in order to be convenient for readers to locate a (or group of) VI(s). The five categories consist of vegetation structure (e.g., LAI, crown closure [CC], green biomass, and plant species, etc.), pigments (e.g., chlorophylls [Chls], carotenoids [Cars], and anthocyanins [Anths]), other biochemical (e.g., lingo-cellulose, N, etc.), water, and stress. Within individual categories, the VIs are arranged in alphabetical order. In the following, a brief summary of these VIs is given. An overview of applications of these VIs in HRS of vegetation is given in Chapter 8 of this book.

Hyperspectral vegetation indices (HVIs), similar to broadband VIs, are designed and developed for estimating and mapping biophysical and biochemical parameters from laboratory/*in situ*, airborne, and spaceborne hyperspectral data. Across a general spectral reflectance curve measured from green vegetation, there are many absorption and reflective features/bands (see some from Figure 5.11). These features/bands construct a spectral basis to design and develop most HVIs in Table 5.1. Many biophysical structurally oriented HVIs rely on some combinations of NIR with red reflectance bands. This is because increases in LAI, CC, and green biomass usually result in higher NIR reflectance and lower red reflectance due to higher multiple cell structure-induced scattering and Chl absorptions. Consequently, this type of HVI is mostly constructed with spectral narrowbands in visible and NIR regions, particularly some typical blue, red, and NIR bands. Pigments and other chemical-related HVIs rely on the absorption and reflective bands/features in visible region, caused by various pigments (Chls, Cars, and Anths), particularly chlorophyll-a and -b content in blue and red narrowbands. Other chemicals, such as lignin and cellulose, are related to some subtle absorption bands in the middle-wave infrared region. Per water-oriented HVIs including some stress VIs, they are all related to several significant water absorption bands (e.g., wavelengths at 975, 1200, 1400, and 1900 nm). Such typical absorption narrow bands are seen in Figure 5.11 to construct various categories of HVIs.

TABLE 5.1
Summary of 82 Spectral Vegetation Indices Extracted from Hyperspectral Data, Collected from Literature

Spectral Index	Characteristics and Functions	Definition	Reference
	Structure (LAI, Crown Closure, Green Biomass, Species, etc.)		
1DL_DGVI, 1st-order derivative green VI derived using local baseline	Quantify variation of LAI and percent green cover of plant canopy	$\sum_{\lambda 626}^{\lambda 795} \lvert R'(\lambda_i) - R'(\lambda_{626}) \rvert \Delta\lambda_i$	Elvidge and Chen 1995
1DZ_DGVI, 1st-order derivative green VI derived using zero baseline	Quantify variation of LAI and percent green cover of plant canopy	$\sum_{\lambda 626}^{\lambda 795} \lvert R'(\lambda_i) \rvert \Delta\lambda_i$	Elvidge and Chen 1995
ARVI, Atmospherically resistant VI	Similar dynamic range to the NDVI, but is, on average, 4 times less sensitive to atmospheric effects than the NDVI	$\{R_{NIR} - [R_{red} - \gamma(R_{blue} - R_{red})]\}/\{R_{NIR} + [R_{red} - \gamma(R_{blue} - R_{red})]\}$	Kaufman and Tanré 1992
ATSAVI, Adjusted transformed soil-adjusted VI	Less affected by soil background and better for estimating homogeneous canopy	$a(R_{800} - aR_{670} - b)/[(aR_{800} + R_{670} - ab + X(1 + a^2)]$, where $X = 0.08$, $a = 1.22$, and $b = 0.03$	Baret and Guyot 1991
EVI, Enhanced VI	Estimate vegetation LAI, biomass and water content and improve sensitivity in high biomass region	$2.5(R_{NIR} - R_{red})/(R_{NIR} + 6R_{red} - 7.5R_{blue} + 1)$	Huete et al. 2002
EVI2, Two-band enhanced VI	Similar to **EVI**, but without blue band and good for atmospherically corrected data	$2.5(R_{NIR} - R_{red})/(R_{NIR} + 2.4R_{red} + 1)$	Jiang et al. 2008
GI, Greenness index	Estimate biochemical constituents and LAI at leaf and canopy levels	R_{554}/R_{677}	Zarco-Tejada et al. 2005
LAIDI, LAI determining index	Sensitive to LAI variation at canopy level with a saturation point > 8	R_{1250}/R_{1050}	Delalieux et al. 2008
MSAVI, Improved soil adjusted vegetation index	A more sensitive indicator of vegetation amount than **SAVI** at canopy level	$0.5[2R_{800} + 1 - ((2R_{800} + 1)^2 - 8(R_{800} - R_{670}))^{1/2}]$	Qi et al. 1994
MSR, Modified simple ratio	More linearly related to vegetation parameters than **RDVI**	$(R_{800}/R_{670} - 1)/(R_{800}/R_{670} + 1)^{1/2}$	Chen 1996; Haboudane et al. 2004
MTVI1, Modified triangular VI 1	More suitable for LAI estimation than **TVI**	$1.2[1.2(R_{800} - R_{550}) - 2.5(R_{670} - R_{550})]$	Haboudane et al. 2004
MTVI2, Modified triangular VI 2	Preserves sensitivity to LAI and resistance to chlorophyll influence	$\{1.5[1.2(R_{800} - R_{550}) - 2.5(R_{670} - R_{550})]\}/\{(2R_{800}+1)^2 - [6R_{800} - 5(R_{670})^{1/2}] - 0.5\}^{1/2}$	Haboudane et al. 2004
NDVI, Normalized difference vegetation index	Responds to change in the amount of green biomass and more efficiently in vegetation with low to moderate density	$(R_{NIR} - R_{red})/(R_{NIR} + R_{red})$	Rouse et al. 1973
NRI, Normalized ratio index	A sensitive indicator of biomass, N concentration and height of crop (wheat)	$(R_{874} - R_{1225})/(R_{874} + R_{1225})$	Koppe et al. 2010

(Continued)

TABLE 5.1 (CONTINUED)
Summary of 82 Spectral Vegetation Indices Extracted from Hyperspectral Data, Collected from Literature

Spectral Index	Characteristics and Functions	Definition	Reference
OSAVI, Optimized soil-adjusted vegetation index	Similar to MSAVI, but more applicable agricultural applications, whereas MSAVI is recommended for more general purposes	$1.16(R_{800} - R_{670})/(R_{800} + R_{670} + 0.16)$	Rondeaux et al. 1996
PSND, Pigment-specific normalized difference	Estimate LAI and Cars at leaf or canopy level	$(R_{800} - R_{470})/(R_{800} + R_{470})$	Blackburn 1998
PVI$_{hyp}$, Hyperspectral perpendicular VI	More efficiently quantify the low amount of vegetation by minimizing soil background influence on vegetation spectrum	$(R_{1148} - aR_{807} - b)/(1 + a^2)^{1/2}$, where $a = 1.17$, $b = 3.37$	Schlerf et al. 2005
RDVI, Renormalized difference VI	Suitable for low to high LAI values	$(R_{800} - R_{670})/(R_{800} + R_{670})^{1/2}$	Roujean and Breon 1995; Haboudane et al. 2004
SAVI, Soil adjusted VI	Similar to NDVI, but more suitable for lower vegetation cover area	$(R_{NIR} - R_{red})(1 + L)/(R_{NIR} + R_{red} + L)$	Huete 1988
sLAIDI, Normalization or standard of the LAIDI	Sensitive to LAI variation at canopy level with a saturation point > 8	$S(R_{1050} - R_{1250})/(R_{1050} + R_{1250})$, where $S = 5$	Delalieux et al. 2008
SPVI, Spectral polygon vegetation index	Estimate LAI and canopy Chls	$0.4[3.7(R_{800} - R_{670}) - 1.2\|R_{530} - R_{670}\|]$	Vincini et al. 2006
SR, Simple ratio	Same as **NDVI**	R_{NIR}/R_R	Jordan 1969
VARI, Visible atmospherically resistant index	Sensitive to variation of vegetation fractionof wheat canopy, but less sensitive to atmopherical effect	$(R_{green} - R_{red})/(R_{green} + R_{red} - R_{blue})$	Gitelson et al. 2002
VARI$_{green}$, Visible atmospherically resistant index for green ref.	Estimate green vegetation fraction (VF) with minimally sensitive to atmospheric effects; better than **NDVI** for moderate to high VF values of VF	$(R_{green} - R_{red})/(R_{green} + R_{red})$	Gitelson et al. 2002
VARI$_{red-edge}$, Visible atmospherically resistant index for red edge ref.	Same as **VARI$_{green}$**	$(R_{red-edge} - R_{red})/(R_{red-edge} + R_{red})$	Gitelson et al. 2002
WDRVI, Wide dynamic range VI	Estimate LAI, vegetation cover, biomass; better than **NDVI**	$(0.1R_{NIR} - R_{red})/(0.1R_{NIR} + R_{red})$	Gitelson 2004
	Pigments (Chls, Cars, and Anths)		
ACI, Anthocyanin content index	Estimate Anths content from reflectance from sugar maple leaves	R_{green}/R_{NIR}	van den Berg and Perkins 2005
ARI, Anthocyanin reflectance index	Estimate Anths content from reflectance changes in the green region at leaf level	$ARI = (R_{550})^{-1} - (R_{700})^{-1}$	Gitelson et al. 2001

(Continued)

TABLE 5.1 (CONTINUED)
Summary of 82 Spectral Vegetation Indices Extracted from Hyperspectral Data, Collected from Literature

Spectral Index	Characteristics and Functions	Definition	Reference
BGI, Blue green pigment index	Estimate Chls and Cars content at leaf and canopy levels	R_{450}/R_{550}	Zarco-Tejada et al. 2005
BRI, Blue red pigment index	Estimate Chls and Cars content at leaf and canopy levels	R_{450}/R_{690}	Zarco-Tejada et al. 2005
CARI, Chlorophyll absorption ratio index	Quantify Chls concentration at leaf level	$(\lvert a\ 670 + R_{670} + b\rvert/(a^2 + 1)^{1/2}) \times (R_{700}/R_{670})$, $a = (R_{700} - R_{550})/150$, $b = R_{550} - (a \times 550)$	Kim et al. 1994
Chl$_{green}$, Chlorophyll index using green reflectance	Estimate Chls content in anthocyanin-free leaves if NIR is set	$(R_{760-800}/R_{540-560}) - 1$	Gitelson et al. 2006
Chl$_{red-edge}$, Chlorophyll index using red edge reflectance	Estimate Chls content in anthocyanin-free leaves if NIR is set	$(R_{760-800}/R_{690-720}) - 1$	Gitelson et al. 2006
CI, Chlorophyll index	Estimate Chls content in broadleaf tree leaves	$(R_{750} - R_{705})/(R_{750} + R_{705})$; $R_{750}/\ [(R_{700} + R_{710}) - 1]$	Gitelson and Merzlyak 1996; Gitelson et al. 2005
CRI, Carotenoid reflectance index	Sufficient to estimate total Cars content in plant leaves	$CRI_{550} = (R_{510})^{-1} - (R_{550})^{-1}$, $CRI_{700} = (R_{510})^{-1} - (R_{700})^{-1}$	Gitelson et al. 2002
DD, Double difference	Estimate total Cars content in plant leaves	$(R_{750} - R_{720}) - (R_{700} - R_{670})$	le Maire et al. 2004
DmSR, Modified simple ratio of derivatives	Quantify Chls content at leaf level	$(DR_{720} - DR_{500})/(DR_{720} + DR_{500})$, where DR_{λ} is 1st derivative of ref. at wavelength λ	le Maire et al. 2004
EPI, Eucalyptus pigment indexes	Correlate best with Chla, Chls and total carotenoid contents	$a \times R_{672}/(R_{550} \times R_{708})^{\beta}$	Datt 1998
LCI, Leaf chlorophyll index	Estimate Chl content in higher plants, sensitive to variation in reflectance caused by Chl absorption	$(R_{850} - R_{710})/(R_{850} + R_{680})$	Datt 1999
mARI, Modified anthocyanin reflectance index	Estimate anthocyanin content from reflectance changes in the green region at leaf level	$mARI = ((R_{530-570})^{-1} - (R_{690-710})^{-1}) \times R_{NIR}$	Gitelson et al. 2006
MCARI, Modified chlorophyll absorption in reflectance index	Respond to Chl variation and estimate Chl absorption	$[(R_{701} - R_{671}) - 0.2(R_{701} - R_{549})] \times (R_{701}/R_{671})$	Daughtry et al. 2000
MCARI1, Modified chlorophyll absorption ratio index 1	Less sensitive to chlorophyll effects; more responsive to green LAI variation	$1.2[2.5(R_{800} - R_{670}) - 1.3(R_{800} - R_{550})]$	Haboudane et al. 2004
MCARI2, Modified chlorophyll absorption ratio index 2	Preserves sensitivity to LAI and resistance to chlorophyll influence	$\{1.5[2.5(R_{800} - R_{670}) - 1.3(R_{800} - R_{550})]\}/\{(2R_{800}+1)^2 - [6R_{800} - 5(R_{670})^{1/2}] - 0.5\}^{1/2}$	Haboudane et al. 2004
mCRI, Modified carotenoid reflectance index	Estimate carotenoid pigment contents in foliage	$mCRI_G = ((R_{510-520})^{-1} - (R_{560-570})^{-1}) \times R_{NIR}$, $mCRI_{RE} = ((R_{510-520})^{-1} - (R_{690-700})^{-1}) \times R_{NIR}$	Gitelson et al. 2006

(Continued)

TABLE 5.1 (CONTINUED)
Summary of 82 Spectral Vegetation Indices Extracted from Hyperspectral Data, Collected from Literature

Spectral Index	Characteristics and Functions	Definition	Reference
mND$_{680}$, Modified normalized difference	Quantify chlorophyll content and sensitive to low content at leaf level	$(R_{800} - R_{680})/(R_{800} + R_{680} - 2R_{445})$	Sims and Gamon 2002
mND$_{705}$, Modified normalized difference	Quantify chlorophyll content and sensitive to low content at leaf level. **mND$_{705}$** performance better than **mND$_{680}$**	$(R_{750} - R_{705})/(R_{750} + R_{705} - 2R_{445})$	Sims and Gamon 2002
mSR$_{705}$, Modified simple ratio	Quantify chlorophyll content and sensitive to low content at leaf level	$(R_{750} - R_{445})/(R_{705} - R_{445})$	Sims and Gamon 2002
MTCI, MERIS terrestrial chlorophyll index	Quantify chlorophyll content of of tree seedlings canopy	$(R_{750} - R_{710})/(R_{710} - R_{680})$	Dash and Curran 2004
NAOC, Normalized area overreflectance curve	Quantify chlorophyll content and sensitive to agricultural vegetation types	$1 - \frac{\int_a^b R\,d\lambda}{R_{max}(a-b)}$	Delegido et al. 2010
NPCI, Normalized pigment chlorophyll ratio index	Assess Cars/Chl ratio at leaf level	$(R_{680} - R_{430})/(R_{680} + R_{430})$	Peñuelas et al. 1994
NPQI, Normalized phaeophytinization index	Detect variation of Chl concentration and Cars/Chl ratio at a leaf level	$(R_{415} - R_{435})/(R_{415} + R_{435})$	Barnes et al. 1992; Peñuelas et al. 1995b
PBI, Plant biochemical index	Retrieve leaf total chlorophyll and nitrogen concentrations from satellite hyperspectral data	R_{810}/R_{560}	Rama Rao et al. 2008
PhRI, Physiological reflectance index	Estimate FPAR, N-stress at crop canopy	$(R_{550} - R_{531})/(R_{550} + R_{531})$	Gamon et al. 1992
PRI, Photochemical/ physiological reflectance index	Estimate carotenoid pigment contents in foliage	$(R_{531} - R_{570})/(R_{531} + R_{570})$	Gamon et al. 1992
PSSR, Pigment specific simple ratio	Estimate carotenoid pigment contents in foliage	R_{800}/R_{675}; R_{800}/R_{650}	Blackburn 1998
PSND, Pigment specific normalized difference	Estimate carotenoid pigment contents in foliage	$(R_{800} - R_{675})/(R_{800} + R_{675})$; $(R_{800} - R_{650})/(R_{800} + R_{650})$	Blackburn 1998
RARS, Ratio analysis of reflectance spectra	Estimate carotenoid pigment contents in foliage	R_{760}/R_{500}	Chappelle et al. 1992
RGR, Red:green ratio	Estimate anthocyanin content with a green and a red band	R_{Red}/R_{Green}	Gamon and Surfus 1999; Sims and Gomon 2002
SIPI, Structural independent pigment index	Estimate carotenoid pigment content change in foliage; related to a ratio, Cars:Chls	$(R_{800} - R_{445})/(R_{800} - R_{680})$	Peñuelas et al. 1995a

(Continued)

TABLE 5.1 (CONTINUED)
Summary of 82 Spectral Vegetation Indices Extracted from Hyperspectral Data, Collected from Literature

Spectral Index	Characteristics and Functions	Definition	Reference
TCARI, Transformed chlorophyll absorption ratio index	Very sensitive to Chls content variations and very resistant to the variations of LAI and solar zenith angle	$3[(R_{700} - R_{670}) - 0.2(R_{700} - R_{550})(R_{700}/R_{670})]$	Haboudane et al. 2002
TCI, Triangle chlorophyll vegetation index	Quantify chlorophyll content of tallgrass at leaf and canopy levels	$[(R_{800} + 1.5 \times R_{550}) - R_{675}]/(R_{800} - R_{700})$	Gao 2006
TVI, Triangular VI	Characterize the radiant energy absorbed by leaf pigments (Chls). Note that the increase of Chls concentration also results in the decrease of the green reflectance.	$0.5[120(R_{750} - R_{550}) - 200(R_{670} - R_{550})]$	Broge and Leblanc 2000; Haboudane et al. 2004
VOG1, Vogelmann red edge index 1	Sensitive to the combined effects of foliage chlorophyll concentration, canopy leaf area, and water content	R_{740}/R_{720}	Vogelmann et al. 1993
	Other Biochemicals		
CAI, Cellulose absorption index	Cellulose & lignin absorption features, discriminates plant litter from soils	$0.5(R_{2020} + R_{2220}) - R_{2100}$	Nagler et al. 2000
DCNI, Double-peak canopy nitrogen index	Quantify variation of N in canopy of crops	$(R_{720} - R_{700})/[(R_{700} - R_{670})(R_{720} - R_{670} + 0.03)]$	Chen et al. 2010
NDLI, Normalized difference lignin index	Quantify variation of canopy lignin concentration in native shrub vegetation	$[\log(1/R_{1754}) - \log(1/R_{1680})]/[\log(1/R_{1754}) + \log(1/R_{1680})]$	Serrano et al. 2002
NDNI, Normalized difference nitrogen index	Quantify variation of canopy N concentration in native shrub vegetation	$[\log(1/R_{1510}) - \log(1/R_{1680})]/[\log(1/R_{1510}) + \log(1/R_{1680})]$	Serrano et al. 2002
RDVI, Renormalized difference vegetation index	Quantify variation of multi-chemical in vegetation	$(R_{NIR} - R_{red})/(R_{NIR} + R_{red})^{1/2}$	Roujean and Breon 1995
	Water		
DWSI, Disease water stress index	Detect water-stressed crops at a canopy level	$(R_{802} + R_{547})/(R_{1657} + R_{682})$	Galvão et al. 2005
LWVI-1, Leaf water VI 1	Estimate leaf water content, an **NDWI** variant	$(R_{1094} - R_{893})/(R_{1094} + R_{893})$	Galvão et al. 2005
LWVI-2, Leaf water VI 2	Estimate leaf water content, an **NDWI** variant	$(R_{1094} - R_{1205})/(R_{1094} + R_{1205})$	Galvão et al. 2005
MSI, Moisture stress index	Detect variation of leaf water content	R_{1600}/R_{819}	Hunt and rock 1989
NDII, Normalized difference infrared index	Detect variation of leaf water content	$(R_{819} - R_{1600})/(R_{819} + R_{1600})$	Hardinsky et al. 1983
NDWI, ND water index	Improving the accuracy in retrieving the vegetation water content at both leaf and canopy levels	$(R_{860} - R_{1240})/(R_{860} + R_{1240})$	Gao 1996

(*Continued*)

TABLE 5.1 (CONTINUED)
Summary of 82 Spectral Vegetation Indices Extracted from Hyperspectral Data, Collected from Literature

Spectral Index	Characteristics and Functions	Definition	Reference
RATIO$_{1200}$, 3-band ratio at 1.2 μm	Estimate relative water content < 60% at leaf level	$2 * R_{1180-1220}/(R_{1090-1110} + R_{1265-1285})$	Pu et al. 2003b
RATIO$_{975}$, 3-band ratio at 975 nm	Estimate relative water content < 60% at leaf level	$2 * R_{960-990}/(R_{920-940} + R_{1090-1110})$	Pu et al. 2003b
RVI$_{hyp}$, Hyperspectral ratio VI	Quantify LAI and water content at canopy level	R_{1088}/R_{1148}	Schlerf et al. 2005
SIWSI, Shortwave infrared water stress index	Estimate leaf or canopy water stress, especially in the semiarid environment	$(R_{860} - R_{1640})/(R_{860} + R_{1640})$	Fensholt and Sandholt 2003
SRWI, Simple ratio water index	Detect vegetation water content at leaf or canopy level	R_{860}/R_{1240}	Zarco-Tejada et al. 2003
WI, Water index	Quantify relative water content at leaf level	R_{900}/R_{970}	Peñuelas et al. 1997
	Stress		
PSRI, Plant Senescence Reflectance Index	Sensitive to the Car/Chl ratio and used as a quantitative measure of leaf senescence and fruit ripening	$(R_{680} - R_{500})/R_{750}$	Merzlyak et al. 1999
RVSI, Red-Edge Vegetation Stress Index	Assess vegetation community stress at canopy level	$[(R_{712} + R_{752})/2] - R_{732}$	Merton and Huntington 1999

Source: Pu, R., and P. Gong, in *Advances in Environmental Remote Sensing: Sensors, Algorithm, and Applications*, series ed. Q. Weng, CRC Press/Taylor & Francis Publishing Group, Boca Raton, FL, 2011, pp. 101–142.
Note: If one index without specifying wavelength in digital, the index can be applied in both multi- and hyper-spectral data.

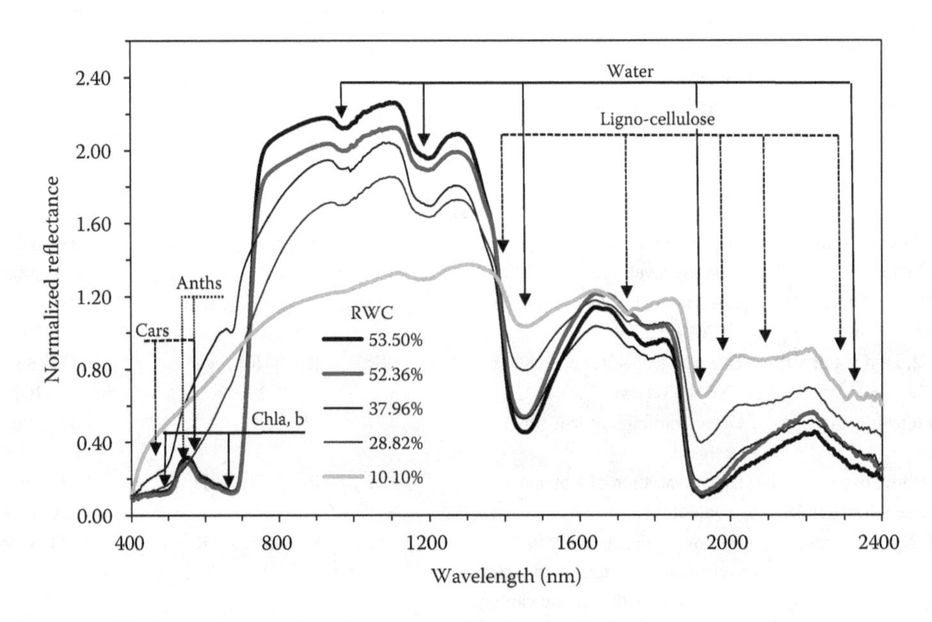

FIGURE 5.11 Spectral reflectance spectra of coast live oak leaves showing the locations of carotenoids (Cars), anthocyanins (Anths), chlorophyll (Chla, b), water, and ligno-cellulose absorptions.

5.6 HYPERSPECTRAL TRANSFORMATION AND FEATURE EXTRACTION

Since hyperspectral data have strong correlations between adjacent spectral bands and high redundancy information contained in hyperspectral data with a few tens to a few hundreds of narrowbands, either data transformations or feature extraction including band selection or other data dimensionality reduction (DR) techniques are suggested before the hyperspectral data are used for thematic information mapping, image classification, and spectral unmixing analysis. To reduce the dimensionality of hyperspectral data, various data transform and feature extraction techniques/algorithms are used. An image data transform generally considers using statistics as a criterion to decorrelate and convert data into a set of uncorrelated data components for analysis. The transforms of this type include two commonly used second-order statistics component transforms, data variance-based principal components analysis (PCA) and signal-to-noise ratio–based maximum noise fraction (MNF) transform (Chang 2013). A feature extraction transform uses a feature extraction-based criterion to produce a set of feature vectors such that data can be represented by these feature vectors. For example, the energy feature of the wavelet decomposition coefficients is computed at each scale for both approximation and details and is used to form an energy feature vector (Pu and Gong 2004). In this section, hyperspectral data transform methods PCA, SNR-based MNF transforms, and independent component analysis (ICA) and feature extraction methods, including canonical discriminant analysis and wavelet transform, are introduced and discussed.

5.6.1 Principal Components Analysis (PCA)

The PCA technique has been extensively applied to reduce the data dimension and feature extraction from hyperspectral data, such as in Gong et al. (2002) and Pu and Gong (2004), for estimating leaf or canopy biophysical and biochemical parameters. PCA makes use of eigenvalues to determine the significance of principal components (PCs) generated by PCA so that DR is accomplished by selecting PCs associated with their greater eigenvalues.

Assume that $\boldsymbol{Z} = \{\boldsymbol{x}_i\}_{i=1}^{N}$ is a set of L-dimensional image pixel vectors and μ is the mean of the sample pool **Z** obtained by $\boldsymbol{u} = \left(\frac{1}{N}\right)\sum_{i=1}^{N}\boldsymbol{x}_i$. We set **X** to be the sample data matrix formed by $\boldsymbol{X} = (\boldsymbol{x}_1, \boldsymbol{x}_2, \ldots, \boldsymbol{x}_N)$. Then a sample covariance matrix of the **Z** can be calculated by $\sum = \left(\frac{1}{N}\right)[\boldsymbol{X}\boldsymbol{X}^T] = \left(\frac{1}{N}\right)\left[\sum_{i=1}^{N}(\boldsymbol{x}_i - \boldsymbol{u})(\boldsymbol{x}_i - \boldsymbol{u})^T\right]$. PC transformation, in fact, is a matter of diagonalizing the sample covariance matrix Σ, a technique that is mathematically performed by determining its eigenvalues $\left\{\sigma_l^2\right\}_{l=1}^{L}$ and corresponding eigenvectors $\{\boldsymbol{\beta}_l\}_{l=1}^{L}$. To calculate the eigenvalues and eigenvectors of the sample covariance matrix, we need to solve the characteristic equation

$$\det(\sigma^2\boldsymbol{I} - \Sigma) = 0 \tag{5.26}$$

for a set solution $\left\{\sigma_l^2\right\}_{l=1}^{L}$, where det(**A**) denotes the determinant of matrix **A**, and **I** is the $L \times L$ identity matrix. If Σ meets full rank assumption, and L nonzero solution exists, where each eigenvalue σ_l^2 represents the variance of data for a particular eigenvector direction. Therefore, to solve Equation 5.26,

$$\boldsymbol{\beta}^T\sum\boldsymbol{\beta} = \boldsymbol{D}_\sigma \text{ and } \boldsymbol{\beta}^T\boldsymbol{\beta} = \mathbf{I} \tag{5.27}$$

are performed, where, $\beta = (\beta_1, \beta_2, \ldots, \beta_L)$ are unit eigenvectors, and $\mathbf{D}_\sigma$ is the diagonal matrix with variances $\{\sigma_l^2\}_{l=1}^L$ (i.e., corresponding eigenvalues) along the diagonal line as

$$D_\sigma = \begin{bmatrix} \sigma_1^2\ 0 \cdots 0 \\ 0\ \sigma_2^2 \cdots 0 \\ \cdots\cdots\cdots\cdots \\ 0\ 0 \cdots \sigma_L^2 \end{bmatrix} \tag{5.28}$$

where $\{\sigma_l^2\}_{l=1}^L$ are eigenvalues and if they are arranged as $\sigma_1^2 \geq \sigma_2^2 \cdots \geq \sigma_L^2$, then we can transform every data sample $\boldsymbol{x}_i$ to a new data sample, $\boldsymbol{y}_i$ ($i = 1, 2, \ldots, N$) by

$$Y = \beta^T Z \tag{5.29}$$

If selecting first q ($q < L$) PC images associated with their greater eigenvalues based on a majority of variance to be accounted for by the q PCs, the hyperspectral DR is achieved. In PCA, if the covariance matrix is replaced by its corresponding correlation coefficient matrix, PCA is called standardized principal components analysis (SPCA). The correlation coefficient matrix **R** can be easily converted from the covariance matrix Σ:

$$R = D_\sigma^{-\frac{1}{2}} \sum D_\sigma^{-\frac{1}{2}} \tag{5.30}$$

From **R** and referring to Equations 5.26 through 5.28, every data sample $\boldsymbol{x}_i$ can be transformed to a new data sample, $\boldsymbol{y}_i^{SPCA}$ ($i = 1, 2, \ldots, N$) by

$$Y^{SPAC} = \left(D_\sigma^{-\frac{1}{2}}\beta\right)^T \mathbf{Z} \tag{5.31}$$

where **Z** is the input data matrix in a standardized form.

5.6.2 Signal-to-Noise Ratio-Based Image Transforms

The goal of PCA is to find principal components in accordance with maximum variance in a hyperspectral image. Because the PCA does not always produce images that show steadily decreasing image quality with increasing component number, Green et al. (1988) developed one transform method called maximum noise fraction (MNF) transform to maximize the signal-to-noise ratio when choosing principal components with increasing component numbers. It was later shown by Lee et al. (1990) that MNF actually performed two stage processes, noise whitening with unit variance, followed by PCA. Because of this, MNF is also referred to as noise-adjusted principal component (NAPC) transform. Then several MNFs (or NAPCs) to maximize the signal-to-noise ratio are selected for further analysis of hyperspectral data, such as for determining end-member spectra for spectral mixture analysis (Pu et al. 2008, Walsh et al. 2008) and hyperspectral mosaic (Hestir et al. 2008), etc.

5.6.2.1 Maximum Noise Fraction (MNF) Transform

In order to address that PCA-ordered PCs are not necessarily ordered by image quality, Green et al. (1988) used an approach similar to PCA, called MNF, based on a different criterion, signal-to-noise ratio, to measure image quality. The idea of MNF is briefly described as follows. Suppose that the lth band image can be represented by an N-dimensional column vector, $\boldsymbol{Z}_l = (\boldsymbol{x}_{l1}, \boldsymbol{x}_{l2}, \cdots, \boldsymbol{x}_{lN})^T$, where $l = 1, 2, \ldots, L$ (bands); N is the image size (total number of image pixels). It assumes that an observation model

$$\mathbf{Z} = \mathbf{S} + \mathbf{N} \tag{5.32}$$

where $\boldsymbol{Z}^T = (\boldsymbol{Z}_1, \boldsymbol{Z}_2, \cdots, \boldsymbol{Z}_L)$ and $\boldsymbol{Z}_l = (x_{l1}, x_{l2}, \cdots, x_{lN})^T$ is the image data matrix of a hyperspectral image with L-bands, and $\boldsymbol{S}^T = (\boldsymbol{S}_1, \boldsymbol{S}_2, \cdots, \boldsymbol{S}_L)$ and $\boldsymbol{N}^T = (\boldsymbol{N}_1, \boldsymbol{N}_2, \cdots, \boldsymbol{N}_L)$ are the uncorrelated signal and noise data matrices. Thus, we have their three corresponding covariance matrices:

$$Cov(\mathbf{Z}) = Cov(\mathbf{S}) + Cov(\mathbf{N}) = \sum = \sum_S + \sum_N \tag{5.33}$$

where Σ_S and Σ_N are the signal and noise covariance matrices of **S** and **N** data matrices. Let us define the noise fraction of the lth band to be

$$\boldsymbol{NF}_l = Var(\boldsymbol{N}_l)/Var(\boldsymbol{Z}_l) \tag{5.34}$$

the ratio of the noise variance to the total variance for band l. Here, $Var(\boldsymbol{Z}_l) = \left(\frac{1}{N}\right)\sum_{i=1}^{N}(x_{li} - u_l)^2$ and $u_l = \left(\frac{1}{N}\right)\sum_{i=1}^{N} x_{li}$. The purpose of maximimizing the noise fraction is to find linear transformations

$$\boldsymbol{Y}_l^{MNF} = \left(\boldsymbol{W}_l^{MNF}\right)^T \mathbf{Z}, 1 = 1, 2, \ldots, \mathrm{L} \tag{5.35}$$

such that the noise fraction for $\boldsymbol{Y}_l^{MNF} = \left(y_{l1}^{MNF}, y_{l2}^{MNF}, \cdots, y_{lN}^{MNF}\right)^T$ is maximum among all linear transformations orthogonal to $\boldsymbol{Y}_j^{MNF}$, $j = 1, 2, \ldots, l$. To obtain the maximum noise fraction (components) for $\boldsymbol{Y}_l^{MNF}$, $l = 1, 2, \ldots, L$, we can first use arguments similar to PCA to find eigenvector matrix (**W**) and a set of corresponding eigenvalues of matrix $\sum_N \sum^{-1}$. Then based on the definition of MNF transform, we can arrange the set of eigenvalues in the order: $\lambda_1^{nf} \geq \lambda_2^{nf} \geq \cdots \geq \lambda_L^{nf}$ so that the MNF components will show steadily increasing image quality (against the usual order of PCA components). To be convenient, the transform vector $\boldsymbol{W}_l^{MNF}$ in Equation 5.35 can be expressed in an MNF transform matrix so that

$$\boldsymbol{Y}^{MNF} = \mathbf{W}^T\mathbf{Z} \tag{5.36}$$

where $\boldsymbol{W} = \left(\boldsymbol{W}_1^{MNF}, \boldsymbol{W}_2^{MNF}, \cdots, \boldsymbol{W}_L^{MNF}\right)$. Since the criterion of NF in Equation 5.34 can be rewritten as

$$\boldsymbol{NF}_l = \frac{Var(\boldsymbol{N}_l)}{Var(\boldsymbol{Z}_l)} = \frac{Var(\boldsymbol{N}_l)}{Var(\boldsymbol{S}_l) + Var(\boldsymbol{N}_l)} = \frac{1}{SNR_l + 1} \tag{5.37}$$

where SNR_l = Var($\mathbf{S}_l$)/Var($\mathbf{N}_l$) is the signal-to-noise ratio of band l. Thus, the MNF components defined by Equation 5.35 will also show SNR steadily increasing with increasing component number.

5.6.2.2 Noise-Adjusted Principal Component Transform

Lee et al. (1990) reinterpreted MNF by Green et al. (1988) as a two-stage process, called noise-adjusted principal component (NAPC) transform. The two-stage process includes that whitened noise variances of each band image to unit variance is first performed, and then PCA transform is carried out on the noise-whitened band images. According to Lee et al. (1990) and Equation 5.34, a fast algorithm developed by Roger (1994) for implementing NAPC transform is summarized as follows.

1. From Σ_N, compute its orthonormalized eigenvector matrix $\mathbf{E}$, and diagonal matrix of its eigenvalues, $\mathbf{D}_n$:

$$\mathbf{E}^T \sum\nolimits_N \mathbf{E} = \mathbf{D}_n \text{ and } \mathbf{E}^T\mathbf{E} = \mathbf{I} \tag{5.38}$$

2. Construct a renormalized matrix $\mathbf{F} = \mathbf{E}\mathbf{D}_n^{-1/2}$ (also called whitening matrix), so that we have

$$\mathbf{F}^T \sum\nolimits_N \mathbf{F} = \mathbf{I} \text{ and } \mathbf{F}^T\mathbf{F} = \mathbf{D}_n^{-1} \tag{5.39}$$

   where $\mathbf{I}$ is the identity matrix and here Σ_N is required to be nonsingular (Roger 1994).
3. Transform the data covariance matrix Σ to the noise-adjusted data covariance matrix Σ_{adj} defined by

$$\sum\nolimits_{adj} = \mathbf{F}^T \sum \mathbf{F}. \tag{5.40}$$

4. From Σ_{adj}, compute its eigenvector matrix, $\mathbf{G}$, such that

$$\Delta_{adj} = \mathbf{G}^T \sum\nolimits_{adj} \mathbf{G} \text{ and } \mathbf{G}^T\mathbf{G} = \mathbf{I} \tag{5.41}$$

   where Δ_{adj} is the diagonal matrix of its eigenvalues.
5. Finally, the NAPC transform matrix, $\mathbf{H}$, can be derived via Equations 5.40 and 5.41.

$$\mathbf{H} = \mathbf{F}\mathbf{G} = \mathbf{E}\mathbf{D}_n^{-1/2}\mathbf{G} \tag{5.42}$$

Now, let $\left\{\mathbf{W}_l^{NAPC}\right\}_{l=1}^{L}$ be the NAPC transform vectors from $\mathbf{H}$ in Equation 5.42, which is $\boldsymbol{H} = \left(\mathbf{W}_1^{NAPC}, \mathbf{W}_2^{NAPC}, \cdots, \mathbf{W}_L^{NAPC}\right)$ that is similar to $\left\{\mathbf{W}_l^{MNF}\right\}_{l=1}^{L}$ in Equation 5.36. Then $\left\{\mathbf{W}_l^{NAPC}\right\}_{l=1}^{L}$ may also be arranged for transformed band images in descending order of SNR. From a point of view of dimensionality reduction (DR), MNF transform in Equation 5.36 and NAPC transform in Equation 5.42 achieve the DR by only retaining q projection vectors, $\left\{\mathbf{W}_l^{NNF}\right\}_{l=L-q+1}^{L}$ and $\left\{\mathbf{W}_l^{NAPC}\right\}_{l=1}^{q}$, corresponding to largest SNRs.

To obtain the MNF transform (Green et al. 1988) or later on NAPC transform (Lee et al. 1990, Roger 1994, Chang and Du 1999), we need to know both Σ and Σ_N. In practice, these covariance matrices are unknown and must be estimated. Usually, Σ is estimated using the sample covariance matrix of $\boldsymbol{X} = (\boldsymbol{x}_1, \boldsymbol{x}_2, \cdots, \boldsymbol{x}_N)$; this approach is adopted for the regular PCA process. However, it is a challenge to estimate Σ_N as random noise exists in every band of the original data but is particularly prominent in those bands with low SNR caused by less sensitivity of detectors. There are different methods to estimate the noise covariance matrix. For example, it is suggested to use sensor dark references that record the systematic noise structure, but these may be unavailable to most end users. Near-neighbor difference method may be applied to estimate the noise covariance in multi- and hyperspectral data (Green et al. 1988, Lee et al. 1990, Chang and Du 1999). This method is generally applicable to spatially homogeneous areas with isolating salt-and-pepper noise, but not for spatially heterogeneous areas where the variation of signals will be mistakenly included as noise. Therefore, applying the near-neighbor difference method on the whole image with spatially heterogeneous areas to search for noises may not be appropriate. Roger (1996) estimated the noise estimates derived from the inverse of the image's covariance matrix in residual-scaled PC or RPC transform, called *residual analysis method.* His experimental result indicated that this was a valid procedure for hyperspectral image data (e.g., AVIRIS). If no separate noise estimates are available or they have to be derived from the image data, then the RPC provides a quick and simple approximation to the MNF or NAPC transform. The computational cost of the RPC transform over and above that of the ordinary PCA is just the inversion of the image covariance matrix. Roger with his colleague developed another method to the noise covariance matrix, called inter/intraband prediction noise estimation method (Roger and Arnold 1996). They used between-band (spectral) and within-band (spatial) correlations to decorrelate the image data via linear regression. They grouped each band into small blocks, each of which is independently decorrelated. Such decorrelation induced noise-like residuals that could be used to estimate the noise. The method worked well in both homogeneous and heterogeneous areas with multiscene AVIRIS images and also performed well for both radiance and reflectance (atmospherically corrected) hyperspectral images.

5.6.3 Independent Component Analysis

The PCA, SNR-based principal component (i.e., MNF and NAPC) transforms are performed based on underlying second-order statistics of data, which implies an assumption that the random vector process underlying hyperspectral imagery is multivariate normal. Regular PC or MNFs or NAPC band images transformed via the second-order statistics of data are uncorrelated. However, if the hyperspectral image data are not normally distributed, they will present higher order statistical moments, and the underlying second-order statistics of data-based component band images will not necessarily be statistically independent. For this case, a computational method called *independent component analysis* (ICA) seeks to search a different linear decomposition of the spectra on a set of normalized basis vectors such that the resulting transformed component images are not only decorrelated but also statistically independent from each other (Eismann 2012). During the last two decades, ICA has received considerable interest because of its extensive applications, ranging from blind source separation and channel equalization to speech recognition, functional magnetic resonance imaging, and hyperspectral imagery. The key idea of ICA assumes that data are linearly mixed by a set of separate independent signal sources and can then be used to unmix these signal sources according to their statistical independency measured by mutual information (Chang 2013). In this section, principle and algorithm of ICA transform will be conceptually introduced and discussed and a numerical implementation algorithm, FastICA, will be briefly described. For more information regarding ICA, the reader can consult more detailed references, such as Hyvärinen and Oja (2000), Hyvärinen et al. (2001), and Falco et al. (2014).

Let us assume **x** as a random variable vector mixed from a source random variable vector **s** via a mixture matrix **A**:

$$\mathbf{x} = \mathbf{A}\mathbf{s} \tag{5.43}$$

where $\mathbf{x} = (x_1, x_2, \cdots, x_n)^T$ is the observed data vector (i.e., n linear mixtures); **A** is the unknown mixture matrix with elements a_{ij}, $i, j = 1, 2, \ldots, n$; and $\mathbf{s} = (s_1, s_2, \cdots, s_n)^T$ is the unknown source vector. By estimating the unmixing matrix of **A**, called **W**, the **s** vector that represents the independent components (ICs) is achieved by

$$\mathbf{s} = \mathbf{W}\mathbf{x} \tag{5.44}$$

where **W** is the inverse of **A**, (i.e., $\mathbf{W} \cong \mathbf{A}^{-1}$), and is called the *separating matrix* or *unmixing matrix*. The purpose of ICA is to find a separating matrix that makes components of the transformed random process **s** statistically independent. To estimate the separating matrix **W**, it is possible if the following assumptions and restrictions are satisfied (Falco et al. 2014): (1) the sources are statistically independent; (2) the independent components must have a non-Gaussian distribution; and (3) the unknown mixture matrix **A** is assumed square and full rank.

The key to estimating the separating matrix **W** for the ICA model is non-Gaussianity. Actually, without non-Gaussianity, the estimation is not possible at all (Hyvärinen and Oja 2000). To use non-Gaussianity in ICA estimation, we need a quantitative measure of non-Gaussianity of a random variable. There are two commonly used measures of non-Gaussianity: The classical measure of non-Gaussianity is kurtosis or the fourth-order cumulant; a second very important measure of non-Gaussianity is negentropy. In the following, the second negentropy measure of non-Gaussianity associated with minimizing mutual information is addressed for estimating the ICA model. To simplify things, let us assume that s is a source random variable that is centered (zero-mean) and has variance equal to one.

Before introducing the estimation algorithm of ICA mode with the negentropy measure of non-Gaussianity, an approach of minimizing mutual information for ICA estimation is first discussed.

Mutual information is a natural measure of the dependence of n random variables. Using the concept of differential entropy, the mutual information $\boldsymbol{I}$ between n (scalar) random variables, s_i, i=1, 2, …, n can be defined as follows:

$$\boldsymbol{I}(\mathrm{s}_1, \mathrm{s}_2, \cdots, \mathrm{s}_n) = \sum_{i=1}^{n} H(\mathrm{s}_i) - H(\boldsymbol{s}) \tag{5.45}$$

where $H(\boldsymbol{s})$ is the differential entropy of random vector $\boldsymbol{s}$ with density $p(\boldsymbol{s})$ defined by

$$H(\boldsymbol{s}) = -\int p(\boldsymbol{s}) \log p(\boldsymbol{s})\, d\boldsymbol{s} \tag{5.46}$$

According to Cover and Thomas (1991), an important property of mutual information may be derived from the linear transformation in Equation 5.44:

$$\boldsymbol{I}(\mathrm{s}_1, \mathrm{s}_2, \cdots, \mathrm{s}_n) = \sum_{i=1}^{n} H(\mathrm{s}_i) - H(\boldsymbol{x}) - \log|det\mathbf{W}| \tag{5.47}$$

If components s_i are chosen in such a way that they are decorrelated and scaled to unit variance, entropy and negentropy differ only by a constant, and the sign. Thus we have

$$\boldsymbol{I}(\mathrm{s}_1, \mathrm{s}_2, \cdots, \mathrm{s}_n) = C - \sum_{i=1}^{n} J(\mathrm{s}_i) \tag{5.48}$$

where C is a constant that is independent of $\boldsymbol{W}$. Equation 5.48 shows the fundamental relation between negentropy and mutual information and the negentropy can be further defined by

$$J(\mathbf{s}) = H(\boldsymbol{s}_{gauss}) - H(\boldsymbol{s}) \tag{5.49}$$

where s_{gauss} is a Gaussian random variable with the same covariance matrix as s. Negentropy is always non-negative and is zero only in the case of Gaussian distribution of random vector s. Per Equation 5.48, it is apparent that defining a set of orthogonal vector directions $\{w_i\}_{i=1}^{n}$ that maximizes the negentropy sum would result in a transformation that minimizes the statistics dependence of the bands of transformed images. To implement the computation (Equation 5.48), in practice, the following approximation is frequently used:

$$J(s) \propto [E\{G(s)\} - E\{G(v)\}]^2 \tag{5.50}$$

where s is a standardized non-Gaussian variable; v is a standardized Gaussian variable with zero mean and unit variance; and G is a nonquadratic function. The accuracy of approximation in Equation 5.50 can be improved by using the following choices of G that have proved very useful:

$$G_1(s) = \frac{1}{a_1} \log \cosh(a_1 s) \tag{5.51}$$

where $1 \le a_1 \le 2$, and

$$G_2(s) = -e^{-s^2/2} \tag{5.52}$$

An algorithm called FastICA searches transform directions $\{w_i\}_{i=1}^{n}$ by efficiently maximizing the negentropy using a Newton's approach solution for the extrema of the expected value $E\{G(\mathbf{w}^T\mathbf{x})\}$ (Eismann 2012). Denote by $g(s)$ the derivatives of $G(s)$:

$$g_1(s) = \tanh(a_1 s) \tag{5.53}$$

and

$$g_2(s) = s e^{-s^2/2} \tag{5.54}$$

The FastICA algorithm proceeds as follows. The algorithm can estimate the ICs in two different ways (Falco et al. 2014): (1) deflationary orthogonalization as shown in Equation 5.55; (2) symmetric orthogonalization as shown in Equation 5.56.

First way:

1. Randomly choose an initial weight vector $\mathbf{w}$.
2. $\mathbf{w}_{i+1} \leftarrow E\left\{\mathbf{x}g\left(\mathbf{w}_i^T\mathbf{x}\right)\right\} - E\left\{g'\left(\mathbf{w}_i^T\mathbf{x}\right)\right\}\mathbf{w}_i$
3. $\mathbf{w}_{i+1} \leftarrow \mathbf{w}_{i+1} - \sum_{j=1}^{i}\left(\mathbf{w}_{i+1}^T\mathbf{w}_j\right)\mathbf{w}_j$ (5.55)
4. Repeat steps 2 and 3 until convergence is reached,

where $E\left\{\mathbf{x}g\left(\mathbf{w}_i^T\mathbf{x}\right)\right\} = \frac{1}{N}\sum_{j=1}^{N}\mathbf{x}_j g\left(\mathbf{w}_i^T\mathbf{x}_j\right)$; $g'(s)$ is the first derivative of $g(s)$.

Second way:
The same steps are taken as the first way except step 3 is replaced by

$$W = (WW^T)^{-\frac{1}{2}} W \tag{5.56}$$

where $W = (w_1, w_2, \cdots, w_n)^T$.

The first way performs orthogonalization using the Gram–Schmidt method, estimating the ICs one by one, while the second way estimates all the ICs in parallel that makes the algorithm faster (Falco et al. 2014). There are many other algorithms for estimating ICs, such as the joint approximate diagonalization of eigenmatrices (JADE), which is a widely used and parameter-free implementation of ICA, and RobustICA, which is a recent method for deflationary ICA in which the kurtosis is the general contrast function to be optimized. The reader can find their detailed information from Falco et al. (2014).

After running the FastICA algorithm (also for other algorithms), as a result a total of n ICs can be generated. However, there are some issues in implementing FastICA to be aware of: (1) FastICA-generated ICs are not necessarily in order of information significance as those produced by PCA or NAPC in the order of decreasing magnitude of eigenvalues or SNRs; (2) ICs generated by FastICA in different runs do not necessarily appear in the same order. These issues are caused primarily by the nature that the initial projection unit vectors used to produce ICs via FastICA are randomly generated. Consequently, an IC generated earlier by FastICA is not necessarily more significant than the one generated later (Chang 2013). Therefore, in order to respond to the issue in using the random initial projection unit vectors by FastICA and make the hyperspectral dimensionality reduction (DR) as PCA and NSPC transforms, Chang (2013) developed three approaches for the ICA transformation: Statistics-prioritized ICA-DR approach, random ICA-DR approach, and initialization-driven ICA-DR approach. See Chang (2013) for detailed information on the three approaches.

5.6.4 Canonical Discriminant Analysis (CDA)

CDA is a dimension-reduction technique like PCA and MNF transforms, equivalent to canonical correlation analysis (e.g., Pu 2012). CDA can be used to determine the relationship between a set of quantitative variables and a set of dummy variables coded from the class variable in a low-dimensional discriminant space (Khattree and Naik 2000, Zhao and Maclean 2000, Pu and Liu 2011). Given a classification variable and multiple quantitative variables, the CDA derives canonical variables, which are linear combinations of the quantitative variables that summarize between-class variation in much the same way as PCA summarizes most variation in the first several principal components. The CDA involves human effort and knowledge derived from training samples, while PCA performs a relatively automatic data transformation and tries to concentrate the majority of data variance in the first several PCs. However, unlike the popularity of PCA and MNF in dimensional reduction preprocessing of hyperspectral image data and feature extraction, the CDA is only occasionally analyzed and tested as a data transformation technique for dimensional reduction and feature extraction from hyperspectral data (e.g., van Aardt and Wynne 2007, Alonzo et al. 2013, Rinaldi et al. 2015). Given that the CDA has a unique capability to extract features and reduce data dimension through data transformation, its principle and algorithm are briefly introduced as follows.

The goal of CDA is to search for a linear combination of independent variables to achieve maximum separation of classes (populations). The new transformed variables are termed canonical variables. Referring to Khattree and Naik (2000) and Johnson and Wichern (2002), CDA assumes g independent populations: $G_1, G_2, \ldots, G_g$ with respective population mean vector $\mu_1, \mu_2, \ldots, \mu_g$ and a common variance–covariance matrix Σ. It is applicable to use a measure of differences of individual population mean vectors $\mu_1, \mu_2, \ldots, \mu_g$ from their overall mean center $\bar{\mu} = \frac{1}{g}\sum_{i=1}^{g} \mu_i$. Let us define a matrix **M**, termed the true between population sums of square and cross-product, as a function of $\mu_1 - \bar{\mu}, \ldots, \mu_g - \bar{\mu}$ as

$$\mathbf{M} = \sum_{i=1}^{g} (\mu_i - \bar{\mu})(\mu_i - \bar{\mu})^T \tag{5.57}$$

and compare it with the common variance-covariance matrix Σ in a meaningful way. It is possible to choose vector **a** to maximize $\mathbf{a}^T\mathbf{M}\mathbf{a}/\mathbf{a}^T\Sigma\mathbf{a}$. To do so, let **x** be a p (dimension) by 1 observation from one of these populations. If **a** is a nonzero, then the choice of **a**, say $\mathbf{u}_1$, resulting from the solution of this optimization problem yields the first canonical variable $Can_1 = \mathbf{u}_1^T\mathbf{x}$, which may be interpreted as the single best linear discriminator of these g populations. The second canonical variable, say $Can_2 = \mathbf{u}_2^T\mathbf{x}$, is chosen in the same way but is subject to the additional restriction that it is orthogonal to the first canonical variable. It may be interpreted as the second-best linear discriminator of these g populations. Proceeding in a similar way, we can obtain a list of $r = \min(p, g - 1)$ canonical variables, uncorrelated with each other.

In practice, independent samples $\{\mathbf{x}_{11}, \mathbf{x}_{12}, \ldots \mathbf{x}_{1n_1}\}, \ldots, \{\mathbf{x}_{g1}, \mathbf{x}_{g2}, \ldots \mathbf{x}_{gn_g}\}$ of sizes $n_1, \ldots, n_g$, are sampled from these g independent populations, and the g population mean vectors $\mu_1, \mu_2, \ldots, \mu_g$ and the common variance–covariance matrix Σ are all unknown. In this case, we estimate μ_i by $\bar{\mathbf{x}}_i = \frac{1}{n_i}\sum_{j=1}^{n_i} \mathbf{x}_{ij}$, $i = 1, 2, \ldots, g$, overall mean center of all populations $\bar{\mu}$ by $\bar{\mathbf{x}} = \frac{1}{\sum_{i=1}^{g} n_i}\sum_{i=1}^{g}\sum_{j=1}^{n_i} \mathbf{x}_{ij} = \frac{1}{\sum_{i=1}^{g} n_i}\sum_{i=1}^{g} n_i\bar{\mathbf{x}}_i$ and the common variance-covariance matrix Σ by the pooled sample variance–covariance matrix $\mathbf{Q}_w = \mathbf{E}_w / \left(\sum_{i=1}^{g} n_i - g\right)$, where $\mathbf{E}_w$ is the pooled within-class sums of squares and cross-product matrix, defined by

$$\mathbf{E}_w = \sum_{i=1}^{g}\sum_{j=1}^{n_i} (\mathbf{x}_{ij} - \bar{\mathbf{x}}_i)(\mathbf{x}_{ij} - \bar{\mathbf{x}}_i)^T \tag{5.58}$$

Also, we estimate the true between-population sums of squares and cross-product **M** by $\mathbf{Q}_b$, a between-class sum of squares and cross-product matrix, given by

$$\mathbf{Q}_b = \sum_{i=1}^{g} n_i(\bar{\mathbf{x}}_i - \bar{\mathbf{x}})(\bar{\mathbf{x}}_i - \bar{\mathbf{x}})^T \tag{5.59}$$

Based on the goal of CDA that makes between-class variance as large as possible and within-class variance as small as possible, choosing β to maximize the objective function λ:

$$\lambda(\beta) = \frac{\beta^T\mathbf{Q}_b\beta}{\beta^T\mathbf{Q}_w\beta} \tag{5.60}$$

To do so, differentiating λ with respect to β and letting it be 0 yield

$$\mathbf{Q}_w^{-1}\mathbf{Q}_b\boldsymbol{\beta} = \lambda\boldsymbol{\beta} \tag{5.61}$$

where βs are eigenvectors corresponding to the r non-zero eigenvalues of $\mathbf{Q}_w^{-1}\mathbf{Q}_b$ and maximizing λ (Xu and Gong 2007). If the r non-zero eigenvalues are ranked from largest to smallest as $\lambda_1 \geq \lambda_2 \geq \ldots \geq \lambda_r$, then $\beta_1 \geq \beta_2 \geq \ldots \geq \beta_r$ are corresponding coefficients of (sample version of) the respective canonical variables. Since the basic interest of conducting CDA is in the reduction of dimensionality of the raw hyperspectral data while keeping maximum separability of different classes, we are interested in evaluating the first several canonical variables corresponding to the first several eigenvectors. There are some ways to test whether the r canonical correlations are significantly different from zero to aid in the selection of important canonical variables (Khattree and Naik 2000), for example, using p-value to test whether a canonical variable is significantly correlated to a set of dummy variables.

5.6.5 Wavelet Transform

Wavelet transform (WT) is a relatively new signal-processing tool that provides a systematic means for analyzing signals at various scales (resolutions) and shifts. WT has proven to be quite powerful in these remote sensing application areas (e.g., dimensionality reduction [DR] and data compression [Mallat 1998, Bruce et al. 2002], texture feature analysis [Fukuda and Hirosawa 1999], and feature extraction [Pittner and Kamarthi 1999, Ghiyamat et al. 2015]). This is attributed to the fact that WT can decompose a spectral signal into a series of shifted and scaled versions of the mother wavelet function, and that the local energy variation (denoted by *peaks and valleys*) of a spectral signal in different bands at each scale can be detected automatically. The local energy variations can provide useful information for further analysis of hyperspectral data (Pu and Gong 2004). With continuous wavelet transform (CWT), one can analyze signals—both single dimensional and multidimensional, such as hyperspectral image cubes—across a continuum of scales. With discrete wavelet transforms (DWT), signals are analyzed over a discrete set of scales, typically being dyadic (2^j, j = 1, 2, 3, …), and the transforms can be realized using a variety of fast algorithms and customized hardware (Bruce et al. 2001). In the following, the WT algorithm and the calculation approach of wavelet transformed energy feature vector are briefly described.

The WT can decompose signals over dilated (scaled) and translated (shifted) wavelets (Rioul and Vetterli 1991, Mallat 1989). Let a set of wavelet basis functions, $\{\psi_{a,b}(\lambda)\}$, be calculated for a hyperspectral pixel spectral signal by shifting and scaling the basis or mother wavelet, $\psi(\lambda)$, by the following:

$$\psi_{a,b}(\lambda) = \frac{1}{\sqrt{a}}\psi\left(\frac{\lambda - b}{a}\right) \tag{5.62}$$

with a zero average condition: $\int_{-\infty}^{+\infty} \psi(\lambda)\,d\lambda = 0$, where $a > 0$ and b are real numbers. The variable a is the scaling factor of a particular basis function and b is the shifting variable along the function's range (Bruce et al. 2001). The WT of $f(\lambda)$ at shifting factor b and scaling factor a can be determined by

$$Wf(a,b) = \langle f, \psi_{a,b} \rangle = \int_{-\infty}^{+\infty} f(\lambda)\frac{1}{\sqrt{a}}\psi\left(\frac{\lambda - b}{a}\right)d\lambda. \tag{5.63}$$

Equation 5.63 is a CWT calculation formula. For CWT, since both scale parameter a and shift parameter b are real numbers, the transform coefficients $Wf(a, b)$ are continuous. The DWT, denoted by $W_{j,k}$, of a function $f(\lambda)$, is defined by the scalar product of $f(\lambda)$ with the scaling function (i.e., the wavelet basis function) $\phi(\lambda)$, which is dilated and translated (Simhadri et al. 1998) as

$$W_{j,k} = \left\langle f(\lambda), \phi_{j,k}(\lambda) \right\rangle, \tag{5.64}$$

where the wavelet basis function $\phi_{j,k}(\lambda)$ can be calculated by

$$\phi_{j,k}(\lambda) = 2^{\frac{-j}{2}} \phi(2^{-j}\lambda - k) \tag{5.65}$$

where j is the jth decomposition level or step and k is the kth wavelet coefficient at jth level. In contrast with the CWT, the scales for the DWT are $a = 2, 4, 8,\ldots,2^j,\ldots,2^J$.

The DWT has been used extensively in the development of fast wavelet algorithms. The decomposition coefficients of a wavelet orthogonal basis can be calculated with a fast algorithm that cascades discrete convolutions with conjugate mirror filters h and g (i.e., low-pass and high-pass filter finite impulse responses, respectively), and subsamples the output. The decomposition formulae are given as follows (Mallat 1998, Hsu et al. 2002):

$$cA_{j+1}[k] = \sum_{n=-\infty}^{+\infty} h[n-2k]cA_j[n] \tag{5.66}$$

$$cD_{j+1}[k] = \sum_{n=-\infty}^{+\infty} g[n-2k]cA_j[n] \tag{5.67}$$

where cA_j are the approximation coefficients at scale 2^j; cA_{j+1} and cD_{j+1} are the approximation and detail components, respectively, at scale 2^{j+1}. In practice, the original signal s is always expressed as coefficient cA_L. Therefore, a multilevel orthogonal wavelet decomposition of cA_L is composed of wavelet detail coefficients of signal s at scales $2^L < 2^j \leq 2^J$, plus the remaining approximation at the largest scale 2^J:

$$[\{cD_j\}_{L<j\leq J}, cA_J]. \tag{5.68}$$

Figure 5.12 illustrates the structure of terminal nodes by the multilevel wavelet decomposition. If the length of cA_j is n, one can see that the downsampling procedure in the wavelet decomposition that reduces the length of cA_{j+1} to $n/2$ achieves the dimension reduction of cA_j. In theory, the maximum number of decomposition levels can be $J = log_2(N)$, where N is the length of the input original signal. In practice, however, the maximum number also depends on which mother wavelet you have chosen (e.g., when the mother wavelet Daubechies 3 is chosen for $N = 167$, maximum number of levels $J = 8$).

In general, we decompose high-dimension pixel signals using the wavelet decomposition and then select the fewest wavelet coefficients required to perform the DR without losing significant useful information. Since each wavelet coefficient directly related to the amount of energy in the signal at different positions and scales, the best wavelet features could be a subset of wavelet coefficients, their energy, and any combination of the two (Zhang et al. 2006). In practice, instead of

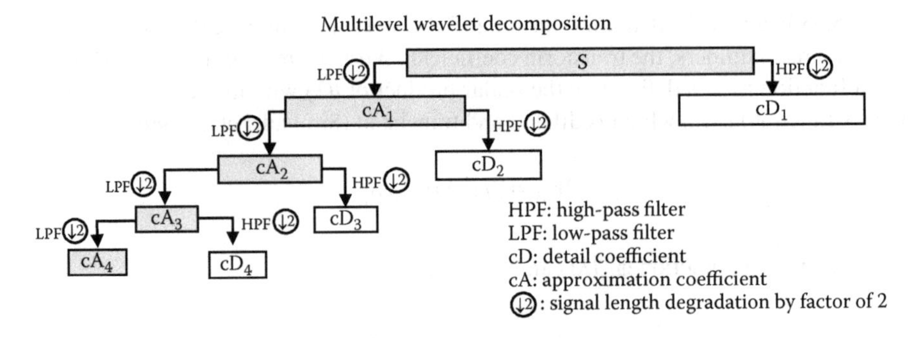

FIGURE 5.12 Illustration of the structure of terminal nodes ($J = 4$) by the multilevel wavelet decomposition.

using all the wavelet coefficients to determine the optimum subset, the wavelet coefficients can also be screened on the basis of their energy. Thus, wavelet coefficients with energy values lower than a predetermined threshold value can be discarded without causing significant loss for signal representations (Peng et al. 2009). For instance, Okamoto et al. (2007) found that only 50 coefficients of 213 contained 99.7% of the total energy in their particular case. To perform a significant DR and feature extraction from hyperspectral data, the retained or all wavelet coefficients derived from each pixel spectrum can be further calculated into an *energy feature vector*. The $1\times(J + 1)$ DWT energy feature vector $\bar{\boldsymbol{F}} = \{F_j\}_{j=1}^{J+1}$ is computed as

$$F_j = \sqrt{\frac{1}{K}\sum_{k=1}^{K} W_{jk}^2} \tag{5.69}$$

where K is number of coefficients at the decomposition level j, while k is the kth coefficient at level j; J is the maximum number of decomposition levels; the length of the feature vector is $(J + 1)$ coming from J levels of detail coefficients and one level of final approximation coefficients. The wavelet energy feature vector provides information about how energy of the hyperspectral spectrum is partitioned according to scales (Bruce et al. 2001, Zhang et al. 2006).

5.7 SPECTRAL MIXTURE ANALYSIS (SMA)

The most widely used methods for extracting surface information from multispectral and hyperspectral images are image classification. With this type of techniques and methods, despite the stochastic concept or statistical property of the methods, each pixel is assigned to one out of several known categories or classes through a statistical or probability decision function. In this way, the final thematic information is obtained disregarding the mostly compositional nature of surface materials. In fact, reflected radiation as recorded in a pixel by a remote sensor has rarely interacted with a volume composed of only a single homogenous material because natural surfaces composed of a single uniform material do not exist in nature (van der Meer et al. 2001). Therefore, most often the electromagnetic radiation observed by the remote sensor as pixel reflectance values results from the spectral mixture of a number of surface materials that have difference spectral natures.

In order to identify various "pure materials" and to determine their spatial proportions or abundances from the remotely sensed data, the spectral mixing process must be understood and properly modeled. Mixture modeling is the forward process of deriving mixed signals from pure end-member spectra while spectral unmixing focuses on doing the reverse, deriving the fractions (or abundance) of the pure end-members from the mixed pixel signal. There are two types of spectral mixing, *linear spectral mixing* (LSM) and *nonlinear spectral mixing.* A standard linear spectral unmixing technique assumes that the spectra collected by the imaging spectrometer can be expressed in the form of a linear combination of end-members weighted by their corresponding proportions or abundances. It should be noted that the linear mixture model assumes minimal secondary reflections or multiple scattering effects in the data-collection procedure, and hence the measured spectra can be expressed as a linear combination of the spectral signatures of materials present in a mixed pixel (Ben-Dor et al. 2013). A nonlinear spectral mixture model assumes that the mixed spectra collected by the imaging instrument are better described by the part of the source radiation being multiplied scattered before being collected at the sensor. Thus, a nonlinear spectral unmixing approach may best characterize the resultant mixed spectra for certain end-member distributions, such as those in which the end-member components are randomly distributed throughout the instrument's FOV (Guilfoyle et al. 2001). Therefore, both linear and nonlinear spectral mixing models are simple tools used to describe spectral mixing processes. Although the nonlinear spectral mixture model can be found in Sasaki et al. (1984), Zhang et al. (1998), and Arai (2013), linear spectral mixing modeling and its inversion have been more widely used since the late 1980s, in which these were extensively applied to extract the proportion or abundance of various components from mixed pixels. Therefore, the relevant spectral mixture modeling and spectral unmixing algorithms reviewed and discussed in the following text are related to LSM, if not all. When using the LSM model to deal with the spectral unmixing issue, the following three assumptions apply:

1. The spectra signals are linearly contributed by a finite number of end-members within each IFOV weighted by their cover percentage (Ichoku and Karnieli 1996).
2. The end-members in a IFOV are homogeneous surfaces and spatially segregated without multiple scattering (Keshava and Mustard 2002).
3. The electromagnetic energy of neighboring pixels does not affect the spectral signal of the target pixel significantly (Miao et al. 2006).

Figure 5.13 illustrates the principle of linear spectral mixing and unmixing that shows the compositional nature of natural surfaces with relative or absolute fractions (or abundance) of a number of spectral components or end-members that together contribute to the observed reflectance of the image. During last three decades, a number of modeling techniques and algorithms have been used for LSM analysis (Ichoku and Karnieli 1996). Unconstrained LSM is the most common form (e.g., Gong et al. 1994, Pu et al. 2008). The sum-to-one constrained method requires the sums of end-member fractions to be one or close to one (e.g., Smith et al. 1990, Adams et al. 1993). The fully constrained method further refines the end-member fractions to be positive (e.g., Adams et al. 1986, Brown et al. 2000). Furthermore, a multiple end-member spectral mixture analysis (MESMA) approach (e.g., Painter et al. 1998, Roberts et al. 1998), a mixture-tuned-matched filtering (MTMF) technique (e.g., Boardman et al. 1995, Andrew and Usting 2008, 2010), and a constrained energy minimization (CEM) approach (e.g., Farrand and Harsanyi 1997) have been developed for spectral unmixing analysis with hyperspectral data. In addition, an artificial neural network algorithm has also been tested to unmix mixed pixels into fraction abundances of end-members in some studies (e.g., Foody 1996, Flanagan and Civco 2001, Pu et al. 2008). Therefore, in the following, we will introduce and discuss these modeling techniques and algorithms.

5.7.1 Traditional Spectral Unmixing Modeling Techniques

Let ρ be an $L \times 1$ reflectance (or digital number, DN) pixel vector in a multispectral or hyperspectral image where L is the number of spectral bands. Assume that $\mathbf{M}$ is an $L \times p$ spectral signature matrix.

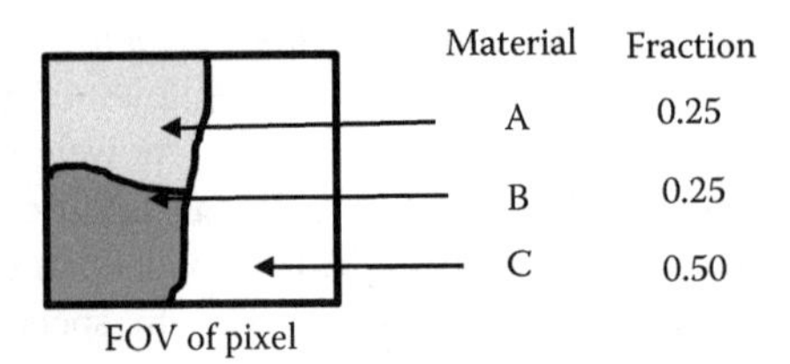

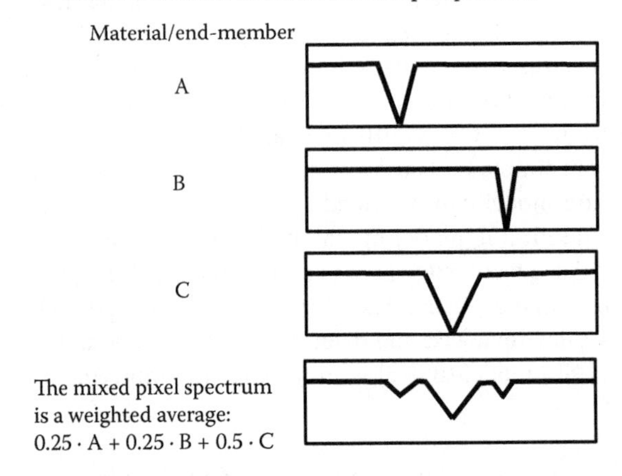

FIGURE 5.13 Principle of linear spectral mixing and unmixing. A single pixel mixed with three materials A, B and C. (Modified from van der Meer, F. D., de Jong, S. M., and Bakker, W., in *Imaging Spectrometry: Basic Principles and Prospective Applications*, eds. F. D. van der Meer and S. M. de. Jong, Springer, The Netherlands, pp. 17–62, 2001.)

Let **F** be a $p \times 1$ fraction or abundance vector associated with a set of end-members of interest in the pixel vector ρ. A linear mixture model assumes that the spectral signature of a pixel vector is linearly superimposed by spectral signatures of image end-members present in the pixel vector and can be described by the relationship with a vector and matrix form:

$$\boldsymbol{\rho} = \mathbf{MF} + \boldsymbol{\varepsilon} \tag{5.70}$$

where

$\boldsymbol{\rho} = \begin{bmatrix} \rho_1 \\ \rho_2 \\ \vdots \\ \rho_L \end{bmatrix}$, $\mathbf{M} = \begin{bmatrix} m_{11}\ m_{12} \cdots m_{1p} \\ m_{21}\ m_{22} \cdots m_{2p} \\ \cdots\cdots\cdots\cdots \\ m_{L1}\ m_{L2} \cdots m_{Lp} \end{bmatrix}$, $\mathbf{F} = \begin{bmatrix} f_1 \\ f_2 \\ \vdots \\ f_p \end{bmatrix}$, $\varepsilon = \begin{bmatrix} \varepsilon_1 \\ \varepsilon_2 \\ \vdots \\ \varepsilon_L \end{bmatrix}$ are the pixel vector in reflectance or DN, spectral signature matrix, fraction or abundance vector of end-members of interest, and an $L \times 1$ additive noise vector, ε, representing a model error (i.e., a variance of ρ not explained by the end-members), respectively. It is well known that inversion of Equation 5.70 (i.e., the spectral unmixing process) can be achieved with a least squares solution when $L > p$, especially in hyperspectral data, $L >> p$ (e.g., Adams et al. 1989, Sohn and McCoy 1997). From an image, end-member spectra signature matrix **M** can be obtained in two ways: (1) The use of spectra from pure pixels, and (2) the use of known fractions of different end-members for a set of training pixels (no less than L) to solve for matrix **M** using the least squares method. For (1), the Pixel Purity Index (PPI) and N-Finder approaches are introduced below. For (2), based on known fractions of different end-members (**F**) for a set of training pixels (ρ) and using Equation 5.70, matrix **M** can be approximated as $\hat{\mathbf{M}}$:

$$\hat{\mathbf{M}} = \boldsymbol{\rho}\mathbf{F}^T(\mathbf{F}\mathbf{F}^T)^{-1}. \tag{5.71}$$

Accordingly, after the spectral signature matrix **M** for a scene of hyperspectral data is obtained, the following four traditional spectral unmixing models can be implemented for estimating fraction vector **F** of end-members from mixed pixels ρ.

1. Model I: Unconstrained Least Squares (ULS) Linear Unmixing

   Using the least squares error as an optimal criterion for solving LSM Equation 5.70, the unconstrained optimal least squares estimate of **F**, $\hat{\mathbf{F}}_{\mathbf{ULS}}$, can be found by minimizing the following least square error function (Heinz and Chang 2001):

   $$\min_{\mathbf{F}}\{(\boldsymbol{\rho}-\mathbf{MF})^T(\boldsymbol{\rho}-\mathbf{MF})\} \tag{5.72}$$

   as

   $$\hat{\mathbf{F}}_{\mathbf{ULS}} = (\mathbf{M}^{\mathbf{T}}\mathbf{M})^{-1}\mathbf{M}^{\mathbf{T}}\boldsymbol{\rho} \tag{5.73}$$

   where $\hat{\mathbf{F}}_{\mathbf{ULS}}$ is a function of the image pixel vector ρ. Hereafter for simplicity, the dependency of $\hat{\mathbf{F}}_{\mathbf{ULS}}$ (or similar LS solutions) on ρ is not included in the notation.
2. Model II: Sum-to-One Constrained LS (SCLS) Linear Unmixing

   The ULS solution of Equation 5.70 is obtained by imposing no constraints on the fraction vector **F**. In order to finally find the fully constrained optimal least squares estimate of **F**, we can first consider the SCLS linear mixing problem (Heinz and Chang 2001, Chang et al. 2004), which can be described as follows:

   $$\min_{\mathbf{F}}\{(\boldsymbol{\rho}-\mathbf{MF})^{\mathbf{T}}(\boldsymbol{\rho}-\mathbf{MF})\} \text{ subject to } \sum_{j=1}^{p} f_j = 1 \text{ or } \mathbf{1}^T\mathbf{F} = 1 \tag{5.74}$$

   where $\mathbf{1}^T = \underbrace{(1,1,\ldots,1)}_{p}$ is a unity vector with all elements equal to one. Then the sum-to-one constrained optimal least squares estimate of F, $\hat{\mathbf{F}}_{\mathbf{SCLS}}$, to Equation 5.74:

   $$\hat{\mathbf{F}}_{\mathbf{SCLS}} = \hat{\mathbf{F}}_{\mathbf{ULS}} + (\mathbf{M}^{\mathbf{T}}\mathbf{M})^{-1}\mathbf{1}\cdot[\mathbf{1}^{\mathbf{T}}((\mathbf{M}^{\mathbf{T}}\mathbf{M})^{-1}\mathbf{1}](\mathbf{1}-\mathbf{1}^{\mathbf{T}}\cdot\hat{\mathbf{F}}_{\mathbf{ULS}}) \tag{5.75}$$

3. Model III: Nonnegativity Constrained LS (NCLS) Linear Unmixing

   Since the SCLS solution derived in Equation 5.75 is only based on the constraint $\sum_{j=1}^{p} f_j = 1$, it does not guarantee that the estimated abundance fractions are nonnegative, i.e., $f_j \geq 0$ for all $1 \leq j \geq p$. However, unlike the SCLS method that produces a closed-form solution, the NCLS method does not have an analytical solution since the abundance nonnegativity constraint (ANC) is formed by a set of linear inequalities rather than equalities

(Heinz and Chang 2001). In general, an NCLS problem can be described by the following optimization problem:

$$\min_{F}\{(\boldsymbol{\rho}-\mathbf{MF})^T(\boldsymbol{\rho}-\mathbf{MF})\} \text{ subject to } \mathbf{F}\geq 0. \tag{5.76}$$

To solve Equation 5.70 under Equation 5.76 for nonnegativity constrained LS estimate of **F**, $\hat{\mathbf{F}}_{\mathbf{NCLS}}$, a Lagrange multiplier vector $\lambda = (\lambda_1, \lambda_2,\ldots, \lambda_p)^T$ needs to be introduced. Finally, two iteration equations may be obtained by differentiating a Lagrangian function associated with the Lagrange multiplier vector (Heinz and Chang 2001, Chang et al. 2004):

$$\hat{\mathbf{F}}_{\mathbf{NCLS}} = (\mathbf{M}^{\mathbf{T}}\mathbf{M})^{-1}\mathbf{M}^{\mathbf{T}}\boldsymbol{\rho} - (\mathbf{M}^{\mathbf{T}}\mathbf{M})^{-1}\boldsymbol{\lambda} = \hat{\mathbf{F}}_{\mathbf{ULS}} - (\mathbf{M}^{\mathbf{T}}\mathbf{M})^{-1}\boldsymbol{\lambda} \tag{5.77}$$

and

$$\boldsymbol{\lambda} = \mathbf{M}^T(\boldsymbol{\rho} - \mathbf{M}\hat{\mathbf{F}}_{\mathbf{NCLS}}) \tag{5.78}$$

The optimal NCLS solution $\hat{\mathbf{F}}_{\mathbf{NCLS}}$ and the Lagrange multiplier vector $\lambda = (\lambda_1, \lambda_2, \ldots, \lambda_p)^T$ can be solved by iterating Equations 5.77 and 5.78. The detailed NCLS iteration algorithm can be found in Chang (2003).

4. Model IV: Fully Constrained LS (FCLS) Linear Unmixing

   A fully constrained least squares (FCLS) problem is set as

$$\min_{F}\{(\boldsymbol{\rho}-\mathbf{MF})^T(\boldsymbol{\rho}-\mathbf{MF})\} \text{ subject to } \sum_{j=1}^{p} f_j = 1 \text{ and } \mathbf{F}\geq 0 \tag{5.79}$$

In order to solve the FCLS problem (i.e., Equation 5.79), Chang et al. (2004) included the abundance sum-to-one constraint (ASC) in the spectral signature matrix **M** by introducing a new signature matrix **N** defined by

$$\mathbf{N} = \begin{bmatrix} \delta\mathbf{M} \\ \mathbf{1}^T \end{bmatrix} \tag{5.80}$$

with $\mathbf{1}^T = \underbrace{(1,1,\ldots,1)}_{p}$, and a vector $\boldsymbol{S}$ denoted by

$$\mathbf{S} = \begin{bmatrix} \delta\boldsymbol{\rho} \\ \mathbf{1} \end{bmatrix} \tag{5.81}$$

The parameter δ in Equations 5.80 and 5.81 is included in order to control the impact of the ASC on the FCLS solution. According to Heinz and Chang (2001) and Chang et al. (2004), with these two Equations 5.80 and 5.81, the FCLS solution can be derived directly from the NCLS solution by replacing **M** and ρ used in the NCLS solution (Equations 5.77 and 5.78) with **N** and **S**. An algorithm of finding the FCLS solution can be summarized as follows:

1. Specify values of the parameter δ (a small value, e.g., 10^{-6}) and the error tolerance *e*.
2. Use Equation 5.73 to generate an unconstrained least squares solution $\hat{\mathbf{F}}_{\mathbf{ULS}}$.
3. Iterate Equations 5.77 and 5.78 with **M** and ρ replaced by **N** and **S** defined in Equations 5.80 and 5.81, respectively, until the algorithm converges within *e*.

In addition, the FCLS solution to the problem Equation 5.79 can be estimated by using a quadratic programming (QP) algorithm (Li 2004).

5.7.2 Artificial Neural Networks Solution to LSM

Apart from the techniques to solve Model IV above, the FCLS solution to the problem Equation 5.79 can also be estimated by a feed-forward artificial neural network (ANN) algorithm, which is a nonlinear solution to the LSM, used for unmixing mixed pixels. The network training mechanism is an error-propagation algorithm (Rumelhart et al. 1986, Pao 1989). The detailed algorithm of ANN is introduced in Section 5.8. In a layered structure, the input to each node is the sum of the weighted outputs of the nodes in the prior layer, except for the nodes in the input layer, which are connected to the feature values of hyperspectral data. The nodes in the last layer output a vector that corresponds to similarities to each fraction of end-members within a mixed pixel. Although more than one layer (hidden layer) between the input and output layers is possible, one hidden layer is usually sufficient for most learning purposes. The network's learning procedure is controlled by a learning rate and a momentum coefficient that need to be specified empirically based on the results of a limited number of tests. The training procedure is iterated by repeatedly presenting training samples (pixels) with known fractions of end-members, and the network training is terminated when the network output meets a minimum error criterion or optimal test accuracy. The trained network can then be applied to estimate the fraction of each end-member in mixed pixels in a scene of hyperspectral image.

5.7.3 Multiple End-Member Spectral Mixture Analysis (MESMA)

This LSM addressed in the previous section has an advantage in which it is relatively simple and provides a physically meaningful measure of abundance in mixed pixels of multi- or hyperspectral data. However, there are a number of limitations (Pu and Gong 2011) to the simple mixing concept: (1) the end-members used in LSM are the same for each pixel, regardless of whether the materials represented by the end-members are present in each pixel; (2) it cannot to account for the fact that the spectral contrast between those materials is variable; (3) it cannot account for subtle spectral differences between materials efficiently; and (4) the maximum number of components that the LSM can map is limited by the number of bands in the image data (Li and Mustard 2003). Therefore, Roberts et al. (1998) introduced MESMA, a technique in which the number of end-members and types are allowed to vary for each pixel in the image, into the solution to LSM. The basic idea is to run unmixing multiple times to each pixel with various subsets of end-members from the total pool of end-members and take the "best fit" results as the final solution to a particular pixel. The best-fit could be the minimum error or most reasonable end-member combination. In general, MESMA overcomes these limitations of the LSM model, such as no total number of end-members limited by the number of bands in the image data.

According to Roberts et al. (1998), the general procedure of the MESMA approach starts with a series of candidate two–end-member models, evaluates each model based on selection criteria then, if required, constructs candidate models that incorporate more end-members (i.e., creating candidate three–end-member models and four–end-member models). Three selection criteria are used to evaluate selection of candidates for two– (or three– or four–) end-member models as follows:

1. A fraction criterion: In considering the 1% error allowed for instrumental noise-generated errors in fractions, a model is selected only if the model produces physically reasonable fractions between −0.01 and 1.01.
2. Root mean square (RMS) error criterion: A model is selected only if the RMS is below the threshold (e.g., 0.025). A lower threshold will reduce the chance that a candidate model is selected, whereas a higher threshold will increase the chance.

3. A residual criterion consisting of a threshold and count: Across hyperspectral bands, the residual threshold is used to evaluate whether any individual residual exceeds an absolute threshold while the residual count is used for counting the number of times the threshold is exceeded contiguously. A contiguous threshold count can be used to distinguish residuals resulting from noise and atmospheric contamination from the residuals resulting from the spectral absorption features caused by end-members in a candidate model associated with a mixture (Roberts et al. 1998).

The result of testing the MESMA by Roberts et al. (1998) with AVIRIS hyperspectral data demonstrates that the technique is capable of discriminating a large number of spectrally distinct types of vegetation while capturing the mosaic-like spatial distribution typical of chaparral. Roessner et al. (2001) made further restrictions of end-member combination and selection in spectral unmixing by considering end-member combination feasibilities based on spatial context in an urban area.

5.7.4 Mixture-Tuned Matched Filtering Technique (MTMF)

MTMF is an advanced spectral unmixing algorithm that does not require that all materials within a scene are known and have identified end-members (Boardman et al. 1995). Therefore, MTMF is initially called partial unmixing that provides a method of solving only that fraction of the data inversion problem that directly relates to the specific goals of the investigation. Figure 5.14 illustrates the concept for a scene with four background materials and two targets of interest. The challenge is how to find the proper projection of the data that hide the background variance while simultaneously maximizing the variance among the targets (Boardman et al. 1995). This method represents an improved alternative to SMA for cases where the number of similar spectra are large or where it is problematic to collect spectra of all potential end-member components within the scene. MTMF treats each end-member independently and, at each pixel for each end-member, models the pixel as a mixture of the end-member and an undefined background material. MTMF outputs a matched filter (MF) score and an infeasibility value for each end-member. The MF score, an estimate of the areal coverage of a pixel by the material of interest, is analogous to the fraction value from simple SMA, and the infeasibility is a measure of how likely a pixel is to contain the material of interest. Pixels are likely to contain materials for which they receive high MF scores and low infeasibilities (Andrew and Ustin 2008). MTMF has proven to be a very powerful tool to detect specific materials that differ slightly from the background. For example, with airborne hyperspectral image data (e.g., AVIRIS and HyMap) MTMF has successfully mapped a variety of invasive species, including Tamarisk (Hamada et al. 2007), perennial pepperweed (Andrew and Usting 2008, 2010), and cheatgrass (Noujdina and Ustin 2008).

The data processing with the MTMF algorithm consists of three steps, according to Boardman et al. (1995): (1) Retrieving apparent (scaled) surface reflectance from hyperspectral raw data (e.g., AVIRIS data); (2) determining pixel purity using MNF transform processing or other methods; and (3) conducting partial unmixing (equivalent to calculating/outputing MF score and an infeasibility value for each end-member in each pixel in an image). In Step 1, the hyperspectral data are first

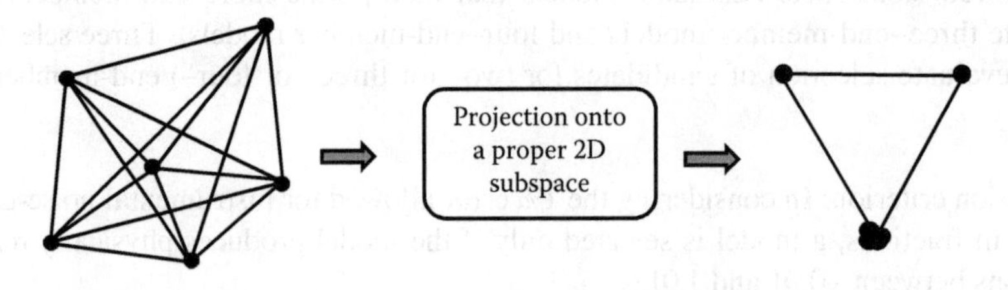

FIGURE 5.14 Schematic diagram of partial unmixing (i.e., MTMF) as a data projection. (Modified from Boardman, J. W., Kruse, F. A., and Green, R. O., *Summaries of the Fifth JPL Airborne Geoscience Workshop* JPL Publication, 95–1, pp. 23–26, NASA Jet Propulsion Laboratory, Pasadena, California, 1995.)

converted to apparent surface reflectance by using an atmospheric correction method (e.g., FLAASH). In Step 2, the data are subjected to a dimensionality analysis and noise-whitening process, using the MNF transform process. The purpose of using the MNF transform process is translating the apparent surface reflectance data to have zero mean. Then the processed data are rotated and scaled so that the noise in every band is uncorrelated and has unit variance. Then the extreme pixels in each projection are identified in the first several MNF images using a PPI index approach. Accordingly, the purest pixels in the scene are rapidly identified. The purest pixels in the scene are then compared against the target spectra. If any are close matches for the target materials, they are identified and separated from the other purest pixels. This allows the method to work on major or low-probability scene components (including targets of interest). For those high-purity pixels that do not closely match a target spectrum, they are used to form a subspace that spans the background and is assigned as a background end-member spectra for a single background end-member, in spite of the complexity of the background. A set of optimal projection vectors is directly calculated for the target-spanning subspace, perpendicular to the background-spanning subspace. In the third step, the spectral unmixing process is applied to the data after the data are projected onto the optimal target subspace. Finally, the abundances or fractions of targets of interest are spatially mapped from the hyperspectral image.

5.7.5 Constrained Energy Minimization (CEM)

The CEM method, developed by Farrand and Harsanyi (1997) and similar to the MTMF method, is an extension of LSM. CEM maximizes the response of a target signature on a pixel-by-pixel basis and suppresses the response of undesired background signatures. CEM technique is based on a linear operator that minimizes the total energy in a hyperspectral image sequence while the response of the operator to the signature of interest is constrained to a desired constant level (Resmini et al. 1997). It is assumed that foreground and background signatures are mixed linearly such as is the case when each photon only interacts with one material. The result of CEM is a vector component image that is comparable to fraction abundance images typically obtained through unmixing. According to Farrand and Harsanyi (1997), the key to the CEM technique is to find a vector $\boldsymbol{w}$ that suppresses the unknown and undesired background signature but enhances the known target signature, $\boldsymbol{d}$. The CEM operator is defined by two constraints. The first constraint is to minimize the total output energy of all pixels. Accordingly, the energy of an individual pixel summed across the wavelength range can be represented by a scalar value y_i, which is a weighted sum of the responses at each spectral band within r_i. More precisely, y_i can be expressed as

$$y_i = \sum_{k=1}^{l} w_k r_{ik} \quad i = 1, 2, \ldots, q \tag{5.82}$$

where $\boldsymbol{w} = (w_1, \ldots w_k, \ldots w_l)^T$ is the vector of weights; l is the number of spectral bands; q is the total number of pixels; and r_{ik} is the radiance recorded in band k for pixel i. Equation 5.82 can be expressed in a vector notation as

$$y_i = \boldsymbol{w}^T \boldsymbol{r}_i \quad i = 1, 2, \ldots, q \tag{5.83}$$

The second constraint is when applied to the target pixel spectrum, the energy of an individual pixel summed across the wavelength range to be 1 (i.e., $y_i = 1$, or in a vector notation, for example, $\boldsymbol{w}^T \boldsymbol{d} = 1$). This constrained minimization problem can be solved by:

$$\boldsymbol{w} = \frac{\boldsymbol{\Phi}^{-1}\boldsymbol{d}}{\boldsymbol{d}^{-1}\boldsymbol{\Phi}^{-1}\boldsymbol{d}} \tag{5.84}$$

where Φ^{-1} is the inverse of the sample correlation matrix of the observation pixel vector ($r_1,\ldots, r_i,\ldots r_q$) from the area of interest. Equation 5.84 defines the CEM operator. In order to calculate Φ^{-1} more accurately, the correlation matrix can be approximated through the use of the first p significant eigenvectors. The equation describing this approximation in a vector notation may be:

$$\hat{\boldsymbol{\Phi}} = \tilde{V}\tilde{\boldsymbol{\Lambda}}\tilde{V}^T \tag{5.85}$$

where $\tilde{\mathbf{V}} = (\tilde{v}_1,\ldots,\tilde{v}_i,\ldots,\tilde{v}_p)$ is the $l \times p$ matrix whose columns are the p significant eigenvectors ($p < l$), and $\boldsymbol{\Lambda} = diag(\lambda_1,\ldots,\lambda_i,\ldots,\lambda_p)$ is a $p \times p$ diagonal matrix. The inverse sample correlation matrix $\hat{\boldsymbol{\Phi}}^{-1}$ can then be estimated by substituting $\boldsymbol{\Lambda}^{-1} = diag\left(\lambda_1^{-1},\ldots,\lambda_i^{-1},\ldots,\lambda_p^{-1}\right)$ for $\tilde{\boldsymbol{\Lambda}}$ in Equation 5.85 (Farrand and Harsanyi 1997).

The inverse of the sample correlation matrix (derived from Equation 5.85) may be considered an optimal estimate of the orthogonal background subspace, and thus the CEM operator is the optimal estimator of the orthogonal subspace projection operator when the background signatures are unknown (Resmini et al. 1997). Therefore, the application of the CEM operator (Equation 5.84) to each pixel in the hyperspectral image sequence will optimally suppress the unknown and undesired background end-members and highlight the spectral signature of interest end-member, and, in fact, thus provide an optimal estimate of the abundance or fraction of the material of interest in each pixel (Resmini et al. 1997).

5.7.6 End-Member Extraction

Appropriate determination of end-members is a key to successful spectral mixture analysis (Gong et al. 1994, Tompkins et al. 1997). Determination of end-members involves identifying the number of end-members and extracting their corresponding spectral signatures. Spectra in the library may be used as end-member spectra, but they must have some correspondence with the sensor characteristics in order to perform the spectral matching and unmixing analysis in HRS. However, such an end-member determination approach relying on spectral library may not be feasible when trying to process large quantities of image data (Veganzones and Graña 2008). Therefore, most spectral unmixing techniques count on image-/scene-based end-member extraction approaches. Such image-based end-member extraction approaches are mostly based on convex geometry assumption of the data distribution. As can be seen in Figure 5.15, there is an assumption that the pixels in an image occupy a space formed by a simplex. A simplex is the simplest geometric shape that can

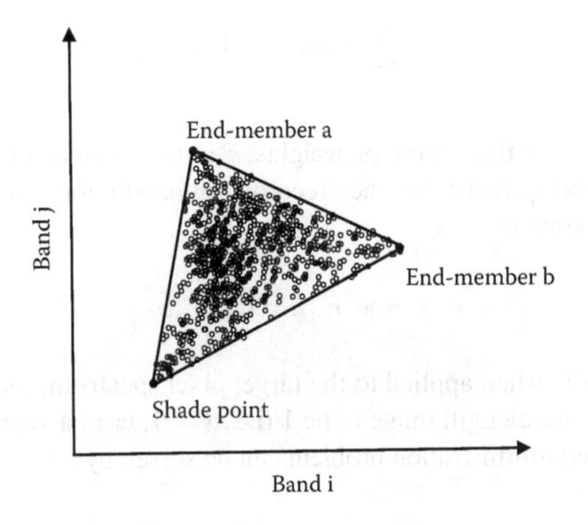

FIGURE 5.15 Scatterplot between two bands that typically shows a triangle shape, a two-dimensional simplex.

contain a space of a given dimension (Winter and Winter 2000). A simplex, for example, can be the line, triangle, and tetrahedron for one dimension, two dimensions, and three dimensions, respectively. The vertices of this simplex are the end-member locations. The pixels located at vertices will be unmixed, or pure, and thus these pixels are composed of only one of the end-members. The set of points located at the vertices of the simplex (i.e., the end-member set) possess the following three properties: (1) They represent the set of points in the data that contain the largest volume; (2) the same pixels are end-members if the entire data cloud is translated or scaled; and (3) the identity of the end-members does not change under any rotations. The third property is particularly important to process hyperspectral data given the fact that rotation of the raw data does not change the identity of the end-members because it allows the use of orthogonal subspace projections (OSPs) for the reduction of the image dimensions. Dimension reduction via OSP (e.g., principal components analysis [PCA] and minimum noise fraction [MNF] transform) is frequently used to reduce the dimensionality of hyperspectral imagery and simplify the further processing (Winter and Winter 2000).

During the last two decades, several techniques/algorithms have been proposed for extracting end-members from hyperspectral scenes. Most algorithms search for pure pixels for end-member spectra in an image cube, which is equivalent to finding the vertices of the simplex containing the data. The assumption that a hyperspectral image can be represented as a simplex forms the basis of the geometrical interpretations of hyperspectral data sets (Boardman 1995). In the following, two techniques will be introduced and reviewed: these techniques are commonly used, and corresponding software is available in most image-processing packages. They are pixel purity index (PPI) (Boardman et al. 1995) and N-Finger technique (Winter 1999). There are many other techniques in current literature, including iterative error analysis (IEA; Neville et al. 1999), optical real-time adaptive spectral identification system (ORASIS; Bowles et al. 1995), convex cone analysis (CCA; Ifarraguerri and Chang 1999), automated morphological end-member extraction (AMEE; Plaza et al. 2002), simulated annealing algorithm (SAA; Bateson et al. 2000), and new virtual dimensionality (VD) concept or the hyperspectral signal identification by minimum error (HySime; Sánchez and Plaza 2014). For those who are interested in these techniques, see the corresponding references.

5.7.6.1 Pixel Purity Index (PPI)

The most famous and extensively used method is the PPI (Boardman et al. 1995). The PPI algorithm is performed through the following three steps. The first step is reducing the data dimensionality and making a noise-whitened process by MNF transform. The next step is repeatedly projecting all data points onto random unit vectors in the N-dimensional space comprising a scatterplot of MNF-transformed data in order to calculate the PPI for each point in the image cube. The extreme pixel in each projection is counted. After many repeated projections to different lines, those pixels with a count meeting a threshold are identified the purest pixels in a scene. The last step is confirming a final set of end-member spectra by loading these potential end-member spectra into an interactive N-dimensional visualization tool and rotating in real time until a desired number of end-members are visually identified as extreme pixels in the data cube. To speed up the PPI process, a fast iterative algorithm to implement PPI is proposed (Chang and Plaza 2006). The Fast Iterative PPI algorithm (FIPPI) improves PPI in several aspects, including (1) producing an appropriate initial set of end-members to speed up the process, and (2) estimating the number of end-members to be generated by virtual dimensionality (VD). Therefore, FIPPI is also an unsupervised and iterative algorithm in which an iterative rule is developed to improve each of the iterations until it reaches an ideal set of end-members (Chang and Plaza 2006).

5.7.6.2 N-Finder

The input data for this process are required to be in a full-spectral image cube, i.e., without either the dimensional reduction or the exemplar thinning process. The N-Finder, developed by Winter (1999), must examine the full data set to find those pure pixels that can be used to describe the various mixed pixels in the scene. The procedure finds the set of pixels that define a simplex with the maximum volume by "inflating" the simplex within the dataset. To do so, first, a dimensionality reduction of the original image is accomplished by using the MNF transform. Second, a random collection of image pixel spectra is used as an initial set of end-members. To refine the estimate of the end-members, every pixel in the image must be evaluated as to its likelihood of being a pure or nearly pure pixel. In the case, the volume must be calculated with each pixel in place of each end-member. Let **E** be defined as

$$\mathbf{E} = \begin{bmatrix} 1 & 1 & \cdots & 1 \\ \mathbf{e}_1 & \mathbf{e}_2 & \cdots & \mathbf{e}_\mathbf{p} \end{bmatrix} \tag{5.86}$$

where $\mathbf{e}_i$ are end-member column vectors and p is the number of end-members used to calculate the simplex volume. The volume of the simplex formed by the end-members is proportional to the determinant of **E** (Plaza et al. 2004):

$$V(\mathbf{E}) = \frac{1}{(p-1)!} abs\left(|\mathbf{E}|\right) \tag{5.87}$$

In order to refine the initial volume estimate, a trial volume is calculated for every pixel in each end-member position by replacing that end-member and recalculating the volume. If the replacement makes the volume increase, the pixel replaces the end-member. This procedure is repeated until there are no replacements of end-members left (Winter 1999).

5.8 HYPERSPECTRAL IMAGE CLASSIFICATIONS

Traditional multispectral classifiers may be used in hyperspectral image classifications and thematic information extraction but they may be less effective than expected with multispectral data. This is because they face the difficulties caused by (1) high dimensionality of hyperspectral data and (2) high correlation of adjacent bands with a relatively limited number of training samples. To solve these problems, image dimension reduction and feature extraction preprocessing methods are necessary before classification. The image dimension reduction processing methods through image transform processing such as PCA, signal-to-noise ratio-based maximum noise fraction (MNF) transforms, and independent component analysis are effective means for performing hyperspectral data classification with traditional multispectral classifiers. Feature extraction schemes such as PCA (or its noise adjusted version, MNF), Fisher's linear discriminant analysis (LDA), canonical discriminant analysis (CDA), or wavelet transform have been applied in transforming and extracting feature variables for efficiently classifying hyperspectral data with traditional classifiers. The feature extraction processing (calculation and selection of feature variables) is realized by maximizing the ordered variance of the whole data set or the ratio of between-class variance to within-class variance of the training samples or transformed features converted into energy features. To date we have developed many segment-based multispectral classifiers and advanced classifiers specially designed for effectively processing hyperspectral data with limited number of training samples, which are appropriate to process hyperspectral data for classification and thematic information extraction. Since some important algorithms and methods associated with hyperspectral image dimension reduction and feature extraction were introduced in Section 5.6, in this section, the image segment-based multispectral classifiers such as segmented PCA, simplified MLC (maximum likelihood classifier), and advanced algorithms including artificial neural networks (ANNs) and support vector machines (SVMs) are introduced and discussed.

5.8.1 Segment-Based Multispectral Classifiers

Jia and Richards (1999) proposed a segmented principal component (PC) transformation (i.e., segmented PCA or segPCA) by selecting subsets of the covariance matrix in a lower segmented dimension. The segPCA can be realized by K subgroups of bands separated based on the "block property" of a correlation matrix of hyperspectral data. Figure 5.16 shows a correlation matrix composed of four blocks (VIS, NIR, MIR1, and MIR2), each with a highly correlated subgroup of bands. According to Jia and Richards (1999), the general procedure of segPCA can be described as follows. The complete data set is first partitioned into subgroups based on a "blocked" correlation matrix of hyperspectral data. Highly correlated bands are selected as subgroups. Let n_1, n_2, ..., n_K be the number of bands in subgroups 1, 2, ..., K, respectively. The PCA is performed separately to each subgroup of data first. Then features can be selected from PCs created from each subgroup of bands (possible total PCs = sum of n_1 through n_K from K subgroups) by either considering making use of variance information of first several PCs from each subgroup for simplicity or considering the separability of classes for each PC calculated from subgroups of bands due to the orthogonal property of transformed PCs. The selected features can be regrouped and transformed again to compress the data further. Generally, the procedure is repeated until the required data reduction ratio is achieved for classification or display purpose of the most informative three features, while important information is essentially preserved. The major advantage of the segPCA is saving computation time as

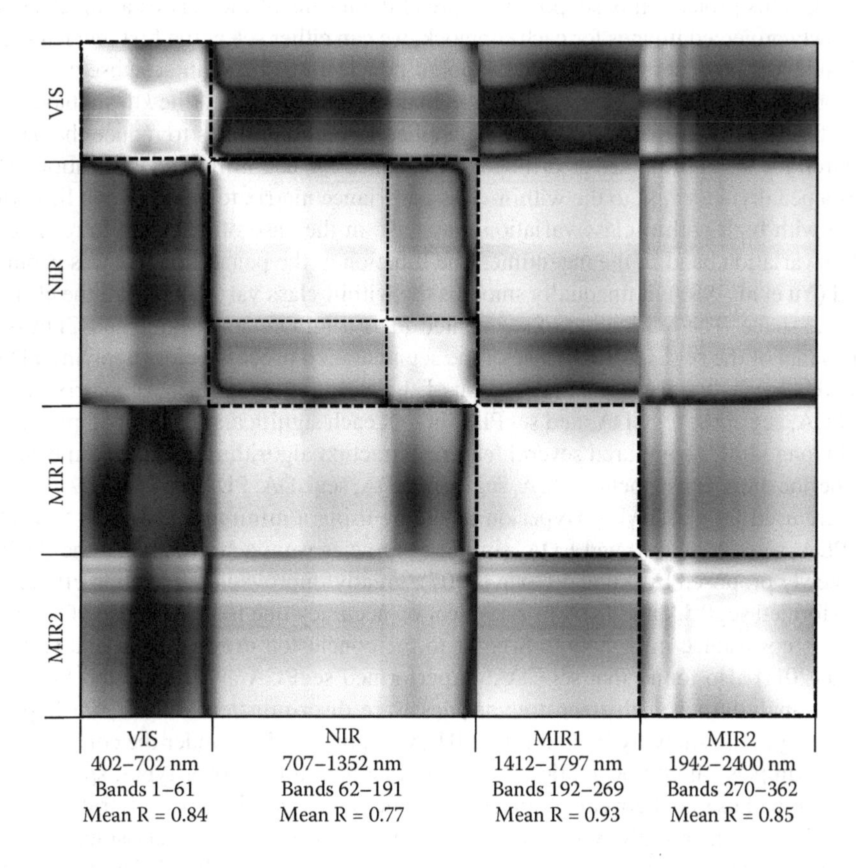

FIGURE 5.16 Band correlation matrix, presented in image grayscale form (in absolute values of correlation: 0–1), which was calculated with 458 hyperspectral measurements. Sub-blocks (VIS: 402–702 nm, NIR: 707–1352 nm, MIR1: 1412–1797 nm, and MIR2: 1942–2400 nm) of the correlation matrix are marked in dashed boxes. The diagonal line in white indicates the highest correlation 1. The darker the tone, the lower is the absolute value of the correlation.

the actual transformation time is reduced. Because of the low dimensionality of finally selected features by the segPCA procedure compared to raw data dimensionality and the orthogonal property, most multispectral classifiers can then be used for classification with a relatively limited number of training samples.

It was suggested that penalized discriminant analysis (PDA) would more efficiently deal with high correlation among the bands by penalizing the high within-class variance and improving the performance of LDA (Yu et al. 1999). Fisher's LDA and CDA search for successive linear combinations of data to maximize the ratio of between-class variance over within-class variance of training samples. They expect spreading the means or the cluster centers of different classes as much as possible while keeping the within-class variation at a similar level for all classes (Yu et al. 1999, Xu and Gong 2007, Pu and Liu 2011). It is based on an assumption of reliable estimation of training statistics.

Segmented LDA (segLDA) first divides the whole spectrum into K subblocks with each block containing a set of continuous highly correlated spectral bands using spectral correlation matrix (Figure 5.16). Denote the dimension of the *k*th sub-block as I_k, and $I_1 + \ldots + I_k + \ldots + I_K = I$ (raw data dimension). For each subblock of spectral bands, estimate the between-class covariance matrix and the within-class covariance matrix in a subspace that has the dimension equal to the number of bands in the subblock. Then, apply LDA to each subblock to generate new component images (features) [with a number of $\min(C-1, I_k)$], where C is a number of classes and I_k is a number of bands in subblock k. This projection is supposed to spread the means of the classes as much as possible. With the newly projected images for each subblock, we can either select the first few feature images from each subblock to generate a combined pool of new features that can be subsequently used for classification or select more feature images [less than $\min(C-1, I_k)$] from the *k*th subblock for k = 1, 2, …, K to form a new subspace. LDA can be applied in multiple times to reduce the data dimension in search for an optimal set of orthogonal subspace for use in final classification. The PDA introduces a penalty matrix Ω to the within-class covariance matrix to penalize and limit the effect that a band with high within-class variation may have in the case of LDA but to reserve the low within-class variation band in the meantime. The function of the penalty matrix was geometrically interpreted (Yu et al. 1999). It unequally smooths the within-class variation for all the classes in the hyperspectral space. The realization of segmented PDA (segPDA) and segmented CDA (segCDA) is similar to that of the segLDA in a sense of the segmentation process before applying PDA except that PDA adds a penalty term to the estimation of the within-class covariance matrix. It is similar to the segPCA, segLDA, segCDA, and segPDA in that each significantly saves computation time.

Xu and Gong (2007) compared several feature extraction algorithms used for dimension reduction of Hyperion data. These include PCA, segPCA, LDA, segLDA, PDA, and segPDA. The selected features were used for classifying Hyperion image by using a minimum distance (MD) classifier. With SegPDA, segLDA, PDA ,and LDA, similar accuracies were achieved while the SegPDA and SegLDA, newly proposed by Xu and Gong (2007), greatly improved computation efficiency. Both also outperformed segPCA and PCA in classification accuracy due to making use of specific intra- and interclass covariance information. Similar to the conclusion drawn by Xu and Gong (2007), Pu and Liu (2011) also found that segCDA outperformed segPCA and segmented SDA (stepwise discriminant analysis) when thirteen tree species were discriminated with *in situ* hyperspectral data. According to the study by Pu and Liu (2011), CDA or segCDA (under the condition with limited training samples) should be applied broadly in mapping forest cover types, species identification, and other land use/land cover classification practices with multi-/hyperspectral remote sensing data because it is superior to PCA and SDA for feature selection for image classification.

Jia and Richards (1994) improved the conventional maximum likelihood classifier as one called simplified maximum likelihood classification (SMLC) that is suitable for classifying hyperspectral data. The SMLC can significantly shorten computation time and effectively use a limited number of training samples. The SMLC method first segments whole spectral space into several subspaces using a spectral correlation matrix (e.g., Figure 5.16) then uses the traditional MLC to classify a hyperspectral image scene. In the following, according to Jia and Richards (1994), the SMLC classifier is introduced.

The conventional MLC is based on the assumption that the probability distribution for each spectral class is expressed in the form of a multivariate normal model with the dimension equaling to the number of spectral bands. Its discriminant function is given by

$$g_i(\boldsymbol{x}) = -ln\left|\sum\nolimits_i\right| - (\boldsymbol{x}-\boldsymbol{m}_i)^T \sum\nolimits_i^{-1} (\boldsymbol{x}-\boldsymbol{m}_i) \quad \mathrm{i} = 1,2,\ldots,\mathrm{C} \tag{5.88}$$

where $\boldsymbol{x}$ is a pixel brightness vector; $\boldsymbol{m}_i$ is the mean pixel spectral vector for class i; Σ_i is its covariance matrix of size N × N; and N is the total number of spectral bands. C is the number of classes available for labeling the pixel. Per the MLC, the decision rule generally is

$$\boldsymbol{x} \in \omega_i \text{ if } g_i(\boldsymbol{x}) > g_j(\boldsymbol{x}) \text{ for all } \mathrm{j}\,(j \neq i) \tag{5.89}$$

where ω_i is spectral class i. The standard classifier is widely employed when the number of bands, N, is small, such as in TM data and SPOT HRV data. With N increasing, however, two problems arise: (1) Computation time increases significantly owing to its quadratic dependence on N, and (2) small classes may not have enough training samples available for reliable estimates of the maximum likelihood statistics $\boldsymbol{m}_i$ and Σ_i. To solve the problems, Jia and Richards (1994) developed an improved version of MLC called simplified maximum likelihood discriminant function (i.e., SMLC).

Considering the global correlation matrix among the complete set of N bands, K subgroups (e.g., four subgroups in Figure 5.16) are formed corresponding to K distinguishable diagonal blocks. There is an assumption of normality made for each class within each subgroup k, $k = 1, 2, \ldots, K$. The corresponding discriminant function of SMLC can be derived in the following manner. In Equation 5.88, let

$$\boldsymbol{y}_i = (\boldsymbol{x} - \boldsymbol{m}_i) \tag{5.90}$$

and partition $\boldsymbol{y}_i$ into its set of independent subvectors: $\boldsymbol{y}_i = \left[\boldsymbol{y}_{i1}^T, \boldsymbol{y}_{i2}^T, \cdots, \boldsymbol{y}_{iK}^T\right]^T$, $\boldsymbol{y}_{ik}^T(\mathrm{k} = 1,2,\ldots,\mathrm{K})$, corresponding to the chosen subgroup k. Therefore, the determinant of its covariance matrix is the product of the determinants of its covariance matrices of diagonal blocks (subgroups):

$$|\Sigma_i| = \prod_{k=1}^{K} |\Sigma_{ik}| \tag{5.91}$$

so that

$$ln|\Sigma_i| = \sum_{k=1}^{K} ln|\Sigma_{ik}|. \tag{5.92}$$

The inverse of a block diagonal matrix can be obtained by inverting individually the blocked matrices, so that

$$\boldsymbol{y}_i^T \Sigma_i^{-1} \boldsymbol{y}_i = \sum_{k=1}^{K} \boldsymbol{y}_{ik}^T \Sigma_{ik}^{-1} \boldsymbol{y}_{ik} \tag{5.93}$$

Considering Equations 5.90 through 5.93, Equation 5.88 can be rewritten as

$$g_i(\boldsymbol{x}) = -\sum_{k=1}^{K} \left\{ ln\left|\Sigma_{ik}\right| + (\boldsymbol{x} - \boldsymbol{m}_i)^T \Sigma_{ik}^{-1} (\boldsymbol{x} - \boldsymbol{m}_i) \right\}, \mathrm{i} = 1,2,\ldots,\mathrm{C}; \mathrm{k} = 1,2,\ldots,\mathrm{K} \quad (5.94)$$

The dimensions of $\boldsymbol{x}$, $\boldsymbol{m}_i$, and Σ_i are reduced to $\mathrm{n_k}$ ($\mathrm{n_k} < \mathrm{N}$), the size of kth subgroup, such that the advantage can be taken of the corresponding quadratic reduction in classification time. Meanwhile, the number of training pixels required for each class, which can be considered the size of the biggest subgroup, is much smaller than that when all bands (N) are used.

5.8.2 Artificial Neural Networks (ANN)

ANN techniques can handle data at any measurement scales although data preprocessing such as normalization needs to be done first. It has been proven that ANN works well even with small training sample size. Therefore, ANN algorithms have been widely used in integration analysis of spatial data from multiple sources including multi-/hyperspectral data for prediction and classification. Rumelhart et al. (1986) developed a frequently used generalized delta rule (GDR) for supervised training of a neural network based on error back propagation. The architecture of a layered network with feed-forward capability is shown in Figure 5.17. The network of elemental processors arranged and connected in a feed-forward manner reminiscent of biological neural nets can be used to classify hyperspectral data or their extracted/transformed features into m classes. In the figure, the basic elements are nodes "o" and links "→." Nodes are arranged in layers and each of them is a processing element. Each input node accepts a single value that corresponds to an element in a set of input variables (e.g., band reflectance). Each node generates an output value. Depending on the layer in which a node is located, its output may be used as the input for all nodes in the next layer. The links between nodes in successive layers are weight coefficients. The number of hidden layers can be greater than one. In the output layer, each node corresponding to a single class generates the membership value [0, 1] of that class.

Each node, except those in the input layer, takes the outputs from all the nodes of its previous layer and uses a linear combination of those outputs from the previous layer as its net input. For a node in layer j, the net input is

$$\mu_j = \sum w_{ji} x_i \quad (5.95)$$

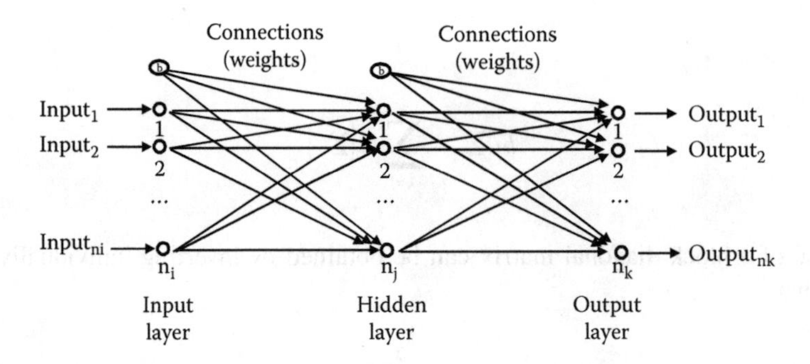

FIGURE 5.17 A three-layer structure feed-forward neural network. "o" denotes nodes of neural network and "→" represents links between the nodes.

where w_{ji} is the link between two nodes from the input layer i to its successive layer j; x_i represent outputs from all the nodes of previous layer i. The output of the node in layer j is

$$o_j = f(\mu_j) \tag{5.96}$$

where f is an activation function that often takes the form of a sigmoidal function:

$$o_j = \frac{1}{1+e^{-(\mu_j+\theta_j)/\theta_0}} \tag{5.97}$$

where parameter μ_j serves as a threshold or bias to shift the activation function along the horizontal axis, and the effect of θ_0 is to modify the shape of the sigmoid. A low value of θ_0 tends to make the sigmoid take on the characteristics of a threshold-logic unit, whereas a high value of θ_0 results in a more gently varying function (Pao 1989). Essentially, functions of a node as expressed by Equations 5.95 and 5.97 work for any node in any layers other than the input layer. For any particular node we replace x_i in Equation 5.95 by an output of a node from the previous layer.

We present the sample $\mathbf{X}_\mathrm{p}$ as input and make the net adjust its weights in all the connecting links and all the thresholds in the nodes in order to achieve the desired output $\mathbf{Y}_\mathrm{p}$ from the output nodes. Once this adjustment has been accomplished by the network, another pair of $\mathbf{X}_\mathrm{p}$ and $\mathbf{Y}_\mathrm{p}$ is presented and the network is asked to learn that association also. In fact, we require that the net find a single set of weights and biases that will satisfy all the input–output pairs presented to it. The process of obtaining the weights and biases is network learning, which is essentially the same as supervised training. During the network training, generally the outputs $\mathbf{O}_p = \{o_{pq}\}$ will not be the same as the desired values $\mathbf{Y}_\mathrm{p}$. Therefore, for each $\mathbf{X}_p$, the sum of squared error is defined as

$$E_p = \sum_{q=1}^{k} (y_{pq} - o_{pq})^2 \tag{5.98}$$

where k is the number of output nodes and the average system error is

$$E = \frac{1}{n}\sum_{p=1}^{n}\sum_{q=1}^{k} (y_{pq} - o_{pq})^2 \tag{5.99}$$

where n is the number of samples (patterns).

With GDR, the optimal set of weights is obtained by changing the weights such that the error E_p can be reduced as rapidly as possible. The purpose of network training is to achieve convergence toward improved weights and thresholds by taking incremental changes $\Delta_p w_{ji}$ proportional to the partial derivatives $-\partial E_p/\partial w_{ji}$ obtainable from Equations 5.95 through 5.98. Starting at the output layer, GDR propagates the error backward to previous layers. This is a process called *error back-propagation* (BP). According to Pao (1989):

$$\Delta_p w_{ji} = \eta \delta_{pj} o_{pi} \tag{5.100}$$

where η is a constant called *learning rate* and the delta is partial derivative $\delta_{pj} = -\partial E_p/\partial \mu_j$. If the jth node is in the output layer, j is replaced by q denoting the qth node in the output layer. We have

$$\delta_{pq} = (y_{pq} - o_{pq})o_{pq}(1 - o_{pq}) \tag{5.101}$$

and if the jth node is in internal layers, then

$$\delta_{pj} = o_{pj}(1 - o_{pj})\sum_{q=1}^{k}\delta_{pq}w_{qj} \tag{5.102}$$

A new weight w_{ji} usually consists of two parts: old weight plus weight change. Since a large η corresponds to rapid learning but might cause system oscillations, Rumelhart et al. (1986) suggested that a new weight change be modified by including some of the weight change in the previous iteration of network learning. Therefore, a new weight w_{ji} is obtained through

$$w_{ji}(n+1) = w_{ji}(n) + \eta(\delta_j o_i) + \alpha\Delta w_{ji}(n) \tag{5.103}$$

where the single-subscript variables apply to any training samples, the quantity $(n + 1)$ denotes the $(n + 1)$th iteration of network learning. α is called a *momentum coefficient.* $\Delta w_{ji}(n)$ is the weight change at which the nth step calculated from all the samples in the training set. Thresholds μ_j are learned in the same manner as are the weights.

In summary, the neural network approach is not a linear procedure. The learning procedure consists of the net starting off with a random set of weights, using one training sample at a time as the network inputs, evaluating the output in a feed-forward manner. Using the error BP procedure that starts by comparing the difference between the network output with the actual value of the sample, the net calculates $\Delta_p w_{ji}$ for every w_{ji} in the net for a particular sample. The procedure is repeated for all the training samples to produce Δw_{ji} for all weights. This iterative training continues until the network output for each training sample is equal to or closes the known output value. However, the learning does not guarantee that the global optimum is achieved (Pao 1989, Eberhart and Dobbins 1990). In general, η and α can be any values within [0, 1] and a number of hidden layers and number of nodes at each hidden layer are variable. Number of iterations is also variable. Consequently, it is difficult to obtain the global optimal solution.

Presently, most methods of ANNs in remote sensing image classification use a BP learning algorithm called multilayer perceptrons (MLPs) for supervised learning classification (Mas and Flores 2008, Xiao et al. 2008). BP is a well-known training algorithm for ANNs and is the easiest to understand and the most commonly used. However, it suffers from several drawbacks. It can be very slow to converge properly on an error minimum, and the parameters that control the training process are difficult to set. There are three other types of ANNs: (1) Radial-basis function (RBF) networks, (2) adaptive resonance theory (ART), and (3) self-organizing maps (SOMs), developed for supervised or unsupervised classification with multi-/hyperspectral data. RBF is a fairly straightforward task to construct a classification ANN if the set of examples is linearly separable. The main idea behind RBFs is to map a classification problem onto a higher-dimensionality space. According to the separability theorem of Cover (1965), a complex pattern-classification problem associated with a high-dimensional space nonlinearly is more likely to be linearly separable than in a low-dimensional space. See the review paper by Mas and Flores (2008) for an introduction to the concept of the algorithm. ART, developed by Carpenter and Grossberg (1995), is a kind of recurrent, self-organized neural network that exhibits unsupervised learning. ART can be used to model and solve the plasticity–stability dilemma (i.e., the ability of the brain to learn quickly and stably without catastrophically forgetting previously acquired knowledge). Therefore, an ART network includes a short-term memory, which captures stimuli, and a long-term memory, which stores learned information (Mas and Flores 2008). Information flows from short-term memory to long-term memory during learning and in the reverse direction during recall. Long-term memory is implemented in the form of weights of connections among neurons in different layers (Carpenter and Grossberg 1995). Based on competitive learning, the neurons in SOMs compete among themselves to be activated. From the different schemes proposed for SOMs, Kohonen's model is the most popular (Kohonen 2001), which captures the main characteristics of cognitive maps in the brain and yet remains computationally tractable. The main goal of an SOM is to map an input signal of arbitrary dimensions to a 1D or 2D output. See Kohonen (2001) for the detailed process that takes place without any supervision.

5.8.3 Support Vector Machines

Support vector machines (SVM) as a new type of classifiers have been successfully applied to the classification of hyperspectral remote sensing data. Traditionally, classifiers first model the density of various classes and then find a separating surface for classification. However, estimation of density for various classes with hyperspectral data suffers from the Hughes phenomenon (Hughes 1968) (i.e., for a limited number of training samples, the classification rate decreases as the dimension increases). The SVM approach does not suffer from this limitation because it directly seeks a separating surface (hyperplane) through an optimization procedure that finds so-called *support vectors* from a set of training samples, which form the boundaries of the classes. This is an especially interesting property for hyperspectral image processing because usually there is only a set of limited training samples available compared to the data dimensionality to define the separating surface for classification. The properties of SVMs make them well suitable for hyperspectral image classification since they can handle data in high dimensionality efficiently, deal with noisy samples in a robust way, and make use of only those most characteristic samples as the support vectors in construction of the classification models. SVMs are considered kernel-based classifiers that are based on mapping data from the original input feature space to a kernel feature space of higher dimensionality and then solving a linear problem in that space (Burges 1998). In the following, according to Burges (1998), Melgani and Bruzzone (2004), Pal and Watanachaturaporn (2004), and Bruzzone et al. (2007), the basic idea of SVMs is introduced and discussed in mathematical formulation.

5.8.3.1 Linear SVM for a Separable Case

In order to be simplified, let us consider a supervised binary classification problem and assume that the training samples consist of N vectors from the *d*-dimensional feature space $\boldsymbol{x}_i \in \mathfrak{R}^d \, (i = 1, 2, \ldots, N)$. Each of the training samples (i.e., each vector $\boldsymbol{x}_i$) belongs to either of the two classes labeled by $y_i \in \{-1, +1\}$. The goal of the SVM classifier is to maximize the separating margin. Let us also assume that the training samples are linearly separable. This means that it is possible to find at least *one linear separating hyperplane* defined by a weighting vector $\boldsymbol{w} \in \mathfrak{R}^d$ (determining the orientation of a discriminating plane) and a scalar $b \in \mathfrak{R}$ (determining the offset of the plane from the origin) that can separate the training samples of two classes without errors (Figure 5.18a). Accordingly, we can use such a hyperplane $\boldsymbol{w} \cdot \boldsymbol{x} + b = 0$ to separate training samples of two classes:

$$\begin{cases} w \cdot \boldsymbol{x}_i + b \geq +1 & \textit{for } y_i = +1, \\ w \cdot \boldsymbol{x}_i + b \leq -1 & \textit{for } y_i = -1. \end{cases} \tag{5.104}$$

The inequalities in Equation 5.104 can be combined into a single inequality as

$$y_i(\boldsymbol{w} \cdot \boldsymbol{x}_i + b) - 1 \geq 0 \quad i = 1, 2, \ldots, N. \tag{5.105}$$

The decision rule is based on the function sign $\{f(x)\}$, where $f(x)$ is the discriminant function associated with the hyperplane and is defined as

$$f(\boldsymbol{x}) = \text{sing}(\boldsymbol{w} \cdot \boldsymbol{x} + b) \tag{5.106}$$

where sing(·) is the signum function. It returns +1 if the element is greater than or equal to zero and −1 otherwise.

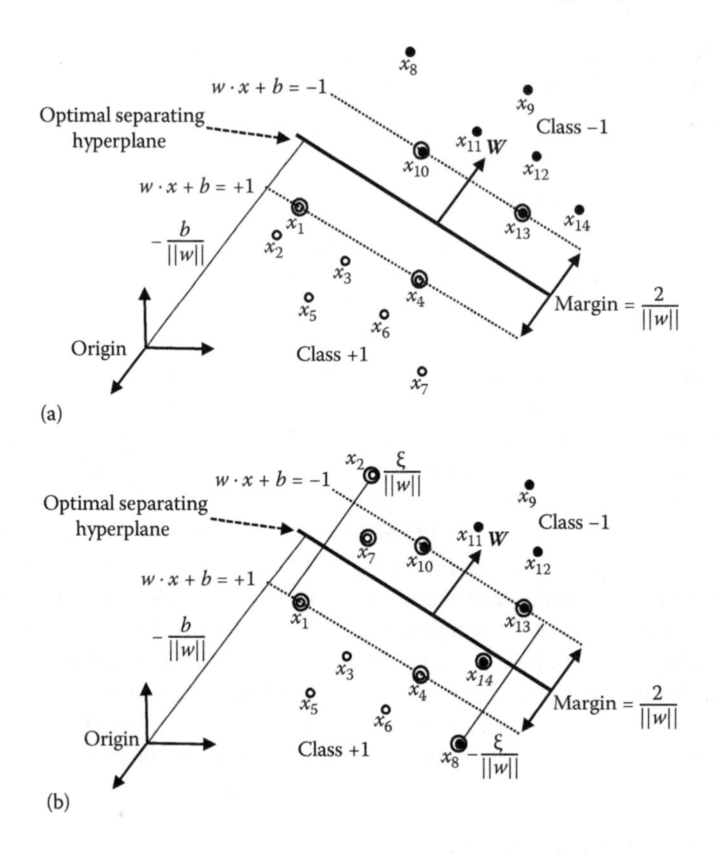

FIGURE 5.18 (a) Representation of the optimal separating hyperplane in a linearly separable case. (b) Representation of the optimal separating hyperplane in a linearly nonseparable case. White and black circles refer to the classes "+" and "−", respectively. Only support vectors (those indicated by an extra circle) are necessary to define the optimal separating hyperplanes.

The goal of SVM classifiers is finding the optimal hyperplane that maximizes the distance between the closest training sample and the separating hyperplane. To do so, we can express this distance as equal to $\frac{1}{\|w\|}$ through a simple rescaling of the hyperplane parameters $\mathbf{w}$ and b such that the geometrical margin between the two classes is given by the quantity $\frac{2}{\|w\|}$ (Figure 5.18). The concept of margin is a key point to study the SVM approach as it is a measure of its generalization capability (Vapnik 1998), which means that the larger the margin, the higher the expected generalization. Thus, the optimal hyperplane can be determined through maximization of the margin $\frac{2}{\|w\|}$ that is equivalent to the minimization of 2-norm $\frac{1}{2}\|\mathbf{w}\|^2$ as the solution of using standard Quadratic Programming (QP) optimization techniques as follows:

$$\begin{cases} \min_{\mathbf{w},b}\left\{\frac{1}{2}\|\mathbf{w}\|^2\right\} \quad \text{subject to the constraints:} \\ y_i(\mathbf{w}\cdot\mathbf{x}_i + b) \geq 1 \quad i = 1,2,\ldots,N \end{cases} \tag{5.107}$$

The solution to the corresponding dual problem to Equation 5.107 is obtained by minimizing the primal Lagrangian formulation in terms of $\boldsymbol{w}$ and b and maximizing with respect to α:

$$\begin{cases} \max\limits_{\boldsymbol{\alpha}} \left\{ \sum_{i=1}^{N} \alpha_i - \frac{1}{2} \sum_{i=1}^{N} \sum_{j=1}^{N} \alpha_i \alpha_j y_i y_j (\boldsymbol{x}_i \cdot \boldsymbol{x}_j) \right\} \text{ subject to} \\ \sum_{i=1}^{N} \alpha_i y_i = 0 \text{ and } \alpha_i \geq 0 \quad i = 1, 2, \ldots, N \end{cases} \tag{5.108}$$

where $\alpha_i (i = 1, 2,\ldots,N)$ are the unknown Lagrange multipliers and can be estimated using QP techniques (Vapnik 1998).

According to the Karush–Kühn–Tucker (KKT) optimality condition (Cristianini and Shawe-Tayor 2000), some of the multipliers α_i will be zeros. The training samples associated with nonzero multipliers α_i values are called the *support vectors* (e.g., $\boldsymbol{x}_1$, $\boldsymbol{x}_4$, $\boldsymbol{x}_{10}$, and $\boldsymbol{x}_{13}$ in Figure 5.18a). These support vectors lie at a distance exactly equal to $\frac{1}{\|w\|}$ from the optimal separating hyperplane (i.e., on the margin bounds), while the remaining training samples are irrelevant for the classification (Bruzzone et al. 2007). The following decision rule associated with the optimal hyperplane is then applied to classify the data vectors into two classes +1 and −1:

$$f(\boldsymbol{x}) = \text{sign}\left\{ \sum_{i \in S} \alpha_i y_i (\boldsymbol{x}_i \cdot \boldsymbol{x}) + b \right\} \tag{5.109}$$

where S is the subset (support vectors) of training samples corresponding to the nonzero Lagrange multipliers α_i.

5.8.3.2 Linear SVM for a Nonseparable Case

The linearly separable case is an ideal case to understand the concept of SVMs. In the case, all training samples are assumed to be separable into two classes with a linear separating hyperplane. However, this ideal case or assumption is rarely met due to noise or mixture during the collection of training data, which means that such an optimistic condition is difficult to satisfy in the classification of real data (Figure 5.18b). In order to tackle nonseparable data, the concept of optimal separating hyperplane has been generalized as the solution that minimizes a cost function that expresses a combination of two criteria: margin maximization (as in the case of linearly separable data) and error minimization (to penalize the wrongly classified samples) (Melgani and Bruzzone 2004). Here, we can introduce *slack variables* $\xi_i \geq 0$, $i = 1, 2,\ldots,N$, to take into account for the noise or error in the dataset due to misclassification so that Equation 5.105 is relaxed to

$$y_i(\boldsymbol{w} \cdot \boldsymbol{x}_i + b) - 1 + \xi_i \geq 0. \tag{5.110}$$

Then the optimization problem for the linearly nonseparable case becomes

$$\begin{cases} \min\limits_{w,b,\xi} \left\{ \frac{1}{2} \|\boldsymbol{w}\|^2 + C \sum_{i=1}^{N} \xi_i \right\} \\ \text{subject to the constraints:} \\ y_i(\boldsymbol{w} \cdot \boldsymbol{x}_i + b) \geq 1 - \xi_i \text{ and} \\ \xi_i \geq 0 \quad i = 1, 2, \ldots, N \end{cases} \tag{5.111}$$

where C is a constant and represents a regularization parameter that allows controlling the penalty assigned to errors. The SVMs with the above-described soft constraints (i.e., ξ_i) are called *soft margin* SVMs, while the ones in Equation 5.107 are called *hard margin* SVMs. In soft margin SVMs, a set of support vectors consists of training samples on and within the upper and lower margins (bounds) and samples falling on the "wrong side" of a margin (e.g., $\boldsymbol{x}_1$, $\boldsymbol{x}_2$, $\boldsymbol{x}_4$, $\boldsymbol{x}_7$, $\boldsymbol{x}_8$, $\boldsymbol{x}_{10}$, $\boldsymbol{x}_{13}$, and $\boldsymbol{x}_{14}$ in Figure 5.18b). From Equation 5.111, it can be seen that when $C \to 0$, the minimization problem is not affected by the misclassification even though $\xi_i > 0$; when $C \to \infty$, and the values of ξ_i approach zeros, the minimization problem reduces to the linear separable case. Therefore, the higher the C value, the greater the penalty associated to misclassified samples. Similar to the objective function in Equation 5.108, the dual optimization problem for a linearly nonseparable case can be obtained as

$$\begin{cases} \max\limits_{\alpha} \left\{ \sum_{i=1}^{N} \alpha_i - \frac{1}{2} \sum_{i=1}^{N} \sum_{j=1}^{N} \alpha_i \alpha_j y_i y_j (\boldsymbol{x}_i \cdot \boldsymbol{x}_j) \right\} \quad \text{subject to} \\ \sum_{i=1}^{N} \alpha_i y_i = 0 \text{ and } 0 \le \alpha_i \le C \quad i = 1, 2, \ldots, N. \end{cases} \tag{5.112}$$

Thus it can be seen that the objective function of the dual optimization problem for the linearly nonseparable case is the same as that of the linearly separable case, except the Lagrange multipliers α_i are bounded by the penalty value C (Pal and Watanachaturaporn 2004). After obtaining the solution of Equation 5.112, the decision rule is also the same one as defined in Equation 5.109.

5.8.3.3 Nonlinear SVM: Kernel Method

There are many cases where a linear separating hyperplane cannot separate classes without misclassification. However, those classes can be separated by a nonlinear separating hyperplane. In fact, we can find that most cases in hyperspectral data classification are nonlinear in nature. For this case, one may think of mapping the data through a proper nonlinear transformation $\Phi(\cdot)$ into a higher dimensional feature space $\Phi(\boldsymbol{x}) \in \Re^{d'}$ $(d' > d)$, where a separating hyperplane between the two classes can be looked at for using the similar method described above, i.e., by means of an optimal hyperplane defined by a weighting vector $\boldsymbol{w} \in \Re^{d'}$ and a bias $b \in \Re$. In the higher transformed dimensional space, data are spread out, and a linear separating hyperplane can be constructed based on *Cover's theorem* on the separability patterns (Cover 1965). For example, Figure 5.19a shows that two-class data in the (relatively lower dimensional) input space may not be separated by a linear hyperplane, but a nonlinear hyperplane can make them separable. However, after mapping (via $\Phi(\cdot)$)

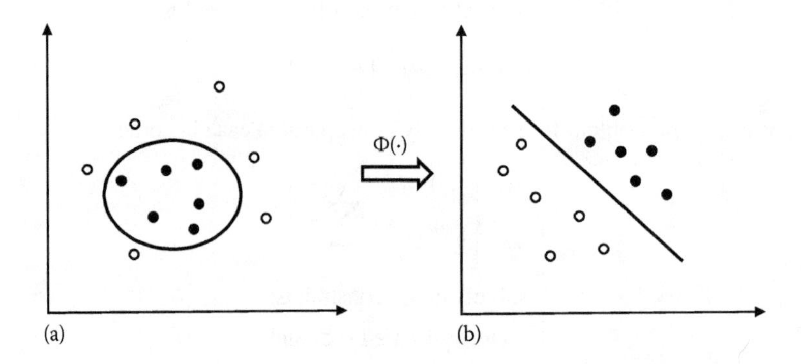

FIGURE 5.19 Representation of the optimal separating hyperplane in a nonlinear case. Mapping nonlinear data (a) to a higher dimension feature space where a linear separating hyperplane can be located (b).

the nonlinear data to a (relatively) higher dimensional feature space, a linear separating hyperplane can be found (Figure 5.19b). For the nonlinear separable data in order to find a linear separating hyperplane as defined in Equation 5.108 for the linearly separable case, one can solve the dual problem by replacing the inner products in the original space $(\boldsymbol{x}_i, \boldsymbol{x}_j)$ with inner products in the transformed space $[\Phi(\boldsymbol{x}_i)\cdot\Phi(\boldsymbol{x}_j)]$. At this point, the main problem consists of the explicit computation of $\Phi(\boldsymbol{x})$, which can prove expensive and at times unfeasible (Melgani and Bruzzone 2004). To deal with this problem, kernel method provides an alternate effective way. The formulation of the kernel function from the inner product is a special case of *Mercer's theorem* (Pal and Watanachaturaporn 2004). Suppose there exists a kernel K such that

$$K(\boldsymbol{x}_i,\boldsymbol{x}_j) = \Phi(\boldsymbol{x}_i)\cdot\Phi(\boldsymbol{x}_j). \tag{5.113}$$

Making use of kernel function allows simplifying the solution of the dual problem considerably, since it avoids the computation of the inner products in the transformed space $[\Phi(\boldsymbol{x}_i)\cdot\Phi(\boldsymbol{x}_j)]$. As a result, the dual optimization problem for a nonlinear case can be expressed as

$$\begin{cases} \max\limits_{\alpha}\left\{\sum_{i=1}^{N}\alpha_i - \frac{1}{2}\sum_{i=1}^{N}\sum_{j=1}^{N}\alpha_i\alpha_j y_i y_j K(\boldsymbol{x}_i,\boldsymbol{x}_j)\right\} \quad \text{subject to} \\ \sum_{i=1}^{N}\alpha_i y_i = 0 \text{ and } 0 \le \alpha_i \le C \quad i = 1,2,\ldots,N. \end{cases} \tag{5.114}$$

Accordingly, in a manner similar to the other two cases above, the dual optimization problem can be solved by using Lagrange multipliers that maximize Equation 5.114. Finally, the corresponding decision rule can be obtained as

$$f(\boldsymbol{x}) = \text{sign}\left\{\sum_{i\in S}\alpha_i y_i K(\boldsymbol{x}_i,\boldsymbol{x}) + b\right\} \tag{5.115}$$

For the kernel-based solution, the selection of a suitable kernel function is essential for a particular problem because the shape of the discriminant function $f(\boldsymbol{x})$ relies on the kind of kernel functions adopted. There are several commonly used kernel functions that are comprised of (i) ~ (v). See the following:

(i) Linear kernel: $\boldsymbol{x}\cdot\boldsymbol{x}_i$;
(ii) Polynomial kernel with order p: $(\boldsymbol{x}\cdot\boldsymbol{x}_i + 1)^p$;
(iii) Radial based function kernel: $\boldsymbol{e}^{\frac{\|x-x_i\|^2}{2\sigma^2}}$;
(iv) Sigmoid kernel: $\tanh(k(\boldsymbol{x}_i, \boldsymbol{x}_j) + \Theta)$; and
(v) Spline kernel: $1 + (\boldsymbol{x}\cdot\boldsymbol{x}_i) + \frac{1}{2}(\boldsymbol{x}\cdot\boldsymbol{x}_i)(\boldsymbol{x}\Lambda\boldsymbol{x}_i) - \frac{1}{6}(\boldsymbol{x}\Lambda\boldsymbol{x}_i)^3$

The performance of sigmoid-, polynomial-, and radial-based function kernels needs the selection of appropriate values of the user-defined parameters (e.g., p, σ, k, and Θ) and the regularization parameter C, which may change from one case to another.

5.8.3.4 SVMs for Multiclass Classification

As discussed in the previous three cases, SVMs are intrinsically developed to solve binary classification problems, where the class labels can be either +1 or −1. However, in general classification of hyperspectral remote sensing data, there are more than two information classes that usually need to be discriminated simultaneously in numerous applications. Therefore, for this case, it is necessary to consider some strategies of combinations of SVMs associated with binary classification problems to address multiclass classification with hyperspectral remote sensing data. There are a number of methods/strategies to generate multiclass SVMs from binary SVMs that have been proposed by researchers (e.g., Melgani and Bruzzone 2004, Bruzzone et al. 2007, Patra and Bruzzone 2012, Sun et al. 2013, Li et al. 2014). The general methods include the parallel-based approach (e.g., one against the rest classification and pairwise classification) and hierarchical tree-based approach (Melgani and Bruzzone 2004, Pal and Watanachaturaporn 2004). Studying and developing multiclass SVMs methods are still a continuing research topic. For a detailed introduction to multiclass SVMs methods/strategies, consult with the relevant references given.

The SVM approach can significantly improve classification accuracy with hyperspectral data. For example, Melgani and Bruzzone (2004) tested four SVM strategies for multiclass discrimination including the one against all, one against one, binary hierarchical tree balanced branches, and binary hierarchical tree one against *all* algorithms. They applied these algorithms to an AVIRIS image acquired over an agricultural area with nine classes and compared their performances with radial basis function neural networks (RBF-NN) and K-nearest neighbor (K-NN) algorithms. They report that greater than 90% overall accuracies by using the multiclass SVMs algorithms can be achieved with an accuracy improvement of 7–12% over the NN and K-NN algorithms. Pal and Mather (2004) and Camps-Valls et al. (2004) used multiclass SVMs methods for hyperspectral data classification compared with NNs and other classifiers. Their results proved that the multiclass SVMs methods outperformed other classifiers. Sun et al. (2013) reported a novel semisupervised method with the pairwise binary SVM classifiers merged as the multiclass classifier for solving the classification problem with hyperspectral data. Their experimental results demonstrated that their proposed method could achieve higher classification accuracy and provided robust solutions for hyperspectral data classification.

5.9 SUMMARY

In this chapter, a relatively complete collection of hyperspectral data analysis and processing techniques and methods were briefly introduced, discussed, and reviewed. In the beginning of this chapter, after discussing the importance and necessity of introducing and reviewing hyperspectral data processing techniques and methods, first-order and second-order spectral derivative analyses were introduced and the sensitivity of derivation spectra to signal-to-noise ratio of hyperspectral data was discussed. To efficiently map and identify surface materials and types, four spectral similarity measures were introduced and discussed in Section 5.3, including three deterministic measures: spectral angle matching (SAM), the Euclidean distance (ED) measure, and the cross-correlogram spectral matching (CCSM) of spectral vectors in the hyperspectral space, and one stochastic measure, spectral information divergence (SID). To extract spectral absorption features (diagnostic features) and some optical position (wavelength) variables from hyperspectral data for estimating and mapping characteristics of geology and vegetation, continuum removal technique, five determination methods of red-edge optical parameters and an approach defining other wavelength position variables using the 10 "slopes" from a hyperspectral spectral curve were reviewed in Section 5.4. The five extraction methods of red-edge optical parameters include the four-point method, fifth-order polynomial fitting, Lagrangian interpolation, invert Gaussian modeling, and linear extrapolation algorithm. In the next section, 82 hyperspectral vegetation indices (HVIs) were defined and summarized in Table 5.1. In Section 5.6, principles and algorithms of the second-order statistics component transforms, data variance-based principal components analysis (PCA), and signal-to-noise ratio (SNR)–based maximum noise fraction transforms: maximum noise fraction (MNF) transform and noise-adjusted principal component (NAPC) transform and independent component analysis (ICA) were introduced and discussed for dimension reduction. In addition, in this section, the canonical discriminant analysis (CDA) and wavelet transform (WT) were discussed for feature extraction from hyperspectral data. In the section for introducing spectral unmixing techniques, algorithms and concepts/principles of the traditional four linear spectral unmixing models—artificial neural networks (ANNs), multiend-member spectral mixture analysis (MESMA), mixture-tuned matched filtering (MTMF), and constrained energy minimization (CEM)—were reviewed and discussed. Two end-member determination techniques, pixel purity index (PPI) and N-finder, were also introduced. In the last section of this chapter, the algorithms and principles/concepts of segment-based traditional classifiers, including segmented PCA, segmented LDA, segmented PDA, segmented CDA, and simplified MLC, and two advanced machine learning techniques, ANNs and support vector machines (SVMs), were introduced and discussed.

REFERENCES

Adams, J. B., M. O. Smith, and A. R. Gillespie. 1989. Simple models for complex natural surfaces: A strategy for the hyperspectral era of remote sensing. *Proceedings of the 1989 International Geoscience and Remote Sensing Symposium*, 16–21. Vancouver, Canada.

Adams, J. B., M. O. Smith, and A. R. Gillespie. 1993. Interpretation based on spectral mixture analysis, in *Remote Geochemical Analysis: Elemental and Mineralogical Composition*, eds. C. M. Pieters and P. Englert, pp. 145–166. New York: Cambridge University Press.

Adams, J. B., M. O., Smith, and P. E. Johnson. 1986. Spectral mixture modeling: A new analysis of rock and soil types at the Viking Lander 1 site. *Journal of Geophysical Research* 91:8098–8112.

Alonzo, M., K. Roth, and D. Roberts. 2013. Identifying Santa Barbara's urban tree species from AVIRIS imagery using canonical discriminant analysis. *Remote Sensing Letters* 4(5):513–521.

Andrew, M.E., and S. L. Ustin. 2008. The role of environmental context in mapping invasive plants with hyperspectral image data. *Remote Sensing of Environment* 112:4301–4317.

Andrew, M.E., and S. L. Ustin. 2010. The effects of temporally variable dispersal and landscape structure on invasive species spread. *Ecological Applications* 20(3):593–608.

Arai, K. 2013. Nonlinear mixing model of mixed pixels in remote sensing satellite images taking into account landscape. *International Journal of Advanced Computer Science and Applications* 4(1):103–109.

Baret, F., and G. Guyot. 1991. Potentials and limits of vegetation indices for LAI and APAR assessment. *Remote Sensing of Environment* 35:161–173.

Barnes, J. D., L. Balaguer, E. Manrique, S. Elvira, and A. W. Davison. 1992. A reappraisal of the use of DMSO for the extraction and determination of chlorophylls a and b in lichens and higher plants. *Environmental and Experimental Botany* 32:85–100.

Bateson, C. A., G. P. Asner, and C. A. Wessman. 2000. End-member bundles: A new approach to incorporating end-member variability into spectral mixture analysis. *IEEE Transactions of Geoscience and Remote Sensing* 38:1083–1094.

Ben-Dor, E., T. Malthus, A. Plaza, and D. Schläpfer. 2013. Hyperspectral remote sensing, in *Airborne Measurements for Environmental Research: Methods and Instruments*, eds. M. Wendisch and J.-L. Brenguier, Wiley-VCH, Weinheim, Germany, pp. 413–456.

Blackburn, G. A. 1998. Quantifying chlorophylls and caroteniods at leaf and canopy scales: An evaluation of some hyperspectral approaches. *Remote Sensing of Environment* 66:273–285.

Boardman, J. W. 1995. Analysis, understanding, and visualization of hyperspectral data as convex sets in n-space. *SPIE Proceedings* 2480:14–22.

Boardman, J. W., F. A. Kruse, and R. O. Green. 1995. Mapping target signatures via partial unmixing of AVIRIS data. *Summaries of the Fifth JPL Airborne Geoscience Workshop* JPL Publication, 95–1, pp. 23–26. Pasadena, CA: NASA Jet Propulsion Laboratory.

Bonhan-Carter, G. F. 1988. Numerical procedures and computer program for fitting an inverted Gaussian model to vegetation reflectance data. *Computational Geosciences* 14(3):339–356.

Bowles, J., P. J. Palmadesso, J. A. Antoniades, M. M. Baumback, and L. J. Rickard. 1995. Use of filter vectors in hyperspectral data analysis. *Proceedings of SPIE* 2553:148–157.

Broge, N. H., and E. Leblanc. 2000. Comparing prediction power and stability of broadband and hyperspectral vegetation indices for estimation of green leaf area index and canopy chlorophyll density. *Remote Sensing of Environment* 76:156–172.

Brown, M., H. G. Lewis, and S. R. Gunn. 2000. Linear spectral mixture models and support vector machines for remote sensing. *IEEE Transactions on Geoscience and Remote Sensing* 38(5):2346–2360.

Bruce, L. M., C. H. Koger, and L. Jiang. 2002. Dimensionality reduction of hyperspectral data using discrete wavelet transform feature extraction. *IEEE Transactions on Geoscience and Remote Sensing* 40:2331–2338.

Bruce, L. M., C. Morgan, and S. Larsen. 2001. Automated detection of subpixel hyperspectral targets with continuous and discrete wavelet. *IEEE Transactions on Geoscience and Remote Sensing* 39:2217–2226.

Bruzzone, L., M. Chi, and M. Marconcini. 2007. Semisupervised support vector machines for classification of hyperspectral remote sensing images, in *Hyperspectral Data Exploitation—Theory and Applications*, ed. C.-I. Chang, pp. 275–311. Hoboken, NJ: John Wiley & Sons, Inc.

Burges, C. J. C. 1998. A tutorial on support vector machines for pattern recognition. *Data Mining and Knowledge Discover* 2(2):121–167.

Camps-Valls, G., L. Gómez-Chova, J. Calpe-Maravilla, J. D. Martín-Guerrero, E. Soria-Olivas, L. Alonso-Chordá, and J. Moreno. 2004. Robust support vector method for hyperspectral data classification and knowledge discovery. *IEEE Transactions on Geoscience and Remote Sensing* 42(7):1530–1542.

Carpenter, G.A., and S. Grossberg. 1995. Adaptive resonance theory (ART), in *The Handbook of Brain Theory and Neural Networks*, ed. M. A. Arbib, pp. 79–82. Cambridge, MA: MIT Press.

Chang, C. I., H. Ren, C. C. Chang, F. D'Amico, and J. Q. Jensen. 2004. Estimation of subpixel target size for remotely sensed imagery. *IEEE Transactions on Geoscience and Remote Sensing* 42(6):1309–1320.

Chang, C.-I. 2000. An information theoretic-based approach to spectral variability, similarity, and discriminability for hyperspectral image analysis. *IEEE Transactions on Information Theory* 46(5):1927–1932.

Chang, C.-I. 2003. *Hyperspectral Imaging: Techniques for Spectral Detection and Classification*. Dordrecht, the Netherlands: Kluwer Academic Publishers.

Chang, C.-I. 2013. *Hyperspectral Data Processing: Algorithm Design and Analysis*. Hoboken, NJ: John Wiley & Sons, Inc.

Chang, C.-I., and Q. Du. 1999. Interference and noise adjusted principal components analysis, *IEEE Trans. on Geoscience and Remote Sensing* 37(5):2387–2396.

Chang, C.-I., and A. Plaza. 2006. A fast iterative algorithm for implementation of pixel purity index. *IEEE Geoscience and Remote Sensing Letters* 3:63–67.

Chappelle, E. W., M. S. Kim, and J. E. McMurtrey, III. 1992. Ratio analysis of reflectance spectra (RARS): An algorithm for the remote estimation of the concentrations of chlorophyll a, chlorophyll b, and carotenoids in soybean leaves. *Remote Sensing of Environment* 39:239–247.

Chen, J. M. 1996. Evaluation of vegetation indices and a modified simple ratio for boreal applications. *Canadian Journal of Remote Sensing* 22:229–242.

Chen, P., D. Haboudane, N. Tremblay, J. Wang, P. Vigneault, and B. Li. 2010. New spectral indicator accessing the efficiency of crop nitrogen treatment in corn and wheat. *Remote Sensing of Environment* 114:1987–1997.

Cho, M. A., and A. K. Skidmore. 2006. A new technique for extracting the red edge position from hyperspectral data: The linear extrapolation method. *Remote Sensing of Environment* 101:181–193.

Clark, R. N., and T. L. Roush. 1984. Reflectance spectroscopy: Quantitative analysis techniques for remote sensing applications. *Journal of Geophysical Research* 89:6329–6340.

Cloutis, E. A. 1996. Hyperspectral geological remote sensing: Evaluation of analytical techniques. *International Journal of Remote Sensing* 17(12):2215–2242.

Cover, T. M. 1965. Geometrical and statistical properties of systems of linear inequalities with applications in pattern recognition. *IEEE Transactions on Electronic Computers* 14:326–334.

Cover, T. M., and J. A. Thomas. 1991. *Elements of Information Theory*. New York: Wiley.

Cristianini, N., and J. Shawe-Tayor. 2000. *An Introduction to Support Vector Machines and Other Kernel-based Learning Methods*. Cambridge, UK: Cambridge University Press.

Dash, J., and P. J. Curran. 2004. The MERIS terrestrial chlorophyll index. *International Journal of Remote Sensing* 25:5003–5013.

Datt, B. 1998. Remote sensing of chlorophyll a, chlorophyll b, chlorophyll a+b, and total carotenoid content in Eucalyptus leaves. *Remote Sensing of Environment* 66:111–121.

Datt, B. 1999. A new reflectance index for remote sensing of chlorophyll content in higher plants: Tests using Eucalyptus leaves. *Journal of Plant Physiology* 154:30–36.

Daughtry, C. S. T., C. L. Walthall, M. S. Kim, E. B. Colstoun, and J. E. McMurtrey. 2000. Estimating corn leaf chlorophyll concentration from leaf and canopy reflectance. *Remote Sensing of Environment* 74:229–239.

Dawson T. P., and P. J. Curran. 1998. A new technique for interpolating the reflectance red edge position. *International Journal of Remote Sensing* 19:2133–2139.

Delalieux, S., B. Somers, S. Hereijgers, W. W. Verstraeten, W. Keulemans, and P. Coppin. 2008. A near-infrared narrow-waveband ratio to determine Leaf Area Index in orchards. *Remote Sensing of Environment* 112:3762–3772.

Delegido, J., L. Alonso, G. González, and J. Moreno. 2010. Estimating chlorophyll content of crops from hyperspectral data using a normalized area over reflectance curve (NAOC). *International Journal Applied Earth Observation and Geoinformation* 12:165–174.

Demetriades-Shah, T. H., M. D. Steven, and J. A. Clark. 1990. High-resolution derivative spectra in remote-sensing. *Remote Sensing of Environment* 33(1):55–64.

Eberhart, R. C., and R. W. Dobbins, eds. 1990. *Neural Network PC Tools: A Practical Guide*. Toronto: Academic Press, Inc.

Eismann, M. 2012. *Hyperspectral Remote Sensing*. Bellingham, WA: SPIE Press.

Eitel, J. U. H., P. E. Gessler, A. M. S. Smith, and R. Robberecht. 2006. Suitability of existing and novel spectral indices to remotely detect water stress in Populus spp. *Forest Ecology and Management* 229(1–3):170–182.

Elvidge, C.D., and Z. Chen. 1995. Comparison of broadband and narrow-band red and near-infrared vegetation indices. *Remote Sensing of Environment* 54:38–48.

Falco, N., J. A. Benediktsson, and L. Bruzzone. 2014. A study on the effectiveness of different independent component analysis algorithms for hyperspectral image classification. *Journal of Selected Topics in Applied Earth Observations and Remote Sensing* 7(6):2183–2199.

Farrand, W. H., and J. C. Harsanyi. 1997. Mapping the distribution of mine tailings in the Coeur d'Alene River Valley, Idaho, through use of a constrained energy minimization technique. *Remote Sensing of Environment* 59:64–76.

Fensholt, R., and I. Sandholt. 2003. Derivation of a shortwave infrared water stress index from MODIS near- and shortwave infrared data in a semiarid environment. *Remote Sensing of Environment* 87(1):111–121.

Flanagan, M., and D. L. Civco. 2001 (April 23–27). Subpixel impervious surface mapping. In *Proceedings of American Society for Photogrammetry and Remote Sensing Annual Convention*. St. Louis, Missouri.

Foody, G. M. 1996. Relating the land-cover composition of mixed pixels to artificial neural network classification. *Photogrammetric Engineering & Remote Sensing* 62:491-499.

Fukuda, S., and H. Hirosawa. 1999. A wavelet-based texture feature set applied to classification of multifrequency polarimetric SAR images. *IEEE Transactions on Geosciences and Remote Sensing* 37: 2282–2286.

Galvão, L. S., A. R. Formaggio, and D. A. Tisot. 2005. Discrimination of sugarcane varieties in Southeastern Brazil with EO-1 Hyperion data. *Remote Sensing of Environment* 94:523–534.

Gamon, J. A., and J. S. Surfus. 1999. Assessing leaf pigment content and activity with a reflectometer. *New Phytologist* 143:105–117.

Gamon, J. A., J. Penuelas, and C. B. Field. 1992. A narrow-waveband spectral index that tracks diurnal changes in photosynthetic efficiency. *Remote Sensing of Environment* 41:35–44.

Gao, B. C. 1996. NDWI—A normalized difference water index for remote sensing of vegetation liquid water from space. *Remote Sensing of Environment* 58:257–266.

Gao, J. 2006. *Canopy Chlorophyll Estimation by Hyperspectral Remote Sensing*. PhD thesis. Kansas State University, Manhattan, Kansas, USA.

Ghiyamat, A., H. Z. M. Shafri, G. A. Mahdiraji, R. Ashurov, A. R. M. Shariff, and S. Mansour. 2015. Airborne hyperspectral discrimination of tree species with different ages using discrete wavelet transform. *International Journal of Remote Sensing* 36(1):318–342.

Gitelson, A. A. 2004. Wide Dynamic Range Vegetation Index for remote quantification of crop biophysical characteristics. *Journal of Plant Physiology* 161:165–173.

Gitelson A. A., and M. N. Merzlyak. 1996. Signature analysis of leaf reflectance spectra: Algorithm development for remote sensing of chlorophyll. *Journal of Plant Physiology* 148:494–500.

Gitelson, A. A., G. P. Keydan, and M. M. Merzlyak. 2006. Three-band model for non-invasive estimation of chlorophyll, carotenoids, and anthocyanin contents in higher plant leaves. *Geophysical Research Letters* 33:L11402. doi: 10.1029/2006GL026457.

Gitelson, A. A., M. N. Merzlyak, and O. B. Chivkunova. 2001. Optical properties and nondestructive estimation of anthocyanin content in plant leaves. *Photochemistry and Photobiology* 74:38–45.

Gitelson, A. A., Y. J. Kaufman, R. Stark, and D. Rundquist. 2002. Novel algorithms for remote estimation of vegetation fraction. *Remote Sensing of Environment* 80:76–87.

Gitelson, A. A., A. Viña, C. Ciganda, D. C. Rundquist, and T. J. Arkebauer. 2005. Remote estimation of canopy chlorophyll content in crops. *Geophysical Research Letters* 32:L08403.

Gong, P., J. R. Miller, and M. Spanner. 1994. Forest canopy closure from classification and spectral unmixing of scene components multisensor evaluation of an open canopy. *IEEE Transactions on Geoscience and Remote Sensing* 32(5):1067–1080.

Gong, P., R. Pu, and R. C. Heald. 2002. Analysis of in situ hyperspectral data for nutrient estimation of giant sequoia. *International Journal of Remote Sensing* 23(9):1827–1850.

Gong, P., R. Pu, and B. Yu. 1997. Conifer species recognition: An exploratory analysis of in situ hyperspectral data. *Remote Sensing of Environment* 62:189–200.

Gong, P., R. Pu, and B. Yu. 2001. Conifer species recognition: Effect of data transformation. *International Journal of Remote Sensing* 22(17):3471–3481.

Gong, P., R. Pu, G.S. Biging, and M. Larrieu. 2003. Estimation of forest leaf area index using vegetation indices derived from Hyperion hyperspectral data. *IEEE Transactions on Geoscience and Remote Sensing* 41(6):1355–1362.

Green, A. A., M. Berman, P. Switzer, and M. D. Craig. 1988. A transformation for ordering multispectral data in terms of image quality with implications for noise removal. *IEEE Transactions on Geoscience and Remote Sensing* 26:65–74.

Guilfoyle, K., M. L., Althouse, and C. Chang. 2001. A quantitative and comparative analysis of linear and nonlinear spectral mixture models using radial basis function neural network. *Remote Sensing of Environment* 39: 2314–2318.

Guyot, G., F. Baret, and S. Jacquemoud. 1992. Imaging spectroscopy for vegetation studies, in *Imaging Spectroscopy: Fundamentals and Prospective Application*, eds. F. Toselli and J. Bodechtel, pp. 145–165. Dordrecht: Kluwer.

Haboudane, D., J. R. Miller, E. Pattery, P. J. Zarco-Tejad, and I. B. Strachan. 2004. Hyperspectral vegetation indices and novel algorithms for predicting green LAI of crop canopies: Modeling and validation in the context of precision agriculture. *Remote Sensing of Environment* 90:337–352.

Haboudane, D., J. R. Miller, N. Tremblay, P. J. Zarco-Tejada, and L. Dextraze. 2002. Integrated narrow-band vegetation indices for prediction of crop chlorophyll content for application to precision agriculture. *Remote Sensing of Environment* 81(2-3):416–426.

Hamada, Y., D. A. Stow, L. L. Coulter, J. C. Jafolla, and L.W. Hendricks. 2007. Detecting Tamarisk species (*Tamarix spp.*) in riparian habitats of Southern California using high spatial resolution hyperspectral imagery. *Remote Sensing of Environment* 109:237–248.

Hardinsky, M. A., V. Lemas, and R. M. Smart. 1983. The influence of soil salinity, growth form, and leaf moisture on the spectral reflectance of Spartina alternifolia canopies. *Photogrammetric Engineering and Remote Sensing* 49:77–83.

He, Y., X. Guo, and J. Wilmshurst. 2006. Studying mixed grassland ecosystems I: Suitable hyperspectral vegetation indices. *Canadian Journal of Remote Sensing* 32(2):98–107.

Heinz, D., and C.-I Chang. 2001. Fully constrained least squares linear mixture analysis for material quantification in hyperspectral imagery. *IEEE Transactions on Geosciences and Remote Sensing* 39: 529–545.

Hestir, E. L., S. Khanna, M. E. Andrew, M. J. Santos, J. H. Viers, J. A. Greenberg, S. S. Rajapakse, and S. L. Ustin. 2008. Identification of invasive vegetation using hyperspectral remote sensing in the California Delta ecosystem. *Remote Sensing of Environment* 112:4034–4047.

Hsu, P.-H., Y.-H. Tseng, and P. Gong. 2002. Dimension reduction of hyperspectral images. *Geographic Information Science* 8:1–8.

Huete, A. R. 1988. A soil adjusted vegetation index (SAVI). *Remote Sensing of Environment* 25:295–309.

Huete, A., K. Didan, T. Miura, E. P. Rodriguez, X. Gao, and L.G. Ferreira. 2002. Overview of the radiometric and biophysical performance of the MODIS vegetation indices. *Remote Sensing of Environment* 83:195–213.

Hughes, G. F. 1968. On the mean accuracy of statistical pattern recognizers. *IEEE Transactions on Information Theory* 14(1):55–63.

Hunt, E. R., and B. N. Rock. 1989. Detection of changes in leaf water content using near and middle-infrared reflectances. *Remote Sensing of Environment* 30: 43–54.

Hyvärinen, A., and E. Oja. 2000. Independent component analysis: Algorithms and applications. *Neural Networks* 13(4–5):411–430.

Hyvärinen, A., J. Karhunen, and E. Oja. 2001. *Independent Component Analysis.* New York: John Wiley & Sons, Inc.

Ichoku, C., and A. Karnieli. 1996. A review of mixture modeling techniques forsub-pixel land cover estimation. *Remote Sensing Reviews* 13:161–186.

Ifarraguerri A., and C.-I. Chang. 1999. Multispectral and hyperspectral image analysis with convex cones. *IEEE Transactions on Geosciences and Remote Sensing* 37:756–770.

Jia, X., and J. A. Richards. 1994. Efficient maximum likelihood classification for imaging spectrometer data sets. *IEEE Transactions on Geoscience and Remote Sensing* 32:274–281.

Jia, X., and J. A. Richards. 1999. Segmented principal components transformation for efficient hyperspectral remote-sensing image display and classification. *IEEE Transactions on Geoscience and Remote Sensing* 37:538–542.

Jia, X., J. A. Richards, and D. E. Ricken. 1999. *Remote Sensing Digital Image Analysis: An Introduction.* Berlin: Springer-Verlag.

Jiang, Z., A. R. Huete, K. Didan, and T. Miura. 2008. Development of a two-band enhanced vegetation index without a blue band. *Remote Sensing of Environment* 112:3833–3845.

Johnson, R. A., and D. W. Wichern. 2002. *Applied Multivariate Statistical Analysis*, 5th ed. Englewood Cliffs, NJ: Prentice Hall.

Jordan, C. F. 1969. Derivation of leaf area index from quality of light on the forest floor. *Ecology* 50:663–666.

Kaufman, Y. J., and D. Tanré. 1992. Atmospherically resistant vegetation index (ARVI) for EOS-MODIS. *Transactions on Geoscience and Remote Sensing* 30(2):261–270.

Keshava, N., and J. F. Mustard. 2002. Spectral unmixing. *IEEE Signal Processing Magazine* 19(1):44–57.

Khattree, R. and D. N. Naik. 2000. *Multivariate Data Reduction and Discrimination with SAS Software.* Cary, NC: SAS Institute Inc.

Kim, M. S., C. S. T. Daughtry, E. W. Chappelle, and J. E. McMurtrey. 1994. The use of high spectral resolution bands for estimating absorbed photosynthetically active radiation (APAR). *Proceedings of The 6th International Symposium on Physical Measurements and Signatures in Remote Sensing*, pp. 299–306. Val d'Isere, France.

Kohonen, T. 2001. Self-organizing maps. *Series in Information Sciences*, 3rd ed. Heidelberg: Springer.

Kong, X., N. Shu, W. Huang, and J. Fu. 2010. The research on effectiveness of spectral similarity measures for hyperspectral image. *IEEE 2010 3rd International Congress on Image and Signal Processing (CISP2010)*, pp. 2269–2273.

Koppe, W., F. Li, M. L. Gnyp, Y. Miao, L. Jia, X. Chen, F. Zhang, and G. Bareth. 2010. Evaluating multispectral and hyperspectral satellite remote sensing data for estimating winter wheat growth parameters at regional scale in the North China Plain. *Photogrammetrie Fernerkundung Geoinformation* 3:167–178.

Kruse, F. A., A. B. Lefkoff, J. W. Boardman, K. B. Heidebrecht, A. T. Shapiro, P. J. Barloon, and A. F. H. Goetz. 1993. The spectral image processing system (SIPS)—Interactive visualization and analysis of imaging spectrometer data. *Remote Sensing of Environment* 44:145–163.

Le Maire, G., C. Francois, and E. Dufrene. 2004. Towards universal broad leaf chlorophyll indices using PROSPECT simulated database and hyperspectral reflectance measurements. *Remote Sensing of Environment* 89:1–28.

Lee, J. B., A. S. Woodyatt, and M. Berman. 1990. Enhancement of high spectral resolution remote sensing data by a noise-adjusted principal components transform. *IEEE Transactions on Geoscience and Remote Sensing* 28(3):295–304.

Li, J. 2004. Wavelet-based feature extraction for improved end-member abundance estimation in linear unmixing of hyperspectral signals. *IEEE Transactions on Geoscience and Remote Sensing* 42(3):644–649.

Li, L., and J. F. Mustard. 2003. Highland contamination in lunar mare soils: Improved mapping with multiple end-member spectral mixture analysis (MESMA). *Journal of Geophysical Research* 108(E6):5053.

Li, W., S. Prasad, and J. E. Fowler. 2014. Decision fusion in kernel-induced spaces for hyperspectral image classification. *IEEE Transactions on Geoscience and Remote Sensing* 52(6):3399–3411.

Mallat, S. G. 1989. A theory for multiresolution signal decomposition: The wavelet representation. *IEEE Transactions on Pattern Analysis and Machine Intelligence* 11:674–693.

Mallat, S. G. 1998. *A Wavelet Tour of Signal Processing.* San Diego, CA: Academic Press.

Mas, J. F., and J. J. Flores. 2008. The application of artificial neural networks to the analysis of remotely sensed data. *International Journal of Remote Sensing* 29(3):617–663.

Melgani, F., and L. Bruzzone. 2004. Classification of hyperspectral remote sensing images with support vector machines. *IEEE Transactions on Geoscience and Remote Sensing* 42(8):1778–1790.

Merton, R., and J. Huntington. 1999 (February 9–11). Early simulation of the ARIES-1 satellite sensor for multi-temporal vegetation research derived from AVIRIS. *Summaries of the Eight JPL Airborne Earth Science Workshop*, JPL Publication 99-17, pp. 299–307. Pasadena, California.

Merzlyak, M. N., A. A. Gitelson, O. B. Chivkunova, and V. Y. Rakitin. 1999. Nondestructive optical detection of pigment changes during leaf senescence and fruit ripening. *Physiologia Plantarum* 106:135–141.

Miao, X., P. Gong, S. Swope, R. Pu, R. Carruthers, G. L. Anderson, J. S. Heaton, and C. R. Tracy. 2006. Estimation of yellow starthistle abundance through CASI-2 hyperspectral imagery using linear spectral mixture models. *Remote Sensing of Environment* 101:329–341.

Miller, J. R., E. W. Hare, and J. Wu. 1990. Quantitative characterization of the vegetation red edge reflectance. 1. An Inverted-Gaussian reflectance model. *International Journal of Remote Sensing* 11:1775–1795.

Miller, J. R., J. Wu, M. G. Boyer, M. Belanger, and E. W. Hare. 1991. Season patterns in leaf reflectance red edge characteristics. *International Journal of Remote Sensing* 12(7):1509–1523.

Nagler, P. L., C. S. T. Daughtry, and S. N. Goward. 2000. Plant litter and soil reflectance. *Remote Sensing of Environment* 71:207–215.

Neville, R. A., K. Staenz, T. Szeredi, J. Lefebvre, and P. Hauff. 1999. Automatic end-member extraction from hyperspectral data for mineral exploration. *Proceedings of the 21st Canadian Symposium on Remote Sensing*. Ottawa, Canada.

Noujdina, N. V., and S. L. Ustin, 2008. Mapping downy brome (Bromus tectorum) using multidate AVIRIS data. *Weed Science* 56:173–179.

Okamoto, H., M. Murata, T. Kataoka, and S. Hata. 2007. Plant classification for weed detection using hyperspectral imaging with wavelet analysis. *Weed Biology and Management* 7:31–37.

Painter, T. H., D. A. Roberts, R. O. Green, and J. Dozier. 1998. The effect of grain size on spectral mixture analysis of snow-covered area from AVIRIS data. *Remote Sensing of Environment* 65:320–332.

Pal, M., and P. M. Mather. 2004. Assessment of the effectiveness of support vector machines for hyperspectral data. *Future Generation Computer Systems* 20:1215–1225.

Pal, M., and P. Watanachaturaporn. 2004. Support vector machines, in *Advanced Image Processing Techniques for Remotely Sensed Hyperspectral Data*, eds. P. K. Varshney and M. K. Aror. Berlin and Heidelberg: Springer-Verlag.

Pao, Y. 1989. *Adaptive Pattern Recognition and Neural Networks*. New York: Addison and Wesley.

Patel, N. K., C. Patnaik, S. Dutta, A. M. Shekh, and A. J. Dave. 2001. Study of crop growth parameters using airborne imaging spectrometer data. *International Journal of Remote Sensing* 22(12):2401–2411.

Patra, S., and L. Bruzzone. 2012. A batch-mode active learning technique based on multiple uncertainty for SVM classifier. *IEEE Geoscience and Remote Sensing Letters* 9(3):497–501.

Peng, Z.K., M. R. Jackson, J. A. Rongong, F.L. Chu, and R. M. Parkin. 2009. On the energy leakage of discrete wavelet transform. *Mechanical Systems and Signal Processing* 23:330–343.

Peñuelas, J., F. Baret, and I. Filella. 1995a. Semi-empirical indices to assess carotenoids/chlorophyll a ratio from leaf spectral reflectance. *Photosynthetica* 31:221–230.

Peñuelas, J., I. Filella, P. Lloret, F. Muñoz, and M. Vilajeliu. 1995b. Reflectance assessment of mite effects on apple trees. *International Journal of Remote Sensing* 16:2727–2733.

Peñuelas, J., J. Piñol, R. Ogaya, and I. Filella. 1997. Estimation of plant water concentration by the reflectance water index WI (R900/R970). *International Journal of Remote Sensing* 18:2869–2875.

Peñuelas, J., J. A. Gamon, A. L. Fredeen, J. Merino, and C. B. Field. 1994. Reflectance indices associated with physiological changes in nitrogen- and water-limited sunflower leaves. *Remote Sensing of Environment* 48:135–146.

Pittner, S., and S. V. Kamarthi. 1999. Feature extraction from wavelet coefficients for pattern recognition tasks. *IEEE Transactions on Pattern Analysis and Machine Intelligence* 21:83–88.

Plaza, A., and C. Chang, eds. 2007. *High Performance Computing in Remote Sensing*. Boca Raton, FL: Taylor & Francis Publishing Group/Chapman & Hall/CRC Press.

Plaza, A., P. Martinez, R. Perez, and J. Plaza. 2002. Spatial/spectral end-member extraction by multidimensional morphological operations. *IEEE Transactions on Geosciences and Remote Sensing* 40: 2025–2041.

Plaza, A., P. Martinez, R. Perez, and J. Plaza. 2004. A quantitative and comparative analysis of end-member extraction algorithms from hyperspectral data. *IEEE Transactions on Geoscience and Remote Sensing* 42:650–663.

Plaza, A., A. Benediktsson, J. Boardamn, J. Brazile, L. Bruzzone, G. CapsValls, J. Chanussot, M. Fauvel, P. Gampa, J. Gulatiri, M. Marconcini, J. Tilton, and G. Trianni. 2009. Recent advances in techniques for hyperspectral image processing. *Remote Sensing of Environment* 113:110–122.

Pu, R. 2009. Broadleaf species recognition with in situ hyperspectral data. *International Journal of Remote Sensing* 30(11):2759–2779.

Pu, R. 2012. Comparing canonical correlation analysis with partial least square regression in estimating forest leaf area index with multitemporal Landsat TM imagery. *Geoscience & Remote Sensing* 49(1):92–116.

Pu, R., and P. Gong. 2004. Wavelet transform applied to EO-1 hyperspectral data for forest LAI and crown closure mapping. *Remote Sensing of Environment* 91:212–224.

Pu, R., and P. Gong. 2011. Hyperspectral remote sensing of vegetation bioparameters, in *Advances in Environmental Remote Sensing: Sensors, Algorithm, and Applications*, series ed. Q. Weng, pp. 101–142. Boca Raton, FL: CRC Press/Taylor & Francis Publishing Group.

Pu, R., and D. Liu. 2011. Segmented canonical discriminant analysis of in situ hyperspectral data for identifying thirteen urban tree species. *International Journal of Remote Sensing* 32(8):2207–2226.
Pu, R., L. Foschi, and P. Gong. 2004. Spectral feature analysis for assessment of water status and health level of coast live oak (*Quercus Agrifolia*) leaves. *International Journal of Remote Sensing* 25(20):4267–4286.
Pu, R., S. Ge, N. M. Kelly, and P. Gong. 2003. Spectral absorption features as indicators of water status in *Quercus Agrifolia* leaves. *International Journal of Remote Sensing* 24(9):1799–1810.
Pu, R., P. Gong, G. S. Biging, and M. R. Larrieu. 2003. Extraction of red edge optical parameters from Hyperion data for estimation of forest leaf area index. *IEEE Transactions on Geoscience and Remote Sensing* 41(4):916–921.
Pu, R., P. Gong, R. Michishita, and T. Sasagawa. 2008. Spectral mixture analysis for mapping abundance of urban surface components from the Terra/ASTER data. *Remote Sensing of Environment* 112:939–954.
Qi, J., A. Chehbouni, A. R. Huete, Y. H. Kerr, and S. Sorooshian. 1994. A modified soil adjusted vegetation index. *Remote Sensing of Environment* 48:119–126.
Rama Rao, N., P. K. Garg, S.K. Ghosh, and V. K. Dadhwal. 2008. Estimation of leaf total chlorophyll and nitrogen concentrations using hyperspectral satellite imagery. *Journal of Agricultural Science* 146:65–75.
Resmini, R. G., M. E. Kappus, W. S. Aldrich, J. C. Harsanyi, and M. Anderson. 1997. Mineral mapping with HYperspectral Digital Imagery Collection Experiment (HYDICE) sensor data at Cuprite, Nevada, USA. *International Journal of Remote Sensing* 18(7):1553–1570.
Rinaldi, M., A. Castrignanò, D. De Benedetto, D. Sollitto, S. Ruggieri, P. Garofalo, F. Santoro, B. Figorito, S. Gualano, and R. Tamborrino. 2015. Discrimination of tomato plants under different irrigation regimes: Analysis of hyperspectral sensor data. *Environmetrics* 26:77–88.
Rioul, O., and M. Vetterli. 1991. Wavelet and signal processing. *IEEE Signal Processing Magazine* 8:14–38.
Roberts, D. A., M. Gardner, R. Church, S. Ustin, G. Scheer, and R. O. Green. 1998. Mapping Chaparral in the Santa Monica Mountains using multiple end-member spectral mixture models. *Remote Sensing of Environment* 65:267–279.
Roessner, S., K. Segl, U. Heiden, and H. Kaufmann. 2001. Automated differentiation of urban surfaces based on airborne hyperspectral imagery. *IEEE Transactions on Geoscience and Remote Sensing* 39(7):1525–1532.
Roger, R. E. 1994. A fast way to compute the noise adjusted principal components transform matrix. *IEEE Transactions on Geosciences and Remote Sensing* 32(1):1194–1196.
Roger, R. E. 1996. Principal components transform with simple, automatic noise adjustment. *International Journal Remote Sensing* 17(14):2719–2727.
Roger, R. E., and J. F. Arnold. 1996. Reliably estimating the noise in AVIRIS hyperspectral imagers. *International Journal of Remote Sensing* 17(10):1951–1962.
Rondeaux, G., Steven, M., and Baret, F. 1996. Optimization of soil-adjusted vegetation indices. *Remote Sensing of Environment* 55:95–107.
Roujean, J. L., and F. M. Breon. 1995. Estimating PAR absorbed by vegetation from bidirectional reflectance measurements. *Remote Sensing of Environment* 51(3): 375–384.
Rouse, J. W., R. H. Haas, J. A. Schell, and D. W. Deering. 1973. Monitoring vegetation systems in the Great Plains with ERTS. *Proceedings, Third ERTS Symposium* 1:48–62.
Rumelhart, D. E., G. E. Hinton, and R. J. Williams. 1986. Learning internal representations by error propagation. *Parallel Distributed Processing-Explorations in the Microstructure of Cognition* 1:318–362. MIT Press, Cambridge, Massachusetts.
Sánchez, S., and A. Plaza. 2014. Fast determination of the number of end-members for real-time hyperspectral unmixing on GPUs. *Journal of Real-Time Image Processing* 9:397–405.
Sasaki, K., S. Kawata, and S. Minami. 1984. Estimation of component spectral curves from unknown mixture spectra. *Applied Optics* 23:1955–1959.
Schaepman, M., S. Ustin, A. Plaza, T. Painter, J. Verrelst, and S. Liang. 2009. Earth system science related imaging spectroscopy–An assessment. *Remote Sensing of Environment* 113:123–137.
Schlerf, M., C. Atzberger, and J. Hill. 2005. Remote sensing of forest biophysical variables using HyMap imaging spectrometer data. *Remote Sensing of Environment* 95:177–194.
Serrano, L., J. Peñuelas, and S. L. Ustin. 2002. Remote sensing of nitrogen and lignin in Mediterranean vegetation from AVIRIS data: Decomposing biochemical from structural signals. *Remote Sensing of Environment* 81:355–364.
Simhadri, K. K., S. S. Iyengar, R. J. Holyer, M. Lybanon, and J. M. Zachary. 1998. Wavelet-based feature extraction from oceanographic images. *IEEE Transactions on Geoscience and Remote Sensing* 36:767–778.
Sims, D. A., and J. A. Gamon. 2002. Relationships between leaf pigment content and spectral reflectance across a wide range of species, leaf structures, and developmental stages. *Remote Sensing of Environment* 81:337–354.

Smith, M. O., S. L. Ustin, J. B. Adams, and A. R. Gillespie. 1990. Vegetation in deserts: 1. A regional measure of abundance from multispectral images. *Remote Sensing of Environment* 31(1):1–26.
Sohn, Y., and R. W. McCoy. 1997. Mapping desert shrub rangeland using spectral unmixing and modeling spectral mixtures with TM data. *Photogrammetric Engineering and Remote Sensing* 63:707–716.
Sun, Z., C. Wang, H. Wang, and J. Li. 2013. Learn multiple-kernel SVMs for domain adaptation in hyperspectral data. *IEEE Geoscience and Remote Sensing Letters* 10(5):1224–1228.
Tompkins, S., J. F. Mustard, C. M. Pieters, and D. W. Forsyth. 1997. Optimization of end-members for spectral mixture analysis. *Remote Sensing of Environment* 59:472–489.
Tsai, F., and W. Philpot. 1998. Derivative analysis of hyperspectral data. *Remote Sensing of Environment* 66(1):41–51.
van Aardt, J. A. N., and R. H. Wynne. 2007. Examining pine spectral separability using hyperspectral data from an airborne sensor: An extension of field-based results. *International Journal of Remote Sensing* 28(2):431–436.
van den Berg, A. K., and T. D. Perkins. 2005. Non-destructive estimation of anthocyanin content in autumn sugar maple leaves. *Horticultural Science* 40(3):685–686.
van der Meer, F., and W. Bakker. 1997a. Cross correlogram spectral matching: Application to surface mineralogical mapping using AVIRIS data from Cuprite, Nevada. *Remote Sensing of Environment* 61(3): 371–382.
van der Meer, F., and W. Bakker. 1997b. CCSM: Cross correlogram spectral matching. *International Journal of Remote Sensing* 18(5):1197–1201.
van der Meer, F. 2006. The effectiveness of spectral similarity measures for the analysis of hyperspectral imagery. *International Journal of Applied Earth Observation and Geoinformation* 8(1):3–17.
van der Meer, F. D., S. M. de Jong, and W. Bakker. 2001. Imaging spectrometry: Basic analytical techniques, in *Imaging Spectrometry: Basic Principles and Prospective Applications*, eds. F. D. van der Meer and S. M. de. Jong, pp. 17–62. Springer, The Netherlands.
Vapnik, V. N. 1998. *Statistical Learning Theory.* New York: Wiley.
Veganzones, M. A., and M. Graña. 2008. End-member extraction methods: A short review, in *Knowledge-Based Intelligent Information and Engineering Systems*, eds. I. Lovrek, R. J. Howlett, and C. Jain, 12th International Conference, *Proceedings*, Part III, pp. 400–407. Zagreb, Croatia. Berlin, Heidelberg: Springer-Verlag.
Vincini, M., E. Frazzi, and P. D'Alessio. 2006. Angular dependence of maize and sugar beet VIs from directional CHRIS/PROBA data. *4th ESA CHRIS PROBA Workshop*, ESRIN, pp. 19–21. Frascati, Italy.
Vogelmann, J. E., B. N. Rock, and D. M. Moss. 1993. Red edge spectral measurements from sugar maple leaves. *International Journal of Remote Sensing* 14:1563–1575.
Walsh, S. J., A. L. McCleary, C. F. Mena, Y. Shao, J. P. Tuttle, A. González, and R. Atkinson. 2008. QuickBird and Hyperion data analysis of an invasive plant species in the Galapagos Islands of Ecuador: Implications for control and land use management. *Remote Sensing of Environment* 112:1927–1941.
Winter, M. E. 1999. N-FINDR: An algorithm for fast autonomous spectral end-member determination in hyperspectral data. *Proceedings of SPIE* 3753:266–275.
Winter, M. E., and E. Winter. 2000. Comparison of approaches for determining end-member in hyperspectral data. In *Proceedings of IEEE International Aerospace Conference*, pp. 305–313.
Xiao, H., X. Zhang, and Y. Du. 2008. A comparison of neural network, rough sets and support vector machine on remote sensing image classification. *7th WSEAS International Conference on Applied Computer and Applied Computational Science (ACACOS '08)*, pp. 597–603. Hangzhou, China.
Xu, B., and P. Gong. 2007. Land use/cover classification with multispectral and hyperspectral EO-1 data. *Photogrammetric Engineering and Remote Sensing* 73(8):955–965.
Yu, B., M. Ostland, P. Gong, and R. Pu. 1999. Penalized linear discriminant analysis for conifer species recognition. *IEEE Transactions on Geoscience and Remote Sensing* 37(5):2569–2577.
Zarco-Tejada, P. J., C. A. Rueda, and S. L. Ustin. 2003. Water content estimation in vegetation with MODIS reflectance data and model inversion methods. *Remote Sensing of Environment* 85:109–124.
Zarco-Tejada, P. J., J. R. Miller, T. L. Noland, G. H. Mohammed, and P. H. Sampson. 2001. Scaling-up and model inversion methods with narrowband optical indices for chlorophyll content estimation in closed forest canopies with hyperspectral data. *IEEE Transactions on Geoscience and Remote Sensing* 39(7):1491–1507.
Zarco-Tejada, P.J., A. Berjón, R. López-Lozano, J. R. Miller, P. Martín, V. Cachorro, M. R. González, and A. Frutos. 2005. Assessing vineyard condition with hyperspectral indices: leaf and canopy reflectance simulation in a row-structured discontinuous canopy. *Remote Sensing of Environment* 99:271–287.

Zhang, J., B. Rivard, A. Sanchez-Azofeifa, and K. Casto-Esau. 2006. Intra- and interclass spectral variability of tropical tree species at La Selva, Costa Rica: Implications for species identification using HYDICE imagery. *Remote Sensing of Environment* 105:129–141.

Zhang, L., D. Li, Q. Tong, and L. Zheng. 1998. Study of the spectral mixture model of soil and vegetation in Poyang Lake area, China. *International Journal of Remote Sensing* 19:2077–2084.

Zhao, G., and A. L. Maclean. 2000. A comparison of canonical discriminant analysis and principal component analysis for spectral transformation. *Photogrammetric Engineering and Remote Sensing* 66(7):841–847.

6 Hyperspectral Data Processing Software

Existing software and image processing packages are specially designed and produced for processing and analyzing multispectral data. When they are used to process hyperspectral data, they are less effective or do not work at all due to the significant characteristics of hyperspectral data, such as huge data volume, high dimensionality, and high correlation among adjacent bands. To increase hyperspectral data processing efficiency and improve the processed results, over the course of the last three decades computer software and tool developers have specifically designed and produced software tools and systems for processing and analyzing hyperspectral data. For user convenience, familiarity, and ability to find hyperspectral processing software and tools/systems currently available in the market, this chapter will introduce the main functionalities and features of a collection of major and minor software tools, programs, and systems for processing and analyzing various hyperspectral datasets.

6.1 INTRODUCTION

Given the fact that hyperspectral data usually characterize huge data volume, high dimensionality, and high correlation among adjacent bands, it is necessary to develop software tools specially for processing hyperspectral data such that the rich and interesting information usually hiding in the data can be easily visualized, extracted, interpreted, and utilized. During the last three decades, numerous software tools, systems, and even packages have been developed for multi-/hyperspectral data analysis. The software tools and systems span the realm of commercial, government, and academic developers (Boardman et al. 2006). Among them, some software and tools developed are for commercial benefit, and the others are public domain and free to be used, but both are for purposes of effective research and applications by using various hyperspectral data collected from laboratory, field, airborne and spaceborne platforms. Some software tools and systems are built in large image and spatial data processing packages (e.g., Exelis Visual Information Solutions, Inc. owned ENVI and Hexagon Geospatial owned ERDAS IMAGINE) as an important component of the packages, and the others are designed and developed as a standalone toolbox/programs for mostly processing and analyzing hyperspectral data (e.g., Spectral Evolution owned DARWin and USGS owned PRISM). Most of the software tools and systems reviewed in this chapter are actively updated and available to access, and some as historically important legacy systems are taken over by other tools/systems using different name by either enhancing existing functionalities and features or adding additional functions and features to the legacy systems. Based on accessible materials, documents, and communications with relevant developers and providers, the major functionalities and features for a specific toolbox/system introduced and reviewed here may not fully reflect the actual capabilities of various software tools and systems for processing and analyzing hyperspectral data. If this is the case, readers should rely on instructions and specifications of software tools and products or directly contact providers or developers.

In this chapter, we focus on an introduction and review of important functions and features of major and minor software tools and systems, rather than the principles and software algorithms themselves for which readers may consult relevant contents introduced in Chapters 4 and 5 as well as other references. The software tools and systems reviewed and introduced in this chapter represent a span of the entire history of relevant software tools/systems development. In the following, a collection of five major software tools/systems (i.e., ENVI, ERDAS, IDRISI, PCI

Geomatics, and TNTmips) in Sections 6.2–6.6 and a collection of twelve minor software tools/programs (i.e., DARWin, HIPAS, ISDAS, ISIS, MATLAB, MuiltSpec, ORASIS, SPECMIN, SPECPR, Tetracorder, and TSG) in Section 6.7, both in alphabetical order of products' names, are introduced and reviewed.

6.2 ENVI

The ENvironment for Visualizing Imagery (ENVI) is a commercial remote sensing image processing software package that was initially focused on processing and analyzing hyperspectral data by using unique and powerful tools/modules (Boardman et al. 2006). ENVI was originally developed by Research Systems Inc. (RSI) in 1977 and is now owned by Exelis Visual Information Solutions Inc. (http://www.exelisvis.com/). ENVI was successful in providing a powerful yet easy-to-use tools/software system for processing and analysis of imaging spectroscopy data. ENVI is now a more widely based and general remote sensing image processing package, having grown beyond its early hyperspectral focus, including various remote sensing data (e.g., multispectral imagery and Radar imagery). Functionally a set of tools/modules for processing and exploring hyperspectral data in ENVI covers the visualization, processing, understanding, and analysis of different hyperspectral data (*in situ*, airborne and spaceborne hyperspectral sensors). In this section, based on ENVI 5.1, the major functions and features of the tools specially designed for processing and analyzing hyperspectral data are grouped and briefly introduced. These tools/features can be easily located from Spectral Toolbox either from the ENVI Image Window Views (Figure 6.1) or from the pull-down menu from ENVI Classic main bar.

6.2.1 Atmospheric Correction

Since correction of atmospheric effects is a critical preprocessing step in analyzing images of surface reflectance, especially analyzing hyperspectral imagery, ENVI provides an atmospheric correction module that has two atmospheric correction modeling tools for retrieving surface spectral reflectance from multispectral and hyperspectral radiance images: Quick Atmospheric Correction (QUAC) and Fast Line-of-Sight Atmospheric Analysis of Spectral Hypercubes (FLAASH®). With the atmospheric correction module (needing a separate license), a user can accurately compensate for atmospheric effects. Since direct measurements of atmospheric properties such as the amount of water vapor, distribution of aerosols, and scene visibility are rarely available, they must be inferred from the image pixels. Hyperspectral images in particular provide enough spectral information within a pixel to independently measure atmospheric water vapor absorption bands (Exelis 2015). Such atmospheric properties are then used to run both QUAC and FLAASH atmospheric correction models to retrieve surface reflectance from multispectral and hyperspectral radiance images.

FLAASH is a first-principles atmospheric correction tool that corrects wavelengths in the visible through near-infrared and shortwave infrared (VNIR–SWIR) regions, up to 3 μm. FLAASH performs with most hyperspectral and multispectral sensors. Usually it is possible to retrieve water vapor and aerosol amount because hyperspectral images contain bands in appropriate wavelength positions. FLAASH can work well for images collected in either vertical (nadir) or slant-viewing geometries. Similar to FLAASH, QUAC also is a VNIR–SWIR atmospheric correction method for multispectral and hyperspectral imagery. QUAC compensates for atmospheric effects relying on atmospheric property parameters estimated directly from the information contained within the scene (observed pixel spectra), without other information. Compared to FLAASH or other physics-based first-principles methods, QUAC performs a more approximate atmospheric correction and generally produces reflectance spectra within approximately +/–15% of the physics-based approaches (Exelis 2015). QUAC also works for any view or solar elevation angles. For example, if a sensor does not have a proper radiometric or wavelength calibration, or the solar illumination intensity is unknown, the model still allows the retrieval of reasonably accurate reflectance spectra

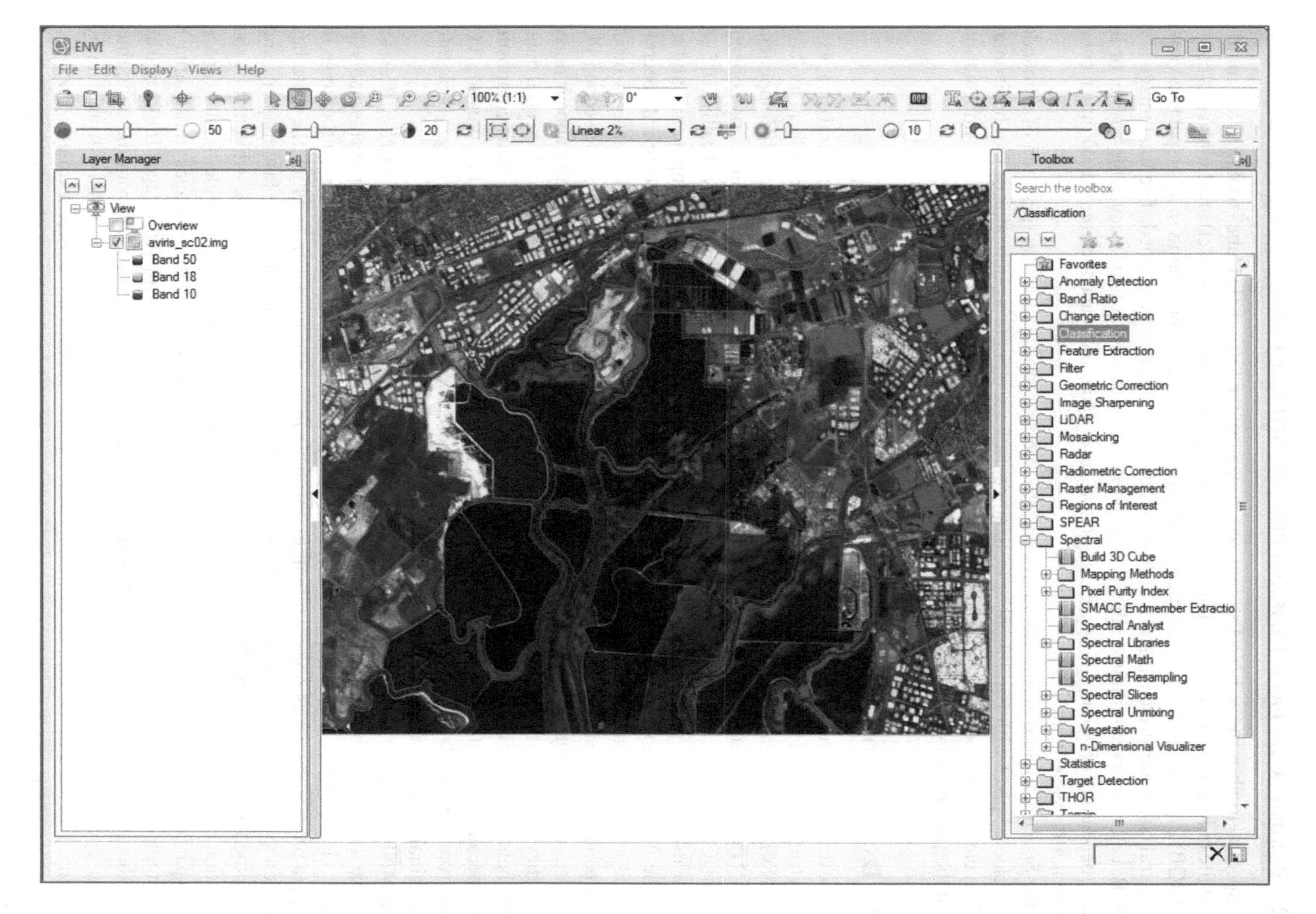

FIGURE 6.1 Image Window views of ENVI (Version 5.1) with Toolbox/Spectral in the left to list most tools/modules used for processing and analyzing hyperspectral data. (Copyright Exelis Visual Information Solutions.)

(Exelis 2015). For substantial operational procedure to perform the atmospheric correction with the two models, see the corresponding user's manual.

6.2.2 Building a 3D Image Cube and Plotting Spectral Curve

In the Spectral toolbox in the ENVI system, a user can use the "Build 3D Cube" tool to take a multispectral file or hyperspectral file (which may be spatially and spectrally subset) and build a red-blue-green (RGB) image with the spectral slice of the top row and far-right column in a perspective view. The spectral slices are stretched and a user-selected color table can be applied. The final image is a 3D, RGB, color-composite image cube. All of the intermediate processing occurs file-to-file, or in memory, depending on your preferences. Therefore, user can either save the created 3D image cube or delete it after viewing it.

In the Spectral toolbox in the ENVI system, the spectral profile plots the spectrum of all bands for the selected pixel with band number or band central wavelength as indicator of plot horizontal axis. You can extract spectra from any multi-/hyperspectral dataset. ENVI uses header information to scale the plot. ENVI plots a spectral profile from each layer.

6.2.3 Data Transformation

To reduce the dimensionality of hyperspectral data and extract features from hyperspectral data, various data transform and feature extraction techniques/algorithms are used. In the Transform toolbox of the ENVI system, three popular data transform techniques are provided: Principal Component Analysis (PCA), Minimum Noise Fraction Transform (MNF), and Independent Components Analysis (ICA), which can be used to transform both multi-/hyperspectral imagery (Exelis 2015).

A user uses PCA to produce uncorrelated output bands, to segregate noise components, and to reduce the dimensionality of data sets. PC bands are linear combinations of the original spectral bands and are uncorrelated. The user can calculate the same number of output PC bands as input spectral bands. The first PC band contains the largest percentage of data variance, the second PC band contains the second-largest data variance, and so on. The last PC bands usually appear noisy because they contain very little variance, much of which is due to noise in the original spectral data. The PC bands may produce more colorful color composite images than spectral color composite images because the data is uncorrelated. ENVI Classic can complete forward (i.e., creating PC images from the raw data space) and inverse PC rotations (i.e., transforming PC images back into their original data space).

The MNF transform, as modified from Green et al. (1988) and implemented in ENVI Classic, is a linear transformation that consists of the following separate principal components analysis rotations: (1) Using the PCs of the noise covariance matrix to decorrelate and rescale the noise in the data (a process known as noise whitening), resulting in transformed data in which the noise has unit variance and no band-to-band correlations (see detailed introduction in the last chapter); and (2) using the PCs derived from the original image data after they have been noise-whitened by the first rotation and rescaled by the noise standard deviation. Here the user can divide the data space into two parts: one part associated with large eigenvalues and coherent eigenimages (most signal), and a complementary part with near-unity eigenvalues and noise-dominated images (most noise). Usually the user only uses the coherent portions and avoids the noise images from the data, thus improving spectral processing results. The MNF tool can help the user remove noise from data by performing a forward transform, determining which bands contain the coherent images, and also run an inverse MNF transform using a spectral subset to include only the good bands, or smoothing the noisy bands before the inverse.

With the transform tool, ICA transformation can be applied to multispectral or hyperspectral datasets to transform a set of mixed, random signals into components that are mutually independent (Exelis 2015). It serves as a tool for blind source separation, where no prior information on the mixing is available. As described in the last chapter, ICA transformation is based on the non-Gaussian assumption of the independent sources, and uses higher-order statistics to reveal interesting features in typically non-Gaussian hyperspectral datasets. ICA transformation can distinguish features of interest even when they occupy only a small portion of the pixels in the image. The applications of ICA transformation in remote sensing in the ENVI system include dimension reduction, extraction of characteristics of the image, anomaly and target detection, feature separation, classification, end-member extraction, noise reduction, and mapping. In the ENVI system, the user can use the ICA transformation tool to calculate forward IC rotations and inverse IC rotations similar to forward PC rotations and inverse PC rotations. The forward transforms can be achieved by calculating new statistics, by using a previously calculated statistics file, or by using an existing transform (Exelis 2015). The ICA transform tool also allows a user to have an option to reduce the dimension of IC bands, and to generate output of only the IC images that a user needs. ICA can also be executed by using a SPEAR tool: SPEAR Independent Components Analysis in the ENVI system.

6.2.4 End-Member Determination and Extraction

With the ENVI system, end-member determination and extraction can be performed by using MNF images created with MNF transformation for running the Pixel Purity Index (PPI) and further performing n-D visualizer tools in the Spectral toolbox. The PPI tool is used to find the most spectrally pure (extreme) pixels in a multispectral or hyperspectral image. The most spectrally pure pixels typically correspond to mixing end-members. The PPI typically runs on an MNF transform result, the images without the noise. The results of the PPI are usually used as input into the n-D Visualizer to further determine end-member pixels (spectra). The PPI is computed by repeatedly projecting n-D scatterplots on a random unit vector. The tool records the extreme pixels in each projection (those pixels that fall onto the ends of the unit vector) and it notes the total number of times each pixel is marked as extreme (most possible as an end-member pixel/spectrum). A pixel purity image (a new output band) is created where each pixel value corresponds to the number of times that pixel was recorded as extreme. In the PPI image, the higher values indicate pixels that are nearer to corners of the n-D data cloud, and thus are relatively purer than pixels with lower values. Pixels with values of 0 were never found to be extreme. Users can choose a disk-based PPI method or a FAST PPI method in ENVI system. The FAST PPI method places the image data into memory and performs the computations in memory, which is much faster than the disk-based PPI method, but requires adequate memory (Exelis 2015). When using FAST PPI, you also have the option to create a new output file or to add to an existing output band (i.e., PPI image).

The n-D Visualizer is an interactive tool in the ENVI system and is used to locate, identify, and cluster the purest pixels and the most extreme spectral responses (end-members) in a dataset in n-D space. The n-D Visualizer tool is designed to help a user visualize the shape of a data cloud that results from plotting image data in spectral space (with image bands as plot axes). The user typically uses the n-D Visualizer with spatially subset MNF images that use only the purest pixels determined from the PPI. A user can use the distribution of these points in n-D space to estimate the number of spectral end-members and their pure spectral signatures. By using the n-D Visualizer, the user can interactively rotate data in n-D space, select groups of pixels into classes, and collapse classes to make additional class selections easier. Then the user can export the selected classes to ROIs and use them as input into classification, and linear spectral unmixing (LSU), etc. The user can view reflectance spectra for specific end-members while the user is selecting end-members and rotating the scatterplot. This allows the user to preview spectra before finalizing spectral classes.

In the Spectral toolbox in the ENVI system, the Spectral Hourglass Wizard is used to guide users step by step through the ENVI Classic hourglass processing flow to find and map image spectral end-members from hyperspectral or multispectral data. The Wizard displays detailed instructions and useful information for each function. The hourglass processing flow uses the spectrally overdetermined nature of hyperspectral data to find the most spectrally pure, or spectrally unique, pixels (end-members) within the dataset and to map their locations and sub-pixel abundances. Each step in the Wizard executes a standalone ENVI Classic function available from the Spectral toolbox. The Sequential Maximum Angle Convex Cone (SMACC) spectral tool is also provided in Spectral toolbox in the ENVI system, which helps find spectral end-members and their abundances throughout an image. This tool is designed for use with previously calibrated hyperspectral data. In comparison to ENVI Classic's Spectral Hourglass Wizard, SMACC provides a faster and more automated method for finding spectral end-members, but it is more approximate and yields less precision (Exelis 2015).

6.2.5 Spectral Unmixing

In the Spectral toolbox, spectral unmixing tools are grouped in Mapping Tools in the ENVI classic system. The ENVI spectral unmixing tools include full end-members unmixing (LSU) and spatial or few end-members unmixing (Matched Filtering or Mixture-Tuned Matched Filtering). Using a LSU tool to determine the relative abundance of materials that are depicted in multispectral or hyperspectral imagery is based on the spectral characteristics of the materials. The reflectance at each pixel of the image is assumed to be a linear combination of the reflectance of each material (or end-member) present within the pixel. The LSU tool in the ENVI system has two constraint options: unconstrained or partially constrained unmixing. In the unconstrained method, abundances may assume negative values and are not constrained to sum-to-unity (one). ENVI Classic also supports an optional, variable-weight, unit-sum constraint in the linear-mixing algorithm. This allows users to define the weight of a sum-to-unity constraint on the abundance fractions. Larger weights in relation to the variance of the data cause the unmixing to honor the unit-sum constraint more closely. To strictly honor the constraint, the weight should be many times the spectral variance of the data (Exelis 2015). LSU tool also permits proper unmixing of MNF transform data, with zero-mean bands.

If not all end-members are known or if a user is only interested in mapping a few end-members, the ENVI system also provides mixture-tuned matched filtering (MTMF) or matched filtering (MF), and constrained energy minimization (CEM) spectral unmixing tools. In the Spectral toolbox, the user can use MTMF or MF to find the abundances of user-defined end-members using a partial unmixing. The MF technique maximizes the response of the known end-member and suppresses the response of the composite unknown background, thus matching the known signature. It provides a rapid means of detecting specific materials based on matches to library or image end-member spectra and does not require knowledge of all the end-members within an image scene. The MF technique may find some false positives for rare materials. The CEM technique is similar to MF in that the only required knowledge is the target spectra to be mapped. Using a specific constraint, CEM uses a finite impulse response (FIR) filter to pass through the desired target while minimizing its output energy resulting from a background other than the desired targets. A correlation or covariance matrix is used to characterize the composite unknown and undesired background. From a mathematical point of view, MF is a mean-centered version of CEM, where the data mean is subtracted from all pixel vectors (Exelis 2015).

6.2.6 Target Detection

In the Spectral toolbox, the ENVI system provides several target detection tools. The Target Detection Wizard guides users through the process to find targets in hyperspectral or multispectral imagery. The targets may be a material or mineral of interest (e.g., mineral alunite) or they may be man-made objects (e.g., military vehicles). Target detection is the process of searching an image for spectra that appear to be a match for a set of spectra from known targets. A detection algorithm can be designed to detect full pixel or subpixel targets. In the ENVI system, the adaptive coherence estimator (ACE) detection algorithm is the mathematically optimal detection algorithm for subpixel targets and is often ranked as a top detection algorithm in comparative tests. The THOR target detection workflow may be used to locate objects within an image that match the signatures of in-scene regions. THOR uses a variety of target detection algorithms to search for targets. After the targets are detected, the workflow will guide a user through combining the rule images from each target detection algorithm into a single detection overlay for each target. The THOR dialog contains help information and allows the user to move forward and back through the workflow. It contains advanced options which a user won't use in this workflow, but the user can read about them in the pull-out panel by clicking the black arrow on the dialog's right side. The Spectral toolbox also provides users with a Finding Targets with BandMax tool. For example, the SAM Target Finder with the BandMax Wizard guides a user through a step-by-step process to find targets in hyperspectral images using the SAM classification and the BandMax algorithm. The BandMax part of the Wizard increases classification accuracy by determining an optimal subset of bands to help a user separate the user's targets from known background materials.

The ENVI system in Spectra toolbox provides an RX Anomaly Detection function that uses the Reed–Xiaoli Detector (RXD) algorithm to detect the spectral or color differences between a region to be tested and its neighboring pixels or the entire dataset (Exelis 2015). This algorithm locates targets that are spectrally distinct from the image background. Results from RXD analysis are unambiguous and have proven very effective in detecting subtle spectral features. ENVI Classic implements the standard RXD algorithm through THOR and SPEAR tools. The THOR and SPEAR anomaly detection workflows/tools are used to locate objects within an image that are spectrally different from the background. After THOR and SPEAR detect the anomalies, the workflows/wizards will guide a user through reviewing each anomaly to determine whether it is of interest or not. The user can also attempt to identify the anomaly by using the Material Identification tool.

6.2.7 Mapping and Discriminant Methods

ENVI provides a variety of spectral mapping methods whose success depends on the data type and quality, and the desired results (Exelis 2015). In addition to the spectral unmixing mapping methods (i.e., LSU, MF/MTMF, and CEM) introduced in Section 6.2.5, there are many other mapping and discriminant methods in the SPEAR toolbox that a user can choose to use. They are spectral angle mapper (SAM), SFF/continuum removal tool, multi-range spectral feature filtering, THOR change detection, etc. SAM is an automated method for comparing image spectra to individual spectra. It determines the similarity between two spectra by calculating the spectral angle between them and treating them as vectors in a space with dimensionality equal to the number of bands. The SAM tool provides a good attempt at mapping the predominant spectrally active material present in a pixel. The spectral feature fitting (SFF) function is used to compare the fit of image spectra to reference spectra using a least-squares technique. The reference spectra are scaled to match the image spectra after the continuum is removed from both datasets. SFF is an absorption-feature–based method for matching image spectra to reference end-members such as end-members (fractions) of an image pixel spectrum. Spectral signatures are typically characterized by multiple absorption features. The multi-range spectral feature filtering (SFF) function allows a user to define multiple and different wavelength ranges around each end-member's absorption features. This SFF method uses the multiple ranges in its comparisons. Each spectral range is interactively defined, and the continuum-removed absorption feature is plotted and extracted. A user can apply optional weights to each spectral range to emphasize more important features. The user can also save the defined wavelength ranges to a file and reuse them. The multi-range SFF executes more slowly than SFF but is likely to produce more accurate results. Both SFF and the multi-range SFF are performed on image spectra after the continuum is removed from both datasets. In the Spectral toolbox, the ENVI system provides a separate Continuum Removal tool to normalize reflectance spectra such that a user can compare individual absorption features from a common baseline. The continuum is a convex hull fit over the top of a spectrum using straight-line segments that connect local spectrum maxima. The first and last spectral data values are on the hull; therefore, the first and last bands in the output continuum-removed data file are equal to 1.0 (Exelis 2015).

ENVI system in the Spectral toolbox provides two change detection tools: THOR and SPEAR tools. The THOR Change Detection workflow is used to identify features that have changed between two images collected over the same area at different times. THOR is designed for use with hyperspectral data, but input files are only required to have at least two bands. The two images without georeference will be coregistered later in the Change Detection workflow. Like the THOR tool, the SPEAR Change Detection tool also provides a way to compare imagery collected over the same area at different times and to highlight features that have changed. The SPEAR tool can create two forms of change detection: absolute and relative. Absolute change detection highlights specifically what has changed, e.g., forest to grassland. Relative change detection shows that something has changed but does not specify what that change is.

6.2.8 Vegetation Analysis and Suppression

ENVI Classic's vegetation analysis tools in the Spectral toolbox classify a scene for agricultural stress, fire fuel distribution, and overall forest health (Exelis 2015). Thus, the user can perform vegetation analysis using tools that guide vegetation index (VI) selection for a specific outcome. With the vegetation analysis tools, the user should first use the Vegetation Index Calculator to calculate the VIs in an image before processing it with a vegetation analysis tool. This allows a user to reuse the data in multiple ways without having to recalculate the VI dataset for each new analysis. Each vegetation analysis tool uses a different set of three VI categories that are combined to produce a map with classifications showing some vegetative property or state in which a user may be interested. The SPEAR vegetation delineation tool in the Spectral toolbox allows a user to quickly identify the presence of vegetation and visualize its level of vigor. The Wizard also provides tools to facilitate creating graphics for use within reports and briefings (Exelis 2015).

The vegetation suppression tool in the Spectral toolbox is used to remove the vegetation spectral signature from multispectral or hyperspectral imagery, using information from red and near-infrared bands. This method helps a user better interpret geologic and urban features and works best with open-canopy vegetation in medium spatial resolution (30 m) imagery. Therefore, the vegetation suppression is most commonly used in lithological mapping and linear feature enhancement in areas with open canopies. For closed canopies in moderate-resolution data, vegetation suppression is primarily used for linear feature enhancement.

6.3 ERDAS IMAGINE

ERDAS IMAGINE is a suite of software that can be used to create, visualize, geo-correct, re-project, model, classify, and compress various remote sensing data for geospatial applications, developed by Hexagon Geospatial (formerly ERDAS, Inc.), initially released in 1978. The latest version was released in 2015 (see http://www.hexagongeospatial.com). As one of the standard ERDAS IMAGINE products, the IMAGINE Spectral Analysis module is used to introduce the concepts, data structures, and initial image processing functionalities of imaging spectrometry (IMAGINE Spectral Analysis 2006). Therefore, the IMAGINE Spectral Analysis module is a set of tools/software to process hyperspectral data. With the set of tools, specific industry-recognized preprocessing techniques have been simplified or automated to provide a rational easy-to-use image preprocessing regimen. There are several powerful analysis algorithms specific to hyperspectral datasets that have been implemented through a simple graphical user interface to produce specific end products for analyses and applications. In the IMAGINE Spectral Analysis module, the software workflow is designed around a set of tasks, including Anomaly Detection, Target Detection, Material Mapping, and Material Identification, and the tools consist of Spectral Analysis Workstation, Minimum Noise

Fraction, and Atmospheric Adjustment, etc. In the following, the functions and features of the tools and tasks in the Spectral Analysis module are briefly introduced and discussed.

6.3.1 IMAGINE Spectral Analysis Workstation

One easy way to access the functions/features of the IMAGINE Spectral Analysis software is the Spectral Analysis Workstation (IMAGINE Spectral Analysis 2006). This is the most comprehensive spectral analysis tool offered within the IMAGINE Spectral Analysis module. This workstation provides a user-friendly interactive interface to all workflow tasks and preprocessing functions. The workstation is started from the same menu, Spectral Analysis, as the workflow tasks from the ERDAS IMAGINE main menu. Figure 6.2 presents a basic layout of the IMAGINE Spectral Analysis Workstation. The Workstation basically consists of Archive libraries, Working libraries, Main view, Zoom view, Overview, Spectrum plot, etc., which are showed in the layout in Figure 6.2. However, the actual layout varies with the selected task or your preferences. The workstation provides the analyst with an environment to work with both hyperspectral imagery and spectral libraries. The interactive interface also allows a user to access viewing specific tools for interactively analyzing the image, spectral signatures, and other data displays.

There are many tools for viewing and processing hyperspectral datasets from the toolbars in the Spectral Analysis Workstation. The analytical tools include anomaly detection, target detection, material mapping, and material identification, etc., while preprocessing tools include atmospheric adjustment, minimum noise fraction, etc.

6.3.2 Anomaly Detection

In the IMAGINE Spectral Analysis module, the anomaly detection tool searches an input hyperspectral image and identifies those pixels that show a spectral signature deviating significantly from the majority of the other pixel spectra (i.e., the background spectra) in the image. In anomaly detection, the user is not required to have knowledge about hyperspectral data preprocessing techniques. After creating the anomaly mask with the tool, the user can display that layer (mask) over the input image for analysis with the viewer swipe tool. Note that Orthogonal Subspace Projection (OSP) is appropriate for this task. OSP can be used for anomaly detection and target detection, and it seeks to simultaneously reduce data dimensionality, suppress the background signal, and maximize the target spectrum signal-to-noise ratio (S/N). In this approach, each pixel vector is projected onto a subspace which is orthogonal to the interfering signals; this minimizes the background signal. Simultaneously, this maximizes the S/N of the desired signature (IMAGINE Spectral Analysis 2006).

6.3.3 Target Detection

The target detection tool is used to search an input multi-/hyperspectral image for a specific material of interest, the target, which is supposedly present in very low concentrations. In a typical analysis scenario, the analyst knows the spectral signature with a low probability material of interest (the target), but does not know background spectra and also does not wish to generate them. Again the OSP technique is very useful for detection of human-made objects in a natural environment. In practice, for target detection with the OSP technique, this assumes that the target spectra are taken from the image under analysis. If the target spectra are from a spectrum library or a different image, the data must be calibrated to reflectance using tools for atmospheric adjustment below.

6.3.4 Material Mapping

In the IMAGINE Spectral Analysis module, the material mapping tool maps an input image for the presence of a specific material or materials based on an input spectrum for the material(s) of interest

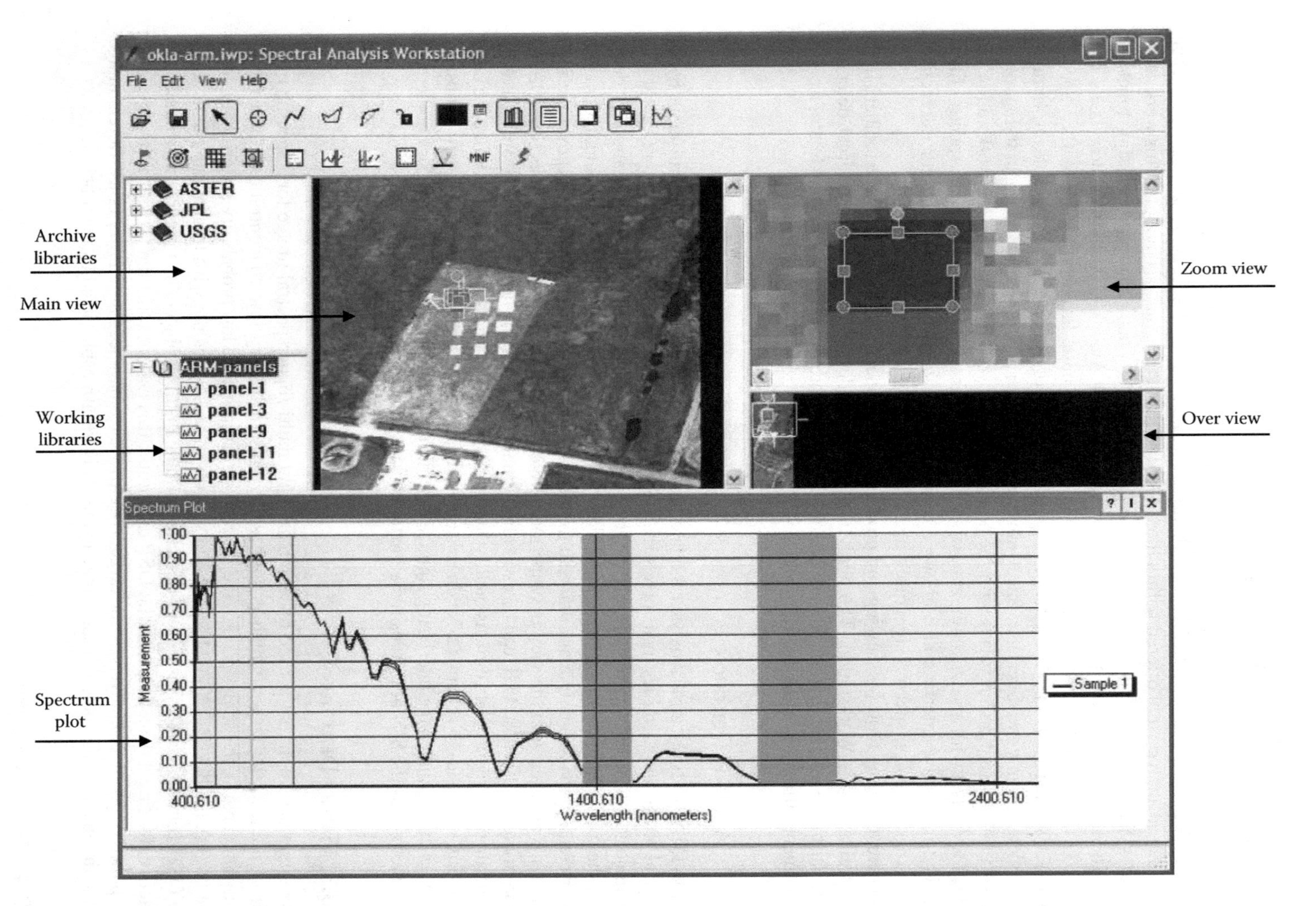

FIGURE 6.2 Layout of IMAGINE Spectral Analysis workstation with a sample image loaded. (Courtesy of ERDAS IMAGINE.)

(IMAGINE Spectral Analysis 2006). The Constrained Energy Minimization (CEM) algorithm in the material mapping tool attempts to maximize the response of a target spectrum and suppress the response of the unknown background signature(s) with hyperspectral imagery. The CEM algorithm is not intended to respond to low-probability signatures (which is the role of the OSP algorithm for target detection). Therefore, this algorithm is appropriate to the phenomenon where the sought material is a significant component of the scene, and thus this matched filter algorithm is optimal for detecting distributed subpixel targets such as mineral occurrences or sparse vegetation. The CEM algorithm maps targets of interest based on their spectra without specific knowledge of the background clutter (IMAGINE Spectral Analysis 2006). Therefore, pixels occurring in sufficient quantity and with sufficient spectral characteristics are used to calculate the covariance of the (background) image statistics. The performance of the CEM technique for subpixel signature detection and mapping is dependent upon the structure of the image correlation matrix. A good estimate of the correlation matrix can be obtained by only considering the first several significant eigenvectors in PCA because the significant eigenvectors are an estimate of the intrinsic dimensionality of the hyperspectral scene under consideration. See IMAGINE Spectral Analysis (2006) for more detailed information on using the CEM tool.

6.3.5 Material Identification

The material identification tool can help identify the material of a pixel or an AOI (area of interest) by comparing the unknown pixel spectrum to a list of candidate materials for which the spectra are known. The algorithm calculates a mathematical similarity index value comparing the unknown spectrum to each of the known spectra. The results are displayed in a ranked list. For the material identification tool, the default analytical metric is Spectral Correlation Mapper (SCM), but SAM is also offered in the IMAGINE Spectral Analysis module. However, SAM commonly gives a far inferior result and is not normally recommended.

The SAM algorithm calculates the angle formed between a reference spectrum and the image spectrum of each pixel (see SAM algorithm detail from Chapter 5). In this presentation, values vary between zero (0) and one (1), with one representing a best match. In the IMAGINE Spectral Analysis module, the reference spectra can be either laboratory-measured or field-measured library spectra, or directly extracted from the image. This method assumes that the data have been reduced to apparent surface reflectance. The output of this tool is a grayscale value representing the angle distance in radians (in N-dimensional space) between the reference spectrum and each pixel spectrum. Darker pixels indicate a smaller angular distance and thus a better match to the reference spectrum. The SCM algorithm is a modified version of the SAM approach in which the data is normalized and centered on the average of the two spectra (de Carvalho Jr. and Meneses 2000). This algorithm function overcomes the two limitations of the SAM: The SAM algorithm basically assumes that both positive and negative correlations are equally acceptable, which is not inherently true; and it is difficult for the SAM to differentiate between a dark material and a shaded area as the SAM quantified only vector direction, not magnitude. In practice, this is not strictly true. The SCM removes these true inconsistencies by normalizing each vector to the vector mean (de Carvalho Jr. and Meneses 2000).

6.3.6 Atmospheric Adjustment

In the IMAGINE Spectral Analysis software, atmospheric adjustment uses empirical algorithms to perform atmospheric correction to hyperspectral image data. The three empirical techniques in the current software are the Internal Average Relative Reflectance (IARR), Modified Flat Field, and Empirical Line technique, and they are all for atmospheric adjustments for hyperspectral imagery (IMAGINE Spectral Analysis 2006). The detailed algorithms and principles of all the three empirical techniques are introduced in Chapter 4 (see corresponding IAR, FFC, and ELC in Section 4.4.2).

The IARR method of atmospheric adjustment is best used when a user has no knowledge of the image that can be used to define spectral control points, while the flat field technique assumes that the user can define a region in the image that is spectrally flat. In the empirical line technique, a user uses several spectral pairs (e.g., from image scene and from spectral library or *in situ* spectral measurements). These spectral pairs are used to plot regression lines that are used to modify the input dataset band by band. Since the goal of this technique is to define the required regression line for each band, the user should use spectra from both bright and dark areas. These points can best constrain and define a regression line for each band.

6.4 IDRISI

The IDRISI Image Processing System in TerrSet software (Geospatial Monitoring and Modeling System) is comprised of an extensive set of procedures for image restoration, enhancement, transformation, and classification of remotely sensed imagery. The IDRISI Image Processing toolset includes a set of routines for working with hyperspectral data (Eastman 2001). Initialized in the mid-1980s, Clark Labs at Clark University has been involved in the development of the IDRISI system, supplying the community with the most comprehensive image processing system in the market. Figure 6.3 shows a screenshot of one of the TerrSet routines that is used to calculate continuum removal and depth analysis using library spectra from the USGS and AVIRIS hyperspectral data. In this section, functions/features of hyperspectral image analysis routines in the TerrSet system are introduced and discussed, mainly including hyperspectral signature collections (from either training sites or spectral libraries), hyperspectral image classifications (both hard and soft classifications), and spectral absorption feature extraction from hyperspectral data.

6.4.1 Hyperspectral Signature Development

In the TerrSet system, the HYPERSIG routine is used to collect spectral signatures from either hyperspectral images (image-based signature) or from the spectral library (spectral library-based signature) for hyperspectral image classification. For the image-based signature development, training sites are delineated in the image and signatures are developed from their statistical characteristics. Because of the large number of bands involved in a hyperspectral image, both the signature development and classification stages make use of different procedures from those used with multispectral data. For an unsupervised procedure, the HYPERAUTOSIG routine discovers signatures based on the concept of signature power. Consult the online help system for specific details for users who hope to experiment with the routine.

Classifying hyperspectral data can also rely upon the use of a library of spectral curves associated with specific Earth surface materials. These spectral curves are measured with very high precision in a lab setting. To effectively use the spectral library-based signatures, however, there are a number of important issues that need to be considered before using these library curves. Since the library curves are developed in a lab setting, the measurements are taken without an intervening atmosphere. You may therefore find it necessary to remove those bands in which atmospheric attenuation is strong. A simple way is first to eliminate the bands that have significant atmospheric attenuation and then to conduct an atmospheric correction procedure to those retained bands that have much less atmospheric effect. As a consequence, the hyperspectral modules in TerrSet assume that the imagery has already been atmospherically corrected if you develop signatures with spectral libraries. For example, the SCREEN module in TerrSet can first be used to screen out bands in which atmospheric scattering has caused significant image degradation, and the ATMOSC module can then be used on the remaining bands to correct for atmospheric absorption and scatterings. Library spectral curves are available from a number of research sites on the web, such as the United States Geological Survey (USGS) Spectroscopy Lab (see http://speclab.cr.usgs.gov).

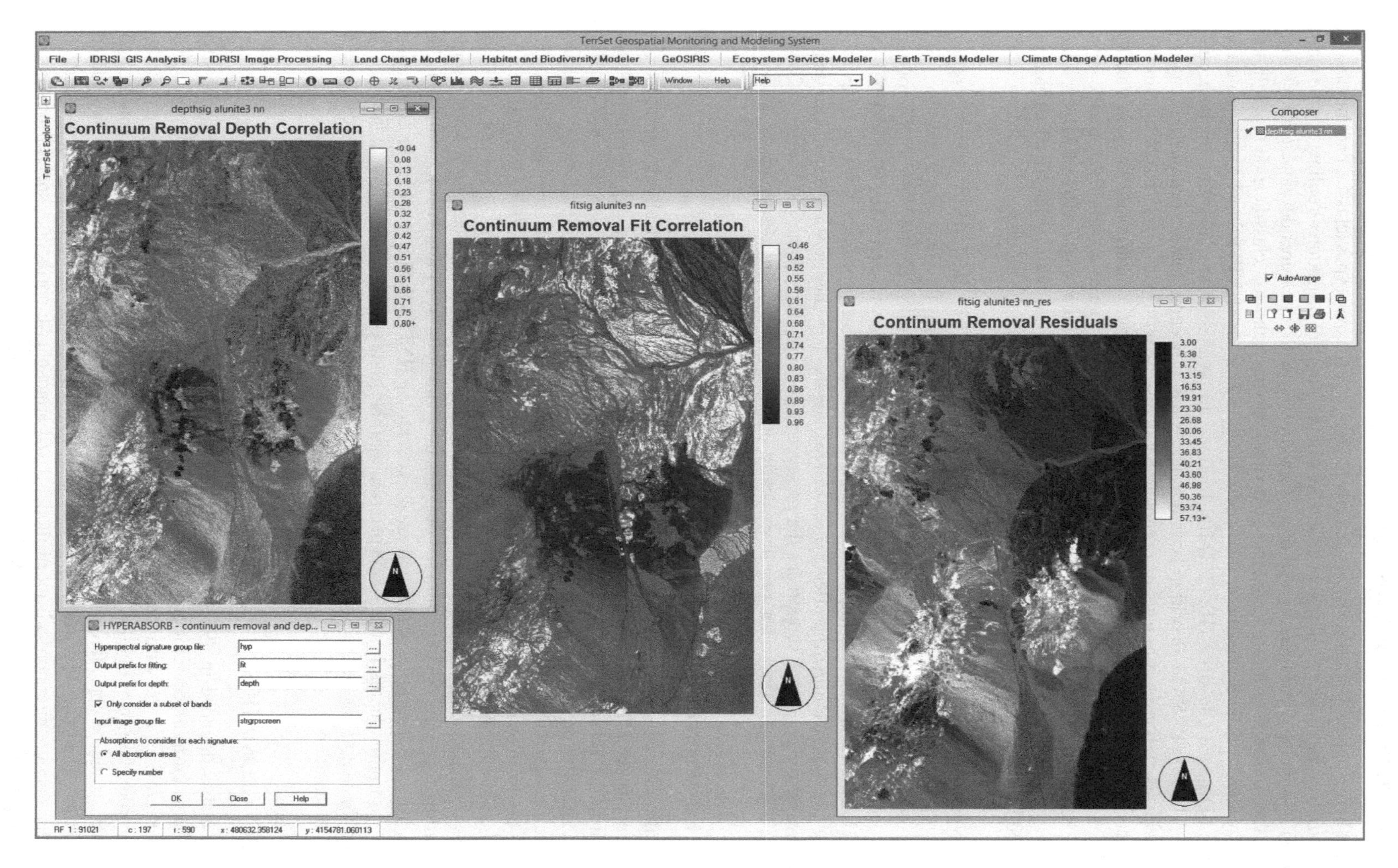

FIGURE 6.3 A screenshot of one of the TerrSet routines that calculates continuum removal and depth analysis using library spectra from the USGS and AVIRIS data. (Courtesy of Clark Labs/Clark University.)

6.4.2 Hyperspectral Image Classification

Once the hyperspectral signatures have been created for surface cover types (classes, end-members), classification of the imagery can proceed. TerrSet offers a range of procedures/routines for the classification of hyperspectral imagery. All routines work well with signatures developed from either training sites or spectral libraries, and can be divided into hard and soft classification types. In the TerrSet system, two hard classifiers, the Spectral Angle Mapper (HYPERSAM routine) and the Minimum-Distance Classifier (HYPERMIN routine), are available for classifying hyperspectral imagery. The HYPERSAM algorithm is a minimum-angle procedure that is specifically designed for use with spectral curve library data (although it can also be used with image-based signatures). In essence, HYPERSAM treats each signature as a vector. Then by comparing the angle formed by an unknown pixel, the origin, and a class mean, and comparing that to all other classes, the class that will be assigned to the unknown pixel is that with the minimum angle. HYPERMIN routine for hyperspectral data classification is specifically intended for use with image-based signatures developed from training sites. It uses logic that is identical to that of the multispectral hard classifier Minimum-Distance Classifier using standardized distances.

For hyperspectral image classification by using soft classifiers in the TerrSet system, there are two supervised soft classifiers: (1) The HYPERUNMIX routine uses the same approach as the LSU for multispectral data, except that it uses hyperspectral signature files; (2) the HYPEROSP routine uses a procedure called Orthogonal Subspace Projection. It is closely related to the HYPERUNMIX routine. However, it attempts to improve the signal-to-noise ratio for a specific cover type by explicitly removing the contaminating effects of mixture elements. The classification result is an image that expresses the degree of support for the presence of the signature of interest (Eastman 2001). Note that this measure is not a fraction intrinsically, but simply a measure of support.

6.4.3 Extraction of Absorption Features

TerrSet also offers a single classifier, the HYPERABSORB routine, specifically designed for use with library spectra for absorption feature extraction and mineral mapping. It is similar in basic operation to the Tetracorder algorithm developed by the USGS (introduced in Section 6.7). The HYPERABSORB procedure specifically looks for the presence of absorption features associated with specific materials. The procedure has proven to be particularly useful in mineral mapping in arid and extraterrestrial environments (Eastman 2001). In the TerrSet, HYPERABSORB can be used to measure the degree of absorption evident in pixels in a hyperspectral image (e.g., AVIRIS image) as compared to a library spectrum (with evident absorption features caused by known minerals) through a process of continuum removal and depth analysis (see Figure 6.3 for example). For instance, after the absorption depths and areas extracted by HYPERSORB routine, the correlation between absorption depths in the pixel spectrum and the library spectrum can give a measure of fit, while the magnitude of the absorption area relative to that in the library spectrum can give a measure of abundance.

6.5 PCI GEOMATICS

PCI Geomatics, founded in 1982, has set the standard in remote sensing and image-processing tools offering customized solutions to the geomatics community in more than 135 countries. PCI Geomatics provides users with tools for remote sensing, digital photogrammetry, geospatial analysis, map production, mosaicking, and more. The hyperspectral image analysis programs in PCI Geomatics should be considered legacy functionality at this point as they have not been updated for some time. The major features of Geomatica hyperspectral image analysis include visualization operations, data compression, advanced model-based atmospheric correction, spectral unmixing, spectral angle mapper, end-member selection, and scatter and spectra plotting. In the following, the features of data visualization, atmospheric correction, and hyperspectral unmixing and mapping analyses are briefly introduced based on PCI Geomatics (2004).

6.5.1 Data Visualization

PCI Geomatics has incorporated three tools to facilitate the visualization of large hyperspectral datasets: the 3D data cube, the thumbnails viewer, and the band cycling tool. A common method for displaying hyperspectral imagery is a 3D data cube. The 3D data cube usually can help users get a sense for the structure of the data they are working with, such as by letting them assess the number and spectral signatures of end-members present in the scene. The 3D data cube is available through PCI Geomatica Focus. This tool allows a user to display, rotate, and excavate the data in three dimensions (see Figure 6.4 for an example). This excavation is useful in visualizing how the spectral response for ground features changes with wavelength.

The thumbnails tool is another effective way to visualize a raster dataset with many bands. The thumbnails tool allows users to display all or a selection of a subset of the bands in a hyperspectral image, simultaneously. This tool is useful because bands that are severely affected by noise or atmospheric effects can be easily discriminated. The band cycling tool in Geomatica Focus is a quick way to cycle through different channel or wavelength ranges in a specified color component to create new color composites. This tool allows a user real time to observe the progressive change that increasing or decreasing the band number being displayed in one of the color display layers has on the composite (PCI Geomatics 2004).

6.5.2 Atmospheric Correction

PCI Geomatica atmospheric correction methods include simple atmospheric correction (i.e., empirical/statistical approaches) and model-based atmospheric transformation model. The FTLOC tool is an image-based correction method that locates spectrally flat targets in the image (see FFC in Section 4.4.2 in Chapter 4 for its detailed introduction). The FTLOC can be used to locate the "flat" targets in a hyperspectral image and to create a radiometric transformation that reduces the presence of atmospheric absorption features in the image data. The transformed image using the FTLOC is more suitable for comparison with ground- or laboratory-measured reflectance spectra than the original image. Another simple atmospheric correction approach is called an empirical line calibration (EMPLINE tool in Geomatica) (see ELC in Section 4.4.2 in Chapter 4 for a detailed introduction) that can be used to convert hyperspectral raw data to apparent surface reflectance values. A reference reflectance spectrum (typically measured on the ground) is required for each of two or more targets that appear in the image data in order to compute a regression line for each band. A set of slope and intercept of the regression line is then applied to every pixel for the band image to calculate the corresponding apparent surface reflectance values (of pixels).

The specific model-based technique that is implemented in Geomatica 9 is based on the technique implemented in the Canada Centre for Remote Sensing (CCRS) Imaging Spectrometer Data Analysis System (ISDAS). The atmospheric transformation technique using an at-sensor radiance look-up table (ATRLUT in Geomatica 9) converts an input radiance dataset (hyperspectral raw data) into a reflectance dataset, or converts an input reflectance dataset into a radiance dataset.

6.5.3 Hyperspectral Unmixing and Mapping

The major image analysis algorithms included in the Geomatica Advanced Hyperspectral package are ENDMEMB (End-member Selection), SPUNMIX (Spectral Linear Unmixing), and SAM (see their detailed introductions in Chapter 5). ENDMEMB tool in Geomatica 9 is used to compute a set of end-member spectra from a hyperspectral image using iterative error analysis (IEA)

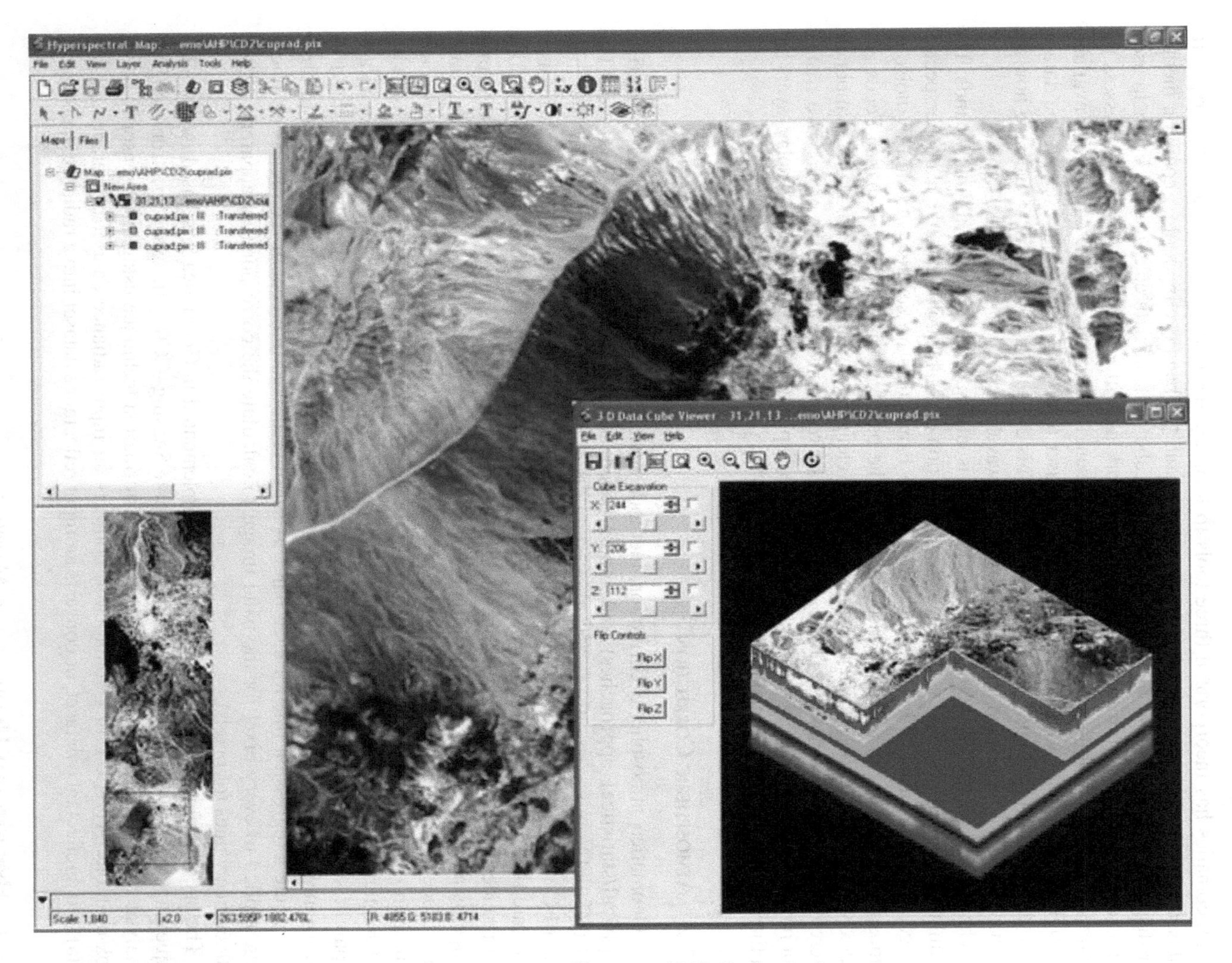

FIGURE 6.4 A screenshot of a 3D hyperspectral data cube, Geomatica 9.1. (Courtesy of PCI Geomatics.)

method. This is the first step for the spectral linear unmixing analysis. Usually, end-members can be thought of as classes or a set of spectra that make up a hyperspectral image. The SPUNMIX tool in Geomatica 9 linearly unmixes a hyperspectral image using a set of end-member spectra created with the ENDMEMB tool. The resulting pixel values in the fraction-map created by the SPUNMIX are estimates of the fractional contribution of each end-member spectrum to the corresponding mixed pixels for the hyperspectral dataset. The SAM in Geomatica 9 is used to classify hyperspectral image data with a set of reference spectra that define classes or materials. The results created with the SAM tool in a raster layer show the smallest spectral angle (to reference spectra) for each pixel, which can be used to label an image pixel into one of number of classes (or materials).

6.6 TNTmips

TNTmips is the software product of MicroImages, Inc., which was founded in 1986 by Lee D. Miller and Michael J. Unverferth with its first software product, the Map and Image Processing System (MIPS). In 1993, MIPS was renamed TNTmips® and made available for all Windows and Mac computers (see http://www.microimages.com/info/index.htm). The prototype hyperspectral analysis process in TNTmips provides specialized interactive analysis tools that allow users to fully exploit the spectral range and spectral resolution provided by hyperspectral datasets. The process allows users to view and save image spectra and to compare image spectra with laboratory spectra stored in spectral libraries (Smith 2013). The functions and features of hyperspectral analysis tools in TNTmips mainly include a hyperspectral explorer tool that allows rapid selection of an informative 3-band set for color display, on-the-fly calibration to reflectance (atmospheric correction) using equal area normalization or flat field correction, integration of the USGS spectral library with more than 500 mineral spectra, integrating spectral plots from image or library spectra, integrating data reduction using PCA or Minimum Noise Fraction (MNF) Transform, identifying spectral end-members using PPI and n-Dimensional Visualizer (n-D Visualizer), hyperspectral image classification using SAM or self-organizing map classifiers, spectral unmixing analysis using linear unmixing or matched filtering, extracting spectral absorption features using continuum removal approach from individual spectra or the entire image, etc. In the following, several major tools/software used to process hyperspectral datasets in TNTmips will be briefly introduced and reviewed.

6.6.1 Hyperspectral Explorer Tool

The hyperspectral explorer tool in TNTmips system is a unique, automated tool for visualizing the hyperspectral image. The tool allows users to automatically create a large number of potential RGB band combinations for a sample area (HA Guides 2015). Users can then preview the combinations as an animated sequence and select the best combination to use for analysis, display, or printing. Figure 6.5 presents a screenshot of the hyperspectral explorer tool.

6.6.2 Atmospheric Correction

The Hyperspectral Analysis process in TNTmips offers several commonly used empirical atmospheric correction methods. These methods are applied to the raw hyperspectral image datasets, and thus you do not need to recompute and store different versions of the image. There are two empirical atmospheric correction methods: The Flat Field Correction (FFC) and the Equal Area Normalization (EAN) in the TNTmips system. The FFC method (see the detailed introduced in Section 4.4.2 in Chapter 4) reduces or compresses atmospheric and solar irradiance effects using the mean spectrum of an area a user designates with the FFC tool. The area the user selects should be topographically and spectrally flat, and it should have a uniform spectral reflectance at all

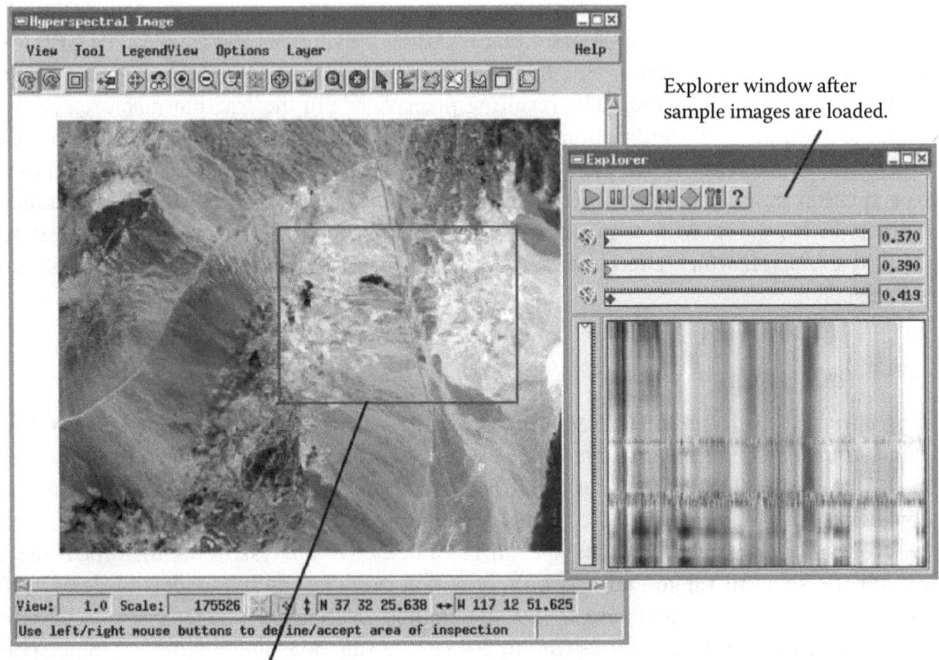

FIGURE 6.5 A screenshot of TNTmips Explorer tool. (From MicroImages, Inc., http://www.microimages.com.)

wavelengths, without significant absorption features across the spectral range. With an appropriate flat field spectrum, this procedure will largely remove atmospheric and solar irradiance effects. If the scene includes large variations in elevation or if the atmosphere was not uniform over the scene, some residual topographic and atmospheric effects will remain in the calibrated image spectra (Smith 2013).

The EAN method is the default method in the hyperspectral analysis tools for atmospheric correction, and thus it is enabled automatically when you start an atmospheric correction process. It is the same as IAR in Section 4.4.2 in Chapter 4 (see there for detailed introduction to the IAR [EAN] method and its drawbacks in correcting atmospheric effects). For a scene of hyperspectral image that includes many different land cover types, averaging process may largely remove spectral features attributed to the surface materials. The resulting spectral values represent reflectance relative to the average spectrum, and in ideal cases should be comparable to true reflectance spectra. When there are large elevation differences or variations in the atmosphere across a scene of hyperspectral image, the EAN method may not be used to remove the atmospheric effects on the image spectra (Smith 2013).

6.6.3 Hyperspectral Image Transformation

There are two image transform methods used to transform hyperspectral images into relatively low-dimensional datasets in TNTmips: PCA and MNF. The PC transform tool in TNTmips is a standard method for deriving a new set of images with reduced spectral redundancy. The MNF procedure first estimates the noise in each image band using spatial variations in brightness values in the TNTmips system (Smith 2013). It then applies two successive principal component transforms. The MNF procedure produces a component set in which noise levels increase consistently with increasing component number. The low-order components should contain most of the image information and little image noise like the result created with MNF processing in the ENVI system.

6.6.4 Hyperspectral Unmixing and Mapping

The hyperspectral analysis process provides the specialized tools in the TNTmips that users need to exploit the spectral range and resolution of hyperspectral images, including classification using spectral angle matching, cross-correlation and self-organizing map classifiers, and subpixel spectral mapping using linear spectral unmixing or matched filtering. For hyperspectral unmixing analysis, the hyperspectral analysis process provides tools for selecting end-members and determining the end-member spectra, which include PPI and the n-Dimensional Visualizer tools. The former is applied to a set of low-order MNF components that provides a "distilled" version of the hyperspectral image that can be used to rapidly identify relatively "pure" image spectra as end-member spectra for spectral unmixing operation. The PPI operation is the first step in identifying these end-member spectra. The latter is used to assess the results of the PPI operation. This tool allows users to create and view an n-dimensional scatterplot of spectra for an area in the hyperspectral image or MNF component set to finally determine the end-member spectra. The hyperspectral analysis process also provides functions to directly use *in situ* (field) spectral measurements and spectral libraries (e.g., using spectra from the USGS spectral library) for spectral unmixing analysis. After end-member spectra for a specific hyperspectral image are ready, users can use two spectral unmixing methods in the hyperspectral analysis process tools: Linear Spectral unmixing Model (LSM) and Matched Filtering (MF) tools, for unmixing the hyperspectral image. See Chapter 5 for detailed introduction to LSM and MF (MTMF) algorithms.

For the image classification analysis based on image spectra (e.g., in DN), the hyperspectral analysis process provides the classification tools including SAM, Cross-Correlation Matching Method (CCM), and the Self-Organizing Map Classifier (SOMC). See detailed introduction to SAM and CCM (CCSM) algorithms in Chapter 5. The SOMC performs an unsupervised classification of the hyperspectral image, which means that users don't have to specify target (training) spectra. The unsupervised procedure uses neural network techniques first to find a best-fit set of 512 class-centers, then assigns all image spectra to one of these classes using spectral angle as a measure of similarity (Smith 2013). Based on Smith (2013), the neural network (in SOMC) used to determine the class center vectors is a square 16 × 32 array of nodes (512), each representing one class. Initially the spectral values assigned to each node are random. Then a large number of image spectra are compared one by one to the full set of nodes. After many sample spectra have been processed and compared, the node values converge to a set of class-center vectors that approximate the distribution of all image spectra in N-dimensional space. The unsupervised classifier is capable of ensuring adequate spectral discrimination of different varieties of common, widespread materials (Smith 2013). If classification with reflectance spectra (e.g., hyperspectral image spectra calibrated to surface reflectance), some spectral matching tools are available in TNTmips, including SAM, CCM, and Band Mapping method. The spectral matching procedures are useful for identifying specific unknown materials in a hyperspectral image. The Band Mapping method compares spectra on the basis of the positions and relative depths of absorption features (local minima in the spectral curves), which are extracted by using a continuum removal approach (see the detailed description to the approach in Chapter 5).

As introduced previously, all major functions and features of the hyperspectral analysis tools included in five major software packages/systems—ENVI, ERDAS IMAGINE, IDRISI, PCI Geomatics, and TNTmips—are summarized in Table 6.1. The functions and features summarized in the table can help readers find what features/functions they may need for processing their hyperspectral data sets. In the following section, there are twelve minor software tools and programs that can also be used to process hyperspectral datasets, including *in situ* and laboratory-based hyperspectral measurements.

TABLE 6.1
A Summary of Functions and Features Within a List of Major Hyperspectral Data Processing and Analysis Tools/Software Packages

Tool/Software Package	Developer (ULR)	Functions/Features
ENVI	Exelis Visual Information Solutions Inc., USA (http://www.exelisvis.com/)	(1) Two atmospheric correction modeling tools for retrieving surface spectral reflectance: QUick Atmospheric Correction (QUAC) and Fast Line-of-sight Atmospheric Analysis of Spectral Hypercubes (FLAASH). (2) Building 3D image cube and plotting spectral curve. (3) Three popular data transform techniques: Principal Component Analysis (PCA), Minimum Noise Fraction Transform (MNF), and Independent Components Analysis (ICA). (4) Endmember determination and extraction using MNF images, Pixel Purity Index (PPI), n-D visualizer, and Sequential Maximum Angle Convex Cone (SMACC) tools. (5) Spectral unmixing tools: Linear Spectral Unmixing (LSU) and Matched Filtering (MF) or Mixture-Tuned Matched Filtering (MTMF). (6) Target detection: Adaptive Coherence Estimator (ACE) detection algorithm and Reed-Xiaoli Detector (RXD) algorithm. (7) Mapping and discriminant methods: Spectral Angle Mapper (SAM), Spectral Feature Fitting (SFF)/Continuum Removal tool, Multi Range Spectral Feature Filtering and THOR and SPEAR Change Detection tools. (8) Vegetation analysis and suppression.
ERDAS	Hexagon Geospatial / ERDAS, USA (http://www.hexagongeospatial.com)	(1) Anomaly detection: Orthogonal Subspace Projection (OSP) technique. (2) Target detection: OSP. (3) Material mapping: Constrained Energy Minimization (CEM) algorithm. (4) Material identification: Spectral Correlation Mapper (SCM) and Spectral Angle Mapper (SAM). (5) Atmospheric adjustment: empirical algorithms (Internal Average Relative Reflectance (IARR), Modified Flat Field, and Empirical Line techniques).
IDRISI	Clark Labs, Clark University, USA (https://clarklabs.org/terrset/)	(1) Hyperspectral signature development: HYPERSIG routine and HYPERAUTOSIG. (2) Hyperspectral image classification: two hard classifiers, the Spectral Angle Mapper (HYPERSAM routine) and the Minimum-Distance Classifier (HYPERMIN routine); two supervised soft classifiers: Linear Spectral Unmixing (HYPERUNMIX routine) and Orthogonal Subspace Projection (HYPEROSP routine). (3) Extraction of absorption features: process of continuum removal and depth analysis (HYPERABSORB routine).
PCI Geomatics	PCI Geomatics, Canada (http://www.pcigeomatics.com/)	(1) Data visualization: three tools for the visualization of large hyperspectral datasets: the 3D Data Cube, the Thumbnails viewer and the Band Cycling tool. (2) Atmospheric correction: empirical/statistical approaches (FTLOC tool for an image-based correction method and EMPLINE tool for an empirical line calibration); and a model-based technique (CCRS Spectrometer Data Analysis System (ISDAS)). (3) Hyperspectral unmixing and mapping: ENDMEMB (Endmember Selection), SPUNMIX (Spectral Linear Unmixing) and SAM (Spectral Angle Mapper) tools for unmixing and mapping.

(Continued)

TABLE 6.1 (CONTINUED)
A Summary of Functions and Features Within a List of Major Hyperspectral Data Processing and Analysis Tools/Software Packages

Tool/Software Package	Developer (ULR)	Functions/Features
TNTmips	MicroImages, Inc., USA (http://www.microimages.com/)	(1) Hyperspectral Explorer tool for visualizing the hyperspectral image. (2) Atmospheric correction: two empirical atmospheric correction methods (the Flat Field Correction (FFC) and the Equal Area Normalization (EAN)). (3) Hyperspectral image transformation: Principal Component Analysis (PCA) and Minimum Noise Fraction (MNF). (4) Hyperspectral unmixing and mapping: classification using spectral angle matching, cross-correlation and self-organizing map classifiers and Band Mapping method with a continuum removal approach, and subpixel spectral mapping using linear spectral unmixing or matched filtering (Pixel Purity Index (PPI) and the n-Dimensional Visualizer tools for determination of endmember spectra).

6.7 OTHER MINOR SOFTWARE TOOLS AND PROGRAMS FOR PROCESSING HYPERSPECTRAL DATA

6.7.1 DARWin

DARWin SP Data Acquisition software was initialized by Spectral Evolution, Inc. (http://spectralevolution.com) in 2010 and is specially used with Spectral Evolution field portable spectrometers and spectroradiometers. It provides an easy-to-use interface for researchers doing field work such as mineral species identification, vegetation research, and soil analysis with spectral measurements taken by using one of the Spectral Evolution instruments. Given the DARWin SP software capable of processing *in situ* hyperspectral data, some tools, including spectral curve smoothing, real-time mineral identification in the field, building spectral library, and vegetation index calculation, are briefly introduced as follows.

6.7.1.1 Set Smoothing Filter Width

In DARWin software, a user can apply a simple smoothing filter to a spectral curve to smooth the curve, if desired. The tool allows the user to adjust the number of samples included in the filter (default = 11 points [bands]). Usually, using a higher number of points provides increased smoothing.

6.7.1.2 EZ-ID Quick Material Identification Tool

The EZ-ID tool in DARWin SP software can help a user quickly identify target samples (e.g., mineral samples) for oreXpress spectrometers. The EZ-ID provides users with real-time mineral identification for field use. When an oreXpress spectrometer takes a spectral measurement from a target, the measured spectrum can immediately match it against a known mineral library, such as the USGS library, or SPECMIN (see its introduction below in this section), or against the user defined spectral library (DARWin SP 2015). EZ-ID provides a Library Builder software module so that users can easily create and add to a custom mineral or other material library based on location, target minerals, or whatever other criteria that fits users' applications. The EZ-ID tool works with samples from a wide range of applications, including vegetation studies, soil research, crop health, raw materials identification, plastics identification, minerals, etc. EZ-ID features include dynamically including or excluding spectral regions of interest for optimal accuracy and a Batch Mode that allows the user to save scans, data, and EZ-ID results in a spreadsheet, etc. (DARWin SP 2015).

6.7.1.3 Vegetation Indices

For vegetation analysis, many broad- and narrowband vegetation indices (VIs) are calculated for the current reflectance spectral scan with DARWin software. Most of these allow the user to vary the broadband wavelength limits used (red, NIR, green, and blue) or the narrowband center wavelengths (DARWin SP Manual 2015), which provides the user with a flexible way to define VIs. In this tool, there are eighteen broadband and narrowband VIs to be calculated from the current spectral scan. Each index can be modified by the user with simple on-screen menu-driven instructions. The eighteen VIs include Normalized Difference Vegetation Index (NDVI), Green Ratio Vegetation Index (GRVI), Simple Ratio Vegetation Index (SR), Difference Vegetation Index (DVI), Soil Adjusted Vegetation Index (SAVI), Red Green Ratio (Red/Green), Atmospherically Resistant Vegetation Index (ARVI), Green Normalized Difference Vegetation Index (Green NDVI), Enhanced Vegetation Index (EVI), Modified Soil Adjusted Vegetation Index Type II (MSAVI2), Infrared Percentage Vegetation Index (IPVI), Summed Green Vegetation Index (Sum Green), Photochemical Reflectance Index (PRI), Red Edge Normalized Vegetation Index (NDVI705), Water Band Index (WBI), Normalized Difference Water Index (NDWI), Photosynthetically Active Radiation (PAR), and Normalized Difference Nitrogen Index (NDNI) (DARWin SP 2015).

6.7.2 Hyperspectral Image Processing and Analysis System (HIPAS)

The HIPAS software system was developed by the Institute of Remote Sensing Applications (now a part of the Institute of Remote Sensing and Digital Earth; RADI), Chinese Academy of Sciences (CAS; see http://www.radi.cas.cn) in 1995. Built on Interactive Data Language (IDL) and implemented on Windows NT workstations, HIPAS meets the requirements for rapidly processing hyperspectral image data and providing prototypes of algorithms (Zhang et al. 2000). HIPAS possesses the capability of processing and analyzing multi-/hyperspectral sensor data, such as Modular Airborne Imaging Spectrometer (MAIS) and Pushbroom Hyperspectral Imager (PHI). With the HIPAS system, many hyperspectral remote sensing application studies were conducted in China, such as mineral identification and mapping, agriculture investigation, urban and wetland studies, etc.

The HIPAS system was designed based on the Object-Oriented method. The system consists of the following six objects: Image object, Lookup table object, Pseudo color table, Ground control points (GCP) object, Filter kernel, Mathematics formula, and Spectral data object. The HIPAS functions can be described with seven main modules: data input and output, data preprocessing, conventional image processing, spectral analysis, interactive analysis tools, spectral database tools, and advanced tools. For example, according to Zhang et al. (2000), the data preprocessing tools include geometric correction, system radiometer calibration, noise removal, image browse, and spectral simulation tools. The spectral analysis tools consist of spectrum filtering and transform, spectral unmixing, spectral matching, quantitative parameter estimation tools, etc. With open storage architecture, HIPAS is fully compatible with some advanced special commercial software such as ENVI, ERDAS, and even the common image-processing system Photoshop® (Yu et al. 2003).

Currently, HIPAS is no longer updated and is functionally taken over by a newly created hyperspectral processing and analysis system named HypEYE in RADI, CAS, in which the hyperspectral processing and analysis functionalities are significantly enhanced. In addition, the forthcoming HypEYE system is built on C++, GDAL, and QT, and it provides users with an easy-to-use platform for hyperspectral image data mining. The main features in the HypEYE system will include data visualization, dimension reduction, spectral unmixing, hyperspectral image classification, and target detection. The enhanced hyperspectral data processing functionalities in the newly created system will include, for example: (1) for spectral unmixing, the whole unmixing chain will be implemented, including a number of end-member estimation, end-member extraction, abundance estimation (multiply algorithms); (2) for hyperspectral image classification, more advanced and machine learning algorithms are included (e.g., SVM and Extraction and Classification of Homogeneous Objects [ECHO]; Landgrebe 1980); and (3) for hyperspectral image target detection, RX, ACE, CEM, and TCIMF algorithms are applied (personal communication with researchers in RADI, CAS, http://www.radi.cas.cn).

6.7.3 Imaging Spectrometer Data Analysis Systems (ISDAS)

As introduced in Section 4.4.3.6 of Chapter 4, ISDAS is an image and spectral analysis system developed by the Canada Centre for Remote Sensing (CCRS) (now called the Canada Centre for Mapping and Earth Observation [CCMEO]) in conjunction with industry for efficiently processing and analyzing hyperspectral imaging data starting late in the 1990s (Staenz et al. 1998). The ISDAS comprises four main categories: data input and output, data preprocessing, data visualization, and data information extraction. These tools provide users with the functionality to handle hyperspectral and associated ancillary data, remove sensor and calibration artifacts, convert at-sensor radiance to surface reflectance, remove BRDF, interactively view and analyze data, evaluate the performance of future sensors, and extract qualitative and quantitative information products (Boardman et al. 2006). An overview of the tools implemented in ISDAS is summarized in Section 4.4.3.6 in Chapter 4. Currently, PCI incorporated a large portion of ISDAS tools in its general purpose system Geomatica, and Vexcel Corporation (Microsoft) used ISDAS tools to build a fast image analysis system based on parallel processing for mineral mapping purposes (Boardman et al. 2006).

6.7.4 Integrated Software for Imagers and Spectrometers (ISIS)

ISIS 2.0 was developed at the beginning of 1989 by the U.S. Geological Survey for NASA and replaced an earlier prototype version (ISIS 2015). ISIS 2.0 was initially developed in the VAX/VMS environment and is implemented in a combination of the Fortran and C languages. The ISIS is a free, specialized digital image processing software package. A key feature of ISIS is an ability to place many types of data in the correct cartographic location. This enables disparate data to be co-analyzed, including standard image processing applications such as contrast, stretch, image algebra, filters, statistical analysis, etc. (ISIS 2015). ISIS can process 2D images and 3D image cubes derived from imaging spectrometers. It can be used to process data from NASA and from international spacecraft missions, including Lunar Orbiter, Apollo, Voyager, Mariner 10, Viking, Galileo, Magellan, Clementine, Mars Global Surveyor, Cassini, Mars Odyssey, Mars Reconnaissance Orbiter, MESSENGER, Lunar Reconnaissance Orbiter, Chandrayaan, Dawn, Kaguya, and New Horizons (ISIS 2015). The ISIS software is a valuable resource for planetary missions that require systematic data processing, products for planning, and research and analysis of derived data products. Therefore, the ISIS can support an instrument for data-processing pipelines, radiometric calibration, photometric calibration, band-to-band registration of multispectral data, ortho-rectification, construction of scientifically accurate and cosmetically pleasing mosaics, generation of control networks solutions, and creation of topographic models (ISIS 2015).

Many functions/features of the ISIS software were designed for processing hyperspectral imaging data (ISIS 2015). Among them, for example, are display tools (e.g., display and analyze cube footprints), filter tools (e.g., filter a cube, smoothing but preserving edges), Fourier Transform tools (e.g., apply a Fourier Transform on a cube), radiometric and photometric correction (e.g., create a flat-field image for line-scan instruments), and mosaicking (e.g., create a mosaic using a list of map projected cubes). The latest version of the ISIS software is version 3.0. This is a completely new software system that is implemented in the C++ language and uses a new data file format and a new user interface. The initial development of the software has concentrated mostly on providing geometry processing, but hyperspectral functionality will be enhanced in the future (Boardman et al. 2006).

6.7.5 MATLAB®

The image processing toolbox in MATLAB (http://www.mathworks.com/products/image/) provides a comprehensive set of reference-standard algorithms, functions, and apps for image processing, analysis, visualization, and algorithm development. With the toolbox, users can perform image analysis, image segmentation, image enhancement, noise reduction, geometric transformations, and image registration. The toolbox is designed mostly for processing multispectral images, but some tools can be used for processing hyperspectral images. However, most researchers are interested in exploiting and utilizing the MATLAB environment to develop their own toolboxes for processing and analyzing hyperspectral images, such as the MATLAB Hyperspectral Image Analysis Toolbox (HIAT) (Arzuaga-Cruz et al. 2004, Rosario-Torres et al. 2015), MATLAB Hyperspectral Toolbox (MHT 2015), and A MATLAB Toolbox for Analysis of Multi-/Hyperspectral Imagery (Ahlberg 2006). In the following, the major features and functions of the MATLAB HIAT toolbox and MATLAB Hyperspectral Toolbox for processing and analyzing hyperspectral images are briefly introduced.

The HIAT is a collection of algorithms that extend the capability of the MATLAB numerical computing environment for processing multi-/hyperspectral imagery. The HIAT was developed at the Center for Subsurface Sensing and Imaging Systems (CenSSIS), which sought to develop a repository of reliable and reusable software tools that could be shared by other researchers and at the Laboratory of Remote Sensing and Image Processing (LARSIP), at the University of Puerto Rico, Mayagüez Campus (Rosario-Torres et al. 2015). HIAT is available for download at http://www.censsis.neu.edu/software/hyperspectral/Hyperspectoolbox.html. MATLAB version 6.5 was used for the implementation of the HIAT (v2.0), but the HIAT toolbox can work with MATLAB version 7.2 to ensure that it is fully upward compatible with different platforms (Rosario-Torres et al. 2015). The processing phases of HIAT (V2.0) can be functionally divided into four groups:

1. Image enhancement (pre-processing)
2. Feature selection/extraction
3. Classification/unmixing
4. Postprocessing

In the HIAT toolbox, the image enhancement pre-processing tools enhance the hyperspectral image in either the spatial domain or the spectral domain, such as by Resolution Enhancement and Principal Component Analysis (PCA) Filtering Enhancement (Rosario-Torres et al. 2015). The feature selection/extraction algorithms provide users the ability to reduce the dimensionality of the hyperspectral image data. The available algorithms include PCA, discriminant analysis, singular value decomposition band selection, optimized information divergence projection pursuit, etc. (Arzuaga-Cruz et al. 2003). In the HIAT toolbox, standard classifiers are included. For example, Fisher's linear discriminant, angle detection, and fuzzy maximum likelihood are included. Spectral unmixing algorithms consist of non-negative sum-to-one, non-negative sum less-than- or equal-to-one, non-negative least-square, and sum-to-one constrained algorithms. To increase classification accuracy, post-processing techniques integrate contextual information of the scene into the resulting classification map, such as supervised and unsupervised extraction and classification of homogeneous objects techniques with window sizes 2×2, 3×3, and 4×4 (Rosario-Torres et al. 2015).

The open source MATLAB Hyperspectral Toolbox includes various hyperspectral data exploitation algorithms. The toolbox is a concise repository of current state-of-the-art processing and analysis algorithms for learning and research purposes. The most important algorithms in the toolbox may include target detection, material abundance mapping with spectral unmixing techniques, automated processing change detection, visualization reading/writing hyperspectral data files, etc. See more detailed information regarding algorithms/tools for processing hyperspectral data from MHT (2015).

6.7.6 MultiSpec

MultiSpec is an image processing system for interactively analyzing Earth's observational multispectral image data (e.g., Landsat series data) and hyperspectral image data from current and future airborne and spaceborne systems (e.g., AVIRIS and Hyperion) (Landgrebe and Biehl 2011). The MultiSpec is a freeware multispectral image data analysis system and its primary objective is as an aid to export the results of our research into devising good methods for analyzing such hyperspectral image data. However, it has also found significant use in other applications such as in multiband medical imagery and in K–12 and university-level educational activities. MultiSpec has its origin in the LARSYS multispectral image data analysis system at Purdue University and has been implemented for Intel- and PowerPC-based Macintosh- and Windows-based PC machines. LARSYS was one of the first remote sensing multispectral data processing systems, originally created during the 1960s. The original purpose for the development of the MultiSpec system in 1988 was to provide an easy-to-use tool that could be used for teaching, research, and especially to provide researchers with the ability to try new techniques without having to program the algorithms (Boardman et al. 2006). MultiSpec's new versions are made available periodically as new algorithms emerge from the present research on hyperspectral processing and analysis. More updated information can be found at https://engineering.purdue.edu/~biehl/MultiSpec/index.html.

The major functions/features of the current version of MultiSpec may include the ability to import data with multi-format, display selected channels of multispectral images and side-by-side channels, calculate and plot histogram, convert image data to different image file formats, create new channels of data from existing channels, cluster data using either a single pass or an iterative (ISODATA) clustering algorithm, enhance statistics for computing training/test statistics (especially for small training sets), select and extract features, classify a designated area in the data file, show a graph of the spectral values of a currently selected pixel or the mean ±SD for a selected area, and show a color presentation of the correlation matrix for a field or class as a visualization tool, especially for hyperspectral data, etc. (Landgrebe and Biehl 2011). To create new bands/features of data from existing bands, the principal component or feature extraction transformation of the existing bands or spectral vegetation indices via calculating a ratio of a linear combination of existing bands divided by a different linear combination of bands are performed. In order to determine the optimal spectral features to be used for a given classification, searching for the best subset of features using any of five statistical distance measures, a method based directly upon decision boundaries defined by training samples, or a second method based directly upon the discriminant functions are used. The methods are especially designed to search for narrow spectral features (e.g., spectral absorption features from hyperspectral data) and for use of projection pursuit as a means of further improving the features extracted (Landgrebe and Biehl 2011). For image classification using multi-/hyperspectral data, six different classification algorithms are available in the MultiSpec system: minimum distance classifier, correlation classifier, matched filter, Fisher linear discriminant, maximum likelihood classifier, and ECHO spectral/spatial classifier.

6.7.7 Optical Real-Time Adaptive Spectral Identification System (ORASIS)

The ORASIS was initially developed at the Naval Research Laboratory (http://www.nrl.navy.mil) in the mid-1990s. The ORASIS is a collection of a number of algorithms that work together to produce a set of end-members that are not necessary from within the data set (Bowles et al. 2003, Bowles and Gillis 2007). It determines end-members by exploiting the scene data using an approach significantly different from either PPI or N-Finder from Section 5.7.5 of the previous chapter. The ORASIS determines the end-members by intelligently extrapolating outside the data set to find end-members that may be closer to pure substances than any of the spectra that exist in the data. According to Bowles and Gillis (2007), hyperspectral data sets provide the ORASIS algorithms with many different mixed spectra of the materials present, and each mixed pixel spectrum gives a clue as to the makeup of the end-members. The collection of algorithms that make up ORASIS can be described as the following components: prescreener, basis selection, end-member selection, and demixing modules. Applications of these algorithms in the ORASIS include automatic target recognition (anomaly detection), data compression, and terrain categorization.

The prescreener module in ORASIS has two main functions: (1) To replace the large original image spectra with a (smaller) representative set (called exemplars), and (2) to associate each image spectrum with exactly one member of the exemplar set. Therefore, the prescreener module could be considered as two-step problem: The first step as the exemplar selection process and the second step as the replacement process. However, in the ORASIS system, the two steps are intimately related. After the prescreener has been performed and the exemplars determined, the next step in the ORASIS system is to project the set of exemplars into an appropriate, lower-dimensional subspace. There are two methods that are available in ORASIS for determining the optimal subspace (Bowles and Gillis 2007). Currently, there are many different algorithms that are used to find the end-members. The end-member selection module in the ORASIS system, in very broad terms, defines the end-members as the vertices of some "optimal" simplex that encapsulates the data. This is pretty similar to some "geometric" end-member algorithms, such as PPI and N-Finder, and is a direct consequence of the linear mixing model. However, unlike PPI and N-Finder, the ORASIS does not have to assume that the end-members must be actual data points (Bowles and Gillis 2007). In the last step after the determination of the end-members, the demixing module in the ORASIS algorithm is to estimate the abundance of each end-member in each mixed pixel in a scene of hyperspectral image. The demixing module allows users to use two methods (i.e., unconstrained demixing and constrained demixing) separately to unmix mixed pixels and to obtain the estimation of the abundance of each end-member in them. The unconstrained demixing is the simplest (and also the fastest) method and doesn't use constraints on the abundance coefficients, while the constrained demixing algorithm applies the non-negativity constraints to the abundance coefficients.

These algorithms in the ORASIS package have been applied in automatic target recognition (ATR), data compression, terrain categorization (TERCAT), etc. One of the more popular and useful applications of hyperspectral imagery is automatic target recognition (ATR), and the ORASIS can be used in ATR, including anomaly detection. Data compression was one of the design goals of ORASIS from the beginning to compress hyperspectral imagery in order to reduce the large amount of space and also to reduce transmission times. Using a mixture model approach, the ORASIS system can be used to categorize terrain (e.g., using a large angle of 10 degrees; Bowles and Gillis 2007).

6.7.8 Processing Routines in IDL for Spectroscopic Measurements (PRISM)

PRISM is a software system that was developed by the U.S. Geology Survey (USGS). It provides a framework to conduct spectroscopic analysis of measurements acquired using laboratory, field, airborne, and spacebased spectrometers. PRISM builds on the legacies of a variety of USGS computer programs that have been created to advance spectroscopy and hyperspectral remote sensing, including SPECPR (Clark 1993) and Tetracorder (Clark et al. 2003). Many capabilities of these programs are available in PRISM routines (Kokaly 2011). PRISM version 1.0 was compiled using a full ENVI 4.8+IDL (Interactive Data Language) 8.0 license running on a computer with a 64-bit version of the Windows 7 operating system, and it is operated correctly in runtime versions of ENVI 4.8 (Kokaly 2011). In general, using the PRISM functions, a user can identify and characterize materials through comparing the spectra of materials of unknown composition with reference spectra of known materials. In addition, PRISM contains more routines for the storage of spectra in database files (SPECPR), import/export of ENVI spectral libraries, importation and processing of field spectra, correction of spectra, arithmetic operations on spectra, interactive continuum removal and comparison of spectral features, calibration of imaging spectrometer data to surface reflectance, and identification and mapping of materials using spectral feature-based analysis of reflectance data (Kokaly 2010, 2011). The PRISM routines can be grouped into four categories:

1. The ViewSPECPR module
2. Spectral analysis functions
3. Image processing functions
4. Material Identification and Characterization Algorithm (MICA)

Starting on the ENVI program, the ENVI menu bar should show a pull-down menu, PRISM, under which there are six items: View Specpr File, Spectral Analysis, Image Processing, MICA, PRISM Help, and About PRISM. The ViewSPECPR module starts by clicking View Specpr File. Once there, the user can list the contents of a SPECPR file and select, plot, and operate the selected SPECPR file (e.g., performing continuum removal on a record; for a detailed description of the ViewSPECPR module, see Kokaly 2005, 2008, 2010, 2011). From Spectral Analysis in PRISM menu, a variety of spectral analysis functions for importing/exporting data to/from SPECPR files, managing and editing SPECPR records, and applying arithmetic, convolution, and interpolation functions to spectra stored in SPECPR data records can be accessed (Kokaly 2011). Many of these routines are available in the ViewSPECPR Module. The additional routines for analyzing spectra measured with ASD spectrometers are also included in PRISM system, including importing spectra in ASD binary files into SPECPR files, converting spectra measured relative to a Spectralon reference panel to reflectance, and correcting for offsets between detectors in ASD spectrometers. With the ENVI spectral library files, spectra stored in SPECPR files and processed using PRISM routines may be linked to ENVI spectral processing functions or other spectral analysis programs that utilize ENVI spectral libraries to store data or utilize the capability to import/export spectra in ENVI spectral library files. The image processing functions allow the user to apply correction factors to imaging spectrometer data. These functions are intended to be used to apply radiative transfer ground calibration (RTGC) corrections of hyperspectral imaging data. The RTGC method employs empirical correction factors derived from field-based spectra measured from calibration areas. The average of field measurements of the ground calibration site need to be converted to reflectance and convolved to the imaging spectrometer spectral characteristics. Usually, the convolved field spectrum is divided by the average of atmospherically corrected reflectance spectra for the pixels that cover the calibration area. The PRISM also includes routines for spectral interpolation and convolution in order to convert spectra stored in USGS spectral libraries to those of imaging spectrometers such as AVIRIS and HyMap (Kokaly 2011).

The Material Identification and Characterization Algorithm (MICA) module in the PRISM system gives the user the ability to identify and map materials by comparing a spectrum, multiple spectra, or even an entire image cube to reference spectra of known materials (Kokaly 2011). A core function of MICA in PRISM is conducting the continuum removal analysis of absorption features so that the extracted absorption features (e.g., wavelength location and absorption depth) can be used to identify and characterize materials. MICA is modeled on the USGS Tetracorder algorithm (Clark et al. 2003), but with additional options for greater user control of feature weightings and identification constraints, and thus the MICA has a significantly greater ability to integrate image results with image processing and GIS software (Kokaly 2011).

6.7.9 SPECMIN

SPECMIN is a reference mineral spectral library and mineral spectral identification system, developed by Spectral International, Inc. (http://www.spectral-international.com). This system works for VIS–NIR–MIR spectroscopy that includes an extensive and dynamic library of reference spectra for minerals, wavelength search/match tables, physical properties of each species in the database, and literature references for the infrared active mineral phases. The automated spectral match tables present matches by major and minor absorption features, mineral species, and mineral groups. The feature analysis tables are available for individual species and diagnostic absorption features for each mineral spectrum selected. The SPECMIN can be used not only to process any specific spectrometer data but also to interface its processed spectra into hyperspectral remote sensing processing as ground truth information.

Spectral International, Inc., now offers two versions (SPECMIN Pro and SPECMIN IP) of its SPECMIN software for Earth science specialists (ESSN 2015). SPECMIN Pro includes a mineral spectral database with over 1,500 spectra of mineral species compiled using a full-range of ASD FieldSpec Pro spectrometer and the PIMA-II shortwave infrared spectrometer and is the most comprehensive version. The SPECMIN Pro allows users to incorporate their proprietary data sets and new collections of spectra in the spectral database using customized features. The spectral data can be searched by species, class, or wavelength. The increasing number of case studies of spectra and information related to analyzed materials includes various types of mineral deposits (e.g., diamonds, base metals, and precious metals), mine and mill waste, natural and man-induced acid drainage, soils, etc. SPECMIN IP is a streamlined version specially designed for users who are mostly interested in image processing, including spectral comparison in remote sensing and GIS database, and comparison of spectra from known minerals with other spectrally responsive materials. This version comes with less background information on the internal spectral databases and does not have all of the wavelength search and match capabilities compared to SPECMIN Pro.

6.7.10 Spectrum Processing Routines (SPECPR)

SPECPR is one of the software tools developed by the Spectroscopy Laboratory, USGS (http://speclab.cr.usgs.gov/software.html), and it is a large-scale interactive program for general one-dimensional array processing. The SPECPR has been optimized for reflectance spectroscopy data and analysis. The program was designed for analysis of laboratory, field, telescopic, and spacecraft spectroscopic data; it processes 1D arrays up to 4,852 data points, and the processing and analysis tools include addition, subtraction, multiplication, division, trigonometric functions, logarithmic and exponential functions, and many more specialized routines (Clark 1993). The detailed list of SPECPR features, analysis routines, and plotting functions can be accessed at http://speclab.cr.usgs.gov/specpr.html.

The SPECPR program was initialized in 1975 at the MIT Wallace Observatory on a Harris 2024 computer. In 1984, the program was moved to the USGS in Denver, Colorado. Hereafter, the SPECPR has been developed with many analysis routines in order to enhance its functionalities of the study of high resolution spectra, absorption band analysis, radiative transfer mixing models, and the desire to manage spectral databases necessitated major changes at USGS. The latest version was updated in 2012, and in the latest analysis area for SPECPR development it is possible to access imaging spectrometer data cubes. This feature allows a user to query large data sets for spectra and then analyze those spectra in detail (Clark 1993). Currently, the SPECPR can access any standard file types common in the terrestrial and planetary remote sensing communities.

6.7.11 Tetracorder

Tetracorder is one of the software tools developed by the Spectroscopy Laboratory, USGS (http://speclab.cr.usgs.gov/software.html), and is a public domain analysis program that is used to identify specific materials and classify image components using hyperspectral data collected from laboratory, field, airborne, and spaceborne platforms (Clark et al. 2003, Livo and Clark 2014). Tetracorder relies on a spectral library (Clark et al. 2007) of materials (including minerals, vegetation, snow, and engineered materials). This library of materials is used to identify species and/or abundance of material from lab- or field-based spectral measurements or airborne or spaceborne hyperspectral images. An important feature of this program is the identification of materials. If the spectral features are diagnostic, the identified material can be robust, whereas in other cases, the identification requires additional verification (Livo and Clark 2014). Tetracorder is a set of software algorithms commanded through an expert system to identify materials based on their spectra (Clark et al. 2003). The results created with the expert system (a set of software algorithms) can be tested and are compared. The algorithms can be trained to compare sets of diagnostic absorption features for a given material according to a set of user-defined rules. To identify and map materials using hyperspectral data, each absorption feature under consideration is normalized against its continuum and then scaled by a multiplicative constant so that its depth matches that of the corresponding reference absorption feature (Dalton et al. 2004). In the Tetracorder system, its primary algorithm is the modified least squares spectral feature matching algorithm based on continuum removed spectral features. Tetracorder can also be used in traditional remote sensing analyses because some of the algorithms are applicable to processing the multispectral data, such as using the matched filter.

Usually, the system first makes each spectrum being partitioned into diagnostic wavelength regions and then the algorithms are tested on different spectral regions. This enables the spectral properties from multiple materials to be analyzed and identified without the need for spectral unmixing analysis (Boardman et al. 2006). The identification step appears to be a unique aspect of the Tetracorder approach. The system ranks the results of the algorithms and selects the best results as the answer. Tetracorder identifies materials through comparing spectrometer measured (including imaging and non-imaging) spectrum (the unknown) to a well-characterized material library spectrum by using techniques that maximize accuracy and performance. Tetracorder software has been used to map many materials on the Earth and throughout the solar system, including ecosystems, disaster response on land and in the ocean, and mineralogical mapping (Livo and Clark 2014).

In addition to the features of Tetracorder, described in Clark et al. (2003), new features have been added, including fuzzy logic and shoulderness (Clark et al. 2010). Therefore, in the current version of the Tetracorder system (Livo and Clark 2014), the principal Tetracorder algorithms include ratio spectrum processing, modified least-squares spectral feature fitting algorithm, red edge or blue edge optical parameter characterization, hooks for future algorithms, etc. And the additional algorithms may include those related to continuum removal and feature extraction algorithms with fuzzy logic, shoulderness with fuzzy logic, constraint analysis, identification analysis algorithms, etc.

6.7.12 The Spectral Geologist (TSG)

The TSG is a specialist processing and analysis software package developed by the CSIRO Earth Science and Resource Engineering Division, Sydney, Australia, in the mid-1990s (http://thespectralgeologist.com). It was developed in support of the PIMA and PIMA-II handheld spectrometers prototyped by CSIRO and successfully commercialized by Integrated Spectronics Pty Ltd. It was designed for analysis of field or laboratory spectrometer data. The TSG is the industry-standard tool for geological analysis of spectral reflectance data of minerals, rocks, and soils, and including drill cores and chips (TSG 2015). It is a fully menu- and icon-driven Windows software program with a series of intuitive commands available through menus and on-screen controls. Since the TSG was initialized in the 1990s, it has grown and added new and advanced processing options, productivity tools, assisted mineralogical interpretation, a batch processing language, new wavelength regions, and new display paradigms (TSG 2015). After 2004, the TSG has become a suite comprised of five scalable programs (e.g., TSG Lite, TSG Pro, TSG Core, and TSG Viewer), offering a hierarchy of options to different types of users. TSG Pro and Lite have five screens and a floater window, whereas TSG Core has seven screens and two floater windows. These screens are designed to allow the user to examine their datasets at different levels of detail, from individual sample spectrum to broad overviews and spatial plots containing thousands of spectra and associated data (TSG 2015).

TABLE 6.2
A Summary of Functions and Features Within a List of Minor Hyperspectral Data Processing and Analysis Software Tools and Programs

Tool/Software	Developer (ULR)	Functions/Features
DARWin	Spectral Evolution, Inc., USA (http://spectralevolution.com)	(1) Spectral curve smoothing with different width filters. (2) Quickly (real-time) material identification tool in the field (materials including vegetation type, soil type, crop health, raw materials and plastics identification, and minerals). (3) Vegetation analysis: there are 18 broadband and narrowband VIs to be calculated from hyperspectral data.
HIPAS	Institute of Remote Sensing Applications (now as a part of Institute of Remote Sensing and Digital Earth, RADI), CAS, China (http://www.radi.cas.cn/)	(1) Data input and output. (2) Data preprocessing (geometric correction, radiometric correction, noise removal tool, image browse tool and spectral simulation tool, etc.). (3) Conventional image processing (image transformation, image filtering, image classification, and registration tools). (4) Spectral analysis (spectrum filtering and transform, spectral unmixing and matching, and quantitative parameter estimation). (5) Interactive analysis tools (spectral/spatial profile and spectral slicing, annotation tool and ROI tools). (6) Spectral database tools. (7) Advanced tools (including image fusion).
ISDAS	Canada Centre for Remote Sensing (CCRS) (now The Canada Centre for Mapping and Earth Observation (CCMEO)	(1) Data input and output. (2) Data preprocessing (major sensor, calibration, and atmospheric modeling artifacts removal tools; the minimum/maximum autocorrelation factor (MAF), the minimum noise fraction (MNF), the principal component analysis (PCA), and the band-moment analysis (BMA); Noise Removal tools in spectral/spatial domains; spectral smile–frown detection and correction tool, keystone detection and correction tool; surface reflectance retrieval tool with CAM5S RT model; and Spectral Simulation tool). (3) Data visualization (1D, 2D and 3D display tools and manipulation environments). (4) Data information extraction (spectral matching, spectral unmixing, automatic endmember selection, interactive endmember selection, quantitative estimation and calculator, etc.).
ISIS	U.S. Geological Survey (https://isis.astrogeology.usgs.gov/)	(1) Display tools (e.g., display and analyze cube footprints). (2) Filter tools (e.g., filter a cube, smoothing but preserving edges). (3) Fourier Transform tools (e.g., apply a Fourier Transform on a cube). (4) Radiometric and photometric correction (e.g., create a flat-field image for line-scan instruments). And (5) Mosaicking (e.g., create a mosaic using a list of map projected cubes).
MATLAB	Mathworks, Inc., USA (http://www.mathworks.com)	Matlab HIAT: (1) Image Enhancement (pre-processing) (resolution enhancement and PCA filtering enhancement). (2) Feature Selection/Extraction (PCA, discriminant analysis, singular value decomposition band selection, and optimized information divergence projection pursuit, etc.). (3) Classification/Unmixing (Fisher's linear discriminant, angle detection, and fuzzy maximum likelihood; for spectral unmixing algorithms: Non Negative Sum to One, Non Negative Sum Less or Equal to One, Non Negative Least Square, and Sum to One Constrained algorithms). (4) Post-Processing. Matlab Hyperspectral Toolbox: Target detection, Material abundance mapping with spectral unmixing, Automated processing change detection, and Visualization reading/writing hyperspectral data files, etc.

(*Continued*)

TABLE 6.2 (CONTINUED)
A Summary of Functions and Features Within a List of Minor Hyperspectral Data Processing and Analysis Software Tools and Programs

Tool/Software	Developer (ULR)	Functions/Features
MultiSpec	Purdue University, USA (https://engineering.purdue.edu/~biehl/MultiSpec/index.html)	Display selected channels of multispectral images and side-by-side channels, calculate and plot histogram, create new channels of data from existing channels, cluster data using either a single pass or an iterative (ISODATA) clustering algorithm, enhance statistics for computing training/test statistics, select and extract features, classify a designated area in the data file (using traditional methods and advanced methods such as ECHO spectral/spatial classifier), show a graph of the spectral values of a currently selected pixel or the mean ± SD for a selected area, and show a color presentation of the correlation matrix for a field or class, etc.
ORASIS	Naval Research Laboratory, USA (http://www.nrl.navy.mil/)	To produce a set of endmembers for spectral unmixing, the four components for the ORASIS: Prescreener (the first step as the exemplar selection process and the second step as the replacement process), basis selection, endmember selection, and demixing modules.
PRISM	U.S. Geological Survey (http://speclab.cr.usgs.gov/software.html)	(1) The ViewSPECPR module (list the contents of a SPECPR file, select, plot and operate the selected SPECPR file (e.g., performing continuum removal on a record)). (2) Spectral analysis functions (a variety of spectral analysis functions for importing/exporting data to/from SPECPR files, managing and editing SPECPR records, and applying arithmetic, convolution and interpolation functions to spectra stored in SPECPR data records). (3) Image processing functions (apply radiative transfer ground calibration (RTGC) approach to correct hyperspectral imaging data). (4) The Material Identification and Characterization Algorithm (MICA) (identify and characterize materials by conducting the continuum removal analysis of absorption features).
SPECMIN	Spectral International, Inc., USA (http://www.spectral-international.com/)	(1) Possess reference mineral spectral libraries (using existing spectral libraries and creating customized ones) and mineral spectral identification system (including identifying and analyzing various types of mineral deposits (e.g., diamonds, base metals and precious metals), mine and mill waste, natural and man-induced acid drainage, and soils, etc.).
SPECPR	U.S. Geological Survey (http://speclab.cr.usgs.gov/software.html)	Analysis of laboratory, field, telescopic, and spacecraft spectroscopic image data, processing of one-dimensional arrays up to 4852 data points, and processing and analyzing tools including addition, subtraction, multiplication, division, trigonometric functions, logarithmic and exponential functions, and many more.
Tetracorder	U.S. Geological Survey (http://speclab.cr.usgs.gov/software.html)	Ratio spectrum processing, modified least-squares spectral feature fitting algorithm, and red edge or blue edge optical parameter characterization, etc. More and additional algorithms: continuum removal and feature extraction algorithms with fuzzy logic, shoulderness with fuzzy logic, constraint analysis, and identification analysis algorithms, etc.
TSG	The Spectral Geologist, the CSIRO Earth Science and Resource Engineering Division, Sydney, Australia (http://thespectralgeologist.com/)	Analysis of field or laboratory spectrometer data of minerals, rocks and soils, including drill cores and chips. TSG Display Tools (not only analyzing the mineral assemblages of each sample and how these vary down a hole or in a project area, but also looking at specific variations in mineral characteristics down hole). Automated Identification tools (interpreting his/her project spectra). The Spectral Maths options (many different processing and analysis approaches, such as the wavelength of a diagnostic absorption associated with a specific mineral composition).

In the TSG suite, TSG Lite is at the entry level TSG for small data sets, with most of the same functions available in TSG Pro. TSG Pro is the most widely used version of TSG, with functionality suitable for the majority of spectral geology applications, including assisted mineralogical interpretation with TSATM. TSG Core is a high-level version of TSG, with advanced functionality for HyLogging datasets with core and chip image handling capabilities, as well as many built-in productivity tools (TSG 2015). And TSG Viewer is a low-cost version of TSG, developed to allow dissemination of already-interpreted spectral data to non-spectral experts within companies and organizations and for training purposes. There are many important features/tools included in the TSG suite. Of them, TSG Display Tools can be used to compare the spectral and ancillary data from all samples/spectra using many different display options available in TSG. These tools allow the user to not only analyze the mineral assemblages of each sample and how these vary down a hole or in a project area, but also look at specific variations in mineral characteristics down hole, between holes, or over a project area. Automated Identification tools are designed to assist a user in interpreting his or her project spectra (TSG 2015). The Spectral Maths options in TSG offer many different processing and analysis approaches to processing a user's spectral data and quantifying specific characteristics of the user's spectra that may relate to a mineral characteristic, such as the wavelength of a diagnostic absorption associated with a specific mineral composition. In addition, the built-in spectral libraries include a reference library of spectra of common minerals and a USGS sample collection.

For the twelve minor tools and programs (i.e., DARWin, HIPAS, ISDAS, ISIS, MATLAB, MultiSpec, ORASIS, PRISM, SPECMIN, SPECPR, Tetracorder, and TSG) designed for processing and analyzing hyperspectral data introduced previously, their major functions and features are summarized in Table 6.2. Like Table 6.1, all functions and features summarized in Table 6.2 can assist readers in finding what tools/programs they may need for processing their hyperspectral data, including imaging and non-imaging datasets.

6.8 SUMMARY

In considering tools that are actively updated, available to access, and used as legacy systems, a collection of major and minor hyperspectral data analysis and processing software tools, programs, and systems were briefly introduced and reviewed in this chapter. In the beginning of this chapter, the importance and necessity of developing software tools and systems specially for processing hyperspectral data were discussed. Then, a collection of five major software systems or packages were introduced and reviewed in Sections 6.2–6.6. They include ENVI, ERDAS, IDRISI, PCI Geomatics, and TNTmips systems and packages, in which functions and features specially for processing hyperspectral data are a major component. The whole systems or packages, in general, also include functionalities and features used for other image (e.g., multispectral) and spatial data (e.g., GIS dataset) processing and analysis. Of them, the ENVI system for processing hyperspectral data has relatively complete processing capabilities, from display image cube and visualization to image preprocessing to hyperspectral information extraction, interpretation, and applications. The five systems and packages are all commercial products and available worldwide. In Section 6.7, a collection of twelve minor software tools and programs in alphabetical order by product name were introduction and reviewed. Their sizes are relatively small, and they serve as standalone products used mostly for processing hyperspectral data only. Most of this type of product, supported by governments, are free to access and reside in the public domain. The minor software tools and programs include DARWin, HIPAS, ISDAS, ISIS, MATLAB, MuiltSpec, ORASIS, SPECMIN, SPECPR, Tetracorder, and TSG. For both collections, major and minor software tools and systems reviewed in this chapter, all major functionalities and features were concisely summarized in Table 6.1 and Table 6.2 so that readers can quickly have a glance at the list of functionalities and features associated with the reviewed software tools and systems.

REFERENCES

Ahlberg, J. 2006. A MATLAB Toolbox for Analysis of Multi/Hyperspectral Imagery. *Sensor Technology, Technical Report*, ISSN 1650-1942. Sweden. Accessed from http://www.foi.se/report?rNo=FOI-R—1962—SE).

Arzuaga-Cruz, E., L. O. Jimenez-Rodriguez, and M. Velez-Reyes. 2003. Unsupervised feature extraction and band subset selection techniques based on relative entropy criteria for hyperspectral data analysis. In *Algorithms and Technologies for Multispectral, Hyperspectral, and Ultraspectral Imagery IX. SPIE Proceedings* 5093:462–473.

Arzuaga-Cruz, E., L. Jimenez-Rodrguez, M. Velez-Reyes, D. Kaeli, E. Rodriguez-Diaz, H. T. Velazquez-Santana, A. Castrodad-Carrau, L. E. Santos-Campis, and C. Santiago. 2004. A MATLAB toolbox for hyperspectral image analysis. In *Proceedings of International Geoscience and Remote Sensing Symposium*, DOI: 10.1109/IGARSS.2004.1370246, pp. 4839–4842.

Boardman, J. W., L. L. Biehl, R. N. Clark, F. A. Kruse, A. S. Mazer, J. Torson, and K. Staenz. 2006 (July 31–August 4). Development and implementation of software systems for imaging spectrometry. In *Proceedings of IGARSS 2006*, Denver, Colorado.

Bowles, J. H., and D. B. Gillis. 2007. An optical real-time adaptive spectral identification system (ORASIS), in *Hyperspectral Data Exploitation: Theory and Applications*, ed. Chein-I Chang, pp. 77–106. Hoboken, NJ: John Wiley & Sons, Inc.

Bowles, J., W. Chen, and D. Gillis. 2003 (July 21–25). ORASIS framework: Benefits to working within the linear mixing model. In *Proceedings of Geoscience and Remote Sensing Symposium*, IGARSS '03, 1:96–98, 10.1109/IGARSS.2003.1293690.

Clark, R.N. 1993. *SPECtrum Processing Routines User's Manual Version 3 (program SPECPR)*: U.S. Geological Survey Open-File Report 93–595. Accessed from http://speclab.cr.usgs.gov.

Clark, R.N., G.A. Swayze, I. Leifer, K. E. Livo, R. Kokaly, T. Hoefen, S. Lundeen, M. Eastwood, R. O. Green, N. Pearson, C. Sarture, I. McCubbin, D. Roberts, E. Bradley, D. Steele, T. Ryan, R. Dominguez, and the Air borne Visible/Infrared Imaging Spectrometer (AVIRIS) Team. 2010. A method for quantitative mapping of thick oil spills using imaging spectroscopy. *U.S. Geological Survey Open-File Report 2010–1167.* Accessed December 20, 2015 from http://pubs.usgs.gov/of/2010/1167.

Clark, R. N., G.A. Swayze, K. E. Livo, R. F. Kokaly, S. J. Sutley, J. B. Dalton et al. 2003. Imaging spectroscopy: Earth and planetary remote sensing with the USGS Tetracorder and expert systems. *Journal of Geophysical Research* 108(E12):5131–5146.

Clark, R. N., G. A. Swayze, R. Wise, E. Livo, T. Hoefen, R. Kokaly, and S. J. Sutley. 2007. USGS digital spectral library splib06a. *U.S. Geological Survey Data Series 231.* Accessed December 20, 2015, from http://speclab.cr.usgs.gov/spectral-lib.html.

Dalton, J. B., D. J. Bove, C. S. Mladinich, and B. W. Rockwell. 2004. Identification of spectrally similar materials using the USGS Tetracorder algorithm: The calcite–epidote–chlorite problem. *Remote Sensing of Environment* 89:455–466.

DARWin SP Manual. 2015. *DARWin SP Application Software Version 1.3 User Manual*, Spectral Evolution, Inc., 1 Canal St. Unit B-1. Lawrence, Massachusetts.

DARWin SP. 2015. *Software for Spectral Evolution Spectrometers, Spectroradiometers, and Spectrophotometers.* Accessed December 6, 2015, from http://spectralevolution.com/software.html.

De Carvalho, O. A., Jr., and P. R. Meneses. 2000. Spectral Correlation Mapper (SCM): An improvement on the spectral angle mapper (SAM). *Paper read at 9th Airborne Earth Science Workshop.* Jet Propulsion Laboratory, 00-18.

Eastman, J. R. 2001. *IDRISI 32.2, Guide to GIS and Image Processing*, vol 2. Worcester, MA: Clark Labs, Clark University.

ESSN. 2015. *Earth Science Software News SPECMIN Suite.* Accessed December 21, 2015, from http://www.canadianminingjournal.com/news/earth-science-software-news-specmin-suite-available.

Exelis, Exelis Visual Information Solutions. 2015. *ENVI Classic Help, ENVI5.1*

Green, A. A., M. Berman, P. Switzer, and M. D. Craig. 1988. A transformation for ordering multispectral data in terms of image quality with implications for noise removal. *IEEE Transactions on Geoscience and Remote Sensing* 26(1):65–74.

HA Guides. 2015. *Hyperspectral Analysis Technical Guides & Other Documentation.* Accessed November 28, 2015, from http://www.microimages.com/documentation/html/Categories/Hyperspectral%20Analysis.htm.

IMAGINE Spectral Analysis. 2006. *IMAGINE Spectral Analysis™ User's Guide.* Switzerland: Leica Geosystems Geospatial Imaging, LLC.

ISIS. 2015. *USGS: Integrated Software for Imagers and Spectrometers*. Accessed December 8, 2015, from https://isis.astrogeology.usgs.gov.

Kokaly, R. F. 2010 (July 25–30). Spectroscopic analysis for material identification and mapping using PRISM, an ENVI/IDL based software package. In *Proceedings of IGARSS 2010*. Honolulu, Hawaii. Accessed December 13, 2015, from http://vigir.missouri.edu/~gdesouza/Research/Conference_CDs/IGARSS_2010/pdfs/2782.pdf.

Kokaly, R. F. 2005. View_SPECPR: Software, installation procedure, and user's guide. *U.S. Geological Survey Open-File Report 2005–1348*.

Kokaly, R. F. 2008. View_SPECPR: Software for plotting spectra (installation manual and user's guide, version 1.2). *U.S. Geological Survey Open-File Report 2008–1183*.

Kokaly, R. F. 2011. PRISM: Processing routines in IDL for spectroscopic measurements (installation manual and user's guide, version 1.0). *U.S. Geological Survey Open-File Report 2011–1155*.

Landgrebe, D., and L. Biehl. 2011. *An Introduction & Reference for MultiSpec©, Version 9.2011*. Purdue University. Accessed from https://engineering.purdue.edu/~biehl/MultiSpec.

Landgrebe, D. A. 1980. Development of a spectral-spatial classifier for Earth observational data. *Pattern Recognition* 12(3):165–175.

Livo, K. E., and R. N. Clark. 2014. The Tetracorder user guide—Version 4.4. *U.S. Geological Survey, Open-File Report 2013–1300*. Accessed from http://dx.doi.org/10.3133/ofr20131300.

MHT. 2015. *MATLAB Hyperspectral Toolbox*. Accessed December 22, 2015, from https://github.com/isaacgerg/matlabHyperspectralToolbox.

PCI Geomatics. 2004. *Hyperspectral Image Analysis, Geomatica 9, Version 9.1*. Accessed November 23, 2015, from http://www.pcigeomatics.com/pdf/Hyperspectral_primer.pdf.

Rosario-Torres, S., M. Vélez-Reyes, and L. O. Jiménez. 2015. *The MATLAB Hyperspectral Image Analysis Toolbox*, DOI: 10.13140/RG.2.1.1220.7203. Accessed from http://www.researchgate.net/publication/279852541.

Smith, R. B. 2013. *Tutorial—Analyzing Hyperspectral Images with TNTmips*. MicroImages, Inc.

Staenz, K., T. Szeredi, and J. Schwarz. 1998. ISDAS—A system for processing/analyzing hyperspectral data. *Canadian Journal of Remote of Sensing* 24:99–113.

TSG. 2015. *The Spectral Geologist (TSG)*. Sydney, Australia: CSIRO Earth Science and Resource Engineering Division. Accessed December 21, 2015, from http://thespectralgeologist.com.

Yu, J., X. Hu, B. Zhang, and S. Ning. 2003 (September 29). A new architecture for hyperspectral image processing and analysis system: Design and implementation. In *Proceedings of SPIE 5286, Third International Symposium on Multispectral Image Processing and Pattern Recognition, 696*. doi:10.1117/12.538648.

Zhang, B., X. Wang, J. Liu, L. Zheng, and Q. Tong. 2000. Hyperspectral image processing and analysis system (HIPAS) and its applications. *Photogrammetric Engineering & Remote Sensing* 66(5): 605–609.

7 Hyperspectral Applications in Geology and Soil Sciences

Compared to multispectral remote sensing, hyperspectral remote-sensing technology is more suitable for application in geology and soils. This is because there are a lot of minerals/rocks and soil types with unique diagnostic absorption features (bands) in the solar-reflected spectrum. Since the advent of imaging spectroscopy technology (i.e., hyperspectral remote sensing) in the early 1980s, using various analysis techniques and methods, the hyperspectral data applied in geology and soil science have been well-documented. Therefore, in this chapter, an overview on application studies of hyperspectral data including laboratory and *in situ* measured hyperspectral data and airborne and spaceborne hyperspectral imaging data in geology and soils is given. Specifically, in Section 7.2, spectral characteristics and properties of various mineral/rock species or classes are summarized and discussed. In Section 7.3, the relevant techniques and methods, including absorption feature extraction, hyperspectral mineral indices, spectral matching methods, spectral unmixing techniques, spectral modeling methods and some advanced techniques and methods, are introduced and their applications with various hyperspectral data for estimating and mapping minerals and rocks are reviewed and discussed. The spectral characteristics of soils and hyperspectral technology applied in estimating and mapping properties of soils are reviewed and discussed in Section 7.4. In addition, three typical geological case studies of hyperspectral remote sensing technology are briefly introduced and summarized in Section 7.5.

7.1 INTRODUCTION

Mineral and rock exploration and mapping by using traditional geological techniques are tedious, expensive, and time-consuming (Ramakrishnan and Bhariti 2015). With multispectral remote sensing techniques (including airborne and spaceborne platforms) sensing a few of the wide bands (usually > 50 nm based on FWHM) in the visible (VIS), shortwave infrared (SWIR), and thermal infrared (TIR) regions, it is possible to discriminate and map lithological and mineral/rocks (e.g., Hewson et al. 2005). However, due to low spectral resolution (and coarse spatial resolution too) with such multispectral sensors, it is difficult or impossible to map detailed mineral/rock species/classes and composition and to estimate relative abundance of mineral constituents within a field of view (FOV) or a pixel. Since the advent of imaging spectroscopy or hyperspectral remote sensing in the early 1980s, the hyperspectral data for geologic applications have been well documented. Why hyperspectral imaging data may be applied to geology and soil identification and mapping is because there is a large number of minerals with unique spectral absorption bands in the solar-reflected spectrum. In the early 1980s, many researchers with geological background had utilized hyperspectral imaging data (e.g., AIS sensor) for mineral identification and geological mapping (e.g., Goetz et al. 1985, Vane and Goetz 1988). Since hyperspectral sensors have an ability to acquire laboratory-like spectra and can provide unique spectrally contiguous imaging data at a high spectral resolution (usually < 10 based on FWHM) that allow precise identification and mapping of minerals and rocks, a variety of hyperspectral imaging sensor data have been used to determine the point, local, and regional distribution of minerals for a range of geology and soil science investigations (Green et al. 1998). To test the potential of geological and soil identification and mapping, the airborne hyperspectral sensors such as AVIRIS, HYDICE, DAIS, HyMap, and spaceborne hyperspectral sensors like EO-1 Hyperion have been extensively attempted and the investigated results and their understanding have been well documented. There are several review articles that provide an overview on most application cases of hyperspectral remote sensing in geologic and soil fields published during the last two decades such as applications of hyperspectral remote sensing in geology (Cloutis 1996), individual mineral exploration using hyperspectral systems (Sabins 1999), the use of remote sensing and GIS in mineral resource mapping (Rajesh 2004), using imaging spectroscopy to study soil properties (Ben-Dor et al. 2009), and hyperspectral geological remote sensing (van der Meer et al. 2012, Ramakrishnan and Bhariti 2015). In addition, how hyperspectral remote sensing (HRS) is applied in geology and soils was also overviewed in Section 1.3.1 in Chapter 1. A lot of research results of geologic remote sensing in the VNIR, SWIR, and TIR parts of the spectrum resulted from USGS scientists' pioneering work (e.g., Hunt 1977, Hunt and Hall 1981, Salisbury et al. 1989), which recorded measured mineral and rock spectra to form a basis for our understanding and testing the feasibility of airborne and spaceborne hyperspectral remote sensing techniques applied in geology and soils sciences.

There is a major advantage of using hyperspectral data in geology and rocks, which is that diagnostic (absorption) spectral features of various minerals and rocks in a spectral region allow us to determine their chemical composition and relative abundance (Crósta et al. 1997). Therefore, the reflectance spectra acquired with fine spectral resolution hyperspectral sensors can be used to identify a large range of surface cover materials and soils that cannot be identified with traditional broadband low spectral resolution data and they can form valuable supplements to more traditional methods and provide information and a perspective not otherwise available. However, there still exist several severe limitations of hyperspectral remote sensing data applied in mineral/rock and soil exploration and mapping. These include the following:

1. Hyperspectral remote sensing data, like multispectral remote sensing data, have a depth penetration of approximately a few micrometers in VNIR–SWIR regions to a few centimeters in the TIR region and some meters (in hyper arid regions) in the microwave region (Rajesh 2004). Therefore, in most cases, users of multi-/hyperspectral data in geology need some indirect clues, such as general geological setting, alteration zones, and associated rock types in a study area. Here, the hyperspectral data are expected to play a greater role in mineral exploration and mapping than other remote sensing data by using their unique diagnostic spectral features along their spectral curves.
2. Since the spectral difference and absorption feature variation are very weak among the various mineral/rock species and classes, to extract subtle and accurate diagnostic spectral information from hyperspectral data for exploring and mapping minerals/rocks and soils, accurate radiometric correction and compensation of atmospheric attenuation are critical to efficiently use hyperspectral data, whereas in multispectral remote sensing, this problem is relatively minor.
3. In addition to the technical obstacle and theoretical difficulty mentioned previously as well as the algorithms and techniques for efficiently processing the hyperspectral data (discussed in Chapter 5), users should remember that the upper mineral layer or soil horizon is not usually fully exposed to the sun's photons. Partial or complete coverage of litter and vegetation also add difficulty to map and identify minerals and rock and soil species or classes, which requires using unique and effective algorithms and techniques to process and analyze various hyperspectral data in order to improve mineral/rock exploration and geological and soil mapping accuracies.

Therefore, in the following, after discussing and summarizing spectral characteristics and signatures of various mineral/rock species or classes, relevant methods and techniques of HRS geological applications are reviewed and evaluated from their applicability, and substantial case studies on applications of hyperspectral data in geology and soil sciences are reviewed and discussed.

7.2 SPECTRAL CHARACTERISTICS OF MINERALS/ROCKS

During the last three decades, scientists have studied the spectral characteristics of minerals and used them for remotely mapping and identifying minerals and determining compositional information of the Earth's surface (Hunt 1977, Clark 1999, Eismann 2012, van der Meer et al. 2012). A great diversity of minerals form the surface of the Earth and their spectral characteristics are quite diverse. Therefore, based on spectra measured in a laboratory and *in situ* with a spectroradiometer, it is necessary to understand the spectral characteristics of minerals (also of rocks and soils). The spectral characteristics of minerals can be described using spectral absorption (diagnostic) bands or spectral reflection minima mostly along the visible-near infrared (VNIR) to middle infrared (MIR) (0.4–2.5 μm) spectral reflectance curves (Hunt 1977, Clark 1999). There are two general processes: electronic and vibrational, which basically cause absorption bands or reflection minima in the spectra of materials. While Burns (1993) examined the details of electronic processes and Farmer (1974) analyzed vibrational processes to provide the fundamentals as well as practical information, Hunt (1977) and Clark (1999) did more work on measuring the spectra of minerals and analyzed and summarized causes of absorption bands or reflection minima in the spectra of minerals, mostly covering the VNIR–MIR spectral range (e.g., the absorption bands or reflection minima in the spectra of minerals shown in Figure 7.1). Therefore, in the following, the spectral (absorption) characteristics of some common minerals (or groups; most presented in Figure 7.1), caused by the two general processes, are reviewed and the possibility of identifying and mapping these common minerals (or groups) from hyperspectral images is discussed. Figure 7.1 serves to locate and identify the origin of all the electronic and vibrational absorption features commonly encountered in the VNIR–MIR spectral range of particulate minerals (Hunt 1977).

7.2.1 Spectral Absorption Characteristics Caused by the Electronic Processes

In the VNIR spectral range, spectral characteristics are influenced primarily by electronic processes such as transition and charge-transfer processes associated with transition-metal ions such as Fe, Cr, Co, and Ni (Eismann 2012). In addition, the electronic phenomena called *color centers* presenting in the visible spectral range also influence the spectral absorption characteristics of a limited number of colored materials.

7.2.1.1 Due to Crystal-Field Effects

Isolated atoms and ions have discrete energy states. Absorption of photons at a specific wavelength causes a change from one energy state to a higher one, while emission of a photon at a specific wavelength occurs as a result of a change in an energy state to a lower one. Both changes associated with absorption and emission of a photon are referred to as transitions (Hunt 1977). In a crystal field, transition-metal ions (e.g., Fe^{2+} and Fe^{3+}) exhibit split orbital energy states and can absorb electromagnetic radiation at frequencies associated with electronic transitions between these energy states (Eismann 2012). Since the crystal field varies with the crystal structure from one mineral to another, the amount of splitting varies and the same ion (e.g., Fe^{2+}) may produce obviously different absorption characteristics, which make specific mineral identification possible from hyperspectral remote sensing data (Clark 1999).

Olivines are an important group of rock-forming minerals and their spectra display a very broad absorption region centered near 1.0 μm, composed of at least three separate features (Hunt 1977, King and Ridley 1987; Figure 7.1). Previous spectral studies also indicate that the wavelength position with the 1 μm absorption feature varies as a function of Mg/Fe ratio with a total shift on the order of 30 nm (King and Ridley 1987, Cloutis 1996). In addition, some minor absorption bands shortwave of ~0·65 μm from some olivine samples were observed (King and Ridley 1987, Sunshine and Pieters 1990). These absorption features results suggest that it would be possible for us to identify and map olivine minerals from high resolution (at least better than 30 nm) spectral

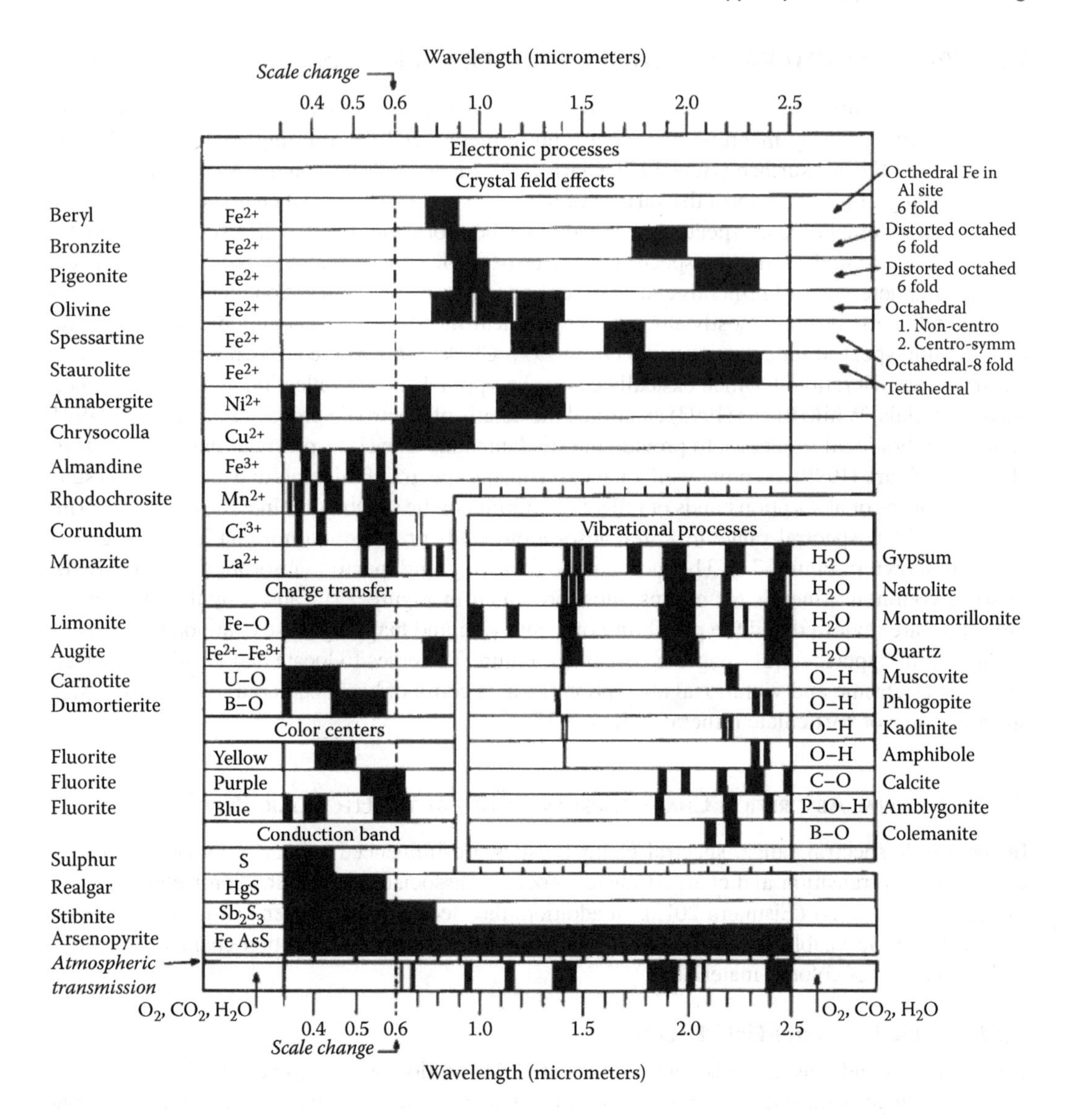

FIGURE 7.1 Major absorption features of minerals. The variety of absorption processes and their wavelength dependence allows us to derive information about the chemistry of a mineral from its reflected or emitted light. (From Hunt, G. R., *Geophysics*, 42, 501–513, 1977. Reprinted with permission from the Society of Exploration Geophysicists.)

data and also to estimate the major and some minor cation abundances associated with olivine minerals (Cloutis 1996).

Both *bronzite* and *pigeonite* belong to mineral group of *pyroxenes* that are potentially important minerals for geological exploration because they are often associated with deposits of platinum group elements and chromite (Cloutis 1996). Both bronzite and pigeonite spectra show an absorption band near 1.0 μm and another at wavelengths longer than 1.8 μm (around 2.0 μm; Figure 7.1). Both features are caused from transitions in the six-fold coordinated ferrous ion located in one of two possible sites (Hunt 1977). The wavelength positions of these bands can vary by up to 500 nm as a function of major element (e.g., Fe, Mg, and Ca) chemistry (Cloutis and Gaffey 1991). Absolute reflectance values of pyroxene spectra depend on grain size. The overall reflectance of pyroxenes increases as the grain size of mineral (and other mafic silicate minerals) decreases. Previous studies suggest that hyperspectral remote sensing data could be used to extract compositional information that in turn could be applied to address various geochemical issues (Cloutis 1996).

The following five minerals, including *annabergite* (Ni^{2+}), *chrysocolla* (Cu^{2+}), *almandine* (Fe^{3+}), *rhodochrosite* (Mn^{2+}), and *corundum* (Cr^{3+}), exemplify features typical of each of the five transition metal ions occasionally present in mineral spectra (Hunt 1977). The features are characteristics of the crystal field transitions in that ion. The annabergite is a sample with transitions from the ground state to three upper states that produce significant absorption bands near 0.4, 0.74, and 1.25 μm because such transitions are spin-allowed. An absorption feature of the chrysocolla associated with cupric ions (Figure 7.1) shows the intense feature near 0.8 μm that is due to the spin-allowed transition from the ground state to the upper state. There are many attempts to study spectral features due to transitions of the ferric ion (Fe^{3+}) in a variety of minerals. The locations of such absorption features can be located in the almandine spectrum (Figure 7.1). Fe^{3+} absorption features are most clearly observed in the spectra of garnets and beryl (Hunt 1977). The absorption features of the mineral rhodochrosite are due to presence of the manganese ion in the blue-green spectral range. For this case, although transitions are spin-forbidden, the features occur in the spectrum at ~0.34, 0.37, 0.41, 0.45, and 0.55 μm (Hunt 1977; Figure 7.1). Transitions derived absorption features of the mineral corundum, displayed in Figure 7.1, are located near 0.4 μm and 0.55 μm, respectively.

7.2.1.2 Due to Charge Transfer

Absorption features of some minerals can also be caused by charge transfers. The charge-transfer, or inter-element electron transition, refers to the process whereby absorbed energy causes an electron to migrate between neighboring ions, or between ions and ligands (Hunt 1977). The transition can also occur between the same metal ions in different valence states, such as between Fe^{2+} and Fe^{3+} and between Mn^{3+} and Mn^{2+}. In general, such absorption bands caused by charge transfers are diagnostic of mineralogy for identifying different minerals in hyperspectral data. Compared to the absorption strength caused by the crystal field effects, the absorption strength caused by the charge transfers is typically hundreds to thousands of times stronger (Hunt 1977, Clark 1999). The location centers of the absorption bands usually occur in the ultraviolet with the wings of the absorption extending into the visible. The red color of iron oxides and hydroxides is mainly resulting from the charge transfer absorptions. It is found that one of the most commonly observed features (absorption bands) in the spectra of terrestrial geologic materials is rapidly decreasing in spectral intensity from the visible into the ultraviolet, and this intensity fall-off phenomenon is particularly evident in the spectra of weathered materials (Hunt 1977, Morris et al. 1985). This phenomenon can be explained by the increased ratio of surface to volume at small-grain–size that results in a greater proportion of grain boundaries where crystal field effects are different, thus resulting in lower magnetic coupling and reduced absorption strength (Morris et al. 1985). Reflectance spectra of iron oxides have such strong absorption bands that the shape changes significantly with grain size (Clark 1999).

For example, Figure 7.1 illustrates the absorption features (0.4 μm) of three minerals: *limonite*, *augite*, and *carnotite*, caused by various types of charge-transfers. The intense absorption feature in the spectrum of limonite is due to the tail of the Fe-O charge-transfer band, while the additional band at longer wavelengths is due to a crystal-field transition. For the augite spectrum, its absorption feature (0.7 μm) is due to the charge-transfer transition between a ferrous and ferric ion, whereas the feature (0.3–0.53 μm) indicated in the carnotite spectrum is due to at least three charge-transfer transitions between the uranium and oxygen atoms (Hunt 1977).

7.2.1.3 Due to Color Centers

A few colored materials display spectral features in the visible that are caused by the presence of electronic phenomena called *color centers*. A color center is caused by irradiation (e.g., by solar UV radiation) of an imperfect crystal. Imperfect crystals in nature have lattice defects that disturb the periodicity of the crystal, caused by impurities (Clark 1999). Such lattice defects can produce discrete energy levels into which excited electrons can fall, and such electrons can become bound to the defects (Hunt 1977). The movement of an electron into the defect requires photon energy. There are many different types of lattice defects that have been studied extensively, but the most commonly encountered color center is referred to as the *F-center.* For example, the yellow, purple, and blue colors of fluorite are typically caused by color centers in natural crystals (Hunt 1977; Figure 7.1).

7.2.1.4 Due to Conduction Band Transitions

In some minerals, there are two energy bands in which the electrons may reside: a higher energy region called the *conduction band*, where electrons move freely throughout the lattice, and a lower energy region called the *valence band*, where electrons are attached to specific ions or bonds (Hunt 1977, Clark 1999). There is a zone of energies between these two energy bands, which is called the *forbidden band* or *band gap*. Metals typically display a very narrow or nonexistent forbidden band gap, while dielectric materials typically have a very wide forbidden band gap. In semiconductors, the forbidden gap is intermediate between metals and dielectrics and corresponds to the energy of visible to near-infrared (VNIR) wavelength photons and the spectrum in this case is approximately a step function (Clark 1999). For this case, the edge of the conduction band is marked by the appearance of an intense absorption edge in the VNIR. The sharpness of this absorption edge is a function of the purity and crystallinity of the material (Hunt 1977). The three minerals, *sulphur* (S), *cinnabar* (HgS), and *stibnite* spectra in Figure 7.1 display clear absorption edges that make the transition from intense absorption in the conduction band at shorter wavelengths to complete transmission in the forbidden gap at longer wavelengths (Hunt 1977). In Figure 7.1, it is interesting that a mineral called *arsenopyrite* shows the conduction (absorption) band covering completely the VNIR range.

7.2.2 Spectral Absorption Characteristics Caused by the Vibrational Processes

Spectral reflectance in the MIR is heavily influenced by vibrational features of water, hydroxyl, carbonate, and sulfate molecules (Hunt 1977, van der Meer 2004). The vibrational features are observed as reflection minima because they are a result of volume scattering (Eismann 2012). According to Clark (1999), the bonds in a molecule or crystal lattice are similar to springs with attached weights: the whole system can vibrate. The strength of each spring (the bond in a molecule) and their masses (the mass of each element in a molecule) determine the frequency of vibration. For a molecule with N atoms, there are $3N-6$ normal modes of vibrations, which are called fundamentals. The additional vibrations are called *overtones* when they include multiples of a single fundamental mode and combinations when they involve different modes of vibrations. When two or more different fundamentals or overtone vibrations occur, a combination tone feature appears located at (or near) the sum of all the fundamentals or overtone frequencies involved (Hunt 1977). Therefore, the number and form of the normal vibrations and the values of the permitted energy levels for any material are determined by the number and type of its constituent atoms, their spatial geometry, and the magnitude of the binding forces between them (Hunt 1977). In addition, it is observed that energy required to excite the fundamental modes of all geologically important materials falls in the mid- and far-infrared regions. What is also observed in the NIR are features due to overtones and combinations of groups that have very high fundamental frequencies. Consequently, the absorption (diagnostic) features by different types of vibrations in MIR region provide a basis for identifying and mapping terrestrial materials from hyperspectral data.

7.2.2.1 Due to Molecular Water

Produced by overtones and combinations of vibrations associated with liquid water that has three fundamental vibrational modes, the diagnostic absorptions in the MIR range in minerals are located near 1.875, 1.454, 1.38, 1.135, and 0. 942 μm. Spectral vibration (absorption) features in some minerals (e.g., *gypsum, natrolite, nontmorillonite*, and *quartz*) due to the vibrational combinations and overtones of molecular water contained in various locations in the minerals shown in Figure 7.1. For example, water may be present in a mineral as individual or clusters of molecules at specific sites essential to its structure, as in hydrates, such as gypsum; water may also be trapped in liquid inclusions in the crystal structure, as in milky quartz (Hunt 1977).

7.2.2.2 Due to Hydroxyl

The hydroxyl ion (OH) group has only one fundamental stretching mode, but its absorption spectral location is dependent on the ion (typically a metal ion) to which it is attached. Exactly where this vibrational feature appears depends upon what the OH is directly attached to, and where it is located in the material (Hunt 1977). For example, there are three sharp bands that occur at 2.719, 2.730, and 2.743 μm in the spectrum of talc. The feature bands due to the fundamental OH-stretching mode locate just outside the MIR range. Thus, what is observed in the MIR are features due to the first overtone of the OH stretch (e.g., the band near 1.4 μm), or due to combination bands of the fundamental vibration stretch with an X–OH bending fundamental (where X is usually Al or Mg), or due to the OH stretch in combination with some lattice or vibrational mode (Hunt 1977). Combination bands in the 2.0 μm region that include the OH fundamental stretching mode generally appear in pairs, such as with the shorter wavelength member typically near either 2.2 or 2.3 μm. The location of the more intense feature appears near 2.2 μm depending upon whether aluminum (Al) is present or near 2.3 μm depending upon whether magnesium (Mg) is present (Hunt 1977, Eismann 2012). In Figure 7.1, minerals *muscovite, phligopite, kaolinite*, and *amphibole* display different vibrational (absorption) features, which are due to the overtone and combination tones of the hydroxyl group appearing in different environments and in various minerals.

7.2.2.3 Due to Carbonate, Borate, and Phosphate

Carbonate, borate, and phosphate also show diagnostic vibrational absorption bands (Clark 1999). Features occur in the MIR spectra between 1.6 and 2.5 μm as a result of overtone or combinations of the internal vibrations of the carbonate ion. Such features are usually quite distinctive (Hunt 1977). The observed absorptions are due to the planar CO_3^{2-} ion. Carbonates are not usually associated with water, so intense water features are frequently not seen to confuse the spectra. In the MIR spectral region, carbonates typically display a series of five very characteristic bands (e.g., *calcite*, Figure 7.1), the two features at the longer wavelengths are quite clearly doubled and are considerably more intense than the three features at shorter wavelengths, which generally have shoulders on their short wavelength sides. The doubling can be explained in terms of lifting of the degeneracies (Hunt 1977). The spectral features indicated in *amblygonite* are due to motions (vibrations) involving the P-O-H group. The features in the spectrum of *colemanite* can be ascribed to overtones of the BO_3^{3-} ion just as similar features are explained in the carbonate spectra by the CO_3^{2-} vibrations (Figure 7.1; Hunt 1977).

7.2.3 Spectral Absorption Characteristics of Alteration Minerals

Hydrothermal alteration characteristics are frequently indicated by iron oxide and sulfate minerals, which produce characteristic signatures throughout the visible and into the NIR range (0.4 to 1.1 μm) as a consequence of electronic processes involving iron (Hunt 1977, Clark 1999). In addition, extremely useful information also exists throughout the MIR range (1.1 to 2.5 μm) that is caused by vibrational processes occurring in some of the molecular groups that constitute alteration minerals and rocks (Hunt and Hall 1981, Hutsinpiller 1988, Kruse 1988). Therefore, the absorption features in the spectra of hydrothermal alteration minerals/rocks are also caused by the electronic processes and vibrational processes and the spectral characteristics may still be explained by the two types of processes associated with metal ions (VNIR range) and water and hydroxyl, etc. (MIR range), as discussed in the last two sections. In this section, a collection of typical alteration minerals and their corresponding positions of reflection minima in the MIR range (1.1 to 2.5 μm) and relative intensities of absorption bands are tabulated (Table 7.1) and a brief discussion is given.

TABLE 7.1
Typical Alteration Minerals and Their Positions of Reflection Minima

Alteration Minerals	Reflection Minima (μm) and Relative Intensities	Sources
Pyrophyllite	1.332 (w); 1.345 (vw); 1.361 (w); 1.392 (vvs); 1.420 (m); 1.92 (vb); 2.062 (m); 2.080 (m); 2.166 (vvs); 2.205 (sh); 2.319 (s); 2.346 (w); 2.390 (ms)	Hunt 1979
Kaolinite	1.330 (sh); 1.357 (m); 1.394 (vs); 1.403 (m); 1.413 (vvs); 1.820 (w); 1.840 (w); 1.914 (vb); 2.090 (sh); 2.120 (sh); 2.162 (vs); 2.194 (sh); 2.209 (vvs); 2.322 (w); 2.357 (w); 2.382 (ms)	
Alunite	1.317 (w); 1.335 (mw); 1.355 (w); 1.375 (sh); 1.424 (vvs); 1.430 (sh); 1.458 (sh); 1.476 (vs); 1.682 (sh); 1.762 (vs); 1.930 (sh); 1.960 (w); 2.010 (vs); 2.060 (w); 2.152 (sb); 2.165 (sb); 2. 178 (s); 2.208 (vvs); 2.317 (vs)	
Muscovite	1.412 (vs); 1.842 (w, b); 1.912 (mb); 2.120 (sh); 2.208 (vvs); 2.240 (sh); 2.348 (s); 2.376 (sh)	
Montmorillonite	1.408 (vs); 1.455 (sh); 1.899 (vvs, b); 1.940 (b, sh); 2.070 (vw, b); 2.090 (w. vb); 2.205 (vvs); 2.232 (sh)	
Jarosite	1.468 (sh); 1.475 (vvs); 1.849 (vs); 1.862 (vs); 2.230 (b, sh); 2.264 (vvs)	
Gypsum	1.375 (sh); 1.443 (vvs); 1.486 (vs); 1.533 (s); 1.745 (vs); 1.772 (sh); 2.075 (vs); 2.125 (w, sh); 2.176 (sh); 2.215 (m); 2.265 (mw)	
Calcite	1.770 (vb); 1.875 (ms); 1.993 (ms); 2.153 (mw); 2.305 (sh); 2.337 (vvs)	
Chlorite	2.23-2.25 (sh); 2.29-2.32 (vs); 2.36-2.38 (w); 2.42-2.43(sh)	Hutsinpiller 1988
Epidote	2.2 (sh); 2.25 (sh); 2.33-2.35 (vs)	
Sericite	2.120 (sh); 2.208 (ws); 2.240 (sh); 2.348 (s); 2.376 (sh)	
White mica	similar to sericite	
Illite	similar to sericite	
Magnesite	2.305; 2.235; 2.137; (only consider 2.1 - 2.5 μm)	Kruse 1988
Dolomite	2.318; 2.265; 2.146; (only consider 2.1 - 2.5 mm)	
Siderite	2.329; 2.252; 2.183; (only consider 2.1 - 2.5 mm)	
Dickite	1.41; 2.20	Hunt and Hall 1981

Note: Code: b, broad; m, medium; s, strong; sh, shoulder; v, very; w, weak.

Table 7.1 lists typical alteration minerals and their positions of reflection minima, based on the published data by Hunt (1979), Hunt and Hall (1981), Hutsinpiller (1988), and Kruse (1988). The reflection minima shown in Table 7.1 are mostly attributed to the vibrational processes related to overtone and combination bands of hydroxyl-group vibrations in the MIR range. Many hydrothermal alteration mineral spectra display well-defined and characteristic features in the regions near 1.4, 1.76, and 2.0 to 2.4 μm. According to Hunt (1979), for remote-sensing purposes, the features in the 2.2 μm region are unique and particularly valuable because they are common to alteration minerals, and they allow discrimination from non-alteration minerals that provide features only as close as to 2.4 μm. In addition, in order to discriminate alteration minerals, at least two filters (one located near 2.166 μm and the other near 2.21 μm) with band width < 0.1 μm would be required.

7.3 ANALYTICAL TECHNIQUES AND METHODS IN GEOLOGICAL APPLICATIONS

In this section, six types of analytical methods with multiple techniques, algorithms, and models specially used for mineralogical mapping and identification from hyperspectral data sets (including laboratory and *in situ* spectral measurements and airborne and spaceborne hyperspectral imaging data) are introduced and reviewed. The principles and algorithms of most of the analytical methods/

techniques are already introduced and discussed previously, so for those readers who are more interested in detailed introduction of algorithms/models, refer to Chapter 5. Here, an emphasis is placed on mineralogical applications of these methods and techniques using hyperspectral data sets. Meanwhile, the characteristics, advantages, and limitations of these methods/techniques during the procedure in their applications will also be briefly evaluated.

7.3.1 Spectral Absorption Feature Extraction in Spectra of Minerals

Since there are diagnostic absorption features in various hyperspectral data (Hunt 1980, Goetz et al. 1985), scientists have developed various techniques and methods to process the hyperspectral image (HSI) data to obtain pixel-based surface compositional information. The techniques developed specially for extracting diagnostic absorption features (e.g., absorption band position, depth, and asymmetry) for identifying and mapping minerals/rocks from HSI data include a direct method extracting absorption features from continuum removal spectra (Clark and Roush 1984), a relative absorption band-depth approach (Crowley et al. 1989), spectral feature fitting (SFF) or least-squares spectral band-fitting techniques (least-squares fitting) (Clark et al. 1990, 1991), a simple linear interpolation technique (van der Meer 2004), and a spectral absorption index (SAI; Huo et al. 2014). In the following, the five techniques/methods and their applications in identifying and mapping minerals/rocks from HSI data are reviewed and discussed.

A continuum corresponds to a background signal unrelated to specific absorption features of interest. Spectra are normalized to a common reference using a continuum formed by defining high points of the spectrum (local maxima) and fitting straight line segments between these points. The continuum is removed by dividing it into the original spectrum. Since the direct absorption feature extraction method with continuum removal technique proposed by Clark and Roush (1984) has been described in Section 5.4 of Chapter 5, readers should refer to that section and its relevant references.

There are many geologists applying analysis skills of HSI data, such as different absorption feature extraction techniques and methods to identify and map various surface materials. Kruse (1988) simulated a continuum spectrum using a second-order polynomial that was fitted to selected channels (channels without known absorption features) in the internal average relative (IAR) reflectance spectra and calculated the continuum removed spectra by dividing the polynomial function into the AIS data. The strongest absorption feature was defined as the wavelength position of the channel with the maximum depth from the continuum removed spectrum, and thus the absorption band position, the band depth, and the band width might be identified and calculated. The absorption features were used to map areas of quartz–sericite–pyrite alteration zone by identifying sericite absorption features at 2.21, 2.25, and 2.35 μm, areas of argillic alteration containing montmorillonite based on a single absorption feature at 2.21 μm, and areas of calcite and dolomite by identifying their sharp diagnostic features at 2.34 and 2.32 μm from AIS images acquired from the northern Grapevine Mountains, Nevada/California. The mapped areas using the AIS data agreed well with the areas identified by field mapping techniques. When using the continuum removal (CR) technique to extract absorption features, the features extracted by the CR could be a result of comprehensive effect of different factors due to a band range containing more than one absorption contribution factors. For this case, Zhao et al. (2015) proposed a new spectral feature extraction method named *reference spectral background removal* (RSBR). Given the reference spectral background, RSBR can eliminate the influence of unwanted contribution factors and extract the absorption feature of the target contribution factor. Using RSBR, the basic absorption feature parameters, including absorption center, absorption width, and absorption depth are extracted. Their experimental results show that RSBR could effectively extract pure absorption features of the target material and that more accurate absorption feature parameters could also be achieved. RSBR can be utilized to eliminate the effects of a wide range of existing background materials, such as vegetation and soil, and extract the absorption features of potential target (altered minerals). However, RSBR has its own limitation when dealing with the multi–end-member issue, and thus the circumstances associated with hyperspectral data are much more complicated. In a vegetated area (with green or dry vegetation), absorption features (e.g., Al-OH feature at 2.20 μm) extracted using the traditional CR technique are usually reduced. To reduce the effect of surface vegetation cover on extracting the depth of the 2.20 μm Al-OH feature, Rodger and Cudahy (2009) proposed a vegetation corrected continuum depth (VCCD) method. The method uses a multiple linear regression model where the coefficients of the linear model are produced via forward modeling, and where the independent variables are continuum removed band depth (CRBD), which are used to detect the presence of green and dry vegetation and the uncorrected Al-OH CRBD. The proposed VCCD method was tested with synthetic datasets as well as hyperspectral data (HyMap) collected at Mount Isa in Queensland, Australia. Their results of using the VCCD method on the uncorrected HyMap data were validated with vegetation free samples collected from the Mount Isa region. Improvements in the R^2 statistics of the corrected 2.20 μm CRBD by using the VCCD method to the vegetation free CRBD were found to be 2–4 times greater than the uncorrected 2.20 μm CRBD.

By using hyperspectral Hyperion image data, covering a mineralized belt in the Noamundi area, eastern India and extracted absorption features in the continuum removed spectra of Hyperion data, Magendrana and Sanjeevi (2014) studied to differentiate iron ores in terms of their grades. They found that (1) spectral curves for iron ore deposits extracted from the Hyperion image pixels display strong absorption at 850–900 nm and 2150–2250 nm wavelengths, which is typical of iron ores, and (2) extracted absorption features from Hyperion data correlate well with the concentration of iron-oxide and alumina (gangue) in the ore samples obtained from the mine face. The experimental results demonstrate that correlations are evident between the concentration of iron oxide and (1) the depth of NIR absorption feature ($R^2 = 0.883$), (2) the width of NIR absorption feature ($R^2 = 0.912$), and (3) the area of the NIR absorption feature ($R^2 = 0.882$). To estimate clay and calcium carbonate ($CaCO_3$) content from laboratory and airborne hyperspectral measurements across VNIR/MIR spectral ranges, Gomez et al. (2008) compared and evaluated performances of the two methods: the continuum removal (CR) and the partial least-squares regression (PLSR). The CR method has been used to correlate spectral absorption bands centered at 2206 nm and 2341 nm with clay and $CaCO_3$ concentrations, while the PLSR method has been used to predict clay and $CaCO_3$ concentrations from VNIR/SWIR full spectrum. Their study shows that when airborne HyMap reflectance data are used, the PLSR technique performs better than the CR approach. In mapping buddingtonite in a study area in the southern Cedar Mountains, Nevada, Baugh et al. (1998) first extracted 2.12 μm ammonium absorption features in buddingtonite for every pixel of AVIRIS imagery by using the CR technique. A linear calibration determined by the laboratory analysis was then applied to convert the band depth at 2.12 μm to ammonium concentration for mapping the mineral in the study area. The analysis result suggests that remote geochemical mapping using imaging spectrometer data would be possible, and presents a methodology that could be extended to quantitatively map other minerals that have absorption features in the MIR range.

The relative absorption band-depth (RBD) technique was developed to detect the presence of diagnostic mineral absorption feature (Crowley et al. 1989) and to create RBD images, like band-ratio images for providing a semi-quantitative measure of mineral absorption intensity from the radiometrically corrected hyperspectral imagery. To produce a RBD image, several data channels from both absorption-band shoulders are summed and then divided by the sum of several channels from the absorption-band minimum (Crowley et al. 1989). Basically, a RBD image can provide a local continuum correction to remove any small band to band radiometric offsets, as well as variable atmospheric effects for each pixel in the data set. Therefore, the resulting RBD image can provide a measure the depth of an absorption feature relative to the local continuum which can be used to identify pixels that have stronger absorption-bands associated with a certain mineral. By using the RBD images, Crowley et al. (1989) demonstrated that a number of rock and soil units were distinguished in the Ruby Mountains, including weathered quartz-feldspar pegmatites, marbles of several compositions, and soils developed over poorly exposed mica schists from AIS imagery in the MIR (1.2–2.4 μm) wavelength range. They reported that (1) the RBD images are both highly specific and sensitive to the presence of particular mineral absorption features, and (2) the RBD technique is especially well suited for detecting weak MIR spectral features produced by soils, which implies that it is possible to improve mapping of subtle lithologic and structural details in semiarid terrains.

Clark et al. (1990) developed a technique called *least-squares fitting* to be used for mapping mineral and vegetation absorption features in imaging spectrometer data. The technique was demonstrated using a least-squares solution of the total band shape of a reference library spectrum to spectra from a hyperspectral data set. The technique first removes a continuum from the both observed and reference (library, L_c) spectra and expands and compresses the reference absorption profile to best fit the observed data. To do so, according to Clark et al. (1990), the contrast in the reference library spectrum absorption feature is modified by adding a simple constant, k:

$$L_c' = (L_c + k)/(1.0 + k) \tag{7.1}$$

where L_c' is the modified continuum removed spectrum that best matches the observed spectrum by adding an appropriate k. This equation can be rewritten in the form:

$$L_c' = a + bL_c \tag{7.2}$$

where $a = k/(1.0 + k)$, and $b = 1.0/(1.0 + k)$. In Equation 7.2, we want to find the a and b that give a best fit to the observed spectrum. The solution can be done using standard least squares fitting using a number of spectral bands available from both continuum removed reference and observed spectra (Clark et al. 1990). The advantage of the technique is that it can map complicated band shapes and fit to all data points comprising the feature. The algorithm can compute the band depth for a particular absorption feature, a goodness of fit, and the reflectance level of the continuum at the band center. Images of the band depth and goodness of fit are used for mapping a specific mineral. The images from this method show minerals' distribution in significantly better detail and with less noise than simpler methods such as band ratios or band depth images computed from two continuum and one band center channels (Clark et al. 1990). Thus, combinations of mineral absorption features may be used to map geologic units.

The least-squares fitting technique was substantially extended to multiple absorption features in the both continuum removed reference and observed spectra (Clark et al. 1991). For the multiple absorption features fitting, the same basic algorithm is still used, and the fits from multiple minerals are compared to determine what mineral absorption features are present in the spectrum. This new version of the technique substantially increases the ability to discriminate between two minerals. The algorithm has been thoroughly tested using mineral spectra where controlled amounts of noise have been added and demonstrated that use of multiple diagnostic absorption features would allow discrimination at even lower signal-to-noise (ratio) levels (Clark et al. 1991).

SFF embedded in the ENVI software (Exelis 2015) is an absorption feature–based method for matching image (observed) spectra to reference end-members, similar to the techniques above developed at the USGS (Clark et al. 1990, 1991). SFF requires users to select reference end-members (minerals) from either an image or a spectral library, to remove the continuum from both the reference and unknown (observed) spectra, and to scale each reference end-member spectrum (by adding a similar constant k as above) to match the unknown spectrum. A "scale" image for each end-member (mineral) selected for analysis is created by first subtracting the continuum-removed spectrum from one end-member spectrum and then inverting it and making the continuum (background) zero. Supposing that a reasonable spectral range has been selected, a large scaling factor is equivalent to a deep spectral feature, while a small scaling factor indicates a weak spectral feature (Exelis 2015). SFF then calculates a least-squares-fit, band-by-band, between each reference end-member and the unknown spectrum. The total root–mean–square (RMS) error is used to form an RMS error image for each end-member. A ratio image of scale image to RMS error image provides a "fit" image that is a measure of how well the unknown spectrum matches the reference spectrum on a pixel-by-pixel basis. In addition, the multi-range SFF function in the ENVI software allows a user to define multiple and different wavelength ranges around each end-member's absorption features (Exelis 2015), similar to the multiple absorption features fitting technique in Clark et al. (1991). Both SFF and multi-range SFF are performed to image spectra after the continuum is removed from both datasets.

By using AVIRIS high spectral resolution data and the least-squares spectral band fitting algorithm (i.e., least-squares fitting technique), Crowley (1993) studied evaporite minerals in Death Valley playas. Evaporite minerals possess spectral absorption features in the AVIRIS spectral range that have been used to map these sedimentary units. In the study, eight different saline minerals were remotely identified, including three borates, hydroboracite, pinnoite, and rivadavite, which have not been previously reported from the Death Valley efflorescent crusts (Crowley 1993). The mapping results demonstrate the potential for using AVIRIS and other imaging spectrometer data to study playa chemistry. Also, using AVIRIS data covering the Bodie and neighboring Paramount mining districts, eastern California and two algorithms, least-squares fitting (from Tricorder) and the spectral angle mapper (SAM), Crósta et al. (1998) mapped hydrothermal alteration minerals and compared the performances of the two algorithms. The experimental results indicated that both algorithms appeared to produce satisfactory results for geologic reconnaissance and mapping applications, but the least-squares fitting generally classified more pixels and identified more mineral species than the SAM. The results also demonstrated that laboratory spectra of rock samples from five localities could match test spectra of minerals identified by the two mapping algorithms for corresponding pixels to some extent.

To calculate the absorption features parameters (absorption band position and depth) from continuum removed image spectra, van der Meer (2004) proposed a simple liner interpolation method (SLIM). Figure 7.2 illustrates the procedure of the SLIM method from a continuum removed spectrum (van der Meer 2004). According to van der Meer (2004) and using continuum removed hyperspectral spectra, the SLIM algorithm is briefly introduced as follows.

To calculate the two distances coefficients, denoted C_1 and C_2, for an absorption feature of interest, the hyperspectral image bands are first determined, which might be served as shoulders of the absorption features by visual inspection using expert knowledge on hyperspectral remote sensing. By definition, Figure 7.2 shows the locations of two shoulder bands: a short wavelength shoulder (denoted S_2) and a long wavelength shoulder (denoted S_1). Next, two bands (denoted A_1 and A_2 in Figure 7.2) located between S_1 and S_2 are selected as the absorption points that may be used in the linear interpolation. Then two distance coefficients C_1 and C_2 are calculated as

$$C_1 = \sqrt{(Depth_1)^2 + (S_1 - A_1)^2} \tag{7.3}$$

and

$$C_2 = \sqrt{(Depth_2)^2 + (S_2 - A_2)^2} \tag{7.4}$$

Referring to Figure 7.2, the absorption wavelength positions (AWP) can be interpolated using parameters C_1, C_2, A_1, and A_2 as

$$AWP = A_1 - \left[\frac{C_2}{C_1 + C_2} \times (A_1 - A_2)\right] \tag{7.5}$$

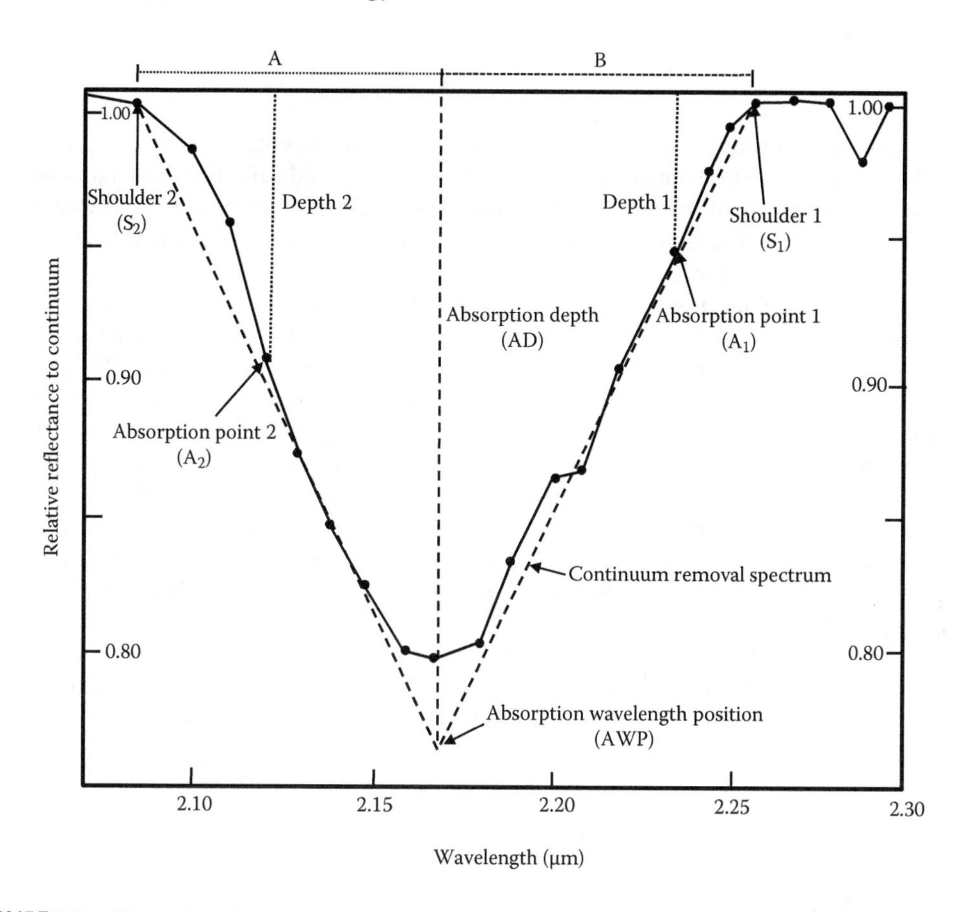

FIGURE 7.2 Illustration of the simple linear interpolation method for extracting absorption feature parameters: absorption wavelength position and depth. (Modified from van der Meer, F., *Journal of Applied Earth Observation and Geoinformation*, 5, 55–68, 2004.)

or

$$AWP = A_2 + \left[\frac{C_1}{C_1 + C_2} \times (A_1 - A_2)\right] \tag{7.6}$$

Therefore, the associated absorption band depth (AD) can be derived as

$$AD = \left[\frac{S_1 - AWP}{S_1 - A_1}\right] \times Depth_1 \tag{7.7}$$

or

$$AD = \left[\frac{AWP - S_2}{A_2 - S_2}\right] \times Depth_2 \tag{7.8}$$

And the asymmetry factor of the absorption feature is calculated as

$$asymmetry = A - B = (AWP - S_2) - (S_1 - AWP) \tag{7.9}$$

By Equation 7.9, this operation returns 0 for a perfect symmetric absorption feature, a negative value for an absorption feature skewing toward the longer wavelength, and a positive value for an absorption feature skewing toward the shorter wavelength. One of the advantages of the linear technique is that it is easily implemented by users (for example using the band math operator of ENVI system) who are not familiar with programming languages (van der Meer 2004).

By using the simple linear interpolation method (SLIM) and AVIRIS data acquired in 1995 over the Cuprite mining area (in Nevada), van der Meer (2004) derived absorption-band position, depth, and asymmetry from the hyperspectral image data. A sensitivity analysis of the method proposed shows that good results can be obtained for estimating the absorption wavelength position; however, the estimated absorption–band–depth is sensitive to the input parameters chosen (e.g., selecting shoulder bands and absorption bands). The testing results demonstrate that the resulting parameter images (including depth, position, and asymmetry of the absorption), after carefully examined and interpreted by an experienced remote sensing geologist, can provide key information on surface mineralogy. In addition, the estimates of the absorption band depth and position can be related to the chemistry of the samples and thus allow to bridge the gap between field geochemistry and hyperspectral remote sensing (van der Meer 2004).

The spectral absorption index (SAI) is defined as the reciprocal of the ratio between the local minimum reflectance of point M and the corresponding value of the spectral absorption baseline, defined with the two shoulder points (bands) P_1 and P_2 (Huo et al. 2014; Figure 7.3), and it can be described as

$$SAI = \frac{S\rho_{\lambda_1} + (1 - S)\rho_{\lambda_2}}{\rho_{\lambda_m}} \tag{7.10}$$

where S is the symmetry defined as $(\lambda_2 - \lambda_m)/(\lambda_2 - \lambda_1)$, and ρ_{λ_1}, ρ_{λ_2}, and ρ_{λ_m} are the corresponding reflectances at points (bands) P_1, P_2, and M. By using HyMap data and based on the SAI and other spectral absorption features, Huo et al. (2014) identified and mapped multiple minerals in a study area in Tudun, eastern Tien Shan, in Northwestern China. Alteration minerals, such as calcite, alumina-rich (Al-rich) muscovite, epidote, and antigorite, were explored, and their relative abundance was depicted. The mapping results demonstrated that the spectral absorption band-depth and the SAI had a higher degree of consistency with the relative abundance of minerals than the other spectral

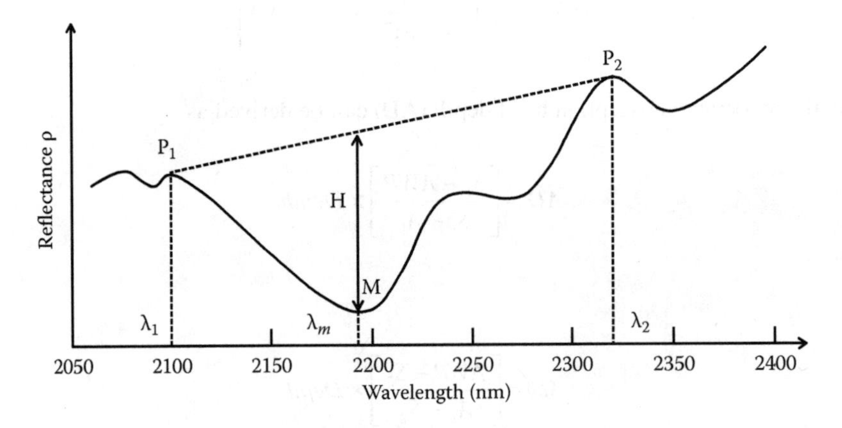

FIGURE 7.3 Spectral absorption features (bands) used for calculating spectral absorption index (SAI). (Modified from Huo, H., Ni, Z., Jiang, X., Zhou, P., and Liu, L., *Remote Sensing*, 6, 11829–11851, 2014.)

absorption features. However, although the SAI and its model put the emphasis on single spectral absorption features, in practical geological applications, it is difficult to distinguish and classify those minerals that have similar spectral absorption features such as epidote and chlorite (Huo et al. 2014).

7.3.2 Identifying and Mapping Minerals Using Hyperspectral Mineral Indices

Hyperspectral mineral indices (HMIs) or hyperspectral band ratios allow features of interest (e.g., absorption features in spectra of minerals) to be accentuated. This band ratioing technique (spectral vegetation indices) has been wildly applied not only to plant spectra for biophysical and biochemical parameter extraction and analysis (e.g., Pu and Gong 2011) but also to mineralogical spectra for discriminating and mapping minerals, soils, and rocks (e.g., van Ruitenbeek et al. 2006, Henrich et al. 2012). Table 7.2 present a list of 23 HMIs extracted from airborne and spaceborne hyperspectral

TABLE 7.2
A List of Hyperspectral Mineral Indices Used for Identifying and Mapping Minerals and Rocks from Hyperspectral Data

Index Name (for Mineral/Rock)	Formula	Suitable for Sensor
Alteration	$R_{1600-1700}/R_{2145-2185}$	AVIRIS, DAIS-7915, HYDICE, HyMAP, Hyperion
Alunite/Kaolinite/Pyrophylite	$(R_{1600-1700}+R_{2185-2225})/R_{2145-2185}$	AVIRIS, DAIS-7915, HYDICE, HyMAP, Hyperion
Amphibole	$R_{2185-2225}/R_{2295-2365}$	AVIRIS, DAIS-7915, HYDICE, HyMAP, Hyperion
Amphibole/MgOH	$(R_{2360-2430}+R_{2185-2225})/R_{2295-2365}$	AVIRIS, DAIS-7915, HYDICE, HyMAP, Hyperion
Carbonate	$R_{10250-10950}/R_{10950-11650}$	DAIS-7915
Carbonate/Chlorite/Epidote	$(R_{2360-2430}+R_{2235-2365})/R_{2295-2365}$	AVIRIS, DAIS-7915, HYDICE, HyMAP, Hyperion
Dolomite	$(R_{2185-2225}+R_{2295-2365})/R_{2235-2365}$	AVIRIS, DAIS-7915, HYDICE, HyMAP, Hyperion
Epidote/Chlorite/Amphibole	$(R_{2185-2225}+R_{2360-2430})/(R_{2235-2365}+R_{2295-2365})$	AVIRIS, DAIS-7915, HYDICE, HyMAP, Hyperion
Ferric Oxides	$R_{1600-1700}/R_{760-860}$	AVIRIS, DAIS-7915, HYDICE, HyMAP, Hyperion
Ferrous Iron	$(R_{2145-2185}/R_{760-860})+(R_{520-600}/R_{630-690})$	AVIRIS, DAIS-7915, HYDICE, HyMAP, Hyperion
Ferrous Silicates	$R_{2145-2185}/R_{1600-1700}$	AVIRIS, DAIS-7915, HYDICE, HyMAP, Hyperion
Host Rock	$R_{2145-2185}/R_{2185-2225}$	AVIRIS, DAIS-7915, HYDICE, HyMAP, Hyperion
Kaolinitic	$R_{2235-2365}/R_{2145-2185}$	AVIRIS, DAIS-7915, HYDICE, HyMAP, Hyperion
Laterite	$R_{1600-1700}/R_{2145-2185}$	AVIRIS, DAIS-7915, HYDICE, HyMAP, Hyperion
Muscovite	$R_{2235-2365}/R_{2185-2225}$	AVIRIS, DAIS-7915, HYDICE, HyMAP, Hyperion
Phengitic	$R_{2145-2185}/R_{2185-2225}$	AVIRIS, DAIS-7915, HYDICE, HyMAP, Hyperion
Quartz Rich Rocks	$R_{10950-11650}/R_{8925-9275}$	DAIS-7915
Sericite/Muscovite/Illite/Smecite	$(R_{2145-2185}+R_{2235-2365})/R_{2185-2225}$	AVIRIS, DAIS-7915, HYDICE, HyMAP, Hyperion
Silica 2	$R_{8475-8825}/R_{8925-9275}$	DAIS-7915
Silica 3	$R_{10250-10950}/R_{8125-8475}$	DAIS-7915
Simple Ratio Eisenhydroxid-Index	R_{MIR}/R_{RED}	AVIRIS, DAIS-7915, HYDICE, HyMAP, Hyperion, CASI1500
Simple Ratio Iron Oxide	R_{RED}/R_{BLUE}	AVIRIS, DAIS-7915, HYDICE, HyMAP, Hyperion, CASI1500
Simple Ratio Ferrous Minerals	R_{SWIR}/R_{NIR}	AVIRIS, DAIS-7915, HYDICE, HyMAP, Hyperion, CASI1500

Source: Henrich, V., G. Krauss, C. Götze, and C. Sandow, *IDB—Index DataBase*, Entwicklung einer Datenbank für Fernerkundungsindizes, 2012. See http://www.indexdatabase.de.

sensors: AVIRIS, CASI-1500, DAIS-7915, HYDICE, HyMap, and Hyperion for estimating and mapping corresponding minerals and rocks (Henrich et al. 2009, 2012). Such mineral indices are mostly named with corresponding minerals/rocks.

Many geologists have used the HMIs or developed their own band ratios/mineral indices from the hyperspectral data sets for their studies. For examples, with high spectral and spatial resolution airborne imaging spectrometer (AIS) data, Feldman and Taranik (1988) utilized three mineral discriminant and mapping methods: band ratios, principal component analysis (PCA), and a signature matching algorithm to map hydrothermal alteration mineralogy in the Tybo mining district, Hot Creek Range, Nevada. Band ratios were created from AIS data in Red Rock Canyon and Music Canyon areas to test their effectiveness for mineralogical discrimination. The band ratios were calculated with a highly reflective band as the numerator and a band within an absorption feature as the denominator. Band ratios are advantageous as they use simple algorithms and consume very little computer time. The highly reflective band at 2095 nm was used as the ratio numerator. Those within the absorption features of kaolinite, montmorillonite, wairakite, clinoptilolite, and calcite were used as ratio denominators. Their comparative results demonstrate that the signature matching algorithm was the most effective method for discriminating alteration minerals and had the potential of identifying mineralogy by matching AIS image spectra with library reference spectra. The PCA method was the second most successful method, and the band ratioing method was the least useful. In addition, although differences in mineralogy can be ascertained using band ratios method, some prior knowledge of mineralogy is necessary for mapping mineral species.

Baissa et al. (2011) proposed to study hand specimens of the carbonate facies of Jurassic age in the Agadir Basin, a part of the Atlantic passive margin, using hyperspectral carbonate facies imagery collected by the camera HySpex SWIR-320m, in the spectrum ranging from 1.3 to 2.5 μm. The hyperspectral images offer the possibility to identify different carbonate minerals by allowing diagenetic facies characterization. They developed a hyperspectral index of carbonate, called the Normalized Difference Carbonate Index (NDCI), to study the overall shape of reflection spectra of carbonates and to map other accessory minerals. The NDCI describes the state of crystallinity by calculating the deep absorption band complementary to the analysis of the abundance of calcite compared to dolomite. The calculation procedure of NDCI begins by finding the minimum reflectance band ρ_{min} of the carbonate absorption band and maximum reflectance band ρ_{max} at the right of the minimum between wavelength 2100 nm and 2400 nm, and the NDCI is defined as

$$NDCI = \frac{\rho_{max} - \rho_{min}}{\rho_{max} + \rho_{min}} \tag{7.11}$$

The experimental result indicates that the NDCI index is effective for studying the phases of crystallization textures without being hindered by individual states or the geometry of the sample surface or by the state of crystallinity of the carbonate mineral species when better than 5 nm spectral resolution hyperspectral data are available (Baissa et al. 2011).

Utilizing Hyperion datasets, the SAM algorithm and the Normalized Hydrocarbon Index, Zhang et al. (2014) extracted end-member spectra of specific minerals, mapped the carbonate, and detected the hydrocarbons in soil. The Hydrocarbon Index (*HI*) algorithm was originally proposed by Kühn et al. (2004). *HI* can be calculated through the following formula:

$$HI = (\lambda_B - \lambda_A)\frac{R_C - R_A}{\lambda_C - \lambda_A} + R_A - R_B \tag{7.12}$$

where λ_A and R_A, λ_B and R_B and λ_C and R_C are the wavelength/reflectance pairs for each "index point" (*cf.* Figure 2 in Kühn et al. [2004]). Hydrocarbon produces an absorption feature that is unique, and prominent at 1730 nm. Therefore, Zhang et al. (2014) took $\lambda_A = 1670$ nm, $\lambda_B = 1730$ nm, $\lambda_C = 1790$ nm, and reflectances R_A, R_B, and R_C at corresponding wavelengths, respectively. Normalized Hydrocarbon Index (*NHI*) was defined as $NHI = HI/R_B$. Their study indicated that the *NHI* index was more useful than *HI* for detecting micro-seeps of coal-bed methane geology. Their experimental results also demonstrate that Hyperion datasets have the ability to identify alterations and micro-seeps of coal-bed methane, and utilizing hyperspectral remote sensing for coal-bed methane exploration is feasible, effective, and low-cost.

Measuring the wavelength position of the absorption feature of white micas in the 2200 nm region is a useful tool for mapping the hydrothermal alteration (van Ruitenbeek et al. 2006). Using HyMap airborne hyperspectral data to study volcanic rocks of the Soansville greenstone belt in the Pilbara region of Western Australia, van Ruitenbeek et al. (2006) developed an alternative method for mapping white micas based on extracted diagnostic spectral information of the mineral of interest from hyperspectral band ratios. To do so, van Ruitenbeek et al. (2006) first estimated the probability of white mica being present in a rock sample using two band ratios (L_{2168}/L_{2185} and L_{2005}/L_{2079}, where L_λ is the radiance at wavelength λ (nm) of HyMap hyperspectral data) as independent variables through a logistic regression equation. They then used a multiple regression analysis approach to determine the relation between band ratios (L_{2220}/L_{2202} and L_{2237}/L_{2220}), calculated from the airborne imaging spectroscopy, and the absorption wavelength of white micas measured in rock samples on the ground. The results of both band–ratio regression analyses can then be combined to show predicted wavelengths of white mica in the areas of highest probability of containing white mica. The experimental results indicate that band ratios calculated from airborne imaging spectroscopy are suitable for mapping subtle shifting spectral features, such as the wavelength position of the absorption feature of white micas in volcanic rock; and the absorption wavelength can be estimated from two band ratios with a multiple linear regression equation (van Ruitenbeek et al. 2006).

7.3.3 Identifying and Mapping Minerals Using Spectral Matching Methods

There are two broad categories of hyperspectral analytical techniques: spectrum matching techniques and spectral unmixing methods, which are developed and used widely for identifying and mapping minerals using hyperspectral data (van der Meer et al. 2012). The matching techniques aim to express the spectral similarity of reference (usually library or field spectra of known materials) to test (usually image pixels) spectra. One of the most common techniques is the SAM (Kruse, Lefkoff, Boardman et al. 1993), which treats the two spectra as vectors in an *n* (band)–dimensional space and calculates the angle between the vectors as a measure of similarity. At small angles the test (pixel) sample is the most likely to be the corresponding reference sample. A matching algorithm based on the cross correlation of image and reference spectra is the cross correlogram spectral matching technique (CCSM; van der Meer and Bakker 1997). A stochastic measure, called the spectral information divergence (SID), was proposed (Chang 2000). For the detailed descriptions to the three spectral similarity measures, see Section 5.3 in Chapter 5. By using synthetic hyperspectral and real hyperspectral data (i.e., AVIRIS data), van der Meer (2006) compared the performances of the four spectral similarity measures: SAM, Euclidean distance measure (ED), CCSM, and SID between a known reference and unknown target spectrum (from hyperspectral data) to discriminate hydrothermal alteration system characterized by the minerals alunite, kaolinite, montmorillonite, and quartz. The comparative results from the study of AVIRIS data show that SID is more effective in mapping the four target minerals than SAM, CCSM, and ED, and that CCSM is better than SAM. Although SID shows the best performance among the four spectral similarity measures in the study by van der Meer (2006), no real application of the measure with hyperspectral data in identifying and mapping geological materials was found in the literature. Therefore, in the following, SAM and its modification versions and CCSM are reviewed based on existing application studies using various hyperspectral data in mineralogical mapping.

7.3.3.1 SAM

There are a lot of studies that successfully applied SAM to identify and map minerals and rock types using either *in situ*, airborne, or spaceborne hyperspectral sensor data. Feldman and Taranik (1988) used airborne imaging spectrometer (AIS) data and three methods: band ratios, principal component analysis (PCA), and a signature-matching algorithm to map hydrothermal alteration mineralogy in the Tybo mining district, Hot Creek Range, Nevada. In their study area, kaolinite and montmorillonite-clinoptilolite alteration zones and limestone formations had been identified from field mapping and laboratory analyses. The spectral matching algorithm locates pixels with spectral signatures that are similar to specified spectra, either reference spectra or other spectra in the image. Both reference and test (pixel) spectra need to be binary-encoded, and then the two binary-encoded vectors are compared to identify and map the test (pixel) spectrum into one of reference spectra by applying a threshold (Feldman and Taranik 1988). The experimental results demonstrate that the spectral matching algorithm was the most effective method for discriminating alteration minerals and had the potential of identifying mineralogy by matching AIS image spectra with library reference spectra. PCA was the second most successful method, and band ratioing was the least useful. In the Bodie and neighboring Paramount mining districts, eastern California, Crósta et al. (1998) used AVIRIS data to map hydrothermal alteration minerals. In their study, two spectral analysis algorithms, SAM and Tricorder, were used, and their performances to identify and map the minerals were compared. The Tricorder was designed by USGS scientists to compare spectra of materials from the USGS Digital Spectral Library to image spectra acquired by hyperspectral sensors, analyzing simultaneously for multiple minerals, using multiple diagnostic spectral features (Clark et al. 2003). The analysis results indicate that both algorithms appear to produce satisfactory results for geologic reconnaissance and mapping applications, but Tricorder generally classified more pixels and identified more mineral species than SAM. Laboratory spectra of rock samples from five locations in the study area were compared with test spectra of minerals identified by the classification algorithms for corresponding pixels, indicating the results matching to some extent. Baugh et al. (1998) also employed AVIRIS data and the SAM approach to map ammonium minerals (buddingtonite) in hydrothermally altered volcanic rocks in the southern Cedar Mountains, Nevada. The SAM approach determines the similarity of a test spectrum (from an AVIRIS pixel) to a reference spectrum (laboratory spectrum) by calculating the “angle” between them, treating each as a vector in a space with dimensionality equal to the number of bands (the actual number of AVIRIS data used). Output from the SAM is a grayscale image where lower values denote a better match with target spectra. Their study indicated that AVIRIS high spectral resolution data could be used for geochemical mapping, and the SAM approach could be extended to quantitatively map other minerals that have absorption features in the MIR range. In comparing the performance of AVIRIS sensor with EO-1 satellite hyperspectral sensor Hyperion for mineral mapping in areas of Cuprite, Nevada, and northern Death Valley, California and Nevada, Kruse et al. (2003) used SAM and mixture-tuned matched filtering (MTMF) mapping techniques. Minerals including carbonates, chlorite, epidote, kaolinite, alunite, buddingtonite, muscovite, hydrothermal silica, and zeolite were successfully estimated and mapped. The comparative results indicate that the Hyperion sensor could provide similar basic mineralogical information as that from the airborne AVIRIS sensor, with the principal limitation being limited mapping of fine spectral detail due to lower signal-to-noise ratios. To assess the influence of different reference spectral sources on SAM mineral mapping results, Hecker et al. (2008) illustrated and quantified the effects that different sources of reference libraries have on SAM classification results by using synthetic images of three mineral end-members (classified by using reference libraries derived from airborne hyperspectral imagery, ground spectra (Portable Infrared Mineral Analyzer), and from a standard library (USGS)). Their results demonstrate that the source of the reference library strongly influences the mineral mapping results using the SAM algorithm if all available wavelengths are used. However, the effect can be partially weakened by using appropriate preprocessing methods, such as spectral subsetting of the data and continuum removal methods. Best results might be achieved by using a feature subset (i.e., limiting the input wavelengths to the diagnostic absorption features). Their experimental results clearly show that the mapping results were improved if all disturbing nondiagnostic bands were excluded for running the SAM mapping procedure (Hecker et al. 2008).

To improve geological mapping results with the SAM method and HSI data, there are a few modification versions of the SAM algorithm (Oshigami et al. 2013, 2015). For example, by using reflectance data obtained in the SWIR regions by the airborne hyperspectral sensor, HyMap, Oshigami et al. (2013) applied the modified spectral angle mapper (MSAM) and continuum-removal methods to map distribution of minerals related to hydrothermal alteration and pegmatite in the Cuprite region, Nevada. The MSAM uses a new spectral parameter (Kodama et al. 2010) to calculate the angle between the difference vectors: a reference spectral vector and a test (pixel) spectral vector. The new spectral parameter (S') is derived through subtracting the average reflectance (S_m) of the spectrum from the spectral reflectance (S) (i.e., $S' = S - S_m$, where S_m is the average reflectance of all band reflectance for a specific sample (pixel), and S and S' are the original spectral reflectance and new spectral parameter at a specific band, respectively). Since the spectral angle calculated by the MSAM method has a range from 0 to π, this spectral angle range is double of that calculated by SAM (0 to $\pi/2$), and thus the MSAM method can potentially improve mineral mapping accuracy (due to wider spectral angle range). In their study, they first carried out continuum removal as a preprocessing step, which results in emphasis of the shape and location of absorption peaks in reflectance spectra, and thus can achieve high-accuracy classification (Oshigami et al. 2013). Their results indicated that the mineral mapping results were consistent with the field survey results and also with X-ray diffraction analyses and spectral measurements of rock samples. Their results further demonstrated that the continuum-removal MSAM method successfully discriminated hydrothermal alteration minerals such as kaolinite and pyrophyllite, and pegmatite-related minerals such as high-aluminium muscovite and lepidolite in southern Namibia for the first time. The main advantages of continuum-removal MSAM method are simplicity and high-precision identification of alteration minerals.

7.3.3.2 CCSM

There are a couple of real application studies that applied the CCSM spectral similarity measure to discriminate and map geological materials with hyperspectral data. By using 1994 AVIRIS data acquired from Cuprite, Nevada, van der Meer and Bakker (1997) used the CCSM mapping technique to map surface mineralogical components. Accurate mapping of kaolinite, alunite, and buddingtonite was achieved using the CCSM technique and AVIRIS data, assessed via three parameters from the cross correlograms that were constructed on a per pixel basis: the correlation coefficient at match position zero, the skewness value, and the significance (based on a Student *t*-test of the validity of the correlation coefficients). The results produced by the CCSM can be tested by decision theory that allows not only the validation of resulting classification, but also a measure of the reliability and accuracy of the results (van der Meer and Bakker 1997). To identify and map hydrothermal alteration zones associated with epithermal gold deposits in the island of Lesvos in Greece, Ferrier et al. (2002) analyzed Landsat TM imagery and ground-based spectroscopy data. For analyzing the field spectroscopy data in order to identify hydrothermal alteration minerals, they used two quantitative techniques' SAM and CCSM, to clearly identify the presence of high grade kaolinite and alunite outcrops within the alteration zones.

7.3.4 Estimating and Mapping the Abundance of Minerals Using Spectral Unmixing Methods

Given the fact that there are a large number of imaging pixels mixed with more than one geological component (mineral), spectral unmixing techniques with hyperspectral data can help improve the accuracy of identifying and mapping various minerals in geological mapping. The spectral unmixing of a pixel is a process of inverting the linear mixture (in some studies also considering non-linear spectral mixture) of end-members to derive proportions of their surface coverage in the pixel. Typically, the first step is to find spectrally unique signatures of pure minerals/end-members (from a spectral library, *in situ* spectra, or directly from imaging spectra), and the second step is to unmix the mixed pixel spectra as (linear or nonlinear) combinations of end-member/material spectra. There are four popularly used spectral unmixing techniques in mineralogical mapping with HSI data, reviewed here, including linear spectral mixture model (LSM), multiple end-member spectral mixture analysis (MESMA), and two partially spectral unmixing models/techniques: MTMF and constrained energy minimization (CEM). For readers interested in a more detailed introduction to the four spectral unmixing techniques/methods, refer to the detailed descriptions of the concepts, algorithms, and principles of the four techniques/methods in Section 5.7 of Chapter 5. Therefore, in the following, the application studies of all the four spectral unmixing techniques and methods are reviewed with various hyperspectral data in geological mapping.

7.3.4.1 Linear Spectral Mixing (LSM)

The LSM model is a traditional and commonly used spectral mixture model (SMM) with various hyperspectral data and has been applied to estimating and mapping abundances of various minerals/rocks starting in the early 1990s. For example, Mustard (1993) used the LSM model and AVIRIS data to investigate an area of soil, grass, and bedrock associated with the Kaweah serpentinite melange in the foothills of the Sierra Nevada Mountains in California. With five spectral end-members plus one for shade in the spectral mixture model, the LSM model was able to account for almost all the spectral variability within the data set and derive information on the areal distribution of several surface cover types. Three of the end-members were shown to accurately model green vegetation, dry grass, and illumination. For the other three end-members, the spatial distributions in mixed pixels were still interpretable and coherent although their spectral distinction was very small. By using AVIRIS data and the linear spectral unmixing program, Bowers and Rowan (1996) examined the alkaline complex and adjacent country rocks in southeastern British Columbia, Canada. The spectral unmixing technique models each pixel spectrum in an AVIRIS image as a linear combination of unique end-member spectra. The end-member spectra from well-exposed and spectrally distinct mineralogical units, vegetation, and snow were determined. Of them, four of the end-members reflect mineralogical variations within the McKay group in the study area, and may represent lateral and vertical variations of sedimentary or metamorphic facies (Bowers and Rowan 1996). The resultant spatial distribution of end-members shows generally close agreement with the published geologic map. However, in several places, the map derived from AVIRIS hyperspectral imagery is more accurate than the published map. Recently, to unmix the spectral mixture of leaf and mineral and to estimate the cover proportions of mineral from the hyperspectral data, Chen et al. (2013) performed a spectral mixture experiment of calcite and green single-layer leaf using a novel SMM where the leaf transmission was taken into account compared with the traditional linear spectral model. The SMM model was thus applied to invert the carbonate minerals from hyperspectral, EO-1 Hyperion, data. The measurement of the leaf cover proportions presents a generally good agreement with the results inversed from Hyperion image. And the mean relative error is less than 10% between the inversion and measurement of the cover proportions. The experimental result indicates that the proposed method was robust to unmix the spectra of leaf and carbonate mineral and to inverse their cover proportions from the hyperspectral data. Further, the results of retrieving the cover proportions of carbonate mineral in the Luanping area in Hebei province, China, show that when different fractions of leaves were overlaid on the calcite, the absorption depths of calcite at the wavelength of 2.33 μm would be lower with the increase of the leaf cover proportions, but the absorption positions kept unchanged (Chen et al. 2013). By using the hyperspectral (EO-1 Hyperion) image data, covering a mineralized belt in the Namenda area, eastern India, Magendran and Sanjeevi (2014) reported the mapping results of a study to differentiate iron ores in terms of their grades. Based on the calibrated hyperspectral spectral data and extracted spectral parameters (e.g., depth, width, area, and wavelength position of absorption features), they used correlation analysis and spectral unmixing analysis methods to estimate and map concentration of iron oxide and alumina (gangue) in ore samples obtained from the mine face. The analysis results indicate that well-defined correlations were evident between the concentration of iron oxide and the extracted spectral parameters, and the linear spectral unmixing resulted in an iron ore abundance map, which, in conjunction with the image- and laboratory-spectra, helped in assessing the grades of iron ores in the study area. Further, the study demonstrates the feasibility of using the Hyperion image data for discriminating various grades of iron ores at specific target locations such as mining areas.

7.3.4.2 MESMA

MESMA is a technique in which the number of end-members and types are allowed to vary for each pixel in the image, into the solution to LSM. The MESMA can overcome limitations of the LSM model, such as no total number of end-members limited by the number of bands in the image data. The basic idea is to run unmixing multiple times to each pixel with various subsets of end-members from the total pool of end-members and take the "best fit" results as the final solution to a particular pixel (Roberts et al. 1998). In the geological mapping application with imaging spectrometer data, usually the situation of multiple minerals (end-members) is normal in a geological mapping area, but for a specific site/pixel, there may exist only a few minerals and thus the MESMA should be an effective technique to estimate and map multiple minerals in the area. There are many researchers who have conducted such application studies of the technique. For instance, using MESMA of SWIR imaging spectrometer data recorded by the HyMap imaging system, Bedini et al. (2009) managed to map the spatial distribution of SWIR-active hydrothermal alteration mineralogy (clays and alunite), clays associated with soils of the quaternary cover and carbonate and andesite lithology in the Rodalquilar caldera complex in southeast Spain. In the study, a total nine end-member set derived from the imagery was used. The MESMA was based on models of two to five end-members evaluated in terms of fraction and RMSE criteria. The MESMA was used to identify SWIR active surface mineralogy and resolve the mixed spectral response from the surface mineralogy and vegetation cover. Based on the verified result with field spectral measurements, the experimental results have the potential to refine the map of hydrothermal alteration zones in the Rodalquilar caldera complex, and also demonstrate that the MESMA could be considered as a very effective unmixing technique in geological applications of imaging spectrometry in semi-arid regions. In mapping the distributions of plagioclase, clinopyroxene, and olivine on the global lunar surface with interference imaging spectrometer (IIM) hyperspectral data based on the MESMA method, Shuai et al. (2013) modified the MESMA technique. The IIM sensor of Chang'E-1 mission acquired hyperspectral data of the global lunar surface within the wavelength of 480–960 nm in which major minerals can be discriminated by faint differences in 32 contiguous hyperspectral bands (Shuai et al. 2013). They modified the MESMA technique to map lunar surface materials because of consideration of the effect of space weathering that produces multiple end-members of lunar minerals by obscuring the pure spectra of minerals at different levels. The major difference between the MESMA and modified MESMA is that the latter contains another physical assumption, which is that all the end-members' spectra in one pixel are affected by the same factor. In their particular study, the assumption specifically states that all the minerals' spectra are affected by the same mass fraction of submicroscopic iron (SMFe) in one pixel of 200-m spatial resolution. As a result, there were six unmixing models in the modified MESMA according to the six space weathering levels, but the number of unmixing models of the original MESMA method would be 216 for the three–end-member model (Shuai et al. 2013). The result demonstrates that the modified MESMA method is an effective approach to quantitative mapping of the lunar minerals and space weathering levels using hyperspectral data.

7.3.4.3 MTMF

MTMF is an advanced spectral unmixing algorithm that does not require that all materials within a scene are known and have identified end-members (Boardman et al. 1995). MTMF has proven to be a very powerful tool to detect specific materials that differ slightly from the background. In the geological mapping with hyperspectral data, the MTMF technique can be used directly to estimate and map a target/interest end-member/mineral rather than considering other background minerals. There are a lot studies that have used the MTMF spectral unmixing technique and airborne and spaceborne hyperspectral image data to identify and map minerals of interest. For example, in a comparative alteration mineral mapping using multi-/hyperspectral sensors (including EO-1 ALI, Hyperion, and Terra ASTER), Hubbard et al. (2003) used spectral mapping methods including the spectral angle mapper, spectral unmixing method MTMF to identify and map hydrothermally altered rocks associated with several young volcanic systems in the Central Andes between Volcan Socompa and Salar de Llullaillaco. The comparative analysis results indicate that the combination of ALI, ASTER, and Hyperion imagery was effective for mapping a variety of minerals characteristic of hydrothermally altered rocks on the South American Altiplano. Among the three sensors, Hyperion image data within the broader band image coverage of ALI and ASTER provided essential leverage for calibrating and improving the mineral mapping accuracy of the multispectral data (i.e., ALI and ASTER). By using the MTMF spectral unmixing method and AVIRIS and EO-1 Hyperion hyperspectral image data, Kruse et al. (2003) identified and mapped minerals including carbonates, chlorite, epidote, kaolinite, alunite, buddingtonite, muscovite, hydrothermal silica, and zeolite in Cuprite, Nevada. Comparison of airborne AVIRIS data with the Hyperion data establishes that Hyperion provides similar basic mineralogical information. The analysis results demonstrate that satellite hyperspectral sensors can produce useful mineralogical information, but also indicate that SNR improvements are required for future spaceborne sensors to allow similar mineral mapping level that is currently possible from airborne hyperspectral sensors such as AVIRIS (Kruse et al. 2003). Using the same spectral unmixing mineral mapping method (i.e., MTMF) as that in Nevada (Kruse et al. 2003), Kruse et al. (2006) identified and mapped minerals including hematite, goethite, kaolinite, dickite, alunite, pyrophyllite, muscovite/sericite, montmorillonite, calcite, and zeolites in the Los Menucos District, Rio Negro, Argentina, from AVIRIS and EO-1 Hyperion hyperspectral image data. The MTMF algorithm provided a consistent way to process multiple AVIRIS flight lines allowing identification and mapping of VNIR- and MIR-active minerals. The mineral mapping results created with both hyperspectral sensors show a good correspondence with the results of field reconnaissance verification and spectral measurements acquired using an ASD field spectrometer. These analysis results illustrate the high potential of hyperspectral remote sensing for geologic mapping and mineral exploration. Recently, Bishop et al. (2011), Kodikara et al. (2012), and Zadeh et al. (2014) also used EO-1 Hyperion hyperspectral data and the MTMF spectral unmixing analysis technique to identify and map various minerals. Bishop et al. (2011) checked two geological mapping techniques, SAM and MTMF, to discriminate and map argillic alteration, iron oxide– and sulphate-bearing minerals in the two target areas in the mountainous region of Pulang, China. The experimental results demonstrate that the MTMF feasibility value would help reduce false positives that are likely with SAM. They therefore believed that the mineral map produced by MTMF was the most appropriate for this region. In addition, the experimental results also demonstrate that the ability to identify alteration minerals and to define their zonation, despite the low SNR of the Hyperion sensor, highlights the advantages of using the hyperspectral data in the context and of using Hyperion in this particular case. Kodikara et al. (2012) analyzed and identified High Magadi beds, chert series, and volcanic tuff in evaporitic lacustrine sediments in Lake Magadi, East African Rift Valley, Kenya. Their analysis results show the usefulness of the hyperspectral remote sensing to map the surface geology of this kind of environment and to locate promising sites for industrial open-pit trona mining in a qualitative and quantitative manner. Zadeh et al. (2014) discriminated and mapped diagnostic alteration minerals around porphyry copper deposits at the Central Iranian Volcano-Sedimentary Complex. The results revealed that Hyperion data proved to be powerful in discriminating and mapping various types of alteration zones, although the data were subjected to adequate preprocessing. With the MTMF technique, the Hyperion sensor was effective for mapping a variety of minerals characteristics of hydrothermally altered rocks, including muscovite, illite, kaolinite, chlorite, pyrophyllite, biotite, hematite, jarosite, and goethite, potassic-biotitic, phyllic, argillic, propylitic alteration zones, etc., in the study area.

7.3.4.4 CEM

The constrained energy minimization (CEM) method developed by Farrand and Harsanyi (1997) is an extension of a traditional spectral mixing method (LSM). Like the MTMF method, the CEM technique is based on a linear operator that minimizes the total energy in a hyperspectral image sequence while the response of the operator to the signature of interest is constrained to a desired constant level (Resmini et al. 1997). CEM maximizes the response of a target signature on a pixel-by-pixel basis and suppresses the response of undesired background signatures. For more detailed introduction to the CEM technique, readers can refer to Section 5.7 in Chapter 5. In geological mapping with HSI data, the "target signature" can be interesting minerals that people hope map while the "undesired background signatures" can be those minerals/alterations that people are not interested in mapping over a specific area. Researchers have successfully applied the CEM technique to estimate and map mineral abundance with hyperspectral data. For examples, Farrand and Harsanyi (1997) developed and applied the CEM technique to estimate and map abundance of ferruginous fluvial sediments deposited on the banks and on the floodplain of the Coeur d'Alene River in northern Idaho where the sediments have been contaminated by trace metals released by mining activities in and around the town of Kellogg, Idaho, from AVIRIS hyperspectral data. The CEM-derived abundance images from the AVIRIS data, produced using both laboratory and image data as the target signatures, were thresholded to create a set of spectra dominated by the ferruginous sediment spectral response (Farrand and Harsanyi 1997). The experimental results demonstrate that the performance of the CEM method used to map the ferruginous sediments has shown to be excellent. However, it is necessary that potential improvements of the CEM method include more rigorous quantitative thresholds for identifying target materials on abundance images. With Hyperspectral Digital Imagery Collection Experiment (HYDICE) sensor's data acquired in an area of the Cuprite mining district, Nevada, Resmini et al. (1997) also applied the CEM technique to map the areal distributions of the minerals alunite, kaolinite, and calcite. Their cross-comparative results indicate that the CEM-derived mineral mapping results agreed with the mapping results created by other studies in the region. Linear spectral unmixing and principal components analysis also produced similar results to those of CEM. Therefore, their experimental results further demonstrate that CEM is a powerful and rapid technique for mineral mapping, which requires only the spectrum of the target mineral to be mapped and no prior knowledge of other background constituents/minerals. Recently, by using EO-1 Hyperion hyperspectral imagery and ground-based spectral measurements, Li et al. (2014) compared six mineral mapping methods of SAM, orthogonal subspace projection (OSP), CEM, adaptive coherence/cosine estimator (ACE), adaptive matched filter (AMF), and elliptically contoured distributions (ECD) for detecting and mapping the small altered rock targets under the covering of vegetation in a forest area. Usually, the outcrop of the altered rocks is small and distributes sparsely, and the altered rocks are difficult to directly identify in the case of dense vegetation coverage. The altered rocks mapping results indicate that the ACE and AMF might be useful in applications of detection of presence of small geological target in vegetation covered area; the CEM was also found to be sensitive to the altered rock; however, the performance of SAM, OSP, and ECD was found to be the poorest. Their results further demonstrate the ability of six target detection algorithms for geological target detection in the forest area from the hyperspectral data.

7.3.5 Estimating and Mapping the Abundance of Minerals Using Spectral Modeling Methods

At VNIR and MIR wavelengths (0.4–2.5 μm), reflectance spectra of minerals contain absorption (diagnostic) features that are of characteristics of the composition and crystal structure of the absorbing species (Hunt 1977, Clark 1999, Sunshine et al. 1990). It is possible to use hyperspectral remote sensing data to extract and fit the absorption features (Sunshine et al. 1990). To fit the absorption features, some techniques have been developed, such as making use of Gaussian in the fitting of absorption bands. During the last two decades, the modified Gaussian model (MGM; Sunshine et al. 1990, Sunshine and Pieters 1993) has been developed and applied to estimate and map the mineralogy of both terrestrial and extraterrestrial surfaces. Compared with other curve-fitting models, the MGM model is more firmly rooted in the crystal field theory and hence expected to provide more valid results (Sunshine et al. 1990). Therefore, in this section, the algorithm and nature of MGM are first introduced and discussed, and then its applications to estimating and mapping mineralogy from hyperspectral data are briefly reviewed. Since the MGM model is a modification version of the traditional Gaussian model (GM), let us consider the GM model first.

The GM model is based on an underlying assumption that absorption features observed in VNIR–MIR spectra are composed of absorption bands that are inherently Gaussian in shape (Sunshine et al. 1990). Under the center limit theorem of statistics, in terms of a center (mean) μ, width (standard deviation) σ, and strength (amplitude) s, a Gaussian distribution, $g(x)$, in a random variable (x), is expressed as

$$g(x) = s \cdot \exp\left\{\frac{-(x-\mu)^2}{2\sigma^2}\right\} \tag{7.13}$$

Based on the consistency of the shape between the Gaussian distribution curve and absorption curve in the VNIR and MIR (0.4–2.5 μm) spectrum, the Gaussian deconvolution can produce a unique fit for a spectral characteristic of a given mineral under certain reasonable restraints (Yang et al. 2010). In addition, the theoretical justification of an absorption band curve being Gaussian distribution curve was also given by many researchers (e.g., Sunshine et al. 1990, Clénet et al. 2011). Thus, the GM method can help estimate and map mineralogy using hyperspectral data. However, according to the experimental results obtained by Sunshine and his colleagues (Sunshine et al. 1990, Sunshine and Pieters 1993), who used the GM model to analyze the absorption spectra of mineral pyroxenes (clinopyroxene and orthopyroxene), it was found that it was not appropriate for the Gaussian model to describe absorption spectra due to electronic transition (e.g., Fe^{2+}) absorptions (e.g., absorption spectra of the mineral pyroxene). For this case, they thought that in a crystal field site, the average bond length will vary due to random thermal vibrations and variations; thus, the random variable (x in Equation 7.13) for electronic transition absorption bands is not the energy of absorption but the average bond length (Sunshine et al. 1990).

Therefore, based on understanding of the average bond length by Sunshine et al. (1990), we not only can apply the central limit theorem of statistics, but also consider the average bond lengths to describe absorption bands of minerals. According to the crystal field theory, the description of electronic transition absorption (Burns 1970, Marfunin 1979), which suggests that a relationship between the absorption energy (e) and average bond length (r), is expressed by a power law: $e \propto r_n$, on a basis, the GM model based on the energy of electronic transition absorption as the random variable is mapped into that based on the average bond length of absorption as the random variable, called the modified Gaussian model (MGM), $m(x)$:

$$m(x) = s \cdot \exp\left\{\frac{-(x^n-\mu^n)^2}{2\sigma^2}\right\} \tag{7.14}$$

Compared to Equation 7.13, changing the exponent of $(x^n - \mu^n)$ varies the symmetry of the distribution, i.e., a relative slope of the left and right wings of the distribution. In this case, it will have more chances to reach the realistic characteristics of the spectral absorption curve. In practice, an appropriate value of the exponent n in Equation 7.14 can be determined empirically. Based on the experimental results (optimal RMS residual error) of Sunshine and his colleagues (Sunshine et al. 1990, Sunshine and Pieters 1993) on the 0.9-μm orthopyroxene absorption feature fitting using different values of n, $n = -1$ (Equation 7.15) could produce the optimal fitting effectiveness (Figure 7.4):

$$m(x) = s \cdot \exp\left\{\frac{-(x^{-1} - \mu^{-1})^2}{2\sigma^2}\right\} \tag{7.15}$$

As illustrated in Figure 7.4, as with the transmission absorption feature, $n \approx -1$ results in the smallest RMS residual error. As a comparison of the fits of various distribution models to the 0.9-μm orthopyroxene reflectance absorption band shown in Figure 7.4, the GM ($n = 1.0$) produces a worst fit, but when using different n values, a progressive improvement can be observed (e.g., $n = -0.2$) until the best fit is achieved ($n = -1.0$). For readers interested in knowing more about the development of the MGM model, modeling parameters' settings, and iterating/adjusting (center, amplitude, and width), please refer to the appendix in Sunshine et al. (1990).

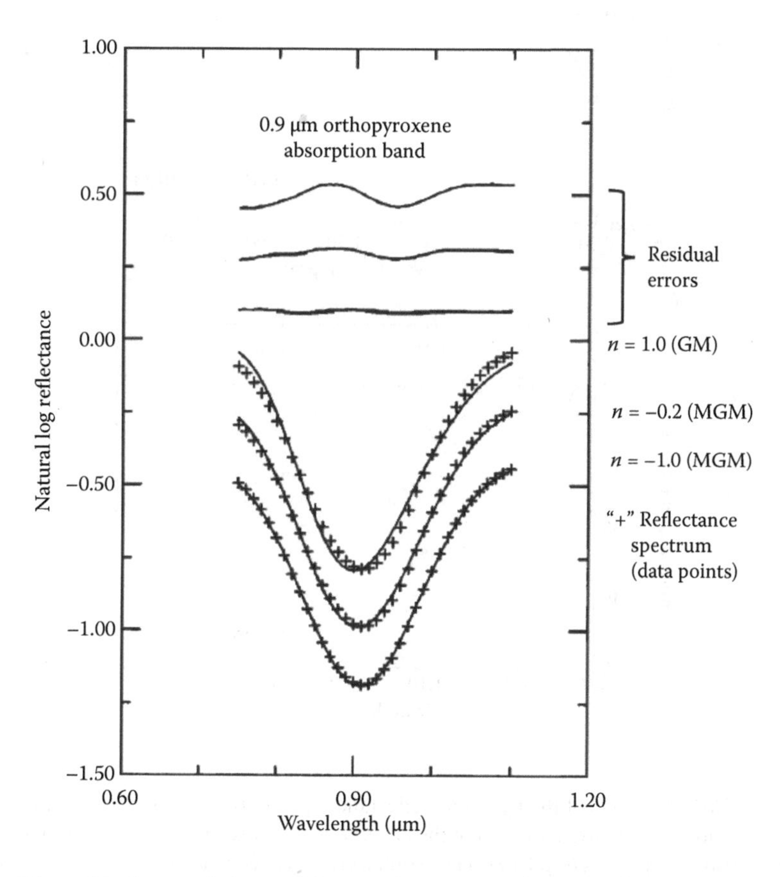

FIGURE 7.4 The residual errors between 0.9 μm orthopyroxene absorption feature models fitted by using different values of n in Equation 7.14. Among all cases, $n = -1$ produces the optimal fit. (©1990 Wiley. Used with permission from Sunshine, J. M., C. M. Pieters, and S. R. Pratt, *Journal of Geophysical Research*, 95, 6955–6966, American Geophysical Union, 1990.)

To test the ability of the MGM deconvolution of mineral absorption bands, Sunshine and his colleagues (Sunshine et al. 1990, Sunshine and Pieters 1993) used the MGM model to unmix and estimate the abundance of two pyroxenes: orthopyroxene (Opx) and clinopyroxene (Cpx). The former came from Webster, North Carolina (Opx), while the latter came from a Hawaiian volcanic bomb (Cpx). Since the Opx and Cpx have different compositions, the wavelength positions of their absorption bands are different. Handpicked mineral separates of these samples (Opx and Cpx) were crushed and wet-sieved with ethanol into 45–75 μm particle size separates. In addition, to conduct a mineral spectral unmixing analysis, several fraction mixtures of the two pyroxenes were created. To do so, the spectral measurements were first taken from different fraction mixtures (of Opx and Cpx) and pure individual pyroxenes such that the MGM and GM models were used to fit and unmix Gaussian distribution curves of absorption bands of pure or mixtures of Opx and Cpx in order to unmix and estimate the abundance of Opx and Cpx.

Figure 7.5 presents spectral curves of pure pyroxenes (Opx and Cpx) and fitting results using the MGM model in order to compare with fitting and unmixing results (i.e., center position, width,

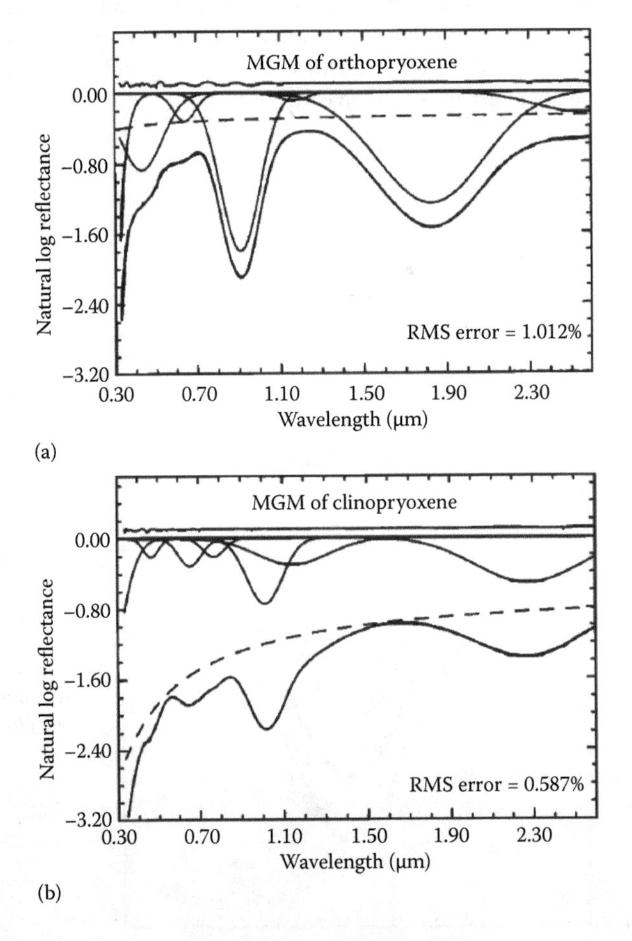

FIGURE 7.5 The MGM models of the pyroxene reflectance spectra (45–75 μm particles): (a) Opx; (b) Cpx. Only single distributions are required for both of the absorption features near 1.0 and 2.0 μm. From top to bottom in the figure (also for Figures 7.6 and 7.7): The residual error between the modeled spectrum and the actual spectrum (offset 10% for clarity), individual MGM distributions representing absorption bands, the continuum or baseline (dashed line), and the modeled spectrum superimposed on to the actual spectrum. (©1990 Wiley. Used with permission from Sunshine, J. M., C. M. Pieters, and S. R. Pratt, *Journal of Geophysical Research*, 95, 6955–6966, American Geophysical Union, 1990.)

and depth of a corresponding absorption band) of different fraction mixtures of two pyroxenes (Opx and Cpx) using the GM model (Figure 7.6) and the MGM model (Figure 7.7). Figure 7.6 shows the results of two different fraction mixtures of the two pyroxenes, fitted and extracted using the GM model under constraints. Figure 7.7 presents the results of two different fraction mixtures of the two pyroxenes, the same mixtures as those in Figure 7.6, fitted and extracted using the MGM model without constraints. From Figure 7.7, the distributions resulting from this unconstrained fit correspond directly to those found in the spectra (Figure 7.5) of the Opx and Cpx used to create these mixtures. Further, as expected from the proportion of Opx in the mixture, the Opx features are significantly stronger than those of the Cpx in the 75/25 Opx/Cpx mixture, while the Cpx features dominate the 25/75 Opx/Cpx mixture. Compared to the results created with the constrained GM model in Figure 7.6, the results shown in Figure 7.7 demonstrate that the unconstrained MGM model can correctly identify and characterize superimposed absorptions in mixture spectra. Consequently, the MGM model may be correctly used to model a reflectance spectrum as a series of modified Gaussian curves with each curve characterized by a central wavelength position, width, and depth.

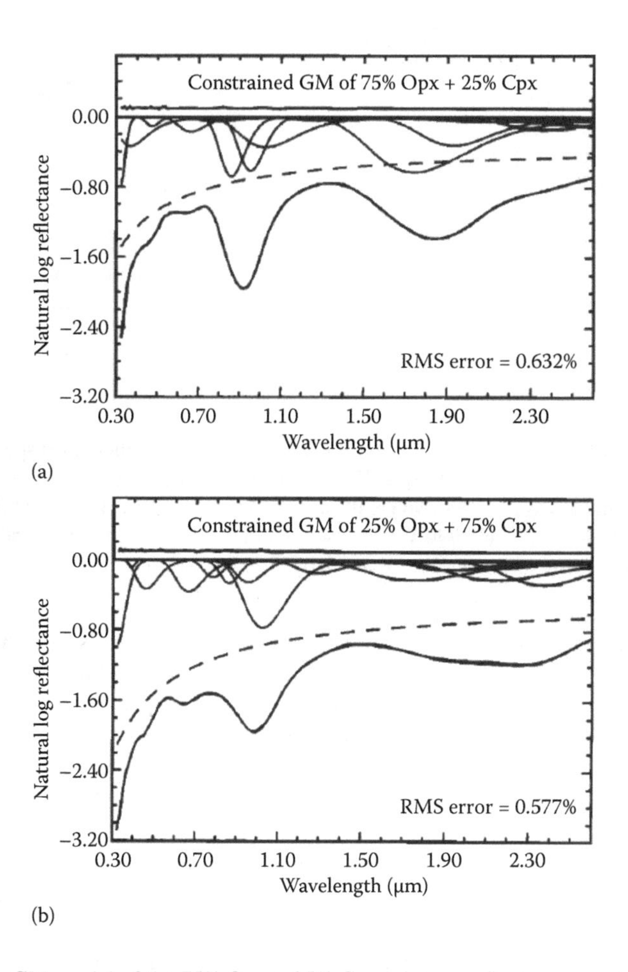

FIGURE 7.6 (a) The GM model of the 75% Opx + 25% Cpx mixture reflectance spectrum (45–75 μm particles). (b) The GM model of the 25% Opx + 75% Cpx mixture reflectance spectrum (45–75 μm particles). These models are constrained to have the Gaussian distributions with centers fixed at the wavelengths determined for the pyroxene end-member spectra. These models of pyroxene mixture spectra work well only under such constraints. (©1990 Wiley. Used with permission from Sunshine, J. M., C. M. Pieters, and S. R. Pratt, *Journal of Geophysical Research*, 95, 6955–6966, American Geophysical Union, 1990.)

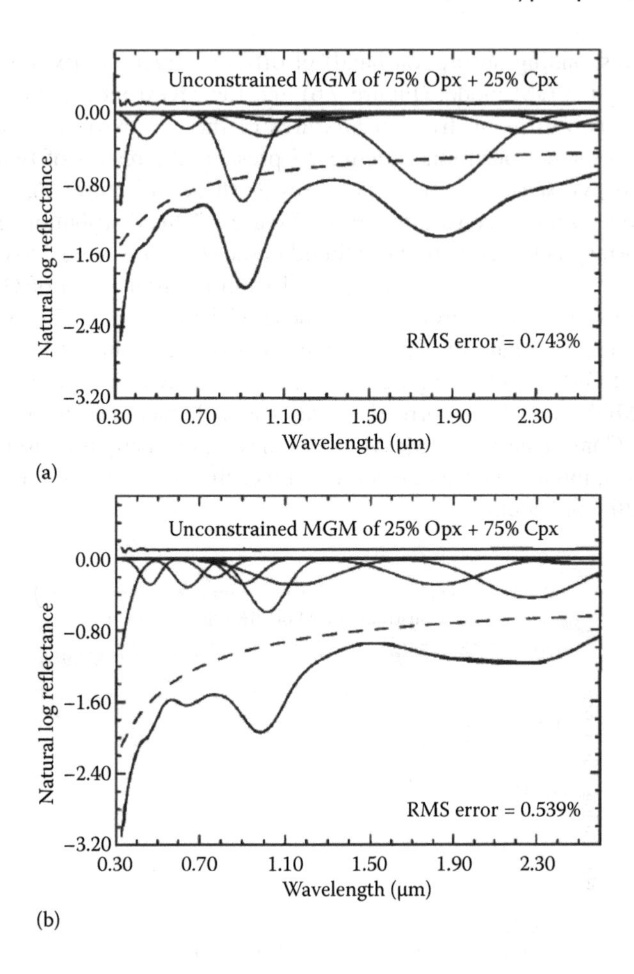

FIGURE 7.7 (a) The MGM model of the 75% Opx + 25% Cpx mixture reflectance spectrum (45–75 μm particles). (b) The MGM model of the 25% Opx + 75% Cpx mixture reflectance spectrum (45–75 μm particles). Unlike the GM model in Figure 7.6, these MGM fits are unconstrained, yet still reflect the proportion of both pyroxene minerals in each sample. (©1990 Wiley. Used with permission from Sunshine, J. M., C. M. Pieters, and S. R. Pratt, *Journal of Geophysical Research*, 95, 6955–6966, American Geophysical Union, 1990.)

Using such a modeling method, the characteristic absorption bands present in mineral reflectance spectra can be identified and quantified in order to estimate and map the mineralogy of both terrestrial and extraterrestrial surfaces. Compared to other spectral analytical approaches, MGM deconvolution analysis possesses a number of advantages (Cloutis 1996):

1. The mathematic basis is more firmly rooted in the physical processes of electronic absorptions.
2. MGM can be used to reduce hyperspectral data dimensions by reducing reflectance spectra to a series of modified Gaussian curves with almost no loss of information.
3. MGM analysis is relatively insensitive to particle size variations.
4. MGM analysis can be directly applied to the data in spectral libraries, resulting in simplified spectral searching.

Given the advantages possessed by the MGM method, there are a large number of studies using the MGM for spectral deconvolution analysis with mineral reflectance spectra collected both in laboratory and on terrestrial and extraterrestrial surfaces. With spectral measurements taken from two experiments, the laboratory case and the field case, Mulder et al. (2013) used the MGM deconvolution method to unmix and estimate mineral abundances of mixtures with at least two constituents using absorption features in the 2.1–2.4 μm wavelength region. The laboratory experiment was conducted on spectra of physical mixtures of almost pure minerals, including kaolinite, dioctahedral mica (illite), smectite, calcite, and quartz, those found to be dominantly present in the study area in the field experiment. The field case study was located in Northern Morocco, centered at around 34.0° N and 4.5° W and covered an area of 15,000 km^2. They first parameterized the absorption behavior of mineral mixtures by using the MGM modeling method, and then predicted mineral abundances by regression tree analysis using these parameters as inputs. Cross validation results demonstrate that the lab spectral samples of kaolinite, dioctahedral mica, smectite, and calcite were predicted with an RMSE less than 9 wt.%, and for the field samples the RMSE was less than 8 wt.% for calcite, dioctahedral mica, and kaolinite abundances. The mineral spectral unmixing method proposed by Mulder et al. (2013) allows the simultaneous quantification of more than two minerals within a complex mixture and thereby enhances the perspectives of spectral analysis for mineral abundances. To process a *priori* unknown mafic mineralogy observed in the VNIR by reflectance spectroscopy in the case of laboratory or natural rock spectra, Clénet et al. (2011) implemented an automatic procedure on the original MGM approach (Sunshine et al. 1990) in considering all the mixture possibilities involving orthopyroxene, clinopyroxene, and olivine. The automatic procedure to characterize magmatic units from olivine/pyroxenes mixtures, according to Clénet et al. (2011), includes initializing the MGM procedure with a proper setting for the Gaussians parameters; performing an automatic analysis of the shape of the spectrum; handling the continuum with a second-order polynomial adjusted on the local maxima along the spectrum and making Gaussians parameters initial settings on the basis of laboratory results available in the literature in the case of simple mixtures of mafic minerals; and assessing the returned MGM solutions on spectroscopic grounds and either validated or discarded, on the basis of a mineralogical sorting. The results derived from their study demonstrate that the MGM outputs are truly informative of the chemical composition of pyroxenes and olivines. And considering some limits on the detection thresholds, these results are quite promising for operationally using MGM in mapping detailed mineralogy with hyperspectral data sets. This version (Clénet et al. 2011) of MGM is fully automated and operational with large amounts of hyperspectral imaging (HSI) data. Later, Clénet et al. (2013) continued to test the performance of the adapted version of the MGM with two natural cases: (1) The Sumail massif in the Oman ophiolite on Earth, based on HyMap data, and (2) the volcanic shield of Syrtis Major on Mars, based on an integrated VNIR–MIR Observatoire pour la minéralogíe, l'eau, les glaces et l'activité (OMEGA)/ Mars Express (MEx) mosaic. By applying the new version of MGM approach, for the first case, they were able to clearly distinguish between the mantle section and the crustal section in a controlled case study on Earth. For the second case, the results are in agreement with previous work, but also olivine appears to be more abundant than previously considered in the central part of the volcanic edifice. Therefore, the results derived from both case studies demonstrate the performance and ability of this new MGM approach to map lithological units.

In order to assess the utility of MGM and SAM spectral mixture modeling approaches to characterize the surface materials over the lunar surface using hyperspectral moon mineralogy mapper data, recently, Kodikara et al. (2016) extracted spectrally pure pixels (end-members) of the area using the pixel purity index (PPI), identified mineralogy of the selected end-member spectrum using the MGM modeling approach, and mapped mineralogically identified end-members using the SAM method. The mapping results demonstrate that the combination of MGM modeling of spectral deconvolution with the SAM method of spectral matching as an effective approach for compositional characterizations of morphological features on the lunar surface. As a modeling approach of spectral deconvolution, MGM was able to identify and characterize both high- and low-Ca pyroxenes along with plagioclase feldspar. The SAM method was able to map identified mineral mixtures output from the MGM. The MGM model could be used to characterize the degree of light absorbed (k) based on a mineral spectrum. Optical constants of minerals are parameters that may be used to describe the degree of light absorbed (k) and refracted (n) in a medium. Trang et al. (2013) parameterized k in the VNIR and MIR (0.6–2.5 μm) spectral range of natural olivine as a function of forsterite number and synthetic pyroxene with respect to the wollastonite and ferrosilite numbers using a diverse and larger sample size, which was expected to lead to robust optical parameters. The olivine spectra were collected at the NASA/Keck Reflectance Experiment Laboratory (RELAB) at Brown University and the USGS Library, while the synthetic pyroxene spectral data were also collected at RELAB (Trang et al. 2013). Trang et al. (2013) characterized each k-spectrum with the MGM that models MIR absorptions due to electronic transitions. They found that their fitting routine characterized the olivine and pyroxene k-spectra in a robust and consistent manner. And they also used regression analysis to characterize each parameter of the Gaussians and the continuum as a function of mineral composition. Their results demonstrate that the developed optical parameters in the study will allow calculations of mineral proportions and compositions on planetary surfaces using spectral data from planetary missions.

7.3.6 Mapping Minerals Using Advanced Techniques and Methods

Some advanced techniques and algorithms, such as machine learning methods, can be directly used for identifying minerals/rocks and mapping geology with laboratory and *in situ* spectral measurements or airborne and satellite hyperspectral imaging data. Artificial neural networks (ANNs), spectral expert systems, and support vector machines (SVMs) methods have been used to identify and map geology by using different hyperspectral data sets. In the following, the application studies of all the three advanced techniques and methods are reviewed with various hyperspectral data in mineralogical mapping.

7.3.6.1 ANNs

The basic ANN algorithm and the other three types of ANNs (radial-basis function networks, adaptive resonance theory, and self-organizing maps) are introduced in Section 5.8.2 of Chapter 5. When the ANN method is applied to identify and map mineralogy, in general, selected hyperspectral band reflectance or extracted absorption features from hyperspectral data using the continuum removal technique or modeling techniques are used as input features; a single hidden layer (or possible to use > one hidden layer) is usually adopted with varying nodes; and output nodes respond abundance of number of mineral/rock end-members. Other network structure parameters, such as learning rate, momentum coefficient, and number of iterations are determined empirically. Using AVIRIS hyperspectral data, Yang et al. (1999) compared the ability of back-propagation neural network (BPN) with SAM for geological mapping. The mapping area is the Cuprite mining district in Nevada. It contains both hydrothermally altered and unaltered rocks. These altered rocks in the mapping area can be subdivided into three mappable units: silicified, opalized, and argillized rocks. Considering the 40 MIR AVIRIS band data, a single hidden-layered neural network containing 40 hidden nodes was adopted. The network nodes used a standard sigmoid transfer function with a gain factor of 1. Training the networks involved 800 iterations of a back-propagation learning algorithm with the learning rate and momentum coefficient as 0.1 and 0.9, respectively. The experimental results show that no misclassification for the training set in the case of BPN was created while 17% of misclassification occurred with the SAM; and the validation accuracy of BPN was also much higher than that created with the SAM (86% vs. 69%). In the study, the better performance by the BPN can be explained by its ability to deal with complex relationships (e.g., 40 dimensions) and the nature of the dataset with the mostly pure pixels of minerals. Therefore, the result demonstrates that BPN has superior the classification ability when applied to imaging spectrometer data for minerology mapping. In mapping rock and alteration minerals with airborne hyperspectral image data, Arvelyna et al. (2011) also utilized, and compared the performance of the BPN and SAM. In the study, HyMap hyperspectral data over the Warmbad district, south of Namibia, were acquired. The MIR data of HyMap from 2.01 μm (band 98) to 2.48 μm (band 126; a total of 29 bands) were used in mapping rock and alteration minerals. A total of 16 spectrally pure end-members were identified from the HyMap image and interpreted by using USGS spectral library data and specific field spectra. In the Warmbad district, the northeastern part of the area mostly consists of meta-gabbro, gabbro norite, and pegmatite intrusions, whereas the southwestern part has a distribution of mostly granodiorite, granite, and diorite. The BPN method was applied using a feed-forward classification technique with a logistic activation function and one hidden layer. The experimental results show that, compared to the SAM method, the BPN method produced better mapping results. The spatial distribution of rocks and alteration minerals is clear on the map created by using the combination of the BPN classification method and input of PPI and field spectra. The characteristics of each observed mineral such as pegmatite, sericite, chlorite-epidote alteration, and amphibole in the study area, have been retrieved using the BPN method, which may be used as the neural network spectra library for the other mapping project in the Warmbad district and its vicinity (Arvelyna et al. 2011).

With laboratory spectra and EO-1 Hyperion satellite hyperspectral data, Patteti et al. (2015) utilized a new feature-tuned ANN model for identification of minerology end-members and geology mapping. The ANN model works on the extracted absorption features from ore and rock spectra, rather than on the original hyperspectral band reflectance as used by a standard classification algorithm. This technique has the additional advantages of reducing the dimensionality of input features to the ANN as well as inhibiting the influences of noisy bands on classification of end-members due to using only the essential absorption bands (features) of mineral spectra. In the study, the ANN model was trained using input features extracted from laboratory spectra of *in situ* bulk ore materials collected from an existing iron ore deposit. The mineral and rock spectra represent iron, manganese, copper, uranium, and bauxite ores and rocks commonly found in India. The input features to the ANN model are spectral features (i.e., absorption band parameters) extracted by either Gaussian or MGM from the laboratory spectra, which include center, width, and strength parameters of each absorption band from each mineral spectrum. The trained ANN model was then applied for the end-member classification of Hyperion hyperspectral image acquired over the iron ore deposit. The ANN model has achieved nearly 97% and 71% classification accuracy in a training set and a test set, respectively, with laboratory spectra of *in situ* bulk ore materials. It was found that the field verification result of end-member classification of Hyperion imagery by using the trained ANN model with laboratory spectra has an overall accuracy of 67%. Therefore, the experimental results indicate that the ANN model is a potential approach for mineral end-member classification and mapping from *in situ* spectral measurements, and this model can be explored for other cases in the future investigation of mineral/ore identification using hyperspectral imaging data.

7.3.6.2 Expert Systems

In this section, two spectral expert bases/systems are briefly introduced and their applications are reviewed. An expert system, the Spectral Expert®, was developed by Horizon GeoImaging, LLC, Frisco, Colorado, for identifying and mapping materials based on extraction of key spectral features from VNIR and MIR reflectance spectra and hyperspectral imagery (HSI; Kruse 2008). In the system, spectral absorption features are automatically extracted from a spectral library, and the extracted spectral diagnostic features and characteristics are developed into "rules." The rules may be used by a non-expert to identify materials by matching individual feature parameters or with a rule-controlled root mean square (RMS) error approach. Based on the spectral system, the identification result of spectral sample is a score between 0.0 (no-match) and 1.0 (perfect-match) for each unknown specific material in the spectral library or for the hyperspectral image data. In addition, a feature-based mixture-index score or image is also created, which may help an analyst understand a possible spectral mixing problem.

According to Kruse (2008), to build the Spectral Expert system, the key components consist of extracting and isolating individual reflectance absorption features, characterizing these features using objective parameters and spectral variability, automatically building rules to describe the spectral features, and identifying unknown materials by matching their absorption features to the defined rules. The facts and rules to construct the system are built on analyzing and extracting of absorption band features (parameters) from a reference spectral library, either from laboratory spectra, field spectral measurements or from airborne/spaceborne HSI data. The refined rules are finally used to analyze unknown spectra by applying the feature extraction and analysis to the unknown spectra, then comparing the analyzed and extracted results to the rules to determine the property of the unknown spectra (pixels). To identify unknown reflectance spectra or HSI datasets, the spectral system matches against the constructed feature-based expert system rules for using an empirical probability. The empirical probability, called *certainty probability* as used in the Spectral Expert, is an empirical measure of the degree of fit of an unknown spectrum to the rules based on the number of rules satisfied for an unknown spectrum against the total number of rules for the reference spectrum. A feature based mixture index (FBMI) can also optionally be calculated to help judge the success of the feature-based analysis approach in the system. FBMI looks at the residual spectral features after matching a given material's features from the rule base. Higher FBMI scores indicate either that the material of interest is not in the rule base, or that there are additional unknown materials influencing the spectral signature. A RMS error method of comparing the known and unknown spectra within the Spectral Expert is also implemented to determine the number of features and the wavelength ranges to use in the fitting. To calculate the RMS error, the continuum is removed for both the known (library) and unknown spectra between the rule-defined continuum endpoints. Finally, the system combines the feature-based Spectral Expert with the RMS-based Spectral Expert in a weighted fashion along with other analysis algorithms including the SAM, binary encoding, and spectral feature fitting.

The spectral expert system was tested by Kruse (2008) using AVIRIS data acquired on June 19, 1997, at the Cuprite, Nevada, site and USGS Spectral Library spectra for mineral mapping. The Cuprite site has been used extensively for nearly 30 years as a test site for remote sensing instrument validation. The tested results demonstrate that the Spectral Expert is of basic success and thus presents an alternative of and supplement to other statistically based HSI analysis methods for analysis of full HSI data cubes. However, high spectral variability and spectral mixing presenting in this test complicated the analysis. Spectral Expert variability and separability analysis tools help deal with these issues. In short, the Spectral Expert works best with unique end-members with well-formed spectral features. Its application prefers working on unique, high-quality reflectance spectra from spectral libraries or culled from the HSI data.

Using the early version of the Spectral Expect, Kruse, Lefkoff, and Dietz (1993) integrated analysis of imaging spectrometer data (AVIRIS) and field spectral measurements, which was used in conjunction with conventional geologic field mapping to characterize bedrock and surficial geology at the northern end of Death Valley, California, and Nevada. In the study, the expert system was used to extract and characterize absorption features by analyzing a suite of laboratory spectra of the most common minerals. Facts and rules defining a generalized knowledge base for analysis of reflectance spectra were built on significant absorption band characteristics from laboratory spectra. The expert system successfully identified minerals using the 224 channel AVIRIS data. Specifically, the expert system along with linear spectral unmixing technique clearly defined the known distributions and abundances (and additional previously unmapped concentrations) of the carbonates, iron oxides, and sericite.

Unlike the spectral expert system introduced by Kruse (2008), which considers the VNIR/MIR (0.4–2.5 μm) spectral region only, Chen et al. (2010) developed a rule-based system that integrates both VNIR/MIR (0.4–2.5 μm) hyperspectral data and TIR (8.0–13 μm) multispectral data and evaluated it with a case study in Cuprite, Nevada. The main objective of developing the rule-based system was for automated identification of minerals and rocks based on their characteristic spectral features in the VNIR/MIR and TIR spectral regions. The rule-based system is related to reflectance spectrum and emissivity spectrum analysis, spectral feature matching algorithms, and decision rules. According to Chen et al. (2010), the knowledge base for the rule-based system was developed from analyses of rock spectra in both VNIR/MIR and TIR regions. The decision rules were constructed based on spectral reflectivity and emissivity features and characteristics of rocks in the VNIR/MIR region and spectral features in the TIR region. These spectral characteristics are related to electronic or vibrational processes associated with the interaction of the atoms and molecules composed by the minerals that make up a rock. For instance, iron dioxide, hydroxyl, and carbonate minerals exhibit absorption features in the VNIR/MIR region. However, most silicate minerals have spectral features in the TIR region (Hunt 1980, Chen et al. 2010). In the study, the reflectance spectra (0.4–2.5 μm) were extracted from AVIRIS data while emissivity spectra (8–13 μm) from MASTER data. The MASTER sensor is of the MODIS/ASTER Airborne Simulator (MASTER). Although the sensor acquires 50 bands of data in the 0.4–13 μm region, only the 10 TIR bands (8–13 μm) were used for the study in order to complement the AVIRIS VNIR/MIR bands. Based on the analyses by Chen et al. (2010), these reflectance spectra from the AVIRIS sensor and emissivity spectra from the MASTER sensor have showed distinctive spectral features that can be related to rock composition.

It was found that the SAM and SFF algorithms (see the detailed introduction to the two algorithms and applications in Sections 7.3.1 and 7.3.3) have shown varying success for identifying and mapping different minerals and rocks using different spectral regions. The SAM shows some advantages over SFF in identifying minerals and rocks when low albedo and relative flat spectral features are utilized, while SFF achieves better performance when it is used for identifying and mapping minerals and rocks from hyperspectral data that exhibit strong diagnostic absorption features. In the study, since the AVIRIS data in the VNIR/MIR range exhibit strong absorption features, the SFF procedure in the USGS Tetracorder System (Clark et al. 2003) based on a least-squares fit is used to compare each diagnostic absorption feature in the AVIRIS pixel spectrum with the continuum-removed reference spectra. For the MASTER data, since the pixel spectra do not show any strong characteristic absorption features, the SAM algorithm is used to estimate and compare the spectral feature from observed spectra with that from reference spectra. For pixels showing strong emittance features in the TIR region (e.g., quartz feature), the wavelength of the minimum emittance is determined using a least-squares fit approach as it is used for the AVIRIS data.

Therefore, the rule-based system employs different spectral feature matching algorithms (SFF and SAM), depending on the nature of the input spectrum. The rule-based system consists of a hierarchy of decision rules that are based on input pixel spectra with individual rock types. In the study, decision rules include spectral reflectance and the presence of diagnostic absorption features in the AVIRIS VNIR/MIR region, and the position of the minimum in the MASTER TIR spectrum is also identified. The decision rules assign each image pixel to just one class of number of classes, or if a pixel fails to reach a predetermined confidence threshold, it is assigned to the class Unknown. In the study, in order to simplify the rule-based system, the confidence threshold is set empirically to 0.1 for SAM (for TIR spectra) and 0.5 for SFF (for VNIR/MIR) (Chen et al. 2010). The rule-based system was tested by Chen et al. (2010) using AVIRIS and MASTER data acquired over the study area, June 19, 1996, and June 9, 1999, respectively. This geological map created with the rule-based system agrees with the previous geological map and provides additional information about unmapped rock units. In comparison with the other mapping methods, such as the SAM, SFF, minimum distance, and maximum likelihood classification methods, the rule-based system was found to achieve a higher overall performance.

In mapping mineralogy, this system demonstrates that integrating different spectral regions (e.g., VNIR/MIR and TIR regions) can result in substantially increased accuracy of identification of rocks compared to using the individual wavelength regions alone.

7.3.6.3 SVMs

The substantial SVM algorithms are introduced in Section 5.8.3 of Chapter 5. Similar to the application of the ANN method to minerology mapping, SVMs can also be directly applied to identifying and mapping minerals/rocks from hyperspectral data sets. Generally, selected hyperspectral band reflectances or extracted intrinsic dimensions of hyperspectral spectra using relevant feature extraction techniques are used as input features; training and test data can be determined either from spectral library or *in situ* spectral measurements or directly from hyperspectral image data; and the trained SVM classifiers are then applied to the hyperspectral image data to identify and map minerals/rock types. There are several different approaches used for classifying multiclass of minerals/rock types using SVMs, which are also briefly addressed in Section 5.8.3 of Chapter 5. In the following, two application cases of SVMs using hyperspectral data sets for identifying and mapping minerology are reviewed. By using independent spectral library and field based hyperspectral imagery, Murphy et al. (2012) evaluated and compared the performance of two classification techniques, SAM and SVM, to identify and map rock types on a vertical mine face at the West Angelas mine in the Hamersley Province of the Pilbara Region of Western Australia. They first compared the relative performance of SAM with SVM for precisely identifying rock classes using libraries of field spectra of common rock types found at the mine site. The spectral libraries were acquired under conditions of direct sunlight, shadow, and variable viewing geometries. They then applied SAM and SVM to hyperspectral imagery acquired from a vertical mine face for identification and mapping of rock types. In the study, the objective was to use library spectra of known rock types (samples) to classify hyperspectral imagery of the whole mine face into ore-bearing and non–ore-bearing rocks. Hyperspectral images were acquired from vertical mine faces with VNIR (400–970 nm) sensor and MIR (971–2516 nm) sensor at 2.22 and 6.35 nm spectral resolution and 6 and 12 cm spatial resolution, respectively. The SVM made use of kernel machine theory to extend the algorithm to perform nonlinear classification, where the transformed data points become linearly separable. The SVM algorithm sought to find a separating decision surface that maximizes the margin of separation between two linearly separable point sets. To perform classification, an alternative approach that makes use of probabilistic estimates of class membership was adopted rather than using the standard procedure by applying a hard decision function to the final SVM outputs. Probabilistic predictions are particularly useful for problems having more than two classes. In the study, the two main approaches were applied to solve the multiclass problems: one against all, where a classifier tests for the presence or absence of each class compared to the rest; and one against one, where the classes are first grouped in pairs and a classifier then processes the classification for each pair. The former was used to classify the spectral libraries and the latter to classify the hyperspectral images. The comparative results demonstrate that, compared to SAM method, SVMs perform better when training spectra are selected from the same type of data that is being classified but not when the training spectra are selected from an independent library acquired with a different sensor and under different conditions. In addition, shadow had a profound impact on classification of rock types using SAM and SVM techniques.

Unlike the work by Murphy et al. (2012) to improve the performance of SVM method for identifying and mapping minerals, Kolluru et al. (2014) developed an SVM-based dimensionality reduction and classification (SVMDRC) framework for hyperspectral data. The proposed SVMDRC unified framework was tested at Los Tollos in the Rodalquilar district of Spain, where alunite, kaolinite, and illite minerals with sparse vegetation cover are predominant. The airborne hyperspectral image was obtained from the HyMap sensor with 126 contiguous spectral bands, covering VNIR/MIR (0.45–2.5 μm) spectral range at spectral resolution 15–20 nm. The methodology in the study is divided into two parts, SVMDRC and SVMC. The SVMDRC was performed with the intrinsic (low) dimensionality of HyMap data, while the SVMC part was directly performed with an SVM classification on the hyperspectral image. Modified broken stick rule (Bajorski 2009) was used to calculate the intrinsic dimensionality of HyMap data that automatically reduce the number of feature bands. With the same training data, the SVMC classified the hyperspectral image into the three minerals with an accuracy of 64.70%, whereas accuracy created with SVMDRC was 82.35%. Therefore, the comparative result clearly suggests that SVM alone was inadequate in producing better mineral classification accuracy from hyperspectral data without considering dimensionality reduction, and incorporation of the intrinsic dimensionality extraction method for hyperspectral data in SVM technique positively enhances the feature separability and provides better classification accuracy.

Table 7.3 presents a summary of the all six types of analytical methods reviewed above from their techniques/algorithms/models, characteristics, advantages and limitations, and major application

TABLE 7.3

A Summary of Techniques/Algorithms, Characteristics, and Advantages/Limitations of the Six Types of Methods Applied in Geology Using Hyperspectral Data Sets

Analytical Method	Technique/Algorithm	Characteristic, Advantage and Limitation	Case Study
Absorption feature extraction	(1) Continuum removal (including a second-order polynomial simulated continuum, reference spectral background removal (RSBR), vegetation corrected continuum depth (VCCD)); (2) a relative absorption band-depth (RBD) approach; (3) spectral feature fitting (SFF) or least-squares spectral band-fitting techniques (least-squares fitting); (4) a simple linear interpolation technique; and (5) spectral absorption index (SAI).	(1) RSBR can eliminate the effects of a wide range of existing background material, such as vegetation and soil, on extracting the absorption features of potential target (minerals). There is a limitation when dealing with multiendmember problem as the circumstances associated with hyperspectral data will be much more complicated. (2) RBD image can provide a local continuum correction to remove any small band to band radiometric offsets, as well as variable atmospheric effects for each pixel in the data set. It is especially well suited for detecting weak SWIR spectral features. (3) Least-squares fitting technique can map complicated band shapes and fit to all data points comprising the feature. (4) Linear technique is easily implemented by users who are not familiar with programming languages. However the estimated absorption-band-depth is sensitive to the input parameters chosen. (5) Although the SAI and its model put the emphasis on single spectral absorption features, in practical geological applications, it is difficult to distinguish and classify those minerals that have similar spectral absorption features.	Clark and Roush 1984; Kruse 1988; Zhao et al. 2015; Rodger and Cudahy 2009; Crowley et al. 1989; Clark et al. 1990, 1991; van der Meer 2004; Huo et al. 2014.
Spectral mineral index	(1) 23 hyperspectral mineral indices (HMIs); (2) Normalized Difference Carbonate Index (NDCI); and (3) Hydrocarbon Index (HI) and Normalized Hydrocarbon Index (NHI).	(1) HMIs use simple algorithms and consume very little computer time. However, for some HMIs, some prior knowledge of mineralogy is necessary for mapping mineral species.(2) NDCI is effective for studying the phases of crystallization textures when hyperspectral data better than 5 nm spectral resolution are available. (3) NHI index was more useful than HI for detecting micro-seeps of coal-bed methane geology, and utilizing hyperspectral data for coal-bed methane exploration is feasible, effective and low at cost.	Henrich et al. 2012; Feldman and Taranik 1988; Van Ruitenbeek et al. 2006; Baissa et al. 2011; Zhang et al. 2014.

(Continued)

TABLE 7.3 (CONTINUED)
A Summary of Techniques/Algorithms, Characteristics, and Advantages/Limitations of the Six Types of Methods Applied in Geology Using Hyperspectral Data Sets

Analytical Method	Technique/Algorithm	Characteristic, Advantage and Limitation	Case Study
Spectral matching methods	(1) Spectral angle mapper (SAM); modified SAM (MSAM); (2) cross correlogram spectral matching technique (CCSM); and (3) spectral information divergence (SID).	(1) SAM method is simple and insensitive to gain factors because the angle between two spectrum vectors is invariant with respect to the lengths of the vectors; the main advantages of MSAM method are of simplicity and high-precision identification of alteration minerals. (2) The CCSM can be tested by decision theory that allows not only the validation of resulting classification, but also a measure of the reliability and accuracy of the results. (3) SID is probablity and stochastic measure.	Kruse et al. 1993a; Crósta et al. 1998; Hecker et al. 2008; Oshigami et al. 2013, 2015; van der Meer and Bakker 1997; Ferrier et al. 2002; Chang 2000.
Spectral mixture methods	(1) Linear spectral mixing model (LSM); (2) multiple endmember spectral mixture analysis (MESMA); (3) mixture tuned matched filtering (MTMF); and (4) constrained energy minimization (CEM).	(1) Easy to understand and simple to use, but (a) fail to account for that the spectral contrast between materials is variable, (b) cannot account for subtle spectral differences between materials efficiently, and (c) the maximum number of endmembers limited by the number of bands in the image data. (2) Although any mixed pixel individual spectrum can be modeled with relatively few endmembers, the number of endmembers and types of endmembers are variable across an image. A subset of all possible models was selected based on optimization for maximal area coverage. (3) MTMF retains the strengths of the MF and linear unmixing techniques, and has the ability to detect subtle changes in spectra. (4) CEM only requires the spectrum of the target mineral to be mapped and no prior knowledge of other background constituents/minerals, but it is necessary that potential improvements are on more rigorous quantitative thresholds for identifying target materials on abundance images.	Mustard 1993; Chen et al. 2013; Roberts et al. 1998; Shuai et al. 2013; Boardman et al. 1995; Bishop et al. 2011; Kodikara et al. 2012; Zadeh et al. 2014; Farrand & Harsanyi 1997; Resmini et al. 1997.

(Continued)

TABLE 7.3 (CONTINUED)
A Summary of Techniques/Algorithms, Characteristics, and Advantages/Limitations of the Six Types of Methods Applied in Geology Using Hyperspectral Data Sets

Analytical Method	Technique/Algorithm	Characteristic, Advantage and Limitation	Case Study
Spectral modeling methods	Modified Gaussian Model (MGM)	Compared to other spectral analytical approaches, MGM deconvolution analysis possesses a number of advantages: (1) The mathematic basis is more firmly rooted in the physical processes of electronic absorptions; (2) MGM can be used to reduce hyperspectral data dimensions by reducing reflectance spectra to a series of modified Gaussian curves with almost no loss of information; (3) the MGM analysis is relatively insensitive to particle size variations; and (4) MGM analysis could be directly applied to the data in spectral libraries, resulting in simplified spectral searching.	Sunshine et al. 1990; Sunshine and Pieters 1993; Clénet et al. 2011, 2013.
Advanced classification methods	(1) Artificial neural networks (ANNs); (2) expert system; and (3) support vector machines (SVMs).	(1) Having ability to deal with complex (nonlinear) relationships between absorption spectra and properties of mineral/rocks, but determining the NN structure parameters emperically. (2) The Spectral Expert works best with unique endmembers with well-formed spectral features; its application prefers working on unique, high-quality reflectance spectra from spectral libraries or imaging data. And (3) SVMs perform better when training spectra are selected from the same type of data that is being classified but not when the same type of data is selected from an independent library acquired with a different sensor and under different conditions.	Yang et al. 1999; Arvelyna et al. 2011; Patteti et al. 2015; Kruse 2008; Chen et al. 2010; Murphy et al. 2012; Kolluru et al. 2014.

references. The summary in the table will give the reader a quick view of the six types of analytical methods using hyperspectral data sets for identifying and mapping minerology.

7.4 HYPERSPECTRAL APPLICATIONS IN SOIL SCIENCES

7.4.1 Spectral Characteristics of Soils

Spectral characteristics of soils are the result of their physical and chemical properties and are often influenced largely by the compositional nature of soils in which the main components are moisture content, organic matter content, texture, structure, iron content, mineral composition, type of clay minerals, and surface conditions of the soil (de Jong and Epema 2001, van der Meer 2001, Eismann 2012). In the visible and NIR regions extending to 1.0 μm electronic transitions related to iron are the main factor determining soil spectral reflectance characteristics. The major absorption diagnostic features for mineral composition appear in the MIR spectral region from 2.0 to 2.5 μm. A strong fundamental OH^- vibration at 2.74 μm also influences the spectral signature of hyrdoxyl-bearing minerals. Furthermore, according to van der Meer (2001), some layered silicates such as clays, micas, and carbonates featured with diagnostic absorption characteristics also occur in the MIR region. Organic matter has a very important influence on the spectral reflectance properties of soils; even amounts only exceeding 2% are known to have a remarkable effect on spectral reflectance so that the overall reflectivity of the soil is reduced and the diagnostic absorption features are sometimes completely obscured. Two significant absorption features near 1.4 and 1.9 μm due to bound and unbound water can be typically seen from soil reflectance curves. A few of the less prominent water absorption features can be found at 0.97, 1.20, and 1.77 μm. In general, increasing moisture content leads to the decrease of overall reflectance of the soil, and a similar effect resulting from increasing the particle size of the soil resulting in a decrease in reflectivity can be observed.

More specifically, Stoner and Baumgardner (1981) examined spectral reflectance characteristics from 485 soil samples collected from the United States and Brazil, which represent 30 suborders of the 10 order of soil taxonomy. The bidirectional reflectance spectra from the 485 soil samples were measured on uniformly moist soils over the 0.52 to 2.32 μm wavelength range with an indoor-used spectrometer. Based on a curve shape, the presence or absence of absorption bands, and the predominance of soil organic matter and iron oxide composition, five distinct soil spectral reflectance curve forms could be identified. These curve forms were further characterized according to genetically homogeneous soil properties in a manner similar to the subdivisions at the suborder level of soil taxonomy. Their analysis results indicate that the five types of soil reflectance spectral curves might represent spectral characteristic variations of a wide range of naturally occurring soil samples. Figure 7.8 presents the five types of soil reflectance spectral curves. In the figure, Type A shows a low overall reflectance with a characteristic concave curve shape ranging from 0.5 to 1.3 μm. There are two strong water absorption bands present at 1.45 and 1.95 μm in this and other types of curves. Type B is characterized by a high overall reflectance and a characteristic convex curve shape from 0.5 to 1.3 μm. In addition to the two strong water absorption bands, a couple of minor water absorption bands can present at 1.2 and 1.77 μm due to the absorption bands observed in transmission spectra of relatively thick water films of the type. The Type C curve is identified here as the iron-affected form, which is distinguished by minor ferric iron absorption at 0.7 μm and a stronger iron absorption band at 0.9 μm. The 2.2 μm hydroxyl absorption band can be seen in this type of curves, but it doesn't exhibit a consistent result. Type D typically has a higher overall reflectance than organic-dominated form, and it slightly shows a concave shape from 0.5 and 0.75 μm with a convex shape from 0.75 and 1.3 μm. The last type (Type E) in the iron-dominated form is unique in that reflectance actually decreases with increasing wavelength beyond 0.75 μm. In this type of curve, due to the strong absorption in the MIR spectral region, the 1.45 and 1.95 μm water absorption phenomena are almost obliterated (Stoner and Baumgardner 1981). The soil spectral

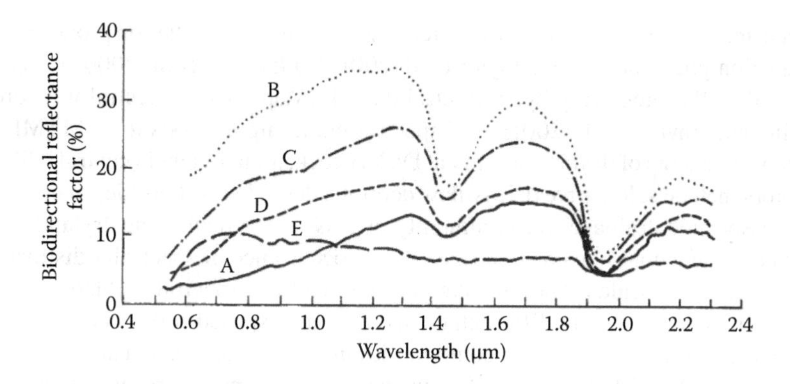

FIGURE 7.8 Characteristic soil bidirectional reflectance spectra. Curve A: Developed, fine-textured soils with high (> 2%) organic matter content. Curve B: Undeveloped soils with low (< 2%) organic matter and low (< 1%) iron oxide content. Curve C: Developed soil with low (< 2%) organic matter and medium (1–4%) iron oxide content. Curve D: Moderately course textured soils with high (> 2%) organic matter content and low (< 1%) iron-oxide content. Curve E: Fine-textured soils with high (> 4%) iron oxide content. (Republished with permission of Soil Science Society of America. From Stoner, E. R., and Baumgardner, M. F., *Soil Science Society of American Journal*, 45, 1161–1165, 1981. Permission conveyed through Copyright Clearance Center, Inc.)

characteristics that describe soils' reflectance properties in the solar part of the spectrum have also been described in many other studies (e.g., Condit 1970, Hunt 1980, Ben-Dor et al. 1999).

7.4.2 Review of Hyperspectral Applications in Soils

There are many techniques and algorithms that have been used in estimating, discriminating, and mapping soil properties with hyperspectral data measured in the laboratory and the field and collected on airborne and spaceborne platforms. They include spectral mixture analysis algorithms (SMA, MESMA, MTMF, and ISMA), SFF, multiple regression and partial least squares regression (PLSR) techniques, and soil moisture Gaussian model (SMGM; Whiting et al. 2004). All spectral unmixing algorithms and SFF are described in Chapter 5, except ISMA, which is the iterative spectral mixture analysis (Rogge et al. 2006). The forms of hyperspectral data input to models of estimating/mapping soil properties may be band reflectance, derivative spectra, ratio indices, features of absorption bands, etc. Visible, near-infrared, and middle infrared (VNIR/MIR, 400–2500 nm), and laboratory, field, and airborne/spaceborne spectroscopy techniques have been proven as good alternatives to costly physical and chemical soil analysis for the estimation of a large range of soil properties. In the following, a review of hyperspectral applications in soils is provided, which covers HRS of a wide range of soil properties.

7.4.2.1 Soil Degradation (Salinity, Erosion, and Deposition)

Soil degradation is a serious global environmental problem and may be exacerbated by climate change and human activities. For example, soil salinization, which is one of the most prevalent land degradation problems, is caused either by climate change in arid and semiarid regions or by human farming activities in agricultural areas. Soil degradation may include physical, chemical, and biological deterioration, such as loss of organic matter; decline in soil fertility; decline in structural condition; erosion; adverse changes in salinity, acidity, or alkalinity; and the effects of toxic chemicals, pollutants, and excessive flooding. Compared to multispectral remote sensing technology, hyperspectral remote sensing technology has shown a promising capability to better study and assess the properties and extents of soil degradation from far distances (e.g., Hill et al. 1995). Using various hyperspectral data (laboratory-based, *in situ*–field based and air/spaceborne-based imaging spectrometers), many researchers have conducted different kinds of studies on properties and extents of soil degradation phenomena (e.g., Taylor et al. 2001, Dutkiewicz et al. 2006, Weng et al. 2010, Malec et al. 2015). By integrating hyperspectral data, HyMap, *in situ* spectral measurements and digital terrain data, Taylor et al. (2001) used spectral unmixing methods (e.g., MTMF method) to map the various indicators of dryland salinity at Dicks Creek, a catchment in central NSW, Australia. These indicators include salt-source debris-flow deposits, degraded soil profiles, halophytic grasses, and a closely associated drainage line community of grasses and reeds. The dryland salinity in the Dicks Creek catchment could be characterized by the occurrence of spectrally distinctive smectite clays around surface salt scalds. Their results indicate that hyperspectral HyMap imagery, coupled with digital terrain data, is able to differentiate and map the exposed soils occurring around saline seeps and also the distinctive vegetation communities in spatially associated areas of water logging. Dehaan and Taylor (2002, 2003) used HyMap imagery and field spectral measurements and spectral unmixing and matching methods (e.g., MTMF and SFF methods) to evaluate saline soils and related vegetation for characterizing and mapping the spatial distribution of irrigation induced soil salinization at the Pyramid Hill test site, Tragowel Plains in Victoria, Australia. Strategies for extracting and mapping spectral end-members from HyMap imagery are assessed. Their experimental results demonstrate that three saline soil end-members could be mapped, which relate well to the surface expressions of soil salinity as measured by ground geophysics; distribution maps created using the SFF method and a restricted wavelength range of field-derived spectra provide an accurate record of the distribution of both vegetation and soil indicators of salinization at the time of image acquisition; the halophytic vegetation could be mapped down to the species level with hyperspectral data, using either field or image-derived spectra; the extent and composition of the surface salts present at Pyramid Hill have changed since the Geoscan imagery (24-band airborne multispectral VNIR/MIR data) had been acquired some eight years earlier and this suggested that the ability to map soil type might be more useful than the capacity to map the salts themselves; and salinized soil and vegetation indicator class maps created with field- and image-derived spectra show a similar spatial distribution to soil salinization as mapped by ground-based geophysical surveys. Dutkiewicz et al. (2006) compared the ability of three hyperspectral sensors: two airborne sensors (HyMap and CASI) and one satellite sensor (Hyperion) to discriminate and map selected symptoms of salinity in a dryland agricultural area in southern Australia. Three cover types with salinity symptoms include the perennial halophytic shrub samphire (*Halosarcia pergranulata*), a salt tolerant grass, sea barley grass (*Hordeum marinum*), and salt-encrusted pans. The three sensors' image spectra were used to map surface salinity symptoms using partial spectral unmixing techniques (i.e., MTMF models). The results indicate that saltpans were discriminated using the gypsum 1750 nm absorption feature, while full-wavelength image spectra were needed to map the halophytic plants. Their study demonstrates that hyperspectral imagery can improve discrimination of vegetation and mineral indicators of surface salinity compared with traditional soil and salinity mapping approach based on aerial photography interpretation; and seasonality of the imagery is important in capturing useful diagnostic spectral differences. Similar studies to those by Dehaan and Taylor (2002, 2003) were also done by Ghosh et al. (2012), Pang et al. (2014), and Moreira et al. (2015). In addition, Weng et al. (2010) used *in situ* spectral measurements and satellite Hyperion spectra directly to estimate and map soil salinization in the Yellow River Delta region of China. In the study, a soil salinity spectral index (SSI) was constructed from continuum-removed reflectance (CR-reflectance) at 2052 and 2203 nm where there exist several spectral absorption features of the salt-affected soils. Based on a strong correlation ($R^2 = 0.83$) between the SSI calculated with *in situ* spectra and soil salt content (SSC), the pixel-based quantitative salinity map was produced from the calibrated Hyperion image data, which was validated with field data successfully with RMSE = 1.921 (SSC) and ($R^2 = 0.63$). The mapping results suggest that the satellite hyperspectral data would have a potential for predicting SSC over a large area.

To describe the status of soil erosion, Hill et al. (1994, 1995) utilized AVIRIS data to parameterize a spectral mixture model with corresponding end-member spectra. Their thematic analysis of

AVIRIS data for mapping soil condition and erosion involved three separate stages: radiometric corrections of AVIRIS data, spectral mixture modeling for spectral decomposition of the original image spectra, and various soil condition classes mapped with a Euclidian minimum distance classifier. Over their study area in the Mediterranean basin in southern France, they estimated the relative abundance of parent material (including rock fragments) and soil particles on the surface. The erosion state of soils (undisturbed, slightly degraded, and severely degraded) was mapped as a function of the relative amounts of developed soil substrates and material of the parent lithology (Figure 7.9). The mapping results indicate that different erosion levels could be mapped with an accuracy of about 80%, which proved superior to applying the approach of Landsat-TM imagery (Hill et al. 1995). The results also suggested that the approach developed in the study would hold some potential for operational applications, including monitoring of erosion processes and changes in vegetation cover which are important elements for desertification monitoring (Hill et al. 1994). With the satellite hyperspectral sensor EnMap data simulated using airborne HyMap data, Malec et al. (2015)

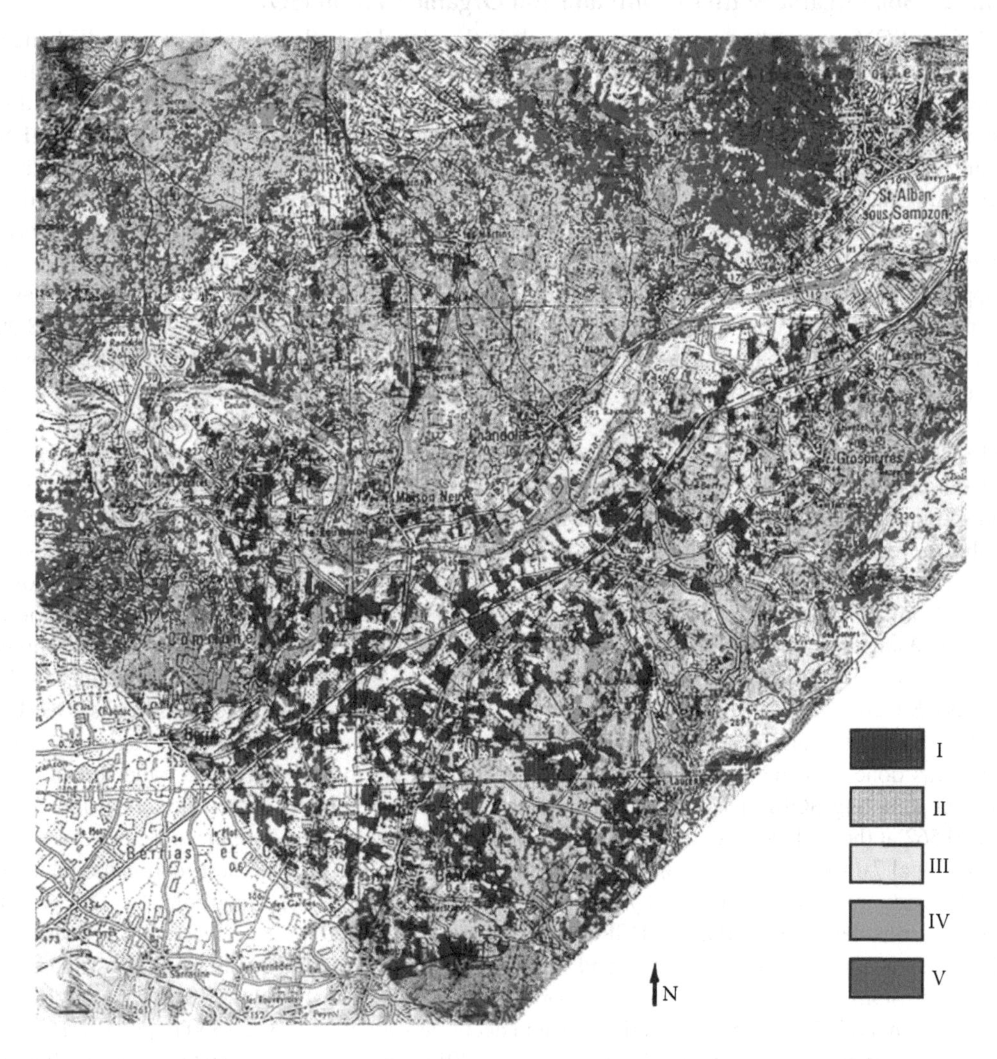

FIGURE 7.9 Map of four soil degradation (erosion) classes derived from AVIRIS data over S-France (I: undisturbed, II: slightly degraded, III: severely degraded; IV and V refer to exposed bedrock of marls and limestone.) Blank areas inside the image region had a vegetation cover of more than 50% and were not interpreted by the spectral mixing model. (After Hill, J., Megier, J., and Mehl, W., *Remote Sensing Reviews*, 12, 107–130, 1995.)

evaluated the capability of simulated EnMAP imagery to map fractional cover in a region near San Jose, Costa Rica, characterized by spatially extensive coffee plantations and grazing in a mountainous terrain. Soil erosion can be linked to relative fractional cover of photosynthetic-active vegetation (PV), non-photosynthetic-active vegetation (NPV), and bare soil (BS), which can be integrated into soil erosion models as the cover-management C-factor (Malec et al. 2015). In the study, the MESMA approach was first used to map the fractions of the three end-members: PV, NPV, and BS with simulated EnMap image data. The C-factor was then calculated based on the estimated fractional covers created by MESMA. The results demonstrate that it is possible to extract quality end-member classes of PV, NPV, and BS with simulated EnMAP imagery, and be able to estimate relative cover fractions for running soil erosion assessment models. From a regional perspective, Malec et al. (2015) thought that the EnMAP data could provide good fractional cover estimates that can be integrated into soil erosion modeling.

7.4.2.2 Soil Organic Matter (SOM) and Soil Organic Carbon (SOC)

SOC from SOM represents one of the major pools in the global C-cycle, in which even small changes in SOC stocks can cause important CO_2 fluxes between terrestrial ecosystems and the atmosphere (Stevens et al. 2006). Since SOM is mainly concentrated on the top A_0 horizon that is exposed to the Sun's radiation, it is a perfect property to be assessed by hyperspectral remote sensing (HRS) technology (Ben-Dor et al. 2009), and thus there are a lot of studies on this assessment. For examples, using laboratory and field-based spectral measurements and airborne hyperspectral data (Compact Airborne Spectrographic Imager [CASI; 405–950 nm] and Shortwave infrared Airborne Spectrographic Imager [SASI; 900–2500 nm] sensors), Stevens et al. (2006, 2008) studied and mapped SOC over cultivated soils in Belgium. Their studies have demonstrated that HRS has a promising potential to map SOM and SOC even with relatively low content. The values of the SOC they assessed ranged from a mean of 1.7% (0.8% min) to 3.0% (5.8% max). They used both stepwise and partial least square regression (PLSR) analysis to relate spectral measurements to SOC contents. The CASI airborne imaging spectroscopy sensor performed poorly, mainly due to its narrow spectral range. However, combing the CASI with SASI showed better results. The degradation of the spectral response to SOC contents with airborne sensor data was due to the variation in soil texture and soil moisture content. Although the RMSE of prediction was found to be 0.17% SOC, which is double the value of the laboratory's accuracy, Stevens et al. (2006, 2008) thought that the processed SOC image was reliable and gave for the first time a spatial overview of the SOM distribution over the study areas. To estimate and map SOM at a regional scale using a relatively rapid and economic method, Wang et al. (2010) used satellite Hyperion hyperspectral data associated with data sources to develop image object (IO)–based SOM estimate model to map SOM in Hengshan County in the northern Shaanxi Province of China. In the study, land degradation spectral response units (DSRUs) were created with various scale parameter values during the image segmentation. Estimating soil SOM was done by using multiple regression and fuzzy logic methods between band reflectances and soil sampling SOM measurements. The determinative coefficient (R^2) of the model increased from 0.562 at the scale level of 25 to 0.722 at the scale level of 100 (see scale meaning in eCognition Professional 7.0 software). The mapping content of SOM based on the DSRU estimation model using the Hyperion image agreed to that by the field survey and by the Kridge interpolation. The experimental results demonstrate that the DSRU estimation model, based on the relationship between the SOM content and features of spectral and regional variables of DSRUs, was valid to estimate the content of SOM and thus the DSRU model provides the potential of mapping SOM content over a large area in a relatively rapid, cost-effective, and accurate way. With CASI-1500 hyperspectral data and *in situ* spectral measurements, Matarrese et al. (2014) also conducted a test to evaluate SOC at a test site in the Apulia Region, Italy. They used CASI-derived derivative spectra to do correlation analysis with SOC field measurements. Their preliminary results have shown a significant response of the airborne hyperspectral sensor in SOC detection, which suggests that HRS technology could be a useful and suitable method for a rapid and efficient SOC monitoring at a local scale.

7.4.2.3 Soil Moisture

Water is considered one of the most important components in the soil system (e.g., Stoner and Baumgardner 1981) and has been estimated with hyperspectral data (e.g., Whiting et al. 2005, Demattê et al. 2006, Finn et al. 2011). For example, using spectral data measured in the laboratory by a spectroradiometer (0.45–2.50 μm), Demattê et al. (2006) developed a model to estimate soil moisture. Spectral reflectance was capable of distinguishing different moisture levels by using the soil line technique (defined by red and NIR bands). In the study, as the soil lost moisture, absorption bands centered at 1.4 and 1.9 μm started to show smaller and narrower concavities, while the band at 2.20 μm increased its absorption as the moisture decreases (Demattê et al. 2006). Their results suggest that soil mineralogy could be simultaneously evaluated with spectra measured from both wet and dry samples. Soil moisture could be differentiated by analyzing band reflectance intensity at the 1.55–1.75 μm wavelengths, but also possible to estimate soil moisture by using a multiple regression model ($R^2 = 0.98$) developed with the intensity of the absorption band features centered at 1.40, 1.90, and 2.20 μm. A robust spectral modeling technique to estimate soil moisture content has been offered by Whiting et al. (2004) by fitting an inverted Gaussian function to the continuum in soil spectra measured in laboratory from a broad range of soil samples. The soil samples were collected in the California Central Valley (high clay content, low carbonate) and La Mancha, Spain (low clay content, high carbonate). The SMGM (similar to the one in Equation 5.18 in Chapter 5) estimates the soil water content, based on the assigned fundamental water absorption region at 2.80 μm (Figure 7.10), which is located beyond the range of common airborne and field hyperspectral instruments. In the study, the SMGM model could accurately estimate the water content within an RMSE of 2.7% using all soil samples with a coefficient of determination (R^2) of 0.94, and an RMSE of 1.7 to 2.5% with R^2 ranging from 0.94 to 0.98 when soil samples were separated between Spain and the United States. The experimental results demonstrate that the SMGM could provide practical water content estimates and also has a potential use in correcting the effects of soil moisture in

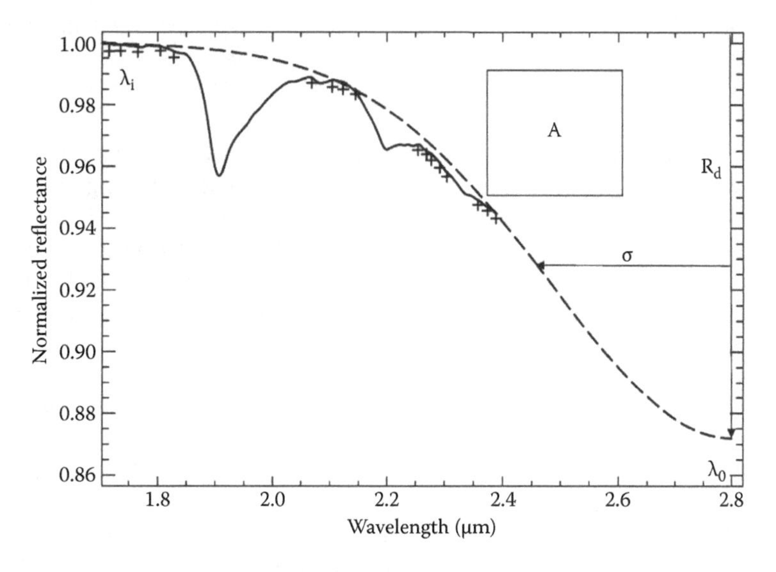

FIGURE 7.10 The inverted Gaussian function (dashed line) fitted to the convex hull points (+) of normalized, natural log spectrum (dashed line) from the point of maximum reflectance (λ_i) to spectrum's longest wavelength (the fundamental water absorption center at 2.8 μm). The functional center (λ_0) is extrapolated to determine the functional amplitude, or depth (R_d), and distant (σ) to the inflection point. The integration along the curve determines the area (A) of one side of the function. The spectra are normalized by dividing the reflectance at each band by this maximum reflectance. (Reprinted from Whiting, M. L., Li, L., and Ustin, S. L., *Remote Sensing Environment*, 89, 535–552. ©2004, with permission from Elsevier.)

hyperspectral images. Using the SMGM modeling method, Whiting et al. (2005) were also able to apply it to the processed AVIRIS and HyMap hyperspectral images to accurately estimate the soil surface moisture content. The improvement made by the SMGM parameter to estimates of clay and carbonate mineral contents is also demonstrated through a simple linear regression analysis of AVIRIS and HyMap images from Kings County, California, and La Mancha, Spain, respectively. Finn et al. (2011) explored whether the reflectance values available in hyperspectral remote sensing datasets can be used in the monitoring of soil water content. They used airborne hyperspectral image data in SWIR region, acquired in 2005 and 2007, to estimate soil moisture at three depth levels (5.08, 20.32, and 30.48 cm) in the Little River Experimental Watershed (LREW), Georgia. A dataset of 85 bands in reflectance in the range of 0.94 to 1.70 μm was used in multiple regression modeling analyses with field-based soil moisture probe data at the three depth levels. The results show that a significant statistical correlation (R^2 value above 0.7 for both sampling years) between the hyperspectral imaging data and the soil moisture probe data at 5.08 cm depth was determined. However, while the multiple regression models for the 20.32 cm and 30.48 cm depths were tested, they were not able to estimate soil moisture to the same degree as that at 5.08 cm depth. The data analysis illustrates that the ability for hyperspectral MIR data to estimate soil moisture deteriorates at the 20.32 and 30.48 cm depths.

Recently, to estimate the soil moisture under plant canopy, Song et al. (2014) proposed a semi-empirical soil moisture model that was expected to be able to estimate soil moisture in vegetation covered areas using Synthetic Aperture Radar and Hyperion data in an agricultural region in Heihe river basin in northwestern China. The model was based on the advanced integrated equation model (AIEM), the Michigan Microwave Canopy Scattering (MIMICS) model (Ulaby et al. 1990) and the Water–Cloud model (Attema and Ulaby 1978). The field-measured data including soil moisture data from the study area were used to confirm the proposed model. The results show that an average absolute deviation and average absolute relative deviation of $0.051 \text{ cm}^3 \cdot \text{cm}^{-3}$ and 19.7% were obtained, respectively, between the estimated soil moisture and the field measurements. The study also suggests that hyperspectral data would have an advantage of retrieving vegetation canopy water content over multispectral remote sensing data, and thus more hyperspectral data should be used in the soil moisture inversion.

7.4.2.4 Soil Contamination

Reflectance spectral properties of soils enable the assessment of various contaminants in them and HRS in laboratory, field and airborne imaging data has proved to be promising for the purpose. For instance, the process of pyrite oxidation at the surface of mine waste may produce acidic water that is gradually neutralized as it drains away from the waste. By considering the process and taking Fe-bearing minerals as indicators of the geochemical conditions, Swayze et al. (2000) used AVIRIS imaging data to map the mineral zones because each of these Fe-bearing secondary minerals is spectrally unique and to evaluate mine waste at the California Gulch Superfund Site near Leadville, Colorado. Based on their preliminary results, the imaging spectroscopy (e.g., AVIRIS sensor) can be used to rapidly screen entire mining districts for potential sources of surface acid drainage and detect acid producing minerals in mine waste or unmined rock outcrops. Following relevant studies (e.g., Swayze et al. 2000) in the United States, the MINEO (assessing and monitoring the environmental impact of mining activities in Europe using advanced Earth Observation techniques) project (Chevrel et al. 2003) investigated six mining areas—five in Europe, including Portugal, United Kingdom, Germany, Austria, and Finland; and one in Greenland—using HyMap airborne hyperspectral data. The HyMap sensor's data were used to map the extent and type of chronicle contamination with heavy metals using primarily trace minerals of pyrite oxidation as an indirect indicator of potential contamination. In the project, promising results have been obtained in combining those resulting maps with other relevant information under GIS for modeling contamination, pollution risk, and site rehabilitation or change detection (Chevrel et al. 2003). These results are important for environmental impact assessment, environmental monitoring of historical mining sites, and remediation planning, etc. Acid mine drainage (AMD) is a key concern for the mining industry due to its impact on the environments (e.g., water and soil) surrounding mine waste deposits. To map regional scale tailings mineral and pinpoints sources of AMD, Zabcic et al. (2014) evaluated the capability of airborne hyperspectral information (HyMap) of the Sotiel-Migollas complex in Spain. They extracted 26 spectral end-members directly from imagery, which represent mostly mineral mixtures. From the 26 spectral end-members, 11 spectral groups were defined each with minor variations in mineral mixtures. The mineral maps were generated through performing the iterative linear spectral unmixing analysis (ISMA) of Rogge et al. (2006). To generate the mineral maps, each pixel was labeled by the most abundant end-member predicted by the ISMA unmixing process. The mineral maps for the detailed investigation of tailings can serve as indicators of the metal, sulphate, and pH levels of the AMD solution at the time of mineral precipitation. Assessed with spectra from samples collected in the field and associated X-ray diffraction measurements, the pH maps tend to be consistent with those observed from the field and laboratory. Consequently, the pH maps are able to offer information on the pH conditions of the tailings thus giving an insight on the different types of oxidation reactions that may occur in mining areas (Zabcic et al. 2014).

7.4.2.5 Soil Classification and Mapping

The use of VNIR/MIR (0.4–2.5 μm) imaging spectrometer can greatly increase the accuracy of the digital soil mapping (e.g., Lagacherie and Gomez 2014). Current literature supports the use of spectral reflectance to determine and map soil properties, mostly using laboratory-based and filed-based soil spectral measurements, but also using airborne/spaceborne hyperspectral imagery in VNIR/MIR (0.4–2.5 μm) spectral regions in many application cases. More spectral information from hyperspectral data indeed advances people's ability to discriminate and map surface properties, including soils. For example, Demattê (2002) used spectral reflectance spectra (covering a spectral range of 0.40–2.50 μm) measured in laboratory from different soil samples. The soil samples were collected in Bauru region, São Paulo State, Brazil, and were identified as Typic Argiudoll (TR), Typic Eutrorthox (LR), Typic Argiudoll (PE), Typic Haplortox (LE), Typic Paleudalf (PV), and Typic Quartzipsamment (AQ). The results show that spectral differences among the different soils were better identified by their spectral curve shapes, reflecting on the intensity of band absorption and angle tendencies; the spectral curve shapes were mainly influenced by soil organic matter, iron, granulometry, and mineralogy constituents; and soils of the same group with different clay textures could be discriminated from their spectral reflectance. Based on the results evaluated by Demattê (2002) using laboratory-based and simulated orbital-based spectral data, it was possible to characterize and discriminate the most soil classes (groups) using laboratory-based spectral measurements. Still using laboratory-based spectral data measured from soil samples, Demattê et al. (2004) also evaluated soil types and soil tillage systems in the same state in Brazil as those in Demattê (2002). The soil samples were located along nine toposequences with basalt and shale. Reflectance spectral curves were evaluated at different soil depths in order to use similar conventional methods of soil classification to determine soil classes. In the study, Demattê et al. (2004) concluded that determination of soil classes was most precise when all spectral curves measured at different depths were analyzed simultaneously; the most important attributes including organic matter, total iron, silt, sand, and mineralogy significantly influenced reflected spectral intensity and spectral features and allowed to characterize and discriminate soil classes; soil line demarcation and number of soil classes detected by spectral analyses were similar to the detailed soil map determined by conventional methods. Their analysis results demonstrate that the laboratory-measured spectral data can be used as a methodology to assist soil surveys and can also group soils under similar tillage systems.

Mapping soil properties with airborne and satellite hyperspectral image data has proved to have a potential (e.g., Hively et al. 2011, Gasmi et al. 2014). By using AVIRIS hyperspectral imagery acquired from the western foothills of the Sierra Nevada near Fresno, California, Mustard (1993) used spectral mixture analysis technique to identify and map the distribution and abundance of five primary surface components (end-members) and also to model the effects of variable illumination.

The five image end-members including dry grass, green vegetation, and three soil types plus a model "shade" of 0.01% reflectance were considered in the study. This spectral unmixing approach successfully separated the transient surface spectral components of illumination, dry grass, and green vegetation from the more stable surface spectral components (end-members) associated primarily with the three soil types. In the study, Mustard (1993) found that the distinction between the three soil image end-members in a spectral context is very subtle, yet the spatial distributions observed in the fractions images reveal coherent and interpretable patterns. Therefore, his study demonstrates that the high spectral and spatial resolution data have the tremendous value in discriminating subtle differences in soil composition, in spite of the complex effects of highly variable transient surface components, such as vegetation. To compare the physical and hydrological properties of the crusts and underlying soil and map Mediterranean soil surface crusts, De Jong et al. (2011) identified the spectral characteristics (0.40–2.5 μm) of the crusted and noncrusted soil surfaces using field spectra and further investigated the potential of mapping crust occurrence using airborne hyperspectral sensor HyMap images. There existed spectral differences between field spectra of non-crusted and crusted soil surfaces in overall reflectance. Stronger absorption features in the clay mineral absorption bands at 2.20 μm than non-crusted soils were shown in 60% of the crusted soil cases. For analyzing airborne hyperspectral imaging data, SFF and linear spectral unmixing algorithms were applied to evaluate the possibilities of mapping surface crusts in the study area and for the case, the field spectra were used as reference spectra. The results show that although the spectral differences between crusted and non-crusted soil surfaces were small, it was possible to identify and map spatial patterns within and between agricultural fields with and without surface crusts, but for the natural areas, it was not possible to map surface crusts due to too fragmented of the landscape in the HyMap imagery. Using satellite Hyperion hyperspectral imagery, Gasmi et al. (2014) estimated two soil properties to study the risk of soil erosion from water: clay and calcium carbonate ($CaCO_3$) in a Mediterranean area of 210 km^2 (plain of the Oued Milyan, Tunisia). The purpose of their study is to determine whether the Hyperion hyperspectral imagery (spatial resolution of 30 m and SNR~50:1) can be used for mapping topsoil properties. They used the PLSR method to map clay and $CaCO_3$ contents with 124 soil samples collected in the study area. The large area (210 km^2) of the study region has allowed analysis of pedological patterns in terms of soil composition and spatial structures. Their results demonstrate that Hyperion data may be used to map clay and $CaCO_3$ contents over bare soils, with corresponding R^2 values of 0.71 and 0.79. And the results also suggest that the Hyperion satellite data might offer an alternative method for digital mapping of soil properties over large areas.

7.5 HYPERSPECTRAL APPLICATIONS IN GEOLOGY: CASE STUDIES

Three case studies are reviewed and summarized here. They are typical application cases of hyperspectral remote sensing (HRS) in identifying and mapping mineralogy using spectral unmixing and spectral matching techniques. Using hyperspectral image (HSI) data (airborne hyperspectral sensors: AVIRIS and HyMap; satellite hyperspectral sensor: Hyperion), Case I is for mapping multiple surficial materials, Case II is for mapping surface hydrothermal alteration minerals, and Case III is for mapping volcanogenic massive sulfide deposits.

7.5.1 Case I: Mapping Multiple Surficial Materials Using HyMap Data–Derived Absorption Features

This case study provides a summary of work by Hoefen et al. (2011) published in *Summaries of Important Areas for Mineral Investment and Production Opportunities of Nonfuel Minerals in Afghanistan* (eds. Peters et al. 2007). In their work, Hoefen et al. (2011) used absorption features derived from HyMap HSI data over the Daykundi area, Afghanistan, to explore huge Sn–W mineralization and associated alteration zones and to detect the presence of selected minerals that may be indicative of past mineralization processes, and they reported results of the imaging spectrometer data analyzed in mineralogy in the study area.

7.5.1.1 Study Area, HSI Data, and Image Preprocessing

The study area, the Daykundi, is approximately 200 km southwest of Kabul, Afghanistan, and is believed to have the potential for tin (Sn) and tungsten (W) deposits likely associated with tertiary felsic intrusive rocks, as well as other potential resources (Peters et al. 2007). The airborne hyperspectral HyMap sensor data were acquired over most of Afghanistan as part of the USGS Oil and Gas Resources Assessment of the Katawaz and Helmand Basins project in 2007. The HSI data were acquired over Afghanistan from August 22 to October 2, 2007. There were a total of 218 flight lines (207 standard data flight lines plus 11 cross-cutting calibration lines) collected over Afghanistan, covering a surface area of 438,012 km^2. HyMap data were calibrated and corrected from radiance to surface reflectance using a multi-step calibration process. The multi-step process consists of three steps:

1. The radiance data were converted to apparent reflectance using the radiative transfer correction program ACORN.
2. The atmospherically corrected apparent reflectance data were further empirically adjusted using ground-based reflectance measurements from ground calibration sites. Five ground-calibration spectra were collected in Afghanistan: Kandahar Air Field, Bagram Air Base, and Mazar-e-Sharif Airport, as well as soil samples from two fallow fields in Kabul.
3. An additional calibration step was taken to address the atmospheric differences caused, in part, by the long distances between the calibration sites and the survey areas. The cross-cutting calibration flight lines collected over the ground calibration areas were used to refine the reflectance calculation for standard data lines, which resulted in reducing residual atmospheric contamination in the HSI data.

7.5.1.2 Mapping Methodology

Calibrated HyMap reflectance data were processed using the Material Identification and Characterization Algorithm (MICA), a module of the USGS Processing Routines for Spectroscopic Measurements (PRISM) software (see its detailed introduction in Chapter 6). The MICA module compared the reflectance spectrum of each pixel of HyMap data to entries in a reference spectral library of minerals, vegetation, water, and other materials, which consists of 97 reference spectra of well-characterized mineral and material standards. The MICA module mapped HyMap data twice to create the two distribution maps of two categories of minerals and other materials that are naturally separated in the wavelength regions of their primary absorption features: MICA was applied using the subset of minerals with absorption features in the VNIR region, creating a 1-μm map of iron-bearing minerals and other materials; and again using the subset of minerals with absorption features in the MIR region, creating a 2-μm map of carbonates, phyllosilicates, sulfates, altered minerals, and other materials. For clarity of presentation, some individual classes in the two maps were grouped by combining selected mineral types (e.g., all montmorillonites or all kaolinites) and representing them with the same color in order to reduce the number of mineral classes. Thus, for the 1-μm map, generic spectral classes, including several minerals with similar absorption features, such as Fe^{3+} Type 1 and Fe^{3+} Type 2, are depicted on the map. For the 2-μm map, since minerals with slightly different compositions but similar spectral features are less easily discriminated, some identified classes consist of several minerals with similar spectra, such as the "chlorite or epidote" class. When a HyMap pixel spectrum cannot be matched with any reference spectra, the pixel was assigned to "not classified" class.

7.5.1.3 Mapping Results

Figures 7.11 and 7.12 present, respectively, the distribution of the iron-bearing minerals (the 1-μm map) and the carbonates, phyllosilicates, sulfates, altered minerals, and other materials (the 2-μm map) created from the HyMap reflectance data for the Daykundi area of interest (AOI). Both maps help identify different lithologies within the dataset, show regional trends and spatial relations of the different minerals, and may be used to improve the accuracy of future maps. Given the large number of classes represented and the subtleties of the distribution patterns represented in the two image maps, it is instructive to show these results as a series of image maps (see Hoefen et al. 2011), each describing a selected group of minerals that are mineralogically related or commonly occur together in specific geologic environments. Comparison results with other mineral data sources indicate that the HyMap data for the Daykundi AOI show good correlation to the various lithologic units of the region.

7.5.1.4 Concluding Remarks

The mineralogical mapping results demonstrate that predominant alteration zones could be mapped using HyMap and field spectroscopy; and for the larger prospects, generally the HyMap mapping results could be correlated back to the gangue minerals, deposit mineralogy, or minerals associated with the host lithology. Therefore, the mapping results with HyMap data analysis and identified numerous sites within the Daykundi study area deserve further investigation, especially detailed geological mapping and geochemical studies.

7.5.2 Case II: Mapping Surface Hydrothermal Alteration Minerals Using Airborne AVIRIS and Satellite Hyperion Imagery

In this case study, a summary of the work by Kruse et al. (2003), published in *IEEE Transactions on Geoscience and Remote Sensing*, is provided. It is regarding the use of airborne and satellite hyperspectral sensor (AVIRIS and Hyperion) data to map surficial hydrothermal alteration minerals. In their work, a general procedure of processing hyperspectral images and mineral mapping method was reported.

7.5.2.1 Study Area and Hyperspectral Data

The study area is located in Cuprite, Nevada. It is an ideal study area for a test of mapping surface hydrothermal alteration minerals using airborne and spaceborne HSI data because it is a relatively undisturbed acid-sulfate hydrothermal system in volcanic rocks that exhibits well-exposed alteration mineralogy, which consists principally of kaolinite, alunite, and hydrothermal silica. This site has been used as a geologic remote sensing test site since the early 1980s, and many test results have been published (e.g., van der Meer 2004, Kruse 2008, Kodama et al. 2010). The hyperspectral data include airborne hyperspectral AVIRIS data acquired on June 19, 1997, and satellite Hyperion data collected on March 1, 2001.

7.5.2.2 Mapping Methodology

Analytical Imaging and Geophysics (AIG), LLC, has developed methods/tools for processing and analysis of HSI data for identifying and mapping minerology, which allow reproducible results with minimal subjective analysis (Kruse et al. 2006; Figure 7.13). All of these approaches and tools are implemented and documented within the ENVI software system (see more details in Chapter 6). The HSI analysis procedure consists of the following steps:

1. HSI data preprocessing, including radiometric correction.
2. Linearly transforming the reflectance data to minimize noise and reduce data dimensionality.
3. Locating the most spectrally pure pixels.

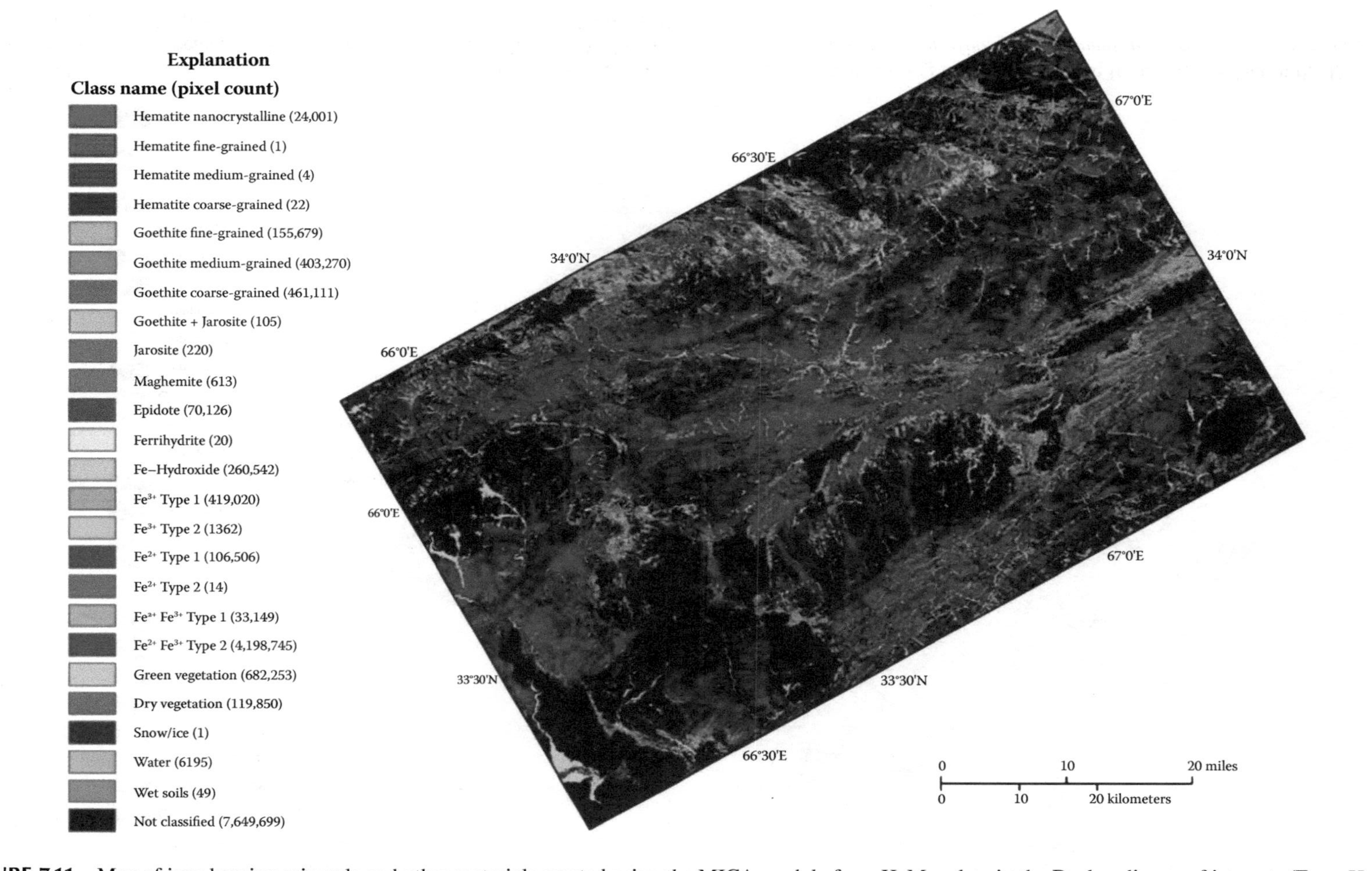

FIGURE 7.11 Map of iron-bearing minerals and other materials created using the MICA module from HyMap data in the Daykundi area of interest. (From Hoefen, T. M., Knepper Jr., D. H., and Giles, S. A., in *Summaries of Important Areas for Mineral Investment and Production Opportunities of Nonfuel Minerals in Afghanistan*, eds. S. G. Peters et al., US Geological Survey, Reston, Virginia, pp. 314–339, 2011.)

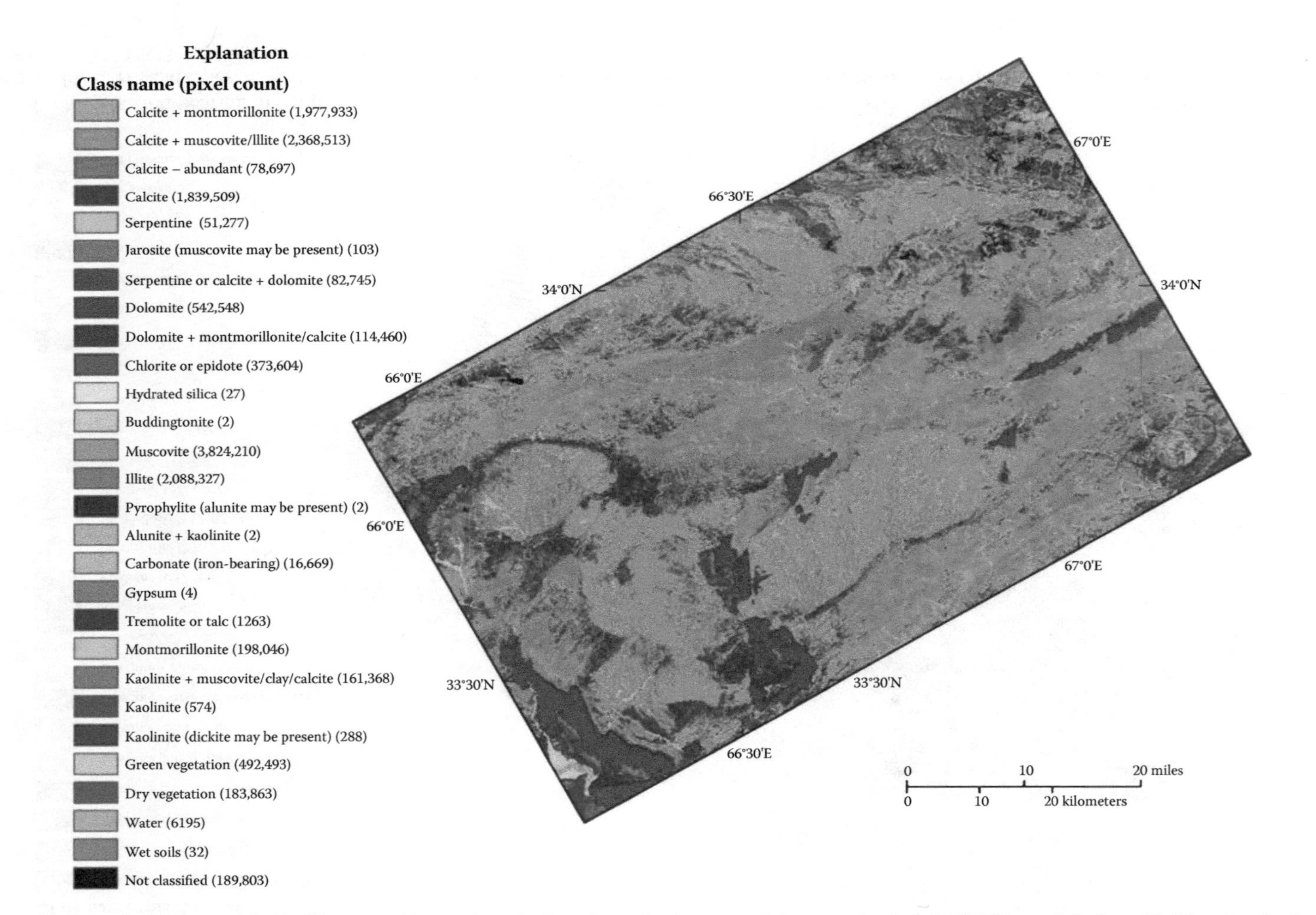

FIGURE 7.12 Map of carbonates, phyllosilicates, sulfates, altered minerals, and other materials created using the MICA module from HyMap data in the Daykundi area of interest. (From Hoefen, T. M., Knepper Jr., D. H., and Giles, S. A., in *Summaries of Important Areas for Mineral Investment and Production Opportunities of Nonfuel Minerals in Afghanistan*, eds. S. G. Peters et al., US Geological Survey, Reston, Virginia, pp. 314–339, 2011.)

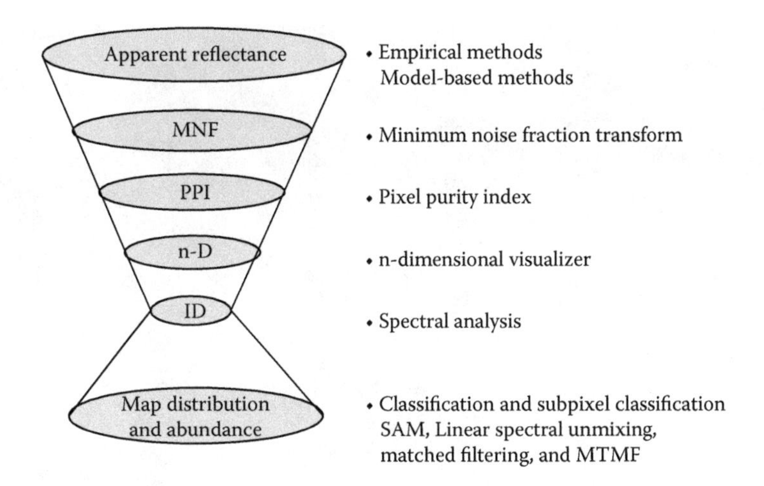

FIGURE 7.13 AIG hyperspectral data processing scheme. Large hyperspectral image (HSI) datasets are reduced to a few key spectra (at the neck of the hourglass) that account for the data using spatial and spectral data reduction techniques. Pixel-based spectral mapping methods are then applied to the full HSI dataset. (©2003 IEEE. Reprinted with permission from Kruse, F. A., Boardman, J. W., and Huntington, J. F., *IEEE Transactions on Geoscience and Remote Sensing*, 41, 1388–1400, 2003.)

4. Identifying and extracting end-member (mineral) spectra.
5. Spatial mapping and abundance estimates for specific image end-members (minerals).

Regarding step 1, the purpose of HSI data preprocessing is to correct and calibrate HSI raw data into apparent reflectance (ground surface scaled reflectance) using either empirical methods (e.g., IAR) or model-based methods (e.g., ACORN; read more about these methods in Chapter 4). The result of this step is that the apparent reflectance of HSI data is retrieved. In step 2, spectral compression, noise suppression, and dimensionality reduction are performed using the minimum noise fraction (MNF) transformation. Then, the low-dimensional MNF feature data (usually first several features) are passed to steps 3 and 4 for automatically locating, extracting pure pixel end-member spectra (e.g., kaolinite, alunite, muscovite, silica, buddingtonite, and calcite) using PPI and n-dimensional scatter plotting tools in the ENVI system. Moreover, end-member spectra are identified also using visual inspection and spectral library comparisons. In step 5, abundance estimate and mineral mapping are done using both the SAM and MTMF. At this step, the spectral analysis/mapping results need to be geometrically corrected using sensor models and aircraft or satellite navigation information.

7.5.2.3 Results

In the study, actual spectral bands covering the shortwave infrared spectral range (2.0–2.4 μm) were selected because these bands are heavily influenced by vibrational features of water, hydroxyl, carbonate, and sulfate molecules (Hunt 1977, van der Meer 2004), and these bands were linearly transformed using the MNF transformation. The MNF results indicate that the AVIRIS data contain significantly more information than the Hyperion data covering approximately the same spatial area and spectral range. For these bands covering the 2.0–2.4 μm range, the MNF analysis indicates dimensionalities of approximately 20 features for AVIRIS data and six features for Hyperion data, respectively, which contain most of spectral information. These low dimensional MNF features were input to PPI and n-dimensional scatterplotting tools in the ENVI system to determine pure mineral end-member spectra for running MTMF mapping algorithm. Therefore, the MTMF, a spectral matching method, was used to estimate and produce minerals maps that show the distribution and abundance of selected minerals (see Figure 7.14). Visually examining and comparing the

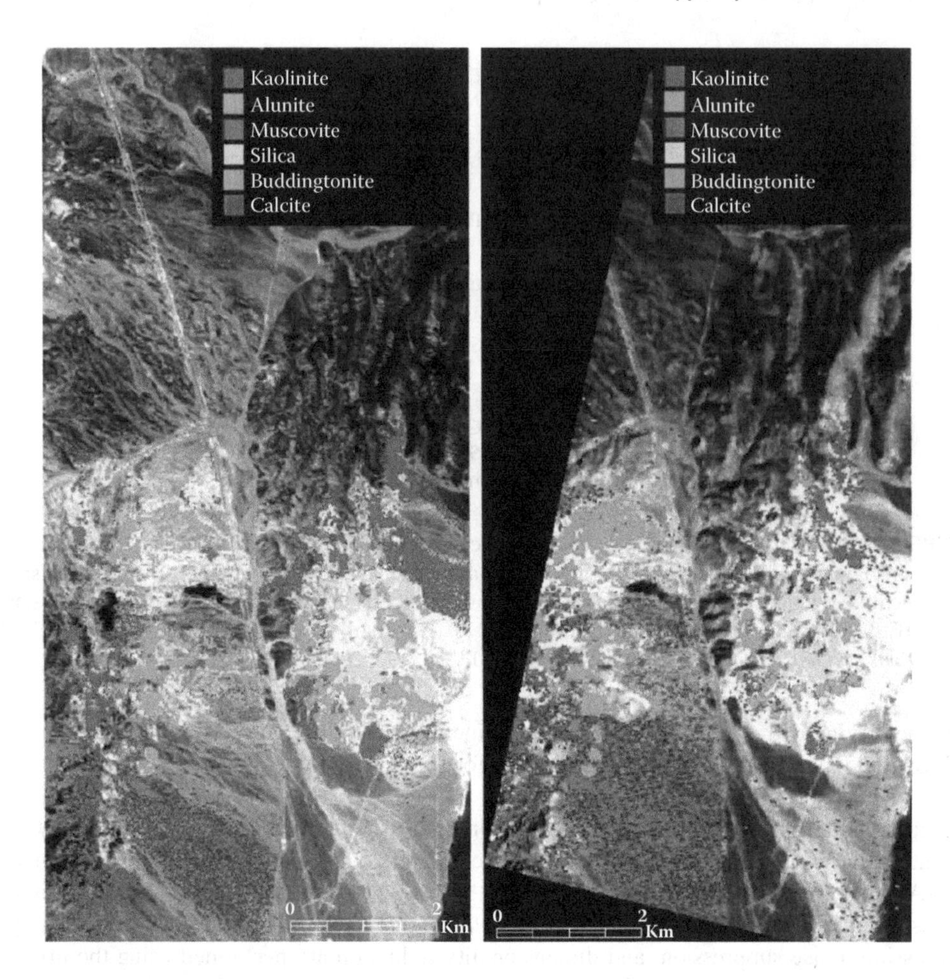

FIGURE 7.14 MTMF mineral maps for (left) AVIRIS and (right) Hyperion produced for the end-members (see legend in the figure) in Cuprite, Nevada. Colored pixels show the spectrally predominant mineral at concentrations greater than 10%. (©2003 IEEE. Reprinted with permission from Kruse, F. A., Boardman, J. W., and Huntington, J. F., *IEEE Transactions on Geoscience and Remote Sensing*, 41, 1388–1400, 2003.)

AVIRIS and Hyperion mineral maps shows that Hyperion generally identified similar minerals and produced similar mineral-mapping results to AVIRIS when using the AVIRIS data as the "ground truth" data and using these mapping methods. However, comparison of the AVIRIS with Hyperion MTMF spectral mapping results also shows that many pixels classified using AVIRIS are unclassified on Hyperion (up to 60%, but variable by minerals). These are errors of omission and are probably explained by the differences in SNR between the two sensors because of Hyperion having lower SNR than the AVIRIS sensor.

7.5.2.4 Concluding Remarks

Mineral mapping results using Hyperion MIR spectrometer (2.0–2.4 μm) data demonstrate that the Hyperion data can be used to produce useful mineralogical information. However, due to lower SNR for Hyperion sensor, fewer end-members with Hyperion data can be identified and mapped than those with AVIRIS. Accuracy assessment and error analysis also indicates that with Hyperion data, in many cases, mineral identification is not possible where some specific minerals are known to exist. In addition, Hyperion often confuses similar minerals that are separable for AVIRIS sensor. Based on the experimental results, it is recommended that future HSI satellite sensors would have significantly higher SNR performance specifications than Hyperion for the MIR (at least 100:1 based on dark current measurements; Kruse et al. 2003).

7.5.3 Case III: Mapping Volcanogenic Massive Sulfide Deposits Using HyMap Imagery

In Case III, the work by van Ruitenbeek et al. (2012) published in *Ore Geology Reviews*, will be briefly introduced and commented. Van Ruitenbeek et al. (2012) used airborne and field hyperspectral techniques to map Archean hydrothermal systems associated with volcanogenic massive sulfide (VMS) deposits, especially mapping the distribution of white mica mineral in the Panorama study area, Western Australia. In their work, relevant results of analyses of white mica distribution maps derived from airborne hyperspectral imagery and their integration with field-based observations and geochemical and oxygen isotope analyses were reported.

7.5.3.1 Study Area and Hyperspectral Data

The Panorama study area is located in the Soansville Greenstone belt in the East Pilbara Terrane in Western Australia. Pilbara Terrane, Australia, is one of the best VMS-type deposits and is extensively studied from the mineralization perspective. The bedrock mineralogy in the area is well-exposed because of the sparse soil and sediment cover, less weathering, and sparse spinifex vegetation cover in the current semiarid climate. To study the bedrock mineralogy in the area, four airborne HyMap hyperspectral images were acquired over the volcanic sequence and adjacent rocks. Each image covered approximately 2.5 by 22 km at a spatial resolution of 5 m and spectral resolution of 20 nm at around 2200 nm. In addition, field sample spectra were measured in laboratory from 223 unweathered volcanic rock sample powders using a portable infrared mineral analyzer with a spectral resolution of 7 nm at 2200 nm, covering spectra range of 1300–2500 nm. The reflectance spectra were used to visually interpret the presence of white mica and chlorite and to calculate spectral absorption features in order to determine mineral abundances.

7.5.3.2 Mapping Methodology

Using the stochastic method described by van Ruitenbeek et al. (2006) for mapping white micas and their absorption wavelength from multiple HyMap band-ratios and field data sets, the HyMap images were further processed and converted into two images: (1) an image representing the probability, ranging from 0 to 1, that an image pixel contains white mica and (2) an image, referred to as absorption wavelength image, showing the wavelength position of the absorption feature near 2200 nm of white micas, ranging approximately from 2195 to 2225 nm. The first image was calculated using a logistic regression model using HyMap band ratios, covering the absorption feature of white micas at 2200 nm and that of spinifex vegetation at near 2050 nm, to estimate the probability for the presence of white micas determined from field spectra of rock samples. The second image was calculated using a multiple linear regression model using HyMap band ratios covering the absorption feature near 2200 nm, which was used to estimate the wavelength position extracted from field spectra of rock samples. Validation of the two images with field measurements showed that mapping accuracy was satisfactory. Next, a template-matching method (van Ruitenbeek et al. 2008) on a pixel-by-pixel basis at various geographic rotation angles was used to classify the white mica probability image and the absorption wavelength image. Finally, the white mica mineral abundance and distribution maps were compared with published hydrothermal alteration maps and differences were interpreted using whole-rock geochemistry and temperature estimates from oxygen isotope geothermometric studies of hydrothermally altered rocks.

7.5.3.3 Results

The white mica species distribution map (Figure 7.15) derived from the 2200 nm absorption feature in the HyMap hyperspectral images shows distinct chemical zoning of white micas within the volcanic sequence, in which the white mica probability image was classified into six spectral groups

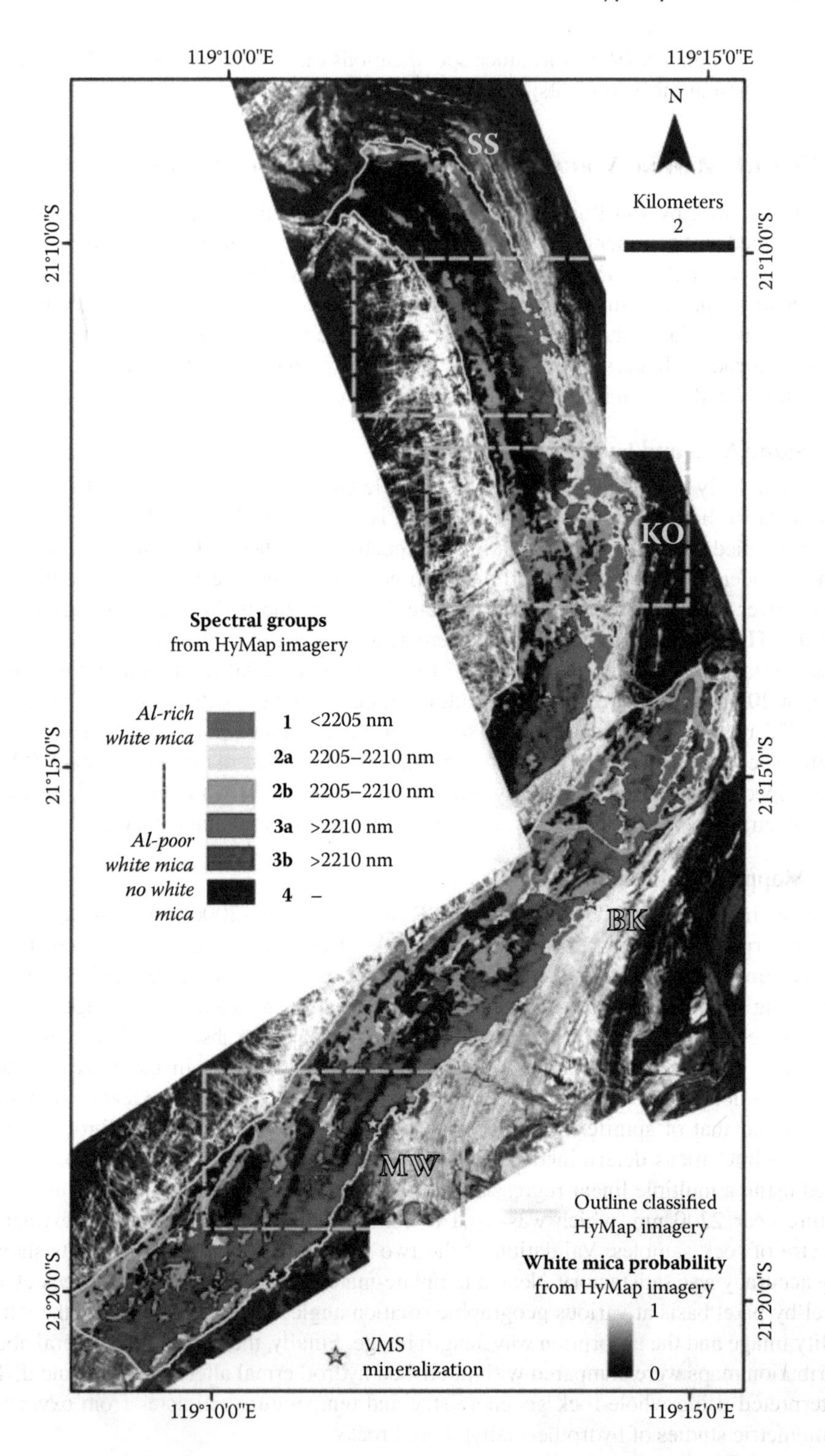

FIGURE 7.15 White mica species distribution map as determined from 2200 nm absorption feature in the HyMap imagery. White mica probability map in grayscale shows sedimentary layering and veining. VMS deposits: SS = Sulphur Springs, KC = Kangaroo Caves, BK = Breakers, MW = Man O'War. (Reprinted from van Ruitenbeek, F. J. A., Cudahy, T. J., van der Meer, F. D., and Hale, M., *Ore Geology Reviews*, 45, 33–46. ©2012, with permission from Elsevier.)

(1, 2a, 2b, 3a, 3b, 4), representing the different types of white mica and their abundances. The resulting map consists of coherent homogenous areas of similar white micas and their abundance. Based on the Al content in Figure 7.15, three different zones were identified, namely (1) Al-rich white mica zones, (2) Al-poor white mica zones predominantly related to K alteration by more evolved hydrothermal fluids, and (3) high to intermediate Al-content white mica zones related to intense alteration by laterally flowing and upwelling evolved fluids.

7.5.3.4 Concluding Remarks

The work by van Ruitenbeek et al. (2012) demonstrated the potential of HRS in mapping the distribution of white mica minerals and characterizing hydrothermal systems and reconstruction of paleo fluid pathways. The use of HRS data allowed for regional-scale mapping of white mica minerals. Therefore, HRS techniques can guide regional-scale exploration for VMS deposits through the identification and mapping of potentially mineralized hydrothermal discharge sites.

7.6 SUMMARY

After introducing the principles, possibility, and advantages and limitations of hyperspectral remote sensing technology in the VNIR–MIR spectral ranges as applied in geology and soil science, the spectral characteristics and properties of minerals/rocks were summarized and discussed. The major spectral characteristics of minerals/rocks are depicted by two general processes: electronic and vibrational, which basically cause absorption bands or reflection minima in the spectra of materials. Next, all six categories of hyperspectral application techniques and methods in geology were introduced and discussed. Their applications in estimating and mapping minerals and rocks with various hyperspectral data are reviewed and their characteristics and advantages/limitations were summarized in Table 7.3. The six categories of hyperspectral analysis techniques and methods include (1) absorption feature extraction methods with continuum removal spectra (including SFF, least-squares fitting, a simple linear interpolation technique, spectral absorption index, etc.), (2) hyperspectral mineral indices, (3) spectral matching techniques (including SAM, CCRM, and SID), (4) spectral unmixing techniques (including LSM, MESMA, MTMF, CEM, etc.), (5) spectral modeling method (MGM model), and (6) three advanced techniques and methods (ANN, expert system, and SVM). In this chapter, hyperspectral remote sensing applied in soil sciences was reviewed in Section 7.4, which involves summary and discussion of spectral characteristics of soils and an overview on hyperspectral remote sensing applications in estimating and mapping soil properties. The estimated and mapped properties of soils are related to soil degradation (salinity, erosion, and deposition), soil organic matter and soil organic carbon, soil moisture, soil contamination, and soil classification and mapping. In the last section (Section 7.5), three typical mineralogical application studies of hyperspectral remote sensing technology were introduced and their results and concluding remarks were summarized.

REFERENCES

Arvelyna, Y., M. Shuichi, M. Atsushi, A. Nguno, K. Mhopjeni, A. Muyongo, M. Sibeso, and E. Muvangua. 2011 (July 24–29). Hyperspectral mapping for rock and alteration mineral with Spectral Angle Mapping and Neural Network classification method: Study case in Warmbad district, south of Namibia. *2011 IEEE International Geoscience and Remote Sensing Symposium*, 1752–1754.

Attema, E. P. W., and F. T. Ulaby. 1978. Vegetation modeled as a water cloud. *Radio Science* 13:357–364.

Baissa, R., K. Labbassi, P. Launeau, A. Gaudin, and B. Ouajhain. 2011. Using HySpex SWIR-320m hyperspectral data for the identification and mapping of minerals in hand specimens of carbonate rocks from the Ankloute Formation (Agadir Basin, Western Morocco). *Journal of African Earth Sciences* 61:1–9.

Bajorski. P. 2009 (April 27). Does virtual dimensionality work in hyperspectral images? In *Proceedings of SPIE 7334, Algorithms and Technologies for Multispectral, Hyperspectral and Ultraspectral Imagery XV*, eds. S. S. Shen and P. E. Lewis, 73341J. doi:10.1117/12.818172. Orlando, Florida.

Baugh, W. M., F. A. Kruse, and W. W. Atkinson. 1998. Quantitative geochemical mapping of ammonium minerals in the southern Cedar Mountains, Nevada, using the Airborne Visible/Infrared Imaging Spectrometer (AVIRIS). *Remote Sensing of Environment* 65:292–308.

Bedini, E., F. van der Meer, and F. van Ruitenbeek. 2009. Use of HyMap imaging spectrometer data to map mineralogy in the Rodalquilar caldera, southeast Spain. *International Journal of Remote Sensing* 30:2327–2348.

Ben-Dor, E., J. A. Irons, and A. Epema. 1999. Soil spectroscopy, in *Manual of Remote Sensing*, 3rd ed, editor A. Rencz, pp. 111–188. New York: John Wiley & Sons, Inc.

Ben-Dor, E., S. Chabrillat, J. A. M. Demattê, G. R. Taylor, J. Hill, M. L. Whiting, and S. Sommer. 2009. Using imaging spectroscopy to study soil properties. *Remote Sensing of Environment* 113:S38–S55.

Bishop, C. A., J. G. Liu, and P. J. Mason. 2011. Hyperspectral remote sensing for mineral exploration in Pulang, Yunnan Province, China. *International Journal of Remote Sensing* 32(9):2409–2426.

Boardman, J. W., F. A. Kruse, and R. O. Green. 1995. Mapping target signatures via partial unmixing of AVIRIS data. *Summaries of the Fifth JPL Airborne Geoscience Workshop JPL Publication*, 95–1, pp. 23–26. Pasadena, CA: NASA Jet Propulsion Laboratory.

Bowers, T. L., and L. C. Rowan. 1996. Remote mineralogic and lithologic mapping of the Ice River alkaline complex, British Columbia, Canada, using ARIVIS data. *Photogrammetric Engineering & Remote Sensing* 62(12):1379–1385.

Burns, R. G. 1970. *Mineralogical Applications of Crystal Field Theory.* New York: Cambridge University Press.

Burns, R. 1993. *Mineralogical Applications of Crystal Field Theory*, 2nd ed. Cambridge: Cambridge University Press.

Chang, C.-I. 2000. An information theoretic-based approach to spectral variability, similarity and discriminability for hyperspectral image analysis. *IEEE Transactions on Information Theory* 46(5):1927–1932.

Chen, S., L. Chen, Y. Liu, and X. Li. 2013. Experimental simulation on mixed spectra of leaves and calcite for inversion of carbonate minerals from EO-1 Hyperion data. *GIScience & Remote Sensing* 50(6):690–703.

Chen, X., T. A. Warner, and D. J. Campagna. 2010. Integrating visible, near-infrared and short-wave infrared hyperspectral and multispectral thermal imagery for geological mapping at Cuprite, Nevada: A rule-based system. *International Journal of Remote Sensing* 31(7):1733–1752.

Chevrel, S., V. Kuosmanen, K. Grösel, S. Marsh, T. Tukiainen, U. Schäffer, L. Quental, P. Vosen, C. Fischer, P. Loudjani, E. Kuronen, and P. Aastrup. 2003. Assessing and monitoring the environmental impact of mining activities in Europe using advanced Earth Observation techniques. *Final Report MINEO Project*, IST-1999-10337.

Clark, R. N. 1999. Spectroscopy of rocks and minerals, and principles of spectroscopy, in *Manual of Remote Sensing, Vol. 3, Remote Sensing for the Earth Sciences*, ed. A. N. Rencz, pp. 3–58. New York: John Wiley & Sons.

Clark, R. N., A. J. Gallagher, and G. A. Swayze. 1990. Material absorption band depth mapping of imaging spectrometer data using a complete band shape least-squares fit with library reference spectra. *Proceedings of the Second Airborne Visible~Infrared Imaging Spectrometer (AVIRIS) Workshop, Jet Propulsion Laboratory Publication 90–54*, 176–186. Pasadena, California.

Clark, R. N., G. A. Swayze, A. J. Gallagher, N. Gorelick, and F. Kruse. 1991. Mapping with imaging spectrometer data using the complete band shape least-squares algorithm simultaneously fit to multiple spectral features from multiple materials. *Proceedings of the Third Airborne Visible~Infrared Imaging Spectrometer (A VIRIS) Workshop, Jet Propulsion Laboratory Publication 91–28*, pp. 2–3. Pasadena, California.

Clark, R. N., and T. L. Roush. 1984. Reflectance spectroscopy: Quantitative analysis techniques for remote sensing applications. *Journal of Geophysical Research* 89:6329–6340.

Clark, R. N., G. A. Swayze, K. E. Livo, R. F. Kokaly, S. J. Sutley, J. B. Dalton, R. R. McDougal, and C. A. Gent. 2003. Imaging spectroscopy: Earth and planetary remote sensing with the USGS Tetracorder and expert systems. *Journal of Geophysical Research* 108(E12):5131. doi:10.1029/2002JE001847.

Clénet, H., P. C. Pinet, Y. Daydou, F. Heuripeau, C. Rosemberg, D. Baratoux, and S. Chevrel. 2011. A new systematic approach using the Modified Gaussian Model: Insight for the characterization of chemical composition of olivines, pyroxenes, and olivine–pyroxene mixtures. *Icarus* 213:404–422.

Clénet, H., P. Pinet, G. Ceuleneer, Y. Daydou, F. Heuripeau, C. Rosemberg, J.-P. Bibring, G. Bellucci, F. Altieri, and B. Gondet. 2013. A systematic mapping procedure based on the Modified Gaussian Model to characterize magmatic units from olivine/pyroxenes mixtures: Application to the Syrtis Major volcanic shield on Mars. *Journal of Geophysical Research Planets* 118:1632–1655.

Cloutis, E. A., and M. J. Gaffey. 1991. Pyroxene spectroscopy revisited: Spectral-corn positional correlations and relationship to geothermometry. *Journal of Geophysical Research* 96:22809–22826.

Cloutis, E. A. 1996. Hyperspectral geological remote sensing: Evaluation of analytical techniques. *International Journal of Remote Sensing* 17:2215–2242.

Condit, H. R. 1970. The spectral reflectance of American soils. *Photogrammetric Engineering* 36:955–966.

Crósta, A. P., and C. R. de F. Souza. 1997 (November 17–19). Evaluating AVIRIS hyperspectral remote sensing data for geological mapping in Laterized Terranes, Central Brazil. *Proceedings of the Twelfth International Conference and Workshops on Applied Geologic Remote Sensing*, Vol. II:II-430–II-437. Denver, Colorado.

Crósta, A. P., C. Sabine, and J. V. Taranik. 1998. Hydrothermal alteration mapping at Bodie, California, using AVIRIS hyperspectral data. *Remote Sensing of Environment* 65:309–319.

Crowley, J. K. 1993. Mapping playa evaporite minerals with AVIRIS data. *Remote Sensing of Environment* 44(2–3):337–356.

Crowley, J. K., D. W. Brickey, and L. C. Rowan. 1989. Airborne imaging spectrometer data of the Ruby mountains, Montana: Mineral discrimination using relative absorption-band-depth images. *Remote Sensing of Environment* 29:121–134.

De Jong, S. M., and G. F. Epema. 2001. Imaging spectrometry for surveying and modeling land degradation: Basic physics of spectrometry, in *Imaging Spectrometry: Basic Principle of Prospective Applications*, eds. F. D. van der Meer and S. M. de Jong, pp. 64–86. The Netherlands: Springer.

De Jong, S. M., E. A. Addink, L. P. H. van Beek, and D. Duijsings. 2011. Physical characterization, spectral response and remotely sensed mapping of Mediterranean soil surface crusts. *Catena* 86:24–35.

Dehaan, R. L., and G. R. Taylor. 2002. Field-derived spectra of salinized soils and vegetation as indicators of irrigation-induced salinization. *Remote Sensing of Environment* 80:406–417.

Dehaan, R., and G. R. Taylor. 2003. Image-derived spectral end-members as indicators of salinity. *International Journal of Remote Sensing* 24:775–794.

Demattê, J. A. M. 2002. Characterization and discrimination of soils by their reflected electromagnetic energy. *Brazilian Journal of Agricultural Research* 37:1445–1458.

Demattê, J. A. M., R. C. Campos, M. C. Alves, P. R. Fiorio, and M. R. Nanni. 2004. Visible–NIR reflectance: A new approach on soil evaluation. *Geoderma* 121:95–112.

Demattê, J. A. M., A. A. Sousa, M. C. Alves, M. R. Nanni. 2006. Determining soil water status and other soil characteristics by spectral proximal sensing. *Geoderma* 135:179–195.

Dutkiewicz, A., M. Lewis, and B. Ostendorf. 2006 (May 8–11). Mapping surface symptoms of dryland salinity with hyperspectral imagery. *The International Archives of the Photogrammetry, Remote Sensing and Spatial Information Sciences* 34, Part XXX.

Eismann, M. 2012. Spectral properties of materials, in *Hyperspectral Remote Sensing*, pp. 133–198. Bellingham, WA: SPIE Press.

Exelis, Exelis Visual Information Solutions. 2015. *ENVI Classic Help, ENVI5.1*

Farmer, V. C. 1974. *The Infra-Red Spectra of Minerals*. London: Mineralogical Society.

Farrand, W. H., and J. C. Harsanyi. 1997. Mapping the distribution of mine tailings in the Coeur d'Alene River Valley, Idaho, through use of a constrained energy minimization technique. *Remote Sensing of Environment* 59:64–76.

Feldman, S. C., and J. V. Taranik. 1988. Comparison of techniques for discriminating hydrothermal alteration minerals with airborne imaging spectrometer data. *Remote Sensing of Environment* 24:67–83.

Ferrier, G., K. White, G. Griffiths, R. Bryant, and M. Stefouli. 2002. The mapping of hydrothermal alteration zones on the island of Lesvos, Greece, using an integrated remote sensing dataset. *International Journal of Remote Sensing* 23(2):341–356.

Finn, M. P., M. Lewis, D. D. Bosch, M. Giraldo, K. Yamamoto, D. G. Sullivan, R. Kincaid, R. Luna, G. K. Allam, C. Kvien, and M. S. Williams. 2011. Remote sensing of soil moisture using airborne hyperspectral data. *GIScience & Remote Sensing* 48(4):522–540.

Gasmi, A., C. Gomez, H. Zouari, A. Masse, and D. Ducrot. 2014 (December). Using Vis-NIR hyperspectral HYPERION data for bare soil properties mapping over Mediterranean area: Plain of the Oued Milyan, Tunisia. *European Academic Research* II(9):11721–11739.

Ghosh, G., S. Kumar, and S. K. Saha. 2012. Hyperspectral satellite data in mapping salt-affected soils using linear spectral unmixing analysis. *Journal of the Indian Society of Remote Sensing* 40(1):129–136.

Goetz, A. F. H., G. Vane, J. E. Solomon, and B. N. Rock. 1985. Imaging spectrometry for earth remote sensing. *Science* 228(4704):1147–1153.

Gomez, C., P. Lagacherie, and G. Coulouma. 2008. Continuum removal versus PLSR method for clay and calcium carbonate content estimation from laboratory and airborne hyperspectral measurements. *Geoderma* 148:141–148.

Green, R. O, M. L. Eastwood, C. M. Sarture, T. G. Chrien, M. Aronsson, B. J. Chippendale, J. A. Faust, B. E. Pavri, C. J. Chovit, M. Solis, M. R. Olah, and O. Williams. 1998. Imaging spectroscopy and the airborne visible/infrared imaging spectrometer (AVIRIS). *Remote Sensing of Environment* 65:227–248.

Hecker, C., M. van der Meijde, H. van der Werff, and F. D. van der Meer. 2008. Assessing the influence of reference spectra on synthetic SAM classification results. *IEEE Transactions on Geoscience and Remote Sensing* 46:4162–4172.

Henrich, V., A. Jung, C. Götze, C. Sandow, D. Thürkow, and C. Gläßer. 2009 (March 16–18). Development of an online indices database: Motivation, concept and implementation. *6th EARSeL Imaging Spectroscopy SIG Workshop Innovative Tool for Scientific and Commercial Environment Applications Tel Aviv*. Israel.

Henrich, V., G. Krauss, C. Götze, and C. Sandow. 2012. *IDB—Index DataBase*. Entwicklung einer Datenbank für Fernerkundungsindizes. Accessed February 4, 2016, from http://www.indexdatabase.de.

Hewson, R. D., T. J. Cudahy, S. Mizuhiko, K. Ueda, and A. J. Mauger. 2005. Seamless geological map generation using ASTER in the Broken Hill-Curnamona province of Australia. *Remote Sensing of Environment* 99:159–172.

Hill, J., J. Mégier, and W. Mehl. 1995. Land degradation, soil erosion and desertification monitoring in Mediterranean ecosystems. *Remote Sensing Reviews* 12:107–130.

Hill, J., W. Mehl, and M. Altherr. 1994. Land degradation and soil erosion mapping in a Mediterranean ecosystem, in *Imaging Spectrometry: A Tool for Environmental Observations*, eds. J. Hill and J. Mégier, pp. 237–260. Dordrecht: Kluwer Academic Publishers.

Hively, W. D., G. W.McCarty, J. B. Reeves III, M. W. Lang, R. A. Oesterling, and S. R. Delwiche. 2011. Use of airborne hyperspectral imagery to map soil properties in tilled agricultural fields. *Applied and Environmental Soil Science*, Vol. 2011, Article ID 358193.

Hoefen, T. M., D. H. Knepper Jr, and S. A. Giles. 2011. Analysis of imaging spectrometer data for the Daykundi area of interest, in *Summaries of Important Areas for Mineral Investment and Production Opportunities of Nonfuel Minerals in Afghanistan*, eds. S. G. Peters et al., pp. 314–339. Reston, VA: US Geological Survey.

Hubbard, B. E., J. K. Crowley, and D. R. Zimbelman. 2003. Comparative alteration mineral mapping using visible to shortwave infrared (0.4–2.4 μm) Hyperion, ALI, and ASTER imagery. *IEEE Transactions on Geoscience and Remote Sensing* 41(6):1401–1410.

Hunt, G. R. 1980. Electromagnetic radiation: The communication link in remote sensing, in *Remote Sensing in Geology*, eds. B. Siegal and A. Gillespia. New York: Wiley.

Hunt, G. R., and R. B. Hall. 1981. Identification of kaolins and associated minerals in altered volcanic rocks by infrared spectroscopy. *Clays and Clay Minerals* 29(1):76–78.

Hunt, G. R. 1977. Spectral signatures of particulate minerals in the visible and near infrared. *Geophysics* 42:501–513.

Hunt, G. R. 1979. Near-infrared (1.3–2.4 μm) spectra of alteration minerals-potential for use in remote sensing. *Geophysics* 44:1974–1986.

Huo, H., Z. Ni, X. Jiang, P. Zhou, and L. Liu. 2014. Mineral mapping and ore prospecting with HyMap data over eastern Tien Shan, Xinjiang Uyghur Autonomous Region. *Remote Sensing* 6:11829–11851.

Hutsinpiller, A. 1988. Discrimination of hydrothermal alteration mineral assemblages at Virginia City, Nevada, using the Airborne Imaging Spectrometer. *Remote Sensing of Environment* 24:53–66.

King, T. V. V., and W. I. Ridley. 1987. Relation of the spectroscopic reflectance of olivine to mineral chemistry and some remote sensing implications. *Journal of Geophysical Research* 92:11457–11469.

Kodama, S., I. Takeda, and Y. Yamaguchi. 2010. Mapping of hydrothermally altered rocks using the Modified Spectral Angle Mapper (MSAM) method and ASTER SWIR data. *International Journal of Geoinformatics* 6:41–53.

Kodikara, G. R. L., P. K. Champatiray, P. Chauhan, and R. S. Chatterjee. 2016. Spectral mapping of morphological features on the moon with MGM and SAM. *International Journal of Applied Earth Observation and Geoinformation* 44:31–41.

Kodikara, G. R. L., T. Woldai, F. J. A. van Ruitenbeek, Z. Kuria, F. van der Meer, K. D. Shepherd, and G. J. van Hummel. 2012. Hyperspectral remote sensing of evaporate minerals and associated sediments in Lake Magadi area, Kenya. *International Journal of Applied Earth Observation and Geoinformation* 14:22–32.

Kolluru, P., K. Pandey, and H. Padalia. 2014. A unified framework for dimensionality reduction and classification of hyperspectral data. *The International Archives of the Photogrammetry, Remote Sensing and Spatial Information Sciences* XL(8):447–453.

Kruse, F. A. 1988. Use of Airborne Imaging Spectrometer data to map minerals associated with hydrothermally altered rocks in the Northern Grapevine Mountains, Nevada and California. *Remote Sensing of Environment* 24:31–51.

Kruse, F. A. 2008 (March 16–20). Expert system analysis of hyperspectral data. *Proceedings, SPIE Defense and Security, Algorithms and Technologies for Multispectral, Hyperspectral, and Ultraspectral Imagery XIV.* Conference DS43, Paper Number: 6966–25. Orlando, Florida.

Kruse, F. A., A. B. Lefkoff, and J. B. Dietz. 1993. Expert system-based mineral mapping in northern Death Valley, California/Nevada using the Airborne Visible/Infrared Imaging Spectrometer (AVIRIS). *Remote Sensing of Environment* 44:309–336.

Kruse, F. A., S. L. Perry, and A. Caballero. 2006. District-level mineral survey using airborne hyperspectral data, Los Menucos, Argentina. *Annals of Geophysics* 49(1):83–92.

Kruse, F. A., J. W. Boardman, and J. F. Huntington. 2003. Comparison of airborne hyperspectral data and EO-1 Hyperion for mineral mapping. *IEEE Transactions on Geoscience and Remote Sensing* 41:1388–1400.

Kruse, F. A., A. B. Lefkoff, J. W. Boardman, K. B. Heidebrecht, A. T. Shapiro, P. J. Barloon, and A. F. H. Goetz. 1993. The spectral image processing system (SIPS)—Interactive visualization and analysis of imaging spectrometer data. *Remote Sensing of Environment* 44:145–163.

Kühn, F., K. Oppermann, and B. Hörig. 2004. Hydrocarbon Index-an algorithm for hyperspectral detection of hydrocarbon. *International Journal of Remote Sensing* 25(12):2467–2473.

Lagacherie, P., and C. Gomez. 2014. What can Global Soil Map expect from Vis-NIR hyperspectral imagery in the near future?, in *Global Soil Map: Basis of the Global Spatial Soil Information System*, eds. D. Arrouays, N. McKenzie, J. Hempel, A. Richer de Forges, and A. B. McBratney, pp. 387–392. Boca Raton, FL: CRC Press.

Li, Q., B. Zhang, L. Gao, L. Lu, and Q. Jiao. 2014. The identification of altered rock in vegetation-covered area using hyperspectral remote sensing. *IEEE Geoscience and Remote Sensing Symposium* (*IGARSS2014*), 2910–2913.

Magendrana, T., and S. Sanjeevi. 2014. Hyperion image analysis and linear spectral unmixing to evaluate the grades of iron ores in parts of Noamundi, Eastern India. *International Journal of Applied Earth Observation and Geoinformation* 26:413–426.

Malec, S., D. Rogge, U. Heiden, A. Sanchez-Azofeifa, M. Bachmann, and M. Wegmann. 2015. Capability of spaceborne hyperspectral EnMap mission for mapping fractional cover for soil erosion modeling. *Remote Sensing* 7:11776–11800.

Marfunin, A. S. 1979. *Physics of Minerals and Inorganic Materials: An Introduction.* New York: Springer-Verlag.

Matarrese, R., V. Ancona, R. Salvatori, M. R. Muolo, V. F. Uricchio, and M. Vurro. 2014(July 13–18). Detecting soil organic carbon by CASI hyperspectral images. *2014 IEEE International Geoscience and Remote Sensing Symposium*, 3284–3287. Quebec, Canada.

Moreira, L. C. J., A. D. S. Teixeira, and L. S. Galvão. 2015. Potential of multispectral and hyperspectral data to detect saline-exposed soils in Brazil. *GIScience & Remote Sensing* 52(4):416–436.

Morris, R. V., H. V. Lauer, C. A. Lawson, E. K. Jr. Gibson, G. A. Nace, and C. Stewart. 1985. Spectral and other physiochemical properties of submicron powders of hematite (-Fe_2O_3), maghemite (-Fe_2O_3), maghemite (Fe_3O_4), goethite (-FeOOH), and lepidochrosite (-FeOOH), *Journal of Geophysical Research* 90:3126–3144.

Mulder, V. L., M. Plötze, S. de Bruin, M. E. Schaepman, C. Mavris, R. F. Kokaly, and M. Egli. 2013. Quantifying mineral abundances of complex mixtures by coupling spectral deconvolution of SWIR spectra (2.1–2.4 µm) and regression tree analysis. *Geoderma* 207–208:279–290.

Murphy, R. J., S. T. Monteiro, and S. Schneider. 2012. Evaluating classification techniques for mapping vertical geology using field-based hyperspectral sensors. *IEEE Transactions on Geoscience and Remote Sensing* 50(8):3066–3080.

Mustard, J. F. 1993. Relationships of soil, grass, and bedrock over the Kaweah serpentinite melange through spectral mixture analysis of AVIRIS data. *Remote Sensing of Environment* 44:293–308.

Oshigami, S., T. Uezato, Y. Yamaguchi, Y. Arvelyna, A. Momose, Y. Kawakami, T. Yajima, S. Miyatake, and A. Nguno. 2013. Mineralogical mapping of southern Namibia by application of continuum-removal MSAM method to the HyMap data. *International Journal of Remote Sensing* 34(15):5282–5295.

Oshigami, S., Y. Yamaguchi, M. Mitsuishi, A. Momose, and T. Yajima. 2015. An advanced method for mineral mapping applicable to hyperspectral images: The composite MSAM. *Remote Sensing Letters* 6(7):499–508.

Pang, G., T. Wang, J. Liao, and S. Li. 2014. Quantitative model based on field-derived spectral characteristics to estimate soil salinity in Minqin County, China. *Soil Science Society of America Journal* 78:546–555.

Patteti, S., B. Samanta, and D. Chakravarty. 2015. Design of a feature-tuned ANN model based on bulk rock-derived mineral spectra for end-member classification of a hyperspectral image from an iron ore deposit. *International Journal of Remote Sensing* 36(8):2037–2062.

Peters, S. G., S. D. Ludington, G. J. Orris, D. M. Sutphin, J. D. Bliss, and J. J. Rytuba, eds., and the U.S. Geological Survey-Afghanistan Ministry of Mines Joint Mineral Resource Assessment Team. 2007. Preliminary non-fuel mineral resource assessment of Afghanistan. *U.S. Geological Survey Open-File Report 2007–1214.*

Pu, R., and P. Gong. 2011. Hyperspectral remote sensing of vegetation bioparameters, in *Advances in Environmental Remote Sensing: Sensors, Algorithm, and Applications*, series ed. Q. Weng, pp. 101–142. Boca Raton, FL: CRC Press/Taylor & Francis Publishing Group.

Rajesh, H. M. 2004. Application of remote sensing and GIS in mineral resource mapping—An overview. *Journal of Mineralogical and Petrological Sciences* 99:83–103.

Ramakrishnan, D., and R. Bharti. 2015. Hyperspectral remote sensing and geological applications. *Current Science* 108(5):879–891.

Resmini, R. G., M. E. Kappus, W. S. Aldrich, J. C. Harsanyi, and M. Anderson. 1997. Mineral mapping with HYperspectral Digital Imagery Collection Experiment (HYDICE) sensor data at Cuprite, Nevada, U.S.A. *International Journal of Remote Sensing* 18(7):1553–1570.

Roberts, D. A., M. Gardner, R. Church, S. Ustin, G. Scheer, and R. O. Green. 1998. Mapping Chaparral in the Santa Monica Mountains using multiple end-member spectral mixture models. *Remote Sensing of Environment* 65:267–279.

Rodger, A., and T. Cudahy. 2009. Vegetation corrected continuum depths at 2.20 μm: An approach for hyperspectral sensors. *Remote Sensing of Environment* 113:2243–2257.

Rogge, D. M., B. Rivard, J. Zhang, and J. Feng. 2006. Iterative spectral unmixing for optimizing per-pixel endmember sets. *IEEE Transactions on Geoscience and Remote Sensing* 44(12):3725–3736.

Sabins, F. F. 1999. Remote sensing for mineral exploration. *Ore Geology Reviews* 14:157–183.

Salisbury, J. W., L. S. Walter, and N. Vergo. 1989. Availability of a library of infrared (2.1–25.0 μm) mineral spectra. *American Mineralogist* 74:938–939.

Shuai, T., X. Zhang, L. Zhang, and J. Wang. 2013. Mapping global lunar abundance of plagioclase, clinopyroxene and olivine with Interference Imaging Spectrometer hyperspectral data considering space weathering effect. *Icarus* 222:401–410.

Song, X., J. Ma, X. Li, P. Leng, F. Zhou, and S. Li. 2014. First results of estimating surface soil moisture in the vegetated areas using ASAR and Hyperion data: The Chinese Heihe River Basin case study. *Remote Sensing* 6:12055–12069.

Stevens, A., B. Wesemael, H. Btholomeus, D. Rosillon, B. Tychon, and E. Ben-Dor. 2008. Laboratory, field and airborne spectroscopy for monitoring organic3 carbon content in agricultural soils. *Geoderma* 144:395–404.

Stevens, A., B. Wesemael, G. Vandenschrick, S. Touré, and B. Tychon. 2006. Detection of carbon stock change in agricultural soils using spectroscopic techniques. *Soil Science Society of America Journal* 70:844–850.

Stoner, E. R., and M. F. Baumgardner. 1981. Characteristic variations in reflectance of surface soils. *Soil Science Society of American Journal* 45:1161–1165.

Sunshine, J. M., and C. M. Pieters. 1990. Extraction of compositional information from olivine reflectance spectra: A new capability for lunar exploration. *Lunar and Planetary Science Conference XXI*, 1223–1224. Houston, TX: Lunar and Planetary Institute.

Sunshine, J. M., C. M. Pieters, and S. R. Pratt. 1990. Deconvolution of mineral absorption bands: An improved approach. *Journal of Geophysical Research* 95:6955–6966.

Sunshine, J. M., and C. M. Pieters. 1993. Estimating modal abundances from the spectra of natural and laboratory pyroxene mixtures using the modified Gaussian model. *Journal of Geophysical Research* 98:9075–9087.

Swayze, G. A., K. S. Smith, R. N. Clark, S. J. Sutley, R. M. Pearson, J. S. Vance, P. L. Hageman, P. H. Briggs, A. L. Meier, M. J. Singleton, and S. Roth. 2000. Using imaging spectroscopy to map acidic mine waste. *Environmental Science and Technology* 34:47–54.

Taylor, G. R., P. Hemphill, D. Currie, T. Broadfoot, and R. L. Dehaan. 2001. Mapping dryland salinity with hyperspectral imagery. *Proceedings IEEE International Geoscience and Remote Sensing Symposium Scanning the Present and Resolving the Future* 1:302–304. Piscataway, NJ: Institute of Electrical and Electronic Engineers.

Trang, D., P. G. Lucey, J. J. Gillis-Davis, J. T. S. Cahill, R. L. Klima, and P. J. Isaacson. 2013. Near-infrared optical constants of naturally occurring olivine and synthetic pyroxene as a function of mineral composition. *Journal of Geophysical Research: Planets* 118:708–732.

Ulaby, F. T., K. Sarabandi, K. McDonald, M. Whitt, and M. C. Dobson. 1990. Michigan microwave canopy scattering model. *International Journal of Remote Sensing* 11:1223–1253.

Van der Meer, F. 2001. Basic physics of spectrometry. *Imaging Spectrometry: Basic Principle of Prospective Applications*, eds. F. van der Meer and S. M. de Jong, pp. 3–18. The Netherlands: Springer.

Van der Meer, F. 2004. Analysis of spectral absorption features in hyperspectral imagery. *International Journal of Applied Earth Observation and Geoinformation* 5:55–68.

Van der Meer, F. 2006. The effectiveness of spectral similarity measures for the analysis of hyperspectral imagery. *International Journal of Applied Earth Observation and Geoinformation* 8(1):3–17.

Van der Meer, F. D., H. M. A. van der Werff, F. J. A. van Ruitenbeek, C. A. Hecker, W. H. Bakker, M. F. Noomen, M. van der Meijde, E. J. M. Carranza, J. B. de Smeth, and T. Woldai. 2012. Multi- and hyperspectral geologic remote sensing: A review. *International Journal of Applied Earth Observation and Geoinformation* 14(1):112–128.

Van der Meer, F., and W. Bakker. 1997. Cross correlogram spectral matching (CCSM): Application to surface mineralogical mapping using AVIRIS data from Cuprite, Nevada. *Remote Sensing of Environment* 61(3):371–382.

Van Ruitenbeek, F. J. A., T. J. Cudahy, F. D. Van der Meer, and M. Hale. 2012. Characterization of the hydrothermal systems associated with Archean VMS-mineralization at Panorama, Western Australia, using hyperspectral, geochemical and geothermometric data. *Ore Geology Review* 45:33–46.

Van Ruitenbeek, F., H. van der Werff, K. Hein, and F. D. van der Meer. 2008. Detection of pre-defined boundaries between hydrothermal alteration zones using rotation-variant template matching. *Computers & Geoscience* 34:1815–1826.

Van Ruitenbeek, F. J. A., P. Debba, F. D. van der Meer, T. Cudahy, M. van der Meijde, and M. Hale. 2006. Mapping white micas and their absorption wavelengths using hyperspectral band ratios. *Remote Sensing of Environment* 102:211–222.

Vane, G., and A. F. H. Goetz. 1988. Terrestrial imaging spectroscopy. *Remote Sensing of Environment* 24:1–29.

Wang, J., T. He, C. Lv, Y. Chen, and W. Jian. 2010. Mapping soil organic matter based on land degradation spectral response units using Hyperion images. *International Journal of Applied Earth Observation and Geoinformation* 12S:S171–S180.

Weng, Y. L., P. Gong, and Z. L. Zhu. 2010. A spectral index for estimating soil salinity in the Yellow River Delta region of China using EO-1 Hyperion data. *Pedosphere* 20(3):378–388.

Whiting, M. L., L. Li, and S. L. Ustin. 2004. Predicting water content using Gaussian model on soil spectra. *Remote Sensing of Environment* 89:535–552.

Whiting, M. L., A. Palacios-Orueta, L. Li, and S. L. Ustin. 2005 (October 23–27). Light absorption model for water content to improve soil mineral estimates in hyperspectral imagery. *Proceedings Pecora 16: Global Priorities in Land Remote Sensing*, Sioux Falls, South Dakota. American Society of Photogrammetry and Remote Sensing.

Yang, C., Y. Han, and P. Han. 2010. The application of modified Gaussian model in hyperspectral image analysis, in *Information Technology in Geo-Engineering*, eds. D. G. Toll, H. Zhu, X. Li, pp. 222–229. Berlin, Germany; Washington. DC: IOS Press.

Yang, H., F. D. van Der Meer, W. Bakker, and Z. J. Tan. 1999. A back-propagation neural network for mineralogical mapping from AVIRIS data. *International Journal of Remote Sensing* 20(1):97–110.

Zabcic, N., B. Rivard, C. Ong, and A. Mueller. 2014. Using airborne hyperspectral data to characterize the surface pH and mineralogy of pyrite mine tailings. *International Journal of Applied Earth Observation and Geoinformation* 32:152–162.

Zadeh, M. H., M. H. Tangestani, F. V. Roldan, and I. Yusta. 2014. Sub-pixel mineral mapping of a porphyry copper belt using EO-1 Hyperion data. *Advances in Space Research* 53:440–451.

Zhang, C., Q. Qin, L. Chen, N. Wang, Y. Bai, and S. Zhao. 2014 (July 13–18). Hyperspectral remote sensing for coal-bed methane exploration. *Proceedings of the 2014 International Geoscience and Remote Sensing Symposium*, 262–265. Québec, Canada.

Zhao, H., L. Zhang, X. Zhang, J. Liu, T. Wu, and S. Wang. 2015. Hyperspectral feature extraction based on the reference spectral background removal method. *Journal of Selected Topics in Applied Earth Observations and Remote Sensing* 8(6):2832–2844.

8 Hyperspectral Applications to Vegetation

Plant and plant ecosystem are important application fields for hyperspectral remote sensing. This is because there exist many unique diagnostic absorption features (bands) in the solar-reflected spectrum from the 0.4 to 2.5 μm spectral range. Since the early 1980s, using various analysis techniques and methods, hyperspectral data applied in vegetation, as in geology and soils, have also been well documented. Thus, in this chapter, an overview on application studies of hyperspectral data including laboratory and *in situ*–measured hyperspectral data and airborne and satellite hyperspectral image data in vegetation is provided. Specifically, spectral characteristics of typical green plants including green leaf structure and plant spectral reflectance curve are introduced in Section 8.2. In Section 8.3, nine types of analytical techniques and methods suitable for extracting and estimating plant biophysical and biochemical parameters and vegetation mapping with hyperspectral data are introduced and reviewed. In the last two sections, an overview on application cases of various hyperspectral data in estimating and mapping a set of plant physical and chemical parameters is offered.

8.1 INTRODUCTION

Imaging spectroscopy (i.e., hyperspectral remote sensing, HRS), as an advanced remote sensing technique, is of growing interest to the remote sensing community. HRS refers to a special type of imaging technology that collects image data in many narrow contiguous spectral bands (<10 nm band width) throughout the visible and solar-reflected infrared portions of the spectrum (Goetz et al. 1985). Imaging spectroscopy systems can produce data with sufficient spectral resolution for direct identification of Earth surface materials with diagnostic spectral (absorption) features, which are from 20 to 40 nm spectral resolution (Hunt 1980). Therefore, the significance of developing hyperspectral remote sensing lies in its ability to acquire a complete reflectance spectrum for each pixel in an image and it is developed for improving identification of materials and quantitative determination of physical and chemical properties of targets of interest, such as minerals, water, vegetation, soils, man-made materials, etc. HRS was developed for mineral mapping in the early 1980s (Goetz et al. 1985). Thus, the initial motivation for the development of imaging spectroscopy was mineral identification, although early experiments were also conducted in botanical remote sensing. However, since 1988, HRS has been successfully applied to a wide range of disciplines, such as geology, vegetation and ecosystem, atmospheric science, hydrology, and oceanography.

Ecosystem and the study of terrestrial vegetation are important application fields for HRS (Green et al. 1998). The spectral properties of plants are strongly determined by their biophysical and biochemical attributes (e.g., plant pigments). Characterizing diagnostic absorption features in plant spectra with hyperspectral data can be done for extraction of bioparameters of plants (e.g., Wessman et al. 1989, Pu and Gong 2004, Cheng et al. 2006, Asner and Martin 2008). A number of such bioparameters as leaf area index (LAI), the amount of live and senesced biomass, absorbed fraction of photosynthetically active radiation (fPAR), moisture content, pigments (e.g., chlorophyll), canopy structure, and community type are correlated with remotely sensed data especially hyperspectral data or their derivatives (Johnson et al. 1994). It is true that major features of peaks and valleys along the spectral reflectance curve of a plant are due to the presence of diagnostic spectral (absorption) features, caused by things such as pigments (e.g., chlorophyll), water, and other chemical constituents. Hyperspectral data have been proven to be more useful in estimating biochemical content and concentration at both leaf and canopy levels (e.g., Peterson et al. 1988, Johnson et al. 1994, Darvishzadeh et al. 2008a, Asner and Martin 2008) and some other ecosystem components such as LAI, plant species composition, and biomass (e.g., Gong et al. 1997, Martin et al. 1998, le Maire et al. 2008) than traditional remotely sensed data. For example, measuring and extracting red-edge optical parameters (e.g., Miller et al. 1990, 1991; Pu et al. 2004) that are related to plant stress or health level is possible with hyperspectral sensors. Therefore, we are able to quantify the phenomena and processes of those bioparameters through testing hypotheses and valuable applications on a variety of terrestrial ecosystems using the spectral property of high resolution of HRS.

Hyperspectral sensors (e.g., AVIRIS, CASI, and Hyperion) on board different types of platforms have made it possible to provide users with high spectral resolution image data. The use of narrow and continuous (1 to 10 nm) instead of broad (50 to 200 nm) spectral bands could offer new potentials for remote sensing applied to quantifying vegetation properties and ecosystem assessment. HRS, in addition to classification and identification of vegetation types, in terrestrial ecosystem study, can be applied to estimating bioparameters and evaluating ecosystem functions. Therefore, in this chapter, an overview of spectral characteristics of typical biophysical and biochemical parameters is offered and applicability and usefulness of HRS techniques and methods suitable for extraction and assessment of plant bioparameters are evaluated. In addition, to a certain extent, a review on extraction and estimation of a set of biophysical and biochemical parameters using HRS techniques and methods and a variety of hyperspectral sensor data is provided.

8.2 SPECTRAL CHARACTERISTICS OF TYPICAL GREEN PLANTS

8.2.1 Green Leaf Structure and Plant Spectral Reflectance Curve

Green plant leaves are the primary photosynthesizing organ. Although plant leaves present various anatomical structures, basic elements are similar, and the variability of the leaf optical properties only results from their arrangement inside the leaf (Verdebout et al. 1994). A typical cross section of a leaf structure is shown in Figure 8.1. Structural elements, such as cell walls and specialized cells, support the leaf. The leaf structure is highly variable depending upon plant species and the environmental conditions in which they are growing. Usually, the plant leaf consists of an outer cuticle, cells stomata or stoma, and intercellular air spaces. Most plant leaves have two distinct layers, composed of a layer of long palisade parenchyma cells in the upper part of the mesophyll and the other layer of more irregularly shaped, loosely arranged spongy parenchyma cells in the lower part of the mesophyll. The photosynthesis occurring inside the green plant leaves relies on two types of food-making cells: palisade parenchyma mesophyll cells and spongy parenchyma mesophyll cells (Jensen 2007). Chlorophyll concentration and content in palisade parenchyma cells mostly determine absorption and reflectance of incidence solar light in visible spectral range while spongy parenchyma mesophyll cells play a key effect on absorption and reflectance of solar light in near-infrared (NIR) range.

Optical properties of a healthy green leaf are influenced by the concentration of plant pigments (e.g., chlorophyll) and other biochemicals, water content, and leaf structure (Lichtenthaler 1987, Gitelson et al. 2001, 2009, Blackburn 2007, Jensen 2007, Ustin et al. 2009). Thus, leaf characteristics all are variable and also the reflectance of vegetation is a function of a very complex changing process within the leaf, the canopy, and the environment in which the vegetation is located. Vegetation is sensitive to most solar radiation from the ultraviolet through short infrared spectral range mostly covering visible (VIS), NIR, and middle infrared (MIR) regions (0.4–2.5 μm), and absorbs plant physiological radiation (solar energy in VIS spectrum) to drive the biological process of photosynthesis necessary for plant growth. The interactions between solar radiation and plants can be allocated into roughly three parts: thermal effects, photosynthetic effects, and photomorphogenic effects of radiation (Kumar et al. 2001). According to Kumar et al. (2001), approximately 70%

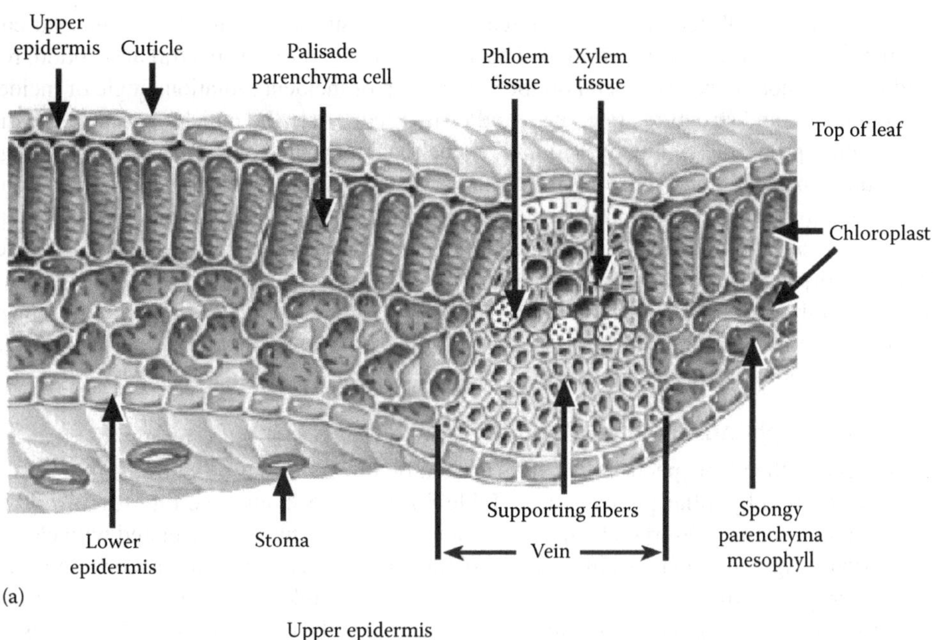

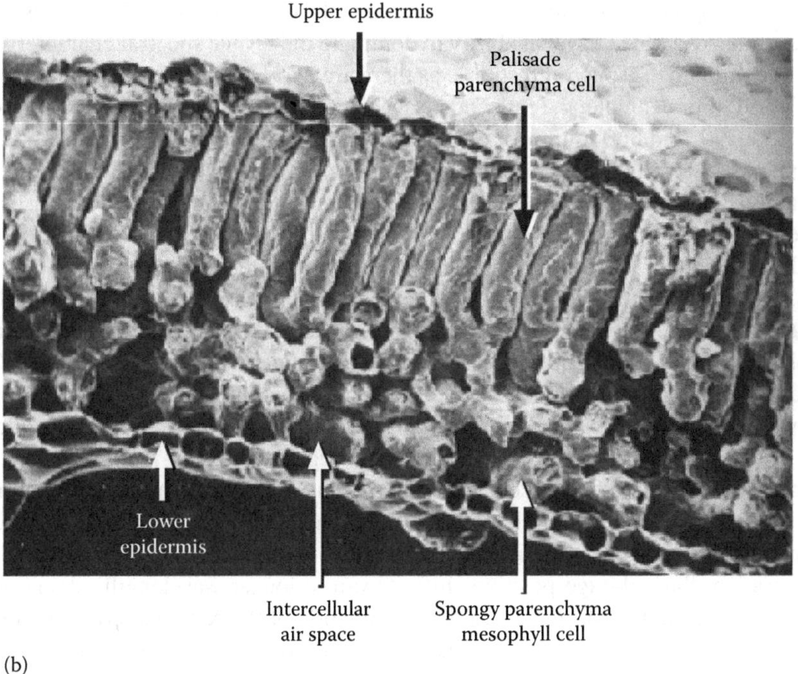

FIGURE 8.1 Hypothetical leaf (a) and a real leaf cross-section (b) of a typical healthy green leaf: Revealing the major structural components that determine the spectral characteristics of vegetation. (From Jensen, J. R., *Remote Sensing of the Environment: An Earth Resource Perspective*, 2nd ed., Pearson Education, New York, 2007. ©2007. Reprinted by permission of Pearson Education, Inc.)

of incoming solar radiation absorbed by plants (e.g., water content) is finally converted into heat and used for maintaining plant temperature and for transpiration (thermal effects); absorption of about 28% total solar energy (called photosynthetically active radiation [PAR]) by plant pigments is used in the photosynthesis process (photosynthetic effects); and absorption of ultraviolet radiation (nominally 2% total energy) is used to produce fluorescence at longer wavelengths (photomorphogenic effects; Eismann 2012). When the solar light reaches the leaf surface, the total incident solar energy is either reflected, absorbed, or transmitted. The amounts and natures of spectral absorption, reflection, and transmission of leaves depend on the wavelength of incident radiation, angle of incidence, surface roughness, and structures (e.g., wax and/or leaf hairs over the cuticle) and also the differences of spectral properties of biophysical and biochemical attributes of the leaves. A typical green leaf spectrum is presented in Figure 1.7 (see Chapter 1) in detail to illustrate major absorption and reflectance features and locations caused by pigments, water, other chemical constituents, and plant cell structure. Leaves from different plant species and environmental conditions may have different strength of absorption and reflectance. From the figure, a typical green leaf spectrum can be divided into three spectral regions: VIS (~0.4–0.7 μm), NIR (~0.7–1.3 μm), and MIR (~1.3–2.5 μm). The plant spectral properties of the three regions associated with absorption, reflectance, and transmittance are discussed as follows.

8.2.1.1 Visible Light Absorbed by Multiple Plant Pigments

In the VIS region, the plant spectrum is characterized by low reflectance and transmittance due to the strong absorption by foliar pigments (see Table 8.1, Figure 8.2; also see Figures 1.7 and 5.11). The visible light is mainly absorbed by multiple plant pigments, which is associated with electronic state transitions. Pigments such as chlorophylls (Chls) and carotenoids (Cars) absorb light of specific light energy, resulting in electron transitions within the molecular structure of the pigment. Although there are many pigments in plant leaves including Chls, Cars, anthocyanins (Anths), phycoerythrin, phycocyanin, etc., which have an ability to dominate the visible spectral characteristics, those of Chls tend to dominate the spectral response as there are 5 to 10 times as much Chls as Cars pigments (Belward 1991). For example, the reflectance spectrum of green vegetation shows absorption peaks around 420, 490, and 660 nm, most of which are caused by strong absorptions of Chls. Moreover, Chl pigments absorb both violet-blue and red light for photosynthesis without absorbing green light, and thus most plants appear green. During photosynthesis, Chls (chloroplasts) use absorbed light energy to convert carbon dioxide and water into carbohydrates. The synthesized organic compounds by the photosynthesis are needed for plants' maintenance and growth.

The absorption characteristics of vegetation in VIS region vary by season. As leaf senesces, Chls usually degrade faster than Cars, which leads to the visible spectral characteristics dominated by Cars and xanthophylls (Kumar et al. 2001). This makes leaves appear yellow because both Cars and xanthophylls absorb blue light and reflect green and red light, which results in the yellow color. As leaves die, brown pigments (tannins) occur, which make the leaf reflectance and transmittance decrease over the VIS region. From red light entering NIR region (660 nm–780 nm), there is a sharp transition from low to high reflectance for healthy green vegetation. The unique feature is called red-edge position (an inflection point between ~700 nm and 725 nm), which can shift when Chls concentration and water content vary in green vegetation. When the plant is healthy and has a high Chls concentration, the red-edge position shifts toward a longer wavelength, otherwise, it shifts toward a shorter wavelength. For an accurate determination of the red-edge position, a large number of spectral measurements in very narrow bands is required and its extraction methods are introduced in Section 5.4 in Chapter 5.

8.2.1.2 NIR Radiation Reflected by Multiscattering of Internal Cellular Structure

The NIR reflectance for a typical healthy green leaf increases dramatically in the spectral range from 0.7 to 1.3 mm (Figures 1.7 and 5.11) and green vegetation NIR spectrum is generally characterized by high reflectance and transmittance and low absorption (Figure 8.2). The actual proportion absorbed, scattered, or reflected and transmitted will vary among species and depends on the internal structure of the leaves. The NIR reflectance spectra of leaves are also influenced by plant development, growth, and senescence. The spongy mesophyll layer in green leaf controls the amount of NIR energy that is reflected (Jensen 2007). The spongy mesophyll layer usually lies below the palisade mesophyll layer and is composed of many cells and intercellular air spaces, as shown in Figure 8.1.

TABLE 8.1
A List of Absorption Features (Wavelengths) Associated With Particular Foliar Biochemical Compounds

Wavelength (nm)	Biochemical	Electron Transition or Bond Vibration
430	**Chlorophyll a**[a]	Electron transition
460	**Chlorophyll b**	Electron transition
640	**Chlorophyll b**	Electron transition
660	**Chlorophyll a**	Electron transition
910	Protein	C-H stretch, 3rd overtone
930	Oil	C-H stretch, 3rd overtone
970	**Water**, starch	O-H bend, 1st overtone
990	Starch	O-H stretch, 2nd overtone
1020	Protein	N-H stretch
1040	Oil	C-H stretch, C-H deformation
1120	Lignin	C-H stretch, 2nd overtone
1200	**Water**, cellulose, starch, lignin	O-H bend, 1st overtone
1400	**Water**	O-H bend, 1st overtone
1420	Lignin	C-H stretch, C-H deformation
1450	Starch, sugar, water, lignin	C-H stretch, 1st overtone, O-H stretch, C-H deformation
1480	Cellulose, water	O-H stretching overtone
1490	Cellulose, sugar	O-H stretch, 1st overtone
1510	**Protein, nitrogen**	N-H stretch, 1st overtone
1530	Starch	O-H stretch, 1st overtone
1540	Starch, cellulose	O-H stretch, 1st overtone
1580	Starch, sugar	O-H stretch, 1st overtone
1690	**Lignin**, starch, protein, nitrogen	C-H stretch, 1st overtone
1730	Protein	C-H stretch
1736	Cellulose	O-H stretch
1780	**Cellulose, sugar**, starch	C-H stretch, 1st overtone/O-H stretch/H-O-H deformation
1820	Cellulose	O-H stretch/C-O stretch, 2nd overtone
1900	Starch	O-H stretch, C-O stretch
1924	Cellulose	O-H stretch, O-H deformation
1940	**Water**, lignin, protein, nitrogen, starch, cellulose	O-H stretch, O-H deformation
1960	Sugar, starch	O-H stretch, O-H rotation
1980	Protein	N-H asymmetry
2000	Starch	O-H deformation, C-O deformation
2060	Protein, nitrogen	N=H rotation, 2nd overtone/N=H rotation/N-H stretch
2080	Sugar, starch	O-H stretch, O-H deformation
2100	**Starch**, cellulose	N=H rotation/C-O stretch/C-O-C stretch, 3rd overtone
2130	Protein	N-H stretch
2180	**Protein, nitrogen**	N-H rotation, 2nd overtone/C-H stretch/C=O stretch/C-N stretch
2240	Protein	C-H stretch,
2250	Starch	O-H stretch, O-H deformation
2270	Cellulose, sugar, starch	C-H stretch/O-H stretch, CH_2 rotation/CH_2 stretch
2280	Starch, cellulose	C-H stretch, CH_2 deformation
2300	Protein, nitrogen	N-H stretch, C=O stretch, C-H rotation, 2nd overtone

(*Continued*)

TABLE 8.1 (CONTINUED)
A List of Absorption Features (Wavelengths) Associated With Particular Foliar Biochemical Compounds

Wavelength (nm)	Biochemical	Electron Transition or Bond Vibration
2310	**Oil**	C-H rotation, 2nd overtone
2320	Starch	C-H stretch, CH_2 deformation
2340	Cellulose	C-H stretch/O-H deformation/C-H deformation/ O-H stretch
2350	Cellulose, nitrogen, protein	CH_2 rotation, 2nd overtone, C-H deforemation, 2nd overtone

Source: Compiled from Williams and Norris (1987), Card et al. (1988), Curran (1989), and Elvidge (1987, 1990).
[a] Chemicals in bold have a wavelength of relatively stronger absorption.

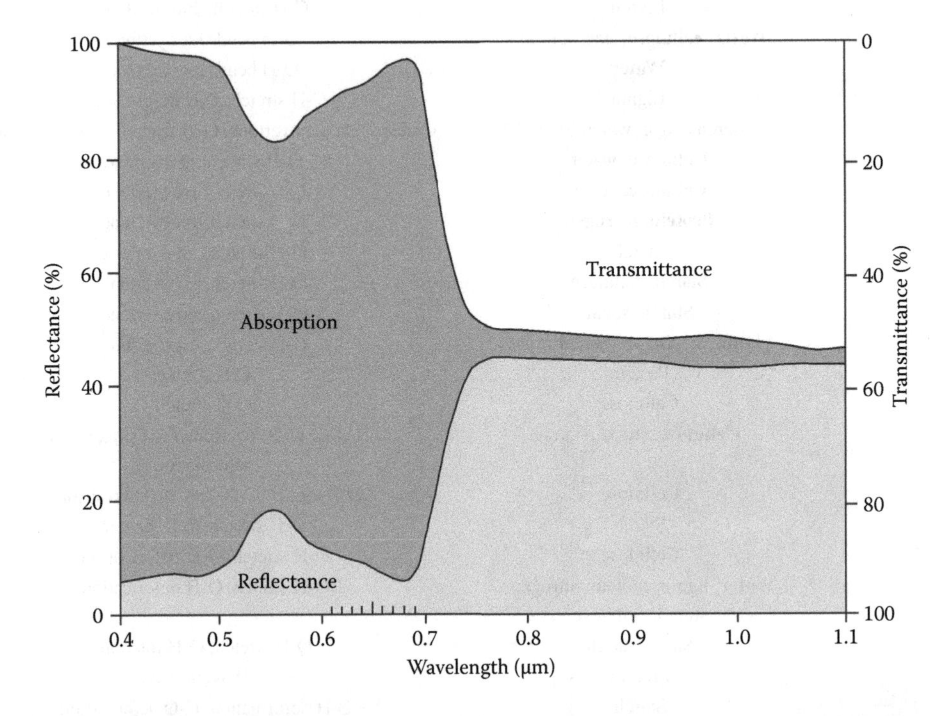

FIGURE 8.2 Hemispherical reflectance, transmittance, and absorption characteristics of big bluestem grass adaxial leaf surface measured using a laboratory spectrometer. The reflectance and transmittance curves are almost mirror images of one another across visible and NIR regions of the spectrum. (From Walter-Shea, E. A., and Biehl, L. L., *Remote Sensing Reviews*, 5(1), 179–205, 1990.)

Since NIR reflectance is due mostly to volume scattering within complex internal cellular structures, the spectral characteristics in the NIR region are mostly determined by the distribution of air spaces and the arrangement, size, and shape of cells in the mesophyll layer. For instance, leaves having a more compact mesophyll layer or higher water content will have fewer air spaces, which allow more transmission and less scattering of radiation (i.e., lower volume reflectance). Inversely, leaves having a spongy mesophyll layer will have more air spaces, and thus more air–water boundaries, and so induce more scattering and less transmission of radiation (Gausman et al. 1970, Kumar et al. 2001). In the NIR region, the reflectance of vegetation canopies usually is much higher than that for single leaves. This is because, as most of the radiation at NIR wavelengths passes through single leaves, the multiple leaf layers in a canopy have an additive effect on reflectance (Belward 1991).

8.2.1.3 MIR Energy Absorbed by Water and Other Biochemical Constituents

In the MIR spectral region, the spectral characteristics are controlled by strong plant liquid–water absorption features and influenced by other foliar biochemical contents such as cellulose, lignin, and pectin (Table 8.1). In Table 8.1, absorption wavelengths of most of all other foliar chemical compounds appear in the MIR region. There is a strong relationship between the reflectance in the MIR region from 1.3 to 2.5 μm and the amount of water present in the leaves of a plant canopy. Figure 5.11 clearly presents the variation of foliar reflectance as the foliar relative water content changes. Water absorptions are caused by transitions in the vibrational and rotational states of the water molecules (Belward 1991), unlike pigments whose absorptions are caused by electron transitions. Usually, the reflectance in this region is much lower than that in the NIR region. In the MIR region, the absorption bands caused by pure liquid water within plant leaves are slightly displaced from those caused by the atmospheric water vapor bands. As the water absorptions in the MIR are fairly strong, they have a carry-over effect such that the spectral ranges between major water absorption bands are also affected. Consequently, increased water content of leaves will not only decrease reflectance in water absorption bands, but they also cause a decrease in reflectance in other regions as well (Kumar et al. 2001). However, as the green leaf dries up, its spectral characteristics are gradually dominated by cellulose, lignin, and pectin contents and have diagnostic features in the 1.8–2.5 μm spectral range.

8.2.2 Spectral Characteristics of Plant Biophysical Parameters

Spectral reflectance properties and characteristics of a list of typical plant biophysical parameters (Table 8.2) have been the subject of studies on plant spectral properties. The typical biophysical parameters include plant canopy leaf area index, specific leaf area, crown closure, vegetation species and composition, biomass, effective fraction of absorbed photosynthetically active radiation (fPAR or fAPAR), and net primary productivity (NPP), which reflect photosynthesis rate, excluding respiration by the plant itself. Such spectral analysis results are useful for determining physicochemical properties of plant directly or indirectly derived from spectral data including multi-/hyperspectral image data in order to assess vegetation and ecosystem conditions and productivity. Some analysis results of spectral properties and characteristics for the list of typical biophysical parameters from hyperspectral data are summarized as follows.

8.2.2.1 Leaf Area Index, Specific Leaf Area, and Crown Closure

LAI, specific leaf area (SLA), and crown closure (CC) are important plant canopy structural parameters for quantifying energy and mass exchange characteristics of ecosystems, especially terrestrial ecosystems (Pu and Gong 2011). Such characteristics involve photosynthesis, respiration, transpiration, carbon and nutrient cycle, and rainfall interception. LAI quantifies the amount of live green leaf material present in the plant canopy per unit ground area while SLA describes the specific leaf area per unit amount of leaf dry mass present in the canopy. CC can only quantify the percentage of area covered by the vertical projection of live green leaf material present in the canopy. The physiological and structural characteristic of plant leaves determine their typically low visible light reflectance except in green light. High NIR reflectance of vegetation allows optical remote sensing to capture detailed information about the live, photosynthetically active plant canopy structure, and thus enhances understanding of the mass exchange between the atmosphere and the plant ecosystem (Zheng and Moskal 2009). As LAI and CC increase, many absorption features become obvious due to increase in their amplitude, width, or location. The diagnostic (absorption) features, including those caused by pigments in the visible region and by water content and other biochemical constituents in the short wave infrared (SWIR) region (Curran 1989,

TABLE 8.2
Typical Plant Biophysical and Biochemical Parameters

Bio-Parameters	Definition and Description	Spectral Properties and Characteristics
Biophysical parameters:		
LAI: leaf area index	A total one-sided area of all leaves in a plant canopy per unit area of ground.	The spectral absorption features caused by pigments in the visible region and by water content and other biochemicals in the SWIR region are useful for extracting and mapping plant LAI and CC.
SLA: specific leaf area	Projected leaf area per unit leaf dry mass (cm^2/g)	Not directly related to water absorption bands but SLA is a leaf structural property linked to the entire constellation of foliar chemicals and photosynthetic processes.
CC: crown closure	Percentage of land area covered by the vertical projection of plant canopies.	As LAI.
Species	Various plant species and species composition	There are spectral differences among different plant species and species composition due to differences and variation in phenology/ physiology, internal leaf structure, bio-chemicals and ecosystem type.
Biomass	A total of absolute amount of vegetation present (often considered in terms of the above-ground biomass) per unit area of ground.	Spectral responses to plant absorption and reflectance, LAI, stand/community structure, species and species composition and image textural information.
NPP: net primary productivity	The rate at which an ecosystem accumulates energy or biomass, elimating the energy used by the process of respiration, which typically corresponds to the rate of photosynthesis, reducing respiration by the plant itself.	Spectra reflect vegetation condition and changes in LAI or canopy light absorption through time in visible and NIR regions.
fPAR or **fAPAR**	Effective fraction of absorbed photosynthetically active radiation in visible region.	In visible spectral region 400–700 nm, most absorbed by plant pigments: Chla and b, Cars and Anths, and leaf water and nitrogen contents for photosynthesis process.
Biochemical parameters:		
Chls (Chla, Chlb): chlorophylls	Green pigments chlorophyll a and b for plant photosynthetically processing and found in green photosynthetic organisms, (mg/m^2 or $nmol/cm^2$)	Chla absorption features are near 430 nm and 660 nm and Chlb absorption features are near 450 nm and 650 nm in vivo (Lichtenthaler 1987; Blackburn 2007). But it is known that *in situ* Chla absorb at both 450 nm and 670 nm.
Cars: carotenoids	Any of a class of yellow to red pigments, including carotenes and xanthophylls, (mg/m^2)	Cars absorption feature in the blue region is near 445 nm *in vivo* (Lichtenthaler 1987). But it is known that *in situ* Cars absorb at 500 nm and even a little bit longer wavelength.
Anths: anthocyanins	Any of various water-soluble pigments that impart to flowers and other plant parts colors ranging from violet and blue to most shades of red (mg/m^2)	Anths absorption feature in the green region is at 530 nm *in vivo*, but *in situ* Anths absorb around 550 nm (Gitelson et al. 2001, 2009; Blackburn 2007).
N: nitrogen	Plant nutrient element (%)	The central wavelengths of N absorption features are near 1.51, 2.06, 2.18, 2.30, and 2.35 μm.

(*Continued*)

TABLE 8.2 (CONTINUED)
Typical Plant Biophysical and Biochemical Parameters

Bio-Parameters	Definition and Description	Spectral Properties and Characteristics
P: phosphorus	Plant nutrient element (%)	No direct and significant absorption features across 0.40–2.50 μm but it does indirectly affect spectral characteristics of other biochemical compounds.
K: potassium	Plant nutrient element (%)	Foliar K concentration has only a slight effect on sclerenhyma cell walls thus on NIR reflectance.
W: water	Leaf or canopy water content or concentration (%)	The central wavelengths of those absorption features are near 0.97, 1.20, 1.40 and 1.94 μm.
Lignin	A complex polymer, the chief noncarbohydrate constituent of wood, which binds to cellulose fibers and hardens and strengthens the cell walls of plants (%)	The central wavelengths of lignin absorption features are near 1.12, 1.42, 1.69, 1.94, 2.05–2.14, 2.27, 2.33, 2.38, and 2.50 μm (Curran 1989; Elvidge 1990).
Cellulose	A complex carbohydrate, which is composed of glucose units, forms the main constituent of the cell wall in most plants (%)	The central wavelengths of cellulose absorption features are near 1.20, 1.49, 1.78, 1.82, 2.27, 2.34, and 2.35 μm.
Pectin	Polymers of galacturonic acid and frequently found in cell walls and in the middle lamella between adjacent cells.	The central wavelengths of cellulose absorption features are near 1.40, 1.82, and 2.27 μm.
Protein	Any of a group of complex organic macromolecules that contain carbon, hydrogen, oxygen, nitrogen, and usually sulfur and are composed of one or more chains of amino acids (%)	The central wavelengths of protein absorption features are near 0.91, 1.02, 1.51, 1.98, 2.06, 2.18, 2.24, and 2.30 μm.

Source: Mostly referring to Pu, R., and P. Gong, in *Advances in Environmental Remote Sensing: Sensors, Algorithm, and Applications*, series ed. Q. Weng, CRC Press/Taylor & Francis Publishing Group, Boca Raton, FL, 2011, pp. 101–142.

Elvidge 1990; Table 8.1), are useful for estimating and mapping LAI and CC. Unlike LAI and CC, the spectral properties of SLA are not directly related to water absorption bands in the full range of a vegetation spectrum. However, SLA has a leaf structural property, which is linked to the entire constellation of foliar chemicals and photosynthetic processes (Wright et al. 2004, Niinemets and Sack 2006). SLA is related to the NIR spectral reflectance that is dominated by the amount of leaf water content and leaf thickness (Jacquemoud and Baret 1990). Thus, at the leaf level, SLA is highly correlated with leaf spectral reflectance (Asner and Martin 2008). Optical remote sensing, especially hyperspectral remote sensing, aims at retrieving the spectral characteristics of plant leaves and canopies through measuring LAI, SLA, and CC biophysical parameters for quantifying vegetation and ecosystems.

8.2.2.2 Species and Composition

Plant foliar spectral variability among different species, or even within single crowns, is attributed not only to differences in internal leaf structure and biochemicals (e.g., water, chlorophyll content, epiphyll cover, and herbivory; Clark et al. 2005) but also to difference and variation in phenology/physiology of plant species. Species-specific biochemical characteristics are related to foliar chemistry (Martin et al. 1998). The relative importance of these biochemical and structural properties among individual species is also dependent upon measured wavelength, pixel-size, and ecosystem type (Asner 1998). For example, sand live oak (*Quercus geminata*) in Tampa, Florida, usually changes its old leaves in April yearly, approximately one month later than all other oak tree species in the area, such that the sand live oak has very different spectral characteristics in April than other oak species (Pu and Landry 2012). In identifying invasive species in Hawaiian forests from native and other introduced species by remote sensing, Asner et al. (2008) confirmed that the observed differences in canopy spectral signatures among the different species are linked to relative differences in measured leaf pigments (Chls and Cars), nutrient (N and P), and structural (SLA) properties, as well as to canopy LAI. Since hyperspectral data allows the identification of vegetation absorption features that may be related to different plant species or varieties (Galvão et al. 2005), it is important to determine the best wavelengths suitable for species recognition in hyperspectral remote sensing.

8.2.2.3 Biomass, NPP, and fPAR or fAPAR

Biomass is basically calculated using the density of unit biomass and the area of plant growth. The unit biomass of plant above-ground biomass (AGB foliage, branch, and stem) can be estimated from optical remote sensing (Zhang and Ni-meister 2014). Biomass is associated with forest components (attributes) that include LAI and canopy structure (CC and height). These components can be directly estimated from optical remotely sensed data, especially from HRS data. Leaf canopy biomass may be calculated as the product of the leaf dry mass per area (LMA, g/m^2, or inverse of SLA) and LAI (le Maire et al. 2008). Therefore, based on the spectral responses to LAI, SLA, and CC, the three biophysical parameters can be estimated from hyperspectral data, and thus AGB and leaf mass of the entire canopy can be estimated. For example, vegetation index (VI) generally enhances vegetation signal as it minimizes the influences from solar irradiance, solar angle, sensor view angle, atmospheric, and soil background effects. Since many narrowbands are available for constructing VIs, selection of the correct wavelengths and bandwidths is important. When some VIs derived from hyperspectral data are used to estimate some biophysical parameters, narrowbands (10 nm) perform better than broadbands (e.g., TM-bands) using standard Red/NIR and Green/NIR and NDVIs ($NDVI_{green}$) (e.g., Gong et al. 2003, Hansen and Schjoerring 2003).

The fPAR quantifies the fraction of the solar radiation absorbed by live leaves for the photosynthesis activity. The fPAR relies on the canopy structure (e.g., LAI and SLA), vegetation element optical properties (e.g., pigment absorption), atmospheric conditions, and angular configuration. In the visible spectral region, most solar radiation energy is absorbed by plant pigments: Chla and Chlb, Cars and Anths, and leaf water and nitrogen contents for photosynthesis process. To measure fPAR or fAPAR over a large area, even at a global scale, optical remote sensing–derived vegetation indices, especially hyperspectral vegetation indices (HVIs), are frequently used to estimate the biophysical parameter fPAR (e.g., Tan et al. 2013, Dong et al. 2015). Although LAI is one of the canopy structure parameters that determine fPAR, in general, VIs are more sensitive to fPAR than LAI, due to the fact that fPAR is a radiation quantity, whereas LAI is nonlinearly related to radiation (Walter-Shea et al. 1997). The NPP can also be indirectly estimated by considering remotely sensed derived LAI and fPAR with an assumption of a light use efficiency (LUE) value (a coefficient describing the conversion of light to organic matter by plants; Gower et al. 1999).

8.2.3 Spectral Characteristics of Plant Biochemical Parameters

Table 8.2 also summarizes a list of typical plant biochemical parameters that include major pigments (chlorophylls [Chls], carotenoids [Cars], and anthocyanins [Anths]), nutrients (nitrogen [N], phosphorus [P], and potassium [K]), leaf or canopy water content (W), and other biochemicals (e.g., lignin, cellulose, pectin, protein, etc.). Analyzing the spectral properties and characteristics of the list of biochemicals is not only for extracting and mapping the list of biochemicals from hyperspectral data, but also for understanding underlying relationships between the list of biophysical parameters and absorption features. In the following, the spectral properties and characteristics of the list of biochemicals are briefly reviewed and discussed.

8.2.3.1 Pigments: Chlorophylls, Carotenoids, and Anthocyanins

Chlorophylls (Chls, Chla, and Chlb) are Earth's most important organic molecules as they are the most important pigments necessary for the photosynthesis process. Chla and Chlb are most common in higher plants. Carotenoids (Cars) are the second major group of plant pigments, composed of carotene and xanthophylls, whereas anthocyanins (Anths) are water-soluble flavonoids which are the third major group of pigments in leaves, but there is no unified explanation for their presence and function (Blackburn 2007). In the VIS spectral region of a spectrum, most of the radiation is absorbed by Chls and other pigments, especially in the red, and blue and converted to chemical energy. When such a pigment absorbs radiant energy, there is a displacement of pi electrons in a resonance system through the pigment molecule (Kumar et al. 2001). Published spectral absorption wavelengths of isolated pigments, such as those shown in Figure 8.3, show that Chla absorption features are near 430 nm and 660 nm and Chlb absorption features are near 450 nm and 650 nm *in vivo* (Lichtenthaler 1987, Blackburn 2007). But it is known that *in situ* Chla absorbs at both 450 nm and 670 nm. Cars absorption feature in the blue region is at 445 nm *in vivo* and β-carotene at 470 nm (Lichtenthaler 1987, Blackburn 2007) *in vivo*. But it is also known that *in situ* Cars absorb at 500 nm and even a little bit longer wavelength. Anths absorption feature in the green region is at 530 nm *in vivo*, but *in situ* Anths absorb around 550 nm (Gitelson et al. 2001, 2009; Blackburn 2007; Ustin et al. 2009). In addition, xanthophylls in higher plants also show their absorption maxima at 425, 450, and 475 nm (Belward 1991). Usually operative pigments in chloroplasts include Chls (65%), Cars (6%), and xanthophylls (29%); percentage distribution is, of course, highly variable (Gates et al. 1965).

8.2.3.2 Nutrients: N, P, and K

The foliage and canopy nitrogen (N) is the most important nutrient element that plants need for growth and is related to a variety of ecological and biochemical processes (Martin et al. 2008). Phosphorus (P) and potassium (K), the second and third most limiting nutrient constituents, are essential at all phases of plant growth. P and K are used in cell division, fat formation, energy transfer, seed germination, and flowering and fruiting (Milton et al. 1991, Jokela et al. 1997). Among the three basic nutrient elements, N can produce several significant absorption features that have been found in the visible, NIR, and MIR regions. In Table 8.1, N absorption features in its isolated form are located near 1.51, 2.06, 2.18, 2.30, and 2.35 μm. Since many biochemical compounds, such as Chls and protein, comprise N, the spectral properties of the compounds are actually characterized

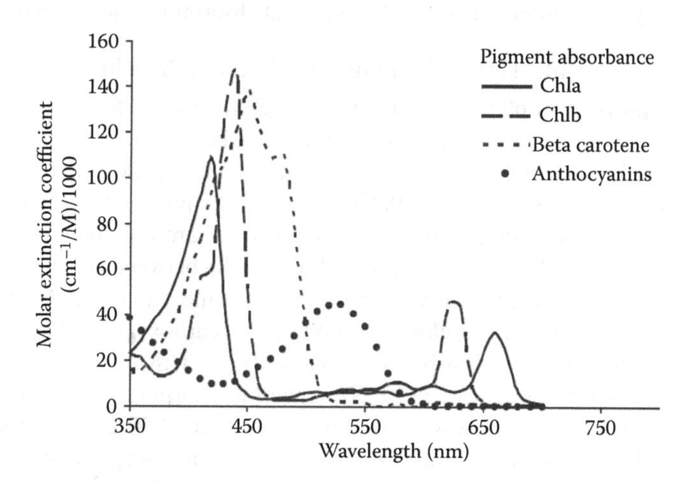

FIGURE 8.3 Absorption spectra of the major plant pigments. (From Blackburn, G. A., *Journal of Experimental Botany*, 58(4), 855–867, 2007. Used by permission of Oxford University Press.)

by N concentration in plant leaves. It is also demonstrated that increases in N concentration have a consistent influence on the overall shape of the 2.1 μm absorption feature due to protein absorptions at 2.054 μm and 2.172 μm, thus supporting the current hypothesis that N-containing protein absorptions represent a sound basis for estimating N concentration using reflectance spectra of dried and ground leaves (Kokaly 2001). Although P has no direct and significant absorption features across the visible, NIR, and MIR regions, and the relative importance of the absorption features and their specific bands for P prediction is hardly known, it does indirectly affect spectral characteristics of other biochemical compounds. The documented spectral changes include higher reflectance in the green and yellow portions of the electromagnetic spectrum in P-deficient plants and a difference in position of the long wavelength edge (the *red edge*) of Chl absorption band centered near 0.68 μm (Milton et al. 1991). Milton et al. (1991) has shown that phosphorus deficiency inhibits a shift of the red edge to longer wavelengths, which was confirmed by Mutanga and Kumar (2007) in estimating foliar phosphorus concentration. Foliar K concentration has only a slight effect on needle morphology and thus on NIR reflectance. This can be explained by the fact that the sclerenchyma cell walls are thicker when the high K concentration is contained in plant leaves, which leads to higher NIR reflectance of leaves (Jokela et al. 1997).

8.2.3.3 Leaf Water Content

Evaluating water status in plant leaves is an important area in hyperspectral remote sensing (Goetz et al. 1985, Curran et al. 1997). Plant leaf water content determines the absorption of infrared radiation. Previous work on assessing the plant water status mainly depends on water spectral absorption bands in the 0.40–2.50 μm region. According to Curran (1989) and Table 8.1, the central wavelengths of the plant water absorption features are near 0.97, 1.20, 1.40, and 1.94 μm. In general, reflectance spectra of green and yellow leaves in those absorption bands are quickly saturated and solely dominated (Elvidge 1990) by changes in leaf water content. The reflectance of dry vegetation shows an absorption feature centered at 1.78 μm by other chemicals (cellulose, sugar, and starch; Curran 1989) rather than by water because pure water does not cause such an absorption feature (Palmer and Williams 1974). However, this absorption feature centered at 1.78 μm is highly correlated with relative water content in leaves (Tian et al. 2001). Since the dominant effect of absorption by leaf water can largely mask the signatures of the biochemical components beyond 1.0 μm (Curran 1989, Elvidge 1990), fresh leaf tissue creates more problems than dried tissue for the spectral analysis of biochemical components. In addition to water being a strong absorber of infrared radiation, the cell structure of fresh plant material scatters light as it passes through multiple air and water surfaces with different refractive indices. These phenomena may also obscure the subtle biochemical absorption features (Kumar et al. 2001).

8.2.3.4 Other Biochemicals: Lignin, Cellulose, Pectin, and Protein

Lignin is a complex polymer of phenylpropanoid, making up 10% to 35% of dry weight of plants, and acts as a barrier to the decomposition of cellulose and hemicellulose, whereas cellulose is a D-glucose polymer and found in the cell walls of all plants, and it makes up one-third to one-half of the dry weight of most plants (Colvin 1980, Crawford 1981, Elvidge 1990). The cellulose's main function is to protect and strengthen plant structures. Pectins are polymers of galacturonic acid and are frequently found in cell walls and in the middle lamella between adjacent cells. Pectin substances are particularly abundant in fruits (Elvidge 1990). The most abundant N bearing compound in plant green leaves is the protein D-ribulose 1-5-diphosphate carboxylase, which is an enzyme that plays a critical role in the fixation of carbon in photosynthesis. This enzyme accounts for 30% to 50% of the N in the green leaves (Elvidge 1990). The spectral absorption features of other biochemicals mostly are located in the short-wave infrared region (1.00–2.50 μm). According to Curran (1989), Table 8.1 and Figure 8.4, the central wavelengths of lignin absorption features are near 1.12, 1.42, 1.69, and 1.94 μm; the central wavelengths of cellulose absorption features are near 1.20, 1.49, 1.78, 1.82, 2.27, 2.34, and 2.35 μm; the central wavelengths of pectin absorption features are near 1.40, 1.82, and 2.27 μm; and the central wavelengths of protein absorption features are near 0.91,

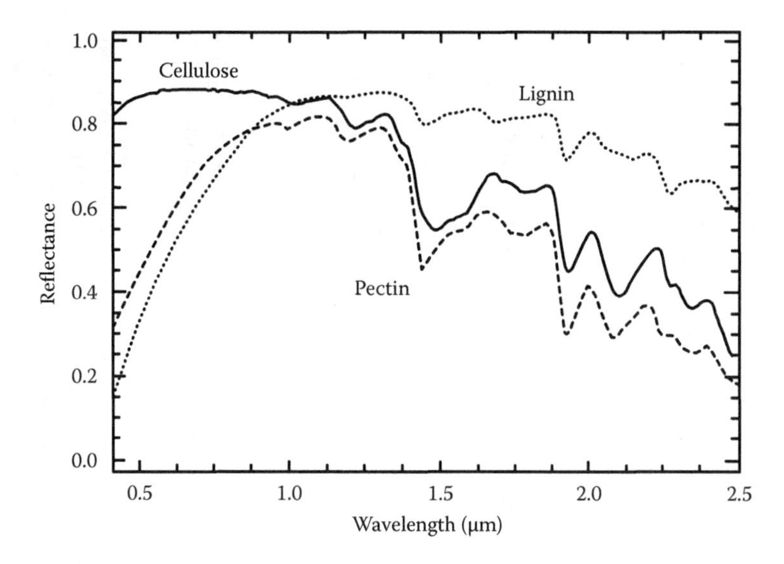

FIGURE 8.4 Spectral reflectance of dry plant constituents: cellulose, lignin, and pectin. (From van der Meer, F. D., *Imaging Spectrometry: Basic Principles and Prospective Applications*, Springer, New York, 2001. With permission of Springer.)

1.02, 1.51, 1.98, 2.06, 2.18, 2.24, and 2.30 μm. Dry plant materials have diagnostic ligno-cellulose absorption features at 2.09 μm in the region (Elvidge 1990). The aforementioned absorptions are mainly due to harmonics and overtones of the peak absorptions (Kumar et al. 2001).

8.3 ANALYTICAL TECHNIQUES AND METHODS NEEDED IN VEGETATION APPLICATIONS

In this section, nine types of analytical techniques and methods specially developed for extracting and estimating plant biophysical and biochemical parameters and vegetation mapping from hyperspectral data sets (including laboratory and *in situ* spectral measurements and airborne and spaceborne hyperspectral image data) are introduced and reviewed. An emphasis is placed on reviewing the principle and applicability of the analytical methods/techniques to vegetation using hyperspectral data sets rather than introducing the development of the techniques and methods themselves. Therefore, for those readers who are more interested in a detailed introduction of the algorithms/models of the methods, please refer to Chapter 5. The nine types of techniques/methods consist of plant spectral derivative analysis, plant spectral absorption feature and wavelength position variable extraction and analysis, spectral vegetation index analysis, spectral unmixing and matching analyses, plant spectral classification analysis, general empirical/statistical analysis methods, physically based modeling methods, and bioparameter mapping methods. In the following, the characteristics, advantages, and limitations of the methods/techniques during the process in their applications to vegetation will also be briefly evaluated. Meanwhile, for the nine types of techniques/methods, the main characteristics; advantages and disadvantages or limitations; major factors affecting detecting, estimating, and mapping results; and some application references are summarized in Table 8.3.

8.3.1 Plant Spectral Derivative Analysis

Using the first- and second-order derivative spectral calculation formulas (Equations 5.1 and 5.2), the corresponding first- and second-order derivative spectra can be calculated from plant spectral

TABLE 8.3
Summary of Techniques/Methods Suitable for Extracting Bioparameters and Mapping Vegetation Using Hyperspectral Data

Technique and Method	Description and Characteristics	Advantages and Limitations	Major Factors	Application Case
1. Plant spectral derivative analysis	Normalized spectral difference of two continuous/neighbor narrow bands with their wavelength interval.	Remove or compress the effect of illumination variations with low frequency on target spectra but sensitive to the SNR of hyperspectral data and higher order spectral derivative processing is susceptible to the noise. Usually, 1st-order derivative spectra are better than higher order derivative spectra.	spectral resolution better than 10 nm and also band continuous.	Gong et al. 1997; Asner et al. 2008; Wang et al. 2008; Pu 2009.
2. Spectral absorption feature and wavelegth position variable analyses	A continuum-removal technique is used to extract these absorption band features: absorption band depth, width, position, area, and asymmetry. Wavelength position variables include red-edge optical parameters and other wavelength-position variables in 0.4–2.5 μm spectral range.	The continuum-removal technique can be performed automatically and manually. The estimated results with the extracted absorption features are usually better than those directly using band reflectance. The manual method is labor intensive, especially for those not very strong absorption bands. For the red-edge optical parameter extraction techniques, the four point method and polynimical fitting method are better than other techniques.	Spectral resolution better than 10 nm and also band continuous; strong absorption bands are required.	Belanger et al. 1995; Pu et al. 2003a; Pu et al. 2003b; Pu et al. 2004; Galvâo et al. 2005; Cho et al. 2008; Huber et al. 2008.
3. Spectral vegetation index analysis	Calculate ratio of images of two bands or normalized difference of two or more than two bands.	Easy to use and reduce impact of sun angle, atmosphere, shadow, topography. However, VI image for different land cover types usually is not normal.	Identify suitable bands to construct VIs.	See application cases from Table 5.1, Chapter 5.
4. Plant spectral mixture analysis	Spectral reflectances from different materials (endmembers) within a pixel are recorded as one spectral response (mixed spectrum). Using linear or nonlinear spectral mixture model to derive fraction (endmember) images from mixed pixels.	A fraction image represents the areal proportion of each endmember, but where the proportioned areas locate within a mixed pixel is not known and sometimes it is difficult to obtain endmember spectra and know all endmembers in a scene.	Identify suitable/pure endmembers and extract their individual spectra.	Roberts et al. 1998; Asner and Heidebrecht 2003; Pu et al. 2008c; Pignatti et al. 2009; Andrew and Ustin 2010.

(Continued)

TABLE 8.3 (CONTINUED)
Summary of Techniques/Methods Suitable for Extracting Bioparameters and Mapping Vegetation Using Hyperspectral Data

Technique and Method	Description and Characteristics	Advantages and Limitations	Major Factors	Application Case
5. Plant spectral matching analysis	With n-dimensional angles (or distances or correlations or discrepancy of probility distributions) to match pixels to reference spectra with smaller angles (shorter distances or higher correlations or lower discrepancy) representing closer matches to the reference spectrum, otherwise representing not matching the reference spectrum.	SAM is a physically based spectral classification and is less sensitive to differences in curve magnitude caused by variation in lighting across a scene, but it is sensitive to noise in any particular band. The accuracy of these spectral matching measures (SAM, ED, CCSM and SID) and techniques is directly affected by geometry of sensors' observations and target size.	Determine a suitable threshold of angle (or distance or correlation or discrepancy) for identification and mapping.	Lass and Prather 2004; Pengra et al. 2007; Pu et al. 2008b; Narumalani et al. 2009.
6. Plant spectral classification analysis	Using supervised/unspervised methods, parameteric/nonparametric algorithms to assign one pixel or image-object into one of classes (species). Dimension reduction and feature extraction are conducted prior to performing hyperspectral data classification.	There is a basis of statistic/probablity or a rule for classification; usually, it is frequently difficult to collect adequate training samples for supervised methods and labeling unsupervised spectral clusters due to relatively high dimensionality of hyperspectral data.	Identify suitable method/ classifier for a specific task and gather adequate training samples.	Camps-Valls et al. 2004; Underwood et al. 2003, 2007; Pu et al. 2008c; Pu and Liu 2011.
7. Empirical/ statistical analysis methods	Univariate and multivariate statistical analyses to correlate and estimate or identify bio-parameters with spectral reflectance, VIs, or derivative spectra or features transformed or extracted from hyperspectral data in the VIS, NIR and MIR wavelengths at leaf- or canopy- or plant community-level	The methods are relatively simple and their modeling results frequently have higher estimation accuracy. However, empirical or statistical relationships are often site-, species- and sensor-specific, and thus cannot be directly applied to other study areas. Multivariate linear regression analysis have three limitations: overfitting of wavebands, intercorrelation of chemicals, and missing of absorption bands. Partial least squares regression can reduce the large number of measured collinear spectral variables to a few non-correlated latent variables.	Avoiding overfitting (using large sample size or low dimen. spectral variables/bands); measuring both bioparameters and spectra accuractly and times closely.	Peterson et al 1988; Johnson et al. 1994; Bolster et al. 1996; Asner and Martin 2008; Huber et al. 2008.

(*Continued*)

TABLE 8.3 (CONTINUED)
Summary of Techniques/Methods Suitable for Extracting Bioparameters and Mapping Vegetation Using Hyperspectral Data

Technique and Method	Description and Characteristics	Advantages and Limitations	Major Factors	Application Case
8. Physically-based modeling methods	The models, including radiative transfer (RT) models and geometric-optical models, rest on the underlying physics and complexity of the leaf internal structure and thus are robust. Basically, there are three types of RT models: leaf-level models (e.g., PROSPECT), canopy-level models (e.g., SAIL) and leaf-canopy coupled models (e.g., PROSAIL). The RT models have been used in the forward mode to calculate leaf or canopy reflectance and transmittance and in inversion to estimate leaf or canopy chemical and physical properties.	Theoretically, they can overcome many limitations held by the empirical/statistical methods, such as case specific, but the structure of the models is complicated and difficult to understand their physical and biological meanings to those who lack plant optical modelling knowledge. In addition, inversion of the models for retrieving vegetation properties frequently is not a easy job. If simplifying the models due to shortage of some structural parameter values, the modeling accuracy may not be ensured.	How to invert the models to retrieve vegetation characteristics from observed reflectance data. Properly applying the three inversion approaches: iterative optimization, LUT and ANN.	Verhoef 1984; Jacquemoud and Baret 1990; Baret et al. 1992; Asner 1998; Dawson et al. 1999; Ganapol et al. 1999; Dash and Curran 2004; Malenovsky et al. 2007; Jacquemoud et al. 2009; Omari et al. 2013.
9. Bio-parameter mapping techniques	Using bioparameter prediction models to estimate or predict pixel-based bioparameters then classifying or density slicing the predicted bioparameter values into several classes to become bioparameter maps.	Hyperspectral image data are easily used to analyze spatial distribution of biophysical and biochemical parameters (bioparameters). It is possible nowadays to estimate and map pixel-based individual bioparameters with the hyperspectral image data.	Establish a realiable and accuracte bioparameter predictive model with pixel-based input spectral variables.	Wessman et al. 1989; Curran et al. 1997; Pu et al. 2003c; Hernández-Clemente et al. 2014.

reflectance curves (laboratory, *in situ*, or imaging hyperspectral data sets). Since *in situ* or imaging hyperspectral data obtained in the field are usually contaminated by illumination variations caused by terrain background, atmosphere, and viewing geometry (Pu and Gong 2011), the spectral derivative analysis has been considered a desirable tool to remove or compress the effect of illumination variations with low frequency (e.g., low frequency or linear effect of soil and atmosphere) on target spectra (e.g., nonlinear plant reflectance curves). However, the spectral derivative technique is sensitive to the signal-to-noise ratio (SNR) of hyperspectral data, and the higher order spectral derivative processing is susceptible to noise (Cloutis 1996). Thus lower-order derivatives (e.g., the first order derivative) are more effective in operational plant spectral analysis as they are less sensitive to the noise. When implementing the spectral derivative analysis, spectral resolution better than 10 nm and spectral bands continuous from wavelength are required.

Derivative spectra have been successfully employed in hyperspectral data analysis for extracting and estimating biophysical and biochemical parameters. For example, Gong et al. (1997) and Pu (2009) reported that the first order derivative of tree spectra could considerably improve the accuracy in recognizing six conifer species commonly found in northern California and eleven urban tree species in Tampa, Florida, respectively, compared to higher order derivative spectra or raw reflectance spectra. Combining *in situ* spectral measurements with airborne imaging spectroradiometer for application data, Wang et al. (2008) mapped an invasive weed (*sericea lespedeza*) in a public grass field in mid-Missouri. The maximal first-order derivative in the red-near infrared region (650–800 nm) was derived to separate the invasive species from the target grass in pastures in Missouri. With a simple threshold approach for the maximum first-order derivative spectrum, *sericea* of various sizes was successfully identified in the study area. Asner et al. (2008) conducted the spectral separability analysis between Hawaiian native and introduced (invasive) tree species with AVIRIS hyperspectral image data for detecting and assessing invasive species. They observed that the spectral differences (measured in reflectance, first-order and second-order derivative spectra) in canopy spectral signatures are linked to relative differences in leaf pigment (Chls and Cars), nutrient (N, P), and structural (specific leaf area) properties, as well as to canopy LAI. These relative differences associated with leaf and canopy properties of trees are helpful to separate invasive species from its background (native) species.

8.3.2 Plant Spectral Absorption Feature and Wavelength Position Variable Analysis

There exist many spectral absorption (i.e., diagnostic) bands across a plant spectral reflectance curve. These absorption bands are caused by plant pigments (e.g., Chls and Cars) in VIS region and by water and other chemical compounds (e.g., lignin, cellulose, pectin, etc.) in the SWIR region. Quantitative measures of these absorption band features from plant spectral reflectance curves may be determined from each absorption peak after the normalization of the raw spectral reflectance curve. A continuum-removal technique proposed by Clark and Roush (1984) can be used to extract these absorption band features, including absorption band depth, width, position, area, and asymmetry. The continuum-removal technique can be performed automatically (see Chapter 6) and manually. These features extracted from plant spectral curves can be used to estimate a list of bioparameters. The estimated result with the extracted absorption features is usually better than that directly using band reflectance. For example, Pu, Ge et al. (2003) reported the effectiveness of these absorption features in correlation with leaf relative water content of oak trees at various stages of disease infection. Galvão et al. (2005) successfully used some absorption features extracted with the continuum-removal technique and other spectral indices from EO-1 Hyperion data to discriminate the five sugarcane varieties in southeastern Brazil. Huber et al. (2008) also successfully estimated foliar biochemistry (concentrations of N and carbon and the content of water) from hyperspectral sensor HyMap data in a mixed forest canopy using the continuum-removal technique.

Spectral wavelength position variables can also be modeled and extracted. Like absorption band features, the wavelength position variables are also useful for estimating a list of plant bioparameters.

The most popularly used position variables are plant red-edge optical parameters. There are five techniques introduced in Chapter 5, including the four-point interpolation method (Guyot et al. 1992), the polynomial fitting (Pu, Gong, Biging, and Larrieu 2003), Lagrangian interpolation (Dawson and Curran 1998), invert Gaussian model fitting (Miller et al. 1990), and linear extrapolation techniques (Cho and Skidmore 2006). All five techniques can help extract red-edge optical parameters (red-edge position and red-well position, etc.). The red-edge optical parameters can be used for estimating Chls concentrations (e.g., Belanger et al. 1995, Curran et al. 1995), nutrient constituent concentrations (e.g., Gong et al. 2002, Cho et al. 2008), leaf relative water content (e.g., Pu, Ge et al. 2003, 2004), and forest LAI (e.g., Pu, Gong, Biging, and Larrieu 2003). In addition, Pu et al. (2004) also proposed to extract 20 spectral variables (10 maximum–first derivatives plus corresponding 10 wavelength-position variables) from "10 slopes" defined across a reflectance curve from 0.4 to 2.5 μm for estimating oak leaf relative water content. Their analysis results indicated that high correlations exist between some maximum–first derivatives and wavelength-position variables and the relative water content of oak leaves.

8.3.3 Spectral Vegetation Index Analysis

The explicit advantages of spectral vegetation indices (VIs) are easy to use, and most VIs can also reduce impact of sun angle, atmosphere, shadow, and topography on target spectra (i.e., plant spectra here). It is worth noting that the probability distribution of VI image for different land cover types is not usually normal. A key factor of determining robustness of a VI depends on identifying suitable wavebands. Compared to using multispectral data to construct spectral VIs, using hyperspectral data to construct spectral VIs (HVIs) can make one more advantage of increasing chance and flexibility to choose appropriate spectral bands. In general, with multispectral data, we may only have one choice of using the only red and NIR band, whereas with hyperspectral data, we can choose many of such red and NIR narrowbands to construct many similar hyperspectral VIs (Zarco-Tejada et al. 2001, Gong et al. 2003, Eitel et al. 2006, He et al., 2006). Table 5.1 (in Chapter 5) summarizes a set of 82 HVIs that reflect newly developed narrowband HVIs from hyperspectral data. These HVIs frequently appear in literature on extracting and evaluating plant biophysical and biochemical parameters from hyperspectral data. To conveniently locate a (or group of) HVI(s) for readers, the total 82 HVIs are grouped into five categories based on the characteristics and functions of the HVIs: vegetation structure (LAI, CC, green biomass, plant species, etc.), pigments (Chls, Cars, Anths, etc.), other biochemicals (ligno-cellulose, N, etc.), water, and stress. For those interested in definitions, functions, characteristics, and references of the 82 HVIs in Table 5.1, see Chapter 5. All the 82 HVIs can be used for estimating and mapping plant bioparameters from various hyperspectral data sets.

For estimating and mapping plant canopy LAI, Gong et al. (2003) and Weihs et al. (2008) used PVI_{hyp} (hyperspectral perpendicular VI), SR, NDVI, RDVI (renormalized difference VI), and RVI_{hyp} (hyperspectral ratio VI), constructed from hyperspectral sensors Hyperion and HyMap, to estimate forest LAI. He et al. (2006) and Darvishzadeh et al. (2008a) estimated LAI of grassland ecosystems with VIs: RDVI, MCARI2 (modified chlorophyll absorption ratio index 2), and NDVI. With LAI difference index (LAIDI), standard of LAI difference index (sLAIDI), and VIs, Delalieux et al. (2008) determined LAI in orchards. And Li et al. (2008) used modified triangular VI 2 (MTVI2) to map LAI over an agricultural area from CASI hyperspectral image data. Some VIs, including ATSAVI (adjusted transformed soil-adjusted VI), LWVI-1 (leaf water VI 1), LWVI-2 (leaf water VI 1), NDVI, SR, TVI (triangular VI), mSR_{705} (modified simple ratio), can be used for identifying and mapping plant species and composition. Galvão et al. (2005) developed and used VIs LWVI-1, LWVI-2, and NDVI to discriminate five sugarcane varieties in southern Brazil with EO-1 Hyperion data. Hestir et al. (2008) used mSR_{705} VI to map invasive species with airborne hyperspectral data (HyMap). Lucas and Carter (2008) assessed vascular plant species richness on Horn Island, Mississippi, with various simple ratio VIs constructed from HyMap hyperspectral image data. For estimating biomass from hyperspectral data, some VIs such as EVI, mND_{705} (modified normalized difference), mSR705, NDVI, SR, and WDRVI (wide dynamic range VI) are very useful. Hansen and Schjoerring (2003) and le Maire et al. (2008) successfully employed various narrow band $NDVI_{like}$ and SR_{like} to estimate wheat crop and broadleaf forest biomass, respectively.

There are many HVIs that were developed for estimating plant pigments, especially for Chls (Chla and Chlb), Cars, and Anths contents. For examples, Peñuelas et al. (1995) proposed using the structural independent pigment index (SIPI) for estimating Cars. Blackburn (1998) used various HVIs—SR, pigment-specific normalized difference (PSND), and SIPI VIs—to quantify Chls and Cars of *Pteridium aquilinum* grass at leaf and canopy scales. Asner et al. (2006) studied the variations in upper-canopy leaf Chls and Cars contents during a climatological transition with the remotely sensed photochemical and carotenoid reflectance indices (PRI, CRI). They found that the PRI and CRI were sensitive to differences in light-use efficiency between invasive and native tree species thus helped separating invasive species (*Myrica faya*) from native species (*Metrosideros polymorpha*). With *in situ* spectral measurements taken from tree leaves, Gitelson et al. (2001, 2006) developed modified carotenoid reflectance index (mCRI), anthocyanin reflectance index (ARI), and modified anthocyanin reflectance index (mARI) VIs to estimate Chls, Cars, and Anths contents. To improve estimation of the plant biochemical from spaceborne hyperspectral image data, Rama Rao et al. (2008) developed a new VI, named plant biochemical index (PBI). PBI is a simple ratio of reflectances at 810 and 560 nm, and has the potential to retrieve leaf total Chls and N concentrations of various crops and at different geographical locations. Using PSND and PRI VIs, Hatfield et al. (2008) correctly estimated pigments of agricultural crops. Le Maire et al. (2008) estimated leaf Chls content of broadleaf forest with NDVI and SR HVIs derived from *in situ* and Hyperion hyperspectral data. Chappelle et al. (1992) recommended using R_{760}/R_{500} as a quantitative measure of Cars. For Anths estimation, Gamon and Surfus (1999) used a ratio of red-to-green reflectances $R_{600-700}/R_{500-600}$ to estimate Anths content at a leaf level. However, according to Sims and Gamon (2002), estimating Cars and Anths contents remains more difficult than estimating Chls content from hyperspectral data.

Many researchers used narrowband VIs to estimate water content at the leaf and canopy levels. For instance, when Peñuelas et al. (1993, 1996) studied the reflectances of gerbera, pepper, bean plants, and wheat in the 950–970 nm region as an indicator of water status, they found that the ratio of the reflectance at 970 nm, one of the water absorption bands, to the reflectance at 900 nm as the reference wavelength (R_{970}/R_{900} or Water Index, WI) closely tracked changes in relative water content (RWC), leaf water potential, stomatal conductance, and cell wall elasticity. Cheng et al. (2006) and Clevers et al. (2008) used normalized difference water index (NDWI), WI, and shortwave infrared water stress index (SIWSI) to estimate vegetation water content for different canopy scenarios with hyperspectral AVIRIS data. Colombo et al. (2008) estimated leaf and canopy water contents in poplar plantation using simple ratio water index (SRWI), normalized difference infrared index (NDII), and moisture stress index (MSI) derived from airborne hyperspectral image data. Pu, Ge et al. (2003) determined water status in coast live oak leaves with $RATIO_{1200}$ (3-band ratio at 1200 nm) and $RATIO_{975}$ (3-band ratio at 975 nm) indices derived from hyperspectral measurements.

A few HVIs were especially designed for estimating nutrient constituents and other chemical compounds such as lignin and cellulose concentration. They are cellulose absorption index (CAI), normalized difference nitrogen index (NDNI), normalized difference lignin index (NDLI), NDVI, PBI, and SR. For example, Serrano et al. (2002) proposed the two indices NDNI and NDLI to assess N and lignin concentrations in chaparral vegetation using AVIRIS hyperspectral image data. Gong et al. (2002) and Hansen and Schjoerring (2003) used narrowband NDVI and SR indices to assess nutrient constituent concentrations (N, P, and K) of a conifer species and N status in wheat crops from hyperspectral data. And with PBI derived from Hyperion hyperspectral data, Rama Rao et al. (2008) estimated leaf N concentration of cotton and rice crops.

8.3.4 Plant Spectral Unmixing Analysis

Unlike laboratory and *in situ* spectral reflectances, which are usually measured from pure materials, a large portion of remotely sensed data is spectrally mixed (Pu and Gong 2011). For this case, the spectral mixing process has to be properly modeled in order to identify various "pure materials" and to determine their spatial proportions from the remotely sensed data, especially hyperspectral image data. In hyperspectral remote sensing of vegetation, pure materials in mixed pixels can be considered pure (homogeneous) vegetation types, individual plant species or specific species groups in which the individual species have a similar spectral signature or plant health levels of the same species. Once the spectral mixing process is modeled, the model can be inverted to obtain the spatial proportions and spectral properties of those pure materials (i.e., vegetation types, plant species, or specific species groups or individual species health levels). Generally, there are two types of spectral mixing: linear spectral mixing and nonlinear spectral mixing. Both spectral mixing models are simple tools used to describe spectral mixing processes. Linear spectral mixing modeling and its inversion have been used widely since the late 1980s in order to extract the abundance of various components within mixed pixels. Nonlinear spectral mixture model can be found in Sasaki et al. (1984) and Zhang et al. (1998). Meanwhile, a multiple end-member spectral mixture analysis (MESMA) approach (Roberts et al. 1998), a mixture-tuned matched filtering (MTMF) technique (Boardman et al. 1995), and a constrained energy minimization (CEM) approach (Farrand and Harsanyi 1997) have been developed for spectral unmixing analysis with hyperspectral data. In addition, an artificial neural network algorithm has also been tested to unmix mixed pixels into fraction abundances of end-members in some studies (e.g., Flanagan and Civco 2001, Pu, Gong, Michishita et al. 2008). For those interested in a detailed introduction to the algorithms and methods of these spectral mixture models and methods/techniques for spectral end-member determination and end-member spectral extraction, consult Chapter 5.

There are many studies on applications of linear spectral mixture model (LSM) to hyperspectral data to estimate the abundance of general vegetation cover or specific vegetation species (e.g., Asner and Heidebrecht 2003, Miao et al. 2006, Judd et al. 2007, Hestir et al. 2008, Walsh et al. 2008, Pignatti et al. 2009). A neural network–based, nonlinear solution of the LSM model also was applied to hyperspectral data to estimate the abundance of specific vegetation species (e.g., Pu, Gong, Tian et al. 2008, Walsh et al. 2008). Several researchers have applied the MESMA technique in a variety of environments for vegetation mapping. Roberts et al. (1998, 2003) used MESMA and hyperspectral image data AVIRIS to map vegetation species and land cover types in southern California chaparral. By using AVIRIS image data and the MESMA approach, Li et al. (2005) and Rosso et al. (2005) mapped coastal salt marsh vegetation in China and marshland vegetation of San Francisco Bay, California, respectively. Fitzgerald et al. (2005) successfully mapped multiple shadow fractions in a cotton canopy with the MESMA approach and hyperspectral imagery. In spectral unmixing analysis, MTMF has also proven to be a very powerful tool to detect specific materials that differ slightly from the background. For example, using airborne hyperspectral image data (e.g., AVIRIS and HyMap) for detecting and mapping invasive plant species, MTMF has been successfully used to map a variety of invasive species, including Tamarisk (Hamada et al. 2007), perennial pepperweed (Andrew and Ustin 2010), and cheatgrass (Noujdina and Ustin 2008).

As a result of spectral unmixing, the fractions represent the areal proportions of end-members in a mixed pixel, but where the proportioned areas locate within the mixed pixel is not known. A key factor to unmix mixed spectra is to identify suitable/pure end-members and extract their individual spectra for training and test purposes.

8.3.5 Plant Spectral Matching Analysis

Like the plant spectral unmixing analysis in the application of hyperspectral remote sensing in vegetation, spectral matching analysis techniques mostly use different similarity measures calculated from hyperspectral data sets to identify and map vegetation types (including health levels) and plant species or specific species groups. The similarity measures include spectral angle mapping (SAM; Kruse et al. 1993), the Euclidean distance (ED; Kong et al. 2010) measure, the cross-correlogram spectral matching (CCSM; van der Meer and Bakker 1997), and spectral information divergence (SID; Chang 2000), which all efficiently utilize the detailed spectral differences in magnitude between two spectra (one used as a reference and the other as a test spectrum). For a more detailed introduction to algorithms of these four measures, see Chapter 5. It is commonly believed that SAM is a physically based spectral classification and is less sensitive to differences in curve magnitude caused by variation in lighting across a scene. In spectral matching, it is worth noting that the accuracy of these spectral matching measures and techniques is directly affected by the geometry of sensor observations and target size. Such effect can be minimized by spectral normalization before conducting spectral matching processing (Pieters 1983). In general, such matching techniques are favorable to change detection of scene components than identifying the unknown scene components (Yasuoka et al. 1990). Spectral matching uses n-dimensional angles (or distances, cross-correlations, or discrepancy of probability distributions) to match pixels to reference spectra with smaller angles (shorter distances, higher cross-correlations, lower discrepancy) representing closer matches to the reference spectrum, otherwise representing not matching the reference spectrum. Usually, a key factor to apply the matching techniques for detecting and mapping vegetation types and species with hyperspectral data is how to determine a suitable threshold of angle (or distance or correlation or discrepancy).

Many researchers have employed spectral similarity measures and matching techniques to identify and map vegetation types and plant species. For example, in the study of the spectroscopic determination of two health levels of the coast live oak leaves, Pu, Kelly et al. (2008) used the CCSM algorithm to discriminate between healthy and infected leaves by matching unknown leaf spectra with known infected leaf spectra associated with water stress. With AISA airborne hyperspectral image data and SAM matching technique, Narumalani et al. (2009) quantified and mapped four dominant invasive plant species, including saltcedar, Russian olive, Canada thistle, and musk thistle, along the floodplain of the North Platte River, Nebraska. Validation procedures confirmed an overall map accuracy of 74%. Also using the SAM technique and airborne hyperspectral image data, Lass and Prather (2004) detected the location of Brazilian pepper trees in the Everglades, and Hirano et al. (2003) mapped wetland vegetation with an invasive species lather leaf (*Colubrina asiatica*). In addition, Pengra et al. (2007) used CCSM measure to map an invasive plant, *Phragmites australis*, in coastal wetland using the Hyperion sensor data, which resulted in a better overall accuracy of 81.4%.

8.3.6 Plant Spectral Classification Analysis

In general, the plant spectral unmixing and matching analyses may be considered spectral classification analyses. However, here the plant spectral classification analysis is more focused on an introduction to and review on contemporary, advanced classifiers and methods that are directly used to identify and map a list of bioparameters and vegetation types. Traditional multispectral classification methods, including supervised and unsupervised classifiers, may be used in hyperspectral image classifications and thematic information extraction in vegetation applications. However, direct use of these traditional classifiers in plant spectral classification analysis with hyperspectral data may be less effective than they are expected with multispectral data. This is due to high dimensionality of hyperspectral data and high correlation of adjacent bands with limited number of training samples. To solve such problems with hyperspectral data for classification analysis of vegetation attributes, two types of hyperspectral data processing techniques may be employed: (1) using commonly used and efficient data transformation and feature extraction techniques and methods to derive low-dimensional variables and features transformed and directly extracted from high-dimensional hyperspectral data; and (2) using segmented versions of existing data transformation algorithms and classifiers directly to conduct plant spectral classification analysis. Once the low-dimensional variables and features are transformed and extracted from hyperspectral data using relevant techniques and methods, most currently available traditional classifiers such as MLC and minimum distance classifier may be directly applied to plant spectral classification analysis with the low-dimensional variables and features as inputs. The data transformation and feature extraction techniques and methods include principal components analysis (PCA), maximum noise fraction (MNF), independent component analysis (ICA), linear discriminant analysis (LDA), canonical discriminant analysis (CDA), and wavelet transform (WT). The segmented version of existing data transformation algorithms and classifiers comprising segmented PCA, segmented LDA, segmented CDA, segmented PDA (penalized discriminant analysis), and simplified MLC may be directly applied to hyperspectral data for plant classification analysis. Usually, the segmentation calculation of hyperspectral data to be suitable for the segmented version of the data transformation algorithms and classifiers is based on a band correlation matrix calculated from individual band digital numbers or reflectances of hyperspectral data (Pu and Liu 2011). All the data transformation and feature extraction techniques and methods and segmented version of the data transformation algorithms and classifiers are introduced and discussed in Chapter 5. Some advanced classifiers, such as artificial neural networks (ANNs) and support vector machines (SVMs), are frequently employed in plant classification analysis with hyperspectral data and their algorithms are also introduced in Chapter 5.

There are many application cases of hyperspectral data in vegetation classification analysis using the plant spectral classification analysis methods and classifiers. For example, Pu and Liu (2011) used segmented CDA, segmented PCA, segmented CDA, and MLC classifier with *in situ* hyperspectral data to discriminate thirteen tree species in Tampa, Florida. They concluded that segmented CDA outperformed segmented PCA and segmented SDA and suggested that CDA (or segmented CDA, under a condition with limited training samples) should be applied broadly in mapping forest cover types, species identification, and in other land use/land cover classification practices with multi-/hyperspectral remote sensing data because it is superior to PCA (or segmented PCA) and SDA (or segmented SDA) for selection of features that are used for vegetation spectral classification analysis. After transforming several MNFs from the AVIRIS hyperspectral image data, Underwood et al. (2003, 2007) used the MLC classifier to classify three nonnative plant species (iceplant, jubata grass, and blue gum) in the coastal areas of California. With several PC images transformed from CASI hyperspectral image data, Pu, Gong, Tian et al. (2008) utilized ANN and LDA algorithms to map the saltcedar invasive species. For a six-class crop classification with hyperspectral data HyMap, Camps-Valls et al. (2004) used SVMs to evaluate their performance in terms of efficiency and robustness, as compared to extensively used ANNs and fuzzy methods. They concluded that SVMs yielded better outcomes than ANNs and fuzzy methods regarding classification accuracy, simplicity, and robustness.

In practice, for the hyperspectral image classification analysis, it is frequently difficult to collect adequate training samples for supervised methods and labeling unsupervised spectral clusters due to relatively high dimensionality of hyperspectral data. The major factors for identifying and classifying vegetation types and plant species with various hyperspectral data are to identify suitable classification methods and gather adequate training samples for a specific task.

8.3.7 Empirical/Statistical Analysis Methods

Empirical or statistical analysis methods include univariate and multivariate regression analyses to correlate biophysical or biochemical parameters with spectral reflectance, vegetation indices (VIs), or derivative spectra or features transformed or extracted from hyperspectral data in the visible, NIR, and MIR wavelengths at leaf- or canopy- or plant community–level (e.g., Peterson et al. 1988, Wessman et al. 1988, Johnson et al. 1994, Grossman et al. 1996, Gitelson and Merzlyak 1997, Gong et al. 1997, Martin and Aber 1997, Martin et al. 1998, Galvão et al. 2005, Colombo et al. 2008, Darvishzadeh et al. 2008a, Hestir et al. 2008, Huber et al. 2008). A multivariate regression approach, the stepwise multiple linear regression (SMLR), is commonly used to develop a calibration equation by selecting a small number of narrowband reflectances or spectral variables (e.g., derivative spectra) that account for a large proportion of the variation in biochemical content or concentration (e.g., Johnson et al. 1994, Curran et al. 1997). The calibrated multivariate regression equation is then used to estimate biochemical concentration or content in additional samples (pixels). Such an approach to estimate biochemical factors shows a high level of reliability and reproducibility when training (calibration) samples are collected under carefully controlled conditions (e.g., using random sampling method to collect training samples).

According to Curran (1989), we need to know that the four main assumptions upon which the multivariate approach is based can be easily violated: (1) the relationships between reflectance or other spectral variable forms and biochemical concentration or content are near-linear; (2) we can extract the vegetation spectra of interest from the hyperspectral data; (3) the relationship between spectra and biochemical composition is not confounded by other factors, such as phenology or canopy geometric structure and variations in solar–Earth surface–sensor geometry; and (4) the biochemical concentrations or contents have been accurately measured.

The first assumption is reasonable whereas the second assumption relies on the success of vegetation pure and interested spectra measured or extracted through spectral unmixing methods (see Section 8.3.4). The third assumption may be reasonable if single-species stands of vegetation are studied. Otherwise, the assumption poses particular problems at the canopy level (e.g., different tree species have different canopy geometry). There are many problems with the fourth assumption. For instance, the *in vivo* state of the biochemicals may significantly differ from their *in vitro* state. The concentration of a biochemical measured in the laboratory may differ significantly from its concentration in the field. Isolated biochemicals may be physically altered by isolation processes such as oxidation, hydrolysis, or denaturation (Kumar et al. 2001). In addition, since the dominant effect of absorption by water largely masks the signatures of the biochemical components beyond 1.0 μm (Tucker and Garratt 1977, Curran 1989, Elvidge 1990), fresh leaf tissue creates more problems than dried (or ground) tissue for biochemical measurement for spectral analysis.

There are many studies on using the MLR approach to estimate biochemical concentrations or contents with band reflectance, derivative spectra, or other spectral features transformed or extracted from hyperspectral data collected from different platforms. These studies generally fall into two categories: the assessment of reflectance for biochemical estimation at leaf scales and those at canopy scales. Studies at the leaf scale include the use of foliage harvested and assessed through the use of either laboratory or field spectrometers, while hyperspectral measurements taken from aircraft or satellites are used to assess the biochemical characteristics at the canopy scale (e.g., at both scales; Peterson et al 1988, Gastellu-Etchegory et al. 1995, Kupiec and Curran 1995). Johnson et al. (1994) established regression models for estimating biochemical concentrations using the MLR approach between chemical compounds of forest canopy and the AVIRIS band reflectance. Using hyperspectral image data acquired by AVIRIS and CASI sensors, Matson et al. (1994) demonstrated that canopy biochemicals carried information about forest ecosystem processes and suggested that some of this chemical information might be estimated remotely using hyperspectral data collected by airborne sensors. They found that the first derivative spectra centered in the range of 1525–1564 nm figured prominently in all nitrogen equations. After correlating vegetation indices of R_{NIR}/R_{700} and R_{NIR}/R_{550} with Chl content, Gitelson and Merzlyak (1996, 1997) demonstrated that the vegetation indices for Chl assessment were important for two deciduous species, maple and chestnut. In spectral feature analysis associated with N, P, and K deficiencies in *Eucalyptus saligna* seedling leaves, Ponzoni and Goncalves (1999) proved that spectral reflectance is better estimated using a combination of nutrient constituents (N, P, and K) as independent variables with the results created from univariate regression and MLR approaches. With MLR models, continuum-removal technique, and normalized HyMap spectra, Huber et al. (2008) estimated foliar concentrations of N and carbon, and content of water in a mixed forest canopy.

Such statistical regression approaches may also be used to estimate and identify a set of biophysical parameters. For example, Martin et al. (1998) determined forest species composition using high spectral resolution remote sensing data with the approach that combines forest species–specific chemical characteristics and previously derived relationships between hyperspectral data (AVIRIS) and foliar chemistry. They classified eleven forest cover types, including pure and mixed stands of deciduous and conifer species with an overall accuracy of 75%. With EO-1 Hyperion hyperspectral image data, Galvão et al. (2005) successfully discriminated five sugarcane varieties in southeastern Brazil using the multiple discriminant analysis method that produced a classification accuracy of 87.5%.

However, although the statistical regression analysis (MLR) has been successfully used in many cases in estimating or identifying plant bioparameters from hyperspectral data, the multivariate regression approach frequently suffers from a list of limitations (Curran 1989) as follows:

- *Overfitting of wavebands.* The large number of wavelengths included in calibration equations is compared with the relatively small training sample size. This occurs when reflectance values in wavebands without having causal relationship with the chemical of interest are selected because their noise patterns fit the data on chemical concentration. This risk becomes obvious with the number of wavebands used.
- *Intercorrelation of biochemicals.* There exists a strong intercorrelation between several biochemicals. For example, starch concentration is often correlated with reflectance at around 0.66 μm. This can be due to starch concentration correlated with Chl concentration, which absorbs waveband strongly at around 0.66 μm.
- *Missing of absorption bands.* In a calibrated MLR equation, some known absorption bands for a particular biochemical are not involved. This may be interference between spectrally close absorption features, but is difficult to prove using current hyperspectral data collected from field and airborne/spaceborne spectrometers (Curran 1989).

An alternative statistical regression technique, partial least squares (PLS) regression, is a technique that reduces the large number of measured collinear spectral variables to a few non-correlated latent variables or principal components (PCs). The PCs represent the relevant structural information present in the measured reflectance spectra and are used to predict the dependent variables (i.e., bioparameters; Darvishzadeh et al. 2008a). PLS regression uses data compression to reduce the number of independent variables, followed by a calibration regression stage consisting of a least-squares fit of the bioparameters to the obtained regression factors. Recently, researchers have increased interest in applying the PLS approach to calibrate relationships between spectral variables derived from hyperspectral data and a set of bioparamters. For example, using spectral measurements taken from leaves and bioparameter data (Chla, Chlb, Cars, Anths, water, N, P, and SLA) collected from 162 Australian tropical forest species, along with PLS technique and canopy radiative transfer modeling, Asner and Martin (2008) concluded that a suite of leaf properties among tropical forest species can be estimated using full-range leaf spectra of fresh foliage collected in the field. Hansen and Schjoerring (2003) used two-band combinations in the normalized difference vegetation indices constructed from *in situ* spectral measurements taken from wheat crop canopy and the PLS technique to estimate canopy green biomass and N status. They concluded that the PLS analysis may provide a useful exploratory and predictive tool when applied in hyperspectral reflectance data analysis. To determine nitrogen, lignin, and cellulose concentrations in dry, ground, green foliage samples of temperate forest woody plants from NIR reflectance data, Bolster et al. (1996) showed that PLS performed consistently better than MLR.

Univariate and multiple regression analysis methods are relatively simple and their modeling results frequently have higher estimation accuracy. However, empirical or statistical relationships are often site-, species-, and sensor-specific, and thus cannot be directly applied to other study areas since the plant canopy structure and sensor viewing geometry may vary from different sites and species.

8.3.8 Physically Processing-Based Modeling Methods

Since the empirical or statistical methods lack robustness and are case specific, during the last two to three decades, physically based modeling approaches, such as radiative transfer modeling approaches, have attracted the attention of a lot of researchers who are interested in retrieving biophysical and biochemical parameters by inversing various physically-based models from simulated spectra or real hyperspectral image data, such as Scattering by Arbitrary Inclined Leaves (SAIL) canopy spectral model (Verhoef 1984, Asner 1998) and PROSPECT leaf spectral model (Jacquemoud and Baret 1990). Such physically based models rest on a theoretical basis of various leaf or canopy scattering and absorption models described by biochemistry, biophysics, canopy structure factors, etc. These models, including radiative transfer (RT) models and geometric-optical (GO) models, consider the underlying physics and complexity of the leaf internal structure and thus are robust and have a potential to replace the statistically based approaches (Zhang et al. 2008a,b). In the context of the remote sensing of vegetation properties, such models have been used in the forward mode to calculate leaf or canopy reflectance and transmittance and in inversion to estimate leaf or canopy chemical and physical properties (Pu and Gong 2011). For example, many researchers employed physically based models at a leaf or canopy level to retrieve biochemical parameters including leaf pigments from either simulated spectra or hyperspectral image data (e.g., Asner and Martin 2008, Feret et al. 2008, Croft et al. 2013, Zou et al. 2015).

Most physically based models are RT models that have been developed at leaf and canopy levels. The RT models mostly simulate leaf reflectance and transmittance spectra between 0.4 μm and 2.5 μm. For considering leaf optical properties, the most important RT models may involve PROSPECT model and its improved versions (Jacquemoud and Baret 1990, Jacquemoud et al. 1996, Fourty et al. 1996, Demarez et al. 1999, le Maire et al. 2004, Zarco-Tejada et al. 2013), the Leaf Incorporating Biochemistry Exhibiting Reflectance and Transmittance Yields (LIBERTY) model (Dawson et al. 1998, Coops and Stone 2005), Leaf Experimental Absorptivity Feasibility MODel (LEAFMOD; Ganapol et al. 1998), etc. For considering canopy optical properties, the most popular RT models are SAIL (Verhoef 1984, Asner 1998) and its improved versions that have been adapted to account for some heterogeneity within the vegetation canopy, such as SAILH (Kuusk 1985), GeoSAIL (Verhoef and Bach 2003), 2M-SAIL (Weiss et al. 2001, le Maire et al. 2008), and 4SAIL2 (Verhoef and Bach 2007). The other important canopy reflectance RT models include Fast Canopy Reflectance (FCR; Kuusk 1994), New Advanced DIscrete Model (NADIM; Jacquemoud et al. 2000, Ceccato et al. 2002), Markov–Chain Canopy Reflectance Model (MCRM; Kuusk 1995) adapted for row crops (Cheng et al. 2006), and the four models used for simulating discontinuous forest canopies: Discrete Anisotropic Radiative Transfer (DART; Demarez and Gastellu-Etchegorry 2000), Spreading of Photons for Radiation INTerception (SPRINT; Zarco-Tejada, Miller, Harron et al. 2004), Forest Light Interaction Model (FLIM; Zarco-Tejada, Miller, Morales et al. 2004), and three-dimensional Forest LIGHT interaction (FLIGHT; Koetz et al. 2004). Moreover, during the last two decades, researchers have also developed some leaf-canopy coupled models, including PROSAIL (Baret et al. 1992, Broge and Leblance 2000), LEAFMOD + CANMOD (Ganapol et al. 1999), LIBERTY + FLIGHT (Dawson et al. 1999), LIBERTY + SAIL (Dash and Curran 2004), PROSPECT + DART (Malenovsky et al. 2007), and PROSPECT + FLAIR (Omari et al. 2013). Among the RT models, based on existing literature review, the most popular and important RT models on leaf, canopy, and leaf–canopy coupled optical properties are PROSPECT, SAIL, and PROSAIL, as well as their improved versions (Jacquemoud et al. 2009).

The PROSPECT leaf optical models, including latest versions PROSPECT-4 and -5 (Feret et al. 2008), can provide specific absorption and scattering coefficients of leaf components. The model is widely used and well validated (e.g., Fourty et al. 1996, Zarco-Tejada et al. 2013). The SAIL canopy-level model is a four-stream radiative transfer model developed by Verhoef (1984). It was later modified by Kuusk (1991) to take the hot spot feature into account. Leaf–canopy coupled optical models PROSAIL, PRODART, PROFLAIR, and PROFLIGHT allow description of both the spectral and directional variation of canopy reflectance as a function of leaf biochemistry (mainly Chls, water, and dry matter contents) and canopy architecture (primarily LAI, SLA, LAD, and relative leaf size). The coupled leaf–canopy and other RT models help us understand the way in which leaf reflectance properties are influenced by the large number of controlling factors at a canopy scale (Demarez and Gastellu-Etchegorry 2000). The coupled models have enabled the development and refinement of spectral indices that are insensitive to dominant factors such as canopy structure, illumination geometry, and soil/litter reflectance (Broge and Leblanc 2000, Daughtry et al. 2000). They have also been used in establishing predictive relationships that have been applied to hyperspectral imagery to map plant pigments (e.g., Chls and Cars contents) (e.g., Haboudane et al. 2002; Zarco-Tejada, Berjón et al. 2005; Hernández et al. 2012, 2014).

The geometric–optical (GO) models also belong to one type of RT models, but they more emphasize the effect of canopy architecture on modeling result, and thus they are very effective in capturing the angular distribution pattern of the reflected radiance and used widely in remote sensing applications (Chen and Leblanc 2001) as aforementioned RT models. There are many different types of GO models. A famous and important model developed by Li and Strahler (1985) described the plant canopy using opaque geometric shapes (cones or cylinders), which cast shadows on the ground. The GO models are frequently used to describe sparse forests or shrublands, where shadowing plays an important role.

The goal of developing the physically based models is inverting them to retrieve vegetation characteristics from observed reflectance data. There are three general types of inversion methods for inversion of physical models, which include iterative optimization methods (e.g., Goel and Thompson 1984, Liang and Strahler 1993, Jacquemoud et al. 2000, Meroni et al. 2004), look-up table (LUT) approaches (e.g., Knyazikhin et al. 1998, Weiss et al. 2000, Combal et al. 2002, Gastellu-Etchegorry et al. 2003, Omari et al. 2013, Ali et al. 2016), and ANNs (e.g., Gong et al. 1999, Walthall et al. 2004, Schlerf and Atzberger 2006). In the iterative optimization approach, a stable and optimum inversion is not guaranteed. In addition, the traditional iterative method is time-consuming and often requires a simplification of the models when processing large datasets. This may result in a decrease of inversion accuracy and make the retrieval of biophysical and biochemical variables unfeasible for large geographic areas (Houborg et al. 2007). LUT approaches can partially overcome the drawback from the iterative method. They operate through a database of simulated canopy reflectance variable in structural and radiometric properties. However, LUT creation can be complicated and requires an extensive set of reliable field measurements. ANN technique, proposed in the forward and inverse modeling of radiative transfer models for retrieving bioparameters, is expected to reduce such complexity of inversion, but its internal structure to finally retrieve the bioparameters is not clear. For proper training (ANN) and representation (LUT), they basically rely on a large database of simulated canopy reflectance spectra to achieve a high degree of accuracy. This increases the computational time for identifying the most appropriate LUT entry and the time required for training an ANN (Kimes et al. 2000, Liang 2004).

8.3.9 Bioparameter Mapping Methods

Imaging spectroscopy techniques have provided an ideal data source to analyze spatial distribution of biophysical and biochemical parameters (bioparameters). Using airborne or spaceborne hyperspectral image data, it is possible nowadays to estimate and map pixel-based individual bioparameters. A general bioparameter mapping method may follow the operational steps as follows:

1. Establish bioparameter prediction models by using either empirical/statistical models or physically processing-based canopy radiative transfer models or geometrical-optical models and using pixel values (spectral variables in various forms; e.g., digital number, band radiance, reflectance, derivative, and vegetation indices) and simulated or measured bioparameters.

2. Estimate and predict pixel-based bioparameter values (outputs) using the calibrated prediction models at step 1 and pixel-based spectral variables (inputs).
3. Adopting clustering or density slicing approaches to classify pixel-based bioparameter predicted values to map spatial distribution of single bioparameters.
4. To map more accurate bioparameters, it is sometimes necessary to perform spectral unmixing processing to obtain pixel-based abundance of end-members. Based on the abundance (fraction) maps of end-members, adopt steps 1 and 2 (pixel-based abundance or faction values of end-members as an input).

During the last two to three decades, many researchers have successfully conducted many studies on using hyperspectral image data to estimate and map pixel based bioparameters with either empirical/statistical methods or physically based modeling techniques. For examples, with high spectral resolution airborne imaging spectrometer (AIS) data, Wessman et al. (1989) estimated and mapped forest canopy lignin concentration and nitrogen (N) mineralization rate across Blackhawk Island, Wisconsin. They used pixel-based AIS derivative spectra as independent variables and lignin concentration and N mineralization rate as dependent variables and simulated stepwise regression models to estimate each pixel the biochemical values. They then classified the pixel-based lignin concentration and N mineralization values predicted by the regression models into eight classes to make corresponding distribution maps of lignin concentration and N mineralization rate (Figure 8.5). In Figure 8.5a, changes in lignin concentration from left to right across the island reflect a continuous change in soil texture resulting from sediment sorting when the island area was an early post-glacial floodplain. Figure 8.5b shows the spatial distribution of annual nitrogen mineralization rates, which was obtained by calculating pixel-based N mineralization rates by using a regression equation with AIS-predicted canopy lignin concentration as an input. Using a similar biochemical parameter mapping technique, Curran et al. (1997) also based on AVIRIS hyperspectral image data to map biochemical—Chls, N, lignin, and cellulose—contents in a slash pine plantation near Gainesville, Florida (Figure 8.6). They first calculated the first derivative spectra of AVIRIS data, then established four five-band multivariate regression equations (corresponding to the four biochemicals) by using a stepwise regression analysis to select five (band) first-derivative spectra. It is clear to see the spatial distribution of the four biochemical contents from Figure 8.6.

Using CASI and AVIRIS airborne hyperspectral sensor data to map forest canopy closure (*Pinus ponderosa*), Gong et al. (1994) adopted spectral unmixing analysis method to obtain five faction images of the five components including the pine fraction image. If the pixel-based continuous pine fraction image is further classified into several classes (i.e., pine crown closure classes), a single forest canopy closure class map can then be obtained. With the AVIRIS data at three different processing stages—the original radiance (OR), corrected radiance (CR), and retrieved surface reflectance (SR)—Pu, Gong, and Biging (2003) followed the following procedure to estimate and map conifer forest LAI at two sites in Argentina:

1. Pixel-based spectra from 15 to 225 homogenous pixels were extracted for each of the 70 LAI measurement plots from the OR, CR, and SR images. An average spectral value was calculated from all pixel spectra extracted from each individual plot.
2. A linear correlation coefficient is calculated between each band and the LAI measurements in order to select those bands with high correlations to establish LAI prediction models.
3. Twenty bands from the AVIRIS data of OR, CR, and SR were selected. The following band combinations were considered: (a) selecting 2–3 bands from each band group in terms of the correlogram (Jia and Richards 1999); (b) selecting those bands with peak values of the correlation curve; and (c) selecting those bands with known absorption features.

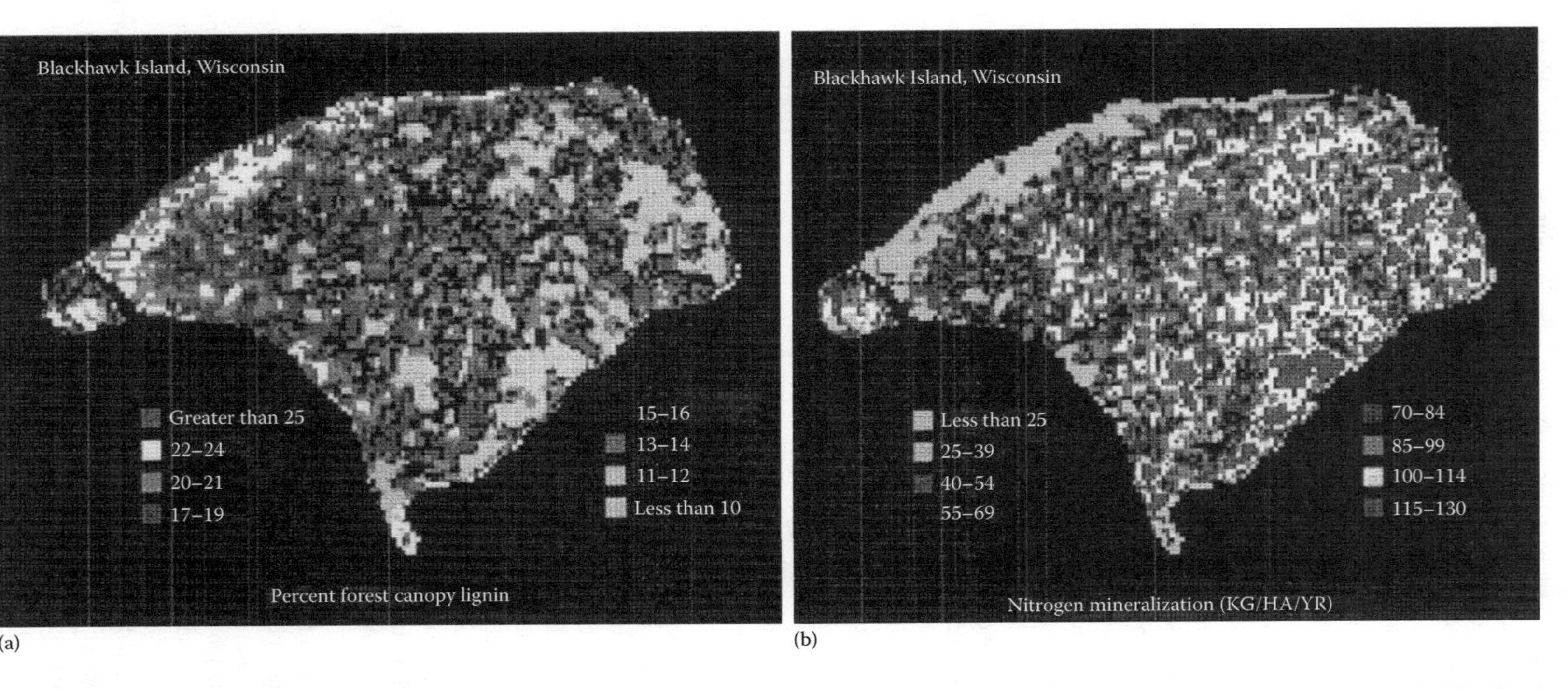

FIGURE 8.5 (a) Spatial distribution of canopy lignin concentration (%) across Blackhawk Island, Wisconsin. (b) Annual nitrogen mineralization rates for Blackhawk Island, Wisconsin, as estimated from AIS-predicted canopy lignin concentrations. (From Wessman, C. A., Aber, J. D., Peterson, D. L., *International Journal of Remote Sensing*, 10, 1293–1316, 1989.)

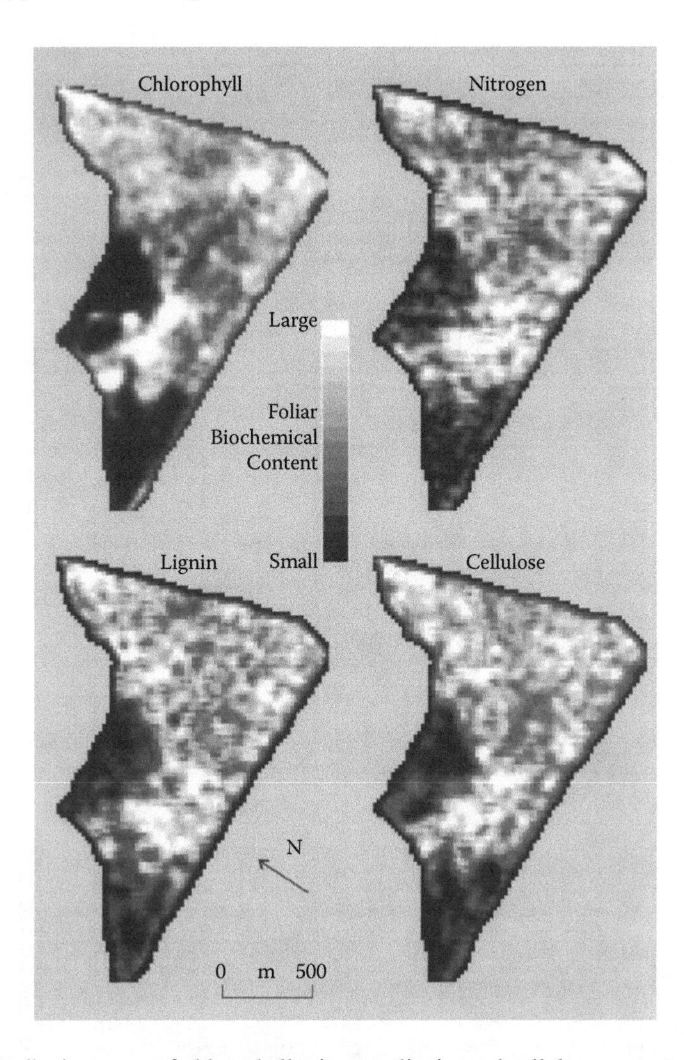

FIGURE 8.6 Distribution maps of chlorophyll, nitrogen, lignin, and cellulose contents for approximately 2 km^2 of slash pine canopy, calculated from AVIRIS hyperspectral imagery. (©1997 IEEE. Reprinted with permission from Curran, P. J., Kupiec, J. A., and Smith, G. A., *IEEE Transactions on Geoscience and Remote Sensing*, 35(2), 415–420, 1997.)

4. Using a piecewise regression procedure, an optimal set of (10) bands from the 20 bands determined at step 3 was selected to develop LAI prediction model. The 10-term prediction model for each type of the AVIRIS data was used to predict pixel-based LAI values for mapping conifer canopy LAI.
5. Pixel-based LAI values estimated from the AVIRIS data were sliced into six classes and colored with a legend referencing to the value ranges for each class.

The six-class LAI mapping results are presented in Figure 8.7. From the LAI maps, the LAI maps created with the AVIRIS data in retrieved surface reflectance (SR) show more reasonable distribution than those created with other AVIRIS data types (OR and CR).

Using hyperspectral imagery (Airborne Hyperspectral Scanner [AHS] with 2 m spatial resolution in 38 bands in the 0.43–12.5 μm spectral range) and model simulation (based on PROSPECT-5+DART model), Hernández-Clemente et al. (2014) drove predictive relationships of Chls and Cars contents (Ca+b and Cx+c) at the canopy level in a conifer forest in a study area consisting

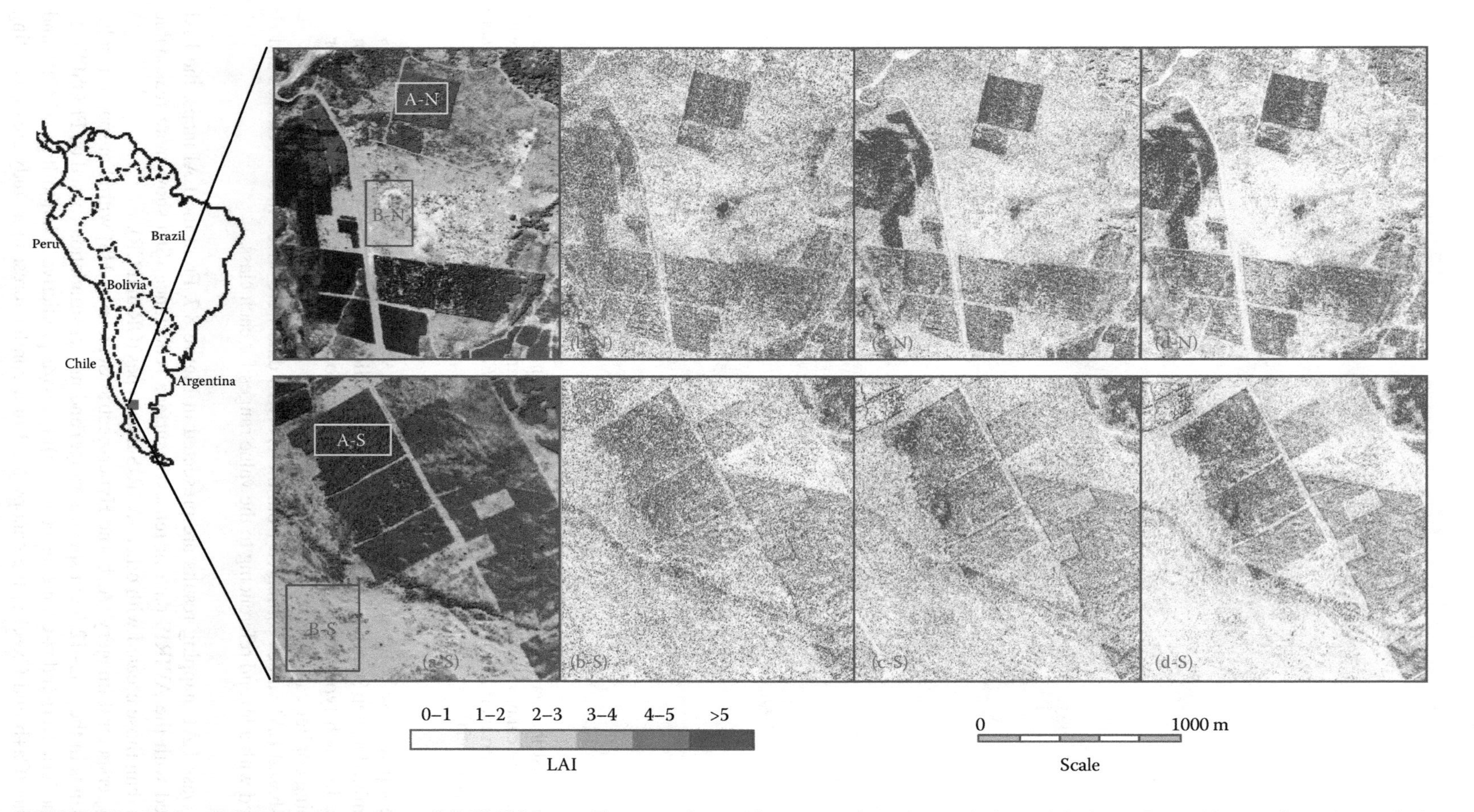

FIGURE 8.7 LAI mapping using three types of AVIRIS images. The upper four subimages are from the north site and the lower four subimages from the south site. (a-N, a-S) False color composite (AVIRIS bands NIR, Red, Green); (b-N, b-S) LAI mapping with original radiance; (c-N, c-S) with corrected radiance; and (d-N, d-S) with retrieved surface reflectance data. (From Pu, R., Gong, P., and Biging, G. S., *International Journal of Remote Sensing*, 24(23), 4699–4714, 2003.)

of a 40-year-old pine afforestation of *Pinus nigra* (Black pine) and *Pinus sylvestris* (Scotch pine). Canopy modeling methods based on infinite reflectance formulations and the discrete anisotropic radiative transfer (DART) model were evaluated in relation to the PROSPECT-5 leaf model for the scaling-up procedure. The PROSPECT-5 model was first used to simulate needle reflectance and transmittance for varying chlorophyll Ca+b (10–60 μg/cm²), carotenoid Cx+c (2–16 μg/cm²), and leaf water (0.01–0.03 cm) content and fixed values of dry matter content and leaf internal structure. They adopted simpler modeling methods to establish predictive relationships between spectral vegetation indices (simulated from PROSPECT-5+DART model) and conifer canopy pigments (Ca+b and Cx+c). The calibrated predictive relationships were then used to predict pixel-based Chls (Ca+b) and Cars (Cx+c) contents from AHS airborne hyperspectral image data. Figure 8.8 presents the mapping results obtained from two samples of *P. sylvestris* and *P. nigra* forest acquired with the AHS hyperspectral imager in a sample area and shows the spatial distribution of two conifer canopy pigments: Ca+b and Cx+c contents over the sampling area.

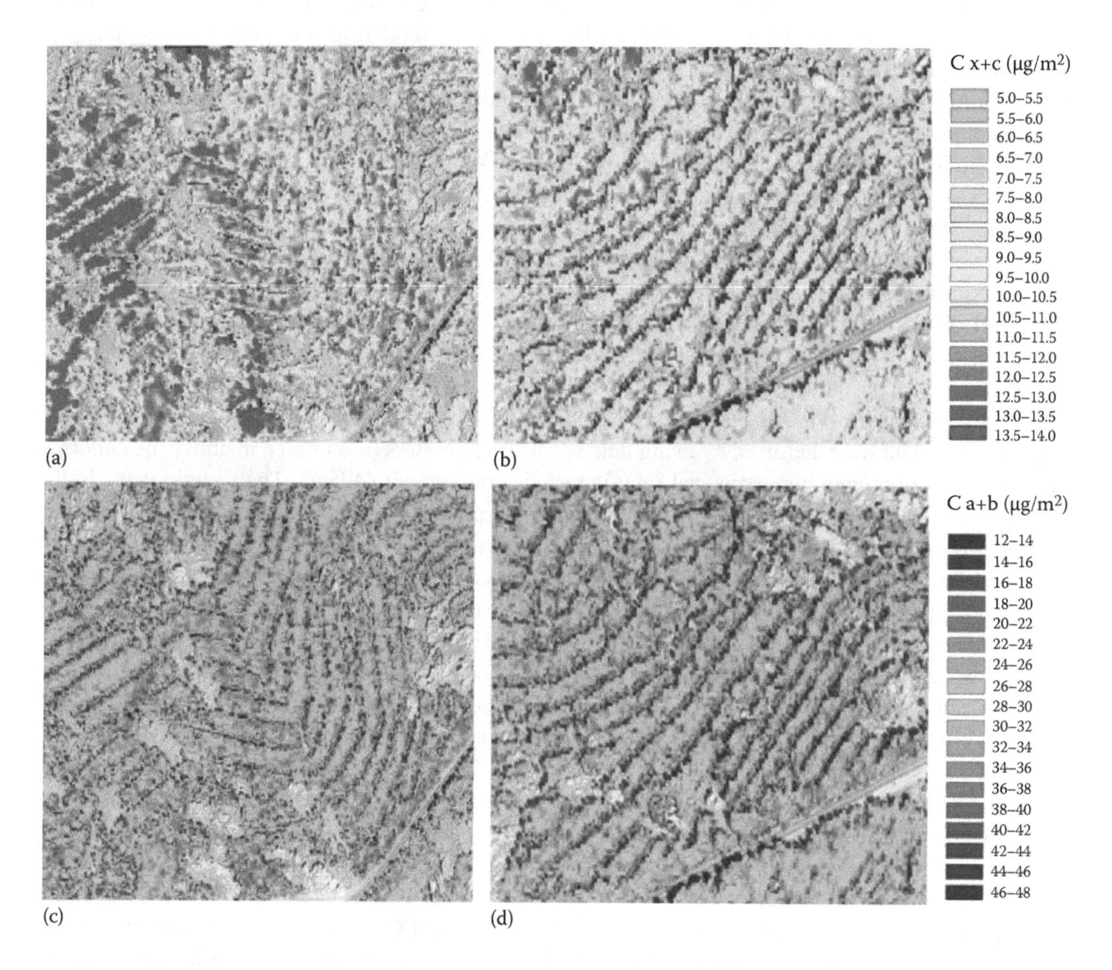

FIGURE 8.8 Mapping results obtained from two samples of *P. sylvestris* and *P. nigra* forest acquired with the AHS hyperspectral imager in a sampling area with (left) high and (right) low concentration of chlorophyll (Ca+b) and carotenoid (Cx+c) pigments. (a and b) Cx+c content estimated from the R_{515}/R_{570} and R_{700}/R_{750} indices simulated from PROSPECT-5+DART model. (c and d) Ca+b content estimated from the R_{700}/R_{750} index simulated PROSPECT-5+DART model. (©2014 IEEE. Reprinted with permission from Hernández-Clemente, R., Navarro-Cerrillo, R. M., and Zarco-Tejada, P. J., *IEEE Transactions on Geoscience and Remote Sensing*, 52(8), 5206–5217, 2014.)

8.4 ESTIMATION OF BIOPHYSICAL PARAMETERS

It has been proven that hyperspectral data are more useful than multispectral data for estimating and mapping plant physical parameters due to subtle spectral information and diagnostic spectral features of hyperspectral data. In this section, a review is given of application cases of various hyperspectral data in estimating and mapping a set of plant physical parameters, including plant canopy LAI, SLA, and crown closure (CC); plant species and composition; and plant canopy biomass, NPP and fPAR, etc. The application approaches are briefly summarized for each group of the biophysical parameters.

8.4.1 Plant Canopy LAI, SLA, and Crown Closure

Since hyperspectral data, including field hyperspectral measurements and airborne and spaceborne hyperspectral image data, were used to estimate and map plant canopy structural parameters LAI, SLA, and CC in the early 1980s, most estimate algorithms and approaches have been using HVIs and simple linear (or nonlinear) and multivariate regression modeling methods. For example, to estimate canopy LAI in a catchment afforested with *Eucalyptus*, *Pinus,* and *Acacia* genera in the KwaZulu-Natal midlands of South Africa, Bulcock and Jewitt (2010) used Hyperion hyperspectral imagery and three vegetation indices, including NDVI, soil adjusted vegetation index (SAVI), and Vogelmann Red Edge Index1 (VOG1). The three HVIs are a combination of reflectance measurements sensitive to the combined effects of foliage Chl concentration, canopy leaf area, foliage clumping, and canopy architecture. Of the three HVIs used in their study, VOG1 was the most robust index with an R^2 and RMSE values of 0.7 and 0.3, respectively. However, both NDVI and SAVI could be used to estimate the LAI of 12-year-old *Pinus patula* accurately. Viña et al. (2011) evaluated different HVIs for the remote estimation of the green LAI of two crop types (maize and soybean) with different canopy architectures and leaf structures. They tested whether the Chl indices (the Chl_{Green}, the $Chl_{Red\text{-}edge}$ and the MERIS terrestrial chlorophyll index, MTCI; see their definitions in Table 5.1) exhibit strong and significant linear relationships with Green LAI. The hyperspectral data were acquired by using field spectroradiometers (mounted 6 m above the canopy) and an aircraft-mounted hyperspectral imaging spectroradiometer (AISA). The tested results indicate that the $Chl_{Red\text{-}edge}$ was the only index insensitive to crop type and produced the most accurate estimations of Green LAI in both crops (RMSE = 0.577 m^2/m^2). Therefore, since the $Chl_{Red\text{-}edge}$ also exhibited low sensitivity to soil background effects, it can constitute a simple yet robust tool for the remote and synoptic estimation of Green LAI. To test HVIs using spaceborne Hyperion hyperspectral data for estimating Chls content and LAI in different canopy structures, Wu et al. (2010) selected wavelengths at the red edge of the vegetation spectrum (705–750 nm) from the Hyperion image to construct different HVIs. Thirty sites were selected in the study area encompassing eight different kinds of vegetation canopy, including forest and crop canopies. The adopted HVIs include SR, modified simple ratio index (mSR), NDVI, mND, TCARI, MCARI, OSAVI, TVI, MSAVI, and MCARI2 (see their definitions in Table 5.1). A linear regression approach was used to simulate relationships between different HVIs and Chls content and LAI. The results show that Chls content and LAI can be successfully estimated by HVIs derived from Hyperion data with a RMSE of 7.20–10.49 mg/cm2 for Chls content and 0.55–0.77 m^2/m^2 for LAI. The HVIs derived from three bands created the best estimation of the Chls content ($MCARI/OSAVI_{705}$) and LAI ($MCARI2_{705}$). The experimental results demonstrate the possibilities for analyzing the variation in Chls content and LAI using Hyperion hyperspectral data derived HVIs with bands from the red edge of vegetation spectrum.

In order to use empirical models to retrieve the structural variables canopy closure (CC), basal area, and timber volume at a plot scale, Fatehi et al. (2015) utilized imaging spectrometer data from the Airborne Prism EXperiment (APEX) in combination with *in situ* measurements of forest structural variables to develop the empirical models. APEX is an airborne pushbroom device with

1000 imaging pixels across track, covering spectral range of 372–2500 nm and spectral resolution from 0.7 nm to 12 nm and 285 bands. The study area is with a complex alpine forest ecosystem located in the Swiss National Park where boreal type forests are distributed among European larch (*Larix decidua L.*) as a dominant tree and with Norway spruce (*Picea abies (L.) Karst*) and Swiss stone pine (*Pinus cembra L.*) as associated species. Simple and stepwise regressions were simulated between HVIs and band depth indices as independent variables and forest structural variables (e.g., CC) as dependent variables. Fatehi et al. (2015) concluded that band-depth indices generated from continuum removed spectra in combination with simple narrowband indices could be used to predict forest structural variables, in particular for CC assessment. Using HyMap airborne hyperspectral images and both statistical and physical models, Darvishzadeh et al. (2011) investigated mapping grassland LAI in the Mediterranean grassland (Majella National Park, Italy). They compared the performance of inversion of the PROSAIL radiative transfer model with HVIs ($NDVI_{like}$ and $SAVI2_{like}$) and partial least squares (PLS) regression for mapping grassland LAI. The inversion of the PROSAIL was carried out by input of HyMap whole spectral range data or spectral subset of the data. A linear regression model was applied between measured LAI and $NDVI_{like}$ and $SAVI2_{like}$, while PLS processed HyMap whole spectral data or spectral subset of the data to produce latent factors. To assess the performance of the investigated models (inversion of PROSAIL, simple and PLS regression models), the normalized RMSE (nRMSE) and R^2 between *in situ* measurements of LAI and estimated LAI values were adopted. The results demonstrate that LAI can be estimated through PROSAIL inversion with accuracies comparable to those of statistical approaches (PLS; $R^2 = 0.89$, nRMSE = 0.22). The accuracy of the PROSAIL model inversion was further increased by using only a spectral subset of the data ($R^2 = 0.91$, nRMSE = 0.18). Similar studies on estimating plant canopy LAI and CC using HVIs and simple or multivariate regression approaches also include Boegh et al. (2002), Gong et al. (2003), He et al. (2006), Delalieux et al. (2008), etc.

A few of the studies used exponential relationship between HVI and LAI to estimate forest canopy LAI from hyperspectral data. For instance, to compare the performance of various HVIs in estimating LAI of structurally different plant species that have different soil backgrounds and leaf optical properties, Darvishzadeh et al. (2009) applied six widely used HVIs that include RVI, NDVI, PVI, TSAVI, and SAVI2 and a red-edge inflection point (REIP), which were extracted from a controlled laboratory experiment. Four different plant species with different leaf shapes and sizes were selected, which include *Asplenium nidus, Halimium umbellatum*, *Schefflera arboricola*, and *Chrysalidocarpus decipiens*. A modified version of Beer's law expressed the variation in vegetation index (VI) as a function of LAI measurements:

$$VI = VI_{\infty} + (VI_g - VI_{\infty})\exp(K_{VI} \cdot LAI),$$

where VI is the HVI, VI_{∞} is the bulk vegetation index, VI_g is the index value corresponding to bare soil, and K_{VI} parameter is equivalent to the extinction coefficient in Beer's law and characterizes the relative increase in HVI due to an increase in LAI. Their experimental results confirmed that bands from the SWIR region contain relevant information for LAI estimation. The study also verified that within the range of LAI studied ($0.3 \le LAI \le 6.1$), linear relationships exist between LAI and the selected HVIs. In an early study conducted by Gong et al. (1995), the Beer's law–based LAI estimation approach actually had been adopted in combining with univariate regression and multiple regression methods, using airborne hyperspectral sensor CASI to estimate and map coniferous forest LAI. Their experimental results indicate that all three LAI estimation methods resulted in LAIs with reasonably low RMSEs.

During the last two to three decades, a large number of studies on estimating plant canopy LAI from *in situ* or airborne/spaceborne hyperspectral data have been conducted through inversion of physically based models. Such studies include retrieval of forest LAI and CC from CASI using FLIM mode (Hu et al. 2000), inversion of leaf–canopy coupled model PROSAIL for retrieving grassland LAI and Chls contents from *in situ* hyperspectral measurements (Darvishzadeh et al. 2008b), inverting PROSPECT and SAIL models using simulated and CASI spectra for estimating crop LAI and leaf Chls content (Li et al. 2008), and mapping forest CC from Hyperion image via inversion of two canopy reflectance models: the Kuusk–Nilson forest reflectance and transmittance model and the Li–Strahler geometric–optical model (Zeng et al. 2008), etc. The modeling input could be band reflectance or wavelet transform coefficients. For example, to improve inversion efficiency of the discrete anisotropy radiative transfer (DART) model using a look-up table (LUT) approach for estimating forest LAI from hyperspectral AVIRIS image data, Banskota et al. (2015) developed a simple and computationally efficient procedure for populating an LUT database with DART simulations over a large number of spectral bands. The inversion procedure of the DART model consists of (1) building a preliminary LUT using model parameters with coarse increments to simulate reflectance for six Landsat TM bands, (2) comparing the preliminary LUT with the TM reflectance to lead to simulations close to Landsat spectra, (3) combining with a sensitivity study and building final LUTs for the full spectrum of AVIRIS narrow bands and six Landsat TM broad bands, and (4) inverting the final LUT to estimate LAI in northern temperate forests from AVIRIS and TM data. The retrieved results indicate that the proposed procedure can be a useful strategy to estimate LAI accurately by DART model inversion from hyperspectral data.

WT can decompose a spectral signal into a series of shifted and scaled versions of the mother wavelet function, and the local energy variation (represented as "peaks and valleys") of a spectral signal in different bands at each scale can be detected automatically and provide some useful information for further analysis of hyperspectral data (Pu and Gong 2004). Banskota et al. (2013) investigated the utility of the discrete wavelet transform (DWT) for estimating forest LAI via inversion of a 3D DART model using AVIRIS hyperspectral image data. They conducted an experiment in the study area comprising a range of broadleaf deciduous forest types within the state of Wisconsin. The DWT transforms the hyperspectral data into wavelet features (coefficients) at a variety of spectral scales. The multiscale features detect and isolate variation in the reflectance continuum not detectable in the original reflectance domain such as amplitude variations over broad and narrow spectral regions, which may be related to LAI magnitude (Pu and Gong 2004). Banskota et al. (2013) employed the DWT technique on reflectance spectra obtained from hyperspectral data to improve estimation of LAI in temperate forests. The DART, combining with the leaf optical model PROSPECT, was inverted with AVIRIS data using an LUT-based inversion approach for inverting to estimate forest LAI. Prior to inversion, model-simulated and AVIRIS hyperspectral spectra were transformed into discrete wavelet coefficients using *Haar* wavelets. Finally, the LUT inversion was performed with three different datasets: (1) the original reflectance bands from AVIRIS data, (2) the full set of wavelet extracted features, and (3) two wavelet subsets of extracted features. The two datasets associated with wavelet-extracted features contained 99.99% and 99.0% of the cumulative energy of the original signal. The results show that the LAI estimated from the full set of wavelet extracted features provided the greatest accuracy, which indicates that the DWT technique can increase the accuracy of LAI estimates by improving the LUT-based inversion of DART using hyperspectral data. Pu and Gong (2004) used mother wavelet *db3* in MATLAB® to transform Hyperion hyperspectral data (167 available bands in their analysis) for extracting features through a dimension reduction for mapping forest LAI and CC at Blodgett Forest Research Station, University of California at Berkeley. Compared with principal component analysis (PCA) and band selection feature extraction methods for mapping forest LAI and CC, the experimental results indicate that the energy features extracted by the WT method are the most effective for mapping forest CC and LAI.

In summary, during the last two to three decades, most application studies of hyperspectral data for estimating and mapping plant canopy LAI and CC are using simple linear (or nonlinear) or multivariate regression approach with spectral variables HVIs (mostly) and spectral derivative (e.g., Gong et al. 1995) and spectral position variable (e.g., Pu, Gong, Biging, and Larrieu 2003). Meanwhile, there are many studies on inversion of various radiative transfer models at leaf, canopy, and coupled leaf–canopy from simulated and hyperspectral image data for retrieving the plant canopy structural parameters, such as LAI and CC, etc.

8.4.2 Plant Species and Composition

The general method to discriminate and map plant species and composition is directly using hyperspectral band reflectance (either all bands or a selected subset of bands). For example, using field spectroradiometer and airborne hyperspectral reflectance spectra (AISA sensor), Mohd Hasmadi et al. (2010) and Zain et al. (2013) discriminated eight species of emergent tree crown in Gunung Stong Forest Reserve and Balah Forest Reserve in Gua Musang, Kelantan, Peninsular Malaysia, with a SAM classifier. The experimental results indicate that high spectral and spatial resolution imagery acquired over a tree canopy of tropical forest has substantial potential for individual tree species mapping.

Given the possibility of using hyperspectral data to estimate foliar and canopy chemicals, some studies on mapping tree species or groups have considered differences of biochemical between different species or species groups, and thus detection of the difference of foliar or canopy biochemicals between species is beneficial to discrimination of plant species and composition. For example, red pine and hemlock were reported to have very similar nitrogen concentration, but very different levels of lignin (Martin et al. 1998). Pu (2009) used 30 selected spectral variables evaluated by ANOVA from *in situ* hyperspectral data to identify 11 broadleaf species in an urban environment. Among the 30 selected spectral variables, most of the spectral variables are directly related to leaf chemistry. For instance, some selected spectral variables are related to water absorption bands near 0.97 μm, 1.20 μm, and 1.75 μm, and the others are related to spectral absorption features of Chls, including red edge optical parameters, simple ratio vegetation index, and reflectance at 680 nm, and the other biochemicals such as lignin (near 1.20 μm and 1.42 μm), cellulose (near 1.20 μm and 1.49 μm), and nitrogen (near 1.51 μm and 2.18 μm) concentrations (Current 1989). Zhao et al. (2016) detected the relationships among the spectral, biochemical, and taxonomic diversity of tree species based on 20 dominant canopy species collected in a subtropical forest study site in China, and found that the number of simulated maximum species, which could be identified based on the eight optimal biochemical components (Chls, Cars, SLA, equivalent water thickness, N, P, cellulose, and lignin) was approximately 15.

Many studies have demonstrated that the combination of hyperspectral data with LiDAR data can lead to improvement of discrimination of tree species and composition. Using a combination of hyperspectral (CASI-1500 imagery) and LiDAR-derived structural parameters in the form of seven predictor datasets with an automated Random Forest modeling approach, Naidoo et al. (2012) classified eight common savanna tree species in the Greater Kruger National Park region, South Africa. The important predictors (i.e., seven predictor datasets) that played an important role in the different classification models included tree height derived from LiDAR, HVIs and tree height, raw band reflectance, continuum removal transformed bands, bands used for SAM classifier, and nutrient and leaf mass–associated bands transformed or selected from CASI-1500 image data. The classification results demonstrate that the hybrid predictor dataset Random Forest model yielded the highest classification accuracy and prediction success for the eight savanna tree species with an overall classification accuracy of 87.68%, which exceeded those achieved in previous tree species classification efforts in South African savannas (Naidoo et al. 2012). Also using hyperspectral AISA imagery and LiDAR data, Zhang and Qiu (2012) developed a neural network based approach to identify urban tree species at the individual tree level. To generate a species-level map for an urban forest with high spatial heterogeneity and species diversity, they conducted a tree top–based species identification from LiDAR data, and then tree species could be recognized by analyzing the hyperspectral data associated with these tree locations. This can avoid problems of double-sided illumination, shadow, and mixed pixels, encountered in the crown-based species classification. Their experimental results indicate that the LiDAR data in conjunction with hyperspectral imagery are capable of not only detecting individual trees and estimating their tree metrics, but also identifying their species types. In addition, Matsuki et al. (2015) also proposed a method to combine CASI-3 hyperspectral image data with LiDAR data for individual tree classification. This new method was applied to classify 16 classes of tree species with hyperspectral and LiDAR data taken over Tama Forest Science Garden in Tokyo, Japan. The method procedure includes (1) removing the influence of shadows in hyperspectral data by applying an unmixing-based correction as the image preprocessing, (2) extracting spectral features of trees by running principal component analysis of the hyperspectral data, (3) deriving individual tree-crown information (sizes and shapes of trees) from the LiDAR data, and (4) classifying pixel-based tree species using SVM classifier with both spectral and tree-crown features. The analysis results indicate that both shadow correction and tree-crown information could improve the classification performance, and thus the hyperspectral data combined with LiDAR data could increase 21.5% of tree species classification accuracy compared to that obtained using hyperspectral data only.

To increase the accuracy of discriminating tree species or species groups, researchers tested several advanced classifiers, such as support vector machines (SVM), random forest (RF), etc., with hyperspectral data. For instance, in order to evaluate the potential of two high spectral and spatial resolution sensors (HySpex-VNIR 1600 and HySpex-SWIR 320i; see their characteristics in Table 3.1 and Chapter 3), Dalponte et al. (2013) considered the following four aspects: (1) three classifiers: SVM, RF, and Gaussian maximum likelihood (MLC); (2) two spatial resolutions (1.5 m and 0.4 m pixel sizes); (3) two sets of spectral bands (all bands and a selected subset of bands); and (4) two spatial levels (pixel and tree/crown levels). There were four tree species and group: Norway spruce (*Picea abies*), Scots pine (*Pinus sylvestris*), scattered Birch (*Betula spp.*), and all other broadleaves (e.g., aspen [*Populus tremula*]). The experimental results show that the HySpex VNIR 1600 sensor was effective in identifying boreal tree species with kappa accuracies over 0.8 (with pine and spruce reaching producer's accuracies higher than 95%); the HySpex-SWIR 320i was limited and was able to properly separate pine and spruce species only; and SVM or RF classifiers had a similar performance in classifying tree species/group. To classify Australian forest species at the leaf, canopy, and community levels in Beecroft Peninsula, New South Wales, Australia, Shang and Chisholm (2014) also assessed the potential of using hyperspectral data and machine-learning classification algorithms. A leaf-level test was conducted using ASD field spectroscopy based on the spectral reflectance of leaf samples collected from the field, and canopy- and community-level studies were conducted based on airborne hyperspectral sensor HyMap where spectral reflectance was derived based on image objects representing the entire (or partial) crown of individual tree or vegetation community patches in imagery. Three machine-learning algorithms, SVM, AdaBoost, and RF, were tested to classify seven native forest species over an Australian eucalyptus forest. A traditional classifier, LDA, was also adopted to compare with the performance of the three machine-learning algorithms. The tree species classification results demonstrate that all three machine-learning algorithms significantly improved the results compared to that produced by the LDA. At the leaf level, RF achieved the best classification accuracy (94.7%), while SVM outperformed the other algorithms at both the canopy (84.5%) and community levels (75.5%). This study suggests that hyperspectral remote sensing and machine-learning classifiers would have substantial potential for the mapping of Australian native forest species. However, in mapping tree species in tropical seasonal semi-deciduous forests with hyperspectral and multispectral data, Ferreira et al. (2016) concluded that the LDA classifier outperformed SVM classifier. This might be related to complexity of tropical forest canopy spectrum and structure and its biodiversity as well.

A few studies have proved that a combination of spectral features with spatial/textural features extracted from hyperspectral data can also improve the discriminant results of tree species and groups. For example, to improve tree species discriminant result with hyperspectral data, Dian et al. (2015) tested the effectiveness of combining spatial with spectral features extracted from CASI airborne hyperspectral image data in discriminating five tree species types—fir, red pine, larch, birch, and willow—in a natural reserve area in Liangshui, Heilongjiang province, Northeast China.

To do so, they (1) reduced the hyperspectral image dimensionality and highlighted variation using the minimum noise fraction (MNF) transform method, (2) extracted textural features of forest tree canopy with a gray-level co-occurrence matrix (GLCM), and (3) classified the five trees species using spectral and textural features of forest canopy and SVM classifier with different kernel functions. The tested results indicated that when using the SVM classifier, MNF and textural features combined with a linear kernel function could result in the best overall accuracy (85.92%). It suggests that combined the spectral with spatial information would improve the accuracy of tree species classification. Laurin et al. (2016) also stressed the importance of textural features and their contribution to the discrimination of tropical forest types and dominant species when using hyperspectral (AISA) data and simulated multispectral Sentinel-2 data and SVM classifier.

In short, hyperspectral data, either *in situ* spectral measurements or airborne/spaceborne hyperspectral imaging data, can be used to discriminate and map plant species and composition. To improve the discriminant results of plant species and composition, researchers have tested and developed techniques and data combination schemes, including identifying tree species based on the differences of leaf and canopy biochemicals between tree species or species groups, combining hyperspectral data with LiDAR data, using advanced classifiers, and combining spectral with spatial/textural features extracted from various hyperspectral data.

8.4.3 Plant Biomass, NPP, fPAR or fAPAR

In general, there are two types of modeling approaches that are used to estimate and map plant biomass using hyperspectral data. First, plant biomass is generally estimated using models that are statistically established in a relationship between spectral responses and field biomass measurements. Second, plant biomass is calculated using multi/hyperspectral data based allometric modeling approaches. The first modeling approaches are generated using either regression analyses or non-parametric computation approaches. The modeling parameters or coefficients are affected by various factors, such as, the atmosphere, sun-view geometry, phenological state of vegetation growth at the time of image acquisition, topography, and imperfections in radiometric calibration and geometric registration.

The first modeling approaches frequently use different hyperspectral data directly to estimate plant canopy biomass. For examples, to investigate the applicability of hyperspectral remote sensing to estimate grassland biomass at the peak of the growing season, Gao et al. (2012) measured hyperspectral data with an ASD Fieldspec3 spectroradiometer and harvested and recorded aboveground net primary productivity (ANPP) simultaneously in Hulunbeier grassland, Inner Mongolia, China. They developed linear and nonlinear regression models to estimate ANPP from the NDVI, constructed from the *in situ* hyperspectral data. Based on coefficients of determination (R^2) and error analysis, optimal regression models were determined for each vegetation type in the study area. The results indicated that a linear equation best fit the arid steppe data; an exponential equation was best suited to wetland vegetation; and power equations were optimal for meadow steppe and sand vegetation. Wang et al. (2011) also used *in situ* spectra measured by an ASD spectrometer to estimate above-ground grass biomass in the Gannan rangelands, Gansu province, China, research work similar to that of Gao et al. (2012). But they tested more spectral features and indices, including original band reflectance, multiple HVIs, and red-, yellow-, and blue-edge optical parameters (Pu et al. 2004) as predictors for developing univariate/multivariate regression models to estimate the grass biomass. The results also indicated that the grassland biomass could be estimated at the canopy level using the *in situ* hyperspectral data. To remotely estimate aboveground biomass (AGB) for a deciduous riparian forest along a river meander bend based on Hyperion hyperspectral band reflectance and other input spectral variables, Filippi et al. (2014) compared three AGB estimate approaches that included multivariate adaptive regression splines (MARS), stochastic gradient boosting (SGB), and Cubist-based AGB approaches. The riparian forest vegetation species mainly include Chinese tallow (*Sapium sebiferum*), American sycamore (*Platanus occidentalis*), hackberry

(*Celtis laevigata*) and Eastern cottonwood (*Populus deltoides*). The input spectral variables, besides band reflectance, consisted of MNF- and ICA-transformed features, HVIs and geomorphometric variables, etc. The results show that although MARS- and SGB-derived estimates are significantly more accurate than Cubist-based AGB, MARS and SGB AGB are not significantly different. Their results also suggest that the three modeling approaches would be applicable in estimating forest AGB using the hyperspectral image data similar to Hyperion sensors across scales and environmental conditions.

Many studies have demonstrated that integrating hyperspectral data with other remote sensing data, such as LiDAR and photogrammetic data, can improve plant biomass estimation. For example, to improve the estimation of forest biomass, Kattenborn et al. (2015) applied interferometric (Tandem-X) and photogrammetric (WorldView-2) based predictors in combination with hyperspectral predictors (EO1-Hyperion) by using four different machine learning algorithms for biomass estimation in temperate forest stands near Karlsruhe, Germany. The study area is representative for European temperate forests and contains both pure and mixed forests of Scots pine (*Pinus sylvestris*), European beech (*Fagus sylvatica*), sessile oak (*Quercus petraea*), red oak (*Quercus rubra*), wildcherry (*Prunus avium*), and hornbeam (*Carpinus betulus*) at different ages. Different machine-learning algorithms, including random forest (RF), generalized additive models (GAM), generalized boosted regression models (GBM), and the boosted version of the GAM (GAMB), have been applied to model complex relationships between remote sensing signals and forest biomass. An iterative model selection procedure was used to identify the optimal combination of predictors, which included selected interferometric and photogrammetric based predictors, such as canopy height models and selected hyperspectral predictors: band reflectances, HVIs, and PCs. This study suggests that a fusion of canopy height and spectral information would allow for accurate estimations of forest biomass from space. Also considering a combination of LiDAR data with hyperspectral data for forest biomass estimation, Swatantran et al. (2011) explored a fusion of structural metrics extracted from LiDAR data and spectral characteristics from hyperspectral sensor AVIRIS data for biomass estimation in the Sierra Nevada, California. LiDAR metrics, AVIRIS HVIs, and MESMA fractions extracted from AVIRIS were compared with field measures of biomass using linear and stepwise regressions at a stand (1 ha) level. AVIRIS derived spectral variables such as water band indices and shade fractions showed strong correlation with LiDAR metric canopy height. LiDAR-derived variables were found to be consistently good predictors of total and species specific biomass, which were better than those extracted from AVIRIS data. This conclusion was consistent with that drawn by Clark et al. (2011) for the estimation of AGB in a Costa Rican tropical rain forest landscape. Species-specific biomass maps and associated errors created from the fusion (Swatantran et al. 2011) were different from those produced without fusion, particularly for hardwoods and pines. The results suggest that, together, the two sensors (LiDAR and AVIRIS) would have many potential applications in carbon dynamics, ecological, and habitat studies. To assess the ability to retrieve biomass in an African tropical forest using both LiDAR and hyperspectral information (AISA sensor), Laurin et al. (2014) used the PLS regression approach to deal with multiple inputs and multicollinearity issues. The input variables include LiDAR metrics and hyperspectral band reflectance and multiple HVIs. The results indicated that if using hyperspectral data only for estimating the tropical AGB, the hyperspectral data had a limited predictive power ($R^2 = 0.36$). However, the efficiency of using PLS with both LiDAR metrics and hyperspectral data in tropical forests for biomass estimation was improved.

In considering both narrowband combinations of the SR and NDVI generated from the Airborne Prism Experiment (APEX) imaging spectroscopy data for estimating grassland and forest AGB, Fatehi et al. (2015) assessed the capability of two strategies to map grassland and forest AGB in a complex alpine ecosystem in the southeastern part of Switzerland. The two mapping strategies consist of a discrete (hard pixel based AGB estimation) and a continuous field (CF; spectral unmixing based AGB estimation) mapping approaches based on APEX hyperspectral image data. *In situ* measurements of grassland and forest AGB were taken in the study area for calibrating empirical models and for validating AGB retrievals. They found a narrowband SR, including spectral bands from the MIR (1689 nm) and NIR (851 nm) as the best regression model to estimate grassland AGB. The highest correlation with an SR for estimating forest AGB was generated from two spectral bands in the MIR (1498, 2112 nm). The results indicate that while both mapping approaches are capable of accurately mapping grassland and forest AGB in complex environments using the hyperspectral image data generally, the CF-based approach yielded higher accuracies due to its incorporating subpixel information (abundances) of different land cover types.

A few researchers used hyperspectral data, with spectral variables HVI, etc., to estimate crop biomass, such as wheat biomass. In order to ascertain the optimal methods for estimating winter wheat biomass, Fu et al. (2014) compared the abilities of univariate techniques involving HVIs and red-edge position (REP) with multivariate calibration techniques including PLS analyses using a combination of band depth parameters with HVIs and REP. The results suggest that PLS analysis, using the combination of optimal $NDVI_{like}$ and band depth parameters, could significantly improve estimation accuracy of winter wheat biomass. In estimating winter wheat biomass, Gnyp et al. (2014) also used *in situ* hyperspectral measurements and satellite hyperspectral sensor Hyperion data to implement a multiscale biomass modeling approach using HVIs for winter wheat in the North China Plain. Their results highlighted that the two HVIs, $(R_{900} \cdot R_{1050} - R_{955} \cdot R_{1220})/(R_{900} \cdot R_{1050} + R_{955} \cdot R_{1220})$ and $(R_{874} - R_{1225})/(R_{874} + R_{1225})$, extracted from *in situ* spectral measurements and Hyperion images could be used for both plot and regional level winter biomass estimation.

The second allometric modeling approaches are physically meaningful because biomass is associated with forest traits (attributes) that include LAI and canopy structure (e.g., crown closure and height). These components can be directly estimated from optical remotely sensed data. Based on literature review, most remote sensing–based allometric model approaches for estimating forest biomass are using multispectral satellite remote sensing images, such as AGB derived from MODIS land data (1 km), foliage-based generalized allometric models (Zhang and Kondragunta 2006), and forest canopy structure retrieved from MODIS and MISR data for biomass calculation (Chopping et al. 2011). Only a few application studies utilized hyperspectral remote sensing data with allometric model approaches to estimate canopy biomass. For example, focusing on estimating AGB for mixed species forests in central southeast Queensland, Australia, Lucas et al. (2008) developed and compared two different approaches for retrieval of biomass from LiDAR and CASI hyperspectral data. The first approach was developed for estimating individual tree biomass, in which stems were located using a LiDAR crown openness index, and each tree crown was delineated and identified to species using CASI data. The individual tree biomass was then estimated using species-specific allometric equations with input of LiDAR-derived height and stem diameter. The second approach was using a jackknife linear regression with six LiDAR strata heights and CASI-derived crown cover information to estimate AGB at the plot-scale that showed a closer correspondence with plot-scale ground data. The study demonstrated that it was feasible to quantify component biomass and AGB through integration of LiDAR and CASI data associated with allometric equations. In a study by combining airborne imaging spectrometer and LiDAR observations with field measurements to assess if the highly invasive nitrogen-fixing tree *Morella faya* alters canopy 3D structure and AGB along a 1500 mm precipitation gradient in Hawaii, Asner et al. (2010) also developed allometric models to relate the combined LiDAR with spectral data to field-based AGB estimates ($R^2 = 0.97$, $P < 0.01$) and to produce a map of biomass stocks throughout native and invaded ecosystems. Their results suggest that the fusion of spectral and LiDAR remote sensing could provide canopy chemical and structural data facilitating a landscape assessment of how biological invasion alters on carbon stocks and other ecosystem properties.

Gross primary production (GPP) and net primary production (NPP) are two important components of the carbon cycle. GPP is defined as the rate at which vegetation captures and stores CO_2 in a given time unit via a photosynthetic process, while NPP is defined as the difference between GPP and autotrophic respiration, and is often measured as net production or accumulation of dry matter in vegetation during a year (Roxburgh et al. 2005). Currently, remote sensing–based approaches applied to estimate GPP at ecosystem and landscape scales are using the principle of radiation-use efficiency (RUE) or light-use efficiency (LUE) with the inputs of remote sensing images and climate data. According to Xiao et al. (2014), GPP and NPP can be estimated as a product of the total amount of APAR and the LUE (i.e., ε_g for GPP and ε_n for NPP) with the formulas GPP = $\varepsilon_g \times$ APAR and NPP = $\varepsilon_n \times$ APAR. Generally, canopy fPAR can be estimated with two approaches: (1) using canopy LAI from *in situ* measurements at a site scale, and (2) using remote sensing data–derived VIs at a large scale. For estimating a large scale GPP, wide-band VIs (e.g., NDVI and SR) are frequently used to estimate canopy fPAR from, for example, MODIS and Landsat TM data. Given that HVIs can improve fPAR estimation due to selected effective narrowbands to construct the HVIs, some researchers have demonstrated this issue. However, compared to using traditional multispectral image data to estimate fPAR, there are only a few of studies on modeling fPAR using hyperspectral data. For example, Tang et al. (2010), Yang et al. (2012), Tan et al. (2013), and Marino and Alvino (2015) used *in situ* ASD spectral measurements–derived HVIs and simple linear or nonlinear models and neural networks to estimate crop (e.g., soybean, corn, and onion) fPAR. Their results indicated that SWIR bands had a great potential for the estimation of fPAR. Using simulated spectra from physically based models (e.g., canopy model FLIGHT and SAIL models), Baret and Guyot (1991) and North (2002) constructed HVIs for estimating vegetation fPAR. In addition, Strachan et al. (2008) and Yu et al. (2014) also successfully utilized hyperspectral image data derived HVIs, such as using Hyperion and CASI sensor data, to estimate and map forest and crop fPAR.

In summary, there are two modeling approaches with hyperspectral data to estimate plant canopy biomass: (1) directly estimate biomass using statistical models associated with relationships between spectral responses and field biomass measurements, and (2) based on allometric modeling approaches to estimate plant biomass. For the first approach, researchers use either different hyperspectral data only or integrate hyperspectral data with other remote sensing data, such as LiDAR and multispectral data, to extract HVIs and LiDAR metrics to estimate plant biomass. In the future, making efficient use of hyperspectral data for the AGB estimations will most probably consider the latter data integration method. The second approaches are not popularly used for estimating plant biomass with hyperspectral data. For estimating NPP and fPAR at a local or a regional scale, most approaches are using multispectral image data derived VIs and simple linear or nonlinear regression models to estimate pixel-based NPP and fPAR, and only a few cases use hyperspectral data derived HVIs to estimate NPP and fPAR.

8.5 ESTIMATION OF BIOCHEMICAL PARAMETERS

Many application studies on estimation of a set of plant biochemical parameters show that only hyperspectral remote sensing technique can be successfully used for estimating and mapping plant chemical components at a leaf or canopy level in a large area. This is determined by the characteristics of hyperspectral data (e.g., diagnostic absorption features available across a spectral reflectance curve of a green plant). This section reviews application cases of various hyperspectral data in estimating and mapping a set of plant biochemical parameters, including plant pigments; nutrient components; water status; and other biochemical factors such as lignin, cellulose, protein, etc. The application techniques and approaches are briefly summarized for each group of the biochemical parameters.

8.5.1 Plant Pigments: Chls, Cars, and Anths

To estimate plant pigments from hyperspectral data (laboratory/field based hyperspectral data and airborne/spaceborne hyperspectral image data), many studies use derivative spectra of hyperspectral data as inputs to regression models to estimate plant pigments and other biochemical parameters (e.g., Curran et al. 1997), but researchers of most application cases also consider hyperspectral vegetation indices (HVIs) constructed from hyperspectral data or spectra simulated from radiative transfer (RT) models. The constructed HVIs are then used to estimate or map pixel-based plant pigments via simple or multivariate regression models (linear or nonlinear form). For example, using

ASD field spectral measurements taken from cotton field, Yi et al. (2014) evaluated cotton Cars in three units of measurement expressed as a mass-per-unit soil surface area (g/m^2), a mass-per-unit leaf area (lg/cm^2), and a mass-per-unit fresh leaf weight (mg/g), respectively. Four methods were used and compared to retrieve amount of Cars: SMLR, existing HVIs, band combination indices, and PLS. The results show that the great sensitivity of reflectance to variation in different units of measurement of Cars was found in the green region at 515–550 nm and at 715 nm and 750 nm bands. When using the mass-per-unit soil surface area (g/m^2) as measurement unit, the best estimation results were obtained based on existing HVI $CI_{red\text{-}edge}$ (i.e., R_{750}/R_{710}). The experimental results also indicated that the Cars expressed on a concentration (mg/g) or content (lg/cm^2) basis at a leaf level could be estimated with the same prediction accuracies as the Cars expressed as a mass-per-unit surface area (g/m^2) at a canopy level using reflectance measurements taken at the canopy level. With satellite hyperspectral sensor (PROBA/CHRIS) image data, Stagakis et al. (2010) studied the potential of monitoring vegetation biophysical and biochemical characteristics through HVIs and different viewing angles for a Mediterranean ecosystem fully covered by the semi-deciduous shrub *Phlomis fruticose*. At each acquisition of image during a two-year period, coincident ecophysiological field measurements were conducted, which included LAI, leaf Chls, Cars, and leaf water potential. The reflectance spectrum of each image was used to calculate a variety of HVIs that involved some published HVIs and additional two-band $NDVI_{like}$ and simple subtraction indices (SSIs). All the HVIs along with raw reflectance and reflectance derivatives were examined using a linear relationship with the ground-measured bioparameters, and the strongest relationships were determined. It is concluded that higher observation angles are better for the extraction of biochemical parameters. HVIs that incorporate red, blue, and IR bands, such as PSRI, SIPI, and mNDVI, present good performance in Chls estimation, and also SSIs that include 701 nm with 511 or 605 nm showed best performance in Chls determination. For Cars estimation, a band on the edge of Car absorption (511 nm) combined with a red band performed best. Also using satellite Hyperion hyperspectral sensor data to estimate Chls content and LAI of eight different kinds of vegetation canopies in southeastern China, Wu et al. (2010) tested HVIs calculated from the red-edge vegetation spectrum (705 and 750 nm) of Hyperion sensor. The results show that Chls content and LAI can be successfully estimated by HVIs derived from Hyperion data. The best HVI for estimating Chls content was $MCARI/OSAVI_{705}$ and for estimating LAI was $MCARI2_{705}$. The results demonstrate the possibilities for quantifying the variation in Chls content and LAI using Hyperion data with bands from the red edge of the vegetation spectrum.

Delegido et al. (2010, 2014) proposed and tested a new spectral index, called normalized area over reflectance curve (NAOC) for remote sensing estimation of the leaf Chls content of heterogeneous areas with different crop canopies, urban vegetation, and different types of bare soil. This index is based on the calculation of the area over the high spectral resolution reflectance curve from the integral of the red–NIR interval, divided by the maximum reflectance in that spectral region. When the NAOC index was applied to PROBA/CHRIS and CASI sensor data to estimate Chls content of different crop types, similar or better results than those created using other methods were obtained. When applying the index to CASI sensor data for mapping detailed Chls content at the individual tree crowns in an urban environment, the result suggests the applicability to identify trees with lowered Chls content due to a suboptimal habitat quality.

Many application cases use simulated spectra from RT models and other *in situ* spectral measurements and airborne/spaceborne hyperspectral images to develop HVIs for estimating plant pigments. Such applications include HVIs calculated from ASD laboratory or *in situ* spectral measurements and simulated spectra via leaf-level RT model, PROSPECT, and leaf-canopy coupled model PROSAIL for developing simple or multivariate regression models to estimate Chls content of crop canopies (e.g., Clevers and Kooistra 2012, Lin et al. 2012, Yu et al. 2014) and Cars content of forest canopy (e.g., Fassnacht et al. 2015); and airborne hyperspectral data and simulated spectra via leaf–canopy coupled models, PROSAIL, PRODART, and PROSAILH for developing Chls estimate models to estimate Chls content or concentration of crop canopy (e.g., Jiao et al. 2014) and oak forests (e.g., Panigada et al. 2010) and for developing Chls and Cars models to estimate Chls and Cars contents of coniferous forests (e.g., Hernández-Clemente et al. 2014). Such spectra simulated from the RT models are ideal to test sensitivity of newly developed HVIs to variation of content or concentration of plant pigments.

The continuum removal spectra and absorption features (e.g., absorption band depth, width, position, area, and asymmetry) extracted from hyperspectral data are also frequently used as input spectral variables for estimating plant pigments. For example, to estimate needle Chls and N concentration in Norway spruce (*Picea abies L. Karst.*) needles from ASD laboratory-measured spectra and airborne HyMap hyperspectral data, Schlerf et al. (2010) systematically compared different types of spectral transformations regarding the accuracy of prediction. There were two types of spectral transformations that included continuum removal (CR) reflectance spectra and band-depth normalized spectra. The best predictive model to estimate Chls was achieved from laboratory spectra using CR transformed data and from hyperspectral HyMap data using band-depth normalized spectra. The wavebands selected into regression models for estimating Chls were typically located in the red-edge region and near the green reflectance peak. The selected wavebands with known absorption features strongly depend on the type of spectral transformation applied, such as the "water removal" approach, which can help locate wavebands directly or indirectly related to known absorption features. The derived Chls and N maps may support the detection and the monitoring of environmental stressors and are also important inputs to many biogeochemical process models. Also to estimate needle Chls content of Norway spruce (*Picea abies L. Karst.*), Malenovský et al. (2013) proposed a new Chls sensitive index located between 650 and 720 nm and termed $ANCB_{650-720}$. The $ANCB_{650-720}$ is defined as the area-under-curve of CR reflectance between 650 and 720 nm normalized by the CR band depth at 670 nm. The CR reflectance curves were calculated from the AISA Eagle airborne imaging spectrometer data and spectra simulated via the PRODART model. The performance of $ANCB_{650-720}$ was validated against ground-measured Chls of ten spruce crowns and compared with Chls estimated by a conventional ANN trained with CR spectra from the RT simulations and also with three previously published chlorophyll optical indices (NDVI between reflectance at 925 and 710 nm, SR between 750 and 710 nm, and the ratio index of TCARI/OSAVI). The ground validation revealed that the $ANCB_{650-720}$ and ANN retrievals were more accurate than the other three published chlorophyll indices. The experimental results indicate that the newly proposed index can not only provide the similar accuracy to ANN, but also computationally efficiently create accurate high spatial resolution airborne Chls maps.

Many researchers have used red-edge optical parameters, extracted directly from hyperspectral data or spectra simulated from RT models to estimate plant canopy Chls content. In many studies, researchers have used classic red-edge optical parameters (see their definitions and extraction techniques in Chapter 5) to estimate plant leaf and canopy Chls content or concentration (e.g., Curran et al. 1995, Cho et al. 2008). Recently, some researchers also developed new red-edge indices extracted from various hyperspectral data for estimating Chls content or concentration at both leaf and canopy levels. For example, to estimate the leaf Chls content of rapeseed (*Brassica napus L.*) and wheat (*Triticum aestivum L.*) crops, Ju et al. (2010) characterized the geometric patterns of the first derivative reflectance spectra in the red-edge region, calculated from ASD spectral measurements. Given that the ratio of the red-edge area less than 718 nm to the entire red-edge area was negatively correlated with leaf Chls, Ju et al. (2010) defined a new red-edge parameter, called red-edge symmetry (RES). The RES was calculated using the reflectance of red-edge boundary wavebands at 675 and 755 nm (R_{675} and R_{755}) and reflectance of red-edge center wavelength at 718 nm (R_{718}), i.e., $RES = (R_{718} - R_{675})/(R_{755} - R_{675})$. The close relationships between the simulated RES and leaf Chls indicated a high feasibility of estimating leaf Chls with simulated RES from AVIRIS and Hyperion data. The simulated results suggest that the RES be readily applicable to airborne and satellite hyperspectral data, such as AVIRIS and Hyperion sensors, as well as ground-based spectral reflectance data. Given that REP (red-edge position) values extracted by using current techniques are relatively insensitive to Chls content at high values, Dash and Curran (2004) proposed a new red-edge index, termed the MERIS terrestrial chlorophyll index (MTCI), which was designed for using spectral data recorded at the standard band settings of the MERIS sensor. MTCI is defined as a ratio of the difference in reflectance between band 10 and band 9 to the difference in reflectance between band 9 and band 8 of the MERIS standard band setting (i.e., MTCI = $(R_{band10} - R_{band9})/(R_{band9} - R_{band8}) = (R_{754} - R_{709})/(R_{709} - R_{681})$). The MTCI index was evaluated with field spectra and MERIS data. The results showed that the index was strongly correlated with REP, but unlike REP, it was sensitive to high values of Chls content of forest canopies.

In order to finally use hyperspectral image data to map plant pigments, many researchers either first use spectra simulated by various RT models to calibrate the plant pigment estimation models, then apply the calibrated models to predict pixel-based plant pigments from hyperspectral image data, or directly retrieve plant pigments from the RT models from hyperspectral image data. For examples, using leaf-level or leaf–canopy coupled RT models (PROSPECT or PRODART) to simulate spectra, Hernández-Clemente et al. (2012) and Wang and Li (2012) developed various HVIs from the simulated spectra for estimating Chls and Cars contents of forest canopies. The developed plant pigments estimate models with selected HVIs were then applied to unmanned aerial vehicle (UAV) image data or ASD *in situ* measurements to predict pixel-based or plot-based forest canopy pigments. Using spectra (e.g., band reflectance) simulated from canopy level RT (SAILH) or GO (4-S) models to directly develop multivariate regression models with measured canopy Chls and Cars contents, Zarco-Tejada et al. (2013) and Yang et al. (2015) then applied the calibrated regression models to predict pixel-based Chls and Cars contents of forest canopies from hyperspectral image data acquired by Hyperion and UAV sensors. The RT (or GO) models, either at a leaf-level or canopy-level or leaf–canopy coupled level, can be directly inverted to retrieve plant pigments. Feret et al. (2008) inverted leaf-level PROSPECTs models to retrieve plant leaf Chls, Cars, and Anths contents. Li and Wang (2013) first directly inverted two leaf-level models to retrieve Chls contents of desert plants, and then applied the calibrated retrieval model to ASD *in situ* spectral measurements to estimate plot-based Chls content of desert plant canopy. Croft et al. (2013) and Omari et al. (2013) also directly inverted 4-S and leaf-level PROSPECT and canopy PROLAIR models by using the LUT approach to retrieve forest canopy Chls content, and then applied the calibrated retrieval models to Hyperion and CASI hyperspectral images to map pixel-based forest canopy Chls content. Usually, the calibrated retrieval models for estimating plant pigments (and also other biochemicals) have a property of generalization for a type of vegetation under a certain condition, and thus such models may have more application potential to other cases outside of the specific study site compared to statistical regression methods.

In short, in order to estimate or map plant pigments from hyperspectral data, although some studies use derivative spectra of hyperspectral data as inputs to regression models to estimate plant pigments and other biochemical parameters, especially in early time of applying hyperspectral techniques in vegetation, most researchers use HVIs to estimate or map pixel-based plant pigments via simple or multivariate regression models. In addition, there are also other forms of spectral variables, extracted from hyperspectral data, which involve continuum removed spectra and absorption features and red-edge optical parameters and indices. Such forms of spectral variables are effective to estimate plant pigments. All such spectral variables and indices are used as inputs for running empirical or statistical models, and thus although such methods and techniques are easy to implement and accurate for a specific case, they have obvious disadvantages, such as being data- and study site–specific. For this case, other researchers invert physically based models to retrieve the plant pigments and apply the calibrated retrieval models to hyperspectral sensor data, especially imaging data.

8.5.2 Plant Nutrients: N, P, and K

Similar to estimating plant pigments using hyperspectral data, most application cases use various HVIs calculated from different hyperspectral data to estimate plant nutrient constitutes using univariate or multivariate regression models. To estimate leaf nitrogen concentration (LNC) in five crop species—rice (*Oryza sativa L.*), corn (*Zea mays L.*), tea (*Camellia sinensis*), gingili (*Sesamum indicum*), and soybean (*Glycine max*)—Shi et al. (2015) analyzed the spectral features, explored the spectral indices, and investigated a suitable simple modeling strategy for estimating the LNC of the five crop species with laboratory-based plant full range spectrum (0.3–2.5 μm). The study results showed that the spectral features [including two-band $NDVI_{like}$ and three-band spectral index (TBSI = $(R_i - R_j + R_l)/(R_i + R_j - R_l)$] for LNC estimation varied among different crop species. TBSI performed better than $NDVI_{like}$ in estimating LNC of crop plants. The study results also indicated that there was no common optimal TBSI and $NDVI_{like}$ for different crop species. By using ASD *in situ* spectral measurements taken from crop sugarcane canopy to estimate the canopy N concentration, Miphokasap et al. (2012) tested two-band $NDVI_{like}$ and simple ration SR_{like} indices using univariate models. They also tested the first derivative spectra using multivariate regression (MLR) models to estimate the plant N concentration. The results showed that the sensitive spectral wavelengths for quantifying the canopy N concentration existed mainly in the visible, red-edge and NIR regions of a spectrum. The results also demonstrated that the estimation model developed by MLR yielded a higher correlation coefficient with N concentration than the model computed by HVIs. Li and Alchanatis (2014) used transformed chlorophyll absorption reflectance index (TCARI) and ratio TCARI/OSAVI (the optimized soil adjusted vegetation index) extracted from AISA airborne hyperspectral image data to estimate crop/potato plant N concentration using a simple regression approach. The ratio TCARI/OSAVI resulted in stronger correlations than TCARI with leaf N concentration from canopy spectral reflectance. The results demonstrated the potential of using HVIs extracted from airborne hyperspectral images for distinguishing spatial variability in leaf N status in potato fields. Similar application studies using ASD *in situ* spectral measurements or airborne or satellite hyperspectral imaging data from crop plants or legume-based pastures were also done by Chen et al. (2010), Li et al. (2010), Kawamura et al. (2011), Tian et al. (2011), and Cilia et al. (2014).

In a way similar (i.e., using various HVIs) as that for quantifying canopy N status from ASD *in situ* spectral measurement, estimating forest canopy nutrient concentration has also been done in some studies. For instance, Cho et al. (2010) and Stein et al. (2014) used the ASD spectra-derived HVIs to estimate loblolly and eucalyptus forests' nutrient constituent, N, P, K, Ca, etc., concentrations. Both studies demonstrated the potential of hyperspectral data for quantifying the nutrient status of the forests.

There are many application cases that use derivative spectra calculated from laboratory-based spectra or airborne and satellite hyperspectral image spectra as input for performing several advanced modeling approaches, such as partial least-squares regression (PLS), ANNs, and SVMs to estimate plant canopy N and P concentration. For instances, using airborne (AVIRIS) and satellite (Hyperion) hyperspectral image data, Coops et al. (2003), Smith et al. (2003), and Townsend et al. (2003) first converted the image data into the first derivative and absorbance spectra as inputs to the PLS modeling approach, and then used the calibrated regression models to estimate forest canopy N content or concentration. The results demonstrated the potential of the satellite Hyperion and airborne AVIRIS image data used for mapping native multispecies eucalyptus forest (Coops et al. 2003), temperate forest (Smith et al. 2003), and mixed oak forest canopy N status. Later, Mitchell et al. (2012) also used airborne HyMap sensor data to calculate the first derivative spectra, and then performed PLS regression model to estimate sagebrush canopy N concentration in dryland ecosystems. Overall, their results were encouraging future landscape scale N estimates and represented a critical step in addressing the confounding influence of bare ground, which was a major influence on quantifying sagebrush canopy N status from an airborne or spaceborne platform. Other studies also used the advanced modeling approaches but other forms of spectral variables, including continuum removal spectra, band absorption features, etc., as inputs to estimate the nutrient constituents. Zhai et al. (2013) used laboratory-based spectra and PLS and SVM modeling approaches to estimate N, P, and K contents from multiple plants, including rice, corn, sesame, soybean, tea, grass, shrub, and arbor plants. The different forms of spectral variables involved in absorbance transformation (log(1/R)) include light scatter and baseline correction, detrending, wavelet, median filter, normalized spectra, and first derivatives of original reflectance (Zhai et al. 2013). The modeling results indicated that the SVM approach coupled with laboratory-based VIS–NIR reflectance data had the potential of estimating the contents of nutrient components. With different forms of spectral variables extracted from airborne hyperspectral sensors (e.g., HyMap) and satellite hyperspectral sensor Hyperion and using PLS and ANN approaches, estimations of P concentration in savanna grass (Mutanga and Kumar 2007), of N, P, and K concentration in boreal forests (Gökkaya et al. 2015), of available N status in eucalyptus trees (Youngentob et al. 2012), and of N, P, and K concentration in tropical forest canopy (Chadwick and Asner 2016) were conducted. The results derived in these studies demonstrate the potential of different types of hyperspectral data for quantifying nutrient status (especially N, P, and K) of different plant canopies.

Extracted from various hyperspectral data, red-edge optical parameters (or indices) are frequently used as input to simple or multiple regression models to estimate canopy N, P, and K, etc. nutrients. For examples, using *in situ* hyperspectral data measured from giant sequoia (*Sequoiadendron giganteum*) plantation at the Blodgett Forest Research Station of the University of California, Berkeley, with a portable spectrometer, Gong et al. (2002) evaluated multiple spectral variables (predictors) including red-edge optical parameters to quantify foliage nutrient constituents' concentration—total nitrogen (TN), total phosphorus (TP), and total potassium (TK)—of giant sequoia using univariate correlation and multivariate regression analysis methods. The spectral variables included original and the first-order derivative spectra, HVI-based, spectral position-based, area-based, and PCA-based predictors. Spectral position-based predictors consisted of parameters extracted from the blue, yellow, and red edges, the green peak and the red well, while area-based variables were calculated as the sum of the first derivative values at each of the three edges (i.e., blue-, yellow-, and red-edges). The results showed that although the spectral position-based predictors did not produce better results than other HVI-based and PCA-based predictors in estimating foliar TN, TP, and TK of giant sequoia, these spectral position predictors had a potential in assessing foliar nutrient status in the hyperspectral data. To estimate N concentration of plant leaves, Cho and Skidmore (2006) developed and tested a new technique for extracting REP parameter from hyperspectral data that aims to mitigate the discontinuity in the relationship between the REP and the N content caused by the existence of a double-peak feature on the derivative spectrum. They used far-red wavebands at 679.65 and 694.30 nm and NIR wavebands at 732.46 and 760.41 nm or at 723.64 and 760.41 nm as the optimal combinations for calculating N-sensitive REPs for three spectral data sets (rye canopy, maize leaf, and mixed grass/herb leaf stack spectra). The experimental results demonstrated that REPs extracted using this new technique showed high correlations with a wide range of foliar N concentrations for both narrow and wider bandwidth spectra. More recently, Li et al. (2014) used ASD *in situ* spectral measurements taken from the crop maize (*Zea mays L.*) canopy to extract red-edge based HVIs (including canopy chlorophyll content index [CCCI], MERIS terrestrial chlorophyll index [MTCI], normalized difference red edge [NDRE] and red-edge chlorophyll index [$CI_{red\text{-}edge}$]) with a univariate regression approach to evaluate red-edge–based HVIs for estimating plant N concentration and uptake of summer maize and to study the influence of bandwidth and crop growth stage changes on the performance of various HVIs. The analysis results showed that the four red-edge–based HVIs (i.e., CCCI, MTCI, NDRE, and $CI_{red\text{-}edge}$) performed similarly better across bandwidths for estimating plant N uptake than NDVI and SR VIs, and, specifically, that the best model with CCCI for assessing summer maize plant N concentration at the early growth stage was found since it could account for ground cover fractions.

Like retrieving plant pigments via inversion of leaf- or canopy-level physically based RT models from hyperspectral data, some researchers also invert RT models to directly assess plant canopy nutrient status. For instances, Wang et al. (2015) estimated leaf N content via relationships between the leaf N content and leaf traits (e.g., leaf chlorophyll, dry matter, and water) that could be retrieved by a physically based model (PROSPECT) inversion. Correlation analysis has shown that the area-based N (content) correlations with leaf traits were better than mass-based N (concentration) correlations. Hence, simple and MLR models were established for area-based N using three leaf traits: leaf Chl content, leaf mass per area (LMA), and equivalent water thickness (EWT). In the study, the three traits were retrieved by the inversion of PROSPECT using an iterative optimization algorithm. The established empirical models and the leaf traits retrieved by PROSPECT were used to estimate leaf N content. The experimental results indicated that a simple linear regression model with EWT as a predictor produced the most accurate estimation of N. The combination of statistical and physically based models provides a moderately accurate estimation of leaf N content. Based on the basic principle of retrieving biochemical parameters by inverting leaf-level RT model PROSPECT, more recently, Yang et al. (2015) extended the original PROSPECT model by replacing the absorption coefficient of Chl in the original PROSPECT model with an equivalent N absorption coefficient to develop an N-based PROSPECT model (N-PROSPECT) for retrieving wheat leaf N content (LNC). N-PROSPECT model was evaluated by comparing the model-simulated reflectance values with the measured leaf reflectance values under the same LNC and a high correlation coefficient ($R = 0.98$) for the spectral range from 400 to 2500 nm was found. The inversion of the N-PROSPECT model was used to retrieve LNC from the measured reflectance with a relatively high accuracy. The modeling result demonstrates that the N-PROSPECT model established in the study can accurately simulate N spectral spectra and retrieve LNC.

Generally, to assess nutrients status at plant foliar or canopy levels, hyperspectral techniques have been explored and their capability is demonstrated. Most popularly used approaches are directly using various HVIs, especially those constructed using absorption bands associated with the nutrient constituents from hyperspectral data and simple and multiple regression models to estimate nutrient content or concentration. There are many researchers who explored to use advanced modeling approaches, such as PLS, ANN, and SVMs, with multiple spectral variables as input, including raw band reflectance, CR spectra, derivative spectra and absorption band features, etc. Such modeling approaches usually result in a better estimation of plant nutrients than the other methods. In addition, red-edge optical parameters and other spectral position variables are good predictors for estimating the plant nutrients. Finally, several researchers have explored the potential of inversion of RT models (e.g., PROSPECT) for retrieving and assessing plant N status and some encouraging retrieval results have been achieved.

8.5.3 Leaf and Canopy Water Content

Compared to other plant biochemical parameters, water has several obvious absorption bands (e.g., at 0.97, 1.20, 1.40, and 1.94 μm in Figure 5.11). Based on the plant water absorption bands, researchers have developed many water sensitive HVIs, and currently the most popularly and commonly used methods and techniques for quantifying plant foliar and canopy water status are related to using various water sensitive HVIs with simple or multivariate regression modeling approaches. For example, to estimate water content of coast live oak leaves (*Quercus agrifolia*) using ASD laboratory-measured spectra, Pu, Ge et al. (2003) used two three-band ratio indices, $RATIO_{975}$ and $RATIO_{1200}$, derived at water absorption bands 975 nm and 1200 nm and water absorption features to correlate relative water content (RWC, %) of oak leaves. The results suggest the two three-band ratio indices might have potential application in assessing water status in vegetation. Using ASD *in situ* spectral measurements taken from *Sirex noctilio*, a pine tree in KwaZulu-Natal, South Africa, Mutanga and Ismail (2010) utilized selected spectral variables consisting of known water absorption features, HVIs and CR absorption features to assess canopy water status for remotely detecting and quantifying symptoms associated with declining forest health in the pine forest. The tested results showed that the variations in foliar water content across the varying levels of *S. noctilio* infestation were strongly linked to the variation in hyperspectral reflectance. Of the spectral variables tested, the water index (WI) provided the strongest linear correlation with foliar water content. Recently, Zhang et al. (2012) studied relationships of equivalent water thickness (EWT), fuel moisture content (FMC), and specific leaf weight (SLW) of cotton leaves with leaf spectra reflectance measured in a laboratory and tried to find sensitive spectral bands and best HVIs (two-band HVIs: $NDVI_{like}$ and SR_{like}) to establish quantitative models for the quick and accurate estimation of EWT, FMC, and SLW in cotton plants under different salinity levels. The sensitive spectral bands for estimating EWT, FMC, and SLW occurred mainly within the NIR and MIR ranges. The high fit results between the measured and estimated values indicated that the EWT and SLW models based on selected two-band HVIs from leaf-level reflectance could be used for the indirect estimation of plant salinity status by monitoring the changes in EWT and SLW caused by soil salinity in cotton plants. Using airborne and satellite hyperspectral image data, many researchers developed water related HVIs to assess foliar or canopy water status. For instance, Casas et al. (2014) assessed a wide range of methods to estimate foliar water content (FWC, g/cm^2), canopy water content (CWC, g/cm^2), fuel moisture content (FMC) of grass, shrub, and forests using multiple HVIs extracted from multitemporal AVIRIS data. The results indicate that standard and recently designed HVIs provided higher estimate accuracy. Dotzler et al. (2015) used water-related HVIs calculated from HySpex airborne hyperspectral image data to assess stress of a deciduous forest. The water-related HVIs included PRI, moisture stress index (MSI), NDWI, and chlorophyll index (CI). The results showed that all four HVIs revealed statistically significant differences in total Chls and water concentration at the canopy level except PRI that worked at the leaf level. Yuan et al. (2010) also used water-related HVIs calculated from satellite Hyperion image to map corn canopy water status in equivalent water thickness (EWT). The results showed that the combined spectral index, (EVI×NDVI)/MSI, has resulted in the best EWT estimation.

There are many application cases in which researchers directly use derivative spectra or water absorption features extracted from continuum-removal (CR) spectra to assess plant water status. For examples, when Al-Moustafa et al. (2012) applied airborne hyperspectral image data (AISA) for estimating live vegetation fuel moisture content (FMC) in a *Calluna vulgaris*–dominated semi-natural upland area in the United Kingdom, they tested spectral variables: the first derivatives and a number of HVIs, calculated from the AISA spectra. The results showed that live FMC exhibited spatial and temporal variations that affected the spectral reflectance measured by the airborne sensor AISA, particularly in the NIR and MIR spectral regions. Using the first derivative and specific HVI improved the estimation of live FMC, but the simple two-wavelength moisture stress index (MSI), based on measurements in the NIR and MIR regions, was effective for FMC estimation. To estimate canopy biochemical parameters in mixed forests, Huber et al. (2008) used CR spectra and absorption features derived from airborne hyperspectral sensor HyMap and MLR regression approach to estimate the foliar concentration of nitrogen, carbon, and water in three mixed forest canopies in Switzerland. The results indicated that the regression models allowed to generate regional maps of biochemical concentration, including water concentration, which might serve as an important tool for monitoring forest health and water status. Moreover, in addition to testing HVIs, Pu, Ge et al. (2003) and Mutanga and Ismail (2010) also tested and demonstrated the potential of water related absorption features from CR spectra for assessing plant water status.

Some hyperspectral data transform and processing techniques can also help improve the estimate accuracy of plant water status. For instance, Cheng et al. (2014) evaluated the application of continuous wavelet analysis (CWA) to airborne imaging spectroscopy AVIRIS data for predicting diurnal and seasonal variation in canopy water content (CWC) for nut tree orchards. They collected CWC measurements and corresponding imagery of AVIRIS twice a day (morning and afternoon) in spring and fall of 2011 in California. Several effective wavelet features were identified and their capability in assessing plant water status was compared with four water sensitive HVIs, including one optimized in this study, for assessing predictive models of CWC. The best performance using CWA associated with a combination of three wavelet features was revealed, which was much better than those by the existing water sensitive HVIs. The results demonstrated the feasibility of applying CWA to airborne imaging spectroscopy data for mapping CWC and its superiority to HVIs for improving prediction of CWC and understanding of spectral–chemical relations. To conduct stress mapping in agriculture, Lelong et al. (1998) used PCA and spectral unmixing techniques to process

MIVIS hyperspectral image data to map water deficiency of wheat in Beauce, France. The result demonstrated the potential of the approach combining of PCA with spectral unmixing for stress detection and mapping of agronomic variables, with a good accuracy compared to HVIs.

Like retrieving most other plant biochemical parameters through the inversion of leaf-, canopy-, or leaf–canopy coupled RT models, plant water information can also be retrieved by using the physically based modeling approaches directly or indirectly. To estimate leaf and canopy water content in poplar plantations, Colombo et al. (2008) investigated the applicability of HVIs including combined indices (e.g., double ratios index) using laboratory-based spectra and airborne MIVIS image data and RT model inversion with leaf and leaf–canopy coupled RT models (PROSPECT and PROSAIL). At both leaf and canopy (landscape) levels, the performance of different HVIs was tested to estimate leaf and canopy equivalent water thickness (EWT) and leaf gravimetric water content (GWC) by using regression models with different hyperspectral data including RT model simulation. The analysis results showed that leaf reflectance was related to changes in EWT rather than GWC, and the comparison results also supported the robustness of hyperspectral spectral indices (HVIs) for estimating water content at both leaf and landscape levels, with lower relative errors compared to those obtained by inversion of leaf and canopy radiative transfer models. Clevers et al. (2008) compared the use of derivative spectra, spectral indices, and continuum removal techniques for 970 nm and 1200 nm spectral regions for estimating canopy water content (CWC) of grasses. Hyperspectral reflectance data representing a range of canopies were simulated using the leaf–canopy coupled RT mode PROSPECT + SAILH. The best results in estimating CWC were obtained by using spectral derivatives at the slopes of the 970 nm and 1200 nm water absorption bands. When using real hyperspectral data of ASD *in situ* spectra and HyMap airborne image data, the best results were also resulted from the derivative spectra near the canopy water absorption feature at 970 nm. Therefore, in order to avoid the potential influence of atmospheric water vapor absorption bands, the derivative spectrum on the right slope of the canopy water absorption feature at 970 nm can best be used for estimating CWC (Clevers et al. 2008). Using simulated spectra via PROSAIL model and ASD *in situ* spectra measured from natural vegetation (pioneer vegetation, grassland, shrubs), Clevers et al. (2010) confirmed the potential of calibrating the relationship between this first derivative of reflectance on the right slope (about 1015 nm up to 1050 nm) of the 970 nm water absorption feature and plant CWC.

In short, to assess plant water status at the leaf and canopy levels using hyperspectral data, most popular methods and techniques are using spectral variables—water sensitive HVIs, derivative spectra, and CR spectra derived water absorption features—and using simple or multivariate regression approaches. To improve plant water estimate accuracy, some data transform techniques (e.g., wavelet transform and spectral unmixing) are adopted to extract effective features from hyperspectral data. RT models at the leaf or canopy or leaf-canopy level are also used to simulate high spectral resolution spectra and invert to retrieve plant water related parameters.

8.5.4 Other Plant Biochemicals: Lignin, Cellulose, and Protein

Compared to successful degree of estimating the biochemical parameters of plant pigments, water, and N, etc., the estimation of lignin, cellulose, and protein using the hyperspectral technique is relatively less successful. This may be attributed to the fact that their absorption spectra have broad and overlapping absorption features and the absorption features are relatively weak (Kokaly et al. 2009). Based on the literature review, most methods used for estimating the biochemicals lignin, cellulose, and protein from hyperspectral data are using reflectance spectra or their derivative spectra (mostly) or logarithmic spectra (a few) from *in situ* spectral measurements and airborne and satellite hyperspectral image data with multivariate linear regression approaches (MLR, mostly in early time) and partial least-squares regression approaches (PLS, most recently) to predict plant lignin, cellulose, and protein in content or concentration at leaf and canopy levels. For example, Wessman et al. (1989) used high spectral resolution AIS data acquired from more than 20 well-studied Wisconsin forest sites to evaluate the potential of remote sensing for estimating forest canopy chemistry including canopy N and lignin content and concentration. They applied stepwise regression approach with input of derivative spectra to determine combinations of wavelengths most highly correlated with canopy chemistry and biomass. The analysis results showed the strong correlations between AIS data and total canopy lignin content in deciduous forests and canopy lignin concentration (total lignin/biomass) in both deciduous and coniferous stands, which indicated that imaging spectrometry could be used to estimate canopy lignin content and, from which, the spatial distribution of annual nitrogen mineralization rates. Also using airborne hyperspectral AIS imagery and stepwise linear regression approach, Peterson et al. (1988) studied to determine optimal wavelengths for predicting biochemical composition, including lignin and starch concentration in forest canopies. The lignin and N concentration of dried and ground deciduous leaves could be predicted from reflectance spectra obtained in the laboratory. The regression analyses were based on raw and smoothed versions of the log(1/R) data and their respective first and second derivatives. The results indicated that absorption features common to the canopy spectra were between 1500 and 1700 nm, which were tentatively attributed to absorption by lignin and starch. To determine whether data from the airborne hyperspectral AVIRIS data could be used to determine forest canopy chemistry at a spatial resolution of 20 m, Martin and Aber (1997) collected foliage and leaf litter ground truthing data at Blackhawk Island, Wisconsin, and Harvard Forest, Massachusetts, to determine canopy-level N and lignin concentrations. They used the MLR approach to develop calibration equations, relating N and lignin concentration to selected first-derivative spectral bands. The calibration equations were then applied to all image pixels to predict canopy N and lignin for both study sites. The results demonstrated that the calibration equations relating AVIRIS spectral data to field-measured canopy chemistry could be used to make spatially explicit estimates of canopy N and lignin within entire AVIRIS scenes. With AVIRIS airborne image data, Curran et al. (1997) also used hyperspectral image data to map a single species pine canopy biochemical concentration and content of Chl, N, lignin, and cellulose. Up to five wavebands in the first derivative were correlated to the concentration and content of the four biochemical components with the stepwise regression modeling approach. The mapping results demonstrated the potential of AVIRIS hyperspectral image data to estimate the four canopy biochemicals.

More recently, many researchers used ASD *in situ* spectral measurements and airborne and satellite hyperspectral image data with PLS regression approaches to estimate the biochemical components. Perbandt et al. (2011) used ASD *in situ* spectra measured from maize to explore the potential of off-nadir field spectral measurements for the non-destructive prediction of dry matter yield (DM), metabolizable energy (ME), and crude protein (CP) in total biomass in the maize canopy. Calibration models were simulated with the PLS approach with Log(1/R) as a predictor spectral variable. The results showed improved prediction accuracies for DM yield and ME using off-nadir measurements, but not for CP, for which nadir measurements were better. Also using ASD *in situ* spectral measurements taken from mixed pastures, Pullanagari et al. (2012) assessed a number of pasture quality parameters including crude protein (CP), acid detergent fiber (ADF), neutral detergent fiber (NDF), lignin, etc., during the fall of 2009. PLS was used to develop the relationship between each of these pasture-quality parameters and spectral reflectance acquired in the 0.5–2.4 μm spectral range. Overall, satisfactory results were produced and high accuracy for pasture quality parameters such as CP, ADF, NDF, and lignin was achieved. The results suggest that the *in situ* canopy reflectance could be used to predict the pasture quality in a timely fashion. Roelofsen et al. (2013) also estimated the canopy concentration of lignin and cellulose in grasses using ASD *in situ* spectra measured from herbaceous plant assemblage and the PLS modeling approach. Using airborne (AVIRIS) and satellite (Hyperion) hyperspectral imagery, Vyas and Krishnayya (2014) and Singh et al. (2015) mapped forest canopy foliar concentration or content of chemicals including lignin and cellulose by calibration models. The calibration models were developed by using the PLS regression approach with all spectral band reflectance or normalized band reflectance.

The results from both studies with imaging spectroscopy demonstrated the applicability of the calibration models across similar forest cover types.

As estimating other biochemical components using the HVIs, the chemical sensitive HVIs are useful to assess the plant status of lignin, cellulose, and protein using hyperspectral data. For instance, to assess protein content in heterogeneous pastures and grain protein content in wheat, Safari et al. (2016) and Jin et al. (2014) collected *in situ* spectral measurements from grasses and a crop. They respectively calculated two-band $NDVI_{like}$ and SR_{like} HVIs and many existing N sensitive HVIs from the *in situ* spectral measurements and then developed calibration models using PLS regression approaches to estimate foliar protein content. Their study results suggest that the hyperspectral data with selected HVIs and MLS modeling technique could provide acceptable prediction accuracies for practical application. Using AVIRIS image data, Serrano et al. (2002) proposed two indices: normalized difference nitrogen index ($NDNI = (\log(1/R_{1510}) - \log(1/R_{1680}))/(\log(1/R_{1510}) + \log(1/R_{1680}))$) and normalized difference lignin index ($NDLI = (\log(1/R_{1754}) - \log(1/R_{1680}))/(\log(1/R_{1754}) + \log(1/R_{1680}))$) to assess N and lignin in native shrub vegetation. And the results showed that the two indices provided good estimates of bulk canopy N and lignin in green continuous canopies.

Only a few studies have attempted to retrieve plant lignin, cellulose, and protein by inversion of RT models and demonstrated the potential of inversion of the physically based models for retrieving the biochemicals using real hyperspectral data. For example, leaf protein and cellulose + lignin for dry leaves could be moderately well estimated through PROSPECT model inversion (e.g., Fourty et al. 1996, Jacquemoud et al. 1996). However, these parameters have not been successfully estimated for fresh leaves using the PROSPECT model (Jacquemoud et al. 1996, Kokaly et al. 2009). For this case, Wang et al. (2015) proposed a new algorithm and demonstrated the applicability of a leaf-level RT model PROSPECT in separating specific absorption coefficients for protein and cellulose + lignin. They evaluated the feasibility in estimating leaf protein and cellulose + lignin content through model inversion with the newly proposed algorithm. In order to alleviate ill-posed problems, inversion needs to be performed across different spectral subsets. The spectral subset of 2.1–2.3 μm yielded the most accurate estimation of leaf cellulose + lignin and protein in fresh leaves. The inversion results indicated that the PROSPECT model with newly calibrated specific absorption coefficients was able to accurately reconstruct leaf reflectance and transmittance, and thus the leaf protein and cellulose + lignin could be estimated at moderate to good accuracies for both fresh and dry leaves.

In general, to estimate plant leaf or canopy concentration or the content of the biochemicals lignin, cellulose, and protein, most techniques and methods use MLR or PLS regression modeling approaches with all-band reflectance, or normalized or logarithmic transforms of the band reflectance and the chemical components sensitive HVIs extracted from various hyperspectral data. Only a few cases were conducted to retrieve the plant biochemicals in both dry and fresh leaves through inversion of RT models (e.g., a PROSPECT leaf optical properties model).

8.6 SUMMARY

After brief introduction to the application of hyperspectral remote sensing to vegetation, spectral characteristics of typical green plants were introduced, including green leaf structure and plant reflectance spectral curve (with a full plant spectral range of 0.4–2.5 μm) and spectral characteristics of typical plant bioparameters. The plant reflectance spectral characteristics can be depicted by three spectral ranges: (1) visible light absorbed by multiple plant pigments, (2) NIR radiation reflected by multiscattering of internal cellular structure, and (3) MIR energy absorbed by water and other biochemical constituents. Spectral characteristics of typical plant bioparameters include biophysical parameters (e.g., plant canopy leaf area index, specific leaf area, crown closure, vegetation species and composition, biomass, effective fraction of absorbed photosynthetically active radiation [fPAR or fAPAR], and net primary productivity [NPP], which reflect photosynthesis rate) and biochemical parameters (e.g., chlorophylls [Chls], carotenoids [Cars], and anthocyanins [Anths]), nutrients (nitrogen [N], phosphorus [P], and potassium [K]), leaf or canopy water content (W), and other biochemicals (e.g., lignin, cellulose, pectin, protein, etc.). The spectral properties and characteristics of the typical plant bioparameters were summarized in Table 8.2. In Section 8.3, nine types of analytical techniques and methods with hyperspectral data specially developed for extracting and estimating plant biophysical and biochemical parameters and vegetation mapping were introduced and reviewed. An emphasis was placed on reviewing the principle and applicability of the analytical methods/techniques to vegetation rather than introducing the development of the techniques and methods (or algorithm) themselves. The nine types of techniques/methods include plant spectral derivative analysis, plant spectral absorption feature, wavelength position variable extraction and analysis, spectral vegetation index analysis, spectral unmixing and matching analyses, plant spectral classification analysis, general empirical/statistical analysis methods, physically based modeling methods, and bioparameter mapping methods. The main characteristics, advantages and disadvantages or limitations, major factors affecting detecting, estimating, and mapping results, and some application cases for the nine techniques and methods were summarized in Table 8.3. In the two remaining sections (8.4 and 8.5) of this chapter, a review of application cases of various hyperspectral data in estimating and mapping a set of plant physical parameters, including plant canopy LAI, SLA, and crown closure, plant species and composition, and plant canopy biomass, NPP and fPAR, etc., and a set of plant biochemical parameters including plant pigments, nutrient components, water status and other biochemicals lignin, cellulose, protein, etc., was provided. The application techniques and approaches were briefly summarized for each group of the biophysical and biochemical parameters.

REFERENCES

Ali, A. M., R. Darvishzadeh, A. K. Skidmore, I. V. Duren, U. Heiden, and M. Heurich. 2016. Estimating leaf functional traits by inversion of PROSPECT: Assessing leaf dry matter content and specific leaf area in mixed mountainous forest. *International Journal of Applied Earth Observation and Geoinformation* 45:66–76.

Al-Moustafa, T., R. P. Armitage, and F. M. Danson. 2012. Mapping fuel moisture content in upland vegetation using airborne hyperspectral imagery. *Remote Sensing of Environment* 127:74–83.

Andrew, M.E., and S. L. Ustin. 2010. The effects of temporally variable dispersal and landscape structure on invasive species spread. *Ecological Applications* 20(3):593–608.

Asner, G. P. 1998. Biophysical and biochemical sources of variability in canopy reflectance. *Remote Sensing of Environment* 64:134–153.

Asner, G. P., and K. B. Heidebrecht. 2003. Imaging spectroscopy for desertification studies: Comparing AVIRIS and EO-1 Hyperion in Argentina drylands. *IEEE Transactions on Geoscience and Remote Sensing* 41:1283–1296.

Asner, G. P., and R. E. Martin. 2008. Spectral and chemical analysis of tropical forests: Scaling from leaf to canopy levels. *Remote Sensing of Environment* 112:3958–3970.

Asner, G. P., R. E. Martin, D. E. Knapp, and T. Kennedy-Bowdoin. 2010. Effects of *Morella faya* tree invasion on aboveground carbon storage in Hawaii. *Biological Invasions* 12:477–494.

Asner, G. P., R. Martin, K. Carlson, U. Rascher, and P. Vitousek. 2006. Vegetation-climate interactions among native and invasive species in Hawaiian rainforest. *Ecosystems* 9:1106–1117.

Asner, G. P., M. O. Jones, R. E. Martin, D. E. Knapp, and R. F. Hughes. 2008. Remote sensing of native and invasive species in Hawaiian forests. *Remote Sensing of Environment* 112:1912–1926.

Banskota, A., R. H. Wynne, V. A. Thomas, S. P. Serbin, N. Kayastha, J. P. Gastellu-Etchegorry, and P. A. Townsend. 2013. Investigating the utility of wavelet transforms for inverting a 3-D radiative transfer model using hyperspectral data to retrieve forest LAI. *Remote Sensing* 5:2639–2659.

Banskota, A., S. P. Serbin, R. H. Wynne, V. A. Thomas, M. J. Falkowski, N. Kayastha, J.-P. Gastellu-Etchegorry, and P. A. Townsend. 2015. An LUT-based inversion of dDART model to estimate forest LAI from hyperspectral data. *Journal of Selected Topics in Applied Earth Observations and Remote Sensing* 8(6):3147–3160.

Baret, F., and G. Guyot. 1991. Potentials and limits of vegetation indices for LAI and APAR assessment. *Remote Sensing of Environment* 35:161–173.

Baret, F., S. Jacquemoud, G. Guyot, and C. Leprieur. 1992. Modeled analysis of the biophysical nature of spectral shifts and comparison with information content of broad bands. *Remote Sensing of Environment* 41:133–142.

Belanger, M. J., J. R. Miller, and M. G. Boyer. 1995. Comparative relationships between some red edge parameters and seasonal leaf chlorophyll concentrations. *Canadian Journal of Remote Sensing* 21(1):16–21.

Belward, A. S. 1991. Spectral characteristics of vegetation, soil and water in the visible, near-infrared and middle-infrared wavelengths, in *Remote Sensing and Geographical Information Systems for Resource Management in Developing Countries*, eds. A. S. Belward, A.S. and C. R. Valenzuela, pp. 31–54. The Netherlands: Kluwer Academic.

Blackburn, G. A. 1998. Quantifying chlorophylls and caroteniods at leaf and canopy scales: An evaluation of some hyperspectral approaches. *Remote Sensing of Environment* 66:273–285.

Blackburn, G. A. 2007. Hyperspectral remote sensing of plant pigments. *Journal of Experimental Botany* 58(4):855–867.

Boardman, J. W., F. A. Kruse, and R. O. Green. 1995. Mapping target signatures via partial unmixing of AVIRIS data. *Summaries of the Fifth JPL Airborne Geoscience Workshop JPL Publication*, 95–1, pp. 23–26. Pasadena, CA: NASA Jet Propulsion Laboratory.

Boegh, E., H. Soegaard, N. Broge, C. B. Hasager, N. O. Jensen, K. Schelde, and A. Thomsen. 2002. Airborne multispectral data for quantifying leaf area index, nitrogen concentration, and photosynthetic efficiency in agriculture. *Remote Sensing of Environment* 81:179–193.

Bolster, K. L., M. E. Martin, and J. D. Aber. 1996. Determination of carbon fraction and nitrogen concentration in tree foliage by near infrared reflectance: A comparison of statistical methods. *Canadian Journal of Forest Research* 26(4):590–600.

Broge, N. H., and E. Leblanc. 2000. Comparing prediction power and stability of broadband and hyperspectral vegetation indices for estimation of green leaf area index and canopy chlorophyll density. *Remote Sensing of Environment* 76:156–172.

Bulcock, H. H., and G. P. W. Jewitt. 2010. Spatial mapping of leaf area index using hyperspectral remote sensing for hydrological applications with a particular focus on canopy interception. *Hydrology and Earth System Sciences* 14:383–392.

Camps-Valls, G., L. Gómez-Chova, J. Calpe-Maravilla, J. D. Martín-Guerrero, E. Soria-Olivas, L. Alonso-Chordá, and J. Moreno. 2004. Robust support vector method for hyperspectral data classification and knowledge discovery. *IEEE Transactions on Geoscience and Remote Sensing* 42(7):1530–1542.

Card, D. H., D. L. Peterson, P. A. Matson, and J. D. Aber. 1988. Prediction of leaf chemistry by the use of visible and near infrared reflectance spectroscopy. *Remote Sensing of Environment* 26:123–147.

Casas, A., D. Riaño, S. L. Ustin, P. Dennison, and J. Salas. 2014. Estimation of water-related biochemical and biophysical vegetation properties using multitemporal airborne hyperspectral data and its comparison to MODIS spectral response. *Remote Sensing of Environment* 148:28–41.

Ceccato, P., N. Gobron, S. Flasse, B. Pinty, and S. Tarantola. 2002. Designing a spectral index to estimate vegetation water content from remote sensing data: Part 1. Theoretical approach. *Remote Sensing of Environment* 82:188–197.

Chadwick, K. D., and G. P. Asner. 2016. Organismic-scale remote sensing of canopy foliar traits in lowland tropical forests. *Remote Sensing* 8:87(1–16).

Chang, C.-I. 2000. An information theoretic-based approach to spectral variability, similarity and discriminability for hyperspectral image analysis. *IEEE Transactions on Information Theory* 46(5):1927–1932.

Chappelle, E. W., M. S. Kim, and J. E. McMurtrey, III. 1992. Ratio analysis of reflectance spectra (RARS): An algorithm for the remote estimation of the concentrations of chlorophyll a, chlorophyll b, and carotenoids in soybean leaves. *Remote Sensing of Environment* 39:239–247.

Chen, J. M., and S. G. Leblanc. 2001. Multiple-scattering scheme useful for geometric optical modeling. *IEEE Transactions on Geoscience and Remote Sensing* 39:1061–1071.

Chen, P., D. Haboudane, N. Tremblay, J. Wang, P. Vigneault, and B. Li. 2010. New spectral indicator assessing the efficiency of crop nitrogen treatment in corn and wheat. *Remote Sensing of Environment* 114:1987–1997.

Cheng, T., D. Riaño, and S. L. Ustin. 2014. Detecting diurnal and seasonal variation in canopy water content of nut tree orchards fromairborne imaging spectroscopy data using continuous wavelet analysis. *Remote Sensing of Environment* 143:39–53.

Cheng, Y.-B., P. J. Zarco-Tejada, D. Riaño, C. A. Rueda, and S. L. Ustin. 2006. Estimating vegetation water content with hyperspectral data for different canopy scenarios: Relationships between AVIRIS and MODIS indexes. *Remote Sensing of Environment* 105:354–366.

Cho, M. A., and A. K. Skidmore. 2006. A new technique for extracting the red edge position from hyperspectral data: The linear extrapolation method. *Remote Sensing of Environment* 101:181–193.

Cho, M. A., A. K. Skidmore, and C. Atzberger. 2008. Towards red-edge positions less sensitive to canopy biophysical parameters for leaf chlorophyll estimation using properties optique spectrales des feuilles (PROSPECT) and scattering by arbitrarily inclined leaves (SAILH) simulated data. *International Journal of Remote Sensing* 29(8):2241–2255.

Cho, M. A., J. van Aardt, R. Main, and B. Majeke. 2010. Evaluating variations of physiology-based hyperspectral features along a soil water gradient in a Eucalyptus grandis plantation. *International Journal of Remote Sensing* 31(12):3143–3159.

Chopping, M., C.B. Schaaf, F. Zhao, Z. Wang, A. W. Nolin, G. G. Moisen, J. V. Martonchik, and, M. Bull. 2011. Forest structure and aboveground biomass in the southwestern United States from MODIS and MISR. *Remote Sensing of Environment* 115:2943–2953.

Cilia, C., C. Panigada, M. Rossini, M. Meroni, L. Busetto, S. Amaducci, M. Boschetti, V. Picchi, and R. Colombo. 2014. Nitrogen status assessment for variable rate fertilization in maize through hyperspectral imagery. *Remote Sensing* 6:6549–6565.

Clark, M. L., D. A. Roberts, and D. B. Clark. 2005. Hyperspectral discrimination of tropical rain forest tree species at leaf to crown scales. *Remote Sensing of Environment* 96:375–398.

Clark, M. L., D. A. Roberts, J. J. Ewel, and D. B. Clark. 2011. Estimation of tropical rain forest aboveground biomass with small-footprint lidar and hyperspectral sensors. *Remote Sensing of Environment* 115:2931–2942.

Clark, R. N., and T. L. Roush. 1984. Reflectance spectroscopy: Quantitative analysis techniques for remote sensing applications. *Journal of Geophysical Research* 89:6329–6340.

Clevers, J. G. P. W, and L. Kooistra. 2012. Using hyperspectral remote sensing data for retrieving canopy chlorophyll and nitrogen content. *Journal of Selected Topics in Applied Earth Observations and Remote Sensing* 5(2):574–583.

Clevers, J. G. P. W., L. Kooistra, and M. E. Schaepman. 2008. Using spectral information from the NIR water absorption features for the retrieval of canopy water content. *International Journal of Applied Earth Observation and Geoinformation* 10:388–397.

Clevers, J. G. P. W., L. Kooistra, and M. E. Schaepman. 2010. Estimating canopy water content using hyperspectral remote sensing data. *International Journal of Applied Earth Observation and Geoinformation* 12:119–125.

Cloutis, E.A. 1996. Hyperspectral geological remote sensing: Evaluation of analytical techniques. *International Journal of Remote Sensing* 17(12):2215–2242.

Colombo, R., M. Meroni, A. Marchesi, L. Busetto, M. Rossini, C. Giardino, and C. Panigada. 2008. Estimation of leaf and canopy water content in poplar plantations by means of hyperspectral indices and inverse modeling. *Remote Sensing of Environment* 112:1820–1834.

Colvin, J. R. 1980. Biosynthesis of cellulose, in *The Biochemistry of Plants: A Comprehensive Treatise, Carbohydrates: Structure and Function*, vol. 3., ed. J. Preiss, pp. 543–570. New York: Academic Press.

Combal, B., F. Baret, and M. Weiss. 2002. Improving canopy variables estimation from remote sensing data by exploiting ancillary information. Case study on sugar beet canopies. *Agronomie* 22(2):205–215.

Coops, N. C., and C. Stone. 2005. A comparison of field-based and modelled reflectance spectra from damaged Pinus radiata foliage. *Australian Journal of Botany* 53:417–429.

Coops, N. C., M.-L. Smith, M. E. Martin, and S. V. Ollinger. 2003. Prediction of Eucalypt foliage nitrogen content from satellite-derived hyperspectral data. *IEEE Transactions on Geoscience and Remote Sensing* 41:1338–1346.

Crawford, R. L. 1981. *Lignin Biodegradation and Transformation*. New York: John Wiley & Sons Inc.

Croft, H., J. M. Chen, Y. Zhang, and A. Simic. 2013. Modelling leaf chlorophyll content in broadleaf and needle leaf canopies from ground, CASI, Landsat TM 5 and MERIS reflectance data. *Remote Sensing of Environment* 133:128–140.

Curran, P. J. 1989. Remote sensing of foliar chemistry. *Remote Sensing of Environment* 30:271–278.

Curran, P. J., J. A. Kupiec, and G. M. Smith. 1997. Remote sensing the biochemical composition of a slash pine canopy. *IEEE Transactions on Geoscience and Remote Sensing* 35(2):415–420.

Curran, P. J., W. R. Windham, and H. L. Gholz. 1995. Exploring the relationship between reflectance red edge and chlorophyll content in slash pine leaves. *Tree Physiology* 15:203–206.

Dalponte, M., H. O. Ørka, T. Gobakken, D. Gianelle, and E. Næsset. 2013. Tree species classification in boreal forests with hyperspectral data. *IEEE Transactions on Geoscience and Remote Sensing* 51(5): 2632–2645.

Darvishzadeh, R., C. Atzberger, A. K. Skidmore, and A. A. Abkar. 2009. Leaf Area Index derivation from hyperspectral vegetation indices and the red edge position. *International Journal of Remote Sensing* 30(23):6199–6218.

Darvishzadeh, R., C. Atzberger, A. Skidmore, and M. Schlerf. 2011. Mapping grassland leaf area index with airborne hyperspectral imagery: A comparison study of statistical approaches and inversion of radiative transfer models. *ISPRS Journal of Photogrammetry and Remote Sensing* 66:894–906.

Darvishzadeh, R., A. Skidmore, M. Schlerf, C. Atzberger, F. Corsi, and M. Cho. 2008a. LAI and chlorophyll estimation for a heterogeneous grassland using hyperspectral measurements. *ISPRS Journal of Photogrammetry and Remote Sensing* 63:409–426.

Darvishzadeh, R., A. Skidmore, M. Schlerf, and C. Atzberger. 2008b. Inversion of a radiative transfer model for estimating vegetation LAI and chlorophyll in a heterogeneous grassland. *Remote Sensing of Environment* 112:2592–2604.

Dash, J., and P. J. Curran. 2004. The MERIS terrestrial chlorophyll index. *International Journal of Remote Sensing* 25(23):5403–5413.

Daughtry, C. S. T., C. L. Walthall, M. S. Kim, E. B. Colstoun, and J. E. McMurtrey. 2000. Estimating corn leaf chlorophyll concentration from leaf and canopy reflectance. *Remote Sensing of Environment* 74:229–239.

Dawson T. P., and P. J. Curran. 1998. A new technique for interpolating the reflectance red edge position. *International Journal of Remote Sensing* 19:2133–2139.

Dawson, T. P., P. J. Curran, and S. E. Plummer. 1998. LIBERTY: Modeling the effects of leaf biochemical concentration on reflectance spectra. *Remote Sensing of Environment* 65:50–60.

Dawson, T. P., P. J. Curran, P. R. J. North, and S. E. Plummer. 1999. The propagation of foliar biochemical absorption features in forest canopy reflectance: A theoretical analysis. *Remote Sensing of Environment* 67:147–159.

Delalieux, S., B. Somers, S. Hereijgers, W. W. Verstraeten, W. Keulemans, and P. Coppin. 2008. A near-infrared narrow-waveband ratio to determine Leaf Area Index in orchards. *Remote Sensing of Environment* 112:3762–3772.

Delegido, J., L. Alonso, G. González, and J. Moreno. 2010. Estimating chlorophyll contentof crops from hyperspectral data using a normalized area over reflectance curve (NAOC). *International Journal Applied Earth Observation and Geoinformation* 12:165–174.

Delegido, J., S. van Wittenberghe, J. Verrelst, V. Ortiz, F. Veroustraete, R. Valcke, R. Samson, J. P. Rivera, C. Tenjoa, and J. Moreno. 2014. Chlorophyll content mapping of urban vegetation in the city of Valencia based on the hyperspectral NAOC index. *Ecological Indicators* 40:34–42.

Demarez, V., and J. P. Gastellu-Etchegorry. 2000. A modeling approach for studying forest chlorophyll content. *Remote Sensing of Environment* 71:226–238.

Demarez, V., J. P. Gastellu-Etchegorry, E. Mougin, G. Marty, C. Proisy, E. Dufrene, and V. Le Dantec. 1999. Seasonal variation of leaf chlorophyll content of a temperate forest inversion of the PROSPECT model. *International Journal of Remote Sensing* 20(5):879–894.

Dian, Y., Z. Li, and Y. Pang. 2015. Spectral and texture features combined for forest tree species classification with airborne hyperspectral imagery. *Journal of the Indian Society of Remote Sensing* 43(1):101–107.

Dong, T., J. Meng, J. Shang, J. Liu, B. Wu, and T. Huffman. 2015. Modified vegetation indices for estimating crop fraction of absorbed photosynthetically active radiation. *International Journal of Remote Sensing* 36(12):3097–3113.

Dotzler, S., J. Hill, H. Buddenbaum, and J. Stoffels. 2015. The potential of EnMAP and Sentinel-2 data for detecting drought stress phenomena in deciduous forest communities. *Remote Sensing* 7:14227–14258.

Eismann, M. 2012. *Hyperspectral Remote Sensing*. Bellingham, WA: SPIE Press.

Eitel, J. U. H., P. E. Gessler, A. M. S. Smith, and R. Robberecht. 2006. Suitability of existing and novel spectral indices to remotely detect water stress in *Populus spp. Forest Ecology and Management* 229(1–3):170–182.

Elvidge, C. D. 1987 (October 26–30). Reflectance characteristics of dry plant materials. *Proceedings of the 21st International Symposium on Remote Sensing Environment*, pp. 721–733. Ann Arbor, Michigan.

Elvidge, C. D. 1990. Visible and near infrared reflectance characteristics of dry plant materials. *International Journal of Remote Sensing* 11:1775–1795.

Farrand, W. H., and J. C. Harsanyi. 1997. Mapping the distribution of mine tailings in the Coeur d'Alene River Valley, Idaho through use of a constrained energy minimization technique. *Remote Sensing of Environment* 59:64–76.

Fassnacht, F. E., S. Stenzel, and A. A. Gitelson. 2015. Non-destructive estimation of foliar carotenoid content of treespecies using merged vegetation indices. *Journal of Plant Physiology* 176:210–217.

Fatehi, P., A. Damm, M. E. Schaepman, and M. Kneubühler. 2015. Estimation of alpine forest structural variables from imaging spectrometer data. *Remote Sensing* 7:16315–16338.

Feret, J.-B., C. François, G. P. Asner, A. A. Gitelson, R. E. Martin, L. P. R. Bidel, S. L. Ustin, G. le Maire, and S. Jacquemoud. 2008. PROSPECT-4 and 5: Advances in the leaf optical properties model separating photosynthetic pigments. *Remote Sensing of Environment* 112:3030–3043.

Ferreira, M. P., M. Zortea, D. C. Zanotta, Y. E. Shimabukuro, and C. R. de S. Filho. 2016. Mapping tree species in tropical seasonal semi-deciduous forests with hyperspectral and multispectral data. *Remote Sensing of Environment* 179:66–78.

Filippi, A. M., İ. Güneralp, and J. Randall. 2014. Hyperspectral remote sensing of aboveground biomass on a river meander bend using multivariate adaptive regression splines and stochastic gradient boosting. *Remote Sensing Letters* 5(5):432–441.

Fitzgerald, G. J., P. J. Pinter Jr., D. J. Hunsaker, and T. R. Clarke. 2005. Multiple shadow fractions in spectral mixture analysis of a cotton canopy. *Remote Sensing of Environment* 97:526–539.

Flanagan, M., and Civco, D. L. 2001 (April 23–27). Subpixel impervious surface mapping. *Proceedings of American Society for Photogrammetry and Remote Sensing Annual Convention*. St. Louis, Missouri.

Fourty, T., F. Baret, S. Jacquemoud, G. Schmuck, and J. Verdebout. 1996. Leaf optical properties with explicit description of its biochmical composition: direct and inverse problems. *Remote Sensing of Environment* 56:104–117.

Fu, Y., G. Yang, J. Wang, X. Song, and H. Feng. 2014. Winter wheat biomass estimation based on spectral indices, band depth analysis and partial least squares regression using hyperspectral measurements. *Computers and Electronics in Agriculture* 100:51–59.

Galvão, L. S., A. R. Formaggio, and D. A. Tisot. 2005. Discrimination of sugarcane varieties in Southeastern Brazil with EO-1 Hyperion data. *Remote Sensing of Environment* 94:523–534.

Gamon, J. A., and J. S. Surfus. 1999. Assessing leaf pigment content and activity with a reflectometer. *New Phytologist* 143:105–117.

Ganapol, B. D., L. F. Johnson, P. D. Hammer, C. A. Hlavka, and D. L. Peterson. 1998. LEAFMOD: A new within-leaf radiative transfer model. *Remote Sensing of Environment* 63:182–193.

Ganapol, B. D., L. F. Johnson, C. A. Hlavka, D. L. Peterson, and B. Bond. 1999. LCM2: A coupled leaf/canopy radiative transfer model. *Remote Sensing of Environment* 70:153–166.

Gao, J.-X., Y.-M. Chen, S.-H. Lü, Ch.-Y. Feng, X.-L. Chang, S.-X. Ye, and J.-D. Liu. 2012. A ground spectral model for estimating biomass at the peak of the growing season in Hulunbeier grassland, Inner Mongolia, China. *International Journal of Remote Sensing* 33(13):4029–4043.

Gastellu-Etchegorry, J. P., F. Gascon, and P. Esteve. 2003. An interpolation procedure for generalizing a look-up table inversion method. *Remote Sensing of Environment* 87(1):55–71.

Gastellu-Etchegorry, J. P., F. Zagolski, E. Mougin, G. Marty, and G. Giordano. 1995. An assessment of canopy chemistry with AVIRIS case study in the Landes Forest, South-West France. *International Journal of Remote Sensing* 16(3):487–501.

Gates, D.M. 1965. Energy, plants, and ecology. *Ecology* 46:1–13.

Gausman, H. W., W. A. Allen, R. Cardenas, and A. J. Richardson. 1970. Relationship of light reflectance to histological and physical evaluation of cotton leaf maturity. *Applied Optics* 9:545–552.

Gitelson, A. A., and M. N. Merzlyak. 1996. Signature analysis of leaf reflectance spectra: Algorithm development for remote sensing of chlorophyll. *Journal of Plant Physiology*. 148:494–500.

Gitelson, A. A., and M. N. Merzlyak. 1997. Remote estimation of chlorophyll content in higher plant leaves. *International Journal of Remote Sensing* 18:2691–2697.

Gitelson, A. A., O. B. Chivkunova, and M. N. Merzlyak. 2009. Nondestructive estimation of anthocyanins and chlorophylls in anthocyanic leaves. *American Journal of Botany* 96(10):1861–1868.

Gitelson, A. A., G. P. Keydan, and M. N. Merzlyak. 2006. Three-band model for non-invasive estimation of chlorophyll, carotenoids and anthocyanin contents in higher plant leaves. *Geophysical Research Letters* 33:L11402.

Gitelson, A. A., M. N., Merzlyak, and O. B. Chivkunova. 2001. Optical properties and nondestructive estimation of anthocyanin content in plant leaves. *Photochemistry and Photobiology* 74:38–45.

Gnyp, M. L., G. Bareth, F. Li, V. I. S. Lenz-Wiedemann, W. Koppe, Y. Miao, S. D. Hennig, L. Jia, R. Laudien, X. Chen, and F. Zhang. 2014. Development and implementation of a multiscale biomass model using hyperspectral vegetation indices for winter wheat in the North China Plain. *International Journal of Applied Earth Observation and Geoinformation* 33:232–242.

Goel, N. S., and R. L. Thompson. 1984. Inversion of vegetation canopy reflectance models for estimating agronomic variables. V. Estimation of leaf area index and average leaf inclination angle using measured canopy reflectance. *Remote Sensing of Environment* 15:69–85.

Goetz, A. F. H., G. Vane, J. E. Solomon, and B. N. Rock. 1985. Imaging spectrometry for earth remote sensing. *Science* 228(4704):1147–1153.

Gökkaya, K., V. Thomas, T. L. Noland, H. McCaughey, I. Morrison, and P. Treitz. 2015. Prediction of macronutrients at the canopy level using spaceborne imaging spectroscopy and LiDAR data in a mixedwood boreal forest. *Remote Sensing* 7:9045–9069.

Gong, P., J. R. Miller, and M. Spanner. 1994. Forest canopy closure from classification and spectral unmixing of scene components multisensor evaluation of an open canopy. *IEEE Transactions on Geoscience and Remote Sensing* 32(5):1067–1080.

Gong, P., R. Pu, and R. C. Heald. 2002. Analysis of in situ hyperspectral data for nutrient estimation of giant sequoia. *International Journal of Remote Sensing* 23(9):1827–1850.

Gong, P., R. Pu, and J. R. Miller. 1995. Coniferous forest leaf area index estimation along the Oregon transect using compact airborne spectrographic imager data. *PE & RS* 61(9):1107–1117.

Gong, P., R. Pu, and B. Yu. 1997. Conifer species recognition: An exploratory analysis on in situ hyperspectral data. *Remote Sensing of Environment* 62:189–200.

Gong, P., D. Wang, and S. Liang. 1999. Inverting a canopy reflectance model using an artificial neural network. *International Journal of Remote Sensing* 20(1):111–122.

Gong, P., R. Pu, G. S. Biging, and M. Larrieu. 2003. Estimation of forest leaf area index using vegetation indices derived from Hyperion hyperspectral data. *IEEE Transactions on Geoscience and Remote Sensing* 41(6):1355–1362.

Gower, S. T., C. J. Kucharik, and J. M. Norman. 1999. Direct and indirect estimation of leaf area index, fAPAR, and net primary production of terrestrial ecosystems. *Remote Sensing of Environment* 70:29–51.

Green, R. O, M. L. Eastwood, C. M. Sarture, T. G. Chrien, M. Aronsson, B. J. Chippendale, J. A. Faust, B. E. Pavri, C. J. Chovit, M. Solis, M. R. Olah, and O. Williams. 1998. Imaging spectroscopy and the airborne visible/infrared imaging spectrometer (AVIRIS). *Remote Sensing of Environment* 65:227–248.

Grossman, Y. L., S. L. Ustin, S. Jacquemoud, E. W. Sanderson, G. Schmuck, and J. Verdebout. 1996. Critique of stepwise multiple linear regression for the extraction of leaf biochemistry information from leaf reflectance data. *Remote Sensing of Environment* 56:182–193.

Guyot, G., F. Baret, and S. Jacquemoud. 1992. Imaging spectroscopy for vegetation studies, in *Imaging Spectroscopy: Fundamentals and Prospective Application*, eds. F. Toselli and J. Bodechtel, pp. 145–165. The Netherlands: Springer.

Haboudane, D., J. R. Miller, N. Tremblay, P. J. Zarco-Tejada, and L. Dextraze. 2002. Integrated narrow-band vegetation indices for prediction of crop chlorophyll content for application to precision agriculture. *Remote Sensing of Environment* 81(2–3):416–426.

Hamada, Y., D. A. Stow, L. L. Coulter, J. C. Jafolla, and L. W. Hendricks. 2007. Detecting Tamarisk species (Tamarix spp.) in riparian habitats of Southern California using high spatial resolution hyperspectral imagery. *Remote Sensing of Environment* 109:237–248.

Hansen, P. M., and J. K. Schjoerring. 2003. Reflectance measurement of canopy biomass and nitrogen status in wheat crops using normalized difference vegetation indices and partial least squares regression. *Remote Sensing of Environment* 86:542–553.

Hatfield, J. L., A. A. Gitelson, J. S. Schepers, and C. L. Walthall. 2008. Application of spectral remote sensing for agronomic decisions. *A Supplement to Agronomy Journal* 100:S-117–S-131.

He, Y., X. Guo, and J. Wilmshurst. 2006. Studying mixed grassland ecosystems I: Suitable hyperspectral vegetation indices. *Canadian Journal of Remote Sensing* 32(2):98–107.

Hernández-Clemente, R., R. M. Navarro-Cerrillo, and P. J. Zarco-Tejada. 2012. Carotenoid content estimation in a heterogeneous conifer forest using narrow-band indices and PROSPECT+DART simulations. *Remote Sensing of Environment* 127:298–315.

Hernández-Clemente, R., R. M. Navarro-Cerrillo, and P. J. Zarco-Tejada. 2014. Deriving predictive relationships of carotenoid content at the canopy level in a conifer forest using hyperspectral imagery and model simulation. *IEEE Transactions on Geoscience and Remote Sensing* 52(8):5206–5217.

Hestir, E. L., S. Khanna, M. E. Andrew, M. J. Santos, J. H. Viers, J. A. Greenberg, S. S. Rajapakse, and S. L. Ustin. 2008. Identification of invasive vegetation using hyperspectral remote sensing in the California Delta ecosystem. *Remote Sensing of Environment* 112:4034–4047.

Hirano, A., M. Madden, and R. Welch. 2003. Hyperspectral image data for mapping wetland vegetation. *Wetlands* 23(2):436–448.

Houborg, R., H. Soegaard, and E. Boegh. 2007. Combining vegetation index and model inversion methods for the extraction of key vegetation biophysical parameters using Terra and Aqua MODIS reflectance data. *Remote Sensing of Environment* 106(1):39–58.

Hu, B., K. Inannen, and J. R. Miller. 2000. Retrieval of leaf area index and canopy closure from CASI data over the BOREAS flux tower sites. *Remote Sensing of Environment* 74:255–274.
Huber, S., M. Kneubühler, A. Psomas, K. Itten, and N. E. Zimmermann. 2008. Estimating foliar biochemistry from hyperspectral data in mixed forest canopy. *Forest Ecology and Management* 256:491–501.
Hunt, G. R. 1980. Electromagnetic radiation: The communication link in remote sensing, in *Remote Sensing in Geology*, eds. B. Siegal and A. Gillespia, pp. 5–45. New York: Wiley.
Jacquemoud, S., and F. Baret. 1990. PROSPECT: A model of leaf optical properties spectra. *Remote Sensing of Environment* 34:75–91.
Jacquemoud, S., C. Bacour, H. Poilve, and J.-P. Frangi. 2000. Comparison of four radiative transfer models to simulate plant canopies reflectance: Direct and inverse mode. *Remote Sensing of Environment* 74(3):471–481.
Jacquemoud, S., S. L. Ustin, J. Verdebout, G. Schmuck, G. Andreoli, and B. Hosgood. 1996. Estimating leaf biochemistry using the PROSPECT leaf optical properties model. *Remote Sensing of Environment* 56:194–202.
Jacquemoud, S., W. Verhoef, F. Baret, C. Bacour, P. J. Zarco-Tejada, G. P. Asner, C. François, and S. L. Ustin. 2009. PROSPECT+SAIL models: A review of use for vegetation characterization. *Remote Sensing of Environment* 113:S56–S66.
Jensen, J. R. 2007. *Remote Sensing of the Environment: An Earth Resource Perspective*, 2nd ed. Upper Saddle River, NJ: Prentice Hall.
Jia, X., and J. A. Richards. 1999. Segmented principal components transformation for efficient hyperspectral remote-sensing image display and classification. *IEEE Transactions on Geoscience and Remote Sensing* 37:538–542.
Jiao, Q., B. Zhang, J. Liu, and L. Liu. 2014. A novel two-step method for winter wheat-leaf chlorophyll content estimation using a hyperspectral vegetation index. *International Journal of Remote Sensing* 35(21):7363–7375.
Jin, X. L., X. G. Xu, H. K. Feng, X. Y. Song, Q. Wang, J. H. Wang, and W. S. Guo. 2014. Estimation of grain protein content in winter wheat by using three methods with hyperspectral data. *International Journal of Agriculture and Biology* 16:498–504.
Johnson, L. F., C. A. Hlavka, and D. L. Peterson. 1994. Multivariate analysis of AVIRIS data for canopy biochemical estimation along the Oregon transect. *Remote Sensing of Environment* 47:216–230.
Jokela, A., T. Sarjala, S. Kaunisto, and S. Huttunen. 1997. Effects of foliar potassium concentration on morphology, ultrastraucture and polyamine concentrations of Scots pine needles. *Tree Physiology* 17:677–685.
Ju, C. H., Y. C. Tian, X. Yao, W. X. Cao, Y. Zhu, and D. Hannaway. 2010. Estimating leaf chlorophyll content using red edge parameters. *Pedosphere* 20(5):633–644.
Judd, C., S. Steinberg, F. Shaughnessy, and G. Crawford. 2007. Mapping salt marsh vegetation using aerial hyperspectral imagery and linear unmixing in Humboldt Bay, California. *Wetlands* 27(4):1144–1152.
Kattenborn, T., J. Maack, F. Faßnacht, F. Enßle, J. Ermert, and B. Koch. 2015. Mapping forest biomass from space—Fusion of hyperspectral EO1-hyperion data and Tandem-X and WorldView-2 canopy height models. *International Journal of Applied Earth Observation and Geoinformation* 35:359–367.
Kawamura, K., A. D. Mackay, M. P. Tuohy, K. Betteridge, I. D. Sanches, and Y. Inoue. 2011. Potential for spectral indices to remotely sense phosphorus and potassium content of legume-based pasture as a means of assessing soil phosphorus and potassium fertility status. *International Journal of Remote Sensing* 32(1):103–124.
Kimes, D. S., Y. Knyazikhin, J. L. Privette, A. A. Abuelgasim, and F. Gao. 2000. Inversion methods for physically-based models. *Remote Sensing Reviews* 18(2–4):381–439.
Knyazikhin, Y., J. V. Martonchik, D. Diner, R. B. Myneni, M. M. Verstraete, B. Pinty, and N. Gobron. 1998. Estimation of vegetation canopy leaf area index and fraction of absorbed photosynthetically active radiation from atmosphere-corrected MISR data. *Journal of Geophysical Research* 103(D24):32239–32256.
Koetz, B., M. Schaepman, F. Morsdorf, P. Bowyer, K. Itten, and B. Allgöwer. 2004. Radiative transfer modeling within heterogeneous canopy for estimation of forest fire fuel properties. *Remote Sensing of Environment* 92:332–344.
Kokaly, R. F. 2001. Investigating a physical basis for spectroscopic estimates of leaf nitrogen concentration. *Remote Sensing of Environment* 75:153–161.
Kokaly, R. F., G. P. Asner, S. V. Ollinger, M. E. Martin, and C. A. Wessman. 2009. Characterizing canopy biochemistry from imaging spectroscopy and its application to ecosystem studies. *Remote Sensing of Environment* 113:S78–S91.

Kong, X., N. Shu, W. Huang, and J. Fu. 2010. The research on effectiveness of spectral similarity measures for hyperspectral image. *IEEE 2010 3rd International Congress on Image and Signal Processing (CISP2010)*, pp. 2269–2273.

Kruse, F. A., A. B. Lefkoff, J. W. Boardman, K. B. Heidebrecht, A. T. Shapiro, P. J. Barloon, and A. F. H. Goetz. 1993. The spectral image processing system (SIPS)—Interactive visualization and analysis of imaging spectrometer data. *Remote Sensing of Environment* 44:145–163.

Kumar, L., K. Schmidt, S. Dury, and A. Skidmore. 2001. Imaging spectrometry and vegetation science, in *Imaging Spectrometry: Basic Principles and Prospective Applications*, eds. F. D. van der Meer and S. M. DeJong, pp. 111–155. Dordrecht: Kluwer.

Kupiec, J., G. M. Smith, and P. J. Curran. 1995 (October 25–29). AVIRIS spectra correlated with the chlorophyll concentration of a forest canopy, in *Summaries of the Fourth Annual JPL Airborne Geoscience Workshop JPL Publication*, ed. R. O. Green, 93-26, pp. 105–108. California.

Kuusk, A. 1985. The hotspot effect of a uniform vegetation cover. *Soviet Journal of Remote Sensing* 3(4):645–658.

Kuusk, A. 1991. The angular distribution of reflectance and vegetation indices in barley and clover canopies. *Remote Sensing of Environment* 37:143–151.

Kuusk, A. 1994. A multispectral canopy reflectance model. *Remote Sensing of Environment* 50:75–82.

Kuusk, A. 1995. A Markov chain model of canopy reflectance. *Agricultural and Forest Meteorology* 76:221–236.

Lass, L. W., and T. S. Prather. 2004. Detecting the locations of Brazilian pepper trees in the everglades with a hyperspectral sensor. *Weed Technology* 18:437–442.

Laurin, G. V., Q. Chen, J. A. Lindsell, D. A. Coomes, F. D. Frate, L. Guerriero, F. Pirotti, and R. Valentini. 2014. Above ground biomass estimation in an African tropical forest with LiDAR and hyperspectral data. *ISPRS Journal of Photogrammetry and Remote Sensing* 89:49–58.

Laurin, G. V., N. Puletti, W. Hawthorne, V. Liesenberg, P. Corona, D. Papale, Q. Chen, and R. Valentini. 2016. Discrimination of tropical forest types, dominant species, and mapping of functional guilds by hyperspectral and simulated multispectral Sentinel-2 data. *Remote Sensing of Environment* 176:163–176.

Le Maire, G., C. Francois, and E. Dufrene. 2004. Towards universal broad leaf chlorophyll indices using PROSPECT simulated database and hyperspectral reflectance measurements. *Remote Sensing of Environment* 89:1–28.

Le Maire, G., C. François, K. Soudani, D. Berveiller, J.-Y. Pontailler, N. Bréda, H. Genet, H. Davi, and E. Dufrêne. 2008. Calibration and validation of hyperspectral indices for the estimation of broadleaved forest leaf chlorophyll content, leaf mass per area, leaf area index and leaf canopy biomass. *Remote Sensing of Environment* 112:3846–3864.

Lelong, C. C. D., P. C. Pinet, and H. Poilvé. 1998. Hyperspectral imaging and stress mapping in agriculture: A case study on wheat in Beauce (France). *Remote Sensing of Environment* 66:179–191.

Li, F., and V. Alchanatis. 2014. The potential of airborne hyperspectral images to detect leaf nitrogen content in potato fields. *Proceedings of International Conference on Material and Environmental Engineering (ICMAEE 2014)*, pp. 103–107.

Li, F., Y. Miao, S. D. Hennig, M. L. Gnyp, X. Chen, L. Jia, and G. Bareth. 2010. Evaluating hyperspectral vegetation indices for estimating nitrogen concentration of winter wheat at different growth stages. *Precision Agriculture* 11:335–357.

Li, F., Y. Miao, G. Feng, F. Yuan, S. Yue, X. Gao, Y. Liu, B. Liu, S. L. Ustin, and X. Chen. 2014. Improving estimation of summer maize nitrogen status with red edge-based spectral vegetation indices. *Field Crops Research* 157:111–123.

Li, L., S. L. Ustin, and M. Lay. 2005. Application of multiple endmember spectral mixture analysis (MESMA) to AVIRIS imagery for coastal salt marsh mapping: A case study in China Camp, CA, USA. *International Journal of Remote Sensing* 26:5193–5207.

Li, P., and Q. Wang. 2013. Retrieval of chlorophyll for assimilating branches of a typical desert plant through inversed radiative transfer models. *International Journal of Remote Sensing* 34(7):2402–2416.

Li, Q., B. Hu, and E. Pattey. 2008. A scale-wise model inversion method to retrieve canopy biophysical parameters from hyperspectral remote sensing data. *Canadian Journal of Remote Sensing* 34(3):311–319.

Li, X., and A. H. Strahler. 1985. Geometric-optical modeling of a conifer forest canopy. *IEEE Transactions on Geoscience and Remote Sensing* GE23:705–721.

Liang, S. 2004. *Quantitative Remote Sensing of Land Surfaces*. Wiley Praxis Series in Remote Sensing. Hoboken, NJ: Wiley & Sons.

Liang, S., and A. H. Strahler. 1993. Calculation of the angular radiance distribution for a coupled atmosphere and leaf canopy. *IEEE Transactions on Geoscience and Remote Sensing* GE31:1081.

Lichtenthaler, H. K. 1987. Chlorophyll and carotenoids: Pigments of photosynthetic biomembranes. *Methods in Enzymology* 148:331–382.
Lin, P., Q. Qin, H. Dong, and Q. Meng. 2012. Hyperspectral vegetation indices for crop chlorophyll estimation: Assessment, modeling and validation. *Proceedings of IGARSS 2012*, pp. 4841–4844.
Lucas, K. L., and G. A. Carter. 2008. The use of hyperspectral remote sensing to assess vascular plant species richness on Horn Island, Mississippi. *Remote Sensing of Environment* 112:3908–3915.
Lucas, R. M., A. C. Lee, and P. J. Bunting. 2008. Retrieving forest biomass through integration of CASI and LiDAR data. *International Journal of Remote Sensing* 29(5):1553–1577.
Malenovsky, Z., L. Homolova, P. Cudlin, R. Z. Milla, M. E. Schaepman, J. G. P. W. Clevers, E. Martin, and J. P. Gastellu-Etchegory. 2007. Physically based retrievals of Norway spruce canopy variables from very high spatial resolution hyperspectral data. In *Proceedings of IEEE IGARSS2007*, vol. 1, pp. 4057–4060. Barcelona, Spain.
Malenovský, Z., L. Homolová, R. Zurita-Milla, P. Lukeš, V. Kaplan, J. Hanuš, J.-P. Gastellu-Etchegorry, and M. E. Schaepman. 2013. Retrieval of spruce leaf chlorophyll content from airborne image data using continuum removal and radiative transfer. *Remote Sensing of Environment* 131:85–102.
Marino, S., and A. Alvino. 2015. Hyperspectral vegetation indices for predicting onion (*Allium cepa L.*) yield spatial variability. *Computers and Electronics in Agriculture* 116:109–117.
Martin, M. E., and J. D. Aber. 1997. Estimation of canopy lignin and nitrogen concentration and ecosystem processes by high spectral resolution remote sensing. *Ecological Applications* 7:431–443.
Martin, M. E., L. C. Plourde, S. V. Ollinger, M.-L. Smith, B. E. McNeil. 2008. A generalizable method for remote sensing of canopy nitrogen across a wide range of forest ecosystems. *Remote Sensing of Environment* 112:3511–3519.
Martin, M. E., S. D. Newman, J. D. Aber, and R. G. Congalton. 1998. Determining forest species composition using high spectral resolution remote sensing data. *Remote Sensing of Environment* 65:249–254.
Matson, P. A., L. F. Johnson, J. R. Miller, C. R. Billow, and R. Pu. 1994. Seasonal changes in canopy chemistry across the Oregon transect: Patterns and spectral measurement with remote sensing. *Ecological Applications* 4:280–298.
Matsuki, T., N. Yokoya, and A. Iwasaki. 2015. Hyperspectral tree species classification of Japanese complex mixed forest with the aid of LiDAR data. *IEEE Journal of Selected Topics in Applied Earth Observations and Remote Sensing* 8(5):2177–2187.
Meroni, M., R. Colombo, and C. Panigada. 2004. Inversion of a radiative transfer model with hyperspectral observations for LAI mapping in poplar plantations. *Remote Sensing of Environment* 92(2):195–206.
Miao, X., P. Gong, S. Swope, R. Pu, R. Carruthers, G. L. Anderson, J. S. Heaton, and C. R. Tracy. 2006. Estimation of yellow starthistle abundance through CASI-2 hyperspectral imagery using linear spectral mixture models. *Remote Sensing of Environment* 101:329–341.
Miller, J. R., E. W. Hare, and J. Wu. 1990. Quantitative characterization of the vegetation red edge reflectance. 1. An Inverted-Gaussian reflectance model. *International Journal of Remote Sensing* 11:1775–1795.
Miller, J. R., J. Wu, M. G. Boyer, M. Belanger, and E. W. Hare. 1991. Season patterns in leaf reflectance red edge characteristics. *International Journal of Remote Sensing* 12(7):1509–1523.
Milton, N. M., B. A. Eiswerth, and C. M. Ager. 1991. Effect of phosphorus deficiency on spectral reflectance and morphology of soybean plants. *Remote Sensing of Environment* 36:121–127.
Miphokasap, P., K. Honda, C. Vaiphasa, M. Souris, and M. Nagai. 2012. Estimating canopy nitrogen concentration in sugarcane using field imaging spectroscopy. *Remote Sensing* 4:1651–1670.
Mitchell, J. J., N. F. Glenn, T. T. Sankey, D. R. Derryberry, and M. J. Germino. 2012. Remote sensing of sagebrush canopy nitrogen. *Remote Sensing of Environment* 124:217–223.
Mohd Hasmadi, I., J. Kamaruzaman, and M. Nyryl Hidayah. 2010. Analysis of crown spectral characteristic and tree species mapping of tropical forest using hyperspectral imaging. *Journal of Tropical Forest Science* 22(1):67–73.
Mutanga, O., and L. Kumar. 2007. Estimating and mapping grass phosphorus concentration in an African savanna using hyperspectral image data. *International Journal of Remote Sensing* 28(21):4897–4911.
Mutanga, O., and R. Ismail. 2010. Variation in foliar water content and hyperspectral reflectance of Pinus patula trees infested by *Sirex noctilio. Southern Forests: A Journal of Forest Science* 72(1):1–7.
Naidoo, L., M. A. Cho, R. Mathieu, and G. Asner. 2012. Classification of savanna tree species, in the Greater Kruger National Park region, by integrating hyperspectral and LiDAR data in a Random Forest data mining environment. *ISPRS Journal of Photogrammetry and Remote Sensing* 69:167–179.
Narumalani, S., D. R. Mishra, R. Wilson, P. Reece, and A. Kohler. 2009. Detecting and mapping four invasive species along the floodplain of North Platte River, Nebraska. *Weed Technology* 23:99–107.

Niinemets, U., and L. Sack. 2006. Structural determinants of leaf light harvesting capacity and photosynthetic potentials. *Progress in Botany* 67:385–419.

North, P. R. J. 2002. Estimation of f_{APAR}, LAI, and vegetation fractional cover from ATSR-2 imagery. *Remote Sensing of Environment* 80:114–121.

Noujdina, N. V., and S. L. Ustin. 2008. Mapping downy brome (Bromus tectorum) using multidate AVIRIS data. *Weed Science* 56:173–179.

Omari, K., H. P. White, K. Staenz, and D. J. King. 2013. Retrieval of forest canopy parameters by inversion of the PROFLAIR leaf-canopy reflectance model using the LUT approach. *Journal of Selected Topics in Applied Earth Observations and Remote Sensing* 6(2):715–723.

Palmer, K. F., and D. Williams. 1974. Optical properties of water in the near infrared. *Journal of the Optical Society of America* 64:1107–1110.

Panigada, C., M. Rossini, L. Busetto, M. Meroni, F. Fava, and R. Colombo. 2010. Chlorophyll concentration mapping with MIVIS data to assess crown discoloration in the Ticino Park oak forest. *International Journal of Remote Sensing* 31(12):3307–3332.

Pengra, B. W., C. A. Johnston, and T. R. Loveland. 2007. Mapping an invasive plant, *Phragmites australis*, in coastal wetlands using the EO-1 Hyperion hyperspectral sensor. *Remote Sensing of Environment* 108:74–81.

Peñuelas, J., F. Baret, and I. Filella. 1995. Semi-empirical indices to assess carotenoids/chlorophyll a ratio from leaf spectral reflectance. *Photosynthetica* 31:221–230.

Peñuelas, J., I. Filella, and L. Sweeano. 1996. Cell wall elastivity and water index (R_{970nm}/R_{900nm}) in wheat under different nitrogen availabilities. *International Journal of Remote Sensing* 17:373–382.

Peñuelas, J., I. Filella, C. Biel, L. Sweeano, and R. Save. 1993. The reflectance at the 950-970 nm region as an indicator of plant water status. *International Journal of Remote Sensing* 14:1887–1905.

Perbandt, D., T. Fricke, and M. Wachendorf. 2011. Off-nadir hyperspectral measurements in maize to predict dry matter yield, protein content and metabolisable energy in total biomass. *Precision Agric.* 12:249–265.

Peterson, D. L., J. D. Aber, D. A. Matson, D. H. Card, N. Swanberg, C. Wessman, and M. Spanner. 1988. Remote sensing of forest canopy and leaf biochemical contents. *Remote Sensing of Environment* 24:85–108.

Pieters, C. M. 1983. Strength of mineral absorption features in the transmitted component of near-infrared reflected light: First results from RELAB. *Journal of Geophysical Research* 88:9534–9544.

Pignatti, S., R. M. Cavalli, V. Cuomo, L. Fusilli, S. Pascucci, M. Poscolieri, and F. Santini. 2009. Evaluating Hyperion capability for land cover mapping in a fragmented ecosystem: Pollino National Park, Italy. *Remote Sensing of Environment* 113:622–634.

Ponzoni, F. J. and J. L. de M. Goncalves. 1999. Spectral features associated with nitrogen, phosphorus, and potassium deficiencies in Eucalyptus saligna seedling leaves. *International Journal of Remote Sensing* 20:2249–2264.

Pu, R. 2009. Broadleaf species recognition with *in situ* hyperspectral data. *International Journal of Remote Sensing* 30(11):2759–2779.

Pu, R., and P. Gong. 2004. Wavelet transform applied to EO-1 hyperspectral data for forest LAI and crown closure mapping. *Remote Sensing of Environment* 91:212–224.

Pu, R., and P. Gong. 2011. Hyperspectral remote sensing of vegetation bioparameters, in *Advances in Environmental Remote Sensing: Sensors, Algorithm, and Applications*, series ed. Q. Weng, pp. 101–142. Boca Raton, FL: CRC Press/Taylor & Francis Publishing Group.

Pu, R., and S. Landry. 2012. A comparative analysis of high resolution IKONOS and WorldView-2 imagery for mapping urban tree species. *Remote Sensing of Environment* 124:516–533.

Pu, R., and D. Liu. 2011. Segmented canonical discriminant analysis of in situ hyperspectral data for identifying thirteen urban tree species. *International Journal of Remote Sensing* 32(8):2207–2226.

Pu, R., L. Foschi, and P. Gong. 2004. Spectral feature analysis for assessment of water status and health level of coast live oak (*Quercus Agrifolia*) leaves. *International Journal of Remote Sensing* 25(20):4267–4286.

Pu, R., P. Gong, and G. S. Biging. 2003. Simple calibration of AVIRIS data and LAI mapping of forest plantation in southern Argentina. *International Journal of Remote Sensing* 24(23):4699–4714.

Pu, R., S. Ge, N. M. Kelly, and P. Gong. 2003. Spectral absorption features as indicators of water status in *Quercus Agrifolia* leaves. *International Journal of Remote Sensing* 24(9):1799–1810.

Pu, R., P. Gong, G. S. Biging, and M. R. Larrieu. 2003. Extraction of red edge optical parameters from Hyperion data for estimation of forest leaf area index. *IEEE Transactions on Geoscience and Remote Sensing* 41(4):916–921.

Pu, R., P. Gong, R. Michishita, and T. Sasagawa. 2008. Spectral mixture analysis for mapping abundance of urban surface components from the Terra/ASTER data. *Remote Sensing of Environment* 112:939–954.

Pu, R., N. M. Kelly, Q. Chen, and P. Gong. 2008. Spectroscopic determination of health levels of Coast Live Oak (*Quercus agrifolia*) Leaves. *Geocarto International* 23(1):3–20.

Pu, R., P. Gong, Y. Tian, X. Miao, and R. Carruthers. 2008. Invasive species change detection using artificial neural networks and CASI hyperspectral imagery. *Environmental Monitoring and Assessment* 140:15–32.

Pullanagari, R. R., I. J. Yule, M. P. Tuohy, M. J. Hedley, R. A. Dynes, and W. M. King. 2012. In-field hyperspectral proximal sensing for estimating quality parameters of mixed pasture. *Precision Agriculture* 13:351–369.

Rama Rao, N., P. K. Garg, S. K. Ghosh, and V. K. Dadhwal. 2008. Estimation of leaf total chlorophyll and nitrogen concentrations using hyperspectral satellite imagery. *Journal of Agricultural Science* 146:65–75.

Roberts, D. A., P. E. Dennison, M. Gardner, Y. Hetzel, S. L. Ustin, and C. Lee. 2003. Evaluation of the potential of Hyperion for fire danger assessment by comparison to the Airborne Visible/Infrared Imaging Spectrometer. *IEEE Transactions on Geoscience and Remote Sensing* 41(6):1297–1310.

Roberts, D. A., M. Gardner, R. Church, S. Ustin, G. Scheer, and R. O. Green. 1998. Mapping Chaparral in the Santa Monica Mountains using multiple endmember spectral mixture models. *Remote Sensing of Environment* 65:267–279.

Roelofsen, H. D., P. M. van Bodegom, L. Kooistra, and J.-P. M. Witte. 2013. Trait estimation in herbaceous plant assemblages from in situ canopy spectra. *Remote Sensing* 5:6323–6345.

Rosso, P. H., S. L. Ustin, and A. Hastings. 2005. Mapping marshland vegetation of San Francisco Bay, California, using hyperspectral data. *International Journal of Remote Sensing* 26:5169–5191.

Roxburgh, S. H., S. L. Berry, T. N. Buckley, B. Barnes, and M. L. Roderick. 2005. What is NPP? Inconsistent accounting of respiratory fluxes in the definition of net primary production. *Functional Ecology* 19(3):378–382.

Safari, H., T. Fricke, and M. Wachendorf. 2016. Determination of fibre and protein content in heterogeneous pastures using field spectroscopy and ultrasonic sward height measurements. *Computers and Electronics in Agriculture* 123:256–263.

Sasaki, K., S. Kawata, and S. Minami. 1984. Estimation of component spectral curves from unknown mixture spectra. *Applied Optics* 23:1955–1959.

Schlerf, M., and C. Atzberger. 2006. Inversion of a forest reflectance model to estimate structural canopy variables from hyperspectral remote sensing data. *Remote Sensing of Environment* 100:281–294.

Schlerf, M., C. Atzberger, J. Hill, H. Buddenbaum, W. Werner, and G. Schüler. 2010. Retrieval of chlorophyll and nitrogen in Norway spruce (*Picea abies L. Karst.*) using imaging spectroscopy. *International Journal of Applied Earth Observation and Geoinformation* 12:17–26.

Serrano, L., J. Peñuelas, and S. L. Ustin. 2002. Remote sensing of nitrogen and lignin in Mediterranean vegetation from AVIRIS data: Decomposing biochemical from structural signals. *Remote Sensing of Environment* 81:355–364.

Shang, X., and L. A. Chisholm. 2014. Classification of Australian native forest species using hyperspectral remote sensing and machine-learning classification algorithms. *IEEE Journal of Selected Topics in Applied Earth Observations and Remote Sensing* 7(6):2481–2489.

Shi, T., J. Wang, H. Liu, and G. Wu. 2015. Estimating leaf nitrogen concentration in heterogeneous crop plants from hyperspectral reflectance. *International Journal of Remote Sensing* 36(18):4652–4667.

Sims, D. A., and J. A. Gamon. 2002. Relationships between leaf pigment content and spectral reflectance across a wide range of species, leaf structures and developmental stages. *Remote Sensing of Environment* 81:337–354.

Singh, A., S. P. Serbin, B. E. Mcneil, C. C. Kingdon, and P. A. Townsend. 2015. Imaging spectroscopy algorithms for mapping canopy foliar chemical and morphological traits and their uncertainties. *Ecological Applications* 25(8):2180–2197.

Smith, M.-L., M. E. Martin, L. Plourde, and S. V. Ollinger. 2003. Analysis of hyperspectral data for estimation oftemperate forest canopy nitrogen concentration: Comparison between an airborne (AVIRIS) and a spaceborne (Hyperion) sensor. *IEEE Transactions on Geoscience and Remote Sensing* 41:1332–1337.

Stagakis, S., N. Markos, O. Sykioti, and A. Kyparissis. 2010. Monitoring canopy biophysical and biochemical parameters in ecosystem scale using satellite hyperspectral imagery: An application on a phlomis fruticosa Mediterranean ecosystem using multiangular CHRIS/PROBA observations. *Remote Sensing of Environment* 114:977–994.

Stein, B. R., V. A. Thomas, L. J. Lorentz, and B. D. Strahm. 2014. Predicting macronutrient concentrations from loblolly pine leaf reflectance across local and regional scales. *GIScience & Remote Sensing* 51(3):269–287.

Strachan, I. B., E. Pattey, C. Salustro, and J. R. Miller. 2008. Use of hyperspectral remote sensing to estimate the gross photosynthesis of agricultural fields. *Canadian Journal of Remote Sensing* 34(3):333–341.

Swatantran, A., R. Dubayah, D. Roberts, M. Hofton, and J. B. Blair. 2011. Mapping biomass and stress in the Sierra Nevada using lidar and hyperspectral data fusion. *Remote Sensing of Environment* 115:2917–2930.

Tan, C., A. Samanta, X. Jin, L. Tong, C. Ma, W. Guo, Y. Knyazikhin, and R. B. Myneni. 2013. Using hyperspectral vegetation indices to estimate the fraction of photosynthetically active radiation absorbed by corn canopies. *International Journal of Remote Sensing* 34(24):8789–8802.

Tang, X., K. Song, D. Liu, Z. Wang, B. Zhang, and F. Yang. 2010 (July 25–30). Application of pca and canopy near, shortwave-infrared bands for soybean and corn fPAR estimation in the Songnen Plain, China. In *Proceedings of GARSS 2010*, pp. 1485–1488. Honolulu, Hawaii.

Tian, Q., Q. Tong, R. Pu, X. Guo, and C. Zhao. 2001. Spectroscopic determination of wheat water status using 1650–1850 nm spectral absorption features. *International Journal of Remote Sensing* 22(12):2329–2338.

Tian, Y. C., X. Yao, J. Yang, W. X. Cao, D. B. Hannaway, and Y. Zhu. 2011. Assessing newly developed and published vegetation indices for estimating rice leaf nitrogen concentration with ground- and space-based hyperspectral reflectance. *Field Crops Research* 120:299–310.

Townsend, P. A., J. R. Foster, R. A. Chastain, Jr., and W. S. Currie. 2003. Application of imaging spectroscopy to mapping canopy nitrogen in the forests of central Appalachian Mountains using Hyperion and AVIRIS. *IEEE Transactions on Geoscience and Remote Sensing* 41:1347–1354.

Tucker, C. J., and M. W. Garratt. 1977. Leaf optical system modelled as a stochastic process. *Applied Optics* 16:635–642.

Underwood, E. C., S. L. Ustin, and D. DiPietro. 2003. Mapping nonnative plants using hyperspectral imagery. *Remote Sensing of Environment* 86:150–161.

Underwood, E. C., S. L. Ustin, and C. M. Ramirez. 2007. A comparison of spatial and spectral image resolution for mapping invasive plants in coastal California. *Environmental Management* 39:63–83.

Ustin, S. L., A. A. Gitelson, S. Jacquemoud, M. Schaepman, G. P. Asner, J. A. Gamon, and P. Zarco-Tejada. 2009. Retrieval of foliar information about plant pigment systems from high resolution spectroscopy. *Remote Sensing of Environment* 113:S67–S77.

Van der Meer, F. D. 2001. Basic physics of spectrometry, in *Imaging Spectrometry: Basic Principles and Prospective Applications*, eds. F. D. van der Meer and S. M. DeJong, pp. 3–16. Dordrecht: Kluwer Academic.

Van der Meer, F., and W. Bakker. 1997. Cross correlogram spectral matching (CCSM): Application to surface mineralogical mapping using AVIRIS data from Cuprite, Nevada. *Remote Sensing Environment* 61(3):371–382.

Verdebout, J., S. Jacquemoud, and G. Schmuck. 1994. Optical properties of leaves: Modelling and experimental studies, in *Imaging Spectrometry: A Tool for Environmental Observations*, eds. J. Hill and J. Mégier, pp. 169–191. Brussels and Luxembourg: ECSC, EEC, EAEC.

Verhoef, W. 1984. Light scattering by leaf layers with application to canopy reflectance modeling: The SAIL model. *Remote Sensing of Environment* 16:125–141.

Verhoef, W., and H. Bach. 2003. Simulation of hyperspectral and directional radiance images using coupled biophysical and atmospheric radiative transfer models. *Remote Sensing of Environment* 87:23–41.

Verhoef, W., and H. Bach. 2007. Coupled soil-leaf-canopy and atmosphere radiative transfer modeling to simulate hyperspectral multi-angular surface reflectance and TOA radiance data. *Remote Sensing of Environment* 109:166–182.

Viña, A., A. A. Gitelson, A. L. Nguy-Robertson, and Y. Peng. 2011. Comparison of different vegetation indices for the remote assessment of green leaf area index of crops. *Remote Sensing of Environment* 115:3468–3478.

Vyas, D., and N. S. R. Krishnayya. 2014. Estimating attributes of deciduous forest cover of a sanctuary in India utilizing Hyperion data and PLS analysis. *International Journal of Remote Sensing* 35(9):3197–3218.

Walsh, S. J., A. L. McCleary, C. F. Mena, Y. Shao, J. P. Tuttle, A. González, and R. Atkinson. 2008. QuickBird and Hyperion data analysis of an invasive plant species in the Galapagos Islands of Ecuador: Implications for control and land use management. *Remote Sensing of Environment* 112:1927–1941.

Walter-Shea, E. A., and L. L. Biehl. 1990. Measuring vegetation spectral properties. *Remote Sensing Reviews* 5(1):179–205.

Walter-Shea, E. A., J. Privette, D. Cornell, M. A. Mesarch, and C. J. Hays. 1997. Relations between directional spectral vegetation indices and leaf area and absorbed radiation in alfalfa. *Remote Sensing of Environment* 61:162–177.

Walthall, C., W. Dulaney, M. Anderson, J. Norman, H. Fang, and S. Liang. 2004. A comparison of empirical and neural network approaches for estimating corn and soybean leaf area index from Landsat ETM+ imagery. *Remote Sensing of Environment* 92(4):465–474.

Wang, C., B. Zhou, and H.L. Palm. 2008. Detecting invasive sericea lespedeza (Lespedeza cuneata) in mid-Missouri pastureland using hyperspectral imagery. *Environmental Management* 41:853–862.

Wang, Q., and P. Li. 2012. Hyperspectral indices for estimating leaf biochemical properties in temperate deciduous forests: Comparison of simulated and measured reflectance data sets. *Ecological Indicators* 14:56–65.

Wang, X., N. Guo, K. Zhang, and J. Wang. 2011. Hyperspectral remote sensing estimation models of aboveground biomass in gannan rangelands. *Procedia Environmental Sciences* 10:697–702.

Wang, Z., A. K. Skidmore, T. Wang, R. Darvishzadeh, and J. Hearne. 2015. Applicability of the PROSPECT model for estimating protein and cellulose + lignin in fresh leaves. *Remote Sensing of Environment* 168:205–218.

Wang, Z., A. K. Skidmore, R. Darvishzadeh, U. Heiden, M. Heurich, and T. Wang. 2015. Leaf nitrogen content indirectly estimated by leaf traits derived from the PROSPECT model. *Journal of Selected Topics in Applied Earth Observations and Remote Sensing* 8(6):3172–3182.

Weihs, P., F. Suppan, K. Richter, R. Petritsch, H. Hasenauer, and W. Schneider. 2008. Validation of forward and inverse modes of a homogeneous canopy reflectance model. *International Journal of Remote Sensing* 29(5):1317–1338.

Weiss, M., F. Baret, R. B. Myneni, A. Pragnere, and Y. Knyazikhin. 2000. Investigation of a model inversion technique to estimate canopy biophysical variables from spectral and directional reflectance data. *Agronomie* 20(1):3–22.

Weiss, M., D. Troufleau, F. Baret, H. Chauki, L. Prévot, A. Olioso, N. Bruguier, and N. Brisson. 2001. Coupling canopy functioning and radiative transfer models for remote sensing data assimilation. *Agricultural and Forest Meteorology* 108:113–128.

Wessman, C. A., J. D. Aber, and D. L. Peterson. 1989. An evaluation of imaging spectrometry for estimating forest canopy chemistry. *International Journal of Remote Sensing* 10:1293–1316.

Wessman, C. A., J. D. Aber, D. L. Peterson, and J. M. Melillo. 1988. Remote sensing of canopy chemistry and nitrogen cycling in temperate forest ecosystems. *Nature* 335:154–156.

Williams, P., and K. Norris, eds. 1987. *Near-Infrared Technology in the Agricultural and Food Industries.* St. Paul, MN: American Association of Cereal Chemists.

Wright, I. J., P. B. Reich, M. Westoby, D. D. Ackerly, Z. Baruch, Bongers, F. et al. 2004. The worldwide leaf economics spectrum. *Nature* 428:821–827.

Wu, C., X. Han, Z. Niu, and J. Dong. 2010. An evaluation of EO-1 hyperspectral Hyperion data for chlorophyll content and leaf area index estimation. *International Journal of Remote Sensing* 31(4):1079–1086.

Xiao, X, C. Jin, and J. Dong. 2014. Gross primary production of terrestrial vegetation, in *Biophysical Applications of Satellite Remote Sensing,* ed. J. M. Hanes, pp. 127–148. Berlin, Heidelberg: Springer Remote Sensing/Photogrammetry, Springer-Verlag.

Yang, F., Y. Zhu, J. Zhang, and Z. Yao. 2012. Estimating fraction of photosynthetically active radiation of corn with vegetation indices and neural network from hyperspectral data. *Chinese Geographical Science* 22(1):63–74.

Yang, G., C. Zhao, R. Pu, H. Feng, Z. Li, H. Li., and C. Sun. 2015. A leaf nitrogen spectral reflectance model of winter wheat (Triticum aestivum) based on PROSPECT: Simulation and inversion. *Journal of Applied Remote Sensing* 9(1):095976.

Yang, X., Y. Yu, and W. Fan. 2015. Chlorophyll content retrieval from hyperspectral remote sensing imagery. *Environmental Monitoring and Assessment* 187:456.

Yasuoka, Y., T. Yokota, T. Miyazaki, and Y. Iikura. 1990 (May 20–24). Detection of vegetation change from remotely sensed images using spectral signature similarity. *Proceedings of the International Geoscience and Remote Sensing Symposium (IGRSS'90),* College Park, Maryland, pp. 1609–1612. Piscataway, NJ: IEEE.

Yi, Q., G. Jiapaer, J. Chen, A. Bao, F. Wang. 2014. Different units of measurement of carotenoids estimation in cotton using hyperspectral indices and partial least square regression. *ISPRS Journal of Photogrammetry and Remote Sensing* 91:72–84.

Youngentob, K. N., L. J. Renzullo, A. A. Held, X. Jia, D. B. Lindenmayer, and W. J. Foley. 2012. Using imaging spectroscopy to estimate integrated measures of foliage nutritional quality. *Methods in Ecology and Evolution* 3:416–426.

Yu, K., V. Lenz-Wiedemann, X. Chen, and G. Bareth. 2014. Estimating leaf chlorophyll of barley at different growth stages using spectral indices to reduce soil background and canopy structure effects. *ISPRS Journal of Photogrammetry and Remote Sensing* 97:58–77.

Yu, Q., S. Wang, R. A. Mickler, K. Huang, L. Zhou, H. Yan, D. Chen, and S. Han. 2014. Narrowband bioindicator monitoring of temperate forest carbon fluxes in northeastern China. *Remote Sensing* 6:8986–9013.

Yuan, J., K. Sun, and Z. Niu. 2010. Vegetation water content estimation using Hyperion hyperspectral data. *The 18th International Conference on Geoinformatics: GIScience in Change, Geoinformatics 2010*. Peking University, Beijing, China.

Zain, R. M., M. H. Ismail, and P. H. Zaki. 2013. Classifying forest species using hyperspectral data in Balah Forest Reserve, Kelantan, Peninsular Malaysia. *Journal of Forest Science* 29(2):131–137.

Zarco-Tejada, P. J., J. R. Miller, A. Morales, A. Berjón, and J. Agüera. 2004. Hyperspectral indices and model simulation for chlorophyll estimation in open-canopy tree crops. *Remote Sensing of Environment* 90:463–476.

Zarco-Tejada, P. J., J. R. Miller, T. L. Noland, G. H. Mohammed, and P. H. Sampson. 2001. Scaling-up and model inversion methods with narrowband optical indices for chlorophyll content estimation in closed forest canopies with hyperspectral data. *IEEE Transactions on Geoscience and Remote Sensing* 39(7): 1491–1507.

Zarco-Tejada, P. J., M. L. Guillén-Climent, R. Hernández-Clemente, A. Catalina, M. R. González, and P. Martín. 2013. Estimating leaf carotenoid content in vineyards using high resolution hyperspectral imagery acquired from an unmanned aerial vehicle (UAV). *Agricultural and Forest Meteorology* 171–172:281–294.

Zarco-Tejada, P. J., A. Berjón, R. López-Lozano, J. R. Miller, P. Martín, V. Cachorro, M. R. González, and A. Frutos. 2005. Assessing vineyard condition with hyperspectral indices: leaf and canopy reflectance simulation in a row-structured discontinuous canopy. *Remote Sensing of Environment* 99:271–287.

Zarco-Tejada, P. J., J. R. Miller, J. Harron, B. Hu, T. L. Noland, N. Goel, G. H. Mohammed, and P. H. Sampson. 2004. Needle chlorophyll content estimation through model inversion using hyperspectral data from Boreal conifer forest canopies. *Remote Sensing of Environment* 89:1989–1999.

Zeng, Y., J. Huang, B. Wu, M. E. Schaepman, S. de Bruin, and J. G. P. W. Clevers. 2008. Comparison of the inversion of two canopy reflectance models for mapping forest crown closure using imaging spectroscopy. *Canadian Journal of Remote Sensing* 34(3):235–244.

Zhai, Y., L. Cui, X. Zhou, Y. Gao, T. Fei, and W. Gao. 2013. Estimation of nitrogen, phosphorus, and potassium contents in the leaves of different plants using laboratory-based visible and near-infrared reflectance spectroscopy: Comparison of partial least square regression and support vector machine regression methods. *International Journal of Remote Sensing* 34(7):2502–2518.

Zhang, C., and F. Qiu. 2012. Mapping individual tree species in an urban forest using airborne lidar data and hyperspectral imagery. *Photogrammetric Engineering & Remote Sensing* 78(10):1079–1087.

Zhang, L., D. Li, Q. Tong, and L. Zheng. 1998. Study of the spectral mixture model of soil and vegetation in Poyang Lake area, China. *International Journal of Remote Sensing* 19:2077–2084.

Zhang, L., Z. Zhou, G. Zhang, Y. Meng, B. Chen, and Y. Wang. 2012. Monitoring the leaf water content and specific leaf weight of cotton (*Gossypium hirsutum L.*) in saline soil using leaf spectral reflectance. *European Journal of Agronomy* 41:103–117.

Zhang, X., and S. Kondragunta. 2006. Estimating forest biomass in the USA using generalized allometric model and MODIS product data. *Geophysical Research Letters* 33:L09402.

Zhang, X., and W. Ni-meister. 2014. Remote sensing of forest biomass, in *Biophysical Applications of Satellite Remote Sensing*, ed. J. M. Hanes, pp. 62–98. New York: Springer, Heidelberg.

Zhang, Y., J. M. Chen, J. R. Miller, and T. L. Noland. 2008a. Leaf chlorophyll content retrieved from airborne hyperspectral remote sensing imagery. *Remote Sensing of Environment* 112:3234–3247.

Zhang, Y., J. M. Chen, J. R. Miller, and T. L. Noland. 2008b. Retrieving chlorophyll content in conifer needles from hyperspectral measurements. *Canadian Journal of Remote Sensing* 34(3):296–310.

Zhao, Y., Y. Zeng, D. Zhao, B. Wu, and Q. Zhao. 2016. The optimal leaf biochemical selection for mapping species diversity based on imaging spectroscopy. *Remote Sensing* 8(216):1–16

Zheng, G., and L. M. Moskal. 2009. Retrieving leaf area index (LAI) using remote sensing: Theories, methods and sensors. *Sensors* 9:2719–2745.

Zou, X., R. Hernández-Clemente, P. Tammeorg, C. L. Torres, F. L. Stoddard, P. Mäkelä, P. Pellikka, and M. Mõttus. 2015. Retrieval of leaf chlorophyll content in field crops using narrow-band indices: Effects of leaf area index and leaf mean tilt angle. *International Journal of Remote Sensing* 36(24):6031–6055.

9 Hyperspectral Applications to Environments

In addition to introductions and reviews of hyperspectral remote sensing (HRS) applied to geology, soil, and vegetation areas in Chapters 7 and 8, HRS can also be applied to other environments and disciplines, such as atmospheric science and urban environment, etc. Hence, in this chapter, an overview of application studies of various hyperspectral data sets to other environments and disciplines is presented. Specifically, they include atmosphere, snow and ice hydrology, coastal environments and inland waters, environmental hazards and disasters, and urban environments.

9.1 INTRODUCTION

As characteristics and number of hyperspectral sensor systems are improved and increased and hyperspectral imaging data sets are understood, a wide range of science research and applications based on spectral imagery recorded by various imaging sensors/systems has taken place during the last three decades. Although HRS is most suitable for research and applications in geological, soil sciences, vegetation, and ecosystem disciplines, based on the nature and characteristics of HRS, HRS has also been used widely in other disciplines and environments, including atmospheric science, snow and ice hydrology, coastal zone and oceanography, inland waters, hazard and disaster monitoring and assessment, and urban environmental studies. Therefore, in addition to introductions to and overviews of HRS research and its applications to geological and soil sciences and vegetation disciplines, an overview of HRS to several other key environments and disciplines is provided in the following. Specifically, in Section 9.2, estimating and mapping techniques/methods and data for key atmospheric parameters including water vapor, clouds, aerosols, and carbon dioxide are reviewed and discussed. Spectral characteristics of snow and ice associated with their hydrological issues are assessed and discussed from various hyperspectral data sets in Section 9.3. Using hyperspectral data, a list of techniques and approaches for extracting and mapping water quality parameters and other components in coastal environments and inland waters are summarized and discussed in Section 9.4. In Section 9.5, HRS studies on mapping and monitoring environmental hazards and disasters, including mining wasters, biomass burning, and landslides are reviewed. In the last section, a review of HRS applied in urban environments, including spectra properties of urban materials and their identification, mapping, and urban thermal environmental characterizing, is presented. In the following review, most studies on HRS in different environmental and discipline areas are discussed in terms of research questions/objectives, principles/mechanisms of methodology/procedures, nature of HRS data sets, key findings and conclusions derived from studies, and any relevant comments.

9.2 ESTIMATION OF ATMOSPHERIC PARAMETERS

Due to scattering and absorption of molecular, gases, and particle constituents, the atmosphere has profound influences on the spectral signal recorded by airborne or spaceborne imaging systems operating in the solar-reflected portion of the spectrum (Vane and Goetz 1993). Many different gases and particles in the atmosphere can absorb and transmit different wavelengths of electromagnetic radiation that passes the atmosphere. Eight major gases, namely water vapor (H_2O), carbon dioxide (CO_2), ozone (O_3), nitrous oxide (N_2O), carbon monoxide (CO), methane (CH_4), oxygen (O_2), and nitrogen dioxide (NO_2) can cause observable absorption features in imaging spectrometer data over the spectral range from 0.4 to 5.0 μm with a spectral resolution between 1 and 20 nm (Gao et al. 2009; Table 9.1). Figure 9.1 presents transmittance spectra of the major atmospheric gases over the spectral range extending to thermal spectral range. From the figure, it is clear that, in the visible range, the absorption caused by H_2O and other gases is very weak and almost transparent in the range, and significant absorption features caused by the major gases are located in the short-wave infrared (SWIR) range. As discussed in Chapter 4, in the SWIR range the absorption substance H_2O has much variability over space and time when compared with other gases (i.e., their contents are relatively stable) in the atmosphere. Nowadays, there are more studies on retrieving atmospheric column water vapor (CWV), cloud, carbon dioxide (CO_2), and aerosol from hyperspectral image data than those on retrieving other gases. Therefore, in the following, estimating the four atmospheric

TABLE 9.1
Main Visible and Short-Wave Infrared Absorption Bands of Atmospheric Gases

Gas	Center Wavelength (mm)	Band Interval (mm)
H_2O	2.70	4.00–2.22
	1.88	2.08–1.61
	1.38	1.56–1.32
	1.14	1.22–1.06
	0.94	0.99–0.88
	0.82	0.85–0.79
	0.72	0.75–0.68
	Visible	0.67–0.44
CO_2	4.30	5.00–4.17
	2.70	2.94–2.60
	2.00	2.13–1.92
	1.60	1.64–1.55
	1.40	1.46–1.43
O_3	4.74	5.00–4.35
	3.30	3.33–3.23
	Visible	0.94–0.44
O_2	1.58	1.59–1.57
	1.27	1.30–1.24
	1.06	1.07–1.06
	0.76	0.78–0.76
	0.69	0.70–0.68
	0.63	0.68–0.63
N_2O	4.50	4.76–4.35
	4.06	4.76–3.57
	2.87	3.03–2.86
CH_4	3.30	4.00–3.13
	2.20	2.50–2.17
	1.66	1.71–1.64
CO	4.67	5.00–4.35
	2.34	2.41–2.30
NO_2	Visible	0.69–0.20

Source: Modified from Petty, G. W., *A First Course in Atmospheric Radiation*, Sundog Publishing, Madison, Wisconsin, 2006.

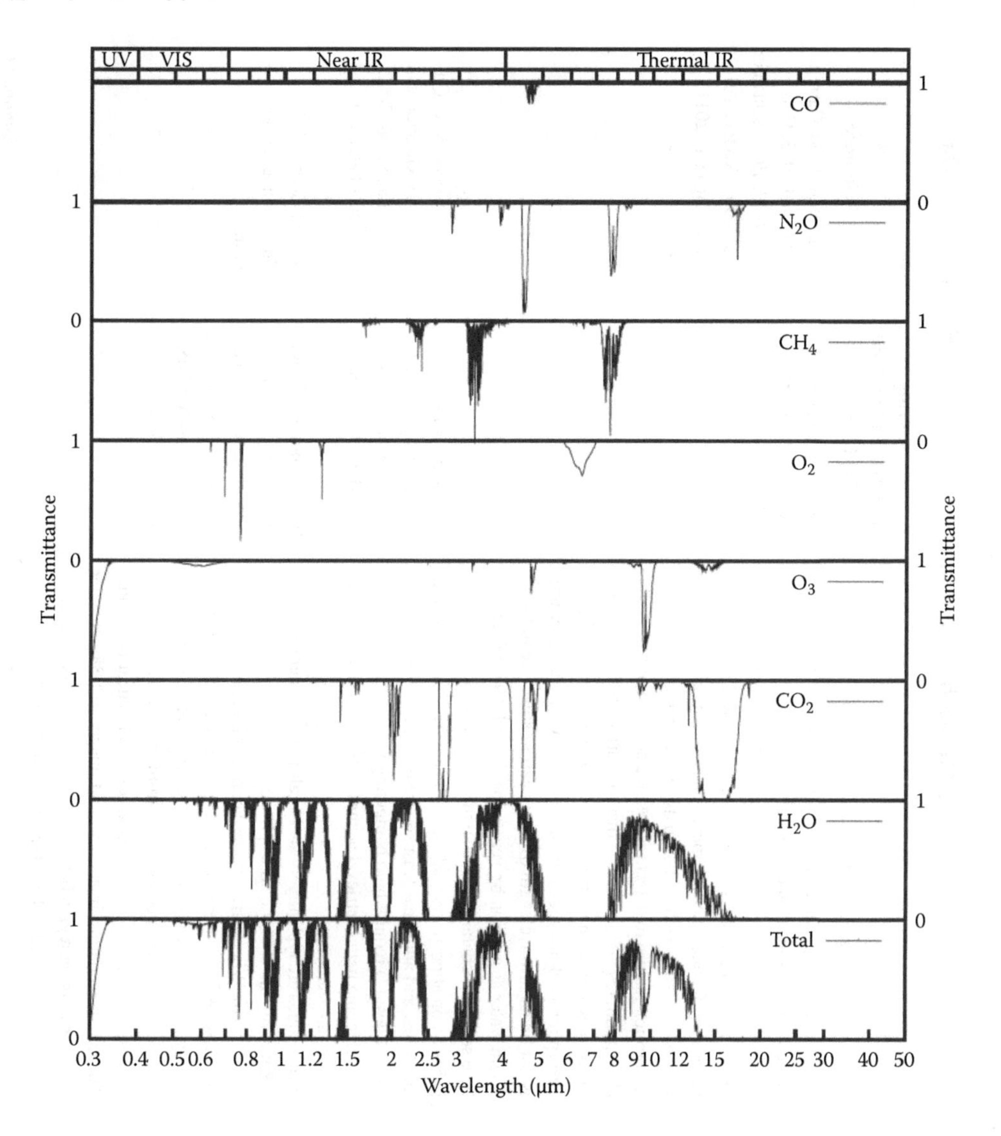

FIGURE 9.1 Transmittance spectra of the major atmospheric gases. (From Petty, G. W., *A First Course in Atmospheric Radiation*, Sundog Publishing, Madison, Wisconsin, 2006.)

parameters, CWV, cloud, aerosol, and CO_2, using hyperspectral remote sensing techniques and data is reviewed separately.

For readers' easy perusal of the major techniques, methods, and hyperspectral sensor data that are currently used for detecting the extent of the four atmospheric parameters (i.e., H_2O, cloud, aerosol, and CO_2) and retrieving their contents or concentrations, Table 9.2 provides a summary to those parameters. The table briefly lists atmospheric parameters/gases, estimating methods and techniques, corresponding characteristics and principles, relevant hyperspectral scattering and absorption bands, appropriate hyperspectral sensor data, and corresponding major references.

9.2.1 Water Vapor (H_2O)

To accurately retrieve atmospheric CWV from water vapor absorption bands in a measured spectrum, e.g., extracted from hyperspectral remote sensing data sets, transmittances of the absorption

TABLE 9.2
A Summary of Methods and Techniques Used for Retrieving and Estimating Atmospheric Parameters/Gases from Hyperspectral Data Sets

Atmospheric Parameter/Gas	Method and Technique	Characteristics and Principles	Scattering and Absorption Bands and Sensors	Reference
Water vapor (H_2O)	Differential absorption techniques: (1) narrow/wide (N/W), (2) continuum interpolated band ratio (CIBR), (3) three-band ratioing (3BR), (4) linear regression ratio (LIRR), (5) atmospheric pre-corrected differential absorption (APDA), and (6) second order derivative algorithm (SODA).	(1) Relatively simple and a practical way to determine WV content from a spectrum of absorption bands at a low computing time cost; (2) a ratioing is performed in this type of techniques between the radiance at bands within the absorption feature (measurement bands) and a reference radiance of bands (reference bands) in the vicinity of the absorption feature to detect the relative strength of the absorption; (3) an assumption that the surface reflectance is linear in wavelength within the spectral region around the absorption peak may lead to inaccuracies when the spectrum is not linear in wavelength over vegetation areas due to the liquid water absorption by vegetation around 0.94 μm, or in bare soils that contain iron in significant proportions.	0.94, 1.14, and 0.82 μm. AVIRIS, CASI, HyMap, HJ-1A HSI.	Frouin et al. 1990; Carrère and Conel 1993; Gao et al. 1993, 1996; Schläpfer et al. 1998; Rodger 2011; and Zhang et al. 2011.
	Spectral fitting techniques: (1) curve fitting algorithm, (2) smoothness test, (3) band fitting technique, and (4) other spectral fitting techniques.	(1) Based on atmospheric radiative transfer, spectral modeling and linear and nonlinear least square fitting theories and techniques; (2) require high computing time cost; (3) achieving better retrieval accuracy of atmospheric column water vapor amount.	0.94, and 1.14 μm; 0.585–0.6 μm; and 0.8–1.25 μm. AVIRIS, CASI-1500, VIRS-200.	Gao and Goetz 1990; Qu et al. 2003; Guanter et al. 2007b; Lang et al. 2002; Barducci et al. 2004.
Clouds	Cirrus mapping approach using narrow spectral channels near the center of the 1.38 μm strong water vapor absorption band.	Cirrus clouds are typically located at altitudes greater than 6 km, while most of the atmospheric water vapor is located below 6 km. When cirrus clouds are present, the narrow channels near 1.38 μm receive a large amount of reflected energy resulting from scattering of solar radiation by the cirrus clouds without contribution from ground surface because of water vapor absorption below 6 km near the band. The radiance contrast in narrow band images in 1.38 μm water vapor absorption region between the cirrus cloud cover areas and cloud-free areas allows the detection of cirrus clouds.	Water vapor absorption bands at 1.38 and 1.85 μm. AVIRIS band images at 1.38 and 1.85 μm and MODIS band at 1.38 μm.	Gao and Goetz 1992; Hutchison and Choe 1996.

(Continued)

TABLE 9.2 (CONTINUED)
A Summary of Methods and Techniques Used for Retrieving and Estimating Atmospheric Parameters/Gases from Hyperspectral Data Sets

Atmospheric Parameter/Gas	Method and Technique	Characteristics and Principles	Scattering and Absorption Bands and Sensors	Reference
	Cloud fraction, shadow and background mapping using (1) band ratioing and (2) physically-based thresholding approach.	(1) Based on a fact that the reflectances of most ground targets vary approximately linearly with wavelength in the 0.94- and the 1.14-μm water vapor band absorption regions, the band ratioing is defined as $BR = (R_{0.94} + R_{1.14})/(2 \times R_{1.04})$, where R represents the apparent reflectance at a wavelength (i.e., 0.94, 1.14, and 1.04 μm). (2) Idendify/classify cloud pixels by applying thresholding approachs to the 1st several principal component images and to selected reflectance narrow bands with a three-step procedure.	0.94, 1.04 and 1.14 μm. AVIRIS and other hyperspectral sensors' data.	Kuo et al. 1990; Gao and Goetz 1991; Feind and Welch 1995; Griffin et al. 2000.
Aerosol	(1) AOT estimated from estimated atmospheric transmittance from hyperspectral image; (2) dark dense vegetation technique (DDV)	(1) Given spectrally uniform surfaces with a reflectance difference of at least 0.5 and based on the relationship between atmospheric AOT (τ_λ) and transmittance (t_λ) ($\tau_\lambda \cong -\ln(t_\lambda)$) to estimate AOT with the relationship; (2) based on the assumption that the aerosol effect is much smaller or negligible at 2.2 μm versus that at the blue and red channels and the atmospheric radiative transfer principle, and the relationships between the surface reflectance at 2.2 μm and those at 0.47 μm and 0.66 μm.	Narrow bands (incl. at 0.47, 0.66 and 2.2 μm) in visible and shortwave spectral ranges. AVIRIS sensor and other hyperspectral sensors' data.	Isakov et al. 1996; Kaufman and Tanré 1996.
	Aerosol optical thickness at 550 nm (AOT at 550 nm) technique	The AOT at 550 nm technique avoids the frequent strategy of using dark targets in the aerosol retrieval. It makes the technique applicable to any kind of land targets, as long as some areas with vegetation and bare soil are present in the window in a hyperspectral image scene. The inversion technique of physically-based model uses a minimization algorithm.	Narrow bands in visible and shortwave spectral ranges. AVIRIS, CASI, MERIS, MIVIS and CHRIS sensors' data.	Guanter et al. 2007a,b; Bassani et al. 2010; Davies et al. 2010.

(Continued)

TABLE 9.2 (CONTINUED)
A Summary of Methods and Techniques Used for Retrieving and Estimating Atmospheric Parameters/Gases from Hyperspectral Data Sets

Atmospheric Parameter/Gas	Method and Technique	Characteristics and Principles	Scattering and Absorption Bands and Sensors	Reference
	(1) L-APOM (LUT-based Aerosol Plume Optical Model); (2) oxygen-dimer (O4) slant column density (SCDs)	(1) L-APOM characterizes the microphysical and optical properties of aerosol plumes from hyperspectral images with high spatial resolution. The inversion process includes three steps: (i) estimating the ground reflectance below the plume (typically for $\lambda > 1.5$ μm); (ii) estimating standard atmospheric terms; and (iii) inverting the L-APOM model. (2) Using oxygen-dimer (O_4) slant column density (SCDs) to test the sensitivities of the O_4 absorption bands at 340, 360, 380, and 477 nm to changes in AEH and its optical properties.	Narrow bands in visible and shortwave spectral ranges; O_4 absorption bands at 340, 360, 380, and 477 nm. AVIRIS and OMI sensors' data.	Alakian et al. 2009; Deschamps et al. 2013; Park et al. 2016.
Carbon dioxide (CO_2)	The joint reflectance and gas estimator (JRGE)	The JRGE method is a two-step algorithm which first estimates the surface reflectance and then the densities of the plume gases. The JRGE first performs an adaptive cubic smoothing spline like estimation of the surface reflectance, and concentrations of several gaseous species are then simultaneously retrieved using a nonlinear procedure based on radiative transfer calculations.	CO_2 narrow absorption bands in 2.0 μm region. AVIRIS sensor's data.	Marion et al. 2004; Deschamps et al. 2013.
	A method combines modeling minimum CO_2 anomalies detectable in AVIRIS data with applying a Cluster-Tuned Matched Filter (CTMF).	Combines modeling minimum CO_2 anomalies detectable in AVIRIS data under different conditions with applying a CTMF for detection of CO_2 plumes in simulated data and in real AVIRIS images acquired over power plants.	CO_2 absorption narrow bands. AVIRIS sensor's data.	Funk et al. 2001; Dennison et al. 2013.

(Continued)

TABLE 9.2 (CONTINUED)
A Summary of Methods and Techniques Used for Retrieving and Estimating Atmospheric Parameters/Gases from Hyperspectral Data Sets

Atmospheric Parameter/Gas	Method and Technique	Characteristics and Principles	Scattering and Absorption Bands and Sensors	Reference
	(1) CIBR; (2) inversion modeling method in fast atmospheric signature code (FASCOD); and (3) sensitivity analysis to the retrieval of CO_2 concentrations in the weak (1.61 and 1.57 μm) and strong (2.01 and 2.05 μm) absorption bands	(1) CIBR was calculated from two Hyperion bands B185 and B186 (2.002 and 2.012 μm). (2) An atmospheric model was first simulated in FASCOD and then the model was inverted with the inputs from hyperspectral remote sensing data (selected Hyperion CO_2 absorption bands) for retrieving CO_2 concentration. (3) Sensitivity analysis to the retrieval of CO_2 concentrations in the weak (1.61 and 1.57 μm) and strong (2.01 and 2.05 μm) absorption bands was performed in order to explore sensors' sensitivity at the weak and strong absorption wavelengths.	Two Hyperion bands 2.002 and 2.012 μm and weak/strong CO_2 absorption bands (PRISMA) at 1.6 and 2.0 μm Hyperion sensor's data; PRISMA simulated data.	Gangopadhyay et al. 2009; Nicolantonio et al. 2015.

bands must be sensitive to the change in the amount of water vapor molecules in the line of sight. Figure 9.2 presents vertical atmospheric transmittances as a function of wavelength between 0.6 and 2.8 μm at four CWV amounts of 0.63, 1.3, 2.5, and 5.0 cm, respectively (Gao and Goetz 1990). In the figure, these transmittance curves indicate that, under typical atmospheric conditions, the transmittances in the 0.94 and 1.14 μm band regions are sensitive to the changes in the amount of water vapor, whereas in the 1.38 and 1.88 μm band regions they are relatively insensitive to the changes in water vapor amount. The other absorption bands centered at about 0.72, 0.82, and 2.18 μm are relatively weak so that it may be difficult to ensure accurate retrieval of CWV from hyperspectral imaging spectra. Therefore, as reviewed in the following text, most existing studies on retrievals of atmospheric CWV amount from hyperspectral data sets are utilization of 0.94 and 1.14 μm water vapor absorption band regions.

To retrieve atmospheric CWV from hyperspectral data sets, there are two general types of estimation techniques: differential absorption techniques and spectral fitting techniques. The differential absorption techniques and algorithms, including narrow/wide (N/W; Frouin et al. 1990), continuum interpolated band ratio (CIBR; Green et al. 1989, Bruegge et al. 1990), three-band ratioing (3BR; Gao et al. 1993), linear regression ratio (LIRR; Schläpfer et al. 1996), and atmospheric precorrected differential absorption (APDA; Schläpfer et al. 1998), were introduced in Section 4.5.1 in Chapter 4.

Based on the study by Frouin et al. (1990), the N/W ratio algorithm has an appropriate exponential relationship with radiosonde total CWV amount measurement with a 10–15% error over land (a commonly acceptable error), but is underestimated up to 20% over water. In order to retrieve total CWV amount from AVIRIS imaging data acquired over Salton Sea, California, using the 0.94 μm water absorption band, Carrère and Conel (1993) compared the performance of two simple techniques (CIBR and N/W). Their comparative results indicated that the CIBR technique was less sensitive to perturbing effects, except for errors in visibility estimate. However, the validation result demonstrated that the CWV amount estimated using the N/W technique matched more closely *in situ* measurements, even after adjusting model parameters for background reflectance, viewing

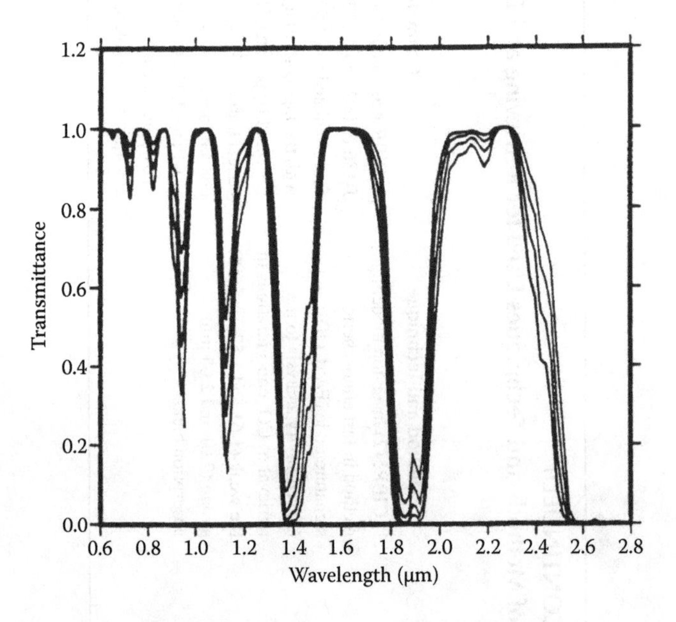

FIGURE 9.2 Vertical atmospheric transmittance as a function of wavelength at different column water vapor amounts under typical atmospheric conditions. The four transmittance curves from the top to the bottom correspond to column water vapor amounts of 0.63, 1.3, 2.5, and 5.0 cm, respectively. (©1990 Wiley. Used with permission from Gao, B., and Goetz, A. F. H., *Journal of Geophysical Research*, 95(D4), 3549–3564, American Geophysical Union, 1990.)

geometry and type of aerosol at the site. According to experimental results of comparing performances of N/W, CIBR, and LIRR techniques for retrieving CWV amount from AVIRIS imaging data, conducted by Schläpfer et al. (1996) over water, only the LIRR method can be evaluated. All the techniques in the 0.94 μm band region showed errors of 30% or more due to the lack of ground reflected radiance (lake water) at that wavelength. However, much better results were achieved over vegetation, and CIBR and LIRR can be quantified with a good accuracy. Using AVIRIS imaging data, their relative noise propagation error over vegetation was 6.7% (CIBR) and 2.6% (LIRR), respectively. Later, Schläpfer et al. (1998) developed a technique called *atmospheric precorrected differential absorption* (APDA) that is derived directly from simplified radiative transfer equations. When the APDA technique was applied to two AVIRIS images acquired in 1991 and 1995, the accuracy of the retrieved total-CWV amounts was within a range of ±5% compared to ground-based radiosonde measurements. By considering a weak water absorption band at 0.82 μm to retrieve the CWV content, Zhang et al. (2011) compared the capability of two water vapor retrieval algorithms, CIBR and APDA, using hyperspectral imagery acquired by the Chinese HJ-1A HyperSpectral Imager (HSI). The HSI sensor has a spectral resolution about 5 nm from 450 nm to 950 nm. Their comparative analysis results indicated that the two algorithms had less difference in the accuracy of water vapor retrieval because of the weak water vapor absorption effect at 0.82 μm. The three-band ratioing (3BR), developed by Gao et al. (1993) by using either 0.94 μm or 1.14 μm absorption band, was considered as the best estimate method of the CWV value from hyperspectral sensor AVIRIS imaging data. Moreover, in practice in implementing the 3BR technique, the center positions and widths of the both side regions and water vapor absorption bands are all allowed to vary in order to reduce any possible errors (Gao et al. 1993).

Rodeger (2011) developed a new method to estimate per-pixel atmospheric CWV and potential band shifts in a hyperspectral sensor. The new method uses variations of a second-order derivative algorithm (SODA) to assess the impact of atmospheric residual features on calculated surface reflectance spectra after atmospheric compensation. The CWV estimated from the HyMap imagery was validated at the two field sites and found a good agreement with *in situ* atmospheric CWV measurements to within 2% and represented a two-fold increase in accuracy over a three band ratio CIBR technique of atmospheric CWV estimation.

Typically, spectral fitting techniques used for retrieving atmospheric CWV amounts from hyperspectral image data are based on atmospheric radiative transfer, spectral molding, and linear and nonlinear least square fitting theories and techniques, and they include a curve-fitting algorithm (Gao and Goetz 1990), smoothness test (Qu et al. 2003), band-fitting technique (Guanter, Estellés et al. 2007), and other spectral fitting techniques. Gao and Goetz (1990) developed a curve-fitting technique to retrieve atmospheric CWV amount. They fitted observed spectra in 1.14 μm and 0.94 μm water vapor band absorption regions using an atmospheric model, a narrowband spectral model, and a nonlinear least squares fitting technique. Spectral curve-fitting techniques have been used extensively in quantitative derivation of information on atmospheric trace gases from solar absorption spectra (Gao and Goetz 1990). The curve-fitting technique is directly applicable for retrieving the CWV amount from AVIRIS imaging spectra measured on clear days with visibilities 20 km or greater. The precision of the retrieved CWV amounts from several AVIRIS data sets was 5% or better and the retrieval results were also independent of the absolute surface reflectance. Their experimental results also suggested that it would be feasible to derive high spatial resolution CWV amounts over land areas from satellite hyperspectral imaging data. Qu et al. (2003) used the smoothness test technique to retrieve CWV from hyperspectral AVIRIS data. The smoothness test technique is based on a principle often used by hyperspectral data analysts that either under- or overestimation of CWV amount results in irregularities in the retrieved surface reflectance. The smoothness test technique is used currently in the HATCH atmospheric correction program for calibrating hyperspectral image data into scaled surface reflectance. A surface reflectance in the 0.8–1.25 μm spectral region is derived for a given amount of water vapor. Then a smoothed reflectance spectrum is constructed accordingly using a truncated cosine series (Qu et al. 2003).

The root-mean-square error (RMSE) difference between the two spectra serves as the smoothness criterion. The results showed that the best water vapor estimation could yield the smoothest retrieved surface reflectance in the water vapor absorbing regions. In practice, once the water vapor amount at the first pixel is derived, the next pixel can use this value as an initial guess and the smoothness testing procedure needs using a few different water vapor amounts to find the proper one. To retrieve atmospheric column water vapor (CWV) amount, Guanter, Estellés et al. (2007) thought band ratio techniques based on a water vapor absorption band at 0.94 μm, such as CIBR (Carrère and Conel 1993) and APDA (Schläpfer et al. 1998), have been shown to perform well with a reasonable computation cost. However, the techniques with an assumption that the surface reflectance is linear in wavelength within the spectral window around the absorption peak may lead to inaccuracies when the spectrum is not linear in wavelength over vegetation areas due to the liquid water absorption by vegetation around 0.94 μm, or in bare soils that contain iron in important proportions. In addition, band ratio methods are very sensitive to sensor spectral and radiometric resolution and calibrations. Therefore, Guater et al. (2007b) developed a new method for CWV retrieval. The new method, called a band-fitting technique, is based on inverting the at-sensor radiance around the 0.94 μm water vapor absorption feature with a simulated spectrum with the adequate CWV value. When the band-fitting technique was tested with CASI-1500 image data for retrieving CWV, a good correlation between ground measurements and CASI-derived CWV was found, with $R^2 = 0.74$.

There are other spectral-fitting techniques that are used to retrieve atmospheric CWV from hyperspectral data. For example, Maurellis et al. (2000) and Lang et al. (2002) investigated the retrieval of atmospheric CWV amount using a new spectral-fitting method applied to Global Ozone Monitoring Experiment (GOME) hyperspectral data. The method called Optical Absorption Coefficient Spectroscopy (OACS) method is well-suited to situations where line widths in the absorption spectrum are much narrower than the instrumental resolution. OACS is an optical absorption spectroscopy technique that was applied to a little-studied visible band between 585 and 600 nm. The OACS was applied to the retrieval of CWV densities from GOME spectral data and achieved high precisions that were better than 0.7% for high CWV and better than 3.4% for low CWV. Barducci et al. (2004) developed a quantitative retrieval algorithm of vertically integrated water vapor content. The basic idea of the proposed algorithm is to use residual line intensity (see Figure 3 in Barducci et al. 2004) that corresponds to the water vapor absorption band centered at 0.94 μm to estimate the line area by spectral integration. The line area is a relevant parameter from an absorption band that is used for estimating the CWV amount. With the new algorithm, the results they obtained from radiometrically calibrated hyperspectral images collected by the AVIRIS sensor over the Cuprite mining district in Nevada suggested that the retrieval algorithm would be accurate and stable and proved that the derived CWV abundance is rather insensitive to changes of surface reflectance. Numerical simulations based on the MODTRAN4 radiative transfer code and an additional empirical H_2O retrieval procedure by using the VIRS-200 imaging spectrometer also demonstrated the better performance of the algorithm.

9.2.2 Clouds

Cirrus clouds play an important role in the global energy balance and climate study because of their covering large area, persistence, and radiative effects. Thin cirrus clouds are difficult to detect in visible images and thermal infrared images in the 10–12 μm atmospheric window, particularly over land, because these clouds are partially transparent (Gao and Goetz 1992).

Hyperspectral sensor data in strong water vapor absorption bands at 1.38 and 1.85 μm provide an operational approach to detect the presence and distribution of cirrus clouds (Gao and Goetz 1992, Hutchison and Choe 1996). For examples, Gao and Goetz (1992) tested the capability of narrow spectral channels near the center of the 1.38 μm strong water vapor band, which was expected to be useful for monitoring cirrus clouds using both AVIRIS image scenes covering land and water areas. They used an approach for the detection of cirrus clouds using channels near the center of the strong

1.38 μm water vapor band from AVIRIS images. Cirrus clouds are typically located at altitudes greater than 6 km over high latitude areas (7 km and 9 km over middle latitude and tropical areas, respectively), while most of the atmospheric water vapor is located below 6 km. Therefore, AVIRIS channels near 1.38 μm receive little reflected energy resulting from scattering of solar radiation by the Earth's surface, because the solar radiation is mostly absorbed by water vapor in the lower atmosphere (<6 km). However, when cirrus clouds are present, the AVIRIS channels near 1.38 μm receive a large amount of reflected energy resulting from scattering of solar radiation by the cirrus clouds. The radiance contrast in AVIRIS images in 1.38 μm water vapor absorption region between the cirrus cloud cover areas and cloud-free areas allows the detection of cirrus clouds. It may be expected for our ability to determine cirrus cloud amounts using space-based remote sensing if channels near the center of the 1.38 pm water vapor band are added to satellite sensors. Hutchison and Choe (1996) also demonstrated the performance of bands near the center of the strong 1.38 μm water vapor band from AVIRIS images for improving thin cirrus detection in daytime imagery collected over land surfaces. They collected both AVIRIS sensor data, which included bands near the center of the strong 1.38 μm water vapor band, and advanced very high resolution radiometer (AVHRR) imagery collected by NOAA operational meteorological satellites. The acquisition time of both sensors' imagery was nearly coincident. The results of thin cirrus detected by both sensors showed that the addition of AVIRIS 1.38 μm imagery significantly improved the detection of optically thin cirrus clouds in these daytime data (the accuracy increased by 30% to 50% compared to those from the AVHRR imagery). However, the 1.38 μm imagery did not mask all solar incident energy reflected by all ground surfaces as suggested in the study. Thus, the accurate use of the 1.38 μm water vapor absorption band data in an automated cloud classification algorithm requires the development of a detection threshold that varies with surface albedo at the 1.38 μm band and atmospheric water vapor concentration (Hutchison and Choe 1996).

Cloud fraction, cloud shadow area, and cloud background analyses were investigated with AVIRIS data in the early 1990s (e.g., Kuo et al. 1990, Gao and Goetz 1991, Feind and Welch 1995). Since the atmospheric CWV concentration decreases rapidly with altitude, the depths of water vapor absorption bands above clouds are usually shallower than those above clear ground surface areas. According to the fact that the reflectances of most ground targets vary approximately linearly with wavelength in the 0.94- and the 1.14-μm water vapor band absorption regions, and the peak absorptions of the water vapor band over cloudy areas are smaller than those over nearby clear ground surface areas, Gao and Goetz (1991) developed a band ratioing (BR) technique to map cloud fraction area in AVIRIS imagery. The BR technique was defined by $BR = (R_{0.94} + R_{1.14})/(2 \times R_{1.04})$, where R represents the apparent reflectance at a wavelength (i.e., 0.94, 1.14, and 1.04 μm), extracted from AVIRIS (or other visible and short wave infrared hyperspectral) imaging data. The BR technique could effectively discriminate among clouds and other clear and high-reflectance surface areas that have similar reflectance values. Otherwise, such discrimination is not possible using standard radiance thresholding techniques. The band ratioing method is also most useful in determining cloud area coverage for partly cloudy conditions. In the band ratioed image, clouds may stand out from a rather uniform background over land. However, in an uneven terrain, it will show a brightness gradient on the band ratioed image because of higher terrains usually with shorter water absorption paths and thus less water absorption and larger ratio values in the ratioed image (Kuo et al. 1990). Such an uneven terrain effect somewhat obscures the distinction between cloud and background and makes the choice of threshold more difficult than on an even terrain. Therefore, Kuo et al. (1990) improved the original BR technique in identifying cloudy areas by using a lower limit of the ratio suggested by the background albedos in the BR ratioing process, and a threshold could then be chosen without ambiguity to map out clouds from the background with uneven terrain. To improve the accuracy of mapping thin cloud and cloud edges, Feind and Welch (1995) applied a technique for registering the thermal infrared multispectral scanner (TIMS) imagery to AVIRIS imagery to applications for studying background and cloud properties and discovered some meaningful results. The registration technique took advantage of the morphology of the fair weather cumulus clouds present in the imagery for estimating intersensor distortions. Compared to the cloud mapping result from AVIRIS data with the BR ratioed imagery, the coregistered TIMS imagery showed that TIMS was superior in detecting thin cloud and cloud-edge pixels, especially over shadowed background.

To map cloud from smoke plumes and biomass burning from hyperspectral imagery, Griffin et al. (2000) used a combination of two approaches to discriminate and classify various features in a smoke/cloud filled scene. They first employed principal component analysis (PCA) to reduce hyperspectral data into first several component images that are orthogonal and decorrelated from each other, and then the different image features could be separated from the component images with appropriate thresholds to the pixel histogram distribution. For instance, the first two principal component images (PC1 and PC2) and fifth principal component image (PC5), transformed from the AVIRIS image scene collected on August 20, 1992, in the foothills east of Linden, California, were useful for discrimination of image features (clouds, smoke plumes, fires, shadows, etc.). The PC1 showed the overall intensities of features such as bright clouds and smoke plumes over backgrounds, while a dark area that appeared to be the source of the thick smoke was apparent in the PC2 image. In the PC5 image, a small fraction of the image pixels was in contrast to the image backgrounds. Since the size of the scatterers in clouds is much larger (scatterer size >> 1 μm) than the sensor's wavelengths, clouds are typically the brightest feature in an AVIRIS image, and also the reflectance from clouds is nearly invariant in the visible and short-wave infrared spectral regions (Griffin et al. 2000). Therefore, they also used physically based thresholding approach (a three-step thresholding approach) to identify different image features to mask out clouds from the hyperspectral image scene. The first step removed dark background objects such as rivers, lakes, and vegetation from consideration by applying a threshold (e.g., 0.20) to the 640 nm reflectance image. The second step took advantage of the spectral invariability of cloud and used a ratio of the 640 to 860 nm reflectance images, and all pixels with ratios < 0.70 were eliminated. This step could remove much of the smoke and vegetation that display distinct changes over the spectral range from 640 to 860 nm. The third step isolated the cloud from the surrounding thick smoke by applying a threshold of 0.35 to the band image with wavelength of 1.60 μm. Those remaining pixels with reflectance values above 0.35 were designated cloud. Their results demonstrated that the combination of the two approaches could provide a useful tool for the characterization and identification of scenes containing clouds, smoke, and active fires from hyperspectral data.

9.2.3 Aerosols

Detection of various aerosols in the atmosphere is very important because they are a type of scattering sources and affect direct and diffuse solar radiation. Knowledge on aerosols is required for the atmospheric correction of imaging spectroscopy and other remotely sensed data. Aerosol also plays an important role in air quality near the surface. For these reasons, observations from airborne and satellite remote sensing, especially hyperspectral remote sensing, have been carried out to investigate aerosol properties at regional and global scales. In the following, an overview on methods, hyperspectral data sets, and results of detecting or estimating aerosol optical properties, such as aerosol optical depth (thickness; AOD or AOT), aerosol effective height (AEH), from various hyperspectral sensor data is provided.

Given sufficiently small sensor errors and spectrally uniform surfaces with a reflectance difference of at least 0.5 and based on the relationship between atmospheric AOT (τ_λ) and transmittance (t_λ) ($\tau_\lambda \cong -\ln(\tau_\lambda)$)(see also Equation 4.8 in Chapter 4), Isakov et al. (1996) employed two scenes of AVIRIS image data to derive aerosol information. At two different study sites covered by the two AVIRIS imagery, high-contrast natural surface and artificial surface were taken as study areas in order to obtain AOT information. Coincident measurements of spectral optical depth from a surface-based sunphotometer also were obtained and used as a validation of the AVIRIS derived retrievals. They first estimated atmospheric transmittance from the AVIRIS image scenes respectively using at-sensor–based and ground-based spectra each with a reflectance difference of at least

0.5, and then used the $\tau_\lambda \cong -\ln(\tau_\lambda)$ relationship to estimate AOT. Comparing the aerosol optical depth (AOT) retrieved from AVIRIS data with the AOT measurements by surface-based sunphotometer, on an average, the two AOT retrievals agreed within an interval of ± 0.1. Their preliminary results suggested that background AOT (i.e., $\tau_{aerosol} < 0.1$) could not be retrieved with adequate accuracy from space; however, the AOT of most polluted atmosphere (i.e., $\tau_{aerosol} > 0.2$) could be retrieved with an adequate accuracy.

Since the aerosol effect is strongest for low background surface reflectance, Kaufman and Tanré (1996) developed a dark pixel approach (called *DDV technique* in Section 4.5.2 of Chapter 4) to directly estimate aerosol loading or AOT from multi-/hyperspectral images with some visible bands (e.g., 0.41 µm, 0.47 µm, 0.66 µm) and MIR band (e.g., 2.2 µm) available. The DDV technique is based on the assumption that the aerosol effect is much smaller or negligible at 2.2 µm versus that at the blue and red channels and the atmospheric radiative transfer principle. Therefore in the DDV technique Kaufman and Tanré (1996), over dark or dense vegetated areas, used the relationships between the surface reflectance at 2.2 µm and those at 0.47 µm and 0.66 µm, derived from Landsat TM and aircraft AVIRIS images over the mid-Atlantic United States to estimate atmospheric AOT. In practice, Kaufman and Tanré (1996) and Kaufman et al. (1997) applied the DDV technique to successfully estimate AOT from EOS–MODIS data for finally correcting atmospheric effect on reflectances of surface materials in visible and shortwave infrared ranges.

The AOT at 550 nm extraction method developed by Guanter, González-Sampedro et al. (2007) and Guanter, Estellés et al. (2007) with hyperspectral image data is an inversion technique of physically based model with a minimization algorithm (see its introduction in Section 4.5.2 of Chapter 4). The AOT at 550 nm avoids the frequent strategy of using dark targets in the aerosol retrieval. It makes the technique applicable to any kind of land targets, as long as some areas with vegetation and bare soil are present in the window in a hyperspectral image scene, which is less restrictive than the dark-targets. The method takes advantage of the large amount of spectral information provided by a hyperspectral sensor with high spectral resolution to better identify the optical atmospheric radiative effects of the aerosol scattering on the at-sensor radiance without any at-ground measurements and working under the usual Lambertian assumption (Bassani et al. 2010). Guanter, González-Sampedro et al. (2007) and Guanter, Estellés et al. (2007) applied the method to the data of two hyperspectral sensors to retrieve AOT at 550 nm and found that the technique was feasible after an extensive validation exercise with ground-based measurements. When using the data of the medium resolution imaging spectrometer (MERIS), Guanter, González-Sampedro et al. (2007) obtained that the values of the RMSE of AOT retrieved from the MERIS data were 0.085 at 440 nm, 0.065 at 550 nm, and 0.048 at 870 nm, compared with the Aerosol Robotic Network ground measurements from two sites. When using CASI-1500 data, Guanter, Estellés et al. (2007) found a good correlation between ground-based AOT and CASI-derived AOT with a Pearson's correlation coefficient R^2 up to 0.71. Bassani et al. (2010) also tested the method (i.e., AOT at 550 nm) for retrieving the AOT at 550 nm for calibrating the surface reflectance from airborne acquired data in the atmospheric window of the visible and VNIR range. They tested the method on five remote sensing images acquired by the multispectral infrared and visible imaging spectrometer (MIVIS) airborne sensor under different geometric conditions to evaluate the reliability of the method for extracting AOT at 550 nm. The results of AOT at 550 nm retrieved from each image were validated with field data contemporaneously acquired by a sun–sky radiometer. A good correlation index, $R^2 = 0.75$, and low root mean square deviation, RMSD = 0.08, were obtained. The tested results indicated that the method was reliable for optical atmospheric studies with airborne hyperspectral images and that the method did not require additional ground-based measurements about AOT.

Using an inversion technique of a physically based model similar to the estimate techniques of AOT at 550 nm from hyperspectral data sets addressed above, Davies et al. (2010) estimated AOD over land surfaces using high spatial resolution, hyperspectral, and multiangle CHRIS sensor data. The CHRIS instrument is mounted aboard the PROBA satellite and provides up to 62 bands. The PROBA satellite allows pointing to obtain imagery from five different view angles (55° and 36° in the backward and forward directions and at nadir) within a short time interval. The inversion technique includes a general physical model of angular surface reflectance and an iterative process that is used to determine the optimum value providing the best fit of the corrected reflectance values for a number of view angles and wavelengths. Guanter et al. (2005) have explored the use of spectral information only for aerosol retrieval from PROBA/CHRIS using the estimate technique of AOT at 550 nm. In the work by Davies et al. (2010), they developed and tested the new inversion technique for estimating AOD at 550 nm also using multiangle PROBA/CHRIS images. Their estimate results were validated with ground-based sunphotometer measurements. Results from 22 CHRIS image sets showed an RMSE of 0.11 in AOD at 550 nm for targets which included homogeneous vegetation, urban, maritime, and bright sandy targets. A quality check based on consistency between three- and five-angle AOD retrievals allowed reduction of the RMSE to 0.06 over 17 CHRIS image sets (Davies et al. 2010). The validation results demonstrated the potential of CHRIS for aerosol retrieval at high spatial resolution, including over bright and urban targets.

Emission measurements of biomass burning derived aerosols and gases have become a major field of interest in regard to radiative budget, atmospheric chemistry, and air quality monitoring for climate change studies. The standard aerosol molding methods may not be suitable for retrieving aerosol optical properties caused by biomass burning using multi-/hyperspectral data sets. For this case, Alakian et al. (2009) developed a retrieval method called LUT-based Aerosol Plume Optical Model (L-APOM) which aims at characterizing the microphysical and optical properties of aerosol plumes from hyperspectral images with high-spatial resolution. The inversion process of L-APOM model for retrieving optical properties of aerosol plumes includes three steps: (1) estimating the ground reflectance below the plume based on the fact that the plume is optically transparent for longer wavelengths (typically for $\lambda > 1.5$ μm); (2) estimating standard atmospheric terms (including the upwelling atmospheric reflectance, the total transmittance, and the atmospheric spherical albedo when considering the atmosphere without a plume) associated with gases and background aerosols; and (3) inverting the L-APOM model for retrieving the plume aerosols properties (Alakian et al. 2009). Given that the inversion process revealed an ill-posed problem, original constraints were added by assuming slow spatial variations of aerosol particles properties within the plume. In the study, the inversion process was validated on a large set of simulated and real hyperspectral images even in the worst cases of noise and the relative estimation errors of aerosol properties remained between 10% and 20% in most cases. Further, when the L-APOM was applied to a real AVIRIS hyperspectral image with a biomass burning plume for which *in situ* measurements were available, the retrieved aerosol properties appeared globally consistent with measurements. The L-APOM model was integrated by Deschamps et al. (2013) into a new method that was used to simultaneously retrieve carbon dioxide (CO_2) and aerosols inside a plume caused by biomass burning, which combines an aerosol retrieval algorithm (i.e., L-APOM model) using visible and NIR wavelengths with a CO_2 estimation algorithm using middle infrared wavelengths hyperspectral data set. After the L-APOM model was applied in retrieving aerosol properties with two scenes of AVIRIS images acquired over vegetation fires to finally estimate CO_2 concentration in smoke plumes, a better agreement between *in situ* measurements and retrieved CO_2 abundances was achieved (Deschamps et al. 2013).

Recently, Park et al. (2016) developed a method to use oxygen-dimer (O_4) slant column density (SCDs) to derive aerosol effective (layer) height (AEH) from a spaceborne UV–visible hyperspectral sensor: the Ozone Monitoring Instrument (OMI). OMI channels are composed of UV-1 (270–314 nm), UV-2 (306– 380 nm), and a visible wavelength range (365–500 nm), with a spectral resolution (FWHM) of 0.63, 0.42, and 0.63 nm. The spatial resolution is 13 km × 24 km at nadir in "global mode." In their study, the sensitivities of the O_4 absorption bands at 340, 360, 380, and 477 nm to changes in AEH and its optical properties were estimated using simulated hyperspectral radiances. They proposed an improved differential optical absorption spectroscopy (DOAS) algorithm for the O_4 absorption bands to retrieve AEH information from the O_4 SCDs based on the sensitivity studies. Then the new algorithm was applied to the O_4 SCD from the OMI data to retrieve the AEH for a real case over East Asia, including error estimates. The OMI spectral data over the visible wavelength range were used to derive the O4I (O_4 Index; see its definition from Park et al. 2016) at 477 nm and the AEH information. The results derived from the real case indicated that about 80% of retrieved AEHs were within the error range of 1 km compared to those obtained from the Cloud-Aerosol LiDAR with Orthogonal Polarization measurements on thick aerosol layer cases (with a correlation coefficient of 0.62 for an AOD > 1.0).

9.2.4 Carbon Dioxide (CO_2)

A more accurate estimation of gas amounts, such as CO_2, is generally required for investigating a variety of phenomena, including biomass burning, volcanoes, combustion of fossil fuels, and other industrial activities–derived pollutions. To estimate a set of atmospheric gas concentrations in an unknown surface reflectance context from hyperspectral sensor data, Marion et al. (2004) developed a new method called joint reflectance and gas estimator (JRGE). The JRGE method is of a two-step algorithm which first estimates the surface reflectance and then the densities of the plume gases. The JRGE first performs an adaptive cubic smoothing spline like estimation of the surface reflectance, and concentrations of several gaseous species are then simultaneously retrieved using a nonlinear procedure based on radiative transfer calculations (Marion et al. 2004). The method takes into account for all the information about gases contained in the data and yields optimized estimates. However, the method is applicable for aerosol-free clear atmospheres in a spectral range from approximately 0.8 to 2.5 μm. The JRGE algorithm has been validated using both simulated data and AVIRIS images acquired during the 1994 Quinault prescribed fire and over the Cuprite mining district in 1997. The validation results indicate that the total CO_2 column has been recovered from AVIRIS data with an accuracy of about 6% to 7% for pixels without any aerosol and with a surface reflectance ≥ 0.1 within gas absorption bands, which was in agreement with *in situ* measurement (345–350 ppmv). However, aerosols in the atmosphere yielded an underestimation of total atmospheric CO_2 content equal to 5.35% about 2 km downwind the fire. The experimental results suggested that if the hyperspectral image was acquired at lower altitude and with a better signal-to-noise ratio, a better estimate accuracy should be expected. To improve the estimate accuracy of atmospheric CO_2 coexisting with aerosols in plumes in the atmosphere, induced by biomass burning from hyperspectral sensor data, Deschamps et al. (2013) reported a new method to simultaneously retrieve CO_2 and AOT inside a plume. The new method combines an aerosol retrieval algorithm (i.e., L-APOM in Section 9.2.3 above) using visible and NIR wavelengths with a CO_2 estimation algorithm (i.e., JRGE) using shortwave infrared (SWIR) wavelengths. To retrieve CO_2 concentration, the JRGE technique (Marion et al. 2004) can take into account the nonlinear spectral variations of the ground reflectance in absorption bands. In general, the JGRE technique is based on an assumption that the soil reflectance slowly varies with the wavelength, whereas the atmospheric transmittance rapidly decreases at the gas absorption band (Deschamps et al. 2013). Finally, CO_2 abundances are estimated using the 2.0 μm absorption band with considering the previously retrieved aerosol properties. To assess the performance of the new method with real hyperspectral data, two AVIRIS images over areas of biomass burning were chosen. The first image was acquired in Quinault, Washington, and the second was acquired in Aberdeen, Washington. The assessment results showed that the new method (i.e., including the aerosol retrieval step before estimating CO_2 abundance) could create a better agreement between retrieved CO_2 abundances and *in situ* measurements, especially for pixels where the plume is not very thick, reduce the standard deviation of estimated CO_2 abundance by a factor of four, and make the spatial distribution of retrieved concentrations coherent.

Imaging spectrometer data that include CO_2 absorption bands in the SWIR range have potential for mapping spatial variation in atmospheric CO_2 concentration. In order to estimate CO_2 that is emitted from the combustion of fossil fuels from point sources such as fossil fuel power plants, high spatial and spectral resolution data from AVIRIS-like sensors may offer a means for detecting plumes and retrieving CO_2 concentrations for point source emissions (Dennison et al. 2013). This is because the AVIRIS-like sensor can provide multiple, contiguous bands covering SWIR absorption features induced by CO_2. For this case, Dennison et al. (2013) conducted a study that combines modeling minimum CO_2 anomalies detectable in AVIRIS data under different conditions with applying a cluster-tuned matched filter (CTMF; Funk et al. 2001, Thorpe et al. 2013) for detection of CO_2 plumes in simulated data and in real AVIRIS images acquired over power plants. For the "classic" AVIRIS sensor (AVIRIS C), its spatial resolution typically ranges between 3 and 20 m depending on platform altitude, while the "next generation" AVIRIS sensor (AVIRIS NG) is capable of spatial resolutions as fine as 1 m with an improved spectral sampling of 5 nm. In the study, to determine the theoretical minimum detectable change in CO_2 concentration, residual radiance values for different CO_2 anomalies were compared to the noise equivalent delta radiance (NEdL) for AVIRIS C and AVIRIS NG. NEdL is the minimum change in radiance distinguishable from sensor noise and is dependent on both wavelength and radiance (Dennison et al. 2013). Thus, NEdL establishes a minimum detection threshold for increased CO_2 concentration. A residual radiance was calculated by subtracting the spectrum from a spectrum with no elevated CO_2 concentration. To assess whether CO_2 plumes from power plants might be detectable in AVIRIS C and NG data sets, the CTMF detection algorithm was applied to the simulated radiance images. The CTMF algorithms were trained with a target spectrum to generate a linear weighting function that produces high values when an unknown spectrum matches the shape of the target spectrum, which contains the residual radiance signal of elevated CO_2 absorption. The modeled residual radiance spectra show that zero to 500 m CO_2 concentration anomalies as low as 100 ppmv for AVIRIS C and 25 ppmv for AVIRIS NG produced residual radiance values that exceeded SWIR NEdL. The experimental results demonstrated the CO_2 plume detection capability based on the CTMF approach with simulated images and all four AVIRIS C images acquired over four U.S. power plants, although the characteristics of the plumes varied according to solar–plume–sensor geometry.

Satellite hyperspectral sensor data sets are good sources used for detecting CO_2 plumes induced by industrial activities in the atmosphere and estimating abundance of CO_2. For example, coalfire has small but significant contributions to CO_2 concentrations in the atmosphere (Gangopadhyay et al. 2009). To estimate the amount of coalfire-induced CO_2 plumes in the atmosphere in a coal-mining region in northern China, Gangopadhyay et al. (2009) applied two methods using simulated and satellite hyperspectral Hyperion spectra. In the first method, a CIBR (Green et al. 1989, Bruegge et al. 1990) band ratioing method was used to estimate atmospheric column CO_2 concentration. As observed, the most significant two CO_2 absorption bands (2.001 and 2.01 μm), which are not considerably influenced by any other atmospheric constituents such as water vapor, were chosen. The two absorption bands are also very close to the central wavelength of Hyperion bands 2.002 and 2.012 μm. After calculating CIBR from two CO_2 absorption bands (2.001 and 2.01 μm) based on the CIBR definition, the columnar abundance of CO_2 can be retrieved by the equation $CIBR = EXP(-\alpha[CO_2]^{\beta})$, where α and β are the two parameters that can be fitted by a calibration relationship between the CIBR and the column CO_2 content (300–10,000 ppmv). To fit the α and β parameters, radiances for three Hyperion bands (1.982, 2.002, and 2.032 μm) could be simulated in fast atmospheric signature code (FASCOD) using the atmospheric conditions under different levels of column CO_2 content as expected in the study area. After the two parameters were fitted, the CO_2 columnar abundance was then retrieved by inverting the equation with the radiance measured at each pixel by Hyperion sensor. In the second method, a relation between CO_2 plume-related radiance and CO_2 concentration was established. To do so, an atmospheric model was first simulated in FASCOD to understand the local radiation transport and then the model was inverted with the inputs from hyperspectral remote sensing data (selected Hyperion CO_2 absorption bands) for retrieving CO_2 concentration for corresponding image pixels. The experimental results indicated that the band ratioing method was much faster and more effective so that the retrieved total atmospheric column CO_2 concentration was adequate for further analysis. Since the inversion method calculates only to a certain level of atmosphere, this method will be much more functional to estimate CO_2 from a particular event. Given that CO_2 emissions from industrial sources are one of the main anthropogenic contributors to the greenhouse effect, Nicolantonio et al. (2015) investigated the potentiality of satellite hyperspectral sensors, such as Hyperspectral Precursor of the Application Mission (PRISMA), to evaluate significant CO_2 concentrations in the lower troposphere. PRISMA, a satellite mission funded by the Italian Space Agency (ASI), is planned to be the next hyperspectral Earth Observation (EO) mission aiming at the development and delivery of hyperspectral products and the qualification of the state-of-art of hyperspectral payload in space (Nicolantonio et al. 2015). In their study, using synthetic hyperspectra in considering designed features required for the PRISMA sensor at the top of atmosphere (TOA) and sensitivity analysis to the retrieval of CO_2 concentrations in the weak (1.61 and 1.57 μm) and strong (2.01 and 2.05 μm) absorption bands was performed in order to explore sensor sensitivity at the weak absorption wavelengths around 1.6 μm and at the strong absorption wavelengths around 2 μm. The spectral and radiometric features designed for the PRISMA sensor, including a spectral resolution of 10 nm or higher (5 and 2 nm) in the spectral range 1.5–2.4 μm and a spatial resolution of 30 m, have been considered in order to evaluate the satellite sensitivity in retrieving CO_2 concentrations up to 20 times the background concentrations. Considering the designed radiometric accuracy of the PRISMA sensor, analysis results showed a relative uncertainty between 3% and 15% on the retrieval of CO_2 concentrations between 390 to 5100 ppmv. With a spectral resolution of 10 nm, the PRISMA sensor would be capable of retrieving a minimal concentration of CO_2 in the low troposphere with an uncertainty below 10%.

9.3 SNOW AND ICE HYDROLOGY

The distribution and characteristics of snow and ice cover on the Earth's surface significantly affect the Earth's radiation budget. Snow grain size, snow impurity and contamination, snow melting, and ice states influence surface albedo and hydrology. Properties of snow and ice recorded in the solar spectral range include fractional cover, grain size, impurities, contaminations, surface liquid water content and phases of snow melting water, and shallow depth (Green et al. 1998).

During the last three decades, there have been many studies on modeling and mapping the properties of snow and ice using a variety of hyperspectral remote sensing techniques. The spectral reflectance of snow varies strongly with wavelength in most solar spectral range (Dozier et al. 1989). Based on the moderate absorptivity of ice in the SWIR wavelengths (0.7–3.0 μm), particularly in the 0.7–1.3 μm spectral region where snow reflectance is most sensitive to the snow grain size, Nolin and Dozier (1993) developed a snow grain size inversion model to derive snow grain size information from AVIRIS spectra. Figure 9.3 presents a relationship between spectral reflectance and snow grain radius in the spectral region of 0.4–2.5 μm. All reflectances in the figure across wavelengths (0.4–2.5 μm) and snow grain sizes (0–1000 μm) were calculated using a discrete-ordinate radiative transfer model (Nolin and Dozier 1993). In the figure, while there is a slight spectral dependence on grain size in the visible range, reflectance decreases rapidly with increasing grain size in the NIR spectral range. To develop a snow grain size inversion model, the wavelength of 1.04 μm was chosen because it is in a spectral region where reflectance is particularly sensitive to grain radius size. To do so for each band image, a model-generated curve of reflectance (R) against snow grain size (GR) was fit with an exponential model of the form $\mathrm{GR} = ae^{bR}$, where a and b are the coefficients simulated using a method of least squares fitting. The exponential model establishes a functional relationship between grain size and surface reflectance for each AVIRIS single band image. Consequently, the model could be inverted with input snowpack reflectance at 1.04 μm from AVIRIS image to retrieve snow grain size. Using the inversion model, estimates of snow grain size for the near-surface snow layer were calculated from the AVIRIS images acquired for the Tioga Pass region and Mammoth Mountain in the Sierra Nevada, California. Note that the Tioga Pass and Mammoth Mountain single-band AVIRIS radiance images were atmospherically corrected to surface reflectance before using the inversion model to estimate the pixel-based snow grain size. Based on the validation result of the inversion technique using a combination of ground-based reflectance measurements

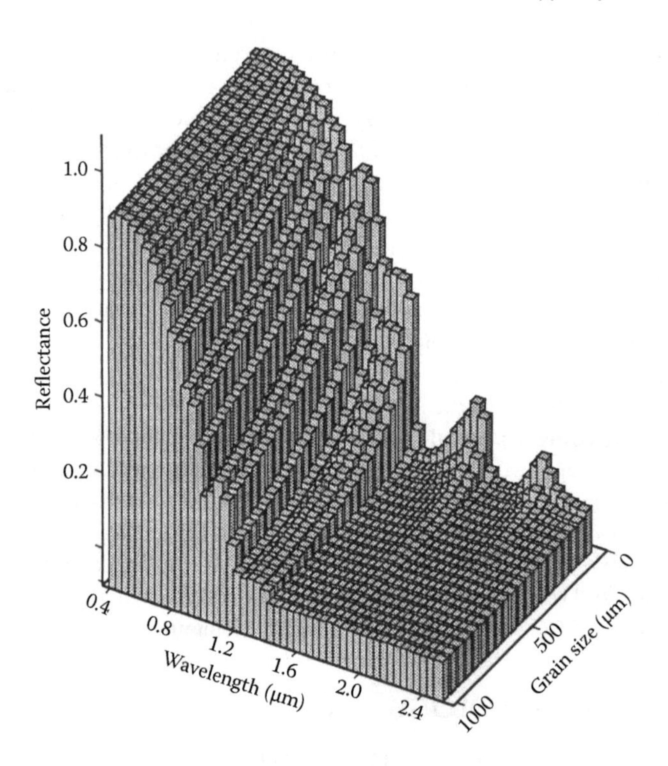

FIGURE 9.3 Spectral reflectance of snow from 0.4 μm to 2.5 μm as a function of snow grain size. Note the steep decrease in reflectance with increasing grain size in the spectral region between 0.7 μm and 1.3 μm. All reflectances across wavelengths (0.4–2.5 μm) and snow grain sizes (0–1000 μm) were made using a discrete-ordinate radiative transfer model. (Reprinted from Nolin, A. W., and Dozier, J., *Remote Sensing Environment*, 44, 231–238, 1993. ©1993, with permission from Elsevier.)

and grain size measurements derived from stereological analysis of snow samples for a wide range of snow grain sizes, for solar incidence angles between 0° and 30°, the inversion technique could provide good estimates of snow grain size from AVIRIS imaging data. In practice, since the technique is limited by the need to know the angle of solar incidence, if in alpine areas, digital elevation models (DEMs) must be available. If these conditions are met, an optically equivalent grain size can be reliably retrieved from reflectance data such as AVIRIS imaging spectra.

However, there exist two limitations of this technique (Nolin and Dozier 1993), which usually are difficult to overcome. The first one is the effect of sensor noise because the inversion model is based on the signal in one wavelength band. The second limitation is the requirements for knowledge of the sun and sensor angles and for high-quality DEMs over many alpine and polar regions in the world. For this case, Nolin and Dozier (2000) developed a new grain size retrieval technique from imaging spectrometer data. Based on a radiative transfer model, the new inversion technique uses an ice absorption feature centered at $\lambda = 1.03$ μm to correlate the optical equivalent snow grain size. There are several prominent snow/ice absorption features over the solar spectrum range, and one of them extends from 0.96 μm to 1.08 μm, centered at $\lambda = 1.03$ μm. Calculated with the continuum removal approach (Clark and Roush 1984), Nolin and Dozier (2000) utilized the area of the ice absorption feature centered at $\lambda = 1.03$ μm rather than the depth of the absorption feature scaled to a relative reflectance value to be related to optically equivalent grain size in a snowpack's surface layer from imaging spectrometer data. The new inversion technique is insensitive to instrument noise and does not require accurate DEM data for a topographic correction. After the new inversion technique was tested using AVIRIS data collected over the eastern Sierra Nevada mountain range in California and the tested result was validated with a combination of ground-based spectrometer data with grain size measurements, the new developed technique was demonstrated to be a robust, accurate inversion technique for estimating the snow grain size in a snowpack's surface layer from hyperspectral imaging data. Figure 9.4 shows a reflectance image of AVIRIS band 70 (1.03 μm) in Mammoth Mountain, California (left), and the corresponding snow grain size image (right) that was produced using the new inversion technique (Nolin and Dozier 2000). Furthermore, in order to accurately estimate optically equivalent snow grain size from observed spectra from AVIRIS data, Painter et al. (2013) and Seidel et al. (2016) developed and utilized another new method derived from the Nolin–Dozier technique (Nolin and Dozier 2000). The new method searches for the best fit between a modeled clean snow spectrum and the AVIRIS observed surface hemispherical–directional reflectance factor ($HDRF_{sfc}$) at the ice absorption features centered at $\lambda \cong 1.03$ μm for retrieving optically equivalent grain radius (snow grain size). The modeled reflectance of the atmosphere at sensor level $HDRF_{sfc}$ was made by using the Atmospheric and Topographic Correction for Airborne Imagery (ATCOR-4) software (Richter and Schläpfer 2014). They retrieved the snow grain size from the spectral range of 1.03–1.06 μm where the reflected light is primarily sensitive to the snow grain size but less by liquid water and water vapor, which usually has severe effects on reflectance retrieval. They used AVIRIS data sets acquired from the Senator Beck Basin Study Area, Colorado, and the Sierra Nevada and Rocky Mountains in the United States, to demonstrate the performance of the new method through ground-based validation.

In order to accurately map snow cover from hyperspectral imaging data, mapping snow-covered pixels can be carried out using the normalized difference snow index: $NDSI = (R_{VIS} - R_{MIR})/(R_{VIS} + R_{MIR})$, where R_{VIS} and R_{MIR} are the $HDRF_{sfc}$ (or just surface reflectance) retrieved from AVIRIS or other hyperspectral sensor data sets at 0.668 μm and 1.502 μm, respectively (Dozier and Marks 1987, Hall et al. 1995). After calculating pixel based NDSI value from the AVIRIS data set to mask full or partially covered snow pixels, Painter et al. (2013) and Seidel et al. (2016) empirically determined NDSI threshold > 0.9 for the Sierra Nevada and 0.93 for Rocky Mountains. Using satellite hyperspectral sensor Hyperion data, Haq (2014) used surface reflectance retrieved from Hyperion bands at similar wavelengths as those from AVIRIS to calculate the NDSI value for masking snow pixels. It is worth noting that using the NDSI thresholding method to map snow cover is

FIGURE 9.4 Reflectance image and snow grain size image of Mammoth Mountain, California. In the left image, vegetation is dark, snow is bright, and ski runs are visible on the north side of the mountain. Because of the anisotropic reflectance of snow, measured reflectance often exceeds unity (i.e., reflectance in some pixels exceeds 1.0). In the right image, pixels with snow cover <90% have been masked to black. (Reprinted from Nolin, A. W., and Dozier, J., *Remote Sensing Environment*, 74, 207–216, 2000. ©2000, with permission from Elsevier.)

not accurate. For accurately mapping snow cover or fractional snow cover from hyperspectral imaging data, spectral unmixing snow-covered area mapping approaches were developed by researchers, especially the multiple end-member snow-covered area and grain size model (MEMSCAG) method (Painter et al. 2003). According to Painter et al. (1998), snow-covered alpine regions often exhibit large surface grain size gradients due to changes in aspect and elevation, and the sensitivity of snow spectral reflectance to the different grain size translates these grain size gradients into spectral gradients. Ice is very transparent in visible wavelengths so that an increase in grain size has little effect on reflectance, but in the NIR region, ice is moderately absorptive. Hence, reflectance is sensitive to grain size, particularly diagnostic ice absorptions at 1.03 μm and 1.26 μm across the wavelength range of 1.0–1.3 μm. To better map snow-covered area and grain size in alpine regions by using multiple snow end-members from hyperspectral imaging data, Painter et al. (1998) developed a spectral unmixing technique to improve spectral mixture analysis of snow properties. They performed numerical simulations to demonstrate the sensitivity of nonlinear mixture analysis to grain size for a range of sizes and snow fractions. They established mixture models with fixed vegetation, rock, and shade with each snow end-member. For each pixel, the snow fraction estimated by the models with least-squares mixing error (RMS) was chosen to produce an optimal map of subpixel snow-covered area. To test the linear spectral unmixing snow-covered area and grain size, they collected AVIRIS data over Mammoth Mountain, California, on April 5, 1994, and a suite of snow image end-members spanning the imaged region's grain size range was extracted. The snow unmixed results were verified with a high spatial resolution aerial photograph and demonstrated to have an equivalent accuracy. However, they experienced that changes in NIR reflectance with snow grain size were the primary source of mixing error in the fixed end-member–suite spectral mixture models. In particular, in regions of grain-size mismatch, snow fractions were overestimated when the grain size was larger than that of the end-member and underestimated when the grain size was smaller than that of the end-member. Furthermore, Painter et al. (1998) demonstrated that subpixel snow mapping would be improved if the snow end-member were allowed to vary to match the spectral shape of the pixel's snow reflectance. Because the spectral reflectance of snow decreases with increasing grain size, varying grain sizes in a scene translate into variability of spectral reflectance, and multiple snow end-members of different grain sizes are necessary to characterize the snow. Therefore, derived from MESMA (Roberts et al. 1998), Painter et al. (2003) developed a MEMSCAG to map snow and its grain size simultaneously using spectral mixture analysis coupled with a radiative transfer model. The MEMSCAG model allows the number of end-members and the end-members themselves to vary pixel-by-pixel and thereby address subscene spatial heterogeneity, and selection of a spectral mixture model at each pixel with the smallest error can produce an optimized map of snow-covered area and a best choice among a range of grain sizes. Painter et al. (2003) tested MEMSCAG with three AVIRIS image scenes acquired over Mammoth Mountain, California, on April 5, 1994, March 29, 1996, and April 29, 1998. The AVIRIS images spanned common snow conditions ranging from fresh, fine-grain snow to coarse, melting snow. The estimates of snow-covered area and grain size were validated with fine-resolution aerial photographs and the stereological analysis of snow samples collected within two hours of the AVIRIS overpasses. The validation results indicated that the RMS error for snow-covered area retrieved using the MEMSCAG model from AVIRIS for the combined set of three images was 4%; and the RMS error for snow grain size retrieved from a 3 × 3 window of AVIRIS data for the combined set of three images was 48 μm.

Recent studies have shown that light-absorbing impurities (e.g., mineral dust, soil, coal, or other organic matter) are crucial in understanding and determining melt rates of snow and ice. Snow and ice spectral reflectance properties are influenced by snow and ice impurities or contaminants. Hyperspectral techniques may be used to identify/discriminate the type and amount of contamination in order to better understand snow spectral reflectance properties and total hemispherical albedo from contaminated or impure snow and ice. Using *in situ* ASD spectral measurements collected from snowpacks with different levels of soil and coal contamination and varying grain size in controlled conditions in the Himalayan region, Singh et al. (2010, 2011) quantitatively analyzed the effects of different levels of soil and coal contamination with varying grain size in controlled conditions on snow spectral reflectance properties and total hemispherical albedo. Relative strength, asymmetry, first derivative, percentage change in reflectance and albedo were used to analyze the *in situ* hyperspectral measurements. Analysis results indicated that the relative strength showed an increase in band depth for higher quantity of contamination but more prominent in coal contamination than soil contamination. And both soil and coal contamination significantly reduced the albedo of snow at a low level of contamination but showed little influence at higher level. However, absorption depth at 1.0 μm was almost unchanged for soil contamination but reduced for coal contamination. Moreover, soil contamination shows a systematic tendency towards left asymmetry of a curve shape with an increase in soil contamination level. Their observations suggest that contamination could initially drastically reduce the albedo but is almost saturated if the snow is further contaminated.

Snow can host an abundant microbial community supported by phytoplankton collectively termed snow algae. *Chlamydomonas nivalis* is the most prevalent alga found in snowfields in the Sierra Nevada of California (Painter et al. 2001a,b). The spectral reflectance of snow containing *C. nivalis* is different from that of snow without algae due to carotenoid (Car) absorption in the spectral range from 0.4 to 0.58 μm and chlorophyll *a* and *b* (Chla,b) absorption in the spectral range from 0.6 to 0.7 μm. In a study on detection and quantification of snow algae with hyperspectral sensor data, Painter et al. (2001a,b) described spectral reflectance measurements of snow containing the snow alga *C. nivalis* and developed a model to retrieve snow algal concentrations from airborne imaging spectrometer AVIRIS data. By analyzing *in situ* spectral reflectance measurements and *C. nivalis* sample data, it was observed that the integral of the scaled Chla, b absorption feature ($I_{0.68}$) varied with algal concentration (C_a), and thus a linear regression model $C_a = 81019.2\ I_{0.68} + 845.2$ (Painter et al. 2001a) and an exponential model $C_a = 922.61EXP\ (13.56\ I_{0.68})$ (Painter et al. 2001b) were developed. Thus, the spatial distribution of snow algal concentrations was mapped by applying the linear and nonlinear models to reflectance data acquired by AVIRIS sensor for the east drainages of Mt. Conness near Tioga Pass in California. Their experimental results suggest that combining this areal biomass concentration retrieved by the simulated models with maps of snow-covered areas created with hyperspectral remote sensing sensors covering larger regions could facilitate broadscale estimates of total algal biomass in snow-covered areas in alpine regions.

Glacier surfaces are not only composed of ice or snow but are heterogeneous mixtures of different materials also, and the latter with light-absorbing impurities strongly influences ice surface spectral characteristics and glacier melt processes (Naegeli et al. 2015). To assess the composition of ice surface materials and their respective impact on surface albedo and glacier melt rates, Naegeli et al. (2015) used the airborne prism experiment (APEX) imaging spectrometer and *in situ* ASD hyperspectral data sets to map the abundances of six predominant surface materials on Glacier de la Plaine Morte, Swiss Alps, which include debris, dirty ice, water, ice, bright ice, and snow. These materials largely differ in their spectral albedo and hence critically impact any model-based calculations of glacier mass balance (Naegeli et al. 2015). APEX is a dispersive pushbroom imaging spectrometer, covering the spectral range from 0.4 μm to 2.5 μm with 313 narrow continuous spectral bands. They applied the spectral angle mapper (SAM) classification algorithm to map the spatial extent and pixel abundance of the six predominant surface materials. The pixel-based classification result showed that about 10% of the ice surface was covered with snow, water, or debris; the remaining 90% of the surface could be divided into three types of glacier ice (i.e., ~7% dirty ice, ~43% pure ice, and ~39% bright ice). They also used APEX reflectance data coupled with *in situ* spectral measurements to map and analyze albedo spatial patterns present on the glacier surface in the study area. The analysis result indicated that a significant amount of light-absorbing impurities resulted in a strong ice-albedo feedback during the ablation season. Their experimental results demonstrated the applicability and potential of imaging spectroscopy in the field of glaciology, contributed to the understanding of spatial variation in glacier surface albedo, and allowed mapping and detecting the occurrence, composition, distribution, and impact of light-absorbing impurities on ice surfaces using hyperspectral data sets.

Measurement of snow melting at regional scales offers the potential to improve measurement, monitoring, and modeling of snow-driven hydrological processes. Hyperspectral remote sensing can provide spectral information regarding snow melting over a regional scale area. Green and Dozier (1995) examined the measurement of the spectral absorption due to liquid water in a melting snowpack with AVIRIS sensor data. They found in the 1.0 μm region there were two phases of water separable based upon their spectral properties. Measurement of these two phases of water requires spectral modeling of the overlapping absorption of the liquid water absorption centered at 0.97 μm and the ice absorption at 1.03 μm. Thus, they developed an equivalent path transmittance model in the 1.0 μm spectral region for quantifying liquid water and ice using the complex refractive indices of water and ice (Kou et al. 1993). Then, the model was inverted using a nonlinear least squares fitting routine to derive the equivalent path length transmittances of liquid water and ice from each spectrum measured by AVIRIS sensor. Validated with field measurements over the AVIRIS imaged area, the retrievals of the equivalent path length transmittances of liquid water and ice were reasonable. Snow grain size and snow melt influence surface albedo and hydrology. Therefore, Green and Dozier (1996) modeled snow reflectance as a function of grain size and liquid water based on the inherent optical properties of ice and liquid water. Their snow reflectance modeling was based on a discrete ordinate radiative transfer approach that uses Mie calculations of snow optical properties, which are based on the complex refractive index of ice and water (Kou et al. 1993, Green and Dozier 1995, Nolin and Dozier 2000). The model was applied to an AVIRIS snow data set acquired over Mammoth Mountain, California. Retrievals of grain size and surface snow melt through inverting the model were generated that were consistent with the expected ranges and distributions (elevations and temperatures) for the data set.

To simultaneously estimate the abundance of the three phases of water in an environment that includes melting snow from hyperspectral imaging data, Green et al. (2006) applied a spectral fitting algorithm to measure the abundance of the three phases of water from a data set acquired by AVIRIS over Mount Rainier, Washington, on June 14, 1996. The fitting algorithm bases on the spectral shift in the absorption coefficient between water vapor, liquid water, and ice at 0.94, 0.98, and 1.03 μm, respectively. To develop the integrated spectral fitting algorithm, the MODTRAN radiative transfer code is linked to the surface liquid water and ice model and integrated into a downhill simplex multidimensional minimization algorithm (Press et al. 1988). During the fitting processing, this fitting algorithm needs iterative adjustment of the model's parameters to match the modeled and measured spectra. When a threshold residual of the sum of the absolute differences is reached, the fitting processing stops and reports the derived abundance of the three phases of water (Green et al. 2006). The output of the fitting algorithm, when applied to the AVIRIS data set, was a spectral fit for each pixel and an image for each phase of water: precipitable water vapor in millimeters and liquid water and ice absorption as equivalent path lengths in millimeters. The results of the three phases of water output by the algorithm in the melting snow from the AVIRIS data set were validated: for water vapor result, with regional radiosonde measurements; and for ice and liquid water results, with the measurements collected with a field spectrometer to show a spectral shift that should occur as snow melts. The validation results demonstrated the capability of spectral fitting algorithm for simultaneously measuring the expressed abundance of water vapor, liquid water, and ice from hyperspectral sensor data. This new method with imaging spectroscopy data offers synoptic information about the abundance and distribution of the three phases of water in snowpacks in alpine regions.

9.4 COASTAL ENVIRONMENTS AND INLAND WATERS

The number and abundance of absorbing and scattering components found in the coastal ocean, lakes, and rivers support the use of spectroscopy to isolate and measure the constituents in these environments. These components or water quality parameters may include chlorophyll, a variety of planktonic species, color-dissolved organic matter, suspended sediments with local and distant sources, turbidity, bottom composition, submerged aquatic vegetation, and Secchi depth (Green et al. 1998). The increased signal-to-noise ratio and absolute calibration of hyperspectral sensors/systems (ground based, airborne, and satellite platforms) in the wavelength range of 0.4–1.0 μm of the spectrum is supporting new investigations and applications in the coastal and inland water environments. In the following inland waters section, the review is arranged in the order of hyperspectral data types: *in situ*, airborne, and satellite hyperspectral remote sensing, while in the coastal environments section, the review is arranged in the order of application targets or components with various hyperspectral data types: seawater quality parameters, benthic substrate type mapping, seagrass characterizing and assessment, and water depth mapping.

9.4.1 Inland Waters

With *in situ* or laboratory hyperspectral measurements taken by spectrometers, such as ASD Field Spec FR spectrometer (ASD Inc.) and SVC HR-1024 spectrometer (Spectra Vista Corporation), researchers have investigated spectral characteristics of a list of water quality parameters [e.g., chlorophyll-a (Chla), cyanobacterial pigment C-phycocyanin (C-PC), colored dissolved organic matter (CDOM), suspended particle matter (SPM), etc.] in inland waters and estimated the parameter values using simple regression models, spectral feature–based retrieval algorithms, and some advanced algorithms. To retrieve Chla and suspended solids (SS) concentrations in Lake Taihu, China, Ma et al. (2007) applied a spectral feature–based method, with which estimation models were developed using features at important and diagnostic absorption bands by Chla and SS, and a spectral derivative method to *in situ* ASD spectra. They identified the absorption trough for Chla located at about 680 nm and the reflectance peak at about 710 nm and the strong scattering due to SS in NIR wavelengths for extracting spectral features used as modeling input for the feature-based method. They also used the first- and second-order derivative spectra to correlate with the Chla and SS concentration measurements. While both methods seemed workable, the comparative analysis indicated that the derivative method was better for estimating Chla and SS concentrations than the spectral feature–based method. Huang et al. (2010) also used *in situ* spectral measurements collected from Tangxun Lake in central China to assess Chla concentration by correlating with spectral variables. To do so, three types of hyperspectral variables, including single-band reflectance, first derivative spectrum, and band ratio, were extracted from the *in situ* spectral measurements and correlated with field-measured Chla concentration. The correlation analysis results indicated that the first derivative spectrum and band ratio variables had a high correlation coefficient ($R^2 > 0.8$, $n = 10$) with the measured Chla concentration.

Colored dissolved organic matter (CDOM) is an important water quality parameter that affects water color and water inherent optical properties. To analyze spectral properties of CDOM and estimate its concentration in Lake Taihu in China, Sun, Li, Wang, Lu et al. (2011) studied the CDOM-retrieval approach with *in situ* spectra collected from the lake. After collecting *in situ* spectral and CDOM measurements, they established a three-layer back propagation (BP) neural network (NN) model. The input layer had four neurons: $R_{rs}(400)$, $R_{rs}(550)$, $R_{rs}(555)$, and $R_{rs}(560)$, and the output layer had only one neuron: CDOM concentration in $a_{CDOM}(440)$, where $R_{rs}(\lambda)$ is the remote sensing reflectance at the wavelength λ and $a_{CDOM}(440)$ is the CDOM absorption coefficient at the wavelength 400 nm. After conducting the analysis of sensitive spectral factors for CDOM absorption retrieval, the four R_{rs} factors at wavelengths 400, 550, 555, and 560 nm were adopted in the NN model. The modeling results showed that in the hidden layer set with 10 nodes, the NN model performed better, resulting in a correlation coefficient of 0.887 and RMSE of 0.156 m^{-1}. Meanwhile, they tested the NN model using other datasets that were collected at different times for applicability analysis, and the derived result also demonstrated a relatively good performance of the NN model (Sun, Li, Wang, Lu et al. 2011). Using *in situ* above-water surface hyperspectral measurements collected in the Mississippi and Atchafalaya River plume regions and the northern Gulf of Mexico, where water types vary from Case 1 to turbid Case 2, Zhu et al. (2011) developed an enhanced inversion model of hyperspectral remote sensing for retrieving the absorption coefficient of CDOM. In the enhanced inversion model, they separated a_g, the absorption coefficient a of CDOM, from a_{dg} (a of CDOM and nonalgal particles) based on two absorption backscattering relationships (i.e., the first one is between a_d [a of nonalgal particles] and b_{bp} [total particulate backscattering coefficient] and the second is between a_p [a of total particles] and b_{bp}). Then, based on Lee's quasi-analytical algorithm (QAA), they developed the so-called extended quasi-analytical algorithm (QAA-E) to decompose a_{dg}, using both a_d-based and a_p-based methods. They tested the absorption-backscattering relationships and the QAA-E using synthetic data from the International Ocean-Colour Coordinating Group as well as their own measured field data. The tested results indicated that the accuracy of CDOM estimation was significantly improved by separating a_g from a_{dg} ($R^2 = 0.81$ and 0.65 for synthetic and *in situ* data, respectively), and that the new enhanced inversion model QAA-E was robust.

In order to retrieve cyanobacterial pigment C-phycocyanin (C-PC) from cyanobacteria-dominated large turbid lakes in China, Sun et al. (2012, 2013) developed and tested an advanced support vector regression (SVR also called SVM) model with *in situ* hyperspectral measurements. They collected both C-PC and *in situ* spectral measurements from three inland lakes in China: Lake Taihu, Lake Chaohu, and Lake Dianchi. Three types of spectral variables, including band reflectance, band ratio, and three-band combination, were compared and, based on better correlation with C-PC measurement, the band ratio was found to be the best candidate to serve for model development. Then Sun et al. (2012) tested two modeling types, a linear regression model and SVR model for retrieving C-PC concentration. The tested results showed that the best-performing model was the SVR model with the highest prediction accuracy and the lowest errors among all modeling methods. The best SVR model took the seven-band ratios (selected from all possible two-band ratios) as inputs to comprehensively characterize the C-PC information, when the C-PC concentration was treated as the only output. To improve the retrieval of C-PC concentration in inland turbid eutrophic waters, with the same data sets of *in situ* spectral and C-PC measurements collected from the three lakes, Sun et al. (2013) first classified the *in situ* spectral samples into three optical types based on the recently developed TD680 optical classification criterion (Sun, Li, Wang, Le et al. 2011) and then applied the same algorithm (SVR) as that used in Sun et al. (2012) to estimate C-PC concentration for each type of samples. TD680 represents the trough depths of remote sensing reflectance [$R_{rs}(\lambda)$] near 680 nm, which usually vary among different spectral cases of different ratios of inorganic suspended matter to total suspended matter (Sun, Li, Wang, Le et al. 2011). In the study, the TD680 was defined as $TD680 = 0.5 \times [R_{rs}(655) + R_{rs}(705)] - R_{rs}(680)$ and according to the proposed TD680 thresholds in Sun, Li, Wang, Le et al. (2011), the investigated turbid waters currently from the three lakes could be divided into three types as follows: Type 1 ($TD680 \geq 0.0082$), Type 2 ($0.0082 > TD680 > 0$), and Type 3 ($TD680 \leq 0$). They then developed three type-specific SVR algorithms and an aggregated SVR algorithm (using total *in situ* spectral and C-PC measurement samples). The performances of the four SVR algorithms were evaluated using validation data sets. The evaluated results showed that the type-specific SVR algorithms significantly outperformed the aggregated SVR algorithm in retrieving C-PC concentration from *in situ* spectral measurements collected from the three lakes.

With airborne hyperspectral imaging data, researchers have studied retrieving and mapping a list of water quality parameters (e.g., Chla, C-PC, CDOM, SPM, turbidity, and Secchi depth) in inland waters through calibrating retrieval models using *in situ* spectral and water quality parameter measurements and then applying the calibrated models to estimate and map the water parameters with airborne hyperspectral imaging data. For examples, Thiemann and Kaufmann (2002) developed models using *in situ* ASD spectral and water trophic parameter measurements for estimating two trophic parameters Secchi disk transparency and Chla concentration from hyperspectral airborne sensors CASI and HyMap data. To determine Secchi disk transparency (water depth), the spectral reflectance between 400 and 750 nm was first used to calculate a spectral coefficient (SpCoef;

Thiemann and Kaufmann 2002), then a regression model between the SpCoef and Secchi disk transparency (SD) as measured with the Secchi disk was developed in an exponential form as follows: $SD = 13.07e^{2.94SpCoef}$. Chla presents two diagnostic absorption bands centered at 435 and 678 nm, and the reflectance peak around 700 nm rises with increasing Chla concentration. Thus, Chla concentration might be linearly quantified using the existing reflectance ratio at 705 and 678 nm (R_{705}/R_{678}). In this study, the wavelengths used for calculating the ratio corresponded to the most frequent maximum and minimum within the *in situ* measured spectra. Hence, a linear relationship could be fitted as follows: $Chla = -52.91 + 73.59Ratio_{(705/678)}$, and applied to derive Chla concentration from the *in situ* reflectance data. To use airborne hyperspectral sensors images (CASI and HyMap), the SD and Chla models were adapted by resampling the *in situ* reflectance spectra to the resolution of CASI and HyMap, respectively. Then the calibrated SD and Chla models were applied to the sensor spectra to estimate and map SD (water depth) and Chla concentration. The validated results by using independent *in situ* reference measurements showed mean standard errors of determination of 1.0–1.5 m for Secchi disk transparency and of 10–11 μg l^{-1} for Chla when using CASI and HyMap sensor data. Sudduth et al. (2015) also used *in situ* spectral and water quality parameter measurements collected in inland waters to calibrate retrieval models and then applied the calibrated models to airborne (AISA) sensor data to retrieve and map pixel-based water quality parameters. In the study, they collected *in situ* ASD spectral spectra and water quality parameters: Chla, turbidity, and N and P species from Mark Twain Lake, a large man-made reservoir in northeastern Missouri. They developed the water quality parameter retrieval models in a spectral index–based method, a modeling method similar to that by Thiemann and Kaufmann (2002), and in a full-spectrum (i.e., partial least squares regression [PLSR]) method using the *in situ* measurements. Then they applied the calibrated retrieval models to AISA imaging data to estimate the pixel-based water quality parameters for the lake. The results showed that most measured water quality parameters were strongly related ($R^2 \geq 0.7$) to *in situ* reflectance spectra across all measurement dates; and that aerial hyperspectral sensing was somewhat less accurate than field method for retrieving the water parameters. Figure 9.5, as a sample, presents the Chla concentration retrieval map created by the PLSR method using AISA hyperspectral image data acquired on August 31, 2005. From the map, the spatial distribution of Chla concentration is clearly presented.

In calibrating water quality parameter retrieval models with *in situ* spectral and water parameter measurements, some researchers investigated physical-based or semi-analytical inversion modeling methods. To retrieve Chla and phycocyanin (C-PC) from hyperspectral imaging data acquired by the CASI-2 and the AISA-Eagle airborne sensors from two shallow lakes in the UK, Hunter et al. (2010) evaluated and compared the performances of four semi-analytical algorithms and some empirical band ratio algorithms. The four semi-analytical algorithms include G05 and G08 algorithms for the retrieval of Chla concentration and S05 and H10 for the retrieval of C-PC concentration. These specific semi-analytical models were chosen because they were specifically developed for pigment retrieval in turbid inland waters (Hunter et al. 2010). Gons et al. (2005) developed the G05 semi-analytical algorithm for Chla retrieval from the medium resolution imaging spectrometer (MERIS) onboard the European Space Agency's Envisat. Gitelson et al. (2008) developed the G08 also for the retrieval of Chla from MERIS data. The semi-analytical algorithm: S05 and H10, developed by Simis et al. (2005) and Hunter et al. (2010) based on Dall'Olmo et al. (2003), are for the retrieval of C-PC from hyperspectral remote sensing data. The G05 and S05 algorithms estimate the values of absorption coefficients $a_{Chl}(665)$ and $a_{C\text{-}PC}(620)$, respectively, and then the retrieved absorption coefficients are converted to pigment concentrations using the specific absorption coefficients for Chla at 665 nm (0.0161 m^{-1}; Gons et al. 2005) and C-PC at 620 nm (0.007 m^{-1}; Simis et al. 2007). The G08 in $Chla = 23.1 + 117.4 \times \{[R_{rs}^{-1}(600-670) - R_{rs}^{-1}(700-730)] \times R_{rs}(740-760)\}$, where $R_{rs}(\lambda_1 - \lambda_2)$ is the remote sensing reflectance from wavelength λ_1 to wavelength λ_2, is used for retrieving Chla from MERIS data and the H10 in $CPC \propto [R_{rs}^{-1}(615) - R_{rs}^{-1}(600)] \times R_{rs}(725)$ is used to estimate C-PC from the sensor's data (Hunter et al. 2010). To assess the performance of the four semi-analytical algorithms, they also used empirical algorithms that were derived by regressing the

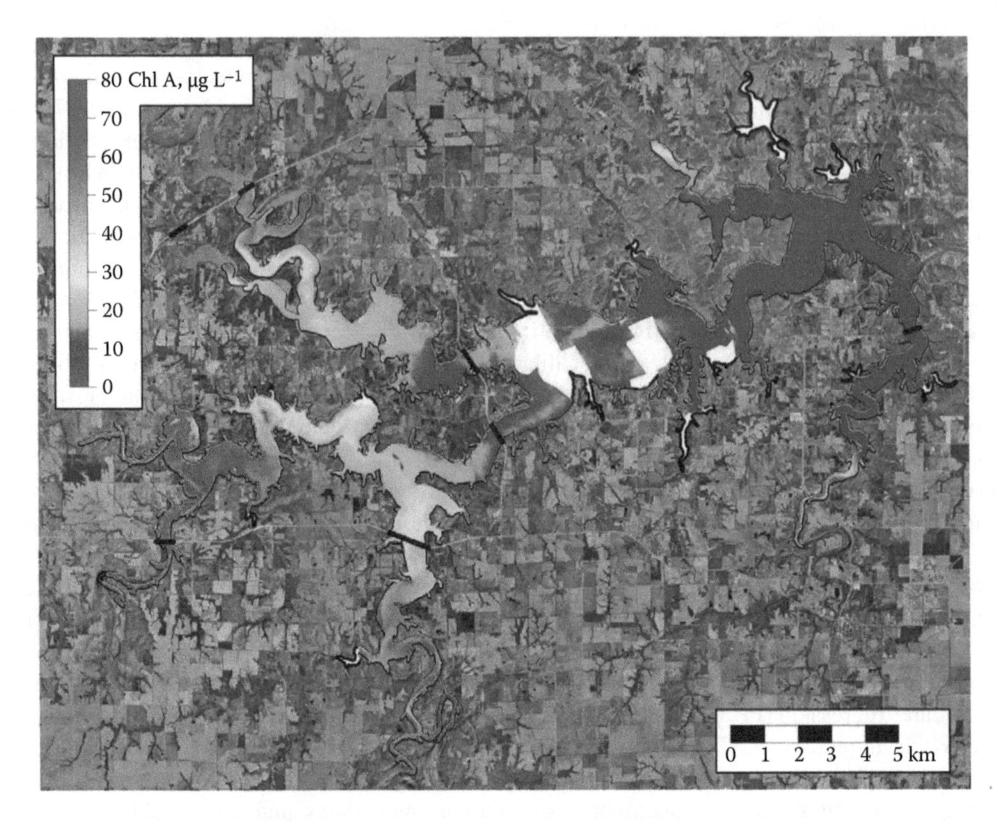

FIGURE 9.5 Map of Chla (Chl A) concentration in Mark Twain Lake created by the PLSR method using AISA hyperspectral aerial image data acquired on August 31, 2005. Black bars across lake arms denote locations of seven bridges where *in situ* water samples and spectrometer data were obtained. White areas in the lake are missing data. (Reprinted with permission of American Society of Agronomy, Crop Science Society of America, and Soil Science Society of America. From Sudduth, K. A., Jang, G.-S., Lerch, R. N., and Sadler, E. J., *Journal of Environmental Quality*, 44, 71–83, 2015. Permission conveyed through Copyright Clearance Center, Inc.)

band ratios [$R_{rs}(705)/R_{rs}(670)$] and [$R_{rs}(705)/R_{rs}(620)$] against the linear and log-transformed Chla and C-PC concentrations using linear and second-order polynomial models for estimating Chla and C-PC from hyperspectral data. The two band ratios have been effectively used in previous studies associated with turbid inland waters (Hunter et al. 2010). After the retrieval accuracies of the four semi-analytical models were compared to those retrieved by optimally calibrated empirical band ratio algorithms, the results indicated that the best-performing algorithm for retrieving Chla was a nonlinear empirical model with the ratio of reflectance at 710 and 670 nm ($R^2 = 0.832$; RMSE = 29.8%). However, this best empirical model was only slightly better in terms of retrieval than the best semi-analytical G05 algorithm that was also strongly correlated with the measured Chla concentration. The best-performing algorithm for retrieving C-PC was the semi-analytical S05 algorithm, because the absorption coefficients and C-PC concentrations returned by the S05 model were strongly correlated with those measured in the lakes. These results demonstrated that the performance of analytically based algorithms could be equally used as more widely used empirical approaches.

By applying a physically based method, Giardino et al. (2015) relied on spectral inversion procedures to simultaneously retrieve water quality parameters (suspended particulate matter [SPM], Chla and CDOM [i.e., $a_{CDOM}(440)$]) in optically deep waters, and water column heights and benthic substrate types in optically shallow waters in Lake Trasimeno, Italy, from airborne imaging spectrometry MIVIS sensor. The spectral inversion procedure implemented in the study was performed with the tool BOMBER (Lee et al. 1998, Giardino et al. 2012). The software tool makes use of bio-optical models for both optically deep and shallow waters. By implementing the inversion procedure in optically deep waters, water quality parameters SPM, Chla, and CDOM were allowed to be retrieved; in optically shallow waters, bottom depth and mapping different types of substrate were achieved as well. To test the physically based inversion procedure, *in situ* spectral and water quality and other field data and airborne MIVIS hyperspectral imaging data were collected. The BOMBER model was parametrized with the *in situ* data. The MIVIS imaging data were first converted to remote sensing reflectance $R_{rs}(\lambda)$ with the atmospheric correction code ATCOR (Richter and Schläpfer 2014) and then used as input to the inversion procedures to retrieve the water quality and other parameters. After being validated with *in situ* data, the MIVIS-derived results indicated that, in optical deep waters, the spectral inversion model provided ranges of SPM, Chla, and $a_{CDOM}(440)$, comparable to *in situ* data collected the day of the airborne campaign; in shallow waters, the water column heights above the substrates derived from MIVIS matching the acoustic soundings and the benthic cover patterns mapped from MIVIS were also comparable to *in situ* observations.

Some researchers directly used airborne sensor data, after converted to water surface remote sensing reflectance or upwelling radiance, and *in situ* water quality parameters to establish linear and nonlinear models to estimate and map water quality parameters. Hamilton et al. (1993) estimated Chla concentration from the AVIRIS data for the surface waters of Lake Tahoe, using the CZCS algorithm for low pigment concentration (Gordon et al. 1983). The CZCS algorithm was given as follows: $log_{10}(Chla) = 0.053 + 1.7\ log_{10}[L_w(550)/L_w(443)]$, where $L_w(550)$ and $L_w(443)$ are the values of upwelled radiance from the surface water provided by the CZCS sensor in bandwidth of 20 nm, which could be approximated using corresponding AVIRIS sensor data in bandwidth of 10 nm. The AVIRIS derived Chla concentration with the CZCS algorithm was found to agree very well with both bottle samples and in-water measurements of upwelling radiance. Olmanson et al. (2013) also used airborne AISA sensor spectra to estimate and map water quality parameters (Chla, volatile suspended solids [VSS], total suspended solids [TSS], turbidity, etc.) in the Mississippi River and its tributaries in Minnesota. They developed simple regression models using single bands and band ratios with *in situ* water quality parameters. The results showed that band ratios of the scattering peak at the red edge (~700 nm) to chlorophyll and other pigment absorption bands at ~670 nm, 592 nm, and 620 nm were strong predictors of VSS and Chla concentrations ($R^2 = 0.73 \sim 0.94$, n = 25), and the scattering peak at ~700 nm was a strong predictor of TSS and turbidity ($R^2 = 0.77 \sim 0.93$, n = 25). The analysis results derived from the study also suggested that hyperspectral imagery could be used to distinguish and map key water quality parameters under complex inherent optical property conditions.

Although there are only a few satellite hyperspectral sensors/systems that are available to be considered for use in inland waters, some researchers did investigate the satellite hyperspectral data sets and applied them for retrieving and mapping water quality parameters (Chla, C-PC, SPM, CDOM, turbidity, Secchi depth, etc.) in inland waters. They used *in situ* spectral measurements to calibrate retrieval models and then applied the calibrated models to satellite hyperspectral sensor data or directly correlated *in situ* water quality parameters to estimate and map pixel-based inland water quality parameters. The satellite hyperspectral sensors or systems may include EO-1/Hyperion, European Space Agency's Envisat/MERIS, and Chinese HJ-1A/HSI (hyperspectral imager). Both Wang et al. (2005) and Yan et al. (2005) explored the Hyperion hyperspectral data with simultaneously collected *in situ* water quality parameters: Chla and suspended sediments (SS) concentrations and the absorption coefficients of yellow substance at 440 nm [$a_{ys}(440)$] data from high turbid inland waters Lake Taihu in eastern China for retrieving and mapping the water quality parameters from the sensor data. Wang et al. (2005) applied empirical correlation analysis methods with three spectral variable forms (i.e., band ratio, band difference, and NDVI) to correlate with water parameters Chla and SS concentrations. The correlation analysis results showed that the band difference technique between R(732~885) (i.e., Hyperion reflectance spectra from wavelength 732 nm to 885 nm) and R(1175~1195) had a stronger correlation with SS ($R > 0.70$, n = 25), while the NDVI technique with R(620~691) and R(722~844) had a close correlation with Chla ($R > 0.90$, $n = 25$). The analysis results suggested that the optimal wavelength and algorithm determined for the retrieval of the water quality parameters would substantially facilitate the water quality monitoring using satellite hyperspectral imaging data. Yan et al. (2005) utilized a physically based method to calibrate spectral analytical models with *in situ* and laboratory-measured data and they inverted the calibrated analytical models to retrieve and map Hyperion pixel based Chla and SS concentrations and $a_{ys}(440)$. The retrieved results from Hyperion data showed that the inversion method of the spectral analytical model, which could account for the mutual influences of the water quality parameters, might be employed to map the spatial distribution of Chla, SS and $a_{ys}(440)$, and the mapped results were comparable with those estimated by the empirical models and the field measurements.

Satellite imaging spectrometer MERIS data have been analyzed for estimating some inland water quality parameters. For example, Gons et al. (2002) applied a calibrated Chla retrieval algorithm to MERIS imaging data to estimate Chla concentration in IJssel Lagoon in the Netherlands. This MERIS algorithm was calibrated for Chla concentration in the range 3~185 mg/m^3 using spectral reflectance calculated from shipboard measurements on the IJssel Lagoon. The MERIS algorithm as follows was adapted from the original equation by Gons (1999): Chla = $\{R_M[a_w(704) + b_b] - a_w(664) - b_b^p\}/a^*(664)$, where R_M is the reflectance ratio at wavelengths 704 nm and 664 nm, i.e., the NIR reflectance peak at 704 nm in eutrophic waters and the red absorption peak at 664 nm of Chla; $a_w(704)$ and $a_w(664)$ are the water absorption coefficients and their values are the 0.630 and 0.402 m^{-1}, respectively; $a^*(664)$ is a mean Chla specific absorption coefficient at 664 nm and its value is 0.0146; b_b is the backscattering coefficient (assumed to be wavelength independent) derived from the reflectance at 776 nm (Gons 1999); and p is an empirical constant close to unity ($p = 1.063$ for MERIS data). The MERIS algorithm was calibrated using data collected in the IJssel Lagoon from 1993 to 1996 and validated using data in the IJssel Lagoon in 1997 and 1999, other Dutch inland waters, the Chinese Lake Taihu, the Scheldt Estuary (Belgium/the Netherlands), the Hudson/Raritan Estuary (New York/New Jersey), and the North Sea off the Belgian coast. In spite of the lower spectral resolution of MERIS as compared with the shipboard spectroradiometer, the standard error of estimate is expected to be similar, i.e., ~9 mg/m^3 of Chla in mesotrophic and eutrophic lakes, rivers, estuaries, and coastal waters (Gons et al. 2002). To classify three water quality parameters: Secchi depth, turbidity, and Chla, Koponen et al. (2002) studied the use of airborne (AISA) and simulated satellite (MERIS) remote sensing data sets. To do so, they used an extensive airborne spectrometer and an *in situ* data set obtained from four lake water quality measurement campaigns in southern Finland during 1996–1998 and developed regression models for each of the three water quality parameters. In their analysis, the best retrieval algorithm for each parameter was found empirically by deriving regression models from all possible bands, band ratios, and band-normalized difference ratios and then selecting the one with the highest R^2. After the three water parameters on each pixel were estimated with their corresponding regression models, Koponen et al. (2002) adopted class limits for the water quality parameters to classify Secchi depth into three classes and turbidity and Chla each into five classes. The class limits for the three water parameters were made by considering two different water quality classification systems: the Water Quality Classification of Inland Waters in Finland and the OECD Lake Classification Scheme (Koponen et al. 2002). Compared to ground truth data, the classification accuracy created with AISA data was 90% for the three Secchi depth classes, 79% for the five turbidity classes, and 78% for the five Chla classes. With simulated MERIS data, the similar classification accuracy as that created with ASIA data was obtained. For monitoring the levels of suspended solids in small and intermediate sized lakes and reservoirs, Tarrant et al. (2010) evaluated the potential application of data from MERIS for estimating total suspended matter (TSM) in four southwestern United States lakes, Roosevelt Lake, Saguaro Lake, Bartlett Lake, and Lake Pleasant, by comparing field data with the image obtained from satellite sensor MERIS. The TSM was estimated from MERIS Level-2 images through the inverse modeling of nonlinear regression procedure or neural network in considering all 13 MERIS bands. The result indicated that the MERIS-derived TSM product was capable of predicting TSM in small to intermediate-sized inland water bodies, although it might be necessary to further validate MERIS-derived TSM product with field data to confirm whether the performance of the inversion modeling method could be repeated on a per case basis.

Recently, sensor data from the Chinese satellite HJ-1A/HSI hyperspectral were used to estimate Chla concentration in Lake Taihu (Li et al. 2010), Lake Dianshan (Zhou et al. 2014), and Lake Qiandao in China (Feng et al. 2015). According to Li et al. (2010), chlorophyll fluorescence properties are effective in detecting Chla concentration in inland waters and provide new optional sensitive bands to retrieve Chla concentration in a complex water. Correlation analysis demonstrated that there was an exponential relationship between Chla and fluorescence line height (FLH). Li et al. (2010) used 50 samples of *in situ* spectral and Chla concentration measurements collected in Lake Taihu from May to August 2008 to establish a regression (exponential) model as follows: $Chla = 7.9755e^{23.797FLH}$. The fitting degree was 0.7302. When the exponential model was applied to two calibrated and georeferenced HJ-1A/HSI images that were acquired on May 9, 2009, to retrieve Chla concentration, the results showed that when compared with empirical MODIS algorithms of Lake Taihu, HSI derived Chla was more similar with that derived from *in situ* reference data. In a similar way as that in Li et al. (2010) using HJ-1A/HSI data to estimate Chla concentration in inland waters, Zhou et al. (2014) based on *in situ* spectral and Chla concentration measurements collected in Lake Dianshan in China and developed a semi-analytical three-band algorithm between remote sensing reflectance and Chla concentration as follows: $Chla = 574.11 \times [R_{rs}^{-1}(653) - R_{rs}^{-1}(691)] \times R_{rs}(748) + 36.796$, where $R_{rs}(\lambda)$ is the remote sensing reflectance at wavelength λ. The ground-based calibrated model could account for 85.6% of variance in Chla concentration measurements. In order to test the performance of the calibrated model with satellite data, HJ-1A/HSI data were analyzed using comparable wavelengths selected from the *in situ* spectral data as $Chla = 328.2 \times [R_{rs}^{-1}(656) - R_{rs}^{-1}(716)] \times R_{rs}(753) + 18.384$, where wavelengths 656, 716, and 753 nm equal HSI bands 67, 80, and 87. The HSI-based model could account for 84.3% of variance in Chla concentration measurements. The high Chla retrieval accuracy indicated that the three-band model could be applied to hyperspectral satellite imagery for the estimation of higher Chla concentration in inland waters. Work similar to that by Li et al. (2010) and Zhou et al. (2014) was done by Feng et al. (2015); they also first used *in situ* spectral and Chla concentration measurements collected from the Lake Qiandao of China to calibrate a four-band model. And then they retuned the calibrated model based on hyperspectral satellite sensor HSI band setting as $Chla = 20.02 \times [R_{rs}^{-1}(661) - R_{rs}^{-1}(706)][R_{rs}^{-1}(717) - R_{rs}^{-1}(683)] + 20.90$, where wavelengths 661, 683, 706, and 717 nm equal HSI bands 68, 73, 78, and 80. The HSI-based model could account for about 80% of variation in Chl- measurements. Finally, the HSI-based model was applied to HSI hyperspectral data for estimating Chla concentration in the Lake Qiandao. The results demonstrated the rationale of the four-band model and the effectiveness of this model for estimating Chla concentration with both *in situ* spectral data and HJ-1A hyperspectral satellite imagery.

For readers' easy perusal of some important algorithms and techniques that are used for retrieving water quality parameters—Chla, C-PC, Secchi depth, suspended matter, etc.—from different hyperspectral data sets, Table 9.3 presents a summary of the important algorithms and techniques. The table briefly summarizes mathematical expressions, characteristics of the algorithms/techniques, and their required hyperspectral sensors/systems, applicable waters, and references.

9.4.2 Coastal Environments

Hyperspectral data (*in situ* spectra and imaging data) have been used to investigate the distribution and concentration (or content) of water quality parameters, benthic substrate mapping, seagrass characterizing and assessing, and water depth retrieval in coastal environments (e.g., for water quality parameters: Carder et al. 1993, Brando and Dekker 2003, Keith et al. 2014; for benthic substrate mapping: Clark et al. 1997, Peneva et al. 2008, Valle et al. 2015; for seagrass mapping and assessment: Williams et al. 2003, Pu and Bell 2013, Dierssen et al. 2015; and for water depth retrieval:

TABLE 9.3
A Summary of Retrieval Algorithms of Water Quality Parameters Using Hyperspectral Data Sets

Water Quality Parameter	Mathmatical Expression	Characteristics: Spectral Variable and Parameter	Hyperspectral Sensor	Suitable Water	Reference
Chlorophyll-a (Chla)	$Chla = 23.1 + 117.4 \times \left\{ \left[\begin{array}{l} R_{rs}^{-1}(600-670) - R_{rs}^{-1}(700-730) \\ \times R_{rs}(740-760) \end{array} \right] \right\}$	$R_{rs}(\lambda_1\text{-}\lambda_2)$ is the remote sensing reflectance from wavelength λ_1 to wavelength λ_2,	MERIS	Inland waters	Gitelson et al. 2008.
	$\log_{10}(Chla) = 0.053 + 1.7 \times \log_{10}[L_w(550)/L_w(443)]$	$L_w(\lambda)$ is the value of upwelled radiance at wavelength λ from the surface water provided by the CZCS sensor.	CACS bands simulated from AVIRIS	Coastal and inland waters	Gordon et al. 1983.
	$Chla = \left\{ R_M [a_w(704) + b_b] - a_w(664) - b_b^p \right\} / a^*(664)$, where R_M is the R_{704}/R_{664}; p=1.063 for MERIS.	$a_w(704) = 0.630$ (m^{-1}) and $a_w(664) = 0.402$ (m^{-1}) are absorptions at 704 nm and at 664 nm, respectively; $a^*(664) = 0.0146$ is a mean Chla specific absorption coefficient at 672 nm; b_b is the backscattering coefficient (assumed to be wavelength independent) = the reflectance at 776 nm; and p is an empirical constant close to unity.	MERIS	Inland waters	Gons 1999; Gons et al. 2002.
	$Chla = 328.2 \times \left[R_{rs}^{-1}(656) - R_{rs}^{-1}(716) \right] \times R_{rs}(753) + 18.384$	Wavelengths 656, 716 and 753 nm equal HSI bands 67, 80 and 87.	HJ-1A/HSI.	Inland waters	Zhou et al. 2014.
	$Chla = 20.02 \times \left[R_{rs}^{-1}(661) - R_{rs}^{-1}(706) \right] \left[R_{rs}^{-1}(717) - R_{rs}^{-1}(683) \right] + 20.90$	Wavelengths 661, 683, 706, and 717 nm equal HSI bands 68, 73, 78, and 80.	HJ-1A/HSI.	Inland waters	Feng et al. 2015.

(*Continued*)

TABLE 9.3 (CONTINUED)
A Summary of Retrieval Algorithms of Water Quality Parameters Using Hyperspectral Data Sets

Water Quality Parameter	Mathmatical Expression	Characteristics: Spectral Variable and Parameter	Hyperspectral Sensor	Suitable Water	Reference
	$\log_{10}(\text{Chla} - a_4) = a_0 + a_1L + a_2L^2 + a_3L^3$, where $L = \log_{10}[R(490)/R(555)]$	The coefficients are $a_0 = 0.3410$, $a_1 = -0.3001$, $a_2 = 2.8110$, $a_3 = -2.0410$, and $a_4 = -0.0400$.	EO-1/Hyperion	Case I sea waters	Liew and Kwoh 2002.
	$\text{Chla} = 17.477 \times \left[R_{rs}^{-1}(686) - R_{rs}^{-1}(703)\right] \times R_{rs}(735) + 6.152$	Estimating Chla concentration ($\mu g\ l^{-1}$).	HICO	Coastal waters	Keith et al. 2014.
Pigment C-phycocyanin (C-PC)	$\text{CPC} \propto \left[R_{rs}^{-1}(615) - R_{rs}^{-1}(600)\right] \times R_{rs}(725)$	$R_{rs}(\lambda)$ is the remote sensing reflectance at wavelength λ.	AISA and CASI	Inland waters	Hunter et al. 2010.
Tubidity	$\text{Turbidity} = 2 \times 10^6 \left[R_{rs}^{2.7848}(646)\right]$	Estimating turbidity (NTU)	HICO	Coastal waters	Keith et al. 2014.
Colored dissolved organic matter (CDOM)	$a_{CDOM}(412) = 0.8426 \times [R_{rs}(670)/R_{rs}(490)] - 0.032$	Estimating absorption coefficient (m^{-1}) of CDOM at wavelength 412 nm.	HICO	Coastal waters	Keith et al. 2014.
Secchi depth	$SD = 13.07e^{2.94SpCoef}$, where SpCoef is the Spectral Coefficient.	The SpCoef is calculated from the spectral reflectance between 400 and 750 nm. SD is used to determine Secchi disk transparency (water depth).	CASI and HyMap	Inland waters	Thiemann and Kaufmann 2002.
Suspended matter	$TD680 = 0.5 \times [R_{rs}(655) + R_{rs}(705)] - R_{rs}(680)$	Trough Depths of remote sensing reflectance $[R_{rs}(\lambda)]$ near 680 nm correlate with ratios of inorganic suspended matter to total suspended matter.	*In situ* ASD spectra.	Inland turbid lake waters.	Sun et al. 2011a.

Sandidge and Holyer 1998, Jay and Guillaume 2015). To assess the feasibility of using both multispectral and hyperspectral approaches for detecting spring blooms of *Phaeocystis globosa*, Lubac et al. (2008) first took *in situ* spectral measurements in the 350–750 nm spectral range with two TriOS radiometers at 3 nm resolution in the inshore and offshore waters of the eastern English Channel and southern North Sea. They considered the two approaches using the *in situ* spectral data to assess the performance of hyperspectral data for detecting spring blooms of *P. globosa*: The first approach was using the two reflectance ratios [$R_{rs}(490)/R_{rs}(510)$ and $R_{rs}(442.5)/R_{rs}(490)$] in the multispectral inversion, while the second approach was using the hyperspectral inversion based on the analysis of the second derivative of $R_{rs}(\lambda)$ ($d\lambda^2 R_{rs}$). The results in the first approach suggested that detection of *P. globosa* blooms was possible from current ocean color sensors, but the effects of Chla and CDOM concentrations and particulate matter composition on the performance of the two-band ratio approach need to be investigated via sensitivity analysis. The results of the hyperspectral approach showed that *P. globosa* blooms might be distinguished from the second derivative of $R_{rs}(\lambda)$ in the 400–540 nm spectral range, and the analysis of $d\lambda^2 R_{rs}$ allowed the establishment of two criteria to discriminate *P. globosa* blooms from diatom blooms. The two criteria were based on the position of the maxima (around 471 nm) and minima (around 499 nm) of $d\lambda^2 R_{rs}$, with shift depending on the presence or absence of *P. globosa* blooms (Lubac et al. 2008). The hyperspectral derivative spectral analysis results confirmed the conclusion drawn by Richardson and Ambrosia (1996) that using airborne AVIRIS derivative spectra could detect algal accessory pigments over leveed salt ponds in the vicinity of Moffett Field, California, especially after the signal-to-noise ratio was improved for AVIRIS spectra in recent years. In order to monitor and better understand algal blooms, Minu et al. (2015) also measured remote sensing reflectance [$R_{rs}(\lambda)$] and absorption coefficients of phytoplankton blooms in coastal waters off Kochi, Southeastern Arabian Sea, using a Satlantic hyperspectral radiometer to investigate differences in the absorption and reflectance of different types of blooms. The Satlantic hyperspectral ocean color radiometer may be used to measure optical data with 255 channels within a spectral range from 300 nm to 1200 nm. The analysis results showed that (1) peaks of the $R_{rs}(\lambda)$ spectra were for *Trichodesmium spp.* bloom at 490 nm and for those of non-bloom areas at 482, 560 and 570 nm; (2) the absorption maxima of phytoplankton were at 435, 437, 438, and 439 nm in the blue region and 632, 674, 675, and 635 nm in the red region, respectively, for *Trichodesmium spp., Chaetoceros spp., Dinophysis spp., and Prorocentrum spp.* blooms; and (3) the variation of chlorophyll specific phytoplankton absorption controlled the behavior of the $R_{rs}(\lambda)$ peaks in these blooms.

To use satellite Hyperion hyperspectral data in retrieving and mapping the water optical parameters (e.g., Chla and CDOM concentrations) of turbid coastal waters (case II waters), Liew and Kwoh (2002) applied a spectral-fitting method that makes use of the spectral information contained in a full spectrum in the retrieval procedure other than traditional methods that are mostly based on band ratios (e.g., mathematic expression 6 for estimating Chla in Table 9.3). This is because the band ratio method works well in case I waters where the water optical properties are influenced only by Chla and co-varying constituents in the ocean. However, in case II waters, the blue absorption bands are often influenced by both Chla and CDOM, which make the band ratio–based algorithm be not able to separate the two optical parameters and thus, the band ratio algorithm often overestimates Chla concentration in coastal waters. Leiw and Kwoh (2002) used the Hyperion spectrum to fit the model for water reflectance and atmospheric transmission by finding a set of fitting parameters that best fits the reflectance curve to the model. Compared to the traditional band ratio method, they have demonstrated the performance of the spectral-fitting technique of inverse modeling to retrieve the two optical parameters (Chla and CDOM) from Hyperion imaging spectra. Also using Hyperion imaging data, Brando and Dekker (2003) first applied a matrix inversion method to satellite hyperspectral imagery to retrieve concentrations of Chla, CDOM, and suspended matter in Eastern Australia, including Moreton Bay in southern Queensland. An analytical approach for water quality retrieval relates the subsurface irradiance reflectance to the water constituent concentrations. A more efficient approach for retrieval of water parameters may be a direct inversion of an analytical model using a linear matrix inversion method (MIM; Lee et al. 2001, Brando and Dekker 2003). In the study, the retrieval results derived from Hyperion data with the MIM method indicated that the MIM method could simultaneously retrieve all constituent concentrations, which were comparable to those estimated in the field on the days of the overpass.

Satellite hyperspectral imager for the coastal ocean (HICO) data were also investigated by Gitelson et al. (2011) and Keith et al. (2014) for estimating Chla, CDOM, and turbidity in coastal and estuarial turbid waters. Gitelson et al. (2011) used red and NIR spectral bands of hyperspectral sensor HICO data to retrieve Chla concentration in productive turbid waters of the Azov Sea, Russia. They analyzed water samples associated with Chla concentration and HICO spectral data, tuned the NIR–red models to optimize the spectral band selections, and retrieved Chla concentrations (mg m^{-3}) from HICO data. An optimal NIR-red model was Chla = 418.88 × [R_{rs}^{-1} (684) – R_{rs}^{-1} (700)] × R_{rs}(720) + 19.275, where $R_{rs}(\lambda)$ is the remote sensing reflectance at wavelength λ. Extracted from HICO data, the NIR–red three-band model with HICO-retrieved remote sensing reflectance at wavelengths 684, 700, and 720 nm could explain more than 85% of Chla concentration variation in the range from 19.67 to 93.14 mg m^{-3} with a RMSE < 10 mg m^{-3} of estimating Chla concentration from the HICO data. The results demonstrated the high potential of HICO data for estimating Chla concentration in turbid productive (Case II) waters in real-time. To estimate Chla concentration and the other two water parameters CDOM and turbidity in coastal case II turbid waters, Keith et al. (2014) also investigated the satellite HICO image data collected from four estuaries along the northwest coast of Florida. They used atmospherically corrected HICO imagery and a comprehensive field validation program. After tuning to HICO spectral data, they developed three regionally specific algorithms to estimate three basic water-quality properties: Chla, CDOM and turbidity from HICO spectral data. The three-band algorithm Chla = 17.477 × [R_{rs}^{-1} (686) – R_{rs}^{-1} (703)] × R_{rs}(735) + 6.152 was used for estimating Chla concentration (μg l^{-1}); the two-band algorithm $a_{CDOM}(412)$ = $0.8426 \times \left[\frac{R_{rs}(670)}{R_{rs}(490)}\right] - 0.032$ was used for estimating absorption coefficient (m^{-1}) of CDOM at wavelength 412 nm; and the one-band algorithm Turbidity = $2 \times 10^6 \left[R_{rs}^{2.7848}(646)\right]$ was used to estimate turbidity (NTU). The validation results indicated that the three-band Chla algorithm performed best ($R^2 = 0.62$) and that CDOM ($R^2 = 0.93$) and turbidity ($R^2 = 0.67$) were also highly correlated with *in situ* measurements. However, they suggested that problems with vicarious calibration of the HICO sensor needed to be resolved and standardized protocols were required for atmospheric correction.

Hyperspectral imaging data have been used to estimate suspended sediments (SS), total suspended matter (TSM), and total inorganic particles (TIP) concentrations in coastal case II turbid waters. For instance, Carder et al. (1993) have used AVIRIS data to study dissolved and particulate constituents in Tampa Bay, Florida. Images of the absorption coefficient at 415 nm and backscatter coefficient at 671 nm were used to map the dissolved and particulate constituents of the Tampa Bay plume. The results were verified at three sites within the bay using *in situ* sampling aboard a ship at the time of the AVIRIS overflight and demonstrated the potential of the AVIRIS imaging data for retrieving the turbid water parameters. Using hyperspectral satellite sensors HICO and Hyperion data, Xing et al. (2012, 2013) have also estimated and mapped water quality parameters SS, TSM, and TIP in the case II turbid waters. In estimating SS with HICO imaging data and *in situ* spectral measurements collected from the Yellow River Estuary and the nearby Bohai Sea, China, Xing et al. (2012) found that the reflectance spectra recorded by HICO sensor were generally consistent with the *in situ*–measured ones. The reflectance peak at 817.4 nm (corresponding band 73 of HICO image) was used as an index of estimating SS concentration. Results also confirmed that the reflectance peak in the spectral region of 450–750 nm shifts toward longer wavelengths with the increase of SS concentration and demonstrates that HICO hyperspectral imagery might be useful for monitoring very turbid coastal waters. In mapping total suspended matter (TSM), total inorganic particles (TIP), and water turbidity with *in situ* spectra measured and Hyperion hyperspectral data collected from Pearl River Estuary, China, Xing et al. (2013) observed that the content of TIP and turbidity was proportional to the concentration of TSM which ranged from 6 mg l^{-1} to 140 mg l^{-1}. They used the band-subtraction algorithms of R_{rs} at 610 nm and 600 nm [$R_{rs}(610) - R_{rs}(600)$] extracted from the *in situ*–measured spectra and of R_{rs} at 609.97 nm and 599.80 nm [$R_{rs}(609.97) - R_{rs}(599.80)$] extracted from the Hyperion spectra in an exponential regression model to estimate the TSM concentration. Compared to field measurements, the RMSEs of estimating TSM from the *in situ* and Hyperion data were 12.6 mg l^{-1} and 5.9 mg l^{-1}, respectively. The results showed that the good performance of the band-subtraction algorithm might be mainly attributed to improvement of the sensitivity of reflectance to suspended sediments by reducing the background influences from water surface reflection and path radiance at the specific wavelengths, a similar function made by a spectral derivative analysis.

Since the advent of hyperspectral remote sensing techniques, researchers have utilized various hyperspectral data to classify and map substrate and bottom types in coastal shallow waters. For example, Clark et al. (1997) mapped tropical coastal environments and most components related to reef, seagrass habitats, coastal wetlands, and mangroves. They demonstrated that the airborne CASI sensor data could provide detailed quantitative information on habitat extent and composition, water depth, seagrass biomass, and mangrove canopy cover, etc. Holden and Ledrew (1999) used *in situ* spectral measurements to identify coral reef features. They collected *in situ* ASD spectral measurements from Fiji and Indonesia and analyzed the spectra data in principal components analysis (PCA) and spectral derivative analysis. The derivative spectra with a three-step process of between 654 and 674 nm, between 582 and 686 nm, and between 506 and 566 nm were used to discriminate coral reef features. Results indicated that the spectra measured in Fiji and those measured in Indonesia were statistically similar, and the proportion of correctly identified spectra of coral features using the three-step process of first derivatives was 75% with the main error source resulting from spectral variability of algae reflectance.

To map benthic and aquatic vegetation types in the complex coastal waters of Saint Joseph's Bay, Florida, Hill et al. (2014) used the spectroscopic aerial mapping system with on-board navigation (SAMSON) hyperspectral imager (band width of 3.2 nm, 156 channels between 400 and 900 nm) to analyze and map several benthic types, including, submerged and floating aquatic vegetation, benthic red algae, bare sand, and optically deep water. They applied stepwise different thresholds to discriminating the benthic types, such as using NIR brightness threshold to separate the imaging area into land and water areas and using NDVI threshold to separate the water area into vegetated and sand areas. They also tested bio-optical methods for the retrieval of absolute seagrass abundance in optically complex coastal waters using the hyperspectral data. The results demonstrated that the ability of hyperspectral imaging data quantified not just areal extent but also productivity of a seagrass meadow in optically complex coastal waters, which could provide information on the capacity of these environments to support marine food webs. Using the *in situ* spectral measurements and CASI airborne hyperspectral imaging data, Valle et al. (2015) applied a supervised classifier, MLC, to map 13 habitat classes in case II waters in the Oka estuary, Basque Country, Spain. See Figure 9.6 for the 13 habitat types. The 13 habitat types were defined along the supralittoral, intertidal, and subtidal zones of an estuary, including *Zostera noltii* seagrass meadows. There were a total of 25 CASI bands in the visible and NIR wavelengths with a ground sampling size of 2 m. Spectral bands were selected for habitat-type discrimination based on the spectral signature of the different habitat classes. Six different band combinations were tested using the MLC algorithm, and the most accurate classification was achieved with a 10-band combination (a mean producer's accuracy 92% and a mean user's accuracy 94%). Figure 9.6 presents the 13-class habitat classification result created with the 10 selected CASI bands. The 13-class habitat result validated based on GPS survey data highlighted the value of CASI data to discriminate and map estuarine habitats, providing key information to be used in supporting the implementation of environmental protection and conservation of coastal habitats.

Mapping seagrass spatial extent and species and assessing seagrass habitats are important because seagrass habitats are characteristic features of shallow waters worldwide and provide a

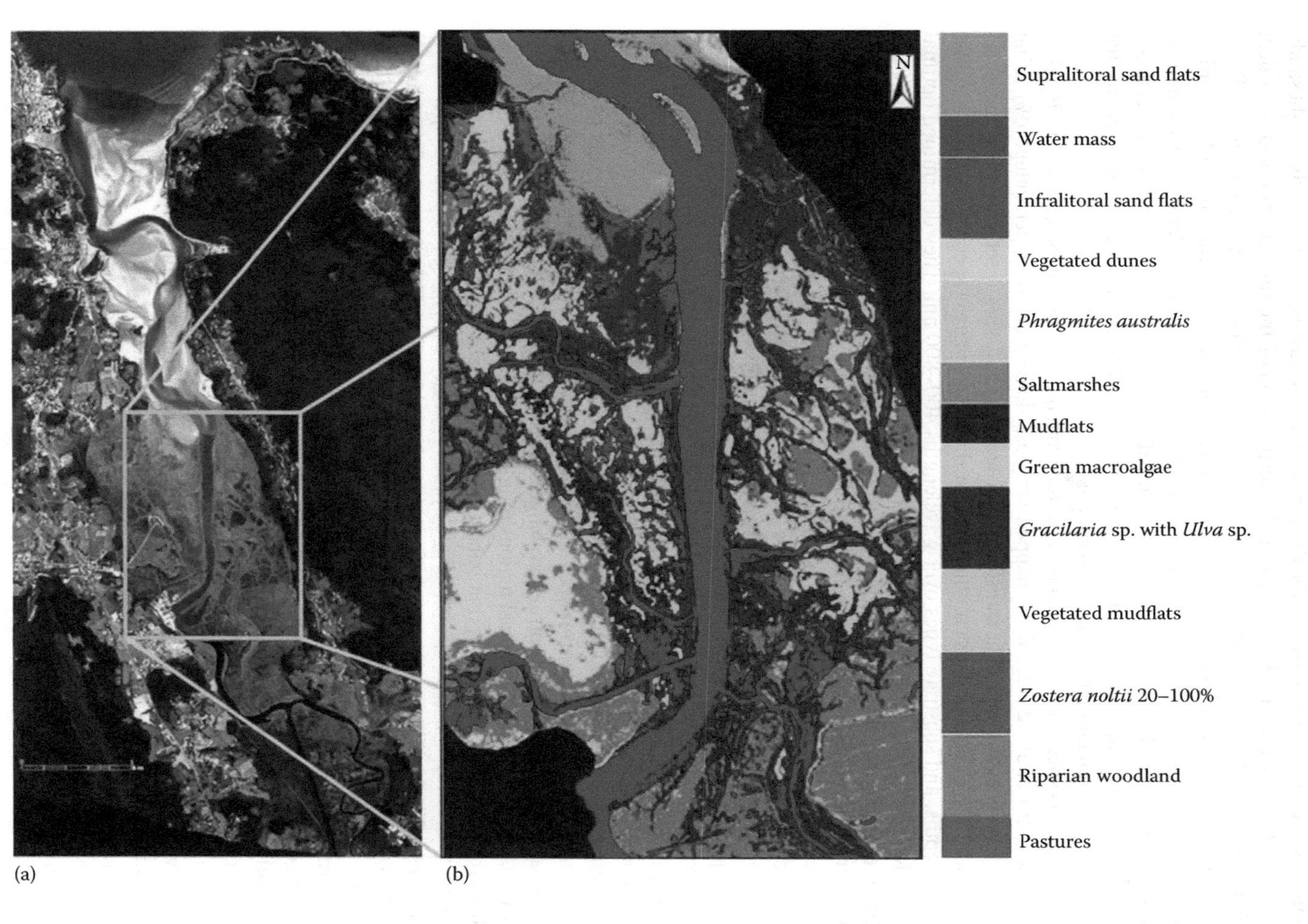

FIGURE 9.6 The 13-habitat classification map with highly discriminated all estuarine habitat classes created using the selected 10 CASI band image. (a) Natural-looking color composite image covering Oka estuary and partial habitat classification image; (b) zoom to San Cristobal and Kanala intertidal areas. (Reprinted from Valle, M. et al., *Estuarine, Coastal and Shelf Science*, 164, 433–442, 2015. ©2015, with permission from Elsevier.)

variety of ecosystem functions. Many researchers have used various hyperspectral data sets to map and assess seagrass characteristics and abundance in coastal shallow waters. Peneva et al. (2008) used HyMap hyperspectral imagery to map seagrass distribution off Horn Island, Mississippi, and estimate its areal coverage. They defined the three basic bottom classes, seagrass beds, bright sandy bottom, and dark sandy bottom, based on visual interpretation of the imagery coupled with field observations respectively in three water-depth zones determined by distance from shore. Three supervised image classification methods, including maximum likelihood classifier (MLC), minimum distance, and spectral angle mapper, were carried out to compare their performances in mapping seagrass coverage. The results showed that the MLC produced the highest overall accuracy of 83%, and that the HyMap-derived seagrass coverage with MLC method was comparable to that created by visual interpretation of aerial photographs.

Using *in situ* ASD spectral measurements collected from the coastal waters off northern Pinellas County, Florida, Pu et al. (2012, 2015) mapped seagrass density or cover percentage (%) using spectral data preprocessing and data transformation techniques. Using a set of 97 field measurements, Pu et al. (2012) compared spectra qualities for different seagrass species, levels of seagrass cover, water depths, and substrate types over wavelengths 400–800 nm. They determined optimal wavelengths for identifying levels of seagrass cover by two-sample t-tests, and also utilized PCA on spectra to evaluate if a set of the first five PCs could be used to effectively discriminate among levels of seagrass cover. The experimental results indicated that the data preprocessing techniques were effective for improving discriminant accuracies of seagrass cover levels. Using the same in situ ASD spectra as in Pu et al. (2012), Pu et al. (2015) also evaluated the use of *in situ* hyperspectral data and hyperspectral vegetation indices (VIs) for distinguishing among levels of percentage submerged aquatic vegetation (%SAV). Analysis procedures included (1) retrieving bottom reflectance, (2) calculating correlation matrices of VIs with %SAV cover, (3) testing the difference of VIs between levels of %SAV cover, and (4) discriminating levels of %SAV cover by using linear discriminant analysis and classification and regression trees classifiers with selected VIs as input. Their experimental results showed that the best VIs for discriminating the four levels of %SAV cover were simple ratio (SR) VI, normalized difference VI (NDVI), modified simple ratio VI, and NDVI × SR; and that the optimal central wavelengths for constructing the best VIs were 460, 500, 610, 640, 660, and 690 nm, with spectral regions ranging from 3 to 20 nm at band width 3 nm, most of which were associated with absorption bands by photosynthetic and other accessory pigments in the visible spectral range. To map seagrass cover percentage, Phinn et al. (2008) compared the ability of two satellite multispectral sensors (Quickbird-2) and Landsat TM and one airborne hyperspectral sensor CASI-2 for mapping benthic types, including four cover classes of seagrass beds (four seagrass cover classes: 0–10%, 10–40%, 40–70%, and 70–100%). Their study was carried out on the Eastern Banks in Moreton Bay, Australia, an area of shallow and clear coastal waters, containing a range of seagrass species, coverage, and biomass levels. The four-class seagrass coverage mapping results demonstrated the higher potential of CASI-2 sensor for mapping seagrass detailed cover levels compared with the other two multispectral satellite sensors.

In order to improve seagrass cover class mapping and assessing seagrass abundance, Pu and Bell (2013) evaluated a protocol that utilized image optimization algorithms followed by atmospheric and sunglint corrections to the three satellite sensors (Landsat TM, EO-1 Advanced Land Imager [ALI], and Hyperion) and a fuzzy synthetic evaluation technique to map and assess seagrass abundance along the northwest coastline of Pinellas County, Florida. After image preprocessed with image optimization algorithms (Zhao et al. 2013) and atmospheric and sunglint correction approaches, the data from the three sensors were used to classify the submerged aquatic vegetation cover (%SAV cover) into 5 classes (i.e., %SAV cover <1%, 1–25%, 25–50%, 50–75%, >75%) with a MLC classifier. Next, the five membership maps were created with the three biometrics (%SAV, leaf area index [LAI], and biomass) along with two environmental factors (water depth and distance-to-shoreline). Based on field measurements, bathymetry, GIS data, and spectral variables derived from the three sensors' data, the pixel-based five membership maps could be created using multiple regression models and expert knowledge (Pu and Bell 2013). Finally, seagrass abundance maps were produced by using the fuzzy synthetic evaluation technique and five membership maps. The comparative results indicated that the Hyperion sensor created the best results of the 5-class classification of %SAV cover (overall accuracy = 87% and Kappa = 0.83 vs. 82% and 0.77 by ALI and 79% and 0.73 by TM) and better multiple regression models for estimating the three biometrics (R^2 = 0.66, 0.62, and 0.61 for %SAV, LAI, and biomass vs. 0.62, 0.61, and 0.55 by ALI and 0.58, 0.56, and 0.52 by TM) for creating seagrass abundance maps along with two environmental factors. The experimental results demonstrated the image optimization algorithms and the fuzzy synthetic evaluation technique were effective in mapping detailed seagrass habitats and assessing seagrass abundance with the 30-m resolution data collected by the three sensors. Figure 9.7 presents the 5-class SAV cover classification map created using Hyperion data and MLC classifier and the seagrass abundance levels mapped using the fuzzy synthetic evaluation technique with the five fuzzy membership maps consisting of the three biometrics and two environmental factors.

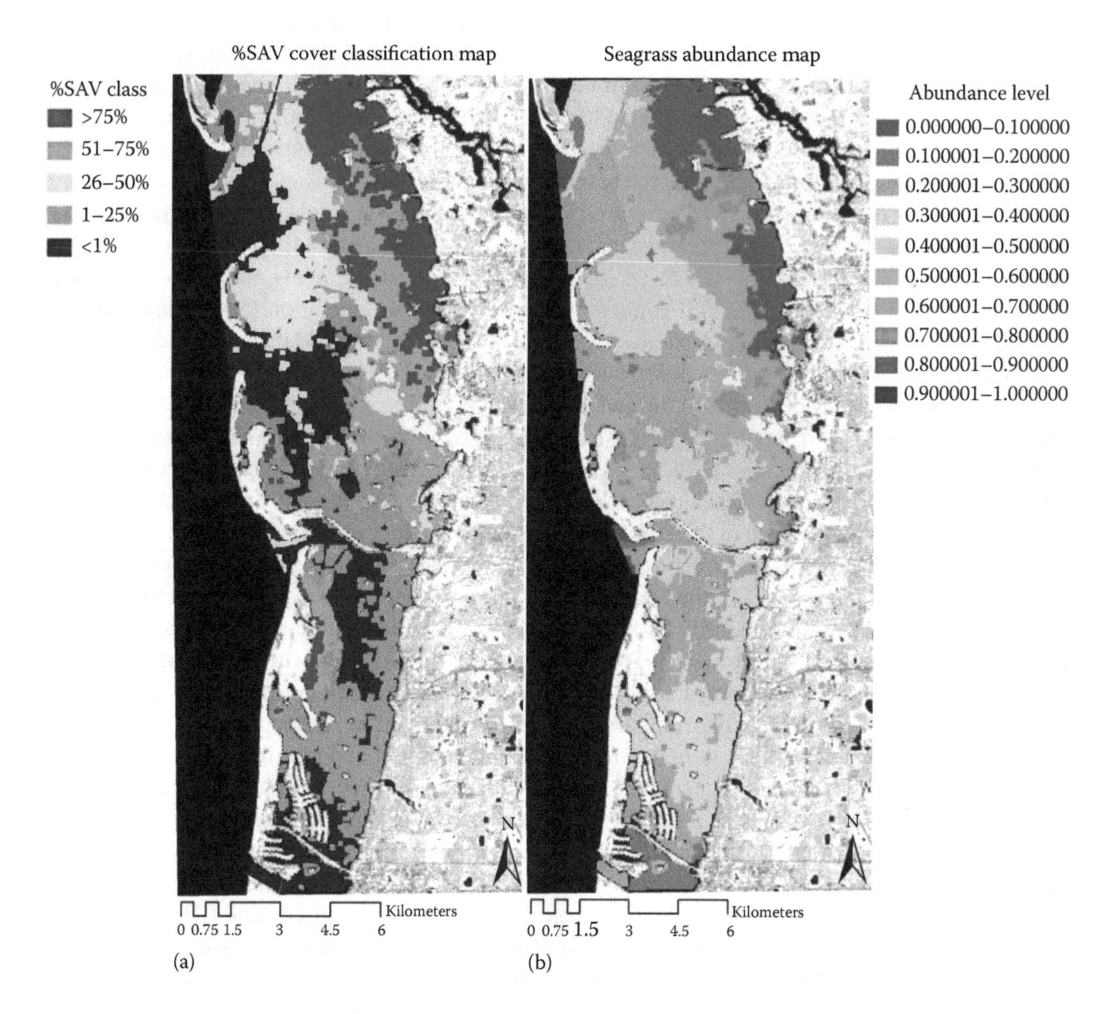

FIGURE 9.7 Results of classification of 5-classes of %SAV cover and seagrass abundance levels created with Hyperion hyperspectral data. (a) The 5-class %SAV cover classification map using supervised classifier MLC with Hyperion image data. (b) The seagrass abundance levels were mapped using the fuzzy synthetic evaluation technique with fuzzy memberships of three biological metrics (i.e., %SAV cover, LAI, and total biomass) and water depth and distance-to-shoreline spatial data layers.

In addition to hyperspectral remote sensing of seagrass cover classes, direct measurements of spectral characteristics of seagrass taxa or species have been conducted in either laboratory or field settings. Researchers have measured the spectral reflectance of seagrass taxa after removal of blades/shoots from sediment and often without the effect of water column (e.g., Fyfe 2003, Thorhaug et al. 2007, Pu et al. 2012). Under laboratory conditions, spectral measurements taken from each of three seagrass species (*Syringodium filiforme*, *Thalassia testudinum*, and *Halodule wrightii*) and five marine algae, and used for comparing among the seagrass species and algae using corresponding first derivative spectra were determined to be useful for distinguishing both among the seagrass species and between seagrass species and green and brown algae (Thorhaug et al. 2007). Using field-measured hyperspectral data, Fyfe (2003) recorded the spectral signatures of the seagrasses *Zostera capricorni, Posidonia australis*, and *Halophila ovalis*, and reported that the seagrass species were spectrally distinct. In a combined laboratory and field-based study, Pu et al. (2012) examined the spectral features of dominant seagrasses characteristic of a subtropical seagrass beds. Using both laboratory and *in situ* spectral measurements, they determined optimal wavelengths (bands) that are effective for identifying seagrass species (i.e., *S. filiforme*, *T. testudinum*, and *H. wrightii*) and evaluated the effectiveness of data preprocessing techniques for determination of optimal wavelengths. The experimental results indicated that the data set of the 2nd derivative normalized spectra could result in the best accuracies for identifying the three species, and the optimal wavelengths were 450, 500, 520, 550, 600, 620, 680, and 700 nm. A set of 5 optimal bands produced higher accuracies for identifying three seagrass species (overall accuracy = 73% and average accuracy = 75%) compared to those from use of PCA.

In addition to hyperspectral data measured in laboratory and *in situ*, many researchers have used airborne and satellite hyperspectral sensor data to test the discrimination of seagrass species and other SAV types. For example, with *in situ* ASD spectral measurements and airborne HyMap imaging data, Williams et al. (2003) studied submerged aquatic vegetation (SAV) stands in the tidal Potomac River at the mouth of Nanjemoy Creek at Blossom Point, Maryland, to identify and map two species of SAV (*Myriophyllum spicatum* and *Vallisneria americana*). After conducting necessary preprocessing of both *in situ* and HyMap hyperspectral data, Williams et al. (2003) adopted a spectral feature fitting (SFF) algorithm (see the introduction to SFF in Section 6.2.7 of Chapter 6) to identify the two SAV species through comparing image spectral data to a set of reference spectra, in this case, the field-measured spectral library database, by a least-squares fit of the continuum removed spectra of the HyMap image. The SAV species identification result derived from hyperspectral HyMap was compared with a product that combined aerial photography with field-based sampling at the end of the SAV growing season. The results indicated that hyperspectral imagery could be used to identify and classify SAV beds in an aquatic environment that could be characterized as optically complex. The hyperspectral data used in the study suggested the presence of epiphytes and sediment coating on the SAV obscured species' biochemical reflectance signatures. When Pan et al. (2016) compared the performance of hyperspectral imagery (CASI-1500) with single-wavelength airborne bathymetric light detection and ranging (LiDAR) for shallow water (<2 m) bathymetry and seagrass mapping, they performed a provisional classification with four supervised classifiers in the study to investigate the capability of airborne bathymetric LiDAR and hyperspectral imagery to identify seagrass species and other bottom types in Redfish Bay State Scientific Area near Corpus Christi, Texas. The four supervised classifiers were SVM, LDA-MLC, PCA-MLC, and SAM, and their performances were compared. The LDA-MLC is an MLC classifier but input with linear discriminant analysis–derived low dimensional data from high-dimensional CASI data while PCA-MLC is also an MLC classifier but input with PCA–derived low-dimensional data (see Chapter 5 for a discussion on the principles and algorithms of LDA-MLC and PCA-MLC). The six different substrate classes, including three seagrass species, *S. filiforme*, *T. testudinum*, and *H. wrightii*, and the other three classes, bare bottom, drift algae, and deep water, were present in the study area. The classification results indicated that hyperspectral imagery alone showed significant capability for substrate classification with over

95% overall accuracy, but features from bathymetric LiDAR alone were not sufficient for substrate classification. Furthermore, a combination of hyperspectral imagery with LiDAR data only marginally improved the overall accuracy of seagrass classification in this study. In differentiating two benthic habitats, seagrass and macroalgae, using satellite hyperspectral HICO data collected over the Indian River Lagoon, Florida, Cho et al. (2014) developed and applied two benthic classification models: $\text{Slope}_{\text{RED}}$ and $\text{Slope}_{\text{NIR}}$. The both $\text{Slope}_{\text{RED}}$ (679 to 790 nm) and $\text{Slope}_{\text{NIR}}$ (696 to 742 nm) with HICO bands are defined as follows: $\text{Slope}_{\text{RED}} = \left|\frac{R_{rs}(679) - R_{rs}(690)}{679 - 690}\right|$ and $\text{Slope}_{\text{RED}} = \left|\frac{R_{rs}(696) - R_{rs}(742)}{696 - 742}\right|$. The slope images were then classified and compared with results obtained from iterative self-organizing data analysis technique and spectral angle mapping classification methods. The validation result indicated that the slope models produced greater overall accuracies (63–64%) and were able to differentiate between seagrass and macroalgae substrates more accurately compared to the results obtained using the other classification methods.

Hyperspectral remote sensing techniques have been studied by many investigators to derive information on the water depth in coastal marine environment. Sandidge and Holyer (1998) used AVIRIS image data in a neural network system to establish quantitative, empirical relationships between water depth and remotely sensed spectral radiance. The AVIRIS data were acquired and analyzed for two areas: the western coast of Florida in the Tampa Bay area and the Florida Keys. The AVIRIS data were calibrated in units of radiance but were not corrected for atmospheric, ocean surface, or illumination effects because retrieval of water depth was attempted from radiance values observed at aircraft altitude. Given the ability of light penetration in water in visible spectral range, only the 36 AVIRIS bands covering the 400–742 nm spectral range were expected to have sufficient water penetration to record the bottom-reflected energy required for depth retrieval. In addition, the five atmospheric bands centered at 867 nm, 943 nm, 1020 nm, 1136 nm, and 1203 nm were considered to reflect the atmospheric origin or to be reflected from the ocean's surface. Consequently, a total of 41 AVIRIS bands (36 water penetrating and 5 atmospheric) were utilized in this study. Finally, the neural network used in the study was a feed-forward, full-connected net with an input layer (41 neurons with 41 AVIRIS bands), one hidden layer (21 neurons), and one output layer with a single neuron representing water depth. Depths in the National Ocean Survey (NOS) Hydrographic Database were used as training and test data sets. A neural network trained on a combination of the two data sets (i.e., the Tampa Bay and Florida Keys data sets) resulted in a combined RMSE of 0.48 m, nearly the same performance as neural networks trained individually for the two areas. The experimental result demonstrated the ability of hyperspectral data with the neural network approach, which operated at a higher level than a more traditional statistical curve-fitting solution for retrieval of the marine environmental information. In mapping water column properties from hyperspectral remote sensing data, Jay and Guillaume (2014) proposed a novel statistical method to improve accuracy of mapping water depth and water quality. Unlike using per-pixel–based traditional inversion methods, the new statistical method considers the spatial correlation between neighboring pixels, because such pixels are often affected by the same water column if the spatial resolution is high enough. The method performs local maximum likelihood (ML) estimations of depth and water quality in large zones with such neighbor-pixels correlation information. It can produce multi-resolution maps depending on local depth. Using the hyperspectral images acquired by a Hyspex VNIR-1600 camera (with different spatial resolutions ranging from 0.4 to 2 m [depending on flight altitude] and a fixed spectral resolution 4.5 nm and a total of 160 spectral bands covering spectral range from 410 nm to 1000 nm), Jay and Guillaume (2014) used the new method to map water depth separately in both shallow and deep waters in the Quiberon Peninsula on the west coast of France. To do so, the entire image was divided into appropriate meshes. In every mesh, water column properties including depth were estimated using both linear and constant depth models (Jay and Guillaume 2014). Final maps were created by combining these modeling results. Results proved that the depth modeling method could improve depth and water quality estimations, especially in shallow water, which indicated that using local information provided by neighboring pixels made this method robust to noise.

9.5 ENVIRONMENTAL HAZARDS AND DISASTERS

Imaging spectroscopy data have been used to determine surface features and compositions that are directly or indirectly related to environmental hazards and disasters. In the following, a review will be conducted on applications of hyperspectral remote sensing techniques (1) to mapping and monitoring impacts of mining wastes and tailings on environments including surface and subsurface soils, vegetation, and wasters in the watershed where mine sites are located, (2) to mapping and characterizing wildland biomass (forest) burning, and (3) to monitoring and assessing vulnerable slopes and landslides.

9.5.1 Mining Wastes and Tailings

There are severe environmental impacts from mining wastes and tailings, which can pollute soils and inland waters within watersheds where mining sites are located and influence vegetation growth and distribution. Such environmental impacts may be characterized and evaluated by using hyperspectral remote sensing techniques through identifying and mapping the extent and abundance of various mining wastes and tailings and estimating pollutant concentration and spatial extent in the polluted streams and lakes surrounding mining sites. For examples, researchers have used different hyperspectral imaging data and mapping and analysis techniques to map and characterize the spatial extent and physical and chemical properties of mining wastes. Farrand and Harsanyi (1997) utilized airborne AVIRIS images to assess the transport of hazardous mine waste downstream through alluvial processes. The trace metals released by mining activities in and around the town of Kellogg, Idaho, have contaminated ferruginous fluvial sediments deposited on the banks and on the floodplain of the Coeur d'Alene River in northern Idaho. Using AVIRIS image with the constrained energy minimization (CEM; see Chapter 5 for a description of the detailed algorithm) mapping technique, they mapped the extent of ferruginous fluvial sediments' spatial distribution. The mapping result indicated that the performance of the CEM mapping technique with AVIRIS hyperspectral data was extremely good for mapping ferruginous sediments. To map and monitor mining wastes and tailings, Mars and Crowley (2003), Shang et al. (2009), Riaza and Carrere (2009), and Zabcic et al. (2014) applied spectral unmixing analysis techniques on airborne sensor images (AVIRIS, HyMap, and Probe-1). Mars and Crowley (2003) used MTMF (see its introduction in Chapter 5) spectral unmixing technique with field data and all spectral bands of AVIRIS image data collected in the Rocky Mountain Range and Forest Region to map 18 mine waste dumps and five vegetation cover types in the southeast Idaho phosphate district. Their analysis results demonstrated that some attributes, including dump morphology, catchment watershed areas associated with individual dumps, stream gradients below dumps, and riparian vegetation densities along outflow pathways could be quantified by using the hyperspectral imagery in conjunction with digital elevation data. However, additional attributes, such as actual dump volumes and compositions in the watershed affected by mining sites, could not be evaluated without additional ancillary information. Using the linear spectral unmixing technique, Riaza and Carrere (2009) monitored mine wastes from sulfide deposits from HyMap hyperspectral image in a small sector of the Odiel River, the river in southwest Spain which drains abandoned sulphide ore mines from the Iberian Pyrite Belt; Zabcic et al. (2014) also mapped regional scale pyritic mine tailings using HyMap imagery to characterize tailings of the Sotiel-Migollas complex in Spain and pinpoints sources of acid mine drainage; and Shang et al. (2009) identified abandoned mine-waste sites and tailing surface and mapped potential sites in northern Ontario, Canada, using airborne hyperspectral Probe-1 sensor data. They all confirmed that the airborne hyperspectral data could be used to map and characterize mining-waste sites and tailings. With Tetracorder's primary subroutine, the modified least squares shape-matching algorithm that compares spectra of unknown materials with hundreds of library reference spectra and identifies the best match, Swayze et al. (2000) used AVIRIS data over the U.S. Environmental Protection Agency (EPA) superfund site at Leadville, Colorado, to map the distribution of acid generating minerals. At Leadville, acidic water and mobilized heavy metals are an environmental hazard. Since each of these Fe-bearing secondary minerals at the site is spectrally unique, imaging spectroscopy AVIRIS data can be used to rapidly screen entire mining districts for potential sources of surface acid drainage and to detect acid-producing minerals in mine waste or unmined rock outcrops. The AVIRIS-derived maps have been employed to aid in the characterization and hazardous waste remediation efforts for the area. Potential sources of acid mine drainage have been identified to protect the Arkansas River (Swayze et al. 2000). The EPA estimated (1998) that the mineral maps generated from AVIRIS data at Leadville accelerated remediation efforts by two years, saving over $2 million in investigation costs. Using satellite hyperspectral Hyperion imagery, Farifteh et al. (2013) identified and mapped mine waste piles and iron oxide by-product minerals in an abandoned mine in southwestern Spain. After the Hyperion image was preprocessed, the Mahalanobis Distance algorithm was first used to differentiate the area covered by mine piles from other main land-use classes, and then the spatial variations of iron oxide and carbonate minerals within the mine area were mapped using the spectral feature fitting (SFF; refer to Chapter 5 for an introduction to SFF) algorithm. The mapping results showed that the mine waste deposits could be easily mapped using available standard algorithms such as Mahalanobis Distance from Hyperion data, and there was an abundance of different minerals such as alunite, copiapite, ferrihydrite, goethite, jarosite, and gypsum within the mine area, derived with the SFF method from Hyperion data.

Mining activities usually cause serious heavy metal pollution in soil and water and affect vegetation growth and distribution around mining sites. Acid sulfate soils (ASSs) are widely spread around the world in coastal areas, inland lakes and rivers, and mine sites. ASSs are potentially harmful to the environment due to their strong acidity production ability and their capability to release trace metals (Shi et al. 2014). Hyperspectral imaging data can be used to map and assess the polluted extent and severity. For example, since secondary iron-bearing minerals, produced by ASSs, have diagnostic spectral features in the VNIR to SWIR spectral range and can be good indicators of the severity of the effects of ASSs, Shi et al. (2014) detected ASSs by using hyperspectral sensor HyMap data for mapping these indicative iron-bearing minerals. They used the spectral features matching method to map iron oxides, hydroxides, and hydroxysulphates, as well as non-iron–bearing minerals from HyMap image data. Coupling with the proximal hyperspectral sensing HyLogger system and field pH measurements, making use of HyMap hyperspectral data could help create comprehensive understanding and estimation of ASSs, both on the surface and in the subsurface. Vegetation growing conditions are an indirect indicator of the environmental problem in mining areas (Zhang et al. 2012). For this case, Zhang et al. (2012) distinguished the stressed and unstressed vegetation growth situations in mining areas using HyMap and Hyperion data in the Mount Lyell mining area in Australia and Dexing copper mining area in China, respectively. Based on analyses of the biogeochemical effect of dominant minerals on the vegetation spectra and vegetation indices, Zhang et al. (2012) developed two hyperspectral indices: vegetation inferiority index (VII) and water absorption disrelated index (WDI) to monitor the environment in the mining area. For the areas with poor vegetation growth (e.g., due to mining effect), based on the index definition, the VII index becomes larger. Due to the strong correlation of the two absorption features (at 0.97 μm and 1.18 μm) in the vegetation areas, the WDI in the region with non-mineral pollution should be close to zero (Zhang et al. 2012). However, the WDI value in bare iron ore areas and vegetation areas contaminated with iron ore is not zero. The experimental results indicated that the VII index could effectively distinguish the stressed and unstressed vegetation growth situations in mining areas while the WDI index was capable of informing whether the target vegetation was affected by a certain mineral. The successful applications of VII and WDI indices demonstrated that hyperspectral remote sensing could provide a good tool to effectively monitor and evaluate the vegetation condition in mining areas.

Active and abandoned various mines (e.g., coal mines) can make a huge impact on the environment. A large environmental problem is caused by acid mine drainage (AMD). Since hyperspectral remote sensing techniques are able to map and characterize mining wastes and tailings, many researchers have used hyperspectral data to analyze, assess, and monitor contaminated waters in the AMD area. For instances, Boine et al. (1999) used CASI hyperspectral images to successfully map different water qualities in the open cast mining residual lakes in the mining area Goitsche, located in Central Germany near Halle and Bitterfeld. They combined CASI data with water sampling of hydrochemical and hydrobiological properties for the retrieval of different water constituents by using an inverse modeling approach. Also in Central Germany near Halle and Bitterfeld and with the same hyperspectral sensor CASI imagery as Boine et al. (1999), Gläßer et al. (2011) successfully tested a new method to further monitor residual lakes using a manifold data set (e.g., CASI hyperspectral data, ground truth data, field and laboratory spectral measurements). They first analyzed optical properties of the mining lakes and defined each of the limnological stages of development and second, according to the lakes' optical properties, developed and utilized algorithms for classification of the hydrochemical parameters. The experimental results showed that the new algorithms enabled the monitoring of mining lakes from acidic to alkaline as well as the quantification of the hydrochemical properties inside the lake waters. By using airborne (HyMap) and satellite (Hyperion) hyperspectral images, Riaza, Buzzi, García-Meléndez, Carrère et al. (2011, 2012, 2015) and Riaza, Buzzi, García-Meléndez, Vázquez et al. (2012) mapped, evaluated, and monitored pyrite mine waste, AMD, and mine waste ponds at a mine site in the Odiel River in the Iberian Pyrite Belt (Spain). The Odiel River is extensively contaminated by sulfide mine wastes, released as sulfuric acid into the waterbodies. Using image processing techniques such as SAM, they could map and evaluate not only mine dams, mill tailings, and mine dumps in a variable state of pyrite oxidation in the AMD area but also local abrupt changes in the water pH in the vicinity of highly contaminated tributaries by means of imaging spectroscopy using HyMap or Hyperion image data. The experimental results confirmed that hyperspectral sensor data were invaluable in giving quick hints on the quality of the rapidly changing state of the pollution produced by sulfide mine wastes, enabling the authorities to activate mitigation procedures.

9.5.2 Biomass Burning

Biomass burning is a major source of trace gases and aerosol particles, with significant changes of atmospheric chemistry, cloud properties, and radiation budget (Kaufman, Justice et al. 1998). In an early study on the smoke, clouds, and radiation–Brazil (SCAR-B) experiment, Kaufman, Hobbs et al. (1998) used multi-/hyperspectral remote sensing data (AVIRIS, NOAA-AVHRR, and MODIS simulator data) to study biomass burning associated with measurements of surface biomass, fires, smoke aerosol and trace gases, clouds, radiation, and their climate effects. They concluded that remote sensing data including high spectral resolution data proved to be useful for monitoring smoke properties, surface properties, and the impacts of smoke on radiation and climate. There are many other researchers who use hyperspectral data to map and monitor forest fire frequency and extent, and to assess the severity. For example, in comparing nearly simultaneous acquisition of Hyperion and AVIRIS to evaluate spaceborne monitoring potential of fire danger in southern California chaparral, Roberts et al. (2003) evaluated the capabilities of Hyperion and AVIRIS for extracting and assessing fire danger parameters derived from sensor data, including surface reflectance, live and dead biomass, canopy water content, species composition, and fuel state. They collected field spectra to support reflectance retrieval and construct a spectral library for vegetation mapping. The conclusions derived from the comparative analysis included that

1. Reflectance spectra retrieved from Hyperion and AVIRIS had similar shape and albedo, but SNR was five times higher in AVIRIS.
2. For fuel condition assessed by spectral mixture analysis, similar fractions and spatial distributions of fuel types were obtained from both Hyperion and AVIRIS imaging data.

3. Hyperion showed a good capability of separating spectral signal of bare soil from that of dry plant litter.
4. Hyperion was capable of retrieving canopy water at 1.20 μm, but had poor performance at 0.98 μm, and the Hyperion sensor noise and instrumental artifacts accounted for poor performance in this spectral region.
5. Species and community mapping showed similar patterns but with better overall accuracy (79%) for AVIRIS compared with that (50%) for Hyperion.

To analyze impacts of fire frequency and severity within recovering forests in the Amazon region using remote sensing, Numata et al. (2011) first used a multispectral Landsat time series dataset to reconstruct the fire history from 1990 to 2002 in a portion of Mato Grosso, Brazil, and then they calculated five narrowband VIs from a satellite hyperspectral Hyperion image for spectral analysis of physiological characteristics of fire-disturbed forests and their recovery. The five VIs included NDVI, carotenoid reflectance index (CRI), photochemical reflectance index (PRI), normalized difference water index (NDWI), and normalized difference infrared index (NDII). In spite of the relatively low SNR of Hyperion sensor, the narrowband-derived indices could provide useful information for monitoring fire disturbed forests, better than currently possible with Landsat. To investigate performances of spectral regions and spectral indices in discriminating burned from unburned areas, and to assess the burn severity of a recent fire in the Kalmthoutse Heide, a heathland area in Belgium, Schepers et al. (2014) used airborne imaging spectroscopy data from the airborne prism experiment (APEX) sensor. The sensor image was acquired with a spatial resolution of 2.4 m and a total of 288 spectral bands covering a spectral range from 0.41 μm to 2.45 μm with a spectral resolution between 5 nm and 10 nm. They used a separability index to estimate the effectiveness of individual bands and spectral indices to discriminate between burned and unburned land, and a modified version of the Geometrically Structured Composite Burn Index (GeoCBI; ref. Schepers et al. 2014) for a field data collection to analyze the burn severity. The experimental results indicated that the normalized burn ratio (NBR) outperformed the other spectral indices and the individual spectral bands in discriminating between burned and unburned areas, and for the burn severity assessment, all spectral bands and indices showed relatively low correlations with the field data GeoCBI index but the variable performance of spectral indices to assess burn severity among vegetation types.

As reviewed above, hyperspectral image data are particularly suitable for mapping burned areas and burn severity but they are also used for monitoring post-fire plant regeneration and ecosystem recovery (Lewis et al. 2011, Mitri and Gitas 2013). For examples, to map forest floor conditions after the 2004 Taylor Complex in Alaska's boreal forest, Lewis et al. (2011) used a combination of pre- and post-fire forest floor depths and post-fire ground cover assessments, measured in the field (e.g., *in situ* ASD spectra), and high-resolution airborne hyperspectral HyMap imagery. They applied a linear spectral unmixing technique with five end-members representing green moss, non-photosynthetic moss, charred moss, ash, and soil to produce fractional cover maps. The study sites spanned low to moderately high burn severity with high cover of green or non-photosynthetic moss representing low consumption while high cover of charred moss, ash, or soil meaning higher consumption. The correlation analysis result between green moss estimated from the HyMap imagery and both post-fire depth and percentage consumption suggested that potential burn severity might be predicted by mapping green (live) moss derived from the remote sensing data, and implied that the method of mapping the condition of the organic forest floor with hyperspectral imagery could be a useful tool to assess the effect of future fires in the boreal region. In mapping post-fire forest regeneration and vegetation recovery in the Mediterranean island of Thasos, Greece, Mitri and Gitas (2013) employed a combination of very high spatial resolution QuickBird, hyperspectral Hyperion imagery, and an object-based image analysis method. The study focused on the mapping of three major post-fire classes (forest regeneration, other vegetation recovery, and unburned vegetation) and of the two main forest regeneration classes, *Pinus brutia* and *Pinus nigra* regeneration.

After the remote sensing images were segmented, the three major post-fire classes and two main forest regeneration classes were classified from the segmented images using a fuzzy rule. The mapping results from both sensors' images were very promising (overall accuracy approximately 84%) compared to ground truthing data.

Hyperspectral image data can be used to accurately detect pixels containing fire and map burned fraction in pixels. Dennison and Roberts (2009) used an AVIRIS image scene acquired over the 2003 Simi Fire in southern California to identify the hyperspectral index that was able to accurately detect pixels containing fire. The most accurate hyperspectral index was found and named the hyperspectral fire detection index (HFDI). The HFDI was constructed with mid-infrared (MIR) bands centered at 2061 and 2429 nm, and the HFDI could detect 1% fire pixel coverage. Consequently, the HFDI was most appropriate for the detection of flaming combustion but might miss lower temperature smoldering combustion at low percent pixel coverage due to low emitted radiance in the MIR spectral range. Compared with two previously proposed hyperspectral fire detection indices, the HFDI could provide improved detection of fire with less variability in background index values (Dennison and Roberts 2009). To evaluate and compare the performance of assessing fire severity of broadband (Landsat OLI) and narrowband hyperspectral data (AVIRIS) in the visible to MIR spectral range (0.35–2.5 μm), Veraverbeke et al. (2014) used an alternative indicator, the burned fraction derived from spectral mixture analysis (SMA). Based on a separability index, they evaluated the separability between the different ground components, or end-members, that comprise post-fire environments (char, green vegetation [GV], non-photosynthetic vegetation [NPV], and substrate) with the multi-/hyperspectral data. The comparative analysis results showed that higher data dimensionality of hyperspectral data resulted in significantly improved post-fire fractional cover mapping and burned fraction estimate compared to multispectral data, and the hyperspectral data, when combined with advanced analysis techniques, significantly improved fire severity assessments.

9.5.3 Landslide Monitoring

Compared with studies on applications of hyperspectral remote sensing techniques to other environment aspects, there are only a few studies on applications of hyperspectral imaging techniques to monitoring landslides and mapping other factors that may lead to landslides. In 1996, using AVIRIS spectral images, Crowley and Zimbelman (1997) investigated a natural hazard associated with the volcanoes of the Cascade Mountains in the western United States. The AVIRIS image spectra were used to map the distribution of specific alteration minerals, which are associated with weak zones in the volcanic slopes and are indicators of regions of slope instability and potential collapse (Crowley and Zimbelman 1997). Two different analysis procedures were applied to analyze the AVIRIS reflectance data set. The first procedure with spectral band–fitting technique compared a series of previously measured mineral or rock reference spectra to each AVIRIS pixel spectrum and identified a single spectrally dominant material that best fits each pixel. The second procedure used linear spectral unmixing to evaluate each AVIRIS spectrum in terms of linear combinations of "pure" spectral end-members (generally minerals). The AVIRIS spectral data could enhance ground-based mapping efforts and should be useful for rapidly identifying hazardous sectors at other volcanoes. As communities grow near these volcanoes, understanding the hazards of instable slopes is of increasing importance. A geomorphological investigation performed by Mondino et al. (2009) for exploiting information derived from airborne hyperspectral sensor MIVIS images included the Cassas landslide, a well-known complex landslide affecting slopes on the southern side of the valley in the Middle Susa Valley (Italian Western Alps). Mondino et al. (2009) proposed a workflow based on neural network algorithms used for both MIVIS image geometric correction and classification. The results indicated that the MIVIS data could be used to identify and characterize major elements of the present-day active unstable slope (including debris-covered areas, fractured/disjointed rock walls, landslide accumulation borders), as well as of individual structural features and landforms which are related to long-term deep-seated gravitational slope deformation. The analysis of MIVIS images for the Cassas landslide also demonstrated that the major geological–geomorphological features (e.g., main scarp, shear structures, transversal minor scarps) could be easily identified through simple remote sensing operations. By integrating LiDAR and hyperspectral AISA data, Sterzai et al. (2010) monitored the Valoria landslide, Modena Province, Italy, a high-risk area with vulnerable elements that are subject to periodic and abrupt reactivations. The multitemporal LiDAR data might be used for calculating a differential surface, highlighting absolute height variations, recognizing the main landslide components, and identifying depletion and accumulation zones, whereas the hyperspectral AISA data could help characterize the landslide terrain roughness. The result derived from both LiDAR and AISA data demonstrated the capabilities of remote sensing techniques to identify and characterize the essential features of an active, rapid earthflow in the study area.

9.6 URBAN ENVIRONMENTS

HRS with high spectral and spatial resolution (for most airborne data) has a great potential for analyzing, quantifying, and mapping complex urban environments. In this section, a review of spectral properties of urban materials and urban spectral libraries, identifying and mapping urban materials and LULC types, and characterizing urban thermal environment using hyperspectral datasets is presented.

9.6.1 Spectral Properties of Urban Materials

Urban areas provide a complex material environment, both in the number of materials present and in the spatial scale of materials variation. Since urban materials consist of natural and artificial materials, their spectral properties are pretty diverse compared with most natural materials. For example, most road materials in an urban area consist of a combination of gravel, concrete, and asphalt. Concrete is a mixture of aggregates such as gravel, sand, water, and Portland cement. Consequently, road surface made from concrete exhibits spectral features characteristic of the minerals composing the aggregate materials, water, and silicates whose spectral properties are introduced in Chapter 7. The dominant spectral features include the quartz reststrahlen doublet, primary water absorption bands, and calcite features near 4.0, 6.5, and 11.3 μm (Eismann 2012). Road surface with asphalt is a mixture of aggregate along with a tar-like oil by-product called petroleum bitumen. The spectra of both fresh and weathered asphalt exhibit different spectral features. The former exhibits a spectral reflectance much like the tar example, which is strongly absorbent over the entire spectral range in VNIR–MIR (0.4–2.5 μm) spectral range, whereas the latter exhibits the spectral characteristics of aggregate material because the tar breaks down (Eismann 2012). In the urban area, many surfaces (road, roof, and building) are painted and coated and thus their spectral properties exhibit the spectral reflectance properties of paints and coatings, which can be very diverse depending on their specific composition. From the basic makeup of paints and coatings, in general, the VNIR spectral characteristics of paints and coatings can be described by the pigment characteristics, while the SWIR, MIR, and LWIR characteristics are influenced by the binder, filler particles, and substrate material (Eismann 2012).

To understand spectral properties of diverse urban materials and efficiently explore the urban complex environment using HRS technology, many researchers have established and applied spectral libraries of natural and artificial materials worldwide (e.g., Ben-Dor et al. 2001a, Roberts et al. 2012, Kotthaus et al. 2014). A spectral library of specific urban targets is made of materials that are found in nature (e.g., water, vegetation, rock, and soil) or artificially manufactured (e.g., plastic, fabrics and metal). For example, Ben-Dor et al. (2001a) built a VIS–NIR spectral library of urban materials using a known spectral library of pure objects (Price 1995), and examined the data against real urban targets sensed by CASI sensor with 48 channels. The spectral library was generated from a collection of more than 3000 different spectra of urban materials, which were then resampled into 48 CASI bands and saved as a database termed *Pure Urban Spectral Library* (PUSL). The relevant urban materials included soil, dry grass, leaf litter, water, shale, plastic sheet, pigments, fiber glass, concrete, limestone, asphalt, rubber, iron cover, metal, bricks, etc. Further, Ben-Dor et al. (2001a) generated an imaging spectroscopy (IS)–based spectral library of corresponding materials from representative targets across the city of Tel-Aviv, Israel. They used atmospherically corrected CASI image as well as *a priori* knowledge of the area, and the spectral database was termed *CASL*. The spectral libraries of urban materials demonstrated that the IS techniques were promising for use in urban environmental studies. For an area-wide identification of urban surface materials using hyperspectral HyMap data, Heiden et al. (2001) developed a spectral library that formed a database. The spectra in the library were measured with a field spectrometer in the wavelength range from 0.35 μm to 2.5 μm. The urban surface materials were categorized in terms of their degree of surface sealing resulting in urban surface cover types, such as the ceramic/mineral group, metallic or synthetic group, and of variations in color or manufacturing types. The created spectral library was used to explore the spectral information content of hyperspectral HyMap data collected in Dresden, Germany. Recently, Kotthaus et al. (2014) created a new online spectral library of urban construction materials including LWIR emissivity spectra of 74 samples of impervious surfaces derived from measurements taken by a portable Fourier Transform InfraRed spectrometer. The online spectral library, included in the London Urban Micromet data Archive (LUMA; http://LondonClimate.info /LUMA/SLUM.html), also contains VIS–SWIR reflectance spectra observed from urban material samples. Given that many urban materials are composed of minerals that exhibit notable absorption and scattering features in the measured spectral range, hyperspectral VIS–SWIR–LWIR information appears especially useful to urban environmental studies. To evaluate the potential in urban science of the forthcoming satellite hyperspectral sensor HyspIRI mission (read the detailed introduction to the mission in Chapter 3), Roberts et al. (2012) developed a spectral library of dominant urban materials (e.g., grass, trees, soil, roof types, and roads) from field and airborne-measured spectra. They used the spectral library to map fractions of impervious, soil, green vegetation (e.g., trees and lawn) and non-photosynthetic vegetation.

9.6.2 Urban Materials and LULC Types

Although one city differs from another in its basic landscape (e.g., street pattern, building architecture, topography, and geomorphology), the main materials that compose urban targets (e.g., asphalt, cement, glass, and vegetation) are similar. Thus, reflectance or emittance from different urban environmental components could have common fingerprints (Ben-Dor 2001). And a variety of absorption and scattering constituents found in the urban complex environment and associated with human infrastructure may form a basis for use of HRS. IS provides a uniform synoptic approach to mapping surface materials in the urban and adjacent environments to support urban characterizing, monitoring, and planning (Green et al. 1998). Therefore, we can use different HRS data sets to identify and map urban materials and land use/land cover (LULC) types. For example, based on a comprehensive set of material spectral library, Kalman and Bassett III (1997) used HYDICE hyperspectral data in classification and material identification in an urban area. They developed a procedure for land cover classification that could be automated and performed with little or no prior knowledge of targets in the scene. Compared to the ground truth observations, the materials identified using the hyperspectral techniques indicated a relatively high agreement. The type of product could potentially provide input to urban analysis models which require accurate assessment of impervious surfaces. Using airborne MIVIS hyperspectral data, Bianchi et al. (1996), and later Fiumie and Marino (1997), demonstrated that it was possible to differentiate between paving materials made with basalt and those made with marble, which are of products found in close proximity of Rome. Mapping the composition of roofs is an important mission for HRS technology. Asbestos, for instance, is a carcinogenic material that should be removed from structures in the populated environment. Since only the SWIR spectral region has a unique spectral fingerprint associated with asbestos, only hyperspectral sensors that have capabilities in the SWIR region are adequate tools for mapping asbestos roofs over an urban environment (Ben-Dor 2001). Based on this evidence, Marino et al. (2000, 2001) utilized MIVIS hyperspectral sensor data integrated with GIS data to map asbestos roofs over two urban areas. In their study, they were able to distinguish between aluminum, tiles, bitumen, and plain concrete using the spectral information provided by the MIVIS with two different approaches of classification strategy, image-based and spectral library–based. Their results showed that hyperspectral images acquired from aircraft platforms could be very useful in order to identify and map asbestos concrete coverings in urban areas.

Due to spectral similarity among some urban materials, a combination of spectral information with shape and context information extracted from high resolution data may improve identification and mapping results of urban materials. For example, to make a discrimination between roofing materials and streets within urban areas, Mueller et al. (2003) integrated hyperspectral information and shape/context information, extracted from HyMap sensor data, for a better differentiation between buildings and open spaces to improve the image analysis result. The hyperspectral image data were acquired for a 2.7 km × 1.7 km north–south transect in the city of Dresden, Germany. Due to the spectral and spatial similarity of roofing bitumen, tar-paper, and asphalt, context knowledge (e.g., shadow information as height information for buildings) and shape knowledge (e.g., long straight objects without shadow for streets) can be used to identify reliable property pixels cover type. The results demonstrated the potential of this method. Segl et al. (2003) also used an approach by considering both spectral and shape features extracted from airborne hyperspectral data of DAIS-7915 to test their potential for automated material-oriented identification of urban surface cover types. This approach was extended by a shape-based classification technique including thermal bands of the DAIS instrument to improve the detection of buildings during the process of identifying seedling pixels. This new approach increased the reliability of differentiation between buildings and open spaces, resulting in more accurate results for the spatial distribution of urban surface cover types. Based on comprehensive field and image spectral libraries of more than 21000 spectra of surface materials widely used in German cities, Heiden et al. (2007) proposed a new approach for the determination and evaluation of such spectral features that are robust against spectral overlap between material classes and within-class variability. There is a two-step process to determine the spectral features. First, image spectra are related to urban surface materials by comparing them with spectra of the field library and by additional field checks. Second, the obtained spectral features are transformed into numerical values for further computer-based analysis using feature functions. The robustness of the interactively defined spectral features was evaluated by a separability analysis. The experimental results indicated that robust spectral features showed the potential for unsupervised detection of end-members in hyperspectral image data.

As reviewed above, it is widely accepted that the HRS technology has a great capability for providing new and useful information for urban environmental studies. Nevertheless, if HRS data are combined with data acquired simultaneously by other advanced sensors, the HRS approach will be able to attain its maximum potential (Ben-Dor 2001). For fine-scale mapping of heterogeneous urban/rural landscapes, Forzieri et al. (2013) explored data fusion strategies that are applied to modern airborne Earth observation systems, including hyperspectral MIVIS, color-infrared ADS40, and LiDAR sensors. The multi-sensor data were collected along a 20-km stretch of the Marecchia River, Italy. Using multiple classification algorithms (maximum likelihood/spectral angle mapper/spectral information divergence) and remote sensing data stack (different multi-sensor data combination), they focused on the identification of the best-performing data fusion configuration and investigated sensor-derived marginal improvements for mapping urban/rural landscapes. The best classification presented a high accuracy (92.57% overall accuracy), and it demonstrated the potential of the proposed approach to define the optimized data fusion and to capture the high spatial variability of natural and human-dominated environments (Forzieri et al. 2013). To compare capabilities among different sensors to identify and map land cover materials within an historical urban area, Cavalli et al. (2008) evaluated the added value of hyperspectral sensors in mapping the complex urban context. They used (1) EO-1 ALI and Hyperion satellite data, (2) LANDSAT ETM+ satellite data, (3) MIVIS airborne data, and (4) the high-resolution IKONOS imagery as reference, acquired over the city of Venice in northeastern Italy. The band-depth (measuring the spectral contrast of the absorption features with respect of its continuum) and sub-pixel analyses applied to subsets of Hyperion and MIVIS hyperspectral imagery were exploited. The results showed that satellite data with a 30 m spatial resolution (ALI, ETM+, and Hyperion) were able to identify only the main urban land cover materials.

From both a spectral and spatial point of view, urban environments are mixed and complex areas, and thus the application of "hard" quantitative classification approaches that use spectral information to map an urban environment can frequently result in a mixed pixel problem. In spite of high spectral resolution for HRS, the spatial resolution of hyperspectral data is frequently relatively low in an urban environment, and a lot of mixed pixels exist in an urban scene of hyperspectral imagery accordingly. To solve the mixed pixel problem using hyperspectral data, many researchers have developed various spectral unmixing ("soft") approaches (including linear and nonlinear ones) to map a complex urban environment where there exist many mixed pixels in an image scene. To map urban materials and LULC types, researchers have explored a multiple end-member approach (MESMA) approach. For example, in testing the potential of Proba/CHRIS satellite hyperspectral sensor data for mapping impervious surfaces in a mixed urban/suburban/rural environment, including part of the city of Leuven, Belgium, Demarchi et al. (2012) evaluated the MESMA approach. MESMA is well-suited for urban environments because it allows the number and types of end-members to vary on a per-pixel basis, which allows control of the large spectral variability in these environments. By modifying the criterion of the RMSE, favoring models with a smaller number of end-members were selected, and the impact of model selection on the accuracy of the unmixing was evaluated using reference fractions derived from 25-cm aerial photography for validation. In spite of the spectral confusion of soil and impervious surface end-members, the unmixed results indicated that average fractional error for impervious surfaces, vegetation, and bare soil was around 15%, which demonstrated the potential of CHRIS hyperspectral data in mapping the major physical components of the urban/suburban environment at the subpixel scale. Lv and Liu (2009) also used satellite Hyperion sensor data to explore MESMA approach for mapping green space for the city of Palo Alto, California. The MESMA approach was applied to generate fraction abundance maps for three types of image end-members (green vegetation, non-photosynthetic vegetation, and impervious surface) from Hyperion imagery. The urban green space mapping results indicated that MESMA was an effective method for extracting information of green space from hyperspectral data in the heterogeneous and spectral mixed urban area. In order to test the potential of HyspIRI mission (read a detailed introduction to the mission in Chapter 3) in mapping urban environment, Roberts et al. (2012) also used MESMA approach to map fractions of impervious, soil, green vegetation, and nonphotosynthetic vegetation from simulated HyspIRI data at the different spatial resolutions and compared the fractional estimates across spatial scales. They also determined and evaluated important surface energy parameters, including albedo, vegetation cover fraction, broadband emissivity, and surface temperature for 14 urban and natural land-cover classes over the Santa Barbara metropolitan region in California. The results demonstrated the utility of HyspIRI data for urban environmental studies and provided an insight of what would be possible at a global scale when HyspIRI data become available (Roberts et al. 2012).

To improve the performance of the MESMA unmixing approach with hyperspectral data, researchers have improved the MESMA approach. Franke et al. (2009) employed a hierarchical approach, in which MESMA was applied to map four levels of complexity ranging from the simplest level consisting of only two classes, impervious and pervious, to 20 classes including differentiated material composition and plant species to improve mapping of urban land cover using HyMap data, which were acquired over the city of Bonn, Germany. In the study, a spectral library containing 1521 end-members was constructed from the HyMap data. In the hierarchical approach with the

MESMA algorithm, lower levels of complexity, mapped at the highest accuracies, were used to constrain spatial models at higher levels of complexity, resulting in lowering spectral confusion between materials. Three end-member selection procedures were used to identify the most representative end-members for each level of complexity, including end-member average RMSE, minimum average spectral angle, and count-based end-member selection. Unmixing mapping accuracies with the approach were achieved from 97.2% for the two lowest complexity levels and a four-class map consisting of vegetation, bare soil, water, and built-up to 75.9% for the highest level, consisting of 20 land cover classes. By combining this with a hierarchical approach that uses spatial information from one level to constrain model selection at a higher level of complexity, the results demonstrated the ability of MESMA to incorporate within-class spectral variability, which was particularly well-suited for urban environments. Fan and Deng (2014) designed an improved MESMA (SASD-MESMA) to enhance the computational efficiency of conventional MESMA. In SASD-MESMA, the parameters of spectral angle (SA) and spectral distance (SD) are used to evaluate the similarity degree between spectral library spectra and image spectra in order to identify the most representative end-member combination for each pixel. The SA is determined by a SAM algorithm while the SD is defined by a sum of absolute differences of reflectances between image spectrum and library (reference) spectrum at each spectral band. Fan and Deng (2014) tested and validated the improved method using Hyperion imagery and field-spectra data collected over the city of Guangzhou, China. The tested results demonstrated that the SA and SD parameters were useful to reduce mismatching in selecting candidate end-members and effective for determining the appropriate end-members for each pixel. Different from the original MESMA that usually selects one end-member spectral signature for each land-cover class, Tan et al. (2014) proposed a modified MESMA approach that allows the selection of multiple end-member signatures for each land-cover class. The MMESMA approach was expected to be able to better accommodate within-class variations and yield better mapping results. Both HySpex and ROSIS sensor (see Chapter 3 for an introduction to both sensors) data were used to test the MMESMA approach. Compared to the unmixing results created with other spectral unmixing approaches, including the original MESMA, LSM, etc., the results created using both airborne hyperspectral sensor data showed the best performance of the MMESMA approach and demonstrated that the proposed MMESMA could generate more reliable abundance fractions from hyperspectral imagery, which tends to contain high within-class spectral variations.

Since urban environments are often characterized by geometrically and spectrally complex scenarios, some researchers have developed nonlinear spectral unmixing techniques to map urban materials and LULC types from hyperspectral data sets. Especially, when considering a 3-D landscape in a mixed pixel, a linear model is no longer valid as irradiated and shadowed areas are present, as well as radiative interactions between facing surfaces (Meganem et al. 2014). For this case, Meganem et al. (2014) introduced a new mixing model adapted to urban environments and aimed to overcome these limitations. This new model was derived from physical equations based on radiative transfer theory, and its analytic expression is linear–quadratic. The nonlinear spectral unmixing model was simplified based on the different radiative components contributing to the signal in a way to make the model easy to use for spectral unmixing. And it was validated using a synthetic but realistic European 3D urban scene. The validated result demonstrated that the quadratic term could not be neglected in urban scenes, essentially when there are many buildings and canyons. To achieve a better knowledge of the effect of the anthropogenic extents over the environment and of hyperspectral unmixing (HSU) architectures to recognize urban materials and structures, Marinoni and Gamba (2016) developed higher order nonlinear mixture models that were used to perform an accurate characterization of the anthropogenic settlements from several Hyperion image scenes recorded over the Marmara Sea area, Turkey. The experimental results showed that the proposed architecture was able to provide an accurate and reliable characterization of urban materials and extents in the considered scenes outperforming other classic HSU models based on a linear or bilinear mixture assumption.

Exploiting imaging spectrometer data with machine learning and partial spectral unmixing algorithms has been demonstrated to be an excellent choice for mapping land cover categories in spectrally complex urban environments. In considering difficulties in deriving quantitative training information that reliably represents pairs of spectral signatures for empirical modeling, Okujeni et al. (2013) presented an approach to overcome this limitation by combining support vector regression (SVR) with synthetically mixed training data to map subpixel fractions of individual urban LULC categories of interest. The approach was tested using HyMap sensor data acquired over Berlin, Germany, and fractions estimated using the approach were validated with reference data and compared to fractions derived from the MESMA approach. The validated results showed the high accuracies of quantitative estimates derived using the proposed approach for four spectrally complex urban land cover types, i.e., fractions of impervious rooftops and pavements and grass- and tree-covered areas. Their findings also demonstrated that the combination of SVR and synthetically mixed training data enabled the use of empirical regression for sub-pixel mapping, and thus the strengths of kernel-based approaches for quantifying urban materials from IS data could be well utilized. Also based on the SVR combined with synthetically mixed training data described in Okujeni et al. (2013), Okujeni et al. (2015) explored the potential of EnMAP data for deriving land covers along the urban–rural gradient of Berlin, Germany. The upcoming hyperspectral satellite mission EnMAP (see Chapter 3 for an introduction to EnMAP) data at 30 m spatial resolution were simulated from HyMap using the EnMAP end-to-end simulator (Okujeni et al. 2015). Land-cover fraction maps derived from the simulated EnMAP data demonstrated that EnMAP imagery would be well suited for mapping impervious, vegetation, and soil surface types according to the V-I-S framework (Ridd 1995). Moreover, the simulated EnMAP data would allow extending the V-I-S framework by more detailed urban materials, such as roof and pavement, and/or low vegetation and tree. However, Okujeni et al. (2015) also advised caution that satellite imaging spectrometer data of improved quality would not completely help overcome well-known phenomena of spectral similarity between materials and spectral confusion caused by the presence of shaded areas within an urban environment. In order to ultimately conduct ecological process studies through a case study of Boulder, Colorado, Golubiewski and Wessman (2010) used convex geometry and partial unmixing algorithms (i.e., MTMF; see Chapter 3) with AVIRIS imagery to identify major landscape elements, including five vegetation end-members that comprised cultivated and natural vegetation, soil, water, and five types of impervious surface. Their study expanded the multispectral unmixing of the V-I-S model with the MTMF algorithms and AVIRIS data, and the experimental results demonstrated a viability of the method for mapping the composition of urban areas.

Urban LULC classes are spectrally heterogeneous and materials from different classes have similar spectral properties. In such a spectrally complex urban environment, although high spectral and spatial resolution hyperspectral data can provide a great potential to characterize urban LULC classes at a material level, the mapping accuracy of urban LULC classes is often limited due to the heterogeneity of urban materials and LULC classes. Hence, it is necessary to exploit object-based image analysis (OBIA) methods when hyperspectral image data with high spatial resolution are available for mapping urban materials and LULC types. Accordingly, researchers are interested in exploring the OBIA methods to map urban environments with hyperspectral images. By training an SVM on pixel information and then applying it to an unsegmented image and segmented images at different levels, van der Linden et al. (2007) investigated the effects of image segmentation on the purely spectral classification of a hyperspectral HyMap data set from a large heterogeneous urban environment. Different effects on mapping the heterogeneous environment were identified with regard to average segment sizes. Highest accuracies for the different urban structure types were achieved at varying segmentation levels. Hence, a straightforward multi-level approach was performed, which combines information from different levels into one final map. The experimental results indicated that the accuracy of the multi-level approach was similar to that of unsegmented data but the multi-level approach comprised the positive effects of more homogeneous segment-based classifications at different levels in one map. In mapping detailed urban LC classes in a portion of Kuala Lumpur, Malaysia, from AISA hyperspectral data, Shafri and Hamedianfar (2015) used the SVM classifier with pixel-based and object-based (OBIA) approaches to test the performance of each method for mapping 12 urban land cover types, including asbestos roofs that are hazardous to human settlements. The OBIA approach could efficiently utilize spatial, spectral, and textural information in a complex urban land cover classification. The mapping result showed that the object-based SVM method could make more accurate characterization of urban land covers from the complex environment compared to the pixel-based SVM method.

9.6.3 Urban Thermal Environment

Urban thermal environment as a result of anthropogenic activities is one of the most interesting topics of remote sensing applications. However, only a few studies were conducted on using hyperspectral sensor data to quantify and map urban thermal environments, such as the urban heat island (UHI) phenomena of a city's climate. This may be due to there being only a few hyperspectral sensors/systems that include thermal spectral range. For example, in order to assess the distribution effect of the UHI in a nontraditional way in Afula city, Israel, Ben-Dor et al. (2001b) used DAIS-7950 hyperspectral sensor data to process and create thermal-based maps. The DAIS sensor consists of 79 bands across VIS–SWIR–MIR–TIR spectral range with seven bands across the TIR region to allow the extraction of parameters such as emissivity and energy fluxes on a per-pixel basis using information from the VIS–NIR–SWIR spectral region. Compared with the traditional UHI mapping way, the spectral-based UHI mapping method could provide additional information about the UHI that cannot be extracted in the traditional way. The mapping result also demonstrated that HRS technology enabled the identification of heat flux parameters and provided a new view of the city's UHI. Microclimate modeling is a powerful tool to quantify the thermal characteristics of urban environments at a local scale. However, the modeling requires mapping urban surface materials at high spatial resolution and knowing object heights. Berger et al. (2015) performed a fusion of airborne hyperspectral CASI and LiDAR remote sensing data, which enables mapping of some of the key input parameters required for urban microclimate modeling. To demonstrate the potential of data-driven microclimate modeling, they presented two case studies for selected test sites in Houston, Texas. The results in the study implied that classification-based microclimate simulations could reveal the thermal properties of urban neighborhoods and thus, facilitate the identification of hot spot areas and critical land cover configurations (Berger et al. 2015).

In assessing surface thermal patterns and their correlation with LC types in Shijiazhuang, Hebei Province, China, Liu et al. (2015) proposed an approach by combining Landsat TM images on the mesoscale level with airborne hyperspectral thermal images (thermal airborne spectrographic imager sensor [TASI], with the spectral range of 8–11.5 μm and 32 bands at a spectral resolution of 0.1095 μm) on the microscale level and found that urban thermal signatures at the two spatial scales could complement each other and the use of airborne imagery data with higher spatial resolution was helpful in revealing more details for understanding urban thermal environments. In their study, land surface temperature (LST) was retrieved from TM thermal band to analyze the thermal spatial patterns and intensity of surface urban heat island (SUHI), and TASI data were utilized to describe more detailed urban thermal characteristics of the city. Surface thermal characteristics were further examined by correlating LST with percentage of imperious surface area (ISA%). The results showed that an obvious surface heat island effect existed in the study area during summer days with an SUHI intensity of 2–4°C, and ISA% could provide an additional metric for the study of SUHI. The results derived from TASI thermal data indicated that among all of the LC types, the diversity of impervious surfaces (rooftops, concrete, and mixed asphalt) contributed most to the SUHI.

9.7 SUMMARY

In this chapter, there are five environmental and discipline areas in which research and applications of HRS were reviewed and summarized. In the atmospheric sciences, after introducing and analyzing spectral properties of major atmospheric gases and their absorption bands, techniques and methods used for retrieving and mapping the four key atmospheric parameters, including water vapor (H_2O), clouds, aerosols, and carbon dioxide (CO_2) using HRS data sets in literature were reviewed and summarized in Table 9.2. In Section 9.3, using different hyperspectral data sets, the spectral characteristics of snow and ice and impacts of snow grain size, snow impurity and contamination, snow melting, and ice states on surface albedo and hydrology were reviewed. HRS techniques and data have been used to estimate and map water quality parameters and other components in coastal environments and inland waters. These components or water quality parameters include chlorophyll, a variety of planktonic species, color dissolved organic matter, suspended sediments with local and distant sources, turbidity, bottom composition, submerged aquatic vegetation, and Secchi depth. The application techniques and algorithms using HRS data sets are reviewed and summarized in Table 9.3. HRS techniques and data can be used in environmental hazard and disaster monitoring. Hence, in Section 9.5, a review is presented on HRS applied (1) for assessing and monitoring impacts of mining wastes and tailings on environments including surface and subsurface soils, vegetation and wasters in the watershed where mine sites are located, (2) for mapping and characterizing wildland biomass burning, and (3) for monitoring and assessing vulnerable slopes and landslides. In Section 9.6, a brief review was given on spectral properties of urban materials and urban spectral libraries, and use of HRS data sets and a variety of analysis techniques and algorithms for identifying and mapping urban materials and land use land-cover classes, as well as characterizing urban thermal environments such as urban heat island phenomena.

REFERENCES

Alakian, A., R. Marion, and X. Briottet. 2009. Retrieval of microphysical and optical properties in aerosol plumes with hyperspectral imagery: L-APOM method. *Remote Sensing of Environment* 113:781–793.

Barducci, A., D. Guzzi, P. Marcoionni, and I. Pippi. 2004. Algorithm for the retrieval of columnar water vapor from hyperspectral remotely sensed data. *Applied Optics* 43(29):5552–5563.

Bassani, C., R. M. Cavalli, and S. Pignatti. 2010. Aerosol optical retrieval and surface reflectance from airborne remote sensing data over land. *Sensors* 10:6421–6438.

Ben-Dor, E. 2001. Imaging spectrometry for urban application, in *Imaging Spectrometry: Basic Principles and Prospective Applications*, eds. F. D. van der Meer and S. M. DeJong, pp. 243–282. Dordrecht: Kluwer Academic.

Ben-Dor, E., N. Levin, and H. Saaroni. 2001a. A spectral based recognition of the urban environment using the visible and near-infrared spectral region (0.4–1.1 μm). A case study over Tel-Aviv, Israel. *International Journal of Remote Sensing* 22(11):2193–2218.

Ben-Dor, E., R. Lugassi, R. Richter, H. Saaroni, and A. Muller. 2001b. Quantitative approach for monitoring the urban heat island effects, using hyperspectral remote sensing. *IEEE 2001 International Geoscience and Remote Sensing Symposium* 6:6 p.

Berger, C., F. Riedel, J. Rosentreter, E. Stein, S. Hese, and C. Schmullius. 2015. Fusion of airborne hyperspectral and LiDAR remote sensing data to study the thermal characteristics of urban environments, in *Computational Approaches for Urban Environments, Geotechnologies and the Environment*, eds. M. Helbich et al., pp. 273–292. Switzerland: Springer International Publishing.

Bianchi, R., R. M. Cavalli, L. Fiumi, C. M. Marino, S. Panuzi, and S. Pignatti. 1996. Airborne remote sensing in urban areas: Examples and considerations on the applicability of hyperspectral surveys over industrial, residential, and historical environments. *Proceedings of the Second International Airborne Remote Sensing Conference and Exhibition*, pp. 439–444. San Francisco, California.

Boine, J., K. Kuka, C. GlaBer, C. Olbert, and J. Fischer. 1999. Multispectral investigations of acid mine lakes of lignite open cast mines in Central Germany. *IGARSS1999 Proceedings*, pp. 855–857. Hamburg, Germany.

Brando, V. E., and A. G. Dekker. 2003. Satellite hyperspectral remote sensing for estimating estuarine and coastal water quality. *Transactions on Geoscience and Remote Sensing* 41(6):1378–1387.

Bruegge, C. J., J. E. Conel, J. S. Margolis, R. O. Green, G. C. Toon, V. Carrere, R. G. Holm, and G. Hoover. 1990. In situ atmospheric water-vapor retrieval in support of AVIRIS validation. *Imaging Spectroscopy of the Terrestrial Environment, SPIE* 1298:150–163.

Carder, K. L., P. Reinersman, R. F. Chen, F. Muller-Karger, C. O. Davis, and M. Hamilton. 1993. AVIRIS calibration and application in coastal oceanic environments. *Remote Sensing of Environment* 44:205–216.

Carrère, V., and J. E. Conel. 1993. Recovery of atmospheric water vapor total column abundance from imaging spectrometer analysis and application to Airborne Visible/Infrared Imaging Spectrometer (AVIRIS) data. *Remote Sensing of Environment* 44:179–204.

Cavalli, R. M., L. Fusilli, S. Pascucci, S. Pignatti, and F. Santini. 2008. Hyperspectral sensor data capability for retrieving complex urban land cover in comparison with multispectral data: Venice City case study (Italy). *Sensors* 8:3299–3320.

Cho, H. J., I. Ogashawara, D. Mishra, J. White, A. Kamerosky, L. Morris, C. Clarke, A. Simpson, and D. Banisakher. 2014. Evaluating Hyperspectral Imager for the Coastal Ocean (HICO) data for seagrass mapping in Indian River Lagoon, FL. *GIScience & Remote Sensing* 51(2):120–138.

Clark, C. D., H. T. Ripley, E. P. Green, A. J. Edwards, and P. J. Mumby. 1997. Mapping and measurement of tropical coastal environments with hyperspectral and high spatial resolution data. *International Journal of Remmote Sensing* 18(2):237–242.

Clark, R. N., and T. L. Roush. 1984. Reflectance spectroscopy: Quantitative analysis techniques for remote sensing applications. *Journal of Geophysical Research* 89:6329–6340.

Crowley, J. K., and D. R. Zimbelman. 1997. Mapping hydrothermally altered rocks on Mount Rainier, Washington, with Airborne Visible/Infrared Imaging Spectrometer (AVIRIS) data. *Geology* 25(6):559–562.

Dall'Olmo, G., A. A. Gitelson, and D. C. Rundquist. 2003. Towards a unified approach for remote estimation of chlorophyll-a in both terrestrial vegetation and turbid productive waters. *Geophysical Research Letters* 30(18):1938.

Davies, W. H., P. R. J. North, W. M. F. Grey, and M. J. Barnsley. 2010. Improvements in aerosol optical depth estimation using multiangle CHRIS/Proba images. *IEEE Transactions on Geoscience and Remote Sensing* 48(1):18–24.

Demarchi, L., F. Canters, J. C.-W. Chan, and T. Van de Voorde. 2012. Multiple endmember unmixing of CHRIS/Proba Imagery for mapping impervious surfaces in urban and suburban environments. *Transactions on Geoscience and Remote Sensing* 50(9):3409–3424.

Dennison, P., and D. Roberts. 2009. Daytime fire detection using airborne hyperspectral data. *Remote Sensing of Environment* 113:1646–1657.

Dennison, P. E., A. K. Thorpe, E. R. Pardyjak, D. A. Roberts, Y. Qi, R. O. Green, E. S. Bradley, and C. C. Funk. 2013. High spatial resolution mapping of elevated atmospheric carbon dioxide using airborne imaging spectroscopy: Radiative transfer modeling and power plant plume detection. *Remote Sensing of Environment* 139:116–129.

Deschamps, A., R. Marion, X. Briottet, and P.-Y. Foucher. 2013. Simultaneous retrieval of CO_2 and aerosols in a plume from hyperspectral imagery: Application to the characterization of forest fire smoke using AVIRIS data. *International Journal of Remote Sensing* 34(19):6837–6864.

Dierssen, H. M., A. Chlus, and B. Russell. 2015. Hyperspectral discrimination of floating mats of seagrass wrack and the macroalgae *Sargassum* in coastal waters of Greater Florida Bay using airborne remote sensing. *Remote Sensing of Environment* 167:247–258.

Dozier, J., and D. Marks. 1987. Snow mapping and classification from Landsat Thematic Mapper data. *Annals of Glaciology* 9:97–103.

Dozier, J., R. E. Davis, and A. W. Nolin. 1989. Reflectance and transmittance of snow at high spectral resolution. *IGARSS '89 Proceedings*, pp. 662–664.

Eismann, M. 2012. *Hyperspectral Remote Sensing*. Bellingham, WA: SPIE Press.

EPA. 1998. U.S. *Environmental Protection Agency Advanced Measurement Initiative Workshop Report EPA-235-R-98-002.*

Fan, F., and Y. Deng. 2014. Enhancing end-member selection in multiple end-member spectral mixture analysis (MESMA) for urban impervious surface area mapping using spectra angle and spectral distance parameters. *International Journal of Applied Earth Observation and Geoinformation* 33:290–301.

Farifteh, J., W. Nieuwenhuis, and E. García-Meléndez. 2013. Mapping spatial variations of iron oxide by-product minerals from EO-1 Hyperion. *International Journal of Remote Sensing* 34(2):682–699.

Farrand, W. H., and J. C. Harsanyi. 1997. Mapping the distribution of mine tailings in the Coeur d'Alene River Valley, Idaho, through the use of a constrained energy minimization technique. *Remote Sensing of Environment* 59:64–76.

Feind, R. E., and R. M. Welch. 1995. Cloud fraction and cloud shadow property retrievals from coregistered TIMS and AVIRIS imagery: The use of cloud morphology for registration. *IEEE Transactions on Geoscience and Remote Sensing* GE-33(1):172–184.

Feng, Q., J. Gong, Y. Wang, J. Liu, Y. Li, A.N. Ibrahim, Q. Liu, and Z. Hu. 2015. Estimating chlorophyll-a concentration based on a four-band model using field spectral measurements and HJ-1A hyperspectral data of Qiandao Lake, China. *Remote Sensing Letters* 6(10):735–744.

Fiumie, L., and C. M. Marino. 1997. Airborne hyperspectral MIVIS data for the characterization of urban historical environments. *Proceedings of the Third International Airborne Remote Sensing Conference and Exhibition II*, pp. 770–771. Copenhagen, Denmark.

Forzieri, G., L. Tanteri, G. Moser, and F. Catani. 2013. Mapping natural and urban environments using airborne multi-sensor ADS40–MIVIS–LiDAR synergies. *International Journal of Applied Earth Observation and Geoinformation* 23:313–323.

Franke, J., D. A. Roberts, K. Halligan, and G. Menz. 2009. Hierarchical multiple endmember spectral mixture analysis (MESMA) of hyperspectral imagery for urban environments. *Remote Sensing of Environment* 113:1712–1723.

Frouin, R., P.-Y. Deschamps, and P. Lecomte. 1990. Determination from space of atmospheric total water vapor, amounts by differential absorption near 940 nm: Theory and airborne verification. *Journal of Applied Meteorology and Climatology* 29:448–459.

Funk, C. C., J. Theiler, D. A. Roberts, and C. C. Borel. 2001. Clustering to improve matched filter detection of weak gas plumes in hyperspectral thermal imagery. *IEEE Transactions on Geoscience and Remote Sensing* 39:1410–1420.

Fyfe, S. K. 2003. Spatial and temporal variation in spectral reflectance: Are seagrass species spectrally distinct? *Limnology and Oceanography* 48:464–479.

Gangopadhyay, P. K., F. van der Meer, and P. van Dijk. 2009. Detecting anomalous CO_2 flux using space borne spectroscopy. *International Journal of Applied Earth Observation and Geoinformation* 11:1–7.

Gao, B., and A. F. H. Goetz. 1990. Column atmospheric water vapor and vegetation liquid water retrievals from airborne imaging spectrometer data. *Journal of Geophysical Research* 95(D4):3549–3564.

Gao, B., and A. F. H. Goetz. 1991. Cloud area determination from AVIRIS data vapor channels near 1 μm. *Journal of Geophysical Research* 96(D2):2857–2864.

Gao, B. C., and A. F. H. Goetz. 1992. Separation of cirrus clouds from clear surface from AVIRIS data using the 1.38 μm water vapor band. In *Summaries of the Third Annual JPL Airborne Geoscience Workshop*, JPL Publ. 92-14, vol. 1:98–100.

Gao, B. C., K. B. Heidebrecht, and A. F. H. Goetz. 1993. Derivation of scaled surface reflectances from AVIRIS data. *Remote Sensing of Environment* 44:165–178.

Gao, B. C., M. J. Montes, C. O. Davos, and A. F. H. Goetz. 2009. Atmospheric correction algorithms for hyperspectral remote sensing data of land and ocean. *Remote Sensing Environment* 113(s1):S17–S24.

George, D. G. 1997. Bathymetric mapping using a compact airborne spectrographic imager (CASI). *International Journal Remote Sensing* 18(10):2067–2071.

Giardino, C., G. Candiani, M. Bresciani, Z. Lee, S. Gagliano, and M. Pepe. 2012. BOMBER: A tool for estimating water quality and bottom properties from remote sensing images. *Computers & Geosciences* 45:313–318.

Giardino, C., M. Bresciani, E. Valentini, L. Gasperini, R. Bolpagni, and V. E. Brando. 2015. Airborne hyperspectral data to assess suspended particulate matter and aquatic vegetation in a shallow and turbid lake. *Remote Sensing of Environment* 157:48–57.

Gitelson, A. A., B.-C. Gao, R.-R. Li, S. Berdnikov, and V. Saprygin. 2011. Estimation of chlorophyll-a concentration in productive turbid waters using a hyperspectral imager for the coastal ocean—The Azov Sea case study. *Environmental Research Letters* 6:024023.

Gitelson, A. A., G. Dall'Olmo, W. Moses, D. C. Rundquist, T. Barrow, T. R. Fisher, D. Gurlin, and J. Holz. 2008. A simple semi-analytical model for remote estimation of chlorophyll a in turbid waters: Validation. *Remote Sensing of Environment* 112:3582–3593.

Gläßer, C., D. Groth, and J. Frauendorf. 2011. Monitoring of hydrochemical parameters of lignite mining lakes in Central Germany using airborne hyperspectral CASI-scanner data. *International Journal of Coal Geology* 86:40–53.

Golubiewski, N. E., and C. A. Wessman. 2010. Discriminating urban vegetation from a metropolitan matrix through partial unmixing with hyperspectral AVIRIS data. *Canadian Journal of Remote Sensing* 36(3):261–275.

Gons, H. J. 1999. Optical teledetection of chlorophyll a in turbid inland waters. *Environmental Science & Technology* 33:1127–1132.

Gons, H. J., M. Rijkeboer, and K. G. Ruddick. 2002. A chlorophyll-retrieval algorithm for satellite imagery (Medium Resolution Imaging Spectrometer) of inland and coastal waters. *Journal of Plankton Research* 24(9):947–951.

Gons, H. J., M. Rijkeboer, and K. G. Ruddick. 2005. Effect of a waveband shift on chlorophyll retrieval from MERIS imagery of inland and coastal waters. *Journal of Plankton Research* 27:125–127.

Gordon, H., D. Clark, J. Brown, O. Brown, R. Evans, and W. Brokenow. 1983. Phytoplankton pigment concentrations in the Middle Atlantic Bight: Comparison of ship determinations and CZCS estimates. *Applied Optics* 22(1):20–36.

Green, R. O., and J. Dozier. 1995. Measurement of the spectral absorption of liquid water in melting snow with an imaging spectrometer. *Summaries of the Sixth Annual JPL Airborne Earth Science Workshop JPL Publication* 95-1, vol. 1:91–94.

Green, R. O., and J. Dozier. 1996. Retrieval of surface snow grain size and melt water from AVIRIS spectra. *Summaries of the Sixth Annual JPL Airborne Earth Science Workshop JPL Publication* 96-4, vol. 1:127–134.

Green, R. O., V. Carrère, and J. E. Conel. 1989. Measurement of atmospheric water vapor using the Airborne Visible/Infrared Imaging Spectrometer. *Workshop Imaging Processing*, American Society for Photogrammetry and Remote Sensing. Sparks, Nevada.

Green, R. O., T. H. Painter, D. A. Roberts, and J. Dozier. 2006. Measuring the expressed abundance of the three phases of water with an imaging spectrometer over melting snow. *Water Resources Research* 42:W10402. doi:10.1029/2005WR004509.

Green, R. O., M. L. Eastwood, C. M. Sarture, T. G. Chrien, M. Aronsson, B. J. Chippendale, J. A. Faust, B. E. Pavri, C. J. Chovit, M. Solis, M. R. Olah, and O. Williams. 1998. Imaging spectroscopy and the airborne visible/infrared imaging spectrometer (AVIRIS). *Remote Sensing of Environment* 65:227–248.

Griffin, M. K., S. M. Hsu, H. K. Burke, and J. W. Snow. 2000. Characterization and delineation of plumes, clouds and fires in hyperspectral images. *Proceedings of the Society of Photo-Optical Instrumentation Engineers (SPIE) Vol. 4049, Algorithms for Multispectral, Hyperspectral, and Ultraspectral Imagery VI*, eds. S. S. Shen and M. R. Descour. Orlando, Florida.

Guanter, L., L. Alonso, and J. Moreno. 2005. First results from the PROBA/CHRIS hyperspectral/multiangular satellite system over land and water targets. *IEEE Geoscience Remote Sensing Letters* 2(3):250–254.

Guanter, L., V. Estellés, and J. Moreno. 2007. Spectral calibration and atmospheric correction of ultra-fine spectral and spatial resolution remote sensing data application to CASI-1500 data. *Remote Sensing of Environment* 109:54–65.

Guanter, L., M. C. González-Sampedro, and J. Moreno. 2007. A method for the atmospheric correction of ENVISAT/MERIS data over land targets. *International Journal of Remote Sensing* 28(3–4):709–728.

Guanter, L., R. Richter, and J. Moreno. 2006. Spectral calibration of hyperspectral imagery using atmospheric absorption features. *Applied Optics* 45:2360–2370.

Hall, D. K., G. A. Riggs, and V. V. Salomonson. 1995. Development of methods for mapping global snow cover using Moderate Resolution Imaging Spectroradiometer data. *Remote Sensing Environment* 54:127–140.

Hamilton, M. K., C. O. Davis, W. J. Rhea, S. H. Pilorz, and K. L. Carder. 1993. Estimating chlorophyll content and bathymetry of Lake Tahoe using AVIRIS data. *Remote Sensing of Environment* 44:217–230.

Haq, M. A. 2014. Comparative analysis of hyperspectral and multispectral data for mapping snow cover and snow grain size. *The International Archives of the Photogrammetry, Remote Sensing and Spatial Information Sciences* XL-8:499–504.

Heiden, U., S. Roessner, K. Segi, and H. Kaufmann. 2001. Analysis of spectral signatures of urban surfaces for their identification using hyperspectral HyMap data. *Proceedings of the IEEE/ISPRS Joint Workshop on Remote Sensing and Data Fusion over Urban Areas*, pp. 173–177.

Heiden, U., K. Segl, S. Roessner, and H. Kaufmann. 2007. Determination of robust spectral features for identification of urban surface materials in hyperspectral remote sensing data. *Remote Sensing of Environment* 111:537–552.

Hill, V. J., R. C. Zimmerman, W. P. Bissett, H. Dierssen, and D. D. R. Kohler. 2014. Evaluating light availability, seagrass biomass, and productivity using hyperspectral airborne remote sensing in Saint Joseph's Bay, Florida. *Estuaries and Coasts* 37:1467–1489.

Holden, H., and E. Ledrew. 1999. Hyperspectral identification of coral reef features. *International Journal of Remote Sensing* 20(13):2545–2563.

Huang, Y., D. Jiang, D. Zhuang, and J. Fu. 2010. Evaluation of hyperspectral indices for chlorophyll-a concentration estimation in Tangxun Lake (Wuhan, China). *International Journal of* Environmental *Research* and *Public Health* 7:2437–2451.

Hunter, P. D., A. N. Tyler, L. Carvalho, G. A. Codd, and S. C. Maberly. 2010. Hyperspectral remote sensing of cyanobacterial pigments as indicators for cell populations and toxins in eutrophic lakes. *Remote Sensing of Environment* 114:2705–2718.

Hutchison, K. D., and N. J. Choe. 1996. Application of 1-38 μm imagery for thin cirrus detection in daytime imagery collected over land surfaces. *International Journal of Remote Sensing* 17(17):3325–3342.

Isakov, V. Y., R. E. Feind, O. B. Vasilyev, and R. M. Welch. 1996. Retrieval of aerosol spectral optical thickness from AVIRIS data. *Int. J. Remote Sensing* 17(11):2165–2184.

Jay, S., and M. Guillaume. 2014. A novel maximum likelihood based method for mapping depth and water quality from hyperspectral remote-sensing data. *Remote Sensing of Environment* 147:121–132.

Kalman, L. S. and E. M. Bassett III. 1997. Classification and material identification in an urban environment using HYDICE hyperspectral data. *Proceedings of SPIE: Imaging Spectrometry III* 3118:57–68.

Kaufman, Y., and D. Tanré. 1996. Strategy for direct and indirect methods for correction the aerosol effect on remote sensing: From AVHRR to EOS-MODIS. *Remote Sensing of Environment* 55:65–79.

Kaufman, Y. J., A. E. Wald, L. A. Remer, B.-C. Gao, R.-R. Li, and L. Flynn. 1997. The MODIS 2.1-μm Channel-Correlation with Visible Reflectance for Use in Remote Sensing of Aerosol. *IEEE Transactions on Geoscience and Remote Sensing* 35:1286–1298.

Kaufman, Y. J., C. O. Justice, L. P. Flynn, J. D. Kendall, E. M. Prins, L. Giglio, D. E. Ward, W. P. Menzel, and A. W. Setzer. 1998. Potential global fire monitoring from EOS-MODIS. *Journal of Geophysical Research* 103(D24):32215–32238.

Kaufman, Y. J., P. V. Hobbs, V. W. J. H. Kirchhoff et al. 1998. Smoke, clouds, and radiation-Brazil (SCAR-B) experiment. *Journal of Geophysical Research* 103(D24):31783–31808.

Keith, D. J., B. A. Schaeffer, R. S. Lunetta, R. W. Gould Jr., K. Rocha, and D. J. Cobb. 2014. Remote sensing of selected water-quality indicators with the hyperspectral imager for the coastal ocean (HICO) sensor. *International Journal of Remote Sensing* 35(9):2927–2962.

Koponen, S., J. Pulliainen, K. Kallio, and M. Hallikainen. 2002. Lake water quality classification with airborne hyperspectral spectrometer and simulated MERIS data. *Remote Sensing of Environment* 79:51–59.

Kotthaus, S., T. E. L. Smith, M. J. Wooster, and C. S. B. Grimmond. 2014. Derivation of an urban materials spectral library through emittance and reflectance spectroscopy. *ISPRS Journal of Photogrammetry and Remote Sensing* 94:194–212.

Kou, L., D. Labrie, and P. Chylek. 1993. Refractive indices of water and ice in the 0.65–2.5 μm spectral range. *Applied Optics* 32(19):3531–3540.

Kuo, K. S., R. M. Welch, B. C. Gao, and A. F. H. Goetz. 1990. Cloud identification and optical thickness retrieval using AVIRIS data. In *Proceedings of the Second AVIRIS Workshop JPL Publication* 90-54, 149–156.

Lang, R., A. N. Maurellis, W. J. van der Zande, I. Aben, J. Landgraf, and W. Ubachs. 2002. Forward modeling and retrieval of water vapor from the Global Ozone Monitoring Experiment: Treatment of narrowband absorption spectra. *Journal of Geophysical Research* 107(D16):10.1029/2001JD001453.

Lee, Z., L. C. Kendall, R. F. Chen, and T. G. Peacock. 2001. Properties of the water column and bottom derived from Airborne Visible Imaging Spectrometer (AVIRIS) data. *Journal of Geophysical Research* 106:11639–11652.

Lee, Z., K. L. Carder, C. D. Mobley, R. G. Steward, and J. S. Patch. 1998. Hyperspectral remote sensing for shallow waters: 1. A semianalytical model. *Applied Optics* 37:6329–6338.

Lewis, S. A., A. T. Hudak, R. D. Ottmar, P. R. Robichaud, L. B. Lentile, S. M. Hood, J. B. Cronan, and P. Morgan. 2011. Using hyperspectral imagery to estimate forest floor consumption from wildfire in boreal forests of Alaska, USA. *International Journal of Wildland Fire* 20:255–271.

Li, L., Q. Yin, H. Xu, C. Gong, and Z. Chen. 2010 (July 25–30). Estimating chlorophyll a concentration in lake water using space-borne hyperspectral data. *Proceeding of 2010 IEEE International Geoscience & Remote Sensing Symposium*, 401–404. Honolulu, Hawaii.

Liew, S. C., and L. K. Kwoh. 2002. Mapping optical parameters of coastal sea waters using the Hyperion imaging spectrometer. *Ocean Remote Sensing and Applications, SPIE Proceedings* 4892. doi:10.1117/12.466828.

Liu, K., H. Su, L. Zhang, H. Yang, R. Zhang, and X. Li. 2015. Analysis of the urban heat island effect in Shijiazhuang, China, using satellite and airborne data. *Remote Sensing* 7:4804–4833.

Lubac, B., H. Loisel, N. Guiselin, R. Astoreca, L. Felipe Artigas, and X. Me´riaux. 2008. Hyperspectral and multispectral ocean color inversions to detect Phaeocystis globosa blooms in coastal waters. *Journal of Geophysical Research* 113:C06026. doi:10.1029/2007JC004451.

Lv, J., and X. Liu. 2009. Sub-pixel mapping of urban green space using multiple end-member spectral mixture analysis of EO-1 Hyperion data. *Proceedings of IEEE 2009 Urban Remote Sensing Joint Event.* doi: 10.1109/URS.2009.5137517.

Ma, R., X. Ma, and J. Dai. 2007. Hyperspectral feature analysis of chlorophyll a and suspended solids using field measurements from Taihu Lake, eastern China. *Hydrological Sciences Journal* 52(4):808–824.

Marino, C. M., C. Panigada, and L. Busetto. 2001. Airborne hyper spectral remote sensing applications in urban areas: Asbestos concrete sheeting identification and mapping. *Proceedings of the IEEE/ISPRS Joint Workshop on Remote Sensing and Data Fusion over Urban Areas*, pp. 212–216.

Marino, C. M., C. Panigada, A. Galli, L. Boschetti, and L. Buseto. 2000. Environmental applications of airborne hyperspectral remote sensing: Asbestos concrete sheeting identification and mapping. *Proceedings of the International Conference on Applied Geologic Remote Sensing*, pp. 607–610. Las Vegas, Nevada.
Marinoni, A., and P. Gamba. 2016. Accurate detection of anthropogenic settlements in hyperspectral images by higher order nonlinear unmixing. *Journal of Selected Topics in Applied Earth Observations and Remote Sensing* 9(5):1792–1801.
Marion, R., R. Michel, and C. Faye. 2004. Measuring trace gases in plumes from hyperspectral remotely sensed data. *IEEE Transactions on Geoscience and Remote Sensing* 42(4):854–864.
Mars, J. C., and J. K. Crowley. 2003. Mapping mine wastes and analyzing areas affected by selenium-rich water runoff in southeast Idaho using AVIRIS imagery and digital elevation data. *Remote Sensing of Environment* 84:422–436.
Maurellis, A. N., R. Lang, W. J. van der Zande, I. Aben, and W. Ubachs. 2000. Precipitable water column retrieval from GOME data. *Geophysical Research Letters* 27(6):903–906.
Meganem, I., P. Déliot, X. Briottet, Y. Deville, and S. Hosseini. 2014. Linear–quadratic mixing model for reflectances in urban environments. *Transactions on Geoscience and Remote Sensing* 52(1):544–558.
Minu, P., S. S. Shaju, V. P. Souda, B. Usha, P. M. Ashraf, and B. Meenakumari. 2015. Hyperspectral variability of phytoplankton blooms in coastal waters off Kochi, south-eastern Arabian Sea. *Fishery Technology* 52:218–222.
Mitri, G. H., and I. Z. Gitasc. 2013. Mapping post-fire forest regeneration and vegetation recovery using a combination of very high spatial resolution and hyperspectral satellite imagery. *International Journal of Applied Earth Observation and Geoinformation* 20:60–66.
Mondino, E. B., M. Giardino, and L. Perotti. 2009. A neural network method for analysis of hyperspectral imagery with application to the Cassas landslide (Susa Valley, NW-Italy). *Geomorphology* 110:20–27.
Mueller, M., K. Segl, and H. Kaufmann. 2003. Discrimination between roofing materials and streets within urban areas based on hyperspectral, shape, and context information. *Proceedings of the 2nd GRSSASPRS Joint Workshop on Data Fusion and Remote Sensing and Data Fusion over Urban Areas*, pp. 196–200.
Naegeli, K., A. Damm, M. Huss, M. Schaepman, and M. Hoelzle. 2015. Imaging spectroscopy to assess the composition of ice surface materials and their impact on glacier mass balance. *Remote Sensing of Environment* 168:388–402.
Nicolantonio, W. D., A. Tiesi, D. Labate, C. Ananasso, L. Candela, and C. Tomasi. 2015 (July 26–31). On the evaluation of prisma hyperspectral satellite sensitivity to significant loadings of carbon dioxide. *International Geoscience and Remote Sensing Symposium (IGARSS 2015)*, pp. 3914–3916. Milan, Italy.
Nolin, A. W., and J. Dozier. 1993. Estimating snow grain size using AVIRIS data. *Remote Sensing of Environment* 44(2–3):231–238.
Nolin, A. W., and J. Dozier. 2000. A hyperspectral method for remotely sensing the grain size of snow. *Remote Sensing of Environment* 74:207–216.
Numata, I., M. A. Cochrane, and L. S. Galvão. 2011. Analyzing the impacts of frequency and severity of forest fire on the recovery of disturbed forest using Landsat time series and EO-1 Hyperion in the southern Brazilian Amazon. *Earth Interactions* 15:Paper #13, 17 pp.
Okujeni, A., S. van der Linden, and P. Hostert. 2015. Extending the vegetation–impervious–soil model using simulated EnMAP data and machine learning. *Remote Sensing of Environment* 158:69–80.
Okujeni, A., S. van der Linden, L. Tits, B. Somers, and P. Hostert. 2013. Support vector regression and synthetically mixed training data for quantifying urban land cover. *Remote Sensing of Environment* 137:184–197.
Olmanson, L. G., P. L. Brezonik, and M. E. Bauer. 2013. Airborne hyperspectral remote sensing to assess spatial distribution of water quality characteristics in large rivers: The Mississippi River and its tributaries in Minnesota. *Remote Sensing of Environment* 130:254–265.
Painter, T. H., D. A. Roberts, R. O. Green, and J. Dozier. 1998. The effect of grain size on spectral mixture analysis of snow-covered area from AVIRIS data. *Remote Sensing of Environment* 65:320–332.
Painter, T. H., J. Dozier, D. A. Roberts, R. E. Davis, and R. O. Green. 2003. Retrieval of subpixel snow-covered area and grain size from imaging spectrometer data. *Remote Sensing of Environment* 85:64–77.
Painter, T. H., F. C. Seidel, A. C. Bryant, S. M. Skiles, and K. Rittger. 2013. Imaging spectroscopy of albedo and radiative forcing by light-absorbing impurities in mountain snow. *Journal of Geophysical Research: Atmospheres* 118:9511–9523.
Painter, T. H., B. Duval, W. H. Thomas, M. Mendez, S. Heintzelman, and J. Dozier. 2001a. Detection and quantification of snow algae with an airborne imaging spectrometer. *Applied and Environmental Microbiology* 67(11):5267– 5272.

Painter, T. H., B. Duval, W. H. Thomas, M. Mendez, S. Heintzelman, and J. Dozier. 2001b. Mapping snow algae concentration in the Sierra Nevada with imaging spectroscopy. *69th Annual Meeting of the Western Snow Conference*. Accessed September 28, 2016, from http://snobear.colorado.edu/WSC/WSC_2001/PDF/PainterEtAl.pdf.

Pan, Z., C. Glennie, J. C. Fernandez-Diaz, and M. Starek. 2016. Comparison of bathymetry and seagrass mapping with hyperspectral imagery and airborne bathymetric LIDAR in a shallow estuarine environment. *International Journal of Remote Sensing* 37(3):516–536.

Park, S. S., J. Kim, H. Lee, O. Torres, K.-M. Lee, and S. D. Lee. 2016. Utilization of O_4 slant column density to derive aerosol layer height from a space-borne UV–visible hyperspectral sensor: Sensitivity and case study. *Atmospheric Chemistry and Physics* 16:1987–2006.

Peneva, E., J. A. Griffith, and G. A. Carter. 2008. Seagrass mapping in the northern Gulf of Mexico using airborne hyperspectral imagery: A comparison of classification methods. *Journal of Coastal Research* 24(4):850–856.

Petty, G. W. 2006. *A First Course in Atmospheric Radiation*. Madison, WI: Sundog Publishing.

Phinn, S., C. Roelfsema, A. Dekker, V. Brando, and J. Anstee. 2008. Mapping seagrass species, cover and biomass in shallow waters: An assessment of satellite multispectral and airborne hyper-spectral imaging systems in Moreton Bay (Australia). *Remote Sensing of Environment* 112:3413–3425.

Press, W. H., B. P. Flannery, S. A. Teukolsky, and W. T. Vetterling. 1988. *Numerical Recipes in C: The Art of Scientific Computing*. Cambridge: Cambridge University Press.

Price, J. C. 1995. Examples of high resolution visible to near-infrared reflectance spectra and a standardized collection for remote sensing studies. *International Journal of Remote Sensing* 16:993–1000.

Pu, R., and S. Bell. 2013. A protocol for improving mapping and assessing of seagrass abundance along the West Central Coast of Florida using Landsat TM and EO-1ALI/Hyperion images. *ISPRS Journal of Photogrammetry and Remote Sensing* 83:116–129.

Pu, R., S. Bell, and D. English. 2015. Developing hyperspectral vegetation indices for identifying seagrass species and cover classes. *Journal of Coastal Research* 31(3):595–615.

Pu, R., S. Bell, L. Baggett, C. Meyer, and Y. Zhao. 2012. Discrimination of seagrass species and cover classes with in situ hyperspectral data. *Journal of Coastal Research* 28(6):1330–1344.

Qu, Z., B. C. Kindel, and A. F. H. Goetz. 2003. The high-accuracy atmospheric correction for Hyperspectral data (HATCH) model. *IEEE Transactions on Geoscience and Remote Sensing* 41:1223–123.

Riaza, A., and V. Carrere. 2009. Monitoring of superficial contamination produced by massive sulphide mine waste along the Odiel River (Andalusia, Spain) using hyperspectral data. *2009 IEEE International Geoscience and Remote Sensing Symposium* 3:1071–1074.

Riaza, A., J. Buzzi, E. García-Meléndez, V. Carrère, A. Sarmiento, and A. Müller. 2012. River acid mine drainage: Sediment and water mapping through hyperspectral Hymap data. *International Journal of Remote Sensing* 33(19):6163–6185.

Riaza, A., J. Buzzi, E. García-Meléndez, V. Carrère, A. Sarmiento, and A. Müller. 2015. Monitoring acidic water in a polluted river with hyperspectral remote sensing (HyMap). *Hydrological Sciences Journal* 60(6):1064–1077.

Riaza, A., J. Buzzi, E. García-Meléndez, V. Carrère, and A. Müller. 2011. Monitoring the extent of contamination from acid mine drainage in the Iberian Pyrite Belt (SW Spain) using hyperspectral imagery. *Remote Sensing* 3:2166–2186.

Riaza, A., J. Buzzi, E. García-Meléndez, I. Vázquez, E. Bellido, V. Carrère, and A. Müller. 2012. Pyrite mine waste and water mapping using Hymap and Hyperion hyperspectral data. *Environmental Earth Sciences* 66:1957–1971.

Richardson, L. L., and V. G. Ambrosia. 1996. Algal accessory pigment detection using AVIRIS image-derived spectral radiance data. *Summaries of the Sixth Annual JPL Airborne Earth Science Workshop JPL Publication* 96-4, 1:189–196.

Richter, R. and D. Schläpfer. 2014. Atmospheric/Topographic Correction for Airborne Imagery. *ATCOR-4 User Guide 6.3.2*. Wessling, Germany: DLR.

Ridd, M. K. 1995. Exploring a V-I-S (vegetation-impervious surface-soil) model for urban ecosystem analysis through remote sensing: Comparative anatomy for cities. *International Journal of Remote Sensing* 16:2165–2185.

Roberts, D. A., D. A. Quattrochi, G. C. Hulley, S. J. Hook, and R. O. Green. 2012. Synergies between VSWIR and TIR data for the urban environment: An evaluation of the potential for the Hyperspectral Infrared Imager (HyspIRI) Decadal Survey mission. *Remote Sensing of Environment* 117:83–101.

Roberts, D. A., M. Gardner, R. Church, S. Ustin, G. Scheer, and R. O. Green. 1998. Mapping chaparral in the Santa Monica Mountains using multiple end-member spectral mixture models. *Remote Sensing of Environment* 65(3):267–279.

Roberts, D. A., P. E. Dennison, M. E. Gardner, Y. Hetzel, S. L. Ustin, and C. T. Lee. 2003. Evaluation of the potential of Hyperion for fire danger assessment by comparison to the airborne visible/infrared imaging spectrometer. *Transactions on Geoscience and Remote Sensing* 41(6):1297–1310.

Rodger, A. 2011. SODA: A new method of in-scene atmospheric water vapor estimation and post-flight spectral recalibration for hyperspectral sensors: Application to the HyMap sensor at two locations. *Remote Sensing of Environment* 115:536–547.

Sandidge, J. C., and R. J. Holyer. 1998. Coastal bathymetry from hyperspectral observations of water radiance. *Remote Sens. Environ.* 65:341–352.

Schepers, L., B. Haest, S. Veraverbeke, T. Spanhove, J. V. Borre, and R. Goossens. 2014. Burned area detection and burn severity assessment of a heathland fire in Belgium using airborne imaging spectroscopy (APEX). *Remote Sensing* 6:1803–1826.

Schläpfer, D., J. Keller, and K. I. Itten. 1996. Imaging spectrometry of tropospheric ozone and water vapor, in *Proceedings of the 15th EARSeL Symposium Basel*, ed. E. Parlow, pp. 439–446. Rotterdam: A. A. Balkema. Accessed from http://www.geo.unizh.ch/dschlapf/paper.html.

Schläpfer, D., C. C. Borel, J. Keller, and K. I. Itten. 1998. Atmospheric precorrected differential absorption technique to retrieve columnar water vapor. *Remote Sensing of Environment* 65: 353–366.

Segl, K., S. Roessner, U. Heiden, and H. Kaufmann. 2003. Fusion of spectral and shape features for identification of urban surface cover types using reflective and thermal hyperspectral data. *ISPRS Journal of Photogrammetry and Remote Sensing* 58:99–112.

Seidel, F. C., K. Rittger, S. M. Skiles, N. P. Molotch, and T. H. Painter. 2016. Case study of spatial and temporal variability of snow cover, grain size, albedo and radiative forcing in the Sierra Nevada and Rocky Mountain snowpack derived from imaging spectroscopy. *The Cryosphere* 10:1229–1244.

Shafri, H. Z. M., and A. Hamedianfar. 2015. Mapping of intra-urban land covers using pixel-based and object-based classifications from airborne hyperspectral imagery. *In Proceedings of the 2015 2nd International Conference on Information Science and Security (ICISS).* doi: 10.1109/ICISSEC.2015.7371017.

Shang, J., B. Morris, P. Howarth, J. Lévesque, K. Staenz, and B. Neville. 2009. Mapping mine tailing surface mineralogy using hyperspectral remote sensing. *Canadian Journal of Remote Sensing* 35(Suppl. 1):S126–S141.

Shi, X.-Z., M. Aspandiar, I. C. Lau, and D. Oldmeadow. 2014. Assessment of Acid Sulphate Soil both on surface and in subsurface using hyperspectral data. *Canadian Journal of Remote Sensing* 39(6):468–480.

Simis, S. G. H., S. W. M. Peters, and H. J. Gons. 2005. Remote sensing of the cyanobacterial pigment phycocyanin in turbid inland water. *Limnology and Oceanography* 50:237–245.

Simis, S. G. H., A. Ruiz-Verdu, J. A. Dominguez-Gomez, R. Pena-Martinez, S. W. M. Peters, and H. J. Gons. 2007. Influence of phytoplankton pigment composition on remote sensing of cyanobacterial biomass. *Remote Sensing of Environment* 106:414–427.

Singh, S. K., A. V. Kulkarni, and B. S. Chaudhary. 2010. Hyperspectral analysis of snow reflectance to understand the effects of contamination and grain size. *Annals of Glaciology* 51(54):83–88.

Singh, S. K., A. V. Kulkarni, and B. S. Chaudhary. 2011. Spectral characterization of soil and coal contamination on snow reflectance using hyperspectral analysis. *Journal of Earth System Science* 120(2):321–328.

Sterzai, P., M. Vellico, M. Berti, F. Coren, A. Corsini, A. Rosi, P. Mora, F. Zambonelli, and F. Ronchetti. 2010. LiDAR and hyperspectral data integration for landslide monitoring: The test case of Valoria landslide. *Italian Journal of Remote Sensing* 42(3):89–99.

Sudduth, K. A. G.-S. Jang, R. N. Lerch, and E. J. Sadler. 2015. Long-term agroecosystem research in the central Mississippi river basin: Hyperspectral remote sensing of reservoir water quality. *Journal of Environmental Quality* 44:71–83.

Sun, D., Y. Li, Q. Wang, C. Le, C. Huang, and K. Shi. 2011. Development of optical criteria to discriminate various types of highly turbid lake waters. *Hydrobiologia* 669:3–104.

Sun, D., Y. Li, Q. Wang, J. Gao, C. Le, C. Huang, and S. Gong. 2013. Hyperspectral remote sensing of the pigment C-phycocyanin in turbid inland waters, based on optical classification. *Transactions on Geoscience and Remote Sensing* 51(7):3871–3884.

Sun, D., Y. Li, Q. Wang, C. Le, H. Lv, C. Huang, and S. Gong. 2012. A novel support vector regression model to estimate the phycocyanin concentration in turbid inland waters from hyperspectral reflectance. *Hydrobiologia* 680:199–217.

Sun, D. Y., Y. M. Li, Q. Wang, H. Lu, C. F. Le, C. C. Huang, and S. Q. Gong. 2011. A neural-network model to retrieve CDOM absorption from in situ measured hyperspectral data in an optically complex lake: Lake Taihu case study. *International Journal of Remote Sensing* 32(14):4005–4022.

Swayze, G.A., K. S. Smith, R. N. Clark, S. J. Sutley, R. M. Pearson, J. S. Vance, P. L. Hageman, P. H. Briggs, A. L. Meier, M. J. Singleton, and S. Roth. 2000. Using imaging spectroscopy to map acidic mine waste. *Environmental Science and Technology* 34:47–54.

Tan, K. X. Jin, Q. Du, and P. Du. 2014. Modified multiple end-member spectral mixture analysis for mapping impervious surfaces in urban environments. *Journal of Applied Remote Sensing* 8:085096–1.

Tarrant, P. E., J. A. Amacher, and S. Neuer. 2010. Assessing the potential of Medium Resolution Imaging Spectrometer (MERIS) and Moderate Resolution Imaging Spectroradiometer (MODIS) data for monitoring total suspended matter in small and intermediate sized lakes and reservoirs. *Water Resources Research* 46:W09532. doi:10.1029/2009WR008709.

Thiemann, S., and H. Kaufmann. 2002. Lake water quality monitoring using hyperspectral airborne data—A semiempirical multisensor and multitemporal approach for the Mecklenburg Lake District, Germany. *Remote Sensing of Environment* 81:228–237.

Thorhaug, A., A. D. Richardson, and G. P. Berlyn. 2007. Spectral reflectance of the seagrasses: Thalassia testudinum, Halodule wrightii, Syringodium filiforme and five marine algae. *International Journal of Remote Sensing* 28(7):1487–1501.

Thorpe, A. K., D. A. Roberts, E. S. Bradley, C. C. Funk, P. E. Dennison, and I. Leifer. 2013. High resolution mapping of methane emissions from marine and terrestrial sources using a Cluster-Tuned Matched Filter technique and imaging spectrometry. *Remote Sensing of Environment* 134:305–318.

Valle, M., V. Palà, V. Lafon, A. Dehouck, J. M. Garmendia, A. Borja, and G. Chust. 2015. Mapping estuarine habitats using airborne hyperspectral imagery, with special focus on seagrass meadows. *Estuarine, Coastal and Shelf Science* 164:433–442.

Van der Linden, S., A. Janz, B. Waske, M. Eiden, and P. Hostert. 2007. Classifying segmented hyperspectral data from a heterogeneous urban environment using support vector machines. *Journal of Applied Remote Sensing* 1(1):013543. doi:10.1117/1.2813466.

Vane, G., and A. F. H. Goetz. 1993. Terrestrial imaging spectrometry: Current status, future trends. *Remote Sensing of Environment* 44:117–126.

Veraverbeke, S., E. N. Stavros, and S. J. Hook. 2014. Assessing fire severity using imaging spectroscopy data from the Airborne Visible/Infrared Imaging Spectrometer (AVIRIS) and comparison with multispectral capabilities. *Remote Sensing of Environment* 154:153–163.

Wang, S., F. Yan, Y. Zhou, L. Zhu, L. Wang, and Y. Jiao. 2005. Water quality monitoring using hyperspectral remote sensing data in Taihu Lake China. P*roceedings of 2005 IEEE International Geoscience and Remote Sensing Symposium* 7:4553–4556.

Williams, D. J., N. B. Rybicki, A. V. Lombana, T. M. O'brien, and R. B. Gomez. 2003. Preliminary investigation of submerged aquatic vegetation mapping using hyperspectral remote sensing. *Environmental Monitoring and Assessment* 81:383–392.

Xing, Q., M. Lou, C. Chen, and P. Shi. 2013. Using in situ and satellite hyperspectral data to estimate the surface suspended sediments concentrations in the Pearl River estuary. *Journal of Selected Topics in Applied Earth Observations and Remote Sensing* 6(2):731–738.

Xing, Q., M. Lou, D. Yu, R. Meng, P. Shi, F. Bragct, L. Zaggia, and L. Tosi. 2012. Features of turbid waters from Hyperspectral Imager for the Coastal Ocean (HICO): Preliminary results at the yellow river delta and the Bohai Sea. *2012 4th Workshop on Hyperspectral Image and Signal Processing (WHISPERS).*

Yan, F., S. Wang, Y. Zhou, L. Zhu, L. Wang, and Y. Jiao. 2005. Inherent optical properties and hyperspectral remote sensing in Taihu Lake China. In *Proceedings of 2005 IEEE International Geoscience and Remote Sensing Symposium* 7:3579–3582.

Zabcic, N., B. Rivard, C. Ong, and A. Mueller. 2014. Using airborne hyperspectral data to characterize the surface pH and mineralogy of pyrite mine tailings. *International Journal of Applied Earth Observation and Geoinformation* 32:152–162.

Zhang, B., D. Wu, L. Zhang, Q. Jiao, and Q. Li. 2012. Application of hyperspectral remote sensing for environment monitoring in mining areas. *Environmental Earth Sciences* 65:649–658.

Zhang, H., Z. Chen, B. Zhang, and D. Peng. 2011 (July 24–29). Comparison of two water vapor retrieval algorithms for hj1a hyperspectral imagery. *IEEE International Geoscience and Remote Sensing Symposium, 2011 (IGARSS 2011)*, pp. 1732–1735. Vancouver, Canada.

Zhao, Y., R. Pu, S. Bell, C. Meyer, L. Baggett, and X. Geng. 2013. Hyperion image optimization in coastal waters. *IEEE Transactions on Geoscience and Remote Sensing* 51(2):1025–1036.

Zhou, L., D. A. Roberts, W. Ma, H. Zhang, and L. Tang. 2014. Estimation of higher chlorophyll a concentrations using field spectral measurement and HJ-1A hyperspectral satellite data in Dianshan Lake, China. *ISPRS Journal of Photogrammetry and Remote Sensing* 88:41–47.

Zhu, W., Q. Yu, Y. Q. Tian, R. F. Chen, and G. B. Gardner. 2011. Estimation of chromophoric dissolved organic matter in the Mississippi and Atchafalaya river plume regions using above surface hyperspectral remote sensing. *Journal of Geophysical Research* 116:C02011. doi:10.1029/2010JC006523.

Index

Page numbers followed by f and t indicate figures and tables, respectively.

D

G

H

Q

R

S

T

X

Y

Z